Calculus
with
Analytic Geometry

Calculus with Analytic Geometry

by Richard H. Crowell
and William E. Slesnick

DEPARTMENT OF MATHEMATICS
DARTMOUTH COLLEGE

W · W · NORTON & COMPANY · INC · NEW YORK

ISBN 0 393 09782 X

Library of Congress Catalog Card No. 67–24281

PRINTED IN THE UNITED STATES OF AMERICA
7 8 9 0

Contents

Preface

This book introduces and develops the differential and integral calculus of functions of one variable. It is intended primarily as a textbook for a one-year course at the college or university level, but would also be suitable for an advanced placement course in secondary school. An alternative possibility would be to use only the first six chapters for an introductory course in calculus lasting one semester or term.

The principal prerequisite for an intelligent reading of this book is a good background in secondary-school mathematics. Although the importance of set theory in the teaching of elementary mathematics may have been exaggerated, it is nonetheless a fact that most modern mathematics is done within the context of intuitive set theory, and we have used it informally. For example, a function is defined on page 13 as a set of ordered pairs. A short summary of some of the common terminology and notations of set theory is given in the third paragraph of Chapter 4 beginning on page 163. Mathematical induction is used occasionally in proofs, e.g., in the proof of the equation $\frac{d}{dx} x^n = nx^{n-1}$ in **(7.5)** on page 61. However, mathematical induction is in no sense an essential prerequisite for using the book. No previous knowledge of trigonometry is actually necessary since the definitions and development of the trigonometric functions as functions of a real variable are self-contained. Nevertheless, an understanding of the elementary trigonometry of angles would certainly be helpful.

As far as we know, there is no single, optimal approach to learning mathematics. One person learns best by slowly and carefully studying the details and logical structures, whereas another prefers to skim at first, work many problems, and return to the theory later—if at all. This book is designed to accommodate a variety of users. One result of this design has been the inclusion of the proofs of a number of theorems whose proofs would probably not be covered in a typical first-year course. They are included for completeness and also to satisfy the reader who, in a given instance, really wants to know WHY. However, to facilitate skipping, the proofs have been indented on the page and set in smaller type than that used in the main body of the text.

In teaching the course out of which this book developed, the authors do not emphasize proofs and do not present many proofs in detail. However, it has been our experience that students can absorb, and do appreciate, a clear and accurate statement of a mathematical proposition even when either no attempt has been made to supply the argument or only the intuitive idea behind the proof has been presented. Some examples of specific theorems whose proofs we would normally omit in teaching a first-year course are:

(7.5), page 127: The $\frac{*}{\infty}$ form of L'Hôpital's Rule.
(3.1), page 258: The Intermediate Value Theorem.
(7.2), page 522: The fact that a power series can be differentiated term by term.
(2.2), page 553: The detailed derivation of the integral formula for the arc length of a parametrized curve.

In addition, the following are some sections of the book that we either skip entirely or mention only briefly in a beginning course:

Section 7, Chapter 2: L'Hôpital's Rule.
All of Chapter 3: Conic Sections. (Sufficient analytic geometry for specific examples in calculus is developed as needed.)
Section 3, Chapter 5: Inverse Function Theorems.
Section 5, Chapter 6: Algebraic and Transcendental Functions.
Section 3, Chapter 8: Numerical Approximations (Continued).
Section 6, Chapter 8: Integration of Discontinuous Functions.
Section 6, Chapter 11: Hyperbolic Functions.

It is our hope that the reader will also find this book useful as a reference work. In his subsequent mathematical training he may wish to return and study more deeply some of the details, which he omitted as a beginning student and which at a later time will appear more meaningful and relevant to his interests.

The format should offer no problems. Definitions are set in boldface type, and theorems in italic. All theorems are numbered with two integers, such as **(2.3)**. The more important ones in addition carry the label **THEOREM,** and the few that are sufficiently famous to have common names are labeled with them. Other propositions and formulas, which are technically also theorems but are not generally referred to as such, are simply numbered. The first integer indicates the section, and the second is a reference within the section. Thus **(2.3)** is the third theorem, proposition, or formula occurring in Section 2. The number of the chapter is not given, but a reference to **(2.3)** is usually accompanied by the number of the page on which it is found. If no page is given, then both the reference and the theorem referred to are in the same chapter. Finally, in individual sections a number, such as (6), may appear to the right of a given equation so that it may be referred to by that number later in the same section.

We are deeply indebted to many people, and it is difficult to make adequate acknowledgments. First, we wish to thank those of our colleagues in the Dartmouth Mathematics Department who have given generously of their time with helpful advice and criticism. Among these, we are especially grateful to Professor Edward M. Brown for his interest and many valuable suggestions. In typing and preparing the manuscript we have had the cooperation from time to time of all the secretaries in the Department. In particular, our thanks are due to Mrs. Margaret P. Carter, Mrs. Helen Hanchett, Mrs. Nancy B. French, Miss Jane Overman, and Mrs. Linda P. Wieselquist. The students who have willingly learned their calculus from unfinished parts of the manuscript also deserve our thanks. We have been fortunate in having the assistance of Mr. Rudd A. Crawford, Jr., of the Concord-Carlisle School District and Harvard University, who prepared the answers to the problems. Finally, although Slesnick is still a bachelor, Crowell is not and owes an inestimable debt to his wife Marilyn for her patience and understanding throughout the writing of this book.

September 1967 R.H.C.
Hanover, N.H. W.E.S.

Calculus
with
Analytic Geometry

CHAPTER 1

Functions, Limits, and Derivatives

1. Real Numbers, Inequalities, Absolute Values. Calculus deals with numerical-valued quantities and, in the beginning, with quantities whose values are real numbers. Some understanding of the basic set R of all real numbers is therefore essential. A **real number** is one that can be written as a decimal: positive or negative or zero, terminating or nonterminating. Examples are

$$\begin{aligned} &1, \quad -5, \quad 0, \quad 14, \\ &\tfrac{2}{3} = 0.666666\ldots, \quad \tfrac{3}{8} = 0.375, \\ &\sqrt{2} = 1.4142\ldots, \\ &-\pi = -3.141592\ldots, \\ &176355.14233333\ldots. \end{aligned}$$

The most familiar subset of R is the set Z of **integers.** These are the numbers

$$\ldots, \quad -3, \quad -2, \quad -1, \quad 0, \quad 1, \quad 2, \quad 3, \ldots. \tag{1}$$

Another subset is the set Q of all rational numbers. A real number r is **rational** if it can be expressed as the ratio of two integers, more precisely, if $r = \dfrac{m}{n}$, where m and n are integers and $n \neq 0$. Since every integer m can be written $\dfrac{m}{1}$, it follows that every integer is also a rational number. A scheme, analogous to (1), which lists all the positive rational numbers is the following:

$$\begin{matrix} \frac{1}{1}, & \frac{2}{1}, & \frac{3}{1}, & \frac{4}{1}, \ldots \\ \frac{1}{2}, & \frac{2}{2}, & \frac{3}{2}, & \frac{4}{2}, \ldots \\ \frac{1}{3}, & \frac{2}{3}, & \frac{3}{3}, & \frac{4}{3}, \ldots \\ \vdots & & & \end{matrix} \tag{2}$$

Of course there are infinitely many repetitions in this presentation since, for example, $\frac{2}{1} = \frac{4}{2} = \frac{6}{3} = \ldots.$ An unsophisticated guess would be that all real numbers are rational. There are, however, many famous proofs that this is not so. For example, a very simple and beautiful argument shows that $\sqrt{2}$ is not rational. (See Problem 13 at the end of this section.) It is not hard to prove that a real number is rational if and only if its decimal expansion beyond some digit consists of a finite sequence of digits repeated forever. Thus the numbers

1.71349213213213213213 . . . (forever),
1.500000000 . . . (forever)

are rational, but

0.101001000100001000001 . . . etc.

is not.

The fundamental algebraic operations on real numbers are addition and multiplication: For any two elements a and b in R, two elements $a + b$ and ab in R are uniquely determined. These elements, called the **sum** and **product** of a and b, respectively, are defined so that the following six facts are true:

(i) ASSOCIATIVE LAWS

$$a + (b + c) = (a + b) + c, \qquad a(bc) = (ab)c.$$

(ii) COMMUTATIVE LAWS

$$a + b = b + a, \qquad ab = ba.$$

(iii) DISTRIBUTIVE LAW

$$(a + b)c = ac + bc.$$

(iv) EXISTENCE OF IDENTITIES
R contains two distinct elements 0 and 1 with the properties that $0 + a = a$ and $1 \cdot a = a$ for every a in R.

(v) EXISTENCE OF SUBTRACTION
For every a in R, there is an element in R denoted by $-a$ such that $a + (-a) = 0$. ["$a - b$" is an abbreviation of "$a + (-b)$."]

(vi) EXISTENCE OF DIVISION
For every $a \neq 0$ in R, there is an element in R denoted by a^{-1} or $\frac{1}{a}$ such that $aa^{-1} = 1$. ("$\frac{a}{b}$" is an abbreviation of "ab^{-1}.")

Addition and multiplication are here introduced as binary operations. However, as a result of the associative law of addition, $a + b + c$ is defined to be the common value of $(a + b) + c$ and $a + (b + c)$. In a like manner we may define the triple product abc and, more generally, $a_1 + \cdots + a_n$ and $a_1 \cdots a_n$. Many theorems of algebra are consequences of the above six facts, and we shall assume them without proof. They are, in fact, frequently taken as part of a set of axioms for R.

Another essential property of the real numbers is that of order. We write $a < b$ as an abbreviation of the statement that a is less than b. Presumably the reader, given two decimals, knows how to tell which one is the smaller. For him the following four facts simply recall the basic properties governing inequalities. On the other hand, they may also be taken as axioms for an abstractly defined relation between elements of R, which we choose to denote by $<$.

(vii) TRANSITIVE LAW
If $a < b$ and $b < c$, then $a < c$.

(viii) LAW OF TRICHOTOMY
For every real number a, one and only one of the following alternatives holds: $a = 0$, or $a < 0$, or $0 < a$.

(ix) *If $a < b$, then $a + c < b + c$.*

(x) *If $a < b$ and $0 < c$, then $ac < bc$.*

Note that each of the above statements except (vi) remains true when restricted to the set Z of integers. Moreover, all the statements are true for the set Q of rational numbers. Hence as a set of axioms for R, they fail to distinguish between two very different sets: R and its subset Q. Later in this section we shall add one more item to the list, which will complete the algebraic description of R.

A real number a is **positive** if $0 < a$ and **negative** if $a < 0$. Since the relation "greater than" is just as useful as "less than," we adopt a symbol for it, too, and abbreviate the statement that a is greater than b by writing $a > b$. Clearly $a > b$ if and only if $b < a$. Statement (x), when translated into English, says that the direction of an inequality is preserved if both sides are multiplied by the same positive number. Just the opposite happens if the number is negative: The inequality is reversed. That is,

(1.1) *If $a < b$ and $c < 0$, then $ac > bc$.*

Proof. Since $c < 0$, statements (iv), (v), and (ix) imply

$$0 = c + (-c) < 0 + (-c) = -c.$$

So $-c$ is positive. Hence by (x), we get $-ac < -bc$. By (ix) again,

$$-ac + (bc + ac) < -bc + (bc + ac).$$

Hence $bc < ac$, and this is equivalent to $ac > bc$.

Two more abbreviations complete the mathematician's array of symbols for writing inequalities:

$$a \leq b \quad \text{means} \quad a < b \quad \text{or} \quad a = b,$$

$$a \geq b \quad \text{means} \quad a > b \quad \text{or} \quad a = b.$$

The geometric interpretation of the set R of all real numbers as a straight line is familiar to anyone who has ever used a ruler, and it is essential to an understanding of calculus. To describe the assignment of points to numbers, consider an arbitrary straight line L, and choose on it two distinct points, one of which we assign to, or identify with, the number 0, and the other to the number 1 (see Figure 1). The rest is automatic. The scale on L is chosen

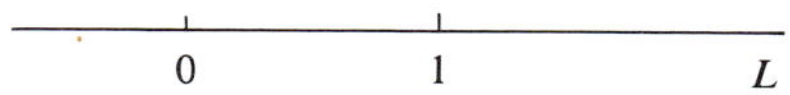

Figure 1

so that the unit of distance is the length of the line segment between the points 0 and 1. Every positive number a is assigned the point on the side of 0 containing 1 which is a units of distance from 0. Every negative number a is assigned the point on the side of 0 not containing 1 which is $-a$ units of distance from 0. Note that if L is oriented so that 1 lies to the right of 0, then *for any two numbers* a *and* b (*positive, negative, or zero*), $a < b$ *if and only if* a *lies to the left of* b. A line which has been identified with R under a correspondence such as the one just described is called a **real number line** (see Figure 2).

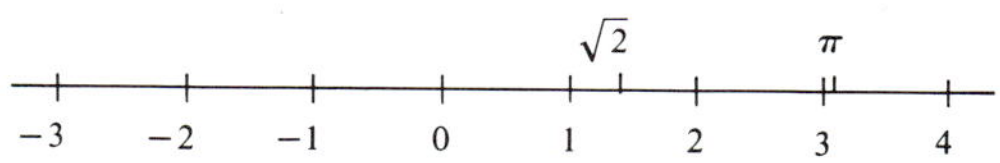

Figure 2

An **interval** is a subset I of R with the property that whenever a and c belong to I and $a \leq b \leq c$, then b also belongs to I. Geometrically an interval is a connected piece of a real number line. A number d is called a **lower bound** of a set S of real numbers if $d \leq s$ for every s in S. It is an **upper bound** of S if $s \leq d$ for every s in S. A given subset of R, and in particular an interval, is called **bounded** if it has both an upper and lower bound. There are four different kinds of bounded intervals:

(a, b), the set of all numbers x such that $a < x < b$;

$[a, b]$, the set of all numbers x such that $a \leq x \leq b$;

$[a, b)$, the set of all numbers x such that $a \leq x < b$;

$(a, b]$, the set of all numbers x such that $a < x \leq b$.

In each case the numbers a and b are called the **endpoints** of the interval. The set $[a, b]$ contains both its endpoints, whereas (a, b) contains neither one. Clearly $[a, b)$ contains its left endpoint but not its right one, and an analogous remark holds for $(a, b]$. It is important to realize that there is no element ∞ (infinity) in the set R. Nevertheless, the symbols ∞ and $-\infty$ are commonly used in denoting unbounded intervals. Thus

(a, ∞) is the set of all numbers x such that $a < x$,

$[a, \infty)$ is the set of all numbers x such that $a \leq x$,

$(-\infty, a)$ is the set of all numbers x such that $x < a$,

$(-\infty, a]$ is the set of all numbers x such that $x \leq a$,

$(-\infty, \infty)$ is the entire set R.

The symbols ∞ and $-\infty$ also appear frequently in inequalities although they are really unnecessary, because, for example,

$$-\infty < x < a \text{ is equivalent to } x < a,$$

$$a \leq x < \infty \text{ is equivalent to } a \leq x,$$

etc. Since ∞ is not an element of R, we shall never use the notations $[a, \infty]$, $x \leq \infty$, etc. An unbounded interval has either one endpoint or none; in each of the above cases it is the number a. We call an interval **open** if it con-

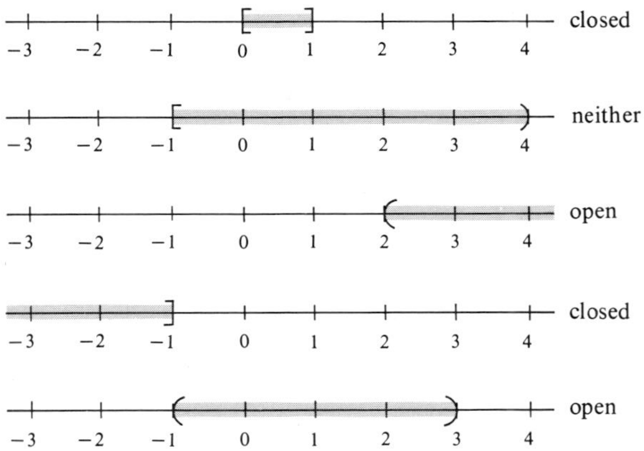

Figure 3

tains none of its endpoints, and **closed** if it contains them all. Thus, for example, (a, b) and $(-\infty, a)$ are open, but $[a, b]$ and $[a, \infty)$ are closed. The intervals $[a, b)$ and $(a, b]$ are neither open nor closed, although they are sometimes called half-open or half-closed. Since $(-\infty, \infty)$ has no endpoints, it vacuously both does and does not contain them. Hence $(-\infty, \infty)$ has the dubious distinction of being both open and closed.

EXAMPLE 1. Draw the intervals $[0, 1]$, $[-1, 4)$, $(2, \infty)$, $(-\infty, -1]$, $(-1, 3)$, and identify them as open, closed, neither, or both (see Figure 3).

It is frequently necessary to talk about the size of a real number without regard to its sign, not caring whether it is positive or negative. This happens often enough to warrant a definition and special notation: The **absolute value** of a real number a is denoted by $|a|$ and defined by

$$|a| = \begin{cases} a & \text{if } a \geq 0, \\ -a & \text{if } a < 0. \end{cases}$$

Thus $|3| = 3$, $|0| = 0$, $|-3| = 3$. Obviously, *the absolute value of a real number cannot be negative.* Geometrically, $|a|$ is the distance between the points 0 and a on the real number line. A generalization that is of extreme importance is the fact that $|a - b|$ *is the distance between the points a and b on the real number line for any two numbers a and b whatsoever.* Probably the best way to convince oneself that this is true is to look at a few illustrations (see Figure 4).

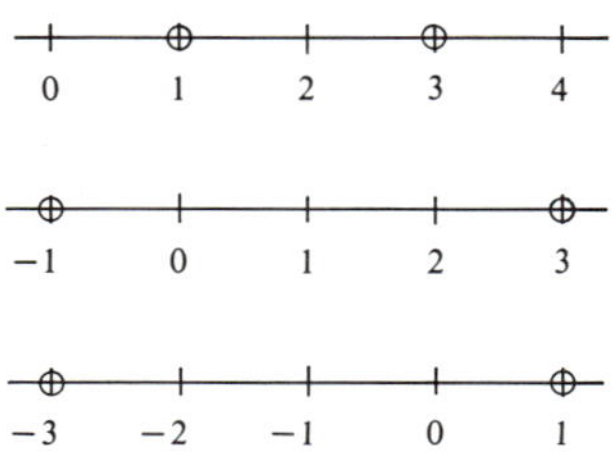

$|3 - 1| = 2 = |1 - 3|$,

$|-1 - 3| = 4$,

$|(-3) - 1| = |1 - (-3)| = 4$.

Figure 4

EXAMPLE 2. Describe the set I of all real numbers x such that $|x - 5| < 3$. For any number x, the number $|x - 5|$ is the distance between x and 5 on a real number line (see Figure 5). That distance will be less than 3 if and only

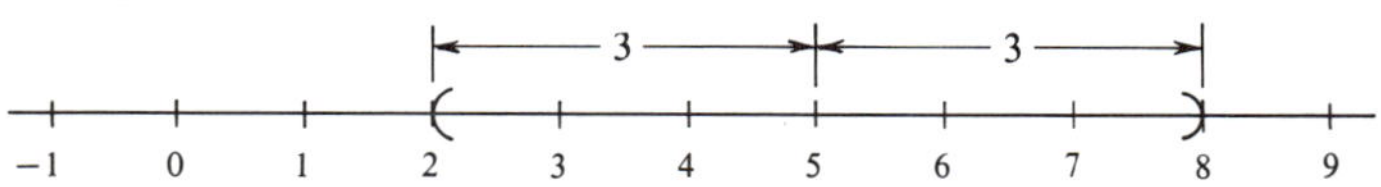

Figure 5

if x satisfies the inequalities $2 < x < 8$. We conclude that I is the open interval $(2, 8)$.

There is an alternative way of writing the definition of the absolute value of a number a which requires only one equation: We do not have to give separate definitions for positive and negative a. This definition uses a square root, and before proceeding to it, we call attention to the following mathematical custom: Although every positive real number a has two square roots, in this book *the expression* $\sqrt{a}$ *always denotes the positive root*. Thus the two solutions of the equation $x^2 = 5$ are $\sqrt{5}$ and $-\sqrt{5}$. Note that the two equations

$$x^2 = a \qquad \text{and} \qquad x = \sqrt{a}$$

are not equivalent. The second implies the first, but not conversely. On the other hand,

$$x^2 = a \qquad \text{and} \qquad |x| = \sqrt{a}$$

are equivalent. Having made these remarks, we observe that

(1.2) $$|a| = \sqrt{a^2}.$$

The formulation (1.2) is a handy one for establishing two of the basic properties of absolute value. They are

(1.3) $$|ab| = |a|\,|b|.$$

(1.4) $$|a + b| \leq |a| + |b|.$$

Proofs. Since $(ab)^2 = a^2b^2$ and since the positive square root of a product of two positive numbers is the product of their positive square roots, we get

$$|ab| = \sqrt{(ab)^2} = \sqrt{a^2b^2} = \sqrt{a^2}\sqrt{b^2} = |a|\,|b|.$$

To prove (1.4), we observe, first of all, that $ab \leq |ab|$. Hence

$$a^2 + 2ab + b^2 \leq a^2 + 2|ab| + b^2 = |a|^2 + 2|a|\,|b| + |b|^2.$$

Thus,

$$|a + b|^2 = (a + b)^2 \leq (|a| + |b|)^2.$$

By taking the positive square root of each side of the inequality (see Problem 9), we get (1.4).

As remarked above, our list of basic facts (i) through (x) about the set R of real numbers is incomplete. One important property of real numbers that together with the others gives a complete characterization is the following:

(xi) LEAST UPPER BOUND PROPERTY
Every nonempty subset of R *which has an upper bound has a least upper bound.*

Suppose S is a nonempty subset of R which has an upper bound. What (xi) says is that there is some number b which (1) is an upper bound, i.e., $s \leq b$ for every s in S, and (2) if c is any other upper bound of S, then $b \leq c$.

It is hard to see at first how such a statement can be so significant. Intuitively it says nothing more than this: If you cannot go on forever, you have to stop somewhere. Note, however, that the rational numbers do *not* have this property. The set of all rational numbers less than the irrational number $\sqrt{2}$ certainly has an upper bound. In fact, each of the numbers 2, 1.5, 1.42, 1.415, 1.4143, and 1.41422 is an upper bound. However, for every rational upper bound, there will always exist a smaller one. Hence there is no rational least upper bound.

PROBLEMS

1. Draw the following intervals and identify them as bounded or unbounded, closed or open, or neither: $(2, 4)$, $[3, 5]$, $(-\infty, -2]$, $[1.5, 2.5)$, $(\sqrt{2}, \pi)$.
2. Draw each of the following subsets of R. For those that are given in terms of absolute values write an alternative description that does not use the absolute value.

   (a) Set of all x such that $4 < x \leq 7.5$.

   (b) Set of all x such that $0 < x < \infty$.

   (c) Set of all x such that $5 \leq x < 8$.

   (d) Set of all x such that $|x| > 2$.

   (e) Set of all y such that $1 < |y| < 3$.

   (f) Set of all z such that $|z - 2| \leq 1$.

   (g) Set of all x such that $|x - a| > 0$.

   (h) Set of all u such that $1 < |u - 1| < 5$.

3. Prove the following facts about inequalities. [*Hint:* Use (viii), (ix), (x), (1.1), and the meanings of $\geq$ and $\leq$. In each problem you will have to consider several cases separately, e.g., $a > 0$ and $a = 0$.]

   (a) If $a \leq b$, then $a + c \leq b + c$.

   (b) If $a \geq b$, then $a + c \geq b + c$.

   (c) If $a \leq b$ and $c \geq 0$, then $ac \leq bc$.

   (d) If $a \leq b$ and $c \leq 0$, then $ac \geq bc$.

4. Prove that a is positive (negative) if and only if $\frac{1}{a}$ is positive (negative).

5. If $0 < a < b$, prove that $\frac{1}{b} < \frac{1}{a}$.

6. If $a > c$ and $b < 0$, prove that $\frac{a}{b} < \frac{c}{b}$.

7. If $a < b < c$, prove that

$$\frac{b}{c} < \frac{b}{a} \quad \text{if } a > 0,$$

$$\frac{b}{c} > \frac{b}{a} \quad \text{if } c < 0.$$

8. Does the set Z of integers have the Least Upper Bound Property? That is, if a nonempty subset of Z has an upper bound, does it have a smallest one?
9. Show that if $0 \leq a \leq b$, then $0 \leq \sqrt{a} \leq \sqrt{b}$.
10. Prove that $a = b$ if and only if $a \leq b$ and $b \leq a$.
11. Show that the Least Upper Bound Property implies the Greatest Lower Bound Property. That is, using (xi), prove that if a nonempty subset of R has a lower bound, then it has a greatest lower bound.
12. Verify the assertion made in the text that if an interval is bounded it must be one of four types: (a, b), $[a, b]$, $(a, b]$, or $[a, b)$. (*Hint:* See Problem 11.)
13. Prove that $\sqrt{2}$ is irrational. (*Hint:* The proof, which is elegant and famous, starts by assuming that $\sqrt{2} = \frac{p}{q}$, where p and q are integers not both even. A contradiction can then be derived.)

2. Ordered Pairs of Real Numbers, the *xy*-Plane, Functions. The set whose members consist of just the two elements a and b is denoted $\{a, b\}$. The notation is not perfect because it suggests that the members a and b have been ordered: a is written first and b second. Actually no ordering is present because $\{a, b\} = \{b, a\}$. Note also that if $a = b$, then $\{a, b\} = \{b, a\} = \{a\}$. It can happen, however, that the ingredient of order is essential. We therefore introduce the notion of an **ordered pair** (a, b) whose first member is a and whose second member is b. The characteristic property of ordered pairs is

$$(a, b) = (c, d) \qquad \textit{if and only if } a = c \textit{ and } b = d.$$

In particular $(a, b) = (b, a)$ *if and only if* $a = b$. In Section 1 we saw that the set R of all real numbers can be thought of as a straight line. We shall now show that every ordered pair (a, b) of real numbers a and b can be identified with a point in a plane. This brings up a notational problem: Is $(5, 7)$ the ordered pair of real numbers or is it the open interval consisting of all x such that $5 < x < 7$? The answer is that it is impossible to tell out of context—just as it is impossible to tell whether the word "well" is the noun or the adverb.

Consider two distinct real number lines drawn in a plane so that they intersect at the number 0 on each line. One of the lines is traditionally drawn horizontal and called the ***x*-axis**, and the other is made perpendicular to it and called the ***y*-axis**. The orientation is chosen so that the number 1 on the x-axis lies to the right of 0, and the number 1 on the y-axis is above 0. It is also customary to use the same scale of distances on both axes. For every ordered pair (a, b) of real numbers, let L_a be the line parallel to the y-axis that cuts the x-axis at a, and let M_b be the line parallel to the x-axis that cuts the y-axis at b. We assign the point of intersection of L_a and M_b to the ordered pair (a, b) (see Figure 6). The numbers a and b are called the **coordinates** of

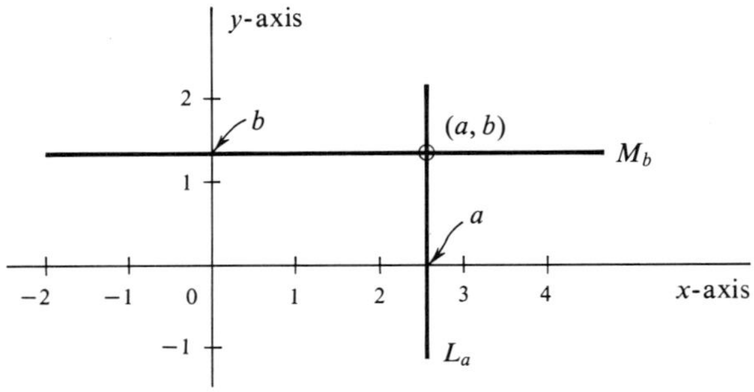

Figure 6

the point. a is the ***x*-coordinate** (or **abscissa**) and b is the ***y*-coordinate** (or **ordinate**).

If the pairs (a, b) and (c, d) are not equal, then the points in the plane assigned to them will be different. In addition, every point in the plane has a number pair assigned to it: Starting with a point, draw the two lines through

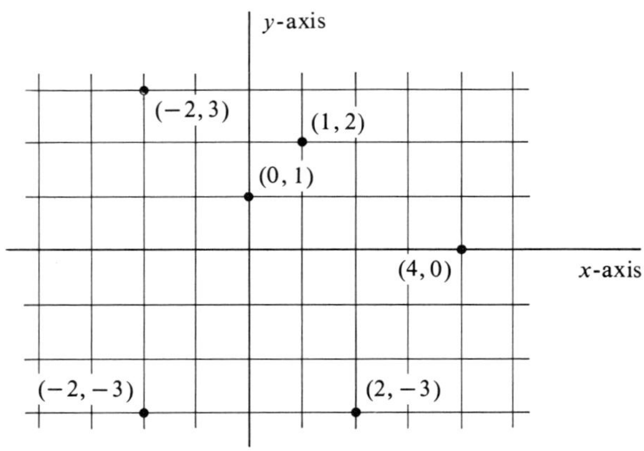

Figure 7

it which are parallel to the x-axis and the y-axis. One line cuts the x-axis at a number a, and the other cuts the y-axis at b. The ordered pair (a, b) has the original point assigned to it. It follows that our assignment

$$\text{pair} \longrightarrow \text{point}$$

is a one-to-one correspondence between the set of all ordered pairs of real numbers, which we denote by R^2, and the set of all points of the plane. It is convenient simply to identify R^2 with the plane together with the two axes.

EXAMPLE 1. Plot the points $(1, 2)$, $(-2, 3)$, $(0, 1)$, $(4, 0)$, $(-2, -3)$, and $(2, -3)$ on the xy-plane (see Figure 7).

The usefulness of the idea of an ordered pair is by no means limited to pairs of real numbers. In plane geometry, for example, we may consider the set of all ordered pairs (T, p) in which T is a triangle and p is the point of intersection of its medians. In the three-dimensional extension of the xy-plane, the set R^3 of all ordered triples (a, b, c) of real numbers is identified with the set of all points in three-dimensional space. The definition of an ordered triple can be reduced to that of an ordered pair by defining (a, b, c) to be $((a, b), c)$.

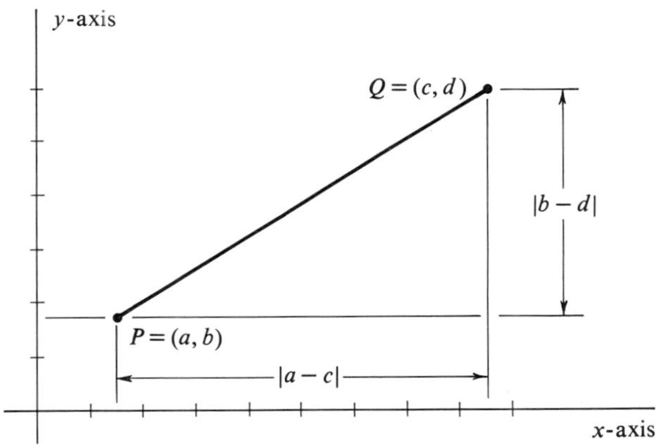

Figure 8

Let $P = (a, b)$ and $Q = (c, d)$ be arbitrary elements in the set R^2 of all ordered pairs of real numbers. We define the **distance** between P and Q by the formula

$$\text{distance}\,(P, Q) = \sqrt{(a - c)^2 + (b - d)^2}. \tag{1}$$

Three simple corollaries of this definition are:

(2.1) *Distance* $(P, Q) \geq 0$; i.e., *distance is never negative.*

(2.2) *Distance* $(P, Q) = 0$ *if and only if* $P = Q$.

(2.3) *Distance* $(P, Q) =$ *distance* (Q, P).

Another consequence of (1) is that it is no longer simply a matter of tradition and convenience that we draw the y-axis perpendicular to the x-axis. It follows from consideration of the Pythagorean Theorem and its converse (see Figure 8) that the above definition of distance between elements of R^2 corresponds with our geometric notion of the distance between points in the Euclidean plane if and only if the two coordinate axes are perpendicular and the scales are the same on both.

EXAMPLE 2. Let C be the subset of the xy-plane consisting of all points whose distance from (1, 1) is equal to 2. Thus C is the circle shown in Figure 9.

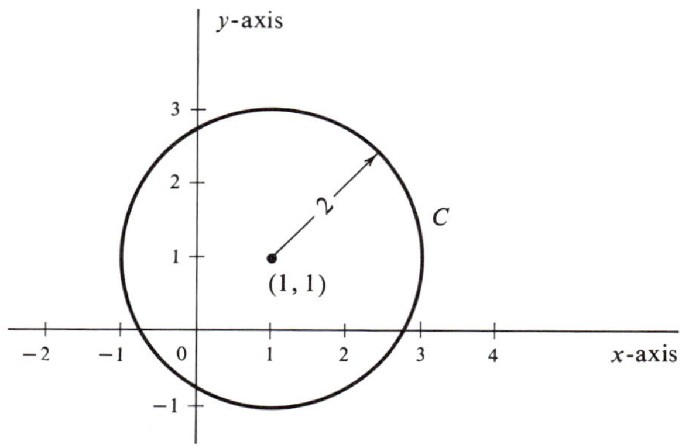

Figure 9

If (x, y) is an arbitrary point in the xy-plane, its distance from (1, 1) is equal to $\sqrt{(x-1)^2 + (y-1)^2}$. Hence, (x, y) belongs to C if and only if

$$\sqrt{(x-1)^2 + (y-1)^2} = 2. \qquad (2)$$

Numbers x and y satisfy (2) if and only if they satisfy

$$(x-1)^2 + (y-1)^2 = 4. \qquad (3)$$

Thus C is the set of all ordered pairs (x, y) that satisfy (3)—or that satisfy (2). Either (2) or (3) is therefore called **an equation of the circle** C.

The set of all points (x, y) in the plane that satisfy a given equation is called the **graph** of the equation. Hence, in the above example, the circle C is the graph of the equation $(x-1)^2 + (y-1)^2 = 4$.

EXAMPLE 3. Let L be the set of all ordered pairs (x, y) such that $y = 2x - 3$. For each real number x, there is one and only one number y such that (x, y) belongs to L: $y = 2x - 3$. To see what L looks like, we plot five of its points (see Table 1). As shown in Figure 10, all these points lie on a straight line. In Section 5 we shall justify the natural conjecture that this straight line is the set L.

TABLE 1

x	$y = 2x - 3$
-1	-5
0	-3
1	-1
2	1
3	3

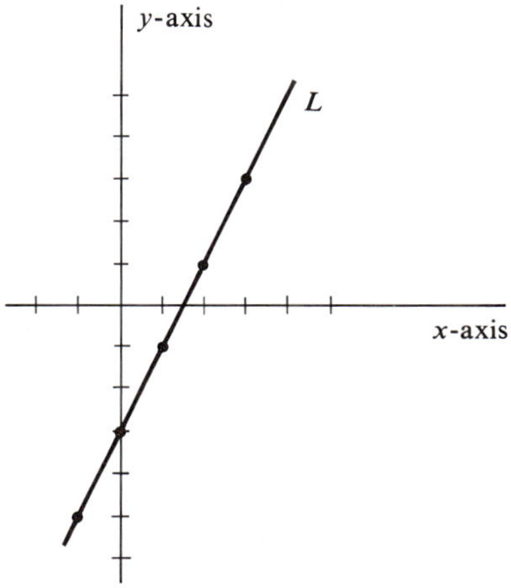

Figure 10

EXAMPLE 4. The set of all pairs (x, y) such that $y^2 = x$ is the curve shown in Figure 11. This curve is a parabola, one of the conic sections, which are studied in greater detail in Chapter 3. At present we shall be satisfied with plotting a few points and connecting them with a smooth curve (see Table 2).

TABLE 2

y	x
0	0
± 1	1
± 2	4

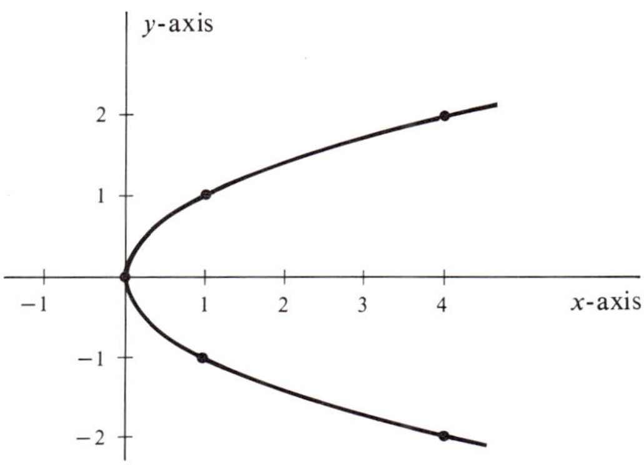

Figure 11

A **function** f is any set f of ordered pairs such that whenever (a, b) and (a, c) belong to f, then $b = c$. Note that every subset of the xy-plane is a set of ordered pairs, but not every subset is a function. In particular, the parabola in Example 4 is not, because it contains both (4, 2) and (4, −2). On the other hand, the straight line in Example 3 is a function. This condition that a function must never contain two pairs (a, b) and (a, c) with $b \neq c$ means geometrically that a subset of the xy-plane is a function if and only if it never intersects a line parallel to the y-axis in more than one point. Hence it is an easy matter to decide which of the following sets are functions and which are not:

(i) The set f of all pairs (x, y) such that $y = x + 1$.

(ii) The set g of all pairs (x, y) such that $x^2 + y^2 = 1$.

(iii) The set F of all pairs (x, y) such that $y = x^2 + 2x + 2$.

(iv) The set h of all pairs (x, y) such that $2x + 3y = 1$.

(v) The set G of all pairs (x, y) such that $y = \sqrt{x + 2}$.

(vi) The set H of all pairs (x, y) such that $y^4 = x$.

The sets f, F, h, and G are functions, but g and H are not.

The **domain** of a function f is the set of all elements a for which there is a corresponding b such that (a, b) belongs to f. Analogously, the **range of** f is the set of all elements b for which there is an a such that (a, b) belongs to f. In (i), the domain of f is the set R of all real numbers and so is the range. On the other hand, in (iii), although the domain of F is equal to R, the range is the interval consisting of all real numbers $y \geq 1$, because we can write $x^2 + 2x + 2 = (x + 1)^2 + 1 \geq 1$.

If a pair (a, b) belongs to a function f, we call b the **value of** f **at** a and write $b = f(a)$. Note that the meaning of $f(a)$ is unambiguous only because the definition of a function forbids having (a, b) and (a, c) both belong to f if $b \neq c$. Therefore the second member of any ordered pair that belongs to f is determined by the first member.

EXAMPLE 5. In (i),

$$f(x) = x + 1, \qquad f(a) = a + 1,$$

$$f(0) = 1, \qquad f(3 + 4) = (3 + 4) + 1 = 8,$$

$$f(-1) = -1 + 1 = 0, \qquad f(a + b) = a + b + 1.$$

In (v),

$$G(x) = \sqrt{x + 2}, \qquad G(2x + y) = \sqrt{2x + y + 2},$$

$$G(0) = \sqrt{2}, \qquad G(-2) = 0,$$

$$G(2) = 2, \qquad G(-3) \text{ is not defined.}$$

To each element a in the domain of a function f there corresponds a value $f(a)$ in the range. This correspondence between domain and range, which is pictured in Figure 12, is the central idea in the definition of a function. Thus the function f that consists of all ordered pairs (x, y) such that $y = x^2$ and $-1 \leq x \leq 2$ is interpreted as the rule of correspondence which assigns to each number in the interval $[-1, 2]$ its square. We can describe f completely and simply by writing

$$f(x) = x^2, \qquad -1 \leq x \leq 2.$$

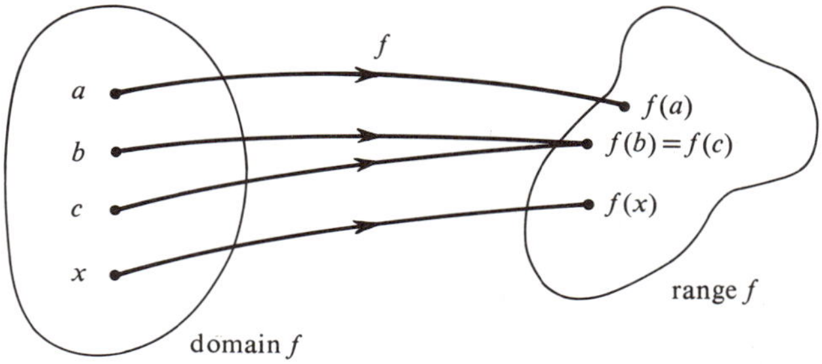

Figure 12

Examples of other functions are

$$g(x) = \sqrt{x+1}, \qquad -1 \leq x < \infty,$$
$$F(x) = x^2, \qquad -\infty < x < \infty,$$
$$h(x) = \frac{x}{x+2}, \qquad x \neq -2.$$

Note that the functions f and F immediately above are *not* equal, although f is a subset of F. Two functions are equal if they are one and the same set of ordered pairs. It follows that

(2.4) *Functions f and g are equal if and only if they have the same domain D and $f(x) = g(x)$ for every element x in D.*

Thus any complete description of a function must include a description of its domain. Sometimes this information is in fact omitted. We shall adopt the convention that if no explicit description of the domain of a function is given, then its domain is assumed to be the largest set of real numbers that makes sense. For example, the domain of the function H defined by

$$H(x) = \frac{1}{x^2 - x - 2} = \frac{1}{(x+1)(x-2)}$$

is assumed to be the entire set of real numbers with the exception of -1 and 2.

It is sometimes helpful to think of a function as a computing machine. Imagine a computing machine, named f, which is provided with an input tape, an output tape, and a button (see Figure 13). One writes a number x on the input tape and pushes the button. If x is one of the inputs which the machine will accept, i.e., if x is in the domain of f, the machine whirs contentedly and prints an output, which we denote $f(x)$, on the output tape. If x is not in the prescribed domain, either nothing happens or a red light flashes.

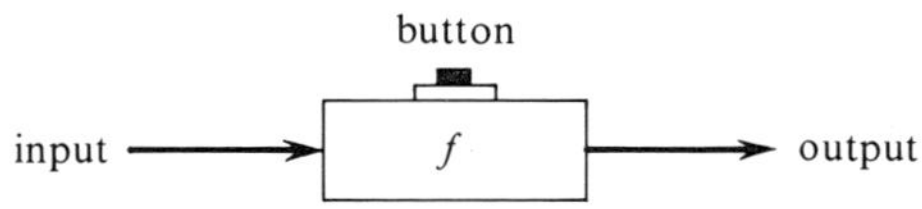

Figure 13

We have already seen that one of the best ways of describing a subset of R^2 is to draw a picture of it. If this subset happens to be a function, we call the picture the graph of the function. More specifically, if a function f is a subset of R^2, its **graph** is the set of all points in the plane that correspond to ordered pairs of the form $(x, f(x))$. Note that the graph of f depends on the correspondence between ordered pairs and points; i.e., it depends on the choice of axes. To illustrate this, in Figure 14 we have drawn the graph of

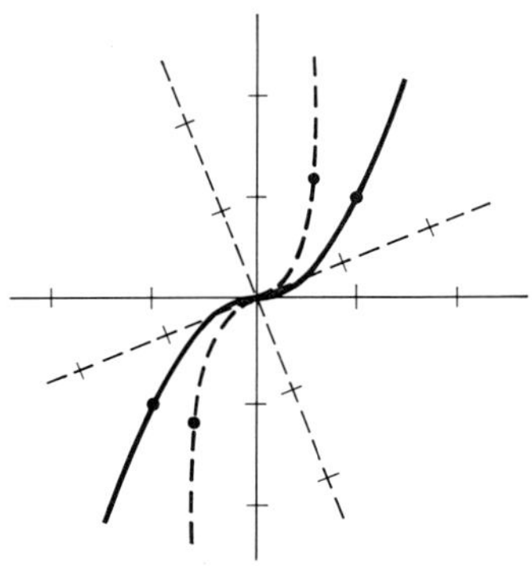

Figure 14

the function f defined by $f(x) = x^3$ for two sets of axes. For a single choice of axes, we simply identify ordered pairs and points, and under this identification a function and its graph become the same thing.

Most of the functions encountered in an introduction to calculus are defined by means of a single equation; e.g., $h(x) = x^3 + 3$. It is a bad mistake, however, to assume that this is always true. The function F given by

$$F(x) = \begin{cases} x^2 + 1 & \text{if } x \geq 0, \\ -\dfrac{x}{2} & \text{if } x < 0, \end{cases}$$

requires two equations for its definition. The graph of F is shown in Figure 15. Another function, which is so wild that it is impossible to draw its graph, is the following:

$$g(x) = \begin{cases} 0 & \text{if } x \text{ is rational,} \\ 1 & \text{if } x \text{ is irrational.} \end{cases}$$

The ordered pairs that comprise a function are not necessarily pairs of numbers. An example is the function, mentioned earlier in this section, which assigns to each triangle the point of intersection of its medians. It is possible for the domain of a function to be a set of ordered pairs. Consider the function f consisting of all ordered pairs $((x, y), z)$, where x, y, and z are

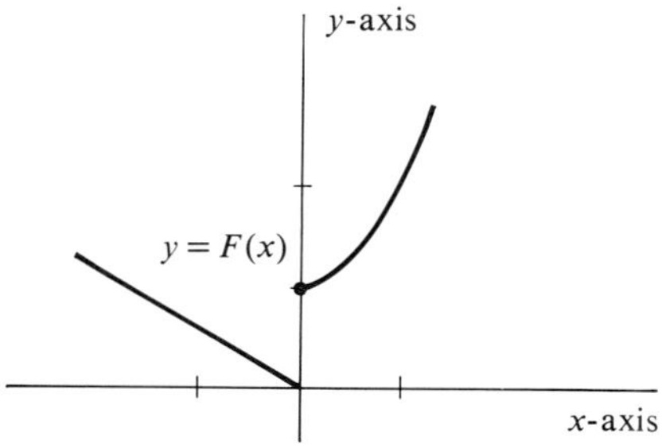

Figure 15

numbers that satisfy $x \geq y$ and $z = 2x^2 + y^2$. We describe this function simply as follows:

$$f(x, y) = 2x^2 + y^2, \qquad x \geq y. \tag{4}$$

As a final example of a function, consider the rule of correspondence that assigns to each person his or her male parent.

As we have indicated, the definition of a function is appallingly general. One of our tasks is to delineate properly the kinds of functions studied in calculus. To begin with, a function f is said to be **real-valued** if its range is a subset of R, the set of real numbers. If the domain of f is a subset of R, we call f a **function of a real variable.** The function $f(x, y)$ defined in (4) has as its domain a subset of R^2. It is a real-valued function of two real variables. For the most part, a first course in calculus is a study of real-valued functions of one real variable.

PROBLEMS

1. Plot the following points in the xy-plane: $(0, -2)$, $(1, 3)$, $(3, 1)$, $(-4, -4)$, and $(5, 0)$.
2. In the xy-plane plot the points $(1, 2)$ and $(2, 1)$, $(-3, 2)$ and $(2, -3)$, $(-2, -3)$ and $(-3, -2)$. Describe the relative positions of the points (a, b) and (b, a) for arbitrary a and b.

3. The x-axis and the y-axis divide R^2 into four quadrants, as shown in Figure 16. Let (a, b) be a point for which neither a nor b is zero. How can you recognize instantly which quadrant (a, b) belongs to?

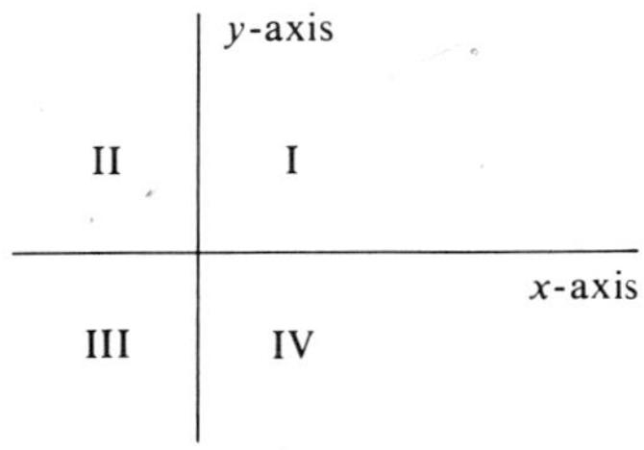

Figure 16

4. Find the distance between

$$\begin{array}{lll} (-1, 2) & \text{and} & (3, 4), \\ (2, 3) & \text{and} & (3, 2), \\ (3, 4) & \text{and} & (-1, 2), \\ (-2, -1) & \text{and} & (2, 1). \end{array}$$

In each case plot the points in R^2.

5. Verify Proposition (2.4).
6. Plot the subsets of the xy-plane defined in (i) through (vi), page 14.
7. In each of the following, plot the subset of R^2 that consists of all pairs (x, y) such that the given equation (or conditions) is satisfied.

(a) $3x + 2y = 3$
(b) $x + y = 1$
(c) $y = |x|$
(d) $y = \sqrt{x}$
(e) $x^2 + y^2 = 4$
(f) $x^2 + 4y^2 = 4$
(g) $x^2 + y^2 = 1$ and $y \geq 0$
(h) $4x^2 - y^2 = 4$
(i) $y = 2x^2 + x - 2$
(j) $y = |x^3|$
(k) $y =$ largest integer less than or equal to x
(l) $y = \begin{cases} 2x + 3, & x \geq 0 \\ \dfrac{x^2}{2}, & x < 0. \end{cases}$

8. In Problem 7, which subsets are functions?
9. Let f and g be two functions defined, respectively, by

$$f(x) = x^2 + x + 1, \qquad -\infty < x < \infty,$$

$$g(x) = \frac{x + 1}{x - 1}, \qquad \text{for every real number } x \text{ except } x = 1.$$

Find:

(a) $f(2), f(0), f(a), f(a + b), f(a - b)$.
(b) $g(0), g(-1), g(10), g(5 + t), g(x^3)$.

10. Give an example of a function f and a function g that satisfy each of the following conditions.

    (a) domain f = domain g, but range $f \neq$ range g.

    (b) domain $f \neq$ domain g, but range f = range g.

    (c) domain f = domain g and range f = range g, but $f \neq g$.

    (d) $f(a) = g(a)$ for every a that belongs to both domains, but $f \neq g$.

11. What is the assumed domain of each of the following functions?

    (a) $f(x) = \dfrac{5}{x-3}$

    (b) $f(x) = \dfrac{x^2+2}{x^2-2}$

    (c) $g(x) = \dfrac{x+3}{x^2+x-12}$

    (d) $f(x) = 5\pi$

    (e) $f(t) = \sqrt{\dfrac{1}{5-t}}$

    (f) $F(x) = \sqrt{x^2 - 8x - 20}$

    (g) The set of all ordered pairs (x, y) such that
    $$\frac{xy - x^2}{x-9} = 7.$$

3. Operations with Functions. If f and g are two functions, a new function $f(g)$, called the **composition** of g with f, is defined by

$$(f(g))(x) = f(g(x)).$$

For example, if $f(x) = x^3 - 1$ and $g(x) = \dfrac{x+1}{x-1}$, then

$$\begin{aligned}(f(g))(x) &= f(g(x)) = (g(x))^3 - 1 \\ &= \left(\frac{x+1}{x-1}\right)^3 - 1 = \frac{2(3x^2+1)}{(x-1)^3}. \end{aligned} \qquad (1)$$

The composition of two functions is the function obtained by applying one after the other. If f and g are regarded as computing machines, then $f(g)$ is the composite machine constructed by feeding the output of g into the input of f as indicated in Figure 17. In general it is not true that $f(g) = g(f)$. In

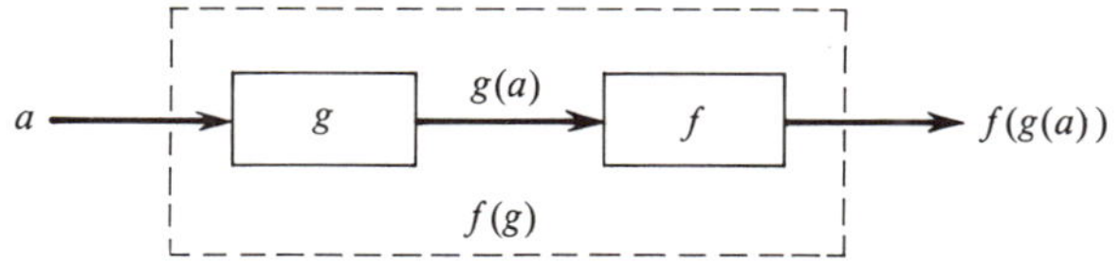

Figure 17

the above example we have

$$\begin{aligned}(g(f))(x) &= g(f(x)) = \frac{f(x)+1}{f(x)-1} \\ &= \frac{(x^3-1)+1}{(x^3-1)-1} = \frac{x^3}{x^3-2},\end{aligned} \tag{2}$$

and the two functions defined by (1) and (2) are certainly not the same. In terms of ordered pairs the composition $f(g)$ of g with f is formally defined to be the set of all ordered pairs (a, c) for which there is an element b such that $b = g(a)$ and $c = f(b)$.

If f and g are two real-valued functions, we can perform the usual arithmetic operations of addition, subtraction, multiplication, and division. Thus for the functions $f(x) = x^3 - 1$ and $g(x) = \dfrac{x+1}{x-1}$, we have

$$\begin{aligned}f(x)+g(x) &= x^3 - 1 + \frac{x+1}{x-1}, \\ f(x)-g(x) &= x^3 - 1 - \frac{x+1}{x-1}, \\ f(x)g(x) &= (x^3-1)\frac{x+1}{x-1} \\ &= (x^2+x+1)(x+1) \qquad \text{if } x \neq 1, \\ f(x)/g(x) &= \frac{x^3-1}{\dfrac{x+1}{x-1}} = \frac{(x^3-1)(x-1)}{x+1}.\end{aligned}$$

Just as with the composition of two functions, each arithmetic operation provides a method of constructing a new function from the two given functions f and g. The natural notations for these new functions are $f+g$, $f-g$, fg, and $\dfrac{f}{g}\cdot$ They are defined by the formulas

$$\begin{aligned}(f+g)(x) &= f(x)+g(x), \\ (f-g)(x) &= f(x)-g(x), \\ (fg)(x) &= f(x)g(x), \\ \frac{f}{g}(x) &= \frac{f(x)}{g(x)} \qquad \text{if } g(x) \neq 0.\end{aligned}$$

The product function fg should not be confused with the composite function $f(g)$. For example, if $f(x) = x^5$ and $g(x) = x^3$, then we have $(fg)(x) = f(x)g(x) = x^5 \cdot x^3 = x^8$, whereas

$$(f(g))(x) = f(g(x)) = (x^3)^5 = x^{15}.$$

We may also form the product af of an arbitrary real number a and real-valued function f. The product function is defined by

$$(af)(x) = af(x).$$

EXAMPLE 1. Let functions f and g be defined by $f(x) = x - 2$ and $g(x) = x^2 - 5x + 6$. Draw the graphs of f, g, $2f$, and $f + g$. We compute the function values corresponding to several different numbers x in Tables 3 and 4. The resulting graphs of f and g are, respectively, the straight line and

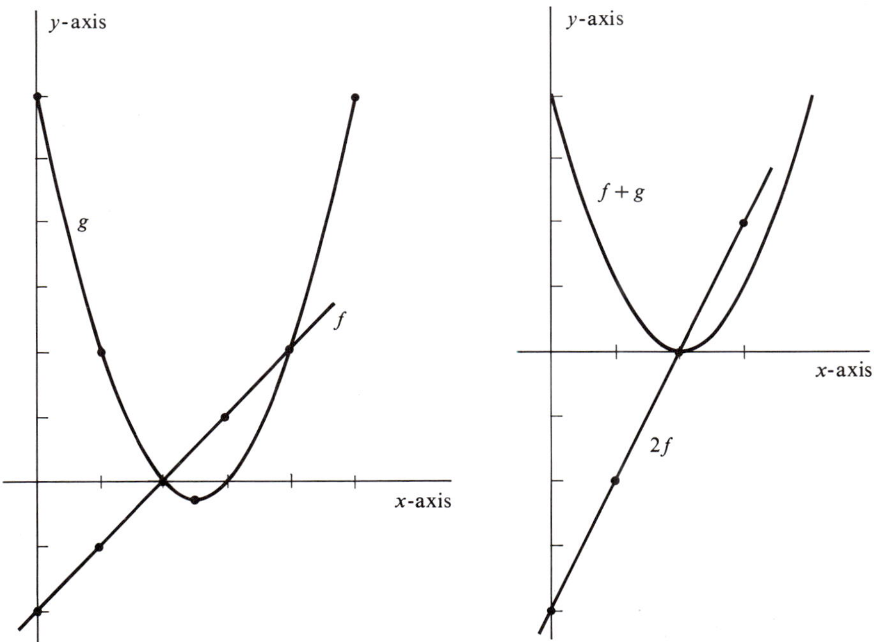

(a) Figure 18 (b)

parabola shown in Figure 18(a). It turns out that the graphs of $2f$ and $f + g$ are also a straight line and a parabola. They are drawn in Figure 18(b). To see why the graph of $f + g$ is a parabola, observe that

$$\begin{aligned}(f + g)(x) &= f(x) + g(x) = (x - 2) + (x^2 - 5x + 6) = x^2 - 4x + 4 \\ &= (x - 2)^2.\end{aligned}$$

It follows that $f + g$ is very much like the function defined by $y = x^2$. Instead of simply squaring a number, $f + g$ first subtracts 2 and then squares.

Its graph will be just like that of $y = x^2$ except that it will be shifted two units to the right.

TABLE 3

x	$f(x)$	$2f(x)$
0	-2	-4
1	-1	-2
2	0	0
3	1	2

TABLE 4

x	$g(x)$
0	6
5	6
$\frac{5}{2}$	$-\frac{1}{4}$
1	2
4	2

Up to this point we have used the letters f, g, h, F, G, and H to denote functions, and the letters x, y, a, b, and c to denote elements of sets—usually real numbers. However, the letters in the second set are sometimes also used as functions. This occurs, for example, when we speak of x as a real variable. As such, it not only is the name of a real number but also can take on many different values: 5, or -7, or π, or .... Thus the variable x is a function. Specifically, it is the very simple function that assigns the value 5 to the number 5, the value -7 to the number -7, the value π to π, ... For every real number a, we have

$$x(a) = a.$$

This function is called the **identity function.**

Suppose, for example, that s is used to denote the distance that a stone falling freely in space has fallen. The value of s increases as the stone falls and depends on the length of time t that it has fallen according to the equation $s = \frac{1}{2}gt^2$, where g is the constant gravitational acceleration. (This formula assumes no air resistance, that the stone was at rest at time $t = 0$, and that distance is measured from the starting point.) Thus s has the value $\frac{9}{2}g$ if t has the value 3, and, more generally, the value $\frac{1}{2}ga^2$ when t has the value a. If we consider t to be another name for the identity function, then s may be regarded as the function whose value is

$$s(a) = \tfrac{1}{2}ga^2 = \tfrac{1}{2}g\big(t(a)\big)^2$$

for every real number a. The original equation $s = \frac{1}{2}gt^2$ then states the relation between the two functions s and t. The fact that s and t take on different values is also expressed by referring to them as variables. A **variable** is simply a name of a function. In our example s is called a dependent variable, and t an independent variable, because the values of s depend on those of t according to $s = \frac{1}{2}gt^2$. Thus an **independent variable** is a name for the identity function, and a **dependent variable** is one that is not independent.

A real variable is therefore a name of a real-valued function. Since the arithmetic operations of addition, subtraction, multiplication, and division have been defined for real-valued functions, they are automatically defined for real variables.

We shall generally use the letter x to denote an independent variable. This raises the question: How does one tell whether an occurrence of x denotes a real number or the identity function? The answer is that the notation alone does not tell, but the context and the reader's understanding should. However, a more practical reply is that it doesn't really make much difference. We may regard $f(x)$ as either the value of the function f at the number x or as the composition of f with the variable x. If x is an independent variable, the function $f(x)$ is then the same thing as f.

EXAMPLE 2. The conventions that we have adopted concerning the use of variables give our notations a flexibility that is both consistent and extremely useful. Consider, for example, the equation

$$y = 2x^2 - 3x.$$

On the one hand, we may consider the subset of R^2, pictured in Figure 19, that consists of all ordered pairs (x, y) such that $y = 2x^2 - 3x$. This subset

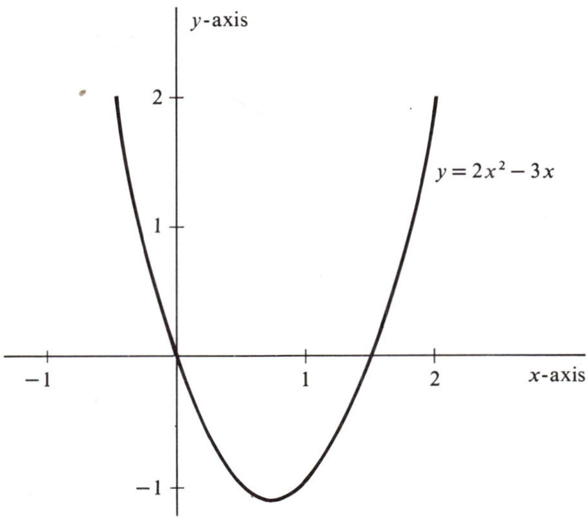

Figure 19

is a function f whose value at an arbitrary real number x is the real number $f(x) = 2x^2 - 3x$. Alternatively, we may regard x as an independent variable, i.e., the identity function. The composition of f with x is then the function $f(x) = 2x^2 - 3x$, whose value at 2, for instance, is

$$\big(f(x)\big)(2) = f\big(x(2)\big) = f(2) = 8 - 6 = 2.$$

A third interpretation is that y is a dependent variable that depends on x according to the equation $y = 2x^2 - 3x$. That is, y is the name of the function $2x^2 - 3x$.

EXAMPLE 3. Let F be the function defined by $F(x) = x^3 + x + 1$. If $u = \sqrt{x - 2}$, then

$$\begin{aligned} F(u) &= u^3 + u + 1 \\ &= (x-2)^{3/2} + (x-2)^{1/2} + 1. \end{aligned}$$

If we denote the function $F(x)$ by w, then

$$\begin{aligned} u + w &= \sqrt{x-2} + x^3 + x + 1, \\ uw &= (x-2)^{1/2}(x^3 + x + 1). \end{aligned}$$

On the other hand, we may let G be the function defined by $G(x) = \sqrt{x - 2}$ for every real number $x \geq 2$. Then $G + F$ and GF are the functions defined, respectively, by

$$\begin{aligned} (G + F)(x) &= G(x) + F(x) \\ &= \sqrt{x-2} + x^3 + x + 1, \\ (GF)(x) = G(x)F(x) &= (x-2)^{1/2}(x^3 + x + 1). \end{aligned}$$

To say that a is a real **constant** means first that it is a real number. Second, it may or may not matter which real number a is, but it is fixed for the duration of the discussion in which it occurs. Similarly, a **constant function**

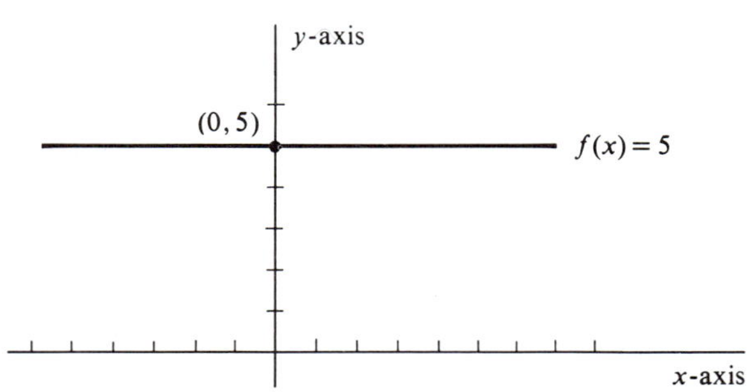

Figure 20

is one which takes on just one value; i.e., its range consists of a single element. For example, consider the constant function f defined by

$$f(x) = 5, \qquad -\infty < x < \infty.$$

The graph of f is the straight line parallel to the x-axis that intersects the y-axis in the point (0, 5); see Figure 20. We shall commonly use lower-case letters at the beginning of the alphabet, e.g., $a, b, c, \ldots$, to denote both constants and constant functions.

EXAMPLE 4. Consider the function $ax + b$, where a and b are constants, $a \neq 0$, and x is an independent variable. The graph of this function is a straight line that cuts the y-axis at b and the x-axis at $-\frac{b}{a}$. It is drawn in Figure 21. This function is the sum of the constant function b and the function which is the product of the constant function a and the identity function x.

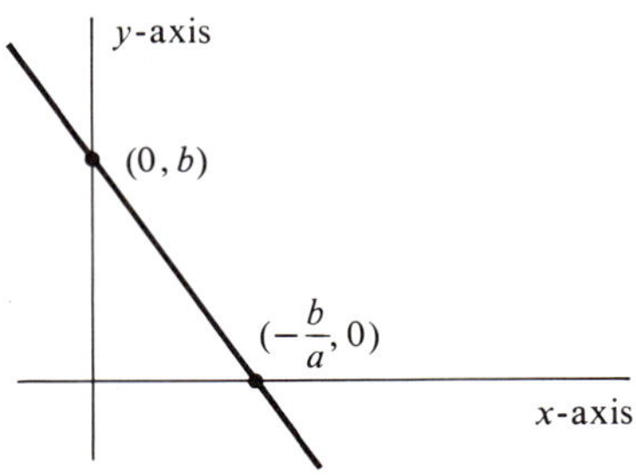

Figure 21

PROBLEMS

1. Let functions f and g be defined by

$$f(x) = x^3 - 4x^2 + 5x - 2 = (x - 2)(x^2 - 2x + 1), \; g(x) = \frac{1}{x}.$$

Find $h(x)$ if

(a) $h = f(g)$ (d) $h = fg$

(b) $h = f + g$ (e) $h = 5fg^2$.

(c) $h = g(f)$

2. What is the domain and range of each of the functions f and g in Problem 1? What is the domain of each of the functions h?

3. If $f(x) = x + 1$ and $g(x) = x - 1$, plot the graph of the function $\frac{f}{g}$.

4. Plot the graph of the composite function $F(g)$, where F and g are the functions defined by $g(x) = x - 2$ and $F(x) = \frac{1}{x}\cdot$
5. If f, g, and h are functions, show that $f(g(h)) = (f(g))(h)$. This is the Associative Law for the Composition of Functions.
6. If f is a real-valued function, how would you define the function $3f$? How would you define $\sqrt{f}$?
7. The velocity v of a freely falling body depends on the distance s that it has fallen according to the equation $v = \sqrt{2gs}$, where g is the constant gravitational acceleration.

   (a) Using an s-axis and a v-axis, plot the dependent variable v as a function of the independent variable s.

   (b) If s depends on the time t according to the equation $s = \frac{1}{2}gt^2$, how does v depend on t?

   Note that the variable v in (a), which depends on s, is not the same function as the variable v in (b), which depends on t. Without knowing which is referred to, the meaning of the value of v at 2 is ambiguous.
8. If $w = u^2 + u + 1$, $u = x^2 + 2$, and $v = x - 1$, what is the value of each of the following functions at an arbitrary real number x?

   (a) $u + v$ (b) $w + v$ (c) wu.

9. If $F(x) = x^3 + x + 2$ and $u = x^2 + 1$ and $w = \dfrac{x+1}{x}$, then

   (a) $(F(u))(x) =$

   (b) $F(w(x)) =$

   (c) $(u + w)(x) =$

10. The equation $y = 2x + 1$ defines y as a function of x. It also defines x as a function of y. Describe the latter function in two ways.
11. Draw the graph of the function $f(x) = ax - 1$ for four different values of the constant a.
12. If f and g are two real-valued functions, give the definitions of the sum $f + g$ and the product fg in terms of ordered pairs.
13. Let f and g be two real-valued functions. In terms of domain f and domain g, what are:

    (a) domain $f(g)$? (b) domain $(f + g)$? (c) domain fg?

4. Limits and Continuity. Consider the function f defined by

$$f(x) = \frac{x^2 - 3x + 2}{x - 2}, \qquad x \neq 2.$$

The domain of f is the set of all real numbers with the exception of the number 2, which has been excluded because substitution of $x = 2$ in the expression for $f(x)$ yields the undefined term $\frac{0}{0}$. On the other hand, $x^2 - 3x + 2 = (x - 1)(x - 2)$ and

$$\frac{(x-1)(x-2)}{x-2} = x - 1, \qquad \text{provided } x \neq 2. \tag{1}$$

The proviso is essential. Without it, (1) is false because, if $x = 2$, the left side is undefined and the right side is equal to 1. We therefore obtain

$$f(x) = x - 1, \qquad x \neq 2.$$

The graph of the function $x - 1$ is a straight line L; so the graph of f is the punctured line obtained from L by omitting the one point (2, 1) (see Figure 22).

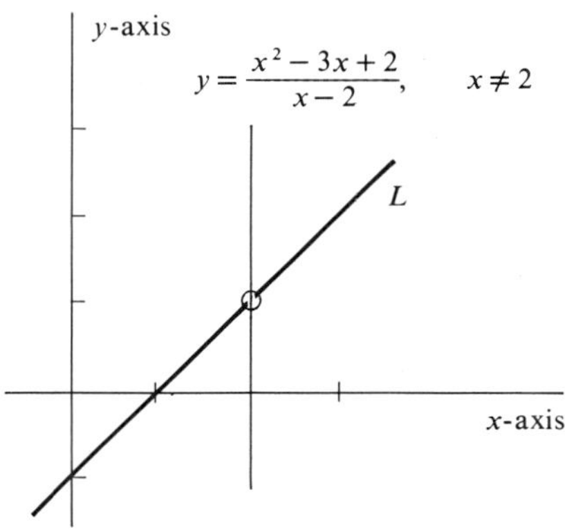

Figure 22

Although the function f is not defined at $x = 2$, we know its behavior for values of x near 2. The graph makes it clear that if x is close to 2, then $f(x)$ is close to 1. In fact, the values $f(x)$ can be brought arbitrarily close to 1 by taking x sufficiently close to 2. We express this fact by writing

$$\lim_{x \to 2} \frac{x^2 - 3x + 2}{x - 2} = 1,$$

which is translated: *The limit of* $\frac{x^2 - 3x + 2}{x - 2}$ *is* 1 *as x approaches* 2.

EXAMPLE 1. Evaluate $\lim_{x \to 3} \frac{\sqrt{x} - \sqrt{3}}{x - 3}$. The function $\frac{\sqrt{x} - \sqrt{3}}{x - 3}$ is not defined at $x = 3$. The following algebraic manipulation puts the function in a form in which its behavior close to 3 can be read off easily:

$$\frac{\sqrt{x} - \sqrt{3}}{x - 3} = \frac{\sqrt{x} - \sqrt{3}}{x - 3} \frac{\sqrt{x} + \sqrt{3}}{\sqrt{x} + \sqrt{3}} = \frac{x - 3}{x - 3} \frac{1}{\sqrt{x} + \sqrt{3}}$$

$$= \frac{1}{\sqrt{x} + \sqrt{3}}, \qquad \text{if } x \neq 3.$$

Again note the proviso $x \neq 3$: When $x = 3$, the last quantity in the preceding equations is equal to $\frac{1}{\sqrt{3}+\sqrt{3}}$, but the first quantity is not defined. However, by taking values of x close to 3, it is clear that the corresponding values of $\frac{1}{\sqrt{x}+\sqrt{3}}$ can be brought as close as we please to $\frac{1}{2\sqrt{3}}$. We conclude that

$$\lim_{x\to 3} \frac{\sqrt{x}-\sqrt{3}}{x-3} = \frac{1}{2\sqrt{3}}.$$

In words: The limit of $\frac{\sqrt{x}-\sqrt{3}}{x-3}$, as x approaches 3, is $\frac{1}{2\sqrt{3}}$.

EXAMPLE 2. If $f(x) = \frac{1}{x}$, evaluate $\lim_{x\to 0} f(x)$. The function f is not defined at 0 (i.e., the number 0 is not in the domain of f). From the graph of f and the list of ordered pairs $(x, f(x))$ shown in Figure 23, it is clear that there are values of x arbitrarily close to 0 for which the corresponding values of $f(x)$ are arbitrarily large in absolute value (see Table 5). We conclude that $\lim_{x\to 0} \frac{1}{x}$ does not exist.

TABLE 5

x	$f(x) = \frac{1}{x}$
1	1
0.1	10
0.01	100
0.001	1000
0.0001	10000
...	...
−1	−1
−0.1	−10
−0.01	−100
−0.001	−1000
−0.0001	−10000
...	...

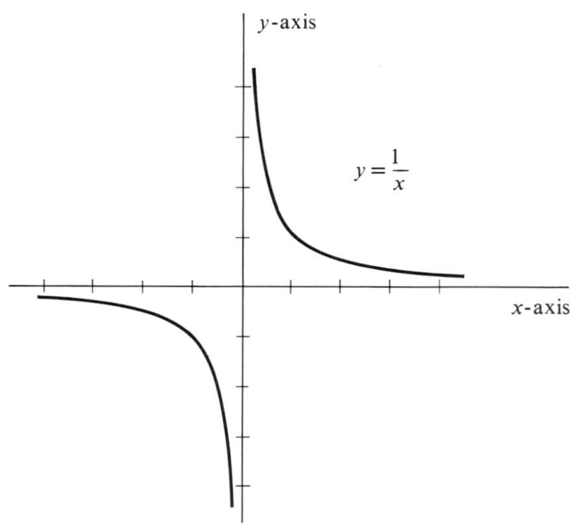

Figure 23

Thus far our examples have been confined to the problem of finding the limit of a function at a number which happens to lie outside the domain of the function. If it happens that the number a is in the domain of f, then it is frequently possible to determine $\lim_{x \to a} f(x)$ at a glance. Consider, for example, the function $f(x) = 2x^2 - x - 2$. As x takes on values closer and closer to 3, the corresponding value of $2x^2$ approaches 18, the value of $-x$ approaches -3, and the constant -2 does not change. We conclude that

$$\lim_{x \to 3} (2x^2 - x - 2) = 13,$$

or that, for this particular function, $\lim_{x \to 3} f(x) = f(3)$.

EXAMPLE 3. It would be incorrect to suppose that if a is in the domain of f, then it always happens that $\lim_{x \to a} f(x) = f(a)$. Consider the two functions f and g defined by

$$f(x) = \begin{cases} x^2 + 1 & \text{if } |x| > 0, \\ 2 & \text{if } x = 0, \end{cases}$$

$$g(x) = \begin{cases} x^2 + 1 & \text{if } x \geq 0, \\ -x^2 - 1 & \text{if } x < 0. \end{cases}$$

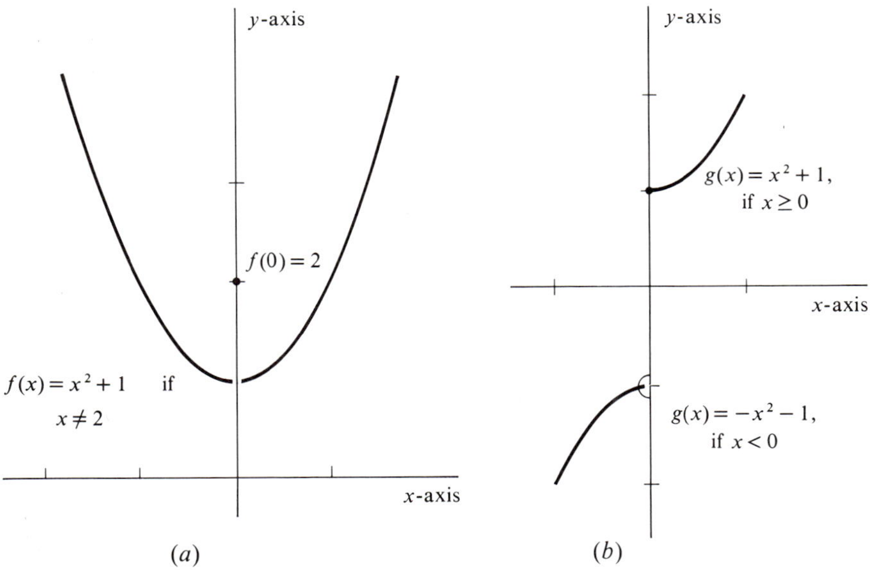

Figure 24

Both these functions are defined on the whole real line; i.e., domain f = domain $g = R$ (see Figure 24). Furthermore,

$$f(0) = 2 \qquad \text{and} \qquad g(0) = 1.$$

As x approaches 0, however, it is clear that $x^2 + 1$ approaches 1 and not 2. Hence

$$\lim_{x \to 0} f(x) = 1 \neq f(0).$$

[Note that in computing $\lim_{x \to a} f(x)$, we consider values of $f(x)$ for all x arbitrarily close to *a but not equal to a*. This point will be made explicit when we give the formal definition.] Turning to g, we see that the value of $g(x)$ near 0 depends on whether x is positive or negative. For any small positive number x, the corresponding number $g(x)$ is close to 1, but if x is small in absolute value and negative, then $g(x)$ is close to -1. Since there is no reason to prefer numbers of one sign to those of the other, we conclude that there is no limit. Thus

$$\lim_{x \to 0} g(x) \text{ does not exist.}$$

The reader may feel that Example 3 loses force because the functions used to make the point were in some sense artificial. There is some truth in the objection. Recall, however, that one of our major objectives is to reduce the class of all functions to those we wish to study in this course. After defining $\lim_{x \to a} f(x)$ precisely, we shall turn our point of view around and use this definition as the major tool in the problem of deciding what does constitute a well-behaved function.

The conceptual problems in trying to give an exact meaning to the expression $\lim_{x \to a} f(x) = b$ revolve around phrases such as "arbitrarily close," "sufficiently near," and "arbitrarily small." After all, there is no such thing in any absolute sense as a *small* positive real number. The number 0.000001 is small in most contexts, but in comparison with 0.000000000001 it is huge. However, we can assert that one number is *smaller than* another. Moreover, the actual closeness of one number x to another number a is just the distance between them: It is $|x - a|$. One way to say that a function f takes on values arbitrarily close to a number b is to state that, for any positive real number ϵ, there are numbers x such that $|f(x) - b| < \epsilon$. We are stating that no matter what positive number ϵ is selected, 10^{17}, or 10^{-17}, or 10^{-127}, there are numbers x so that the distance between $f(x)$ and b is smaller than ϵ. Thus the difficulty inherent in the phrase "arbitrarily close" has been circumvented by the prefix "for any." To finish the definition, we want to be able to say that $f(x)$ is arbitrarily close to b whenever x is sufficiently close, but not equal, to a. What does "sufficiently close" mean? The answer is this: If an arbitrary $\epsilon > 0$ is chosen with which to measure the distance between $f(x)$ and b, then it must be the case that there is a number $\delta > 0$ such that whenever x is in the domain of f and within a distance δ of a, but not equal to a, then the distance between $f(x)$ and b is less than ϵ. The situation is pictured in Figure 25. First $\epsilon > 0$ is chosen arbitrarily. There must then exist a number $\delta > 0$ such that whenever x lies in the interval $(a - \delta, a + \delta)$, and $x \neq a$ then the point $(x, f(x))$ lies in the shaded rectangle. We summarize by giving

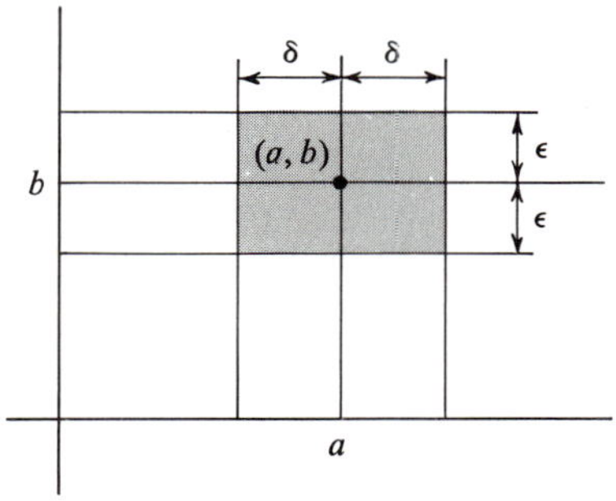

Figure 25

the definition: Let f be a real-valued function of a real variable. Then **the limit as x approaches a of $f(x)$ is b**, written

$$\lim_{x \to a} f(x) = b,$$

if, for any $\epsilon > 0$, there exists $\delta > 0$ such that whenever x is in the domain of f and $0 < |x - a| < \delta$, then $|f(x) - b| < \epsilon$. (There is a strong tradition for using the Greek letters ϵ and δ in the definition of limit. In fact, the part of calculus that deals with rigorous proofs of the various properties of limits is sometimes referred to as "epsilonics.")

EXAMPLE 4. The idea behind a formal definition can sometimes be grasped most easily by looking at an example where the condition is *not* satisfied. Consider the function g defined in Example 3 whose graph is drawn in Figure 24(b). We shall prove that $\lim_{x \to 0} g(x) \neq 1$. To do this, we must establish the negation of the limit condition: There is an $\epsilon > 0$ such that, for any $\delta > 0$, there is a number x in the domain of g such that $0 < |x| < \delta$ and $|g(x) - 1| \geq \epsilon$. There are many possible choices for ϵ. To be specific, take $\epsilon = \frac{1}{2}$. We must now show that for every positive number δ, there is a non-zero number x in the open interval $(-\delta, \delta)$ such that the distance between $g(x)$ and 1 is greater than or equal to $\frac{1}{2}$ (Figure 26). Take $x = -\dfrac{\delta}{2}$. This number is non-zero, lies in $(-\delta, \delta)$, and furthermore

$$g(x) = g\left(-\frac{\delta}{2}\right) = -\frac{\delta^2}{4} - 1 < -1.$$

Hence $|g(x) - 1| > 2 \geq \frac{1}{2}$.

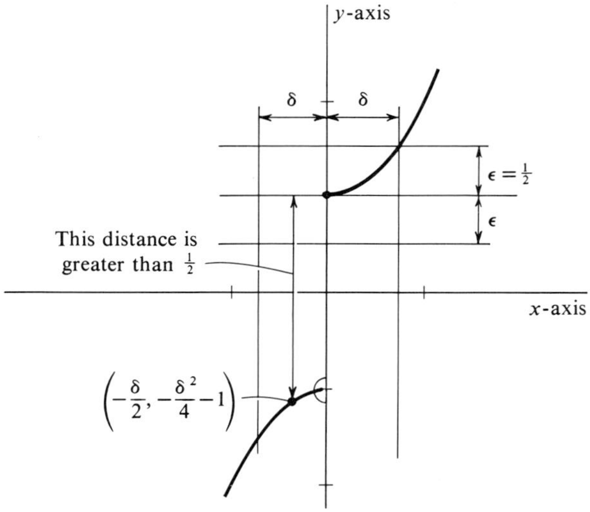

Figure 26

The basic limit theorem is the following:

(4.1) THEOREM. *If* $\lim_{x \to a} f(x) = b_1$ *and* $\lim_{x \to a} g(x) = b_2$, *then*

(i) $\lim_{x \to a} [f(x) + g(x)] = b_1 + b_2$.

(ii) $\lim_{x \to a} cf(x) = cb_1$.

(iii) $\lim_{x \to a} f(x)g(x) = b_1 b_2$.

(iv) $\lim_{x \to a} \dfrac{f(x)}{g(x)} = \dfrac{b_1}{b_2}$ *provided* $b_2 \neq 0$.

The proofs are given in Appendix A. They are not difficult, and (i) and (ii) especially follow directly from the definition of limit and the properties of the absolute value. Some ingenuity in algebraic manipulation is required for (iii) and (iv). Note that we have already assumed that this theorem is true. For example, the assertion that $\lim_{x \to 3} (2x^2 - x - 2) = 13$ is a corollary of (i), (ii), and (iii).

If a function f is defined for every x in R and if its graph contains no breaks, then it is apparent from looking at the graph that $\lim_{x \to a} f(x) = f(a)$. Logically, however, this intuitive point of view is backward. So far, we have constructed the graph of a function f by plotting a few isolated points and then joining them with a smooth curve. In so doing we are assuming that if

x is close to a, then $f(x)$ is close to $f(a)$. That is, we are assuming that $\lim_{x \to a} f(x) = f(a)$. Now that we have given a formal definition of limit, we shall reverse ourselves and use it to say precisely what is meant by a function whose graph has no breaks. Such a function is called continuous. The definitions are as follows: A real-valued function f of a real variable is **continuous at** a if a is in the domain of f and $\lim_{x \to a} f(x) = f(a)$. The function f is simply said to be **continuous** if it is continuous at every number in its domain.

A continuous function is one whose graph has no breaks, but the graph need not be a smooth curve. For example, the function with the saw-tooth graph shown in Figure 27 is continuous. The reader should also realize

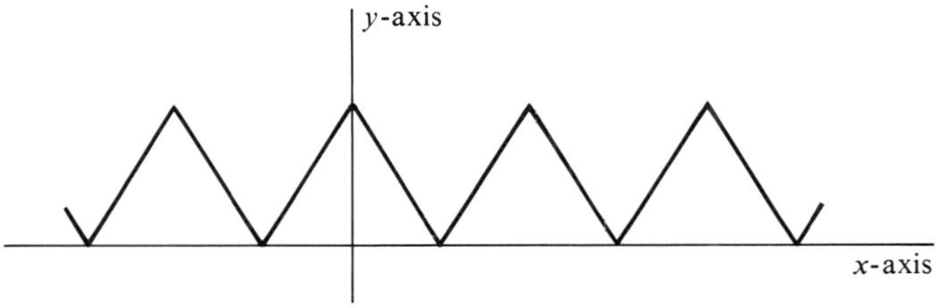

Figure 27

that many functions that are not continuous fail to be so at only a few isolated places. Thus the function f in Example 3, whose graph is drawn in Figure 24(a), has its only discontinuity at 0. It is continuous everywhere else. Finally, we emphasize the fact that there are two conditions in the definition of continuity. Even though $\lim_{x \to 1} \frac{x^3 - 1}{x - 1} = 3$, the function $\frac{x^3 - 1}{x - 1}$ is not continuous at $x = 1$ simply because it is not defined there.

If two functions f and g are continuous at a, then it is not difficult to prove that the sum $f + g$ is also continuous at a. To begin with, a is in the domain of $f + g$ since we have $(f + g)(a) = f(a) + g(a)$. Furthermore, we know that $\lim_{x \to a} f(x) = f(a)$ and that $\lim_{x \to a} g(x) = g(a)$. It follows by Theorem (4.1)(i) that

$$\lim_{x \to a} [f(x) + g(x)] = f(a) + g(a).$$

Since $f(x) + g(x) = (f + g)(x)$, we get

$$\lim_{x \to a} (f + g)(x) = (f + g)(a),$$

which proves the continuity of $f + g$ at a. The other parts of the basic limit theorem (4.1) imply similar results about the products and quotients of continuous functions. We summarize these in

(4.2) THEOREM. *If two functions f and g are continuous at a, then so are*

(i) $f + g$.

(ii) cf, *for any constant c.*

(iii) fg.

(iv) $\frac{f}{g}$, *provided* $g(a) \neq 0$.

A real-valued function f of one real variable is called a **polynomial** if there exist a nonnegative integer n and real numbers $a_0, a_1, \ldots, a_n$ such that, for every real number x,

$$f(x) = a_0 + a_1x + \cdots + a_nx^n.$$

The following functions are all examples of polynomials:

$$f(x) = 2 - 4x + 3x^2,$$
$$f(y) = 117y^{239} + \tfrac{3}{2}y + \pi,$$
$$f(x) = x,$$
$$f(x) = 5,$$
$$g(s) = (s^2 + 2)(s^5 - 1) = s^7 + 2s^5 - s^2 - 2.$$

It is equally important to be able to recognize that a given function is not a polynomial. Examples of functions which are not polynomials are

$$f(x) = |x|,$$
$$f(x) = \frac{1}{x},$$
$$f(x) = x^2 + x + 3x^{-2},$$
$$f(x) = \sqrt{x},$$
$$F(y) = (y^2 - 1)^{3/2}.$$

Algebraically the set of all polynomials is much like the set of integers: The sum, difference, and product of any two polynomials is again a polynomial, but, in general, the quotient of two polynomials is not a polynomial. Moreover, the algebraic axioms (i) through (v) listed in Section 1 also hold.

Just as a rational number is one which can be expressed as the ratio of two integers, a **rational function** is one which can be expressed as the ratio of two polynomials. Examples are the functions

$$f(x) = \frac{x^3 + 2x + 2}{x^4 + 1},$$

$$g(x) = x^{-3} = \frac{1}{x^3},$$

$$f(x) = x^2 + 2x + 1 = \frac{x^2 + 2x + 1}{1},$$

$$g(x) = \pi.$$

The domain of every polynomial is the entire set R of real numbers. Similarly, the domain of a given rational function $\frac{p(x)}{q(x)}$, where $p(x)$ and $q(x)$ are polynomials, is the whole set R with the exception of those numbers x for which $q(x) = 0$. Furthermore, we have

(4.3) *Every polynomial is a continuous function, and every rational function* $\frac{p(x)}{q(x)}$ *is continuous except at those values of x for which $q(x) = 0$.*

Proof. The identity function x is clearly continuous, and so is every constant function. Since every polynomial can be constructed from the identity function x and from constants using only the sums and products of these and the resulting functions, it follows from Theorem (4.2) that every polynomial is continuous. The assertion about the continuity of rational functions then follows from part (iv) of Theorem (4.2).

It is occasionally useful to modify the definition of $\lim_{x \to a} f(x)$ to allow x to approach a from only one side:

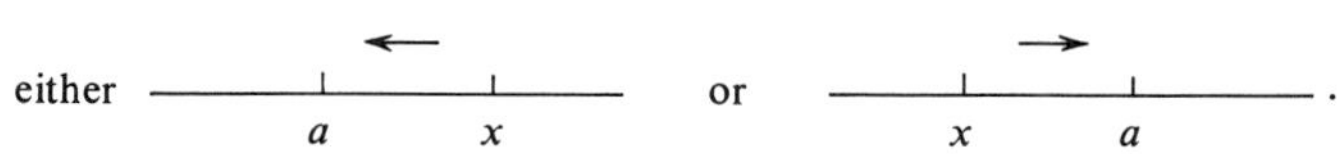

When this is done, we speak of either the limit from the right or the limit from the left and write either

$$\lim_{x \to a+} f(x) \qquad \text{or} \qquad \lim_{x \to a-} f(x),$$

according as the additional condition is $x > a$ or $x < a$. Thus for the function

$$f(x) = \begin{cases} x - 1, & x \geq 2, \\ x^2 - 1, & x < 2, \end{cases}$$

whose graph is shown in Figure 28, the limit of $f(x)$ as x approaches 2 does not exist. Nevertheless, we obtain

$$\lim_{x \to 2+} f(x) = 1, \qquad \lim_{x \to 2-} f(x) = 3.$$

Similarly, for the function g in Figure 24(b), we have $\lim_{x \to 0+} g(x) = 1$, $\lim_{x \to 0-} g(x) = -1$.

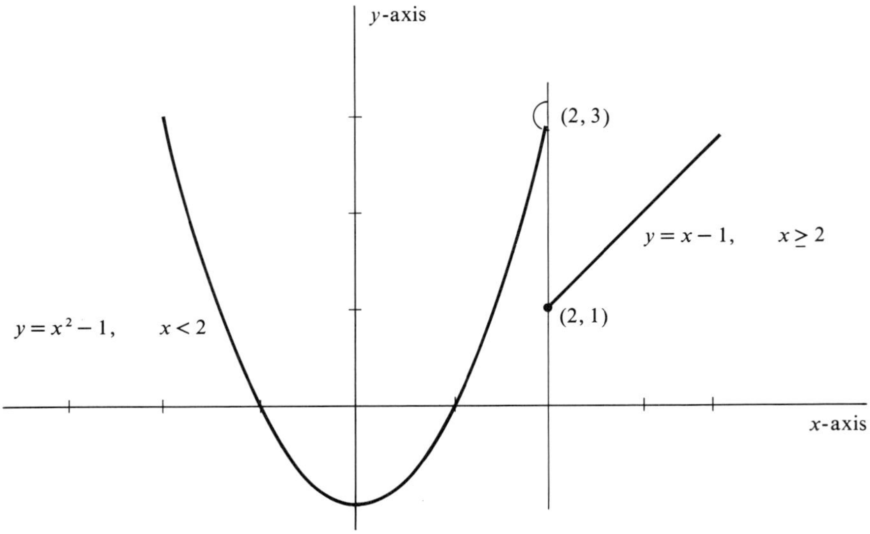

Figure 28

The graph of the rational function

$$f(x) = \frac{x+1}{x} = 1 + \frac{1}{x}, \qquad x \neq 0,$$

together with a list of some of the ordered pairs $(x, f(x))$ that comprise f is shown in Figure 29. From both Figure 29 and Table 6 it is clear that as x increases without bound, $f(x)$ becomes arbitrarily close to 1. We express this fact by writing

$$\lim_{x \to +\infty} \frac{x+1}{x} = 1.$$

TABLE 6

x	$f(x)$
1	2
10	1.1
100	1.01
1,000	1.001
1,000,000	1.000001
⋮	

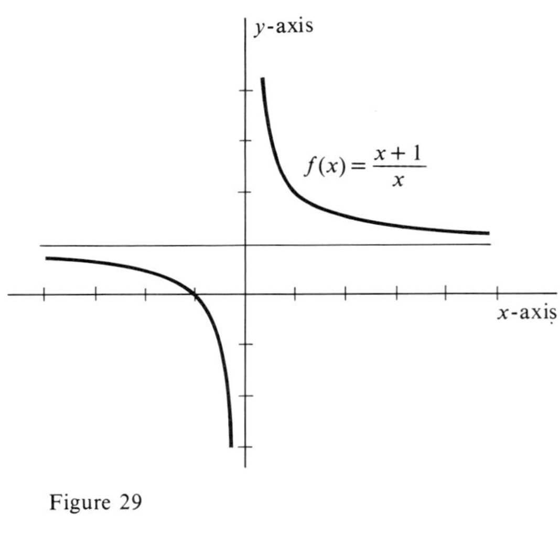

Figure 29

Since $f(x)$ also becomes arbitrarily close to 1 as x decreases without bound, i.e., as $-x$ increases without bound, we write

$$\lim_{x \to -\infty} \frac{x+1}{x} = 1.$$

The definition is as follows: Let f be a real-valued function of a real variable. Then **the limit of $f(x)$ is b as x increases without bound**, written

$$\lim_{x \to +\infty} f(x) = b,$$

if, for any $\epsilon > 0$, there exists $\delta > 0$ such that whenever x is in the domain of f and $\delta < x$, then $|f(x) - b| < \epsilon$. The analogous definition for $\lim_{x \to -\infty} f(x) = b$ is obvious.

The symbols $+\infty$ and $-\infty$ can also be used to refer to the behavior of the values of the function as well as the independent variable. If, as x approaches a, the corresponding value $f(x)$ of the function increases without bound, we may express the fact by writing

$$\lim_{x \to a} f(x) = +\infty.$$

The reader should be able to attach the correct meanings to the various other possibilities:

$$\lim_{x \to a} f(x) = -\infty,$$

$$\lim_{x \to -\infty} f(x) = +\infty,$$

$$\lim_{x \to a+} f(x) = -\infty, \text{ etc.}$$

It is essential to keep in mind that $+\infty$ *and* $-\infty$ *are not numbers*. They are not elements of R. They are used simply as convenient abbreviations for describing the unbounded characteristics of certain functions. The symbol $+\infty$ (or simply ∞) in an expression for a bound will always mean that the quantity referred to increases without limit in the positive direction. Similarly, $-\infty$ always indicates the negative direction. Thus we shall not say $\lim_{x \to 0} \frac{1}{x} = \infty$. But we do say

$$\lim_{x \to 0+} \frac{1}{x} = \infty,$$

$$\lim_{x \to 0-} \frac{1}{x} = -\infty,$$

$$\lim_{x \to 0} \left|\frac{1}{x}\right| = \infty.$$

PROBLEMS

1. Compute the following limits.

(a) $\lim_{x\to 1} \dfrac{x^3 - 1}{x - 1}$

(b) $\lim_{x\to 2} \dfrac{x^2 - 5x + 6}{x - 2}$

(c) $\lim_{x\to 3} \dfrac{x^2 - 5x + 6}{x - 3}$

(d) $\lim_{x\to 1} \left(\dfrac{x^2}{x - 1} - \dfrac{1}{x - 1}\right)$

(e) $\lim_{x\to 0} x^{3/2}$

(f) $\lim_{x\to 0} \dfrac{\sqrt{x + 1} - 1}{x}$

(g) $\lim_{x\to 0} |x|$

(h) $\lim_{h\to 0} \left(\dfrac{1 - h^2}{h^2} + \dfrac{6h^2 - 1}{h^2}\right)$

(i) $\lim_{x\to 0} \dfrac{(a + x)^3 + 2(a + x) - a^3 - 2a}{x}$

(j) $\lim_{h\to 0} \dfrac{2(x + h)^2 - (x + h) - 2x^2 + x}{h}$.

2. For each of the following functions, find those numbers (if any) at which the function is not continuous.

(a) $x^3 + 3x - 1$

(b) $f(x) = |x|$

(c) $\dfrac{x^3 + x + 1}{x^2 - x - 2}$

(d) $g(x) = \dfrac{x^3 - 3x - 2}{x - 2}$

(e) $\sqrt{x + 3}$

(f) $h(x) = \dfrac{|x|}{x}$

(g) $f(x) = \begin{cases} |x|, & |x| \le 1 \\ 2 - x^2, & |x| > 1 \end{cases}$

(h) $F(x) = \begin{cases} 1 & \text{if } x \text{ is rational} \\ 0 & \text{if } x \text{ is irrational} \end{cases}$

(i) $f(x) = \begin{cases} x^2 + 2x + 1 & \text{if } x \ne 1 \\ 1 & \text{if } x = 1 \end{cases}$

3. A function f is said to have a **removable discontinuity** at a if it is not continuous at a, but can be assigned a value $f(a)$ [or possibly reassigned a new value $f(a)$] such that it becomes continuous there.

(a) Locate the removable discontinuities in Problem 2.

(b) Show that the only discontinuities a rational function can have are either removable or infinite. That is, if $r(x)$ is a rational function that is not continuous at a, show that either a is a removable discontinuity or $\lim_{x\to a} |r(x)| = +\infty$.

4. Using Theorem (4.1), prove that if $\lim_{x\to a} f(x) = b_1$ and $\lim_{x\to a} g(x) = b_2$, then

$$\lim_{x\to a} [f(x) - g(x)] = b_1 - b_2.$$

5. Show that

(a) $\lim_{x \to +\infty} f(x) = b$ if and only if $\lim_{t \to 0+} f\left(\frac{1}{t}\right) = b$.

(b) $\lim_{x \to -\infty} f(x) = b$ if and only if $\lim_{t \to 0-} f\left(\frac{1}{t}\right) = b$.

6. Using Problem 5, compute

(a) $\lim_{x \to \infty} \frac{1}{1 + x}$ (c) $\lim_{t \to -\infty} \frac{4t^2 - 3t + 1}{t^2}$

(b) $\lim_{x \to \infty} \frac{3x + 1}{x}$ (d) $\lim_{t \to \infty} \frac{3t^3 + 7t^2 - 2}{t^3 + 1}$.

7. True or false?

(a) If $\lim_{x \to a} f(x) = b$, then $\lim_{x \to a+} f(x) = b$ and $\lim_{x \to a-} f(x) = b$.

(b) If $\lim_{x \to a+} f(x) = b$ and $\lim_{x \to a-} f(x) = b$, then $\lim_{x \to a} f(x) = b$.

8. Define a function f and draw its graph such that $\lim_{x \to 2+} f(x) = 2$ and $\lim_{x \to 2-} f(x) = 0$.

9. Compute

(a) $\lim_{x \to 2} \frac{1}{x^2 - 4x + 4}$ (c) $\lim_{x \to 3+} \frac{|x| - 3}{x - 3}$

(b) $\lim_{x \to 2} \frac{x}{x^2 + 3x - 10}$ (d) $\lim_{x \to 1+} \frac{\sqrt{x - 1}}{x - 1}$.

10. Does the set of rational functions satisfy axioms (i) through (vi) in Section 1? (*Hint:* Be careful; note Problem 3.)
11. Give the formal definition in terms of inequalities of $\lim_{x \to a+} f(x) = b$.
12. Define a function f and draw its graph such that $\lim_{x \to \infty} f(x) \neq \lim_{x \to -\infty} f(x)$, although both limits exist.
13. Prove that it is impossible to choose a rational function in Problem 12.
14. Give the formal definition in terms of inequalities and absolute values of

(a) $\lim_{x \to a} f(x) = \infty$ (b) $\lim_{x \to -\infty} f(x) = +\infty$.

5. Straight Lines and Their Equations. We shall define a **straight line** in R^2 to be any subset L consisting of all ordered pairs (x, y) such that

$$\boxed{ax + by + c = 0, \qquad \text{where } a^2 + b^2 > 0.} \tag{1}$$

The inequality $a^2 + b^2 > 0$ simply says that the constants a and b are not both equal to zero. Of course two different equations can define the same line. For example, the set of all ordered pairs (x, y) such that $4x - 3y + 5 = 0$ is the same line as the set of pairs for which $28x = 21y - 35$. For this reason, we speak of *an* equation of a straight line and not *the* equation.

(5.1) *Suppose that straight lines L_1 and L_2 are defined, respectively, by*

$$a_1x + b_1y + c_1 = 0, \qquad a_1^2 + b_1^2 > 0,$$
$$a_2x + b_2y + c_2 = 0, \qquad a_2^2 + b_2^2 > 0.$$

Then $L_1 = L_2$ if and only if there is a nonzero constant k such that

$$a_2 = ka_1,$$
$$b_2 = kb_1,$$
$$c_2 = kc_1.$$

Proof. If such a k exists, then the two equations are equivalent, and so $L_1 = L_2$. Conversely, suppose that $L_1 = L_2$. We may assume without loss of generality that $b_1 \neq 0$. Then the point $\left(0, -\frac{c_1}{b_1}\right)$ lies on L_1 since it satisfies the first equation; i.e.,

$$a_1 \cdot 0 + b_1\left(-\frac{c_1}{b_1}\right) + c_1 = 0.$$

Because the two lines are equal, the point also lies on L_2, and so

$$a_2 \cdot 0 + b_2\left(-\frac{c_1}{b_1}\right) + c_2 = 0.$$

Hence

$$c_2 = \left(\frac{b_2}{b_1}\right) c_1.$$

In addition, the point $\left(1, -\frac{a_1 + c_1}{b_1}\right)$ lies on L_1 because

$$a_1 + b_1\left(\frac{-a_1 - c_1}{b_1}\right) + c_1 = 0.$$

This point then also lies on L_2, and this fact means that

$$a_2 + b_2\left(\frac{-a_1 - c_1}{b_1}\right) + c_2 = 0.$$

Hence

$$a_2 = \frac{b_2}{b_1}a_1 + \frac{b_2}{b_1}c_1 - c_2 = \left(\frac{b_2}{b_1}\right) a_1.$$

Since $b_2 = \left(\frac{b_2}{b_1}\right) b_1$ trivially, we obtain the desired conclusion by setting $k = \frac{b_2}{b_1}$. Note that $k \neq 0$, for if it were zero, we would get $a_2 = b_2 = 0$, contrary to assumption.

One consequence of Theorem (5.1) is that it enables us to recognize at a glance whether or not different equations define the same straight line. Another corollary arises in connection with the following definitions: A line L defined by an equation $ax + by + c = 0$ with $a^2 + b^2 > 0$ will be called **vertical** if $b = 0$ and **horizontal** if $a = 0$. It follows from the theorem that b must equal zero for every such equation which defines a vertical line and that a must equal zero for every such equation which defines a horizontal line. Thus the definitions are not dependent on the particular equation which defines L.

If $P = (a, b)$ and $Q = (c, d)$ are two points in R^2 and $a \neq c$, the **slope of the line segment** joining P to Q is, by definition,

$$\boxed{m(P, Q) = \frac{d - b}{c - a}.}$$

Note that

$$m(P, Q) = \frac{d - b}{c - a} = \frac{b - d}{a - c} = m(Q, P).$$

The absolute value of $m(P, Q)$ is the ratio of the vertical to horizontal distance between P and Q (see Figure 30). It is simply a measure of steepness. A

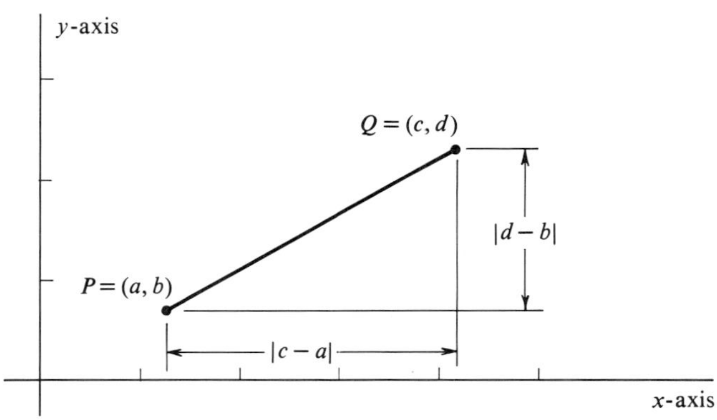

Figure 30

segment with positive slope goes up as it goes to the right; one with negative slope goes down as it goes to the right (Figure 31). If $a = c$, the segment is vertical, and the slope is not defined.

(5.2) *Let L be the straight line defined by the equation $ax + by + c = 0$, where $b \neq 0$. If P and Q are any two distinct points on the line, then*

$$m(P, Q) = -\frac{a}{b}.$$

Proof. Let $P = (x_1, y_1)$ and $Q = (x_2, y_2)$. An equation equivalent to the original one is

$$y = -\left(\frac{a}{b}\right)x - \frac{c}{b}. \tag{1}$$

It follows that $x_1 \neq x_2$, since, otherwise, substitution in this equation would yield $y_1 = y_2$, which would then imply $P = Q$. We obtain

$$m(P, Q) = \frac{y_2 - y_1}{x_2 - x_1} = \frac{-\frac{a}{b}x_2 - \frac{c}{b} + \frac{a}{b}x_1 + \frac{c}{b}}{x_2 - x_1} = -\frac{a}{b},$$

and this completes the proof.

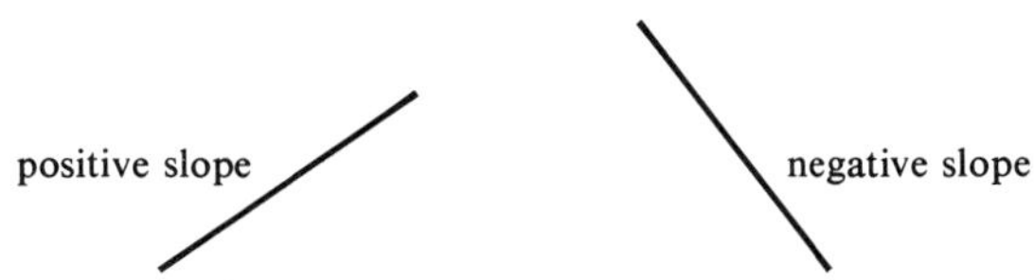

Figure 31

As a result of Theorems (5.1) and (5.2), we can unambiguously define the **slope of a nonvertical line** L, which we shall denote by m_L, as follows: For any pair of distinct points P and Q on L, we define

$$\boxed{m_L = m(P, Q).}$$

It follows at once that m_L depends only on the line L. For if P' and Q' are any other two distinct points on the line, then

$$m(P, Q) = -\frac{a}{b} = m(P', Q').$$

(Since L is not vertical, $b \neq 0$.) Furthermore, any other equation defining L can be written $kax + kby + kc = 0$ with $k \neq 0$, and, of course, $-\dfrac{ka}{kb} = -\dfrac{a}{b}$. We note that the slope of a vertical line is not defined.

EXAMPLE 1. Find an equation of the straight line L through the point (a, b) and with slope m. If (x, y) is any other point on the line, then

$$m = \frac{y - b}{x - a},$$

which implies

$$\boxed{y - b = m(x - a).} \tag{2}$$

This is an equation of the line. For suppose L were defined by some equation $a_1x + b_1y + c_1 = 0$. An equivalent equation is

$$y = -\left(\frac{a_1}{b_1}\right)x - \frac{c_1}{b_1},$$

or, since $m = -\dfrac{a_1}{b_1}$,

$$y = mx - \frac{c_1}{b_1}. \tag{3}$$

Since we are given that (a, b) lies on L, we get $b = ma - \dfrac{c_1}{b_1}$, or

$$\frac{c_1}{b_1} = ma - b.$$

Substitution in (3) yields $y = mx - ma + b$, which is equivalent to (2).

Suppose that S is an arbitrary subset of R^2 with the following three properties:

(i) S contains a point (a, b); i.e., S is a nonempty set.

(ii) The slope $m(P, Q)$ is defined and is equal to the same fixed number m, for every pair of distinct points P and Q in S.

(iii) S contains every point (x, y) in R^2 which is connected to (a, b) by a line segment of slope m.

These are certainly the geometric properties of a nonvertical straight line. It follows from (i) and (ii) that the coordinates of every point (x, y) in S satisfy the equation

$$y - b = m(x - a). \tag{4}$$

Conversely, it follows from (iii) that, for every pair of real numbers x and y which satisfy (4), the point (x, y) must lie in S. Thus the set S is the graph of (4), and, as such, it is a straight line. Since nonvertical straight lines, as we have defined them, have the above three properties, we conclude that our definition coincides with the natural geometric one.

We define two lines L_1 and L_2 to be **parallel** if they are both vertical or if they have the same slope.

The following fact, which we shall prove later using trigonometry, can also be deduced from Figure 32 by the methods of plane geometry.

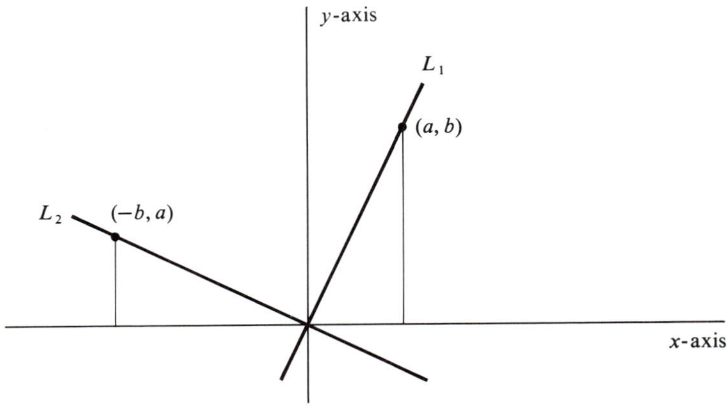

Figure 32

(5.3) *Two nonvertical lines L_1 and L_2 with slopes m_1 and m_2, respectively, are perpendicular if and only if $m_1m_2 = -1$.*

EXAMPLE 2. (a) Write an equation of the straight line L_1 that passes through $(-2, 4)$ and $(3, 7)$. (b) Write an equation of the line L_2 passing through $(5, -2)$ and parallel to L_1. (c) Write an equation defining the line L_3 that passes through $(-1, -3)$ and is perpendicular to L_1.

The slope of the segment joining $(-2, 4)$ and $(3, 7)$ is $\dfrac{7-4}{3+2} = \dfrac{3}{5}$. An arbitrary point (x, y) other than $(3, 7)$ belongs to L_1 if and only if

$$\frac{y-7}{x-3} = \frac{3}{5}.$$

Hence an equation defining L_1 is $5(y - 7) = 3(x - 3)$, or, equivalently,

$$3x - 5y + 26 = 0.$$

The line L_2 also has slope $\frac{3}{5}$. Since it passes through $(5, -2)$, it is defined by

$$\frac{y+2}{x-5} = \frac{3}{5} \qquad \text{if } x \neq 5,$$

or, more generally, by $5(y + 2) = 3(x - 5)$, which is equivalent to

$$3x - 5y - 25 = 0.$$

The slope of the perpendicular is $-\frac{5}{3}$. Hence we obtain the equation

$$\frac{y+3}{x+1} = -\frac{5}{3}, \qquad x \neq -1,$$

or $3y + 9 = -5x - 5$, as an equation of L_3.

What functions have graphs that are straight lines? The answer is an easy one. If f is defined by

$$f(x) = ax + b, \qquad -\infty < x < \infty,$$

then its graph, which is the set of all ordered pairs (x, y) such that $y = ax + b$, is certainly a straight line. Conversely, if the graph of an arbitrary function f is a straight line, then the equation $y = f(x)$ is equivalent to one of the form

$$a_1x + b_1y + c_1 = 0, \qquad a_1^2 + b_1^2 > 0. \tag{5}$$

If b_1 were zero, both points $\left(-\frac{c_1}{a_1}, 0\right)$ and $\left(-\frac{c_1}{a_1}, 1\right)$ would satisfy (5), but the definition of function makes this impossible for the equation $y = f(x)$. We conclude that $b_1 \neq 0$ and that (5) is therefore equivalent to

$$y = -\left(\frac{a_1}{b_1}\right)x - \frac{c_1}{b_1}.$$

It follows [see Theorem (2.4), page 15] that the functions $f(x)$ and $-\left(\frac{a_1}{b_1}\right)x - \frac{c_1}{b_1}$ are equal. Thus the functions whose graphs are straight lines are precisely those of the form $ax + b$. These are the polynomials of degree less than 2, the **linear functions.**

PROBLEMS

1. For each of the following lines, find an equation that defines it.

   (a) The line through (2, 3) with slope 1.

   (b) The line through (0, 1) with slope 1.

   (c) The line through (0, 1) with slope -2.

   (d) The line through $(-1, -3)$ with slope $-\frac{1}{2}$.

   (e) The line through $(-2, 1)$ and $(-1, -1)$.

   (f) The line containing the points (1, 0) and (0, 1).

   (g) The line through the origin containing the point $(1, -19)$.

(h) The line with slope 0 that passes through (3, 4).

(i) The line through (2, 5) and (2, 8).

2. Draw the line defined by each of the following equations, and find the slope.

(a) $x + y = 1$
(b) $x = -y$
(c) $2x - 4y = 3$
(d) $7x = 3$
(e) $7y = 3$
(f) $4x + 3y = 10$.

3. Determine whether P, Q, and R lie on a line. If they do, draw the line and write an equation for it.

(a) $P = (0, 0)$, $Q = (-1, 3)$, $R = (3, -4)$.
(b) $P = (\frac{1}{2}, \frac{3}{2})$, $Q = (\frac{5}{2}, -\frac{7}{2})$, $R = (-\frac{3}{2}, \frac{13}{2})$.
(c) $P = (a_1, a_2)$, $Q = (b_1, b_2)$, $R = (c_1, c_2)$.

4. Draw the set of all ordered pairs (x, y) such that

(a) $4x^2 + 4xy + y^2 + 12x + 6y + 9 = (2x + y + 3)^2 = 0$.
(b) $5x^2 + 7xy + 2y^2 + 3x + 3y = (5x + 2y + 3)(x + y) = 0$.

5. The x-coordinate of a point where a curve intersects the x-axis is called an **x-intercept** of the curve. The definition of a **y-intercept** is analogous.

(a) Find the x- and y-intercept of the line defined by $y - 3x = 10$. Draw the line.
(b) Write an equation for the line with slope m and y-intercept equal to b.

6. For each of the following equations, define the function $f(x)$ whose graph is the set of ordered pairs that satisfy the equation. Which ones are linear functions?

(a) $3x - y = 7$
(b) $5y = 3$
(c) $2|x| + 3y = 4$
(d) $x - y = 1$
(e) $y^2 + 2x + 3 = 0$ (two functions)
(f) $x^2 - 2xy + y^2 = 0$
(g) $y = 3x^2 + 4x + 2$
(h) $5x + 3y = 1$.

7. Among the lines defined by the following equations, which pairs are parallel and which perpendicular?

(a) $4x + 2y = 13$
(b) $3x - 6y = 0$
(c) $3x + 2y = 6$
(d) $y = -2x$
(e) $4x = 13$
(f) $4y = 13$.

8. (a) Write an equation of the straight line L_1 that contains the points (1, 3) and (3, −2).

(b) Write an equation of the line with x-intercept 1 that is parallel to L_1.

(c) Write an equation of the line perpendicular to L_1 that passes through (1, 3).

9. Prove that the two lines L_1 and L_2 in Figure 32 are perpendicular. (*Hint:* Use congruent right triangles or the converse of the Pythagorean Theorem.)

6. The Derivative. The concept of the line tangent to a curve at a point is an important one in geometry. However, it is not so simple an idea as it may first appear. Consider the graph of a function f and a point $P = (a, f(a))$ on the graph, as illustrated in Figure 33. Many people who would have little difficulty drawing the line tangent to the graph at P would not find it easy to give an accurate definition of the tangent line. For example, to say that the tangent line at P is the line which cuts the graph at the single point P, although true for some curves, is obviously not correct in general (in particular, see Figure 33). We shall show that the problem of defining the tangent line to the graph of f at P can be expressed in purely analytic terms involving the function f. In fact, the problem leads directly to the definition of the derivative of a function, the central idea in differential calculus.

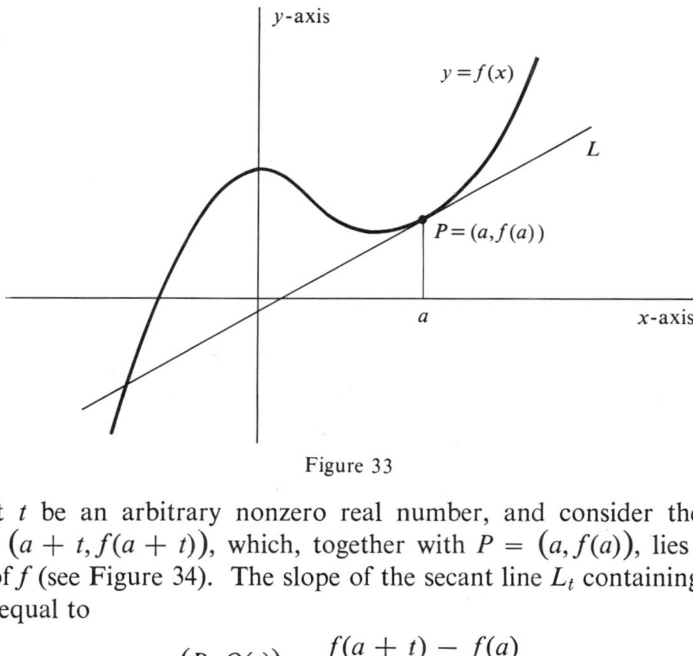

Figure 33

Let t be an arbitrary nonzero real number, and consider the point $Q(t) = (a + t, f(a + t))$, which, together with $P = (a, f(a))$, lies on the graph of f (see Figure 34). The slope of the secant line L_t containing P and $Q(t)$ is equal to

$$m(P, Q(t)) = \frac{f(a + t) - f(a)}{t}. \tag{1}$$

If t is small in absolute value, then L_t is an approximation to what we shall define to be the tangent line. The smaller the value of $|t|$, the better the approximation will be. In some sense, therefore, we would like to define the tangent line L to be the limit, as t approaches zero, of the lines L_t. We can do this, for although we have not defined a limit of lines, we have defined limits for functions, and hence we can express the limit of the slope of L_t. According to equation (1), it is given by

$$\lim_{t \to 0} m(P, Q(t)) = \lim_{t \to 0} \frac{f(a + t) - f(a)}{t}. \tag{2}$$

We shall define the **tangent line** to the graph of f at P to be the line through P having this limit as its slope, provided the limit exists.

Leaving the geometric interpretation aside for the moment, we observe that the value of the limit in (2) depends only on the function f and on the number a. Hence we give the following definitions: An arbitrary real-valued

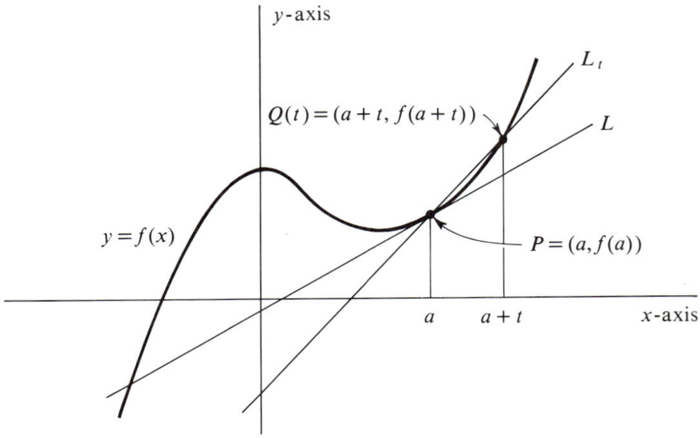

Figure 34

function f of a real variable is **differentiable at** a number a in its domain if

$$\lim_{t \to 0} \frac{f(a+t) - f(a)}{t}$$

exists (i.e., is finite). The **derivative of f at a**, denoted $f'(a)$, is this limit. Thus

$$f'(a) = \lim_{t \to 0} \frac{f(a+t) - f(a)}{t}.$$

If f is differentiable at every number in its domain, it is simply called a **differentiable function**.

Thus the slope of the line tangent to the graph of f at the point $(a, f(a))$ is equal to the derivative $f'(a)$. It follows that an arbitrary point (x, y) lies on this line if and only if

$$y - f(a) = f'(a)(x - a),$$

and we therefore obtain the following equation of the tangent line:

$$y = f(a) + f'(a)(x - a).$$

Note that the only variables that appear in this equation are x and y, and these occur with exponent 1. The equation therefore defines y as a linear function of x.

EXAMPLE 1. Find the derivative of the function $f(x) = x^2 + 2$ at $x = 2$, and write an equation of the line tangent to the graph of f at the point $(2, 6)$. As we have seen above, the slope of the tangent line is the derivative $f'(2)$, and

$$f'(2) = \lim_{t \to 0} \frac{f(2 + t) - f(2)}{t}.$$

We have $f(2) = 6$, and $f(2 + t) = (2 + t)^2 + 2 = t^2 + 4t + 6$. Hence

$$\frac{f(2 + t) - f(2)}{t} = \frac{t^2 + 4t}{t} = t + 4, \qquad \text{if } t \neq 0.$$

So

$$f'(2) = \lim_{t \to 0} (t + 4) = 4.$$

The tangent line passes through $(2, 6)$ and has slope 4. Hence (x, y) lies on the tangent if

$$\frac{y - 6}{x - 2} = 4, \qquad x \neq 2,$$

and we therefore obtain

$$y - 6 = 4(x - 2) \qquad \text{or} \qquad 4x - y - 2 = 0,$$

as an equation of the line.

EXAMPLE 2. Consider the function g defined by

$$g(x) = \frac{1}{x + 2}, \qquad x \neq -2.$$

Compute the derivative $g'(3)$. By definition,

$$g'(3) = \lim_{t \to 0} \frac{g(3 + t) - g(3)}{t}.$$

We have $g(3) = \frac{1}{5}$, and $g(3 + t) = \dfrac{1}{t + 5}$.

$$\begin{aligned} \frac{g(3 + t) - g(3)}{t} &= \frac{1}{t}\left(\frac{1}{t + 5} - \frac{1}{5}\right) = \frac{5 - (t + 5)}{5t(t + 5)} \\ &= -\frac{t}{5t(t + 5)} \\ &= -\frac{1}{5(t + 5)}, \qquad \text{if } t \neq 0. \end{aligned}$$

We conclude that

$$g'(3) = \lim_{t \to 0} \left(- \frac{1}{5(t+5)} \right) = - \frac{1}{25} \cdot$$

EXAMPLE 3. Find $F'(a)$, where $a > 0$ and F is the function

$$F(x) = \frac{1}{\sqrt{x}}, \qquad 0 < x < \infty,$$

and write an equation of the line tangent to the graph of F at the point $(4, \frac{1}{2})$. By the definition of the derivative,

$$F'(a) = \lim_{t \to 0} \frac{F(a+t) - F(a)}{t} \cdot$$

In this case,

$$\frac{F(a+t) - F(a)}{t} = \frac{1}{t} \left(\frac{1}{\sqrt{a+t}} - \frac{1}{\sqrt{a}} \right) \cdot$$

The problem in computing any derivative from the definition is always the same. We set up the fraction $\dfrac{F(a+t) - F(a)}{t}$ and then compute the limit. To begin with, we are faced with a fraction both the numerator and denominator of which approach zero. The limit we seek is the relative rate at which numerator and denominator go to zero. With most examples it is not possible to tell from a cursory glance just what that relative rate is. So we experiment, performing various algebraic manipulations that hopefully will finally change the fraction into a form from which we can tell what the limit is. In the present example the following manipulation will do the trick:

$$\begin{aligned} \frac{1}{t} \left(\frac{1}{\sqrt{a+t}} - \frac{1}{\sqrt{a}} \right) &= \frac{1}{t} \frac{\sqrt{a} - \sqrt{a+t}}{\sqrt{a}\sqrt{a+t}} \frac{\sqrt{a} + \sqrt{a+t}}{\sqrt{a} + \sqrt{a+t}} \\ &= \frac{1}{t} \frac{a - (a+t)}{\sqrt{a}\sqrt{a+t}(\sqrt{a} + \sqrt{a+t})} \\ &= \frac{-1}{\sqrt{a^2 + at}(\sqrt{a} + \sqrt{a+t})}, \qquad \text{if } t \neq 0. \end{aligned}$$

It is now possible to see what happens as $t \to 0$.

$$F'(a) = \lim_{t \to 0} \frac{-1}{\sqrt{a^2 + at}(\sqrt{a} + \sqrt{a+t})} = \frac{-1}{2a\sqrt{a}} = - \frac{1}{2a^{3/2}} \cdot$$

Our principal interpretation of the derivative $F'(a)$ is that it is the slope of the line tangent to the graph of F at the point $(a, F(a))$. For this particular function F, an equation of the tangent line at $(4, \frac{1}{2})$ is therefore found by writing

$$\frac{y - \frac{1}{2}}{x - 4} = F'(4) = -\frac{1}{2 \cdot 4^{3/2}} = -\frac{1}{16}.$$

Hence an equation of the tangent is

$$y - \tfrac{1}{2} = -\tfrac{1}{16}(x - 4).$$

The notation $f'(a)$ for the derivative suggests that we regard f' as a new function whose value at a is the number $f'(a)$. The domain of f' is the set of all real numbers a for which $\lim_{t \to 0} \frac{f(a + t) - f(a)}{t}$ exists. With this point of view, it is natural to think of the derivative evaluated not only at an arbitrary, but fixed, number a but also at a variable x. In so doing, we are admitting the same dual interpretations that were discussed in Section 3. That is, we can interpret $f'(x)$ either as the value of the function f' at the number x, whence

$$f'(x) = \lim_{t \to 0} \frac{f(x + t) - f(x)}{t},$$

or as the composition of the variable x with the function f'.

EXAMPLE 4. If $f(x) = x^3 - 1$, plot the graph of the derived function f'. For any real number x,

$$f'(x) = \lim_{t \to 0} \frac{f(x + t) - f(x)}{t}.$$

We have

$$\begin{aligned} f(x + t) - f(x) &= \big((x + t)^3 - 1\big) - (x^3 - 1) \\ &= 3x^2 t + 3xt^2 + t^3, \end{aligned}$$

and so

$$\frac{f(x + t) - f(x)}{t} = 3x^2 + 3xt + t^2, \qquad \text{if } t \neq 0.$$

Consequently,

$$f'(x) = \lim_{t \to 0} (3x^2 + 3xt + t^2) = 3x^2.$$

The graph of the function $f'(x) = 3x^2$ is the parabola shown in Figure 35, on which the graph of the original function $f(x) = x^3 - 1$ has also been drawn.

x	$3x^2$	$x^3 - 1$
0	0	−1
1	3	0
2	12	7
−1	3	−2
−2	12	−9

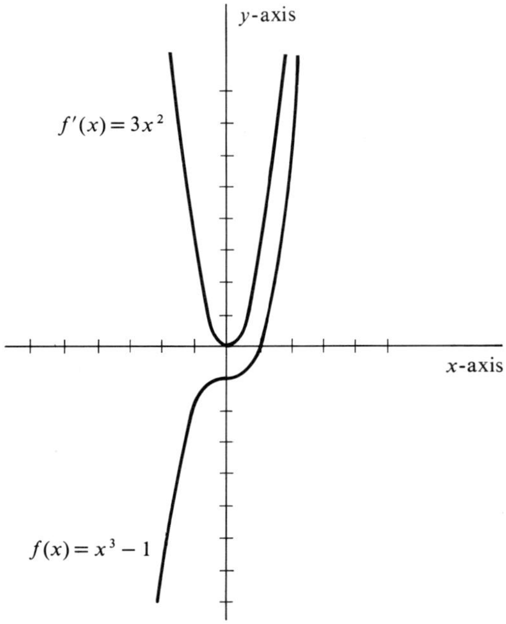

Figure 35

It is not surprising that there are several common notations for the derivative of a function. One strong tradition reflects the basic fact that the derivative is the limit of a ratio by writing it as a ratio. Thus

$$\frac{df}{dx} = f'.$$

This way of writing the derivative is called the differential notation. Using it, we denote the derivative of f at a by

$$\frac{df}{dx}(a) = f'(a).$$

EXAMPLE 5. Let $f(x) = x^3 - 1$. It was shown in Example 4 that $f'(x) = 3x^2$. Each of the following equations is an example of acceptable notation.

$$\frac{df}{dx}(2) = 3 \cdot 2^2 = 12,$$

$$\frac{df}{dx}(a) = 3a^2,$$

$$\frac{df}{dx} = 3x^2,$$

$$\frac{d}{dx}(x^3 - 1) = 3x^2.$$

One could also write $\dfrac{df}{dx}(x) = f'(x) = 3x^2$. There is no need for it, however, since $f'(x)$ becomes identified with f' when it is regarded as the composition of the independent variable x with the function f'.

It should be emphasized that although the notation $\dfrac{df}{dx}$ suggests a ratio, the derivative as we have defined it is *not* a ratio—even though it is the limit of one. $\dfrac{df}{dx}$ is simply an abbreviation of f'.

There are a few variations on the two notations that we have given for the derivative which we shall also use frequently. If $y = f(x)$, we may write any one of

$$y' = \frac{dy}{dx} = f' = \frac{df}{dx}$$

for the derivative. Similarly, for the derivative at a real number a, we have

$$y'(a) = \frac{dy}{dx}(a) = f'(a) = \frac{df}{dx}(a).$$

Still other notations for the derivative, which we shall seldom use, but which the reader may encounter in other books are

$$Df = D_x f = Dy = D_x y = \dot{y},$$

where it is assumed that $y = f(x)$.

EXAMPLE 6. It follows from the computation in Example 3 that if $F(x) = x^{-1/2}$, $x > 0$, then the derivative is given by $F'(x) = -\frac{1}{2}x^{-3/2}$. If we write $y = x^{-1/2}$, $x > 0$, the derivative is also written

$$y' = \frac{dy}{dx} = -\frac{1}{2x^{3/2}}\cdot$$

The value of the derivative at 4 is

$$y'(4) = \frac{dy}{dx}(4) = -\frac{1}{2 \cdot 4^{3/2}} = -\frac{1}{16}\cdot$$

The slope of a straight line is the ratio of a change in y to a change in x. It therefore measures the rate of change of y per unit change in x for the ordered pairs (x, y) that make up the line. Consider the two lines defined by $y = 10x - 3$ and $y = x - 3$ respectively. The rate of change of y to x is 10 for the first and 1 for the second. For a function whose graph is not a straight line, however, the concept of the rate of change of y, or $f(x)$, with respect to x is more profound. There is the problem that the change in functional values $f(x)$ per unit change in x will not be constant along the graph. More basic, however, is the question of the precise meaning or definition of the rate of change. The answer is provided by the derivative. Since $f'(a)$ is the slope of the line tangent to the graph of f at the point $(a, f(a))$, it measures the rate of change of $f(x)$ with respect to x at that point. In

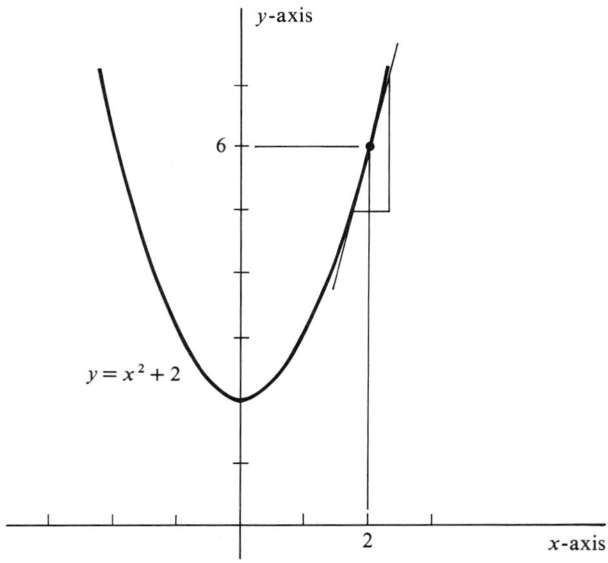

Figure 36

Example 1 we showed that if $f(x) = x^2 + 2$, then $f'(2) = 4$. We interpret the number 4 not only as the slope of the line tangent to the graph of f at $(2, 6)$ but also as the rate of change of $f(x)$ with respect to x there. From the picture of the graph in Figure 36 it is apparent that at $(2, 6)$ a small change in x produces a corresponding change four times as great in $f(x)$.

In Section 4 the idea of limit was introduced by examples and by exploiting the reader's intuitive understanding of continuity and continuous curves. We then gave a formal definition and proceeded in terms of it to go back and define continuity precisely. We shall do an analogous thing here and now define the **slope of the graph of f at the point $(a, f(a))$**, or more simply the **slope of the curve** $y = f(x)$ **at** $(a, f(a))$, to be the derivative $f'(a)$.

We conclude this section with the theorem

(6.1) *If a function f is differentiable at a, then it is continuous there.*

Proof. The hypothesis that

$$\lim_{t \to 0} \frac{f(a + t) - f(a)}{t}$$

exists implies tacitly that a is in the domain of f. If a quotient approaches a finite limit as the denominator approaches zero, then the numerator must also approach zero. This fact is a consequence of the theorem that the limit of a product is the product of the limits [see part (iii) of Theorem (4.1), page 32]. In this case, we have

$$\begin{aligned} \lim_{t \to 0} [f(a + t) - f(a)] &= \lim_{t \to 0} \left[\frac{f(a + t) - f(a)}{t} \cdot t \right] \\ &= \lim_{t \to 0} \left[\frac{f(a + t) - f(a)}{t} \right] \cdot \lim_{t \to 0} t \\ &= f'(a) \cdot 0 = 0. \end{aligned}$$

The equation $\lim_{t \to 0} [f(a + t) - f(a)] = 0$ is equivalent to

$$\lim_{t \to 0} f(a + t) = f(a). \qquad (3)$$

If we set $x = a + t$, then x approaches a as t approaches 0, and conversely. So (3) becomes

$$\lim_{x \to a} f(x) = f(a),$$

and the proof is complete.

PROBLEMS

1. Let $f(x) = 3x^2 + 4$. Using the definition of the derivative, compute

   (a) $f'(1)$ (b) $f'(a)$, for an arbitrary real number a.

2. Write an equation of the line tangent to the graph of the function f in Problem 1 at the point

   (a) $(1, 7)$ (b) $(a, f(a))$.

3. If $F(x) = \dfrac{3}{2x + 1}$, compute $F'(3)$ using the definition of the derivative.

4. Using the definition of the derivative, compute $f'(a)$ for each of the following functions.

(a) $f(x) = x^3$

(b) $f(x) = x^2 + 3x + 5$

(c) $f(x) = 7$

(d) $f(x) = \sqrt{x}, \quad a > 0$

(e) $f(x) = x + \dfrac{1}{x^2}, \quad x \neq 0$

(f) $f(x) = x^3 + 3x^2 + 3x + 1$

(g) $f(x) = \sqrt{x^2 + 1}$

(h) $f(x) = \dfrac{1}{\sqrt{x^2 + 1}}$

(i) $f(x) = x^{1/3}$.

5. Using the results of Problem 4, find an equation of the line tangent to the graph of f at the point $(a, f(a))$, where

(a) $f(x) = x^3$ and $a = 0$.

(b) $f(x) = x^2 + 3x + 5$ and $a = 1$.

(c) $f(x) = 7$ and a is arbitrary.

(d) $f(x) = x + \dfrac{1}{x^2}$ and a is not zero.

6. (a) If $F(x) = x^2$, use the definition of the derivative to find $F'(x)$.

(b) Plot the graphs of F and F' on the same xy-plane.

7. (a) Show that the function $|x|$ is not differentiable at 0 and interpret this fact geometrically.

(b) Compute the derivative at -1 and at 1 of the function $|x|$.

8. Show that the function $\sqrt{x}$ is not differentiable at 0. Draw the graph and interpret the nondifferentiability geometrically.

9. Using the results of Problem 4, find

(a) $\dfrac{df}{dx}(-1)$ if $f(x) = x^2 + 3x + 5$.

(b) $\dfrac{df}{dx}(3)$ if $f(x) = x^3$.

(c) $\dfrac{df}{dx}(b)$ if $f(x) = x^3 + 3x^2 + 3x + 1$.

(d) $\dfrac{d\sqrt{x^2 + 1}}{dx}$.

(e) $\dfrac{d(x^2 + 3x + 5)}{dx}(a)$.

(f) $\dfrac{d}{dx}\left(x + \dfrac{1}{x^2}\right)$.

10. (a) If $y = 2x + 1$, find $\dfrac{dy}{dx}(a)$.

(b) If $s = 16t^2$, find $\dfrac{ds}{dt}(2)$.

(c) If $s = 16t^2$, find $\dfrac{ds}{dt}$.

11. Using the definition of the derivative, prove that if $y = ax^2 + bx + c$, then $\dfrac{dy}{dx} = 2ax + b$.
12. Give an example of a continuous function that fails to have a derivative at some point.

7. Derivatives of Polynomials and Rational Functions. Computing $f'(x)$ from the definition of the derivative by evaluating $\lim_{t\to 0} \dfrac{f(x + t) - f(x)}{t}$ can be quite a job. In this section we shall develop a set of theorems from which the derivatives of many functions, including all polynomials and rational functions, can be found easily and, what is more important, in a completely routine way.

(7.1) *If f and g are differentiable functions, then their sum $f + g$ is differentiable. Moreover, $(f + g)' = f' + g'$.*

Proof. Let a be a number in the domain of $f + g$. Recall that by the definition of the sum of two functions

$$(f + g)(a) = f(a) + g(a),$$

$$(f + g)(a + t) = f(a + t) + g(a + t).$$

Hence, by the definition of the derivative,

$$\begin{aligned}(f + g)'(a) &= \lim_{t\to 0} \frac{(f + g)(a + t) - (f + g)(a)}{t}\\ &= \lim_{t\to 0} \frac{f(a + t) + g(a + t) - \big(f(a) + g(a)\big)}{t}\\ &= \lim_{t\to 0} \left(\frac{f(a + t) - f(a)}{t} + \frac{g(a + t) - g(a)}{t}\right).\end{aligned}$$

It follows from the existence of $f'(a)$ and $g'(a)$ and the fact that the limit of a sum is the sum of the limits [see the basic limit theorem (4.1)(i)] that we may continue the above sequence of equalities, writing

$$\begin{aligned} &= \lim_{t\to 0} \frac{f(a+t) - f(a)}{t} + \lim_{t\to 0} \frac{g(a+t) - g(a)}{t} \\ &= f'(a) + g'(a) \\ &= (f' + g')(a). \end{aligned}$$

This completes the proof.

(7.2) *If f is a differentiable function and c is a constant, then cf is differentiable and $(cf)' = cf'$.*

Proof. For any number a in the domain of f, we have $(cf)(a) = cf(a)$. Hence

$$\begin{aligned} (cf)'(a) &= \lim_{t\to 0} \frac{(cf)(a+t) - (cf)(a)}{t} \\ &= \lim_{t\to 0} \frac{cf(a+t) - cf(a)}{t} \\ &= \lim_{t\to 0} \left(c \cdot \frac{f(a+t) - f(a)}{t}\right). \end{aligned}$$

By the basic limit theorem (4.1)(ii) and the assumed existence of $f'(a)$, we can continue the chain of equalities, writing

$$\begin{aligned} &= c \lim_{t\to 0} \frac{f(a+t) - f(a)}{t} \\ &= cf'(a) = (cf')(a). \end{aligned}$$

This completes the proof.

By taking $c = -1$, we get as a corollary of (7.1) and (7.2) that

$$(f - g)' = f' - g'.$$

(7.3) *The derivative of any constant function is the constant function zero;* i.e.,

$$c' = 0.$$

Proof. Recall that we allow ourselves the liberty of denoting a real number and the constant function whose value is that real number by the same letter. Doing so here, we have

$$c(a) = c(a+t) = c,$$

for any numbers a and t. Hence

$$c'(a) = \lim_{t\to 0} \frac{c(a+t) - c(a)}{t} = \lim_{t\to 0} \frac{0}{t} = 0.$$

EXAMPLE 1. Let $f(x) = x^3$, and $g(x) = \sqrt{x+1}$ $(x \geq -1)$, and $h(x) = x^2 + 3$, and suppose we are given the information that

$$f'(x) = 3x^2,$$
$$g'(x) = \frac{1}{2\sqrt{x+1}}, \qquad x > -1,$$
$$h'(x) = 2x.$$

It follows from the three theorems developed so far in this section that the derivatives of the functions

(a) $5x^3 - 2\sqrt{x+1}$,

(b) x^2,

(c) $3x^3 + 13x^2 + 7$,

are, respectively,

(a′) $15x^2 - \dfrac{1}{\sqrt{x+1}}$,

(b′) $2x$,

(c′) $9x^2 + 26x$.

For example, to get (b′), we write x^2 in the form $(x^2 + 3) - 3$. Then

$$(x^2)' = (x^2+3)' - 3' = 2x - 0 = 2x.$$

The others are equally routine.

The next theorem deals with the derivative of the product of two functions and its conclusion is perhaps unexpected. Note that it does *not* turn out that the derivative of a product is the product of the derivatives.

(7.4) *If f and g are differentiable functions, then their product fg is differentiable. Moreover,* $(fg)' = f'g + g'f$.

Proof. Let a be a number in the domain of fg. By the definition of the product of two functions we have

$$(fg)(a) = f(a)g(a),$$
$$(fg)(a+t) = f(a+t)g(a+t).$$

Hence

$$(fg)'(a) = \lim_{t\to 0} \frac{(fg)(a+t) - (fg)(a)}{t}$$
$$= \lim_{t\to 0} \frac{f(a+t)g(a+t) - f(a)g(a)}{t}.$$

The following algebraic manipulation will enable us to put the above fraction into a form in which we can see what the limit is:

$$\begin{aligned} f(a + t)g(a + t) - f(a)g(a) &= f(a + t)g(a + t) - f(a)g(a + t) \\ &\quad + f(a)g(a + t) - f(a)g(a) \\ &= [f(a + t) - f(a)]g(a + t) \\ &\quad + [g(a + t) - g(a)]f(a). \end{aligned}$$

Thus

$$(fg)'(a) = \lim_{t \to 0} \left[\frac{f(a + t) - f(a)}{t} g(a + t) + \frac{g(a + t) - g(a)}{t} f(a) \right].$$

The limit of a sum of products is the sum of the products of the limits. [Again, see the limit theorem (4.1).] Moreover, $f'(a)$ and $g'(a)$ exist by hypothesis. Finally, since g is differentiable at a, it is continuous there [see Theorem (6.1)]; and so $\lim_{t \to 0} g(a + t) = g(a)$. We conclude that

$$\begin{aligned} (fg)'(a) &= \left[\lim_{t \to 0} \frac{f(a + t) - f(a)}{t} \right] \lim_{t \to 0} g(a + t) \\ &\quad + \left[\lim_{t \to 0} \frac{g(a + t) - g(a)}{t} \right] f(a) \\ &= f'(a)g(a) + g'(a)f(a) = (f'g + g'f)(a). \end{aligned}$$

This completes the proof of the product rule for differentiation.

EXAMPLE 2. Suppose we are given the information that the functions $f(x) = (x^2 + 2)^3$ and $g(x) = (x^2 + 2)^5$ have derivatives

$$f'(x) = 6x(x^2 + 2)^2,$$

$$g'(x) = 10x(x^2 + 2)^4.$$

Find the derivative of $f(x)g(x) = (x^2 + 2)^8$. Theorem (7.4), which is sometimes called Leibnitz's Rule, states that

$$(f(x)g(x))' = f'(x)g(x) + g'(x)f(x).$$

Hence

$$\begin{aligned} ((x^2 + 2)^8)' &= 6x(x^2 + 2)^2(x^2 + 2)^5 + 10x(x^2 + 2)^4(x^2 + 2)^3 \\ &= 16x(x^2 + 2)^7. \end{aligned}$$

The graph of the identity function x is the straight line defined by the equation $y = x$, which passes through the origin and has constant slope 1. It follows that the derivative of the identity function is the constant function 1. Thus

$$x' = 1. \tag{1}$$

We can apply the product (Leibnitz's) rule and obtain

$$(x^2)' = (xx)' = x'x + x'x = 1x + 1x = 2x.$$

Since $x^3 = xx^2$, and we have just found the derivative of each factor, we can use the product rule again to get

$$\begin{aligned}(x^3)' &= (xx^2)' = x'x^2 + (x^2)'x = 1x^2 + 2x \cdot x \\ &= 3x^2.\end{aligned}$$

Again,

$$\begin{aligned}(x^4)' &= (xx^3)' = x'x^3 + (x^3)'x = 1x^3 + 3x^2 \cdot x \\ &= 4x^3.\end{aligned}$$

These results suggest not only the statement of the next theorem, but also how to prove it.

(7.5) *If x is the identity function and n is a positive integer, then* $(x^n)' = nx^{n-1}$.

Proof. We have already proved the theorem for $n = 1$. (Actually we have also proved it for $n = 2, 3$, and 4, but for the moment this is irrelevant.) Suppose we had proved it for all positive integers up to and including k. In particular, we would know that $(x^k)' = kx^{k-1}$. We could then use the product rule to derive $(x^{k+1})' = (xx^k)' = x'x^k + (x^k)'x = 1x^k + kx^{k-1} \cdot x = (k + 1)x^k$. Thus the theorem is true for $n = 1$, and if it is true for $n = k$, it is then also true for $n = k + 1$. We conclude that the theorem holds for every positive integer n. This is an example of a proof by mathematical induction. The reasoning can be paraphrased like this: Suppose I know that I can get on the bottom rung of a ladder. Suppose further that, if I am standing on any rung, then I can reach the next rung. It follows that I can climb the ladder.

EXAMPLE 3. Find the derivatives of the polynomials:

$$\begin{aligned}f(x) &= x^3 - 2, \\ g(x) &= 3x^2 + 7x - 13, \\ y &= 4x^4 + 3x^3 + 2x^2 + x, \\ s &= \tfrac{1}{2}gt^2 \qquad \text{(g is a constant, and t is an independent variable).}\end{aligned}$$

The answers are immediate:

$$\begin{aligned}f'(x) &= 3x^2, \\ g'(x) &= 6x + 7, \\ y' &= 16x^3 + 9x^2 + 4x + 1, \\ s' &= gt.\end{aligned}$$

It should be clear that, as a result of the rules developed so far, the derivative of any polynomial function can be computed immediately and in a purely mechanical way. We turn next to the derivative of a ratio.

(7.6) *If f and g are differentiable functions, then the quotient $\frac{f}{g}$ is differentiable* [if $g(a) = 0$, then $\left(\frac{f}{g}\right)(a)$ is not defined]. *Moreover,*

$$\left(\frac{f}{g}\right)' = \frac{gf' - fg'}{g^2}.$$

Proof. We first prove that the function $\frac{1}{g}$ is differentiable at a number a in its domain provided $g(a) \neq 0$. By definition,

$$\left(\frac{1}{g}\right)'(a) = \lim_{t\to 0} \frac{\left(\frac{1}{g}\right)(a+t) - \left(\frac{1}{g}\right)(a)}{t}$$

$$= \lim_{t\to 0} \frac{\frac{1}{g(a+t)} - \frac{1}{g(a)}}{t}.$$

Note that since g is continuous at a [see Theorem (6.1)] and $g(a) \neq 0$, we know that $g(a + t) \neq 0$ for sufficiently small values of t. Since

$$\frac{\frac{1}{g(a+t)} - \frac{1}{g(a)}}{t} = \frac{g(a) - g(a+t)}{tg(a)g(a+t)}$$

$$= -\left(\frac{1}{g(a)g(a+t)}\right)\left(\frac{g(a+t) - g(a)}{t}\right),$$

we have

$$\left(\frac{1}{g}\right)'(a) = \lim_{t\to 0}\left[-\left(\frac{1}{g(a)g(a+t)}\right)\left(\frac{g(a+t) - g(a)}{t}\right)\right].$$

The derivative $g'(a)$ exists by hypothesis, and $\lim_{t\to 0} g(a + t) = g(a) \neq 0$. The basic limit theorem (4.1) therefore implies that

$$\left(\frac{1}{g}\right)'(a) = -\left(\frac{1}{g(a)\lim_{t\to 0} g(a+t)}\right)\left(\lim_{t\to 0}\frac{g(a+t) - g(a)}{t}\right)$$

$$= -\frac{1}{(g(a))^2}g'(a) = -\frac{g'(a)}{(g(a))^2}.$$

This proves the differentiability of the function $\frac{1}{g}$ and also establishes the following special case of the quotient rule:

$$\left(\frac{1}{g}\right)' = -\frac{g'}{g^2}. \qquad (2)$$

The general form of (7.6) can now be obtained using the product rule:

$$\left(\frac{f}{g}\right)' = \left(f\cdot\frac{1}{g}\right)' = f'\cdot\frac{1}{g} + f\cdot\left(\frac{1}{g}\right)'$$

$$= \frac{f'}{g} - f\cdot\frac{g'}{g^2} = \frac{gf' - fg'}{g^2}.$$

This completes the proof.

EXAMPLE 4. Find the derivatives of the following rational functions:

$$f(x) = \frac{x^2+1}{x},$$

$$g(y) = \frac{y^2 - 3y + 1}{y^3 - 1},$$

$$h(s) = \frac{1}{s^3},$$

$$w = \frac{u-a}{u-b}$$ (a and b are constants and u is an independent variable).

Applying our six rules, we get

$$f'(x) = \frac{x\cdot 2x - (x^2+1)\cdot 1}{x^2} = \frac{x^2-1}{x^2},$$

$$g'(y) = \frac{(y^3-1)(2y-3) - (y^2-3y+1)3y^2}{(y^3-1)^2}$$

$$= \frac{-y^4 + 6y^3 - 3y^2 - 2y + 3}{(y^3-1)^2},$$

$$h'(s) = \frac{-3s^2}{s^6} = -\frac{3}{s^4},$$

$$w' = \frac{(u-b)\cdot 1 - (u-a)\cdot 1}{(u-b)^2} = \frac{a-b}{(u-b)^2}.$$

It is important to realize that the symmetry present in the product rule is missing in the quotient rule. For the former, order is immaterial: The prime appears once on one factor and once on the other, and that is all there is to remember. This is not so for the quotient rule, however, where the wrong order will result in the wrong sign in the answer. There is no help for it but to memorize the formula precisely.

The formula for the derivative of x^n has been proved only if n is a non-negative integer. (It holds for $n = 0$ because $x^0 = 1$.) The next theorem enlarges the scope of the formula to include all integers.

(7.7) *If x is the identity function and n is an integer (positive, negative, or zero), then $(x^n)' = nx^{n-1}$.*

Proof. We shall assume that n is a negative integer, since the theorem is known to be true otherwise. Then $m = -n$ is a positive integer, and $x^n = \dfrac{1}{x^m}$. Using (2) and (7.5), we get

$$\begin{aligned}(x^n)' &= \left(\frac{1}{x^m}\right)' = -\frac{(x^m)'}{x^{2m}} = -\frac{mx^{m-1}}{x^{2m}} \\ &= (-m)x^{-m-1} = nx^{n-1}.\end{aligned}$$

This completes the proof.

Thus, for example, if $f(x) = x^{-7}$, then $f'(x) = -7x^{-8}$. In Section 8 we shall show that the formula is actually valid, not only for integers, but for any rational number n. Finally, in Chapter 5 we shall prove that $(x^a)' = ax^{a-1}$, for any real number a.

Let us summarize in a single list the theorems that we have developed for finding derivatives. To provide practice, we shall this time employ the alternative $\dfrac{d}{dx}$ notation. Let u and v be differentiable functions of x, and c a constant. Then

(i) $$\frac{d(u+v)}{dx} = \frac{du}{dx} + \frac{dv}{dx},$$

(ii) $$\frac{d(cu)}{dx} = c\frac{du}{dx},$$

(iii) $$\frac{dc}{dx} = 0,$$

(iv) $$\frac{d(uv)}{dx} = u\frac{dv}{dx} + v\frac{du}{dx},$$

(v) $$\frac{dx^n}{dx} = nx^{n-1}, \quad \textit{where } n \textit{ is any integer},$$

(vi) $$\frac{d}{dx}\left(\frac{u}{v}\right) = \frac{v\dfrac{du}{dx} - u\dfrac{dv}{dx}}{v^2}.$$

Note that we have proved these theorems for arbitrary differentiable functions u and v, not just for polynomials and rational functions.

EXAMPLE 5. Let

$$y = 2x^3 + 7x + 1,$$
$$u = x^7 + \frac{1}{x^5},$$
$$s = \frac{3t^2 + 2t + 1}{t - 4}.$$

Then

$$\frac{dy}{dx} = 6x^2 + 7,$$

$$\frac{du}{dx} = 7x^6 - \frac{5}{x^6},$$

$$\frac{ds}{dt} = \frac{(t - 4)(6t + 2) - (3t^2 + 2t + 1) \cdot 1}{(t - 4)^2} = \frac{3t^2 - 24t - 9}{(t - 4)^2}.$$

We have seen in this section that the derivative of a polynomial is another polynomial, and the derivative of a rational function is a new rational function. Once we have found the derivative f' of any function f, we can go on and find the derivative of f'. The new function, denoted f'', is called the **second derivative** of f. Clearly,

$$f''(a) = \lim_{t \to 0} \frac{f'(a + t) - f'(a)}{t}.$$

The **third derivative,** written f''', is the derivative of the second derivative, and, in principle, we can go on forever and form derivatives of as high order as we like. It would obviously be absurd to write the seventeenth derivative with seventeen primes, so we adopt the alternative notation $f^{(n)}$ for the nth derivative of f.

The differential notation for the higher derivatives is based on the idea that $\dfrac{d}{dx}$ is a function, sometimes called an operator, which assigns to a function its derivative with respect to x. Hence we write

$$\frac{d}{dx}\left(\frac{df}{dx}\right) = \frac{d^2f}{dx^2} = f'',$$

$$\frac{d}{dx}\left(\frac{d^2f}{dx^2}\right) = \frac{d^3f}{dx^3} = f''',$$

$$\frac{d^nf}{dx^n} = f^{(n)},$$

$$\frac{d^2f}{dx^2}(a) = f''(a), \text{ etc.}$$

In addition, if a variable is used to denote a function, for example, if $y = f(x)$, we also use the expressions

$$\frac{d^2y}{dx^2} = y'' = f'', \qquad \frac{d^ny}{dx^n} = y^{(n)} = f^{(n)}, \text{ etc.}$$

EXAMPLE 6. Let $f(x) = x^3 + 3x^2 + 1$. Then

$$\begin{aligned} f'(x) &= 3x^2 + 6x, \\ f''(x) &= 6x + 6, \\ f'''(x) &= 6, \\ f^{(n)}(x) &= 0, \qquad \text{if } n > 3. \end{aligned}$$

As another example, let $y = \dfrac{1}{x+1}$. Then

$$\begin{aligned} \frac{dy}{dx} &= -\frac{1}{(x+1)^2}, \\ \frac{d^2y}{dx^2} &= \frac{2}{(x+1)^3}, \\ &\vdots \\ \frac{d^ny}{dx^n} &= \frac{(-1)^n n!}{(x+1)^{n+1}}, \qquad n! = 1 \cdot 2 \cdot 3 \cdots n. \end{aligned}$$

PROBLEMS

1. With the aid of the rules for differentiation given in this section, compute $f' = \dfrac{df}{dx}$ for each of the following functions.

(a) $f(x) = 3x^2 + 4x + 1$

(b) $f(x) = x^2(x + 1)$

(c) $f(x) = x^3(x + 2)^2$

(d) $f(x) = (x^2 - 4)(x^2 + 2x + 3)$

(e) $f(x) = 2x^2 + \dfrac{1}{3x^3}$

(f) $f(x) = \dfrac{2x}{2 - x}$

(g) $f(x) = \dfrac{2x}{(2 - x)^2}$

(h) $f(x) = \dfrac{x^3}{x^5 + 1}$

(i) $f(x) = \left(\dfrac{3 - x}{3 + x}\right)^2$

(j) $f(x) = (x^2 + 1)^3$

(k) $f(x) = \dfrac{2x + 1}{x^2 + x}$ (m) $f(x) = (x + x^{-1})^2$

(l) $f(x) = (x^2 + 1)^{-1}$ (n) $f(x) = (x - a)(x - b)(x - c)$.

2. Determine an equation of the line tangent to the parabola $y = x^2 - 4x + 5$ at the point $(1, 2)$. Draw the parabola and the tangent line.
3. The parabola $y = ax^2 + bx + c$ passes through $(0, 4)$ and is tangent to the line $2x + y = 2$ at the point $(1, 0)$. Find the coefficients a, b, and c for the parabola.
4. Show that if f, g and h are differentiable functions, then

$$(fgh)' = f'gh + fg'h + fgh'.$$

5. What is the correct product rule for differentiation, analogous to the one in Problem 4, for (a) four factors, (b) n factors?
6. Obtain an equation of the tangent line to the graph of the function $f(x) = \dfrac{x^3}{x^2 + 1}$ at the point where $x = 2$.
7. (a) If $f(z) = 2z^2 + 2 + \dfrac{2}{z^2}$, then $f'(2) = \cdots$.

   (b) If $f(z) = 2z^2 + 2 + \dfrac{2}{z^2}$, then $f'(x) = \cdots$.

   (c) If $y = \dfrac{x + 1}{x - 1}$, then $\dfrac{dy}{dx} = \cdots$.

   (d) If $y = \dfrac{1}{x}$, then $\dfrac{dy}{dx}(2) = \cdots$.

   (e) If $f(x) = \dfrac{x^2 + 1}{x^2}$, then $\dfrac{df}{dx}(a) = \cdots$.

   (f) If $w = 3u^2 + 4u + 2$, then $\dfrac{dw}{du} = \cdots$.
8. The parabola $y = ax^2 + bx + c$ is tangent to the line $y = 4x + 7$ at the point $(-1, 3)$. In addition, $\dfrac{dy}{dx}(-2) = 0$. Find the coefficients a, b, and c.
9. For each of the following functions f, compute the derivative f' and the second derivative f''.

   (a) $f(x) = 3x^2 + 2x + 1$ (d) $f(t) = t^3(t^2 - 1)$

   (b) $f(x) = 5x + 1$ (e) $f(x) = x^3 + \dfrac{1}{x^2}$

   (c) $f(x) = \dfrac{x^4}{12} + \dfrac{x^3}{6} + \dfrac{x^2}{2} + x + 1$ (f) $f(s) = \dfrac{s^2 - 1}{s^2 + 1}$.

10. The line $y = 3x - 1$ is tangent to the graph of the function $f(x) = ax^3 + bx^2 + c$ at the point $(1, 2)$. Furthermore, $\frac{d^2f}{dx^2}(1) = 0$. Compute a, b, and c.

11. (a) If $f(x) = x^3 - x^2 + x - 1$, then $\frac{d^2f}{dx^2} = \cdots$.

(b) If $y = \frac{x-1}{x+1}$, then $\frac{d^2y}{dx^2} = \cdots$.

(c) If $s = at^3 + bt^2 + ct + d$, where a, b, c, and d are constants, compute $\frac{d^3s}{dt^3}$.

(d) If $y = \frac{1}{x^2}$, then $\frac{d^3y}{dx^3}(a) = \cdots$.

12. Find all points on the graph of the function $\frac{x^3}{3} - x^2$ at which the tangent line is perpendicular to the tangent line at $(1, -\frac{2}{3})$.

13. There are many examples of a function f and a number a such that $f(a)$ is defined (a is in the domain of f) but $f'(a)$ does not exist. Another way of saying the same thing is that the domain of f' can be a *proper* subset of the domain of f. It is equally possible for $f'(a)$ to be defined and $f''(a)$ not to be. Let f be the function defined by

$$f(x) = \begin{cases} \frac{x^2}{2} & \text{if } x \geq 0, \\ -\frac{x^2}{2} & \text{if } x \leq 0. \end{cases}$$

(a) Compute f'.

(b) Is f a differentiable function? [That is, does $f'(a)$ exist for every real number a?]

(c) Show that $f''(0)$ does not exist, and compute $f''(x)$ for $x \neq 0$.

14. Same as Problem 13 except that $f(x) = x^{4/3}$.

15. (a) Draw the graph of the function g defined by

$$g(x) = \begin{cases} x^2, & x \leq 1, \\ 2x - 1, & x > 1. \end{cases}$$

(b) Compute g' and g''.

(c) Are g and g' differentiable functions?

8. The Chain Rule. The theorems in Section 7 were concerned with finding the derivatives of functions that were constructed from other functions using the algebraic operations of addition, multiplication by a constant, multiplication, and division. In this section we shall derive a similar formula, called the **Chain Rule,** for the derivative of the composition $f(g)$ of a differentiable function g with a differentiable function f. Before giving the theorem, we remark that an alternative way of writing the definition of the derivative of a function f is

$$f'(a) = \lim_{x \to a} \frac{f(x) - f(a)}{x - a}. \tag{1}$$

The substitution $x = a + t$ will transform (1) into the expression that we have heretofore used for the derivative. An equation equivalent to (1) is

$$\lim_{x \to a} \left[\frac{f(x) - f(a)}{x - a} - f'(a) \right] = 0.$$

If we set

$$r(x) = \frac{f(x) - f(a)}{x - a} - f'(a), \tag{2}$$

then $\lim_{x \to a} r(x) = 0$. The quantity $r(x)$ is the difference between the slope of the line segment joining $(a, f(a))$ to $(x, f(x))$ and the slope of the line tangent to the curve $y = f(x)$ at $(a, f(a))$. Two equations equivalent to (1) and (2), and which together simply constitute another way of saying that $f'(a)$ is the derivative of f at a, are

$$\begin{cases} f(x) - f(a) = [f'(a) + r(x)](x - a), \\ \lim_{x \to a} r(x) = 0. \end{cases} \tag{3}$$

We shall use these in proving the following:

(8.1) THE CHAIN RULE. *If f and g are differentiable functions, then so is the composite function $f(g)$. Moreover, $[f(g)]' = f'(g)g'$.*

Proof. Let a be a number in the domain of g such that $g(a)$ is in the domain of f. By definition

$$\begin{aligned} [f(g)]'(a) &= \lim_{x \to a} \frac{(f(g))(x) - (f(g))(a)}{x - a} \\ &= \lim_{x \to a} \frac{f(g(x)) - f(g(a))}{x - a}. \end{aligned}$$

The intuitive idea behind the Chain Rule can be seen by writing

$$\begin{aligned} [f(g)]'(a) &= \lim_{x \to a} \left[\frac{f(g(x)) - f(g(a))}{g(x) - g(a)} \frac{g(x) - g(a)}{x - a} \right] \\ &= \left[\lim_{x \to a} \frac{f(g(x)) - f(g(a))}{g(x) - g(a)} \right] \left[\lim_{x \to a} \frac{g(x) - g(a)}{x - a} \right]. \end{aligned}$$

Setting $y = g(x)$ and $b = g(a)$ and noting that y approaches b as x approaches a, we have

$$\begin{aligned}[f(g)]'(a) &= \lim_{y\to b} \frac{f(y) - f(b)}{y - b} \lim_{x\to a} \frac{g(x) - g(a)}{x - a} \\ &= f'(b)g'(a) \\ &= f'\big(g(a)\big)g'(a) = \big(f'(g)g'\big)(a),\end{aligned}$$

which is the desired result. This argument fails to be a rigorous proof primarily because there is no reason to suppose that $g(x) - g(a) \neq 0$ for all x sufficiently close to a. To overcome this difficulty, we use equations (3). These equations, with a typical element in the domain of f denoted by y instead of x and with the derivative evaluated at b, become

$$\begin{cases} f(y) - f(b) = [f'(b) + r(y)](y - b), \\ \lim_{y\to b} r(y) = 0. \end{cases}$$

Substituting $y = g(x)$ and $b = g(a)$, we get

$$f\big(g(x)\big) - f\big(g(a)\big) = [f'\big(g(a)\big) + r\big(g(x)\big)][g(x) - g(a)].$$

Hence

$$\frac{f\big(g(x)\big) - f\big(g(a)\big)}{x - a} = [f'\big(g(a)\big) + r\big(g(x)\big)]\frac{g(x) - g(a)}{x - a}.$$

We know that $\lim_{x\to a} \frac{g(x) - g(a)}{x - a} = g'(a)$. In addition, since g is differentiable at a, it is continuous there [see Theorem (6.1)], and so $\lim_{x\to a} g(x) = g(a) = b$. Since $\lim_{y\to b} r(y) = 0$, it follows that $|r(y)|$ can be made arbitrarily small by taking y sufficiently close to b. Because $\lim_{x\to a} g(x) = b$, we may therefore conclude that $\lim_{x\to a} r\big(g(x)\big) = 0$. The basic limit theorem (4.1) asserts that the limit of a sum or product is the sum or product, respectively, of the limits. Hence

$$\begin{aligned}[f(g)]'(a) &= \lim_{x\to a} \frac{f\big(g(x)\big) - f\big(g(a)\big)}{x - a} \\ &= \left[\lim_{x\to a} f'\big(g(a)\big) + \lim_{x\to a} r\big(g(x)\big)\right] \lim_{x\to a} \frac{g(x) - g(a)}{x - a} \\ &= [f'\big(g(a)\big) + 0]g'(a) = f'\big(g(a)\big)g'(a) \\ &= \big(f'(g)g'\big)(a),\end{aligned}$$

and the proof of the Chain Rule is complete.

EXAMPLE 1. If $F(x) = (x^2 + 2)^3$, compute $F'(x)$. One way to do this problem is to expand $(x^2 + 2)^3$ and use the differentiation formulas developed in Section 7.

$$F(x) = (x^2 + 2)^3 = x^6 + 6x^4 + 12x^2 + 8,$$
$$F'(x) = 6x^5 + 24x^3 + 24x.$$

Another method uses the Chain Rule. Let g and f be the functions defined, respectively, by $g(x) = x^2 + 2$ and $f(y) = y^3$. Then

$$f(g(x)) = (x^2 + 2)^3 = F(x),$$

and, according to the Chain Rule,

$$F'(x) = [f(g(x))]' = f'(g(x))g'(x).$$

Since $g'(x) = 2x$ and $f'(y) = 3y^2$, we get $f'(g(x)) = 3(x^2 + 2)^2$ and

$$\begin{aligned} F'(x) &= 3(x^2 + 2)^2(2x) \\ &= 6x(x^4 + 4x^2 + 4), \end{aligned}$$

which agrees with the alternative solution above.

EXAMPLE 2. Find the derivative of the function $(3x^7 + 2x)^{128}$. In principle, we could expand by the binomial theorem, but with the Chain Rule at our disposal that would be absurd. Let $g(x) = 3x^7 + 2x$ and $f(y) = y^{128}$. Then $g'(x) = 21x^6 + 2$ and $f'(y) = 128y^{127}$. Setting $y = 3x^7 + 2x$, we get

$$\begin{aligned} ((3x^7 + 2x)^{128})' &= [f(g(x))]' = f'(g(x))g'(x) \\ &= 128(3x^7 + 2x)^{127}(21x^6 + 2). \end{aligned}$$

The above two examples are instances of the following corollary of the Chain Rule: If f is a differentiable function, then

$$(f^n)' = nf^{n-1}f', \qquad \text{for any integer } n.$$

To prove it, let $F(y) = y^n$. Then $F(f) = f^n$, and we know that $F'(y) = ny^{n-1}$. Consequently, $(f^n)' = [F(f)]' = F'(f)f' = nf^{n-1}f'$. A significant generalization of this result is

(8.2) *If f is a positive differentiable function and r is any rational number, then $(f^r)' = rf^{r-1}f'$.*

The requirement that f is positive assures that f^r is defined. A nonpositive number cannot be raised to an arbitrary rational power. However, as we shall show on page 267, the requirement that r be a rational number is unnecessary. Theorem (8.2) is actually true for any real number r.

Proof. Let $r = \frac{m}{n}$, where m and n are integers, and set $h = f^r = f^{m/n}$. Then $h^n = (f^{m/n})^n = f^m$, which implies that $(h^n)' = (f^m)'$. Using the above formula for the derivative of an integral power of a function, we get

$$nh^{n-1}h' = mf^{m-1}f'.$$

Solving for h', we obtain

$$\begin{aligned} h' &= \frac{m}{n} h^{1-n} f^{m-1} f' = \frac{m}{n} (f^r)^{1-n} f^{m-1} f' \\ &= r f^{r-rn+m-1} f' \\ &= r f^{r-1} f'. \end{aligned}$$

This completes the proof—almost. Note that we have in the argument tacitly assumed that h, the function whose derivative we are seeking, is differentiable. Is it? If it is, how do we know it? The answer to the first question is yes, but the answer to the second is not so easy. The problem can be reduced to a simpler one: *If n is a positive integer and g is the function defined by $g(x) = x^{1/n}$, for $x > 0$, then g is differentiable.* If we know this fact, we are out of the difficulty because the Chain Rule tells us that the composition of two differentiable functions is differentiable. Hence $g(f)$ is differentiable, and $g(f) = f^{1/n}$. From this it follows that $(f^{1/n})^m$ is differentiable, and $(f^{1/n})^m = f^{m/n}$. (When we express r as a ratio $\frac{m}{n}$, we can certainly take n to be positive.) A proof that $x^{1/n}$ is differentiable, if $x > 0$, is most easily given as an application of the Inverse Function Theorem (3.4), Chapter 5 (see page 261). However, the intuitive reason is simple: If $y = x^{1/n}$ and $x > 0$, then $y^n = x$, and by interchanging x and y we obtain the equation $x^n = y$. The latter equation defines a smooth curve whose slope at every point is given by the derivative $\frac{dy}{dx} = nx^{n-1}$. Interchanging x and y amounts geometrically to a reflection about the line $y = x$. We conclude that the original curve $y = x^{1/n}$, $x > 0$, has the same intrinsic shape and smoothness as that defined by $y = x^n$, $y > 0$. It therefore must have a tangent line at every point, which means that $x^{1/n}$ is differentiable.

EXAMPLE 3. If $y = x^{1/n}$, then

$$\frac{dy}{dx} = \frac{1}{n} x^{(1/n)-1} = \frac{1}{nx^{1-1/n}}, \qquad x > 0.$$

EXAMPLE 4. Find the derivative of the function $F(x) = (3x^2 + 5x + 1)^{5/3}$. If we let $f(x) = 3x^2 + 5x + 1$, then Theorem (8.2) implies that

$$\begin{aligned} F'(x) &= \tfrac{5}{3} f(x)^{2/3} f'(x) \\ &= \tfrac{5}{3}(3x^2 + 5x + 1)^{2/3}(6x + 5). \end{aligned}$$

With the $\frac{d}{dx}$ notation for the derivative, the Chain Rule can be written in a form that is impossible to forget. Let f and g be two differentiable functions. The formation of the composite function $f(g)$ is suggested by writing $u = g(x)$ and $y = f(u)$. Thus x is transformed by g into u, and the resulting u is then transformed by f into $y = f(u) = f(g(x))$. We have

$$\frac{du}{dx} = g'(x),$$

$$\frac{dy}{du} = f'(u),$$

$$\frac{dy}{dx} = [f(g(x))]'.$$

By the Chain Rule, $[f(g(x))]' = f'(g(x))g'(x) = f'(u)g'(x)$, and so

$$\boxed{\frac{dy}{dx} = \frac{dy}{du}\frac{du}{dx}.} \tag{4}$$

The idea that one can simply cancel out du in (4) is very appealing and accounts for the popularity of the notation. It is important to realize that the cancellation is valid because the Chain Rule is true, and *not* vice versa. Thus far, du is simply a part of the notation for the derivative and means nothing by itself. Note also that (4) is incomplete in the sense that it does not say explicitly at what points to evaluate the derivatives. We can add this information by writing

$$\frac{dy}{dx}(a) = \frac{dy}{du}(u(a))\frac{du}{dx}(a).$$

EXAMPLE 5. If $w = z^2 + 2z + 3$ and $z = \frac{1}{x}$, find $\frac{dw}{dx}(2)$. By the Chain Rule,

$$\begin{aligned} \frac{dw}{dx} &= \frac{dw}{dz}\frac{dz}{dx} \\ &= (2z + 2)\left(-\frac{1}{x^2}\right). \end{aligned}$$

When $x = 2$, we have $z = \frac{1}{2}$. Hence

$$\frac{dw}{dx}(2) = (2 \cdot \tfrac{1}{2} + 2)(-\tfrac{1}{4}) = -\tfrac{3}{4}.$$

EXAMPLE 6. Two functions, which we shall define in Chapter 11, are the hyperbolic sine and the hyperbolic cosine, denoted by $\sinh x$ and $\cosh x$ respectively. These functions are differentiable and have the interesting property that

$$\frac{d}{dx}\sinh x = \cosh x,$$

$$\frac{d}{dx}\cosh x = \sinh x.$$

Furthermore, $\sinh(0) = 0$ and $\cosh(0) = 1$. Compute the derivatives at $x = 0$ of

(a) $(\cosh x)^2$,

(b) the composite function $\sinh(\sinh x)$.

By (8.2), we obtain for (a)

$$\frac{d}{dx}(\cosh x)^2 = 2\cosh x \frac{d}{dx}\cosh x = 2\cosh x \sinh x,$$

and so

$$\frac{d}{dx}(\cosh x)^2(0) = 2\cosh 0 \sinh 0 = 0.$$

Part (b) requires the full force of the Chain Rule: Setting $u = \sinh x$, we obtain

$$\begin{aligned}\frac{d}{dx}\sinh u &= \frac{d}{du}\sinh u \frac{du}{dx} \\ &= \cosh u \cosh x,\end{aligned}$$

or

$$\frac{d}{dx}\sinh(\sinh x) = \cosh(\sinh x)\cosh x.$$

Hence

$$\begin{aligned}\frac{d}{dx}\sinh(\sinh x)(0) &= \cosh(\sinh 0)\cosh 0 \\ &= \cosh 0 \cosh 0 = 1.\end{aligned}$$

PROBLEMS

1. In each of the following problems find $[f(g)]'(x)$.

   (a) $f(y) = y^5$ and $g(x) = x^2 + 1$.

   (b) $f(y) = y^2 + 2y$ and $g(x) = x^2 - 2x + 2$.

   (c) $f(y) = y^3$ and $g(x) = \dfrac{x}{x^2 + 1}$.

   (d) $f(u) = \dfrac{u}{u + 1}$ and $g(x) = x^2$.

   (e) $f(x) = x^{-2}$ and $g(x) = x^{1/3}$.

   (f) $f(x) = x^4$ and $g(t) = \dfrac{t^2 - 1}{t^2 + 1}$.

   (g) $f(x) = g(x) = x^2 + 3x + 2$.

2. Find f' given that

   (a) $f(x) = (1 + x^2)^{10}$

   (b) $f(x) = (x^4 + 3x^3 + 2x^2 + x + 4)^6$

   (c) $f(t) = (t^2 + 1)^4(2t^2 - 3)^3$

   (d) $f(x) = \sqrt{x^3 - 1}$

   (e) $f(x) = \left(\dfrac{x - 1}{x + 1}\right)^3$

   (f) $f(s) = \dfrac{1}{\sqrt{s^2 + 1}}$

   (g) $f(y) = \dfrac{y^2}{(y^2 + 1)^{3/2}}$

   (h) $f(u) = \dfrac{5}{\left(u + \dfrac{1}{\sqrt{u}}\right)^4}$.

3. If $f(y) = y^{-2}$ and $g(x) = \dfrac{x^{1/2}}{\sqrt{5x^3 + 6x^2 + 4x}}$, compute the derivative of the composite function $f(g)$ in two ways:

   (a) By finding $f(g(x))$ first and then taking its derivative.

   (b) By the Chain Rule.

4. If $z = 5y^7 + 2y^2 + 1$ and $y = 2x^2 - 6$, find $\dfrac{dz}{dx}$ and $\dfrac{dz}{dx}(2)$.

5. If $y = x^3$ and $x = \dfrac{1}{\sqrt{t^2 + 5}}$, compute $\dfrac{dy}{dt}$ and $\dfrac{dy}{dt}(2)$ using the Chain Rule.

6. Let $y = x^2 + 3x + 2$ and $x = \dfrac{t - 1}{t + 1}$. Compute $\dfrac{dy}{dt}(2)$ in two ways:

   (a) By evaluating the composite function $y(t)$ and then by taking its derivative.

   (b) By the Chain Rule.

7. Prove directly by induction on n without using the Chain Rule that if f is a differentiable function and n is a positive integer, then $(f^n)' = nf^{n-1}f'$.

8. Prove as a corollary of the Chain Rule that

$$[f(g(h))]' = f'(g(h))g'(h)h'.$$

9. Using Problem 8, show that if $w = f(z)$ and $z = g(y)$ and $y = h(x)$, then

$$\frac{dw}{dx} = \frac{dw}{dz}\frac{dz}{dy}\frac{dy}{dx}.$$

10. Let $w = z - \frac{1}{z}$, $z = \sqrt{y^3 + 1}$, and $y = 2x^3 - x + 1$. Find $\frac{dw}{dx}(1)$.

11. Using Example 6, compute

(a) $\frac{d}{dx}(\sinh x)^2$

(b) $\frac{d^2}{dx^2}\sinh x$

(c) $\frac{d^2}{dx^2}\cosh x\,(0)$

(d) $\frac{d}{dx}\sinh(\cosh x)$.

12. If $z = f(y)$ and $y = g(x)$, show that $\frac{d^2z}{dx^2} = \frac{d^2z}{dy^2}\left(\frac{dy}{dx}\right)^2 + \frac{dz}{dy}\frac{d^2y}{dx^2}$.

13. If $z = 2y^3 - 3y + 1$ and $y = x^2 - 1$, compute $\frac{d^2z}{dx^2}(2)$ in two ways:

(a) By evaluating the composite function $z(x)$ and finding $z''(2)$.

(b) Using the result of Problem 12.

14. Let $f(x)$ be a differentiable function with the property that $f'(x) = \frac{1}{x}$. If $g(x)$ is a differentiable function with the property that its composition with f is the identity function, i.e., $f(g(x)) = x$, prove that $g' = g$.

9. Implicit Differentiation. The subset C of the xy-plane consisting of all ordered pairs (x, y) that satisfy the equation

$$\frac{x^2}{9} - \frac{y^2}{4} = 1 \qquad (1)$$

is the hyperbola shown in Figure 37. It is apparent from the figure that the whole set C is not a function, since it is easy to find instances of ordered pairs (a, b) and (a, c) in C with $b \neq c$. For example, both $(6, 2\sqrt{3})$ and $(6, -2\sqrt{3})$ lie on the curve. On the other hand, many subsets of C are functions. For instance, the set of all ordered pairs (x, y) in C for which $x \geq 3$ and $y \geq 0$, which is drawn with a heavy curve in Figure 37, is a function $f(x)$. Central to the ideas that follow is the fact that since the points $(x, f(x))$ that comprise

f belong to C, they satisfy the equation of the hyperbola. That is,

$$\frac{x^2}{9} - \frac{(f(x))^2}{4} = 1, \tag{2}$$

for every $x \geq 3$. We say that the function f is defined **implicitly** by (1).

It is geometrically obvious that the hyperbola has a tangent line at every point, and we therefore conclude that the function $f(x)$ is differentiable except at $x = 3$, where the tangent is vertical. We can compute $f'(x)$ most easily by observing that since (2) holds for every x in the domain of f, it is an

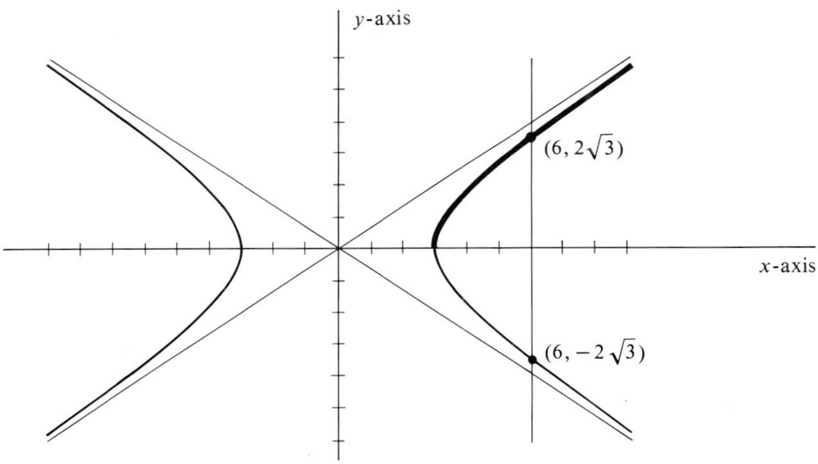

Figure 37

equality between two functions. Specifically, the composite function $\frac{x^2}{9} - \frac{(f(x))^2}{4}$ is equal to the constant function 1. Equal functions have equal derivatives. Hence

$$\left[\frac{x^2}{9} - \frac{(f(x))^2}{4}\right]' = 1'.$$

The rules of differentiation yield

$$\frac{2x}{9} - \frac{2}{4} f(x) f'(x) = 0,$$

and solving for $f'(x)$, we obtain

$$f'(x) = \frac{4x}{9f(x)}. \tag{3}$$

In particular, if $x = 6$, then $f(x) = 2\sqrt{3}$, and

$$f'(6) = \frac{4 \cdot 6}{9 \cdot 2\sqrt{3}} = \frac{4}{3\sqrt{3}} \cdot$$

It is important to realize that there is no single function f defined implicitly by equation (1). The set of all points (x, y) of C for which $y \leq 0$ is another such function, and it includes the point $(6, -2\sqrt{3})$. Note that if this were the function that we denoted by f, we would still obtain equations (2) and (3). For this f, however, we have $f(6) = -2\sqrt{3}$. Hence, this time,

$$f'(6) = \frac{4 \cdot 6}{9(-2\sqrt{3})} = -\frac{4}{3\sqrt{3}} \cdot$$

EXAMPLE 1. The set of all points (x, y) that satisfy the equation

$$5x^2 - 6xy + 5y^2 = 8 \tag{4}$$

can be shown to be the ellipse shown in Figure 38. What is the slope of the line tangent to the ellipse at $(0, 2\sqrt{\frac{2}{5}})$? It is clear from the figure that the set

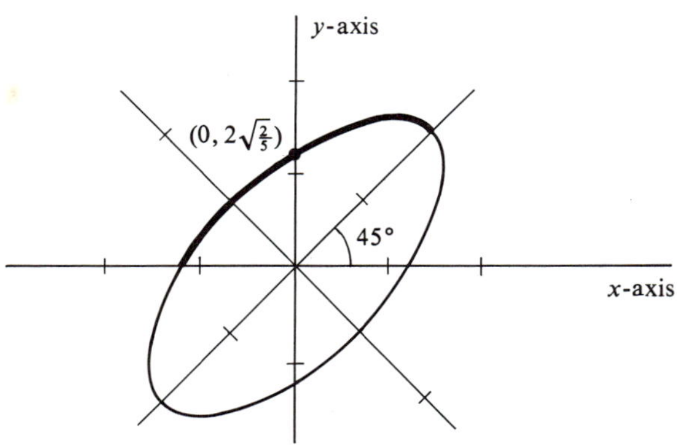

Figure 38

of all pairs (x, y) on the ellipse for which $y > 0$ and $y > x$ (drawn with a heavy curve in the figure) is a differentiable function $f(x)$. This function is implicitly defined by equation (4). Thus

$$5x^2 - 6xf(x) + 5\big(f(x)\big)^2 = 8,$$

for every x in the domain of f. Since this is an equality between two functions we obtain by differentiating both sides,

$$10x - 6f(x) - 6xf'(x) + 10f(x)f'(x) = 0.$$

Solving for $f'(x)$, we get

$$f'(x) = \frac{3f(x) - 5x}{5f(x) - 3x}.$$

This problem deals with an implicitly defined function whose graph passes through the point $(0, 2\sqrt{\frac{2}{5}})$. Hence $f(0) = 2\sqrt{\frac{2}{5}}$ and therefore $f'(0) = \frac{3}{5}$, which is the slope of the desired tangent line.

The definition, which we have thus far illustrated with two equations, is the following: A function $f(x)$ is **defined implicitly by an equation** $F(x, y) = c$, where c is a constant, if $F(x, f(x)) = c$ for every x in the domain of f. We emphasize that, in general, an equation in x and y defines y as a function of x in many ways. The most we can hope for in the way of uniqueness is that, for a given point (a, b) such that $F(a, b) = c$, we can choose an open interval containing a which is the domain of precisely one continuous function $f(x)$ defined implicitly by $F(x, y) = c$ with $f(a) = b$.

Note that in both our examples the derivative f' of the implicitly defined function was computed without solving the original equation for f. The fact that this is always possible is almost too good to be true—especially for an equation where first solving for y in terms of x is either impractical or even impossible (except by numerical techniques). This method of finding the derivative of an implicitly defined function by differentiating both sides of the equation that defines the function is called **implicit differentiation.**

EXAMPLE 2. The point (2, 1) lies on the curve defined by the equation

$$x^3y + xy^3 = 10.$$

Assuming that this equation implicitly defines a differentiable function $f(x)$ whose graph passes through (2, 1), compute $f'(2)$. Letting y stand for $f(x)$, we obtain by implicit differentiation

$$3x^2y + x^3\frac{dy}{dx} + y^3 + x3y^2\frac{dy}{dx} = 0.$$

Hence

$$\frac{dy}{dx} = -\frac{3x^2y + y^3}{3xy^2 + x^3}.$$

At the point $x = 2$, $y = 1$, we therefore get

$$\left.\frac{dy}{dx}\right|_{\substack{x=2\\y=1}} = -\frac{13}{14}.$$

EXAMPLE 3. The set of all pairs (x, y) that satisfy the equation

$$y^3 + yx^2 + ax^2 - 3ay^2 = 0 \tag{5}$$

is the curve, called a **trisectrix,** shown in Figure 39. Find $\frac{dy}{dx}$ when $x = a$. So stated, the problem is impossible. There are three distinct points on the

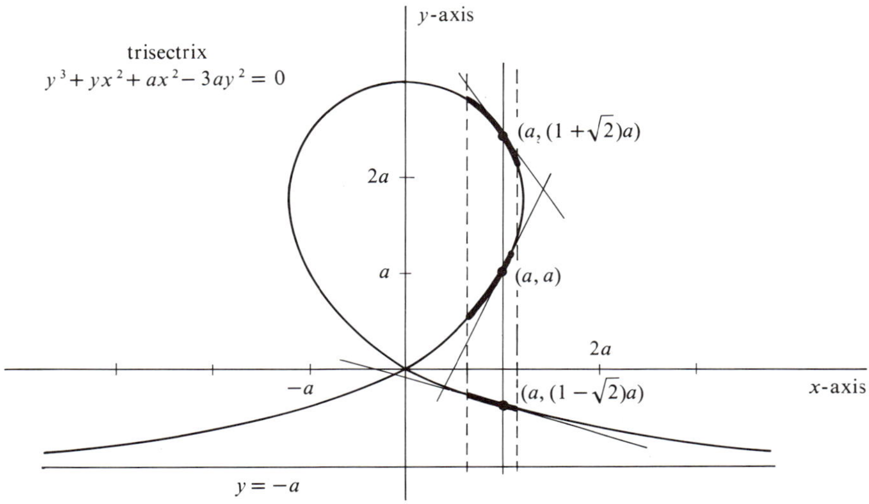

Figure 39

curve with x-coordinate equal to a, which may be found by substituting $a = x$ in equation (5) and then solving for y. The points are (a, a), $(a, (1 + \sqrt{2})a)$, and $(a, (1 - \sqrt{2})a)$. As shown in the figure, we may select a small interval about a to serve as the domain of three different implicitly defined functions. To find the derivative of each one at $x = a$, we proceed by implicit differentiation:

$$3y^2 \frac{dy}{dx} + x^2 \frac{dy}{dx} + 2xy + 2ax - 6ay \frac{dy}{dx} = 0.$$

Hence

$$\frac{dy}{dx} = \frac{2xy + 2ax}{6ay - x^2 - 3y^2}.$$

Thus the derivatives at a of the three differentiable functions defined implicitly by equation (5) are, respectively,

$$\left.\frac{dy}{dx}\right|_{\substack{x=a \\ y=a}} = 2,$$

$$\left.\frac{dy}{dx}\right|_{\substack{x=a \\ y=a(1+\sqrt{2})}} = -1 - \frac{\sqrt{2}}{2},$$

$$\left.\frac{dy}{dx}\right|_{\substack{x=a \\ y=a(1-\sqrt{2})}} = -1 + \frac{\sqrt{2}}{2}.$$

The reader should note that in each of the above examples of implicit differentiation the existence of an implicitly defined differentiable function has either been assumed outright or justified geometrically from a picture. The problem of giving analytic conditions which ensure that an equation $F(x, y) = c$ implicitly defines y as a differentiable function of x in the neighborhood of a point (a, b) is the subject of the Implicit Function Theorem. A discussion and proof of this famous theorem may be found in any standard text in advanced calculus.

PROBLEMS

1. The equation $(x^2 + y^2)^2 = 2(x^2 - y^2)$ (see Figure 40) implicitly defines a differentiable function $f(x)$ whose graph passes through the point $\left(\frac{\sqrt{3}}{2}, \frac{1}{2}\right)$. Compute $f'\left(\frac{\sqrt{3}}{2}\right)$.

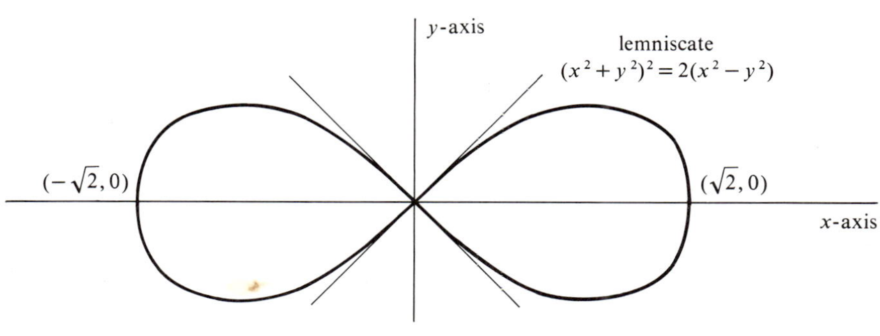

Figure 40

2. Compute the slope of the tangent line to the circle $x^2 + y^2 = 4$ at the point $(1, \sqrt{3})$ and at the point $(1, -\sqrt{3})$.
3. (a) The equation $x^3 + y^3 - 6xy = 0$ (see Figure 41) implicitly defines a differentiable function $f(x)$ whose graph passes through (3, 3). Compute $f'(3)$.

   (b) How many differentiable functions $f(x)$ having a small interval about the number 3 as a common domain are implicitly defined by the equation in (a)?

   (c) Compute $f'(3)$ for each of them.

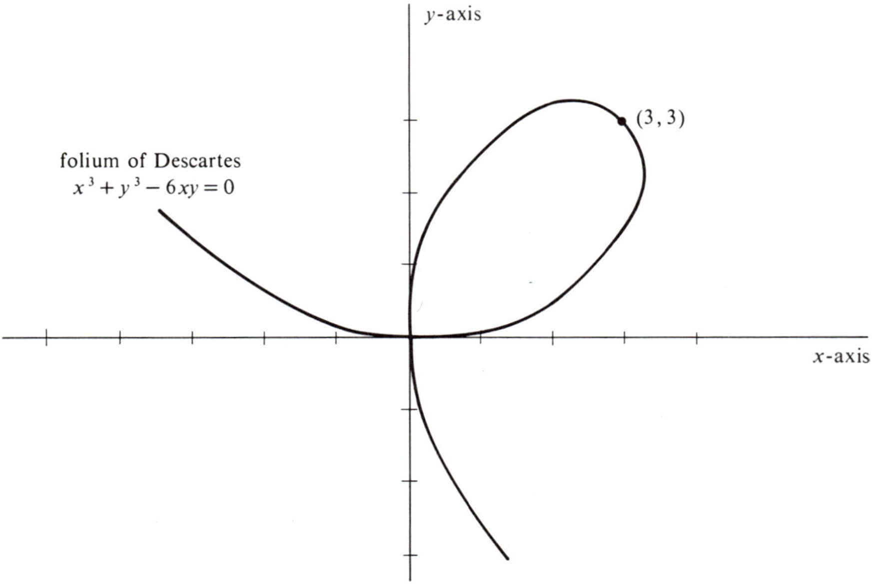

Figure 41

4. For each of the following equations calculate $\frac{dy}{dx}$ at the point specified.
   (a) $4x^2 + y^2 = 8$, at the point (1, 2).
   (b) $y^2 = x$, at the point (4, 2).
   (c) $y^2 = x^5$, at one point (1, 1).
   (d) $y^2 = \frac{x - 1}{x + 1}$, at the point (a, b).
   (e) $y^2 = \frac{x^2 - 1}{x^2 + 1}$, at the point (a, b).
   (f) $x^2y + xy^2 = 6$, at the point (1, 2).
   (g) $x^2 + 2xy = 3y^2$, at the point (1, 1).
   (h) $5y^2 = x^2y + \frac{2}{xy^2}$, at the point (2, 1).
   (i) $x^{3/2} + y^{3/2} = 2$, at the point (1, 1).
   (j) $x^5 + 3x^2y^3 + 3x^3y^2 + y^5 = 8$, at the point (1, 1).

5. What is the slope of the line tangent to the graph of $y^3x^2 = 4$ at the point (2, 1)? Calculate $y''(2)$.

6. Each of the following equations implicitly defines y as a differentiable function of x in the vicinity of the point (a, b). Compute $\frac{dy}{dx}(a)$ and $\frac{d^2y}{dx^2}(a)$.

(a) $x^2 - y^2 = 1, \quad (a, b) = (\sqrt{2}, 1).$

(b) $y^2 = 1 - xy, \quad (a, b) = (0, 1).$

(c) $xy^2 = 8, \quad (a, b) = (2, -2).$

(d) $x^2y^3 = 1, \quad (a, b) = (-1, 1).$

CHAPTER 2

Applications of the Derivative

1. Curve Sketching. The slope of the tangent line to the graph of a function is one interpretation of the derivative, and the rate of change of y with respect to x is another. Both interpretations aid us in the sketching of graphs. A little practice will show that we need plot relatively few points for a sketch if we know the slope of the graph at each of these points. Let us consider the function f defined by

$$f(x) = \tfrac{1}{3}x^3 - 4x^2 + 12x - 5.$$

Its domain is R, and, for each real value of x, we find the corresponding value $f(x)$. To help us make the sketch, we look at the derivative:

$$f'(x) = x^2 - 8x + 12 = (x - 2)(x - 6).$$

If $x < 2$, each of the factors of $f'(x)$ is negative, and hence their product is positive. Thus the first derivative is positive for each value of x less than 2. With the rate-of-change interpretation, this means that the rate of change of f with respect to x is positive or that $f(x)$ increases whenever x does. Thus, as x increases from $-\infty$ to 2, $f(x)$ increases. The graph goes up as one moves to the right until $x = 2$.

If $x = 2$, then $f'(x) = 0$, and the tangent, having a slope of 0, is horizontal. If $2 < x < 6$, the first factor of $f'(x)$ is positive, the second factor is negative, and their product is negative. With a negative rate of change, $f(x)$ must decrease as x increases. Thus, as x increases from 2 to 6, $f(x)$ decreases. The graph goes down as one moves to the right from $x = 2$ to $x = 6$.

If $x = 6$, then $f'(x) = 0$, and the tangent to the graph is again horizontal.

If $x > 6$, both factors of $f'(x)$ are positive, and hence their product is positive. Thus $f(x)$ increases as one goes to the right beyond $x = 6$. Since $f(2) = \frac{1}{3} \cdot 8 - 4 \cdot 4 + 12 \cdot 2 - 5 = 5\frac{2}{3}$ and $f(6) = \frac{1}{3} \cdot 216 - 4 \cdot 36 + 12 \cdot 6 - 5 = -5$, we plot the points $(2, 5\frac{2}{3})$ and $(6, -5)$. At each of these points we sketch a horizontal line segment. An additional point may be found by inspection: Since $f(0) = -5$, the point $(0, -5)$ is also plotted on Figure 1.

We now know that the graph comes up from lower left through $(0, -5)$ to $(2, 5\frac{2}{3})$, goes down from $(2, 5\frac{2}{3})$ to $(6, -5)$, and then goes up to the right from (6, −5). But we do not know its shape. Further information on this may be obtained from the second derivative:

$$f''(x) = 2x - 8 = 2(x - 4).$$

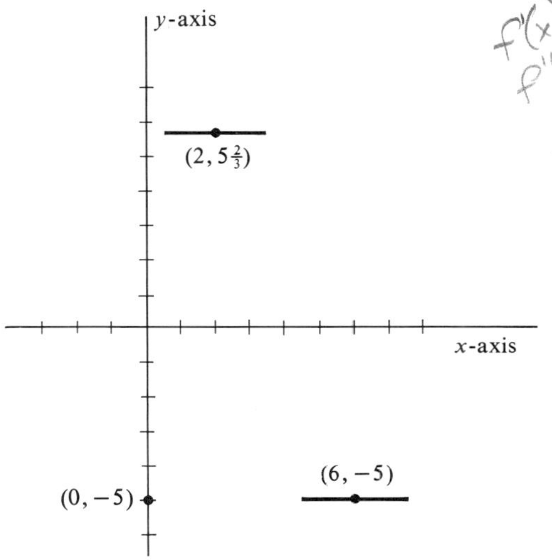

Figure 1

If $x < 4$, then $f''(x)$ is negative. Since f'' is the rate of change of f' with respect to x, this means that f' is decreasing as x is increasing from $-\infty$ to 4. If we interpret f' as the slope of the tangent, then this means that the slope of the tangent decreases as x increases. We can get some idea of shape here if we plot three points on Figure 2(a), the middle one the highest and with a horizontal tangent drawn through it. Note that the tangent through the middle point has slope less than that of the tangent through the left point

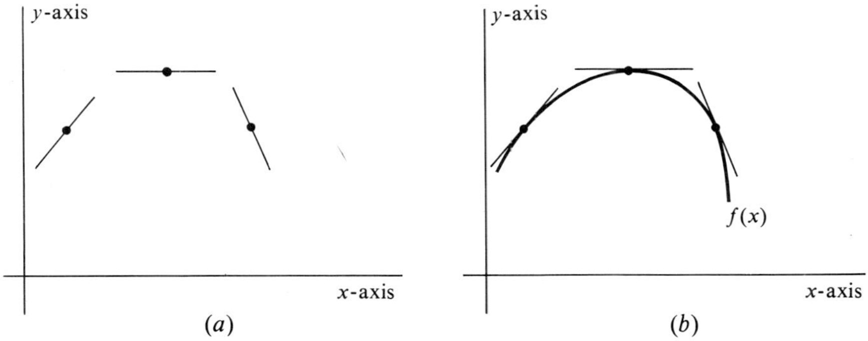

Figure 2

and that the tangent through the right point has a slope which is still less. The slopes of tangents at intermediate points will take on intermediate values, and thus a curve passing through these three points with these three tangents must be concave downward or must "bend" down. Whenever $f''(x) < 0$, the graph of $f(x)$ will be bending down. The part of the curve through the three points of Figure 2(a) with the appropriate tangents is drawn in Figure 2(b).

If $x = 4$, then $f''(x) = 0$, and the rate of change of f' with respect to x is 0. Thus the slope of the tangent has ceased decreasing.

If $x > 4$, then $f''(x) > 0$, and f' increases as x increases. Thus the slope of the tangent increases as x increases. Again we plot three points, the middle one the lowest and with a horizontal tangent drawn through it. These are shown in Figure 3(a). Since the slope increases as x increases, the tangent

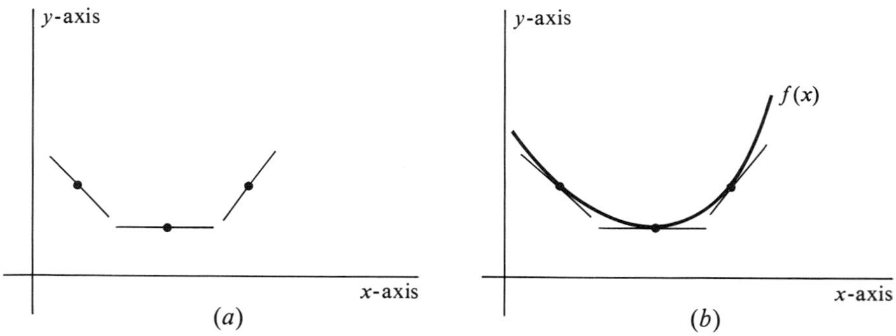

Figure 3

through the middle point has slope greater than that of the tangent through the left point, and the tangent through the right point has a slope which is still greater. The slopes of tangents at intermediate points take on intermediate values, and thus a curve passing through these three points with these tangents must be concave upward or must "bend" up. Whenever $f''(x) > 0$, the graph of $f(x)$ will be bending up. The part of the curve through the three points of Figure 3(a) with the appropriate tangents is drawn in Figure 3(b).

After finding that $f(4) = \frac{1}{3} \cdot 64 - 4 \cdot 16 + 12 \cdot 4 - 5 = \frac{1}{3}$, we are ready to sketch the graph in Figure 4. The graph is concave downward from the far left through $(0, -5)$ to a high point at $(2, 5\frac{2}{3})$ and on to $(4, \frac{1}{3})$. It is then concave upward from $(4, \frac{1}{3})$ to a low point at $(6, -5)$ and on upward to the right. The graph is, of course, incomplete, since it continues indefinitely both downward to the left and upward to the right. The point $(2, 5\frac{2}{3})$, being higher than any nearby point on the graph, is called a local, or relative, maximum point. It is certainly not the highest point on the graph, hence the word "local," or "relative." Similarly, the point (6, −5) is a local, or relative, minimum.

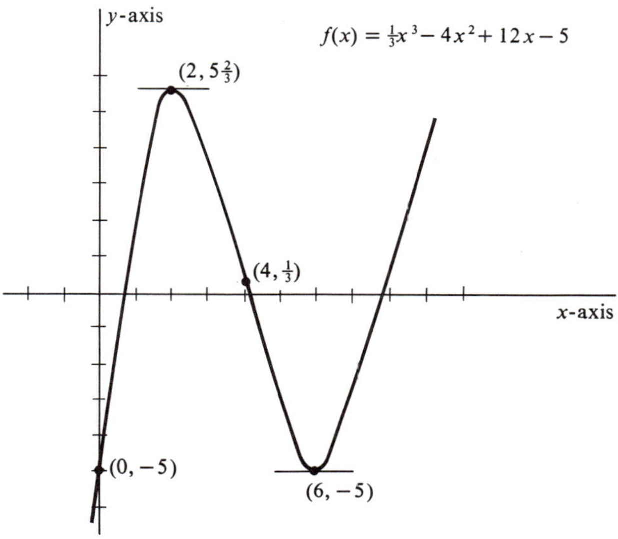

Figure 4

In summary, the graph of a function is concave downward when the second derivative of the function is negative and concave upward when the second derivative of the function is positive. The graph has horizontal tangents when the first derivative is 0. The points where the tangent is horizontal may be local maximum points or local minimum points or, as we shall see in Example 1, points of horizontal inflection.

It is important to understand clearly the definitions of the various expressions used in sketching graphs. The ordered pair $(a, f(a))$ is a **local maximum point** or a **local minimum point** of the function f if there is an open interval of the x-axis containing a such that, for every number x in that interval,

$$f(a) \geq f(x) \qquad \text{or} \qquad f(a) \leq f(x),$$

respectively. As we have indicated, the words relative maximum and relative minimum are also used. On the other hand, the pair $(a, f(a))$ is an **absolute maximum point** if, for every x in the domain of f,

$$f(a) \geq f(x),$$

and an **absolute minimum** point if

$$f(a) \leq f(x).$$

An **extreme point** is one that is either a maximum or minimum point (local or absolute). If $(a, f(a))$ is an extreme point, we shall call $f(a)$ the **extreme**

value of the function and shall say that the function has the extreme value **at** a. For example, we say that the function f in Figure 4 has a local minimum value of -5 which occurs at $x = 6$. However, this function has no absolute maximum or minimum points.

Any point $(a, f(a))$, where $f'(a) = 0$, is called a **critical point** of f.

A point of inflection is a point where the concavity changes sign. Thus $(a, f(a))$ is a **point of inflection** of the function f if there is an open interval on the x-axis containing a such that, for any numbers x_1 and x_2 in that interval,

$$f''(x_1)f''(x_2) < 0 \tag{1}$$

whenever $x_1 < a$ and $x_2 > a$. The inequality (1) simply says that $f''(x_1)$ and $f''(x_2)$ are of opposite sign. A characteristic of a point of inflection of a function is that its tangent line crosses the graph of the function at that point. The different possibilities are illustrated in Figure 5. Of the points P, Q, and

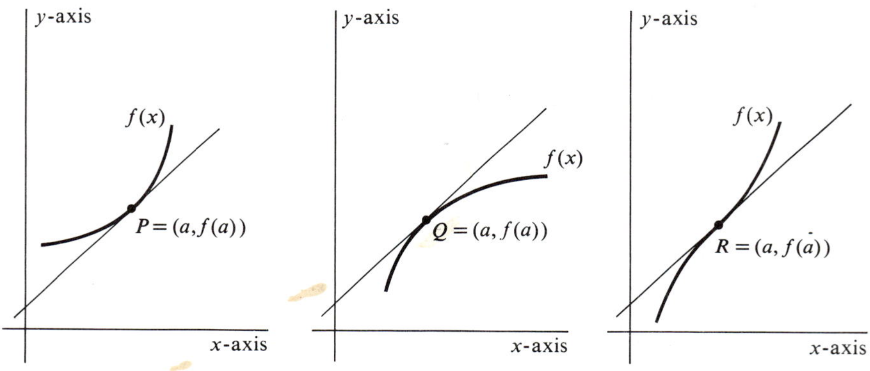

Figure 5

R, only R is a point of inflection. *If a function has a point of inflection $(a, f(a))$ and the second derivative $f''(a)$ exists, then $f''(a) = 0$.* However, it is possible to have a point of inflection at a point where there is no second derivative (see Problem 10 at the end of this section).

EXAMPLE 1. Sketch the graph of $f(x) = (x + 1)^3(x - 1)$. We first compute the derivatives:

$$f'(x) = (x + 1)^3 + (x - 1)3(x + 1)^2 = (x + 1)^2(4x - 2),$$

$$f''(x) = (x + 1)^2 4 + (4x - 2)2(x + 1) = 12x(x + 1).$$

Setting $f'(x) = 0$, we obtain $x = -1$ and $x = \frac{1}{2}$. Thus $(-1, 0)$ and $(\frac{1}{2}, -\frac{27}{16})$ are critical points and the tangents through these points are horizontal. Setting $f''(x) = 0$, we get solutions $x = -1$ and $x = 0$. Thus, if there are any points of inflection, they must occur at these two places. It is easy to see that the sign of the second derivative for values of x along the x-axis follows the pattern

positive	negative	positive	
-1	0		x-axis

We conclude that $(-1, 0)$ and $(0, -1)$ are in fact points of inflection. The point $(-1, 0)$, being both a critical point and a point of inflection, is a point of horizontal inflection. The graph is shown in Figure 6. Note that the graph

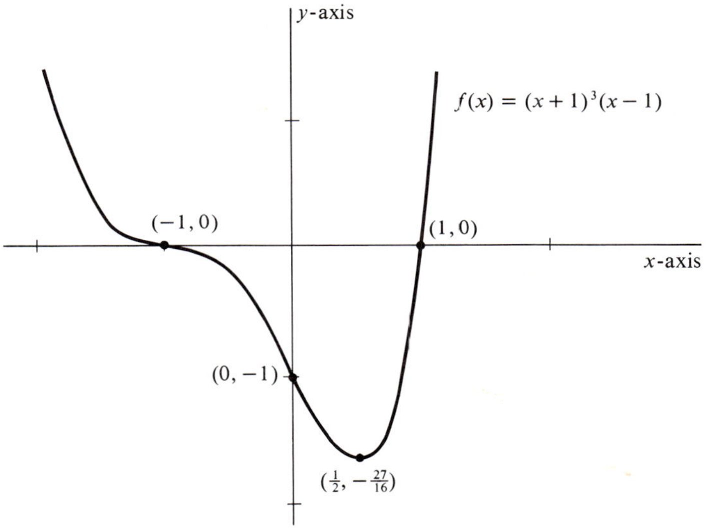

Figure 6

crosses the tangent at the point of horizontal inflection and that the slope of the tangent line does not change sign at that point. The first derivative (hence the slope of the tangent) increases to zero, as we go from the left to $x = -1$ and then decreases again through negative values, from $x = -1$ to $x = 0$. The point $(\frac{1}{2}, -\frac{27}{16})$, being the lowest point on the graph, is not only a local minimum point but also an absolute minimum point. $-\frac{27}{16}$ is the absolute minimum value of this function.

In the plotting of graphs of polynomial functions, we are frequently helped by knowing that a straight line can cut the graph of a polynomial function of the nth degree in at most n points. This is a consequence of the algebraic fact that, if $p(x)$ is a polynomial function of degree n, then the equation $p(x) = 0$ can have at most n distinct real roots. The function in Example 1 may be expanded to show that it is a polynomial of degree 4. It is possible to draw a straight line which will cut its graph in four points, but no straight line which will cut it in as many as five points.

EXAMPLE 2. Sketch the graph of $f(x) = 1 - x^{2/3}$. As before, we find derivatives:

$$f'(x) = -\tfrac{2}{3}x^{-1/3},$$
$$f''(x) = \tfrac{2}{9}x^{-4/3}.$$

For no values of x will we have $f'(x) = 0$, and so there are no critical points. On the other hand, $f'(0)$ is not defined and the graph has a vertical tangent at (0, 1). The first derivative is defined for all other values of x and is positive for $x < 0$ and negative for $x > 0$. Thus, at each point on the graph to the left of the vertical axis, the slope of the tangent is positive, increasing without limit as $x \to 0-$. At each point on the graph to the right of the vertical axis, the slope of the tangent is negative, increasing from negative numbers large in absolute value as x increases from 0. The second derivative $f''(x)$ is positive for all values of x except for $x = 0$, where it is not defined. It follows that there are no points of inflection. The graph is nowhere concave downward.

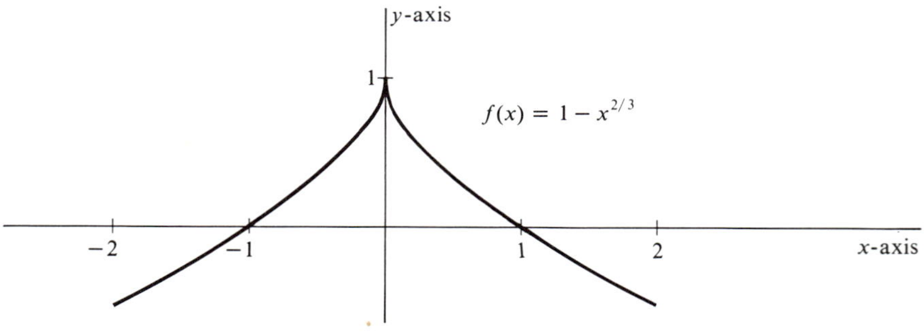

Figure 7

Note in Figure 7 that the portion of the graph which lies to the right of the vertical axis is the "mirror reflection" across that axis of the portion which lies to the left of the vertical axis. Note also that $f(-x) = f(x)$. Such a function, where $f(-x) = f(x)$, is called an **even function** and its graph will always contain two halves which can be brought into coincidence with each other by folding the graph along the vertical axis (Figure 8). The problem of graphing an even function is simplified by graphing it for positive values of x and drawing, for negative values of x, a reflection over the vertical axis of the right half of the graph.

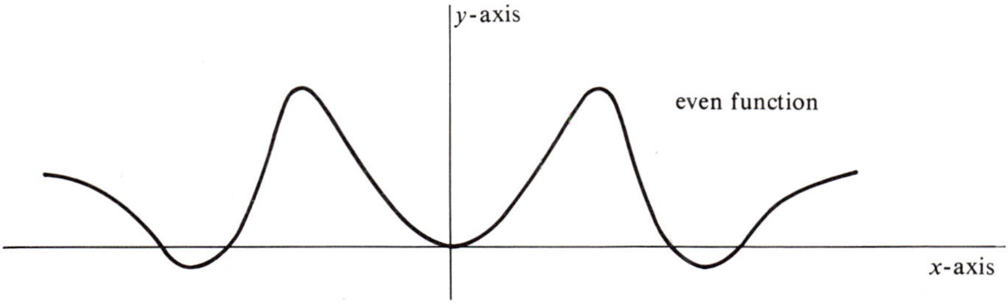

Figure 8

EXAMPLE 3. Sketch the graph of $f(x) = x + \frac{4}{x}$. The derivatives are

$$f'(x) = 1 - \frac{4}{x^2},$$

$$f''(x) = \frac{8}{x^3}\cdot$$

The first derivative vanishes for $x = 2$ and $x = -2$, and thus we see that $(2, 4)$ and $(-2, -4)$ are critical points. The second derivative is negative when x is negative, so the curve is bending down at $(-2, -4)$ and that point must be a local maximum point. Similarly, $f''(2) > 0$ and $(2, 4)$ is a local

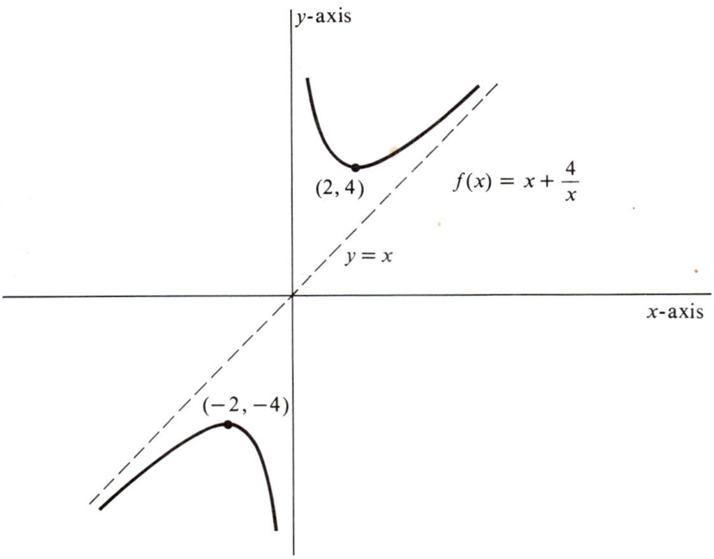

Figure 9

minimum point. $f(0)$ is undefined and $|f(x)|$ increases as $x \to 0$. Behavior for large values of $|x|$ can be seen, since $f(x)$ approaches x as x increases or decreases without bound.

From Figure 9, one can see the graph approaching the vertical axis as $x \to 0$ from either side, and also approaching the graph of the equation $y = x$ as $|x|$ increases without bound. Note that the two parts of the graph are reflections of each other across the origin, and also that $f(-x) = -f(x)$. Any function f for which $f(-x) = -f(x)$ is called an **odd function.** The graph of an odd function may be obtained by first drawing the graph of $f(x)$, where $x \geq 0$ (see Figure 10). We may then obtain the remainder of the

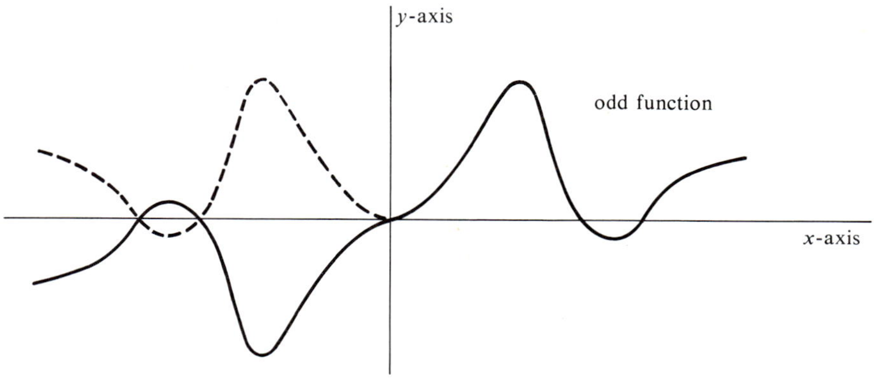

Figure 10

graph by first reflecting this positive part about the y-axis and then about the x-axis. The result of reflecting first about one axis and then about the other we shall call reflection about the origin.

Summarizing the techniques of curve sketching, we find the first and second derivatives of the function with respect to x, we find the points where either derivative vanishes, and we determine critical points and points of inflection. Points of general use in graphing $f(x)$ include:

1. The tangent is horizontal if $f'(x) = 0$.

2. The curve is concave downward if $f''(x)$ is negative, concave upward if $f''(x)$ is positive.

3. $(a, f(a))$ is a local maximum point if $f'(a) = 0$ and $f''(a) < 0$, a local minimum point if $f'(a) = 0$ and $f''(a) > 0$.

4. $(a, f(a))$ is a point of inflection if $f''(x)$ changes sign as x increases through a.

5. Even and odd functions need be investigated carefully only for $x \geq 0$. The rest of the graph of an even function is found by reflection across the vertical axis, of an odd function by reflection about the origin.

PROBLEMS

1. Sketch the graph of each of the following functions, carefully labeling all extreme points and all points of inflection. Classify each extreme point as to type.

(a) $x^2 - 5x + 6$

(b) $3 - 2x - x^2$

(c) $2x^2 - 3x - 1$

(d) $5 - 2x^2$

(e) $x^3 - 3x$

(f) $(x + 1)(x^3 - x^2 - 5x + 13)$

(g) $\dfrac{x + 2}{x}$

(h) $\dfrac{16}{x} + x^2$

(i) $x^3 - x$

(j) $x^4 - x$

(k) $(a^{2/3} - x^{2/3})^{3/2}$

(l) $|x|$

(m) $|x - 7|$

(n) $3 + \dfrac{6}{x - 2}$

(o) $x^4 - 8x^2 + 3$

(p) $\dfrac{x^2 - 5x + 4}{8x}$

(q) $(x - 1)(x - 2)(x - 3)$

(r) $1 + 6x - \frac{1}{2}x^3$.

2. Show that a polynomial function of x which consists only of even powers of x is an even function.
3. Show that a polynomial function of x which consists only of odd powers of x is an odd function.
4. (a) Show that the graph of the function $ax^2 + bx + c$, $a \neq 0$, always has an absolute extreme point.
   (b) Which of the constants a, b, and c determines the type of extreme point of the graph?
   (c) What is the extreme value of $ax^2 + bx + c$?
   (d) Write $ax^2 + bx + c$ as $a\left(x^2 + \frac{b}{a}x\right) + c$, complete the square on $x^2 + \frac{b}{a}x$ without changing the function, and find the result of (c) algebraically.
5. Show that the graph of $x^3 - 12x$ has a local maximum point but no absolute maximum point and that it also has a local minimum point which is strictly local.
6. Sketch the graph of $f(x)$, if $f(0) = 3$ and $f'(x) = -1$ for all real values of x.

7. Sketch the graph of $f(x)$, if $f(-1) = 2$ and $f'(x) = \frac{1}{2}$ for all real values of x.
8. Sketch the graph of $f(x)$, if $f(0) = 0$ and $f'(x) = x$ for all real values of x.
9. Construct a function which has a local maximum point, with local maximum point defined as in this section, but would not have a local maximum if the definition were changed to demand $f(a) > f(x)$, $x \neq a$, instead of $f(a) \geq f(x)$.
10. Graph the function $x^{1/3}$ and show that it has a point of inflection where neither the first nor the second derivative exists.

2. Maximum and Minimum Problems. In sketching the graph of a function, we spent some time looking for maximum and minimum points, both local and absolute. This idea suggests that we can use the same technique to find that value (or those values) of a variable which maximize or minimize a length, an area, or a profit. For example, what should be the dimensions of the rectangular field which can be enclosed with a fixed length of fencing but has the greatest area? Or, what are the dimensions of the quart can which can be made from the least amount of tin? Or, if the telephone company were to reduce the rate per instrument for each new instrument over a certain number, what number of telephones would give them the greatest profit?

These are all problems which can be solved by calculus and, more specifically, by the technique developed in Section 1. However, before we tackle them, we shall consider the theorems which justify the methods which we shall use in solving them.

(2.1) THEOREM. *If a belongs to an open interval in the domain of f, if $f'(a)$ exists, and if $(a, f(a))$ is a local extreme point* (either a maximum or a minimum), *then* $f'(a) = 0$.

Proof. Geometrically this theorem is obvious. We shall prove it, only in the case that $(a, f(a))$ is a local maximum point, since a similar proof (with the inequalities reversed) is valid for a local minimum point. Since $(a, f(a))$ is a local maximum point, $f(a + t) \leq f(a)$ for all t in some open interval containing 0. Thus $f(a + t) - f(a) \leq 0$. If t is negative, $\dfrac{f(a + t) - f(a)}{t} \geq 0$ and

$$\lim_{t \to 0-} \frac{f(a + t) - f(a)}{t} \geq 0.$$

If t is positive,

$$\frac{f(a + t) - f(a)}{t} \leq 0 \quad \text{and}$$

$$\lim_{t \to 0+} \frac{f(a + t) - f(a)}{t} \leq 0.$$

Since $f'(a)$ exists, the two limits above must have a common value which is $f'(a)$. Thus $f'(a)$ is both greater than or equal to zero and also less than or equal to zero. The only number which satisfies both of these conditions is zero, hence $f'(a) = 0$. This completes the proof.

In our sketches we found points where the first derivative vanished. If the curve was concave downward at that point, we identified a local maximum point; if the curve was concave upward at that point, we identified a local minimum point. We summarize this result in the following theorem.

(2.2) THEOREM. *Let f be a function with a continuous second derivative and with $f'(a) = 0$. Then $(a, f(a))$ is a local maximum point if $f''(a) < 0$ and is a local minimum point if $f''(a) > 0$.*

This theorem is easily proved when we have more mathematics at our command. Specifically, it follows quickly from Taylor's Formula with the Remainder (see Problem 13, page 540). For the work at hand, it will be sufficient to understand the theorem and to be able to use it.

If the domain of f is restricted to a closed interval, we may find an absolute extreme point which lies on the boundary of the interval. Consider the function graphed in Figure 11. This function is defined on the closed interval

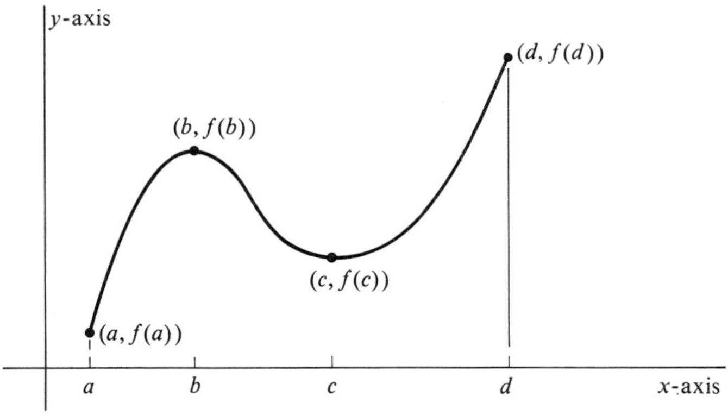

Figure 11

$[a, d]$ and has a local minimum point at $(c, f(c))$. However, there are several points in $[a, d]$ which are lower than $(c, f(c))$, and the absolute minimum point is $(a, f(a))$. Similarly, $(b, f(b))$ is a local maximum point but $(d, f(d))$ is the absolute maximum point. This suggests the following theorem.

(2.3) THEOREM. *Let f be a differentiable function whose domain is restricted to a closed interval containing a. If $(a, f(a))$ is an extreme point then $f'(a) = 0$ or a is an endpoint of the interval.*

Proof. This theorem is an immediate corollary of Theorem (2.1). Let the domain be $[c, d]$. If $a \neq c$ and $a \neq d$, then a lies in the open interval (c, d) and, by Theorem (2.1), $f'(a) = 0$. If $a = c$ or $a = d$, then a is an endpoint of the interval.

For many functions, Theorem (2.3) has the virtue of reducing an apparently impossible problem to a simple one. In principle, the problem of finding the maximum values of a function over a closed interval involves the examination of $f(x)$ for every x in the interval, i.e., for an infinite number of points. This theorem tells us that we need look only at those values of x at which the first derivative vanishes and those which are endpoints. For most functions there is only a small finite number of such points.

Theorem (2.3) tells us where to look for the extreme points of a differentiable function defined on a closed interval, but it does not guarantee the existence of any. To complete the theory, we add a statement of the following fundamental existence theorem.

(2.4) THEOREM. *Every real-valued continuous function whose domain is a closed bounded interval has at least one absolute maximum point and at least one absolute minimum point.*

We omit the proof. The result sounds perfectly obvious, of course, and it is obvious in the sense that if continuity means what we want it to mean, then (2.4) must be true. To see whether it, in fact, follows logically from our definitions, of course, requires proof. Further insight into the theorem may be found in Problems 21 and 22, where it is shown that functions that do not satisfy the hypotheses of (2.4) can fail to have absolute extreme points.

Let us now look at the problems suggested at the beginning of the section.

EXAMPLE 1. What are the dimensions of the largest rectangular field which can be enclosed with 200 feet of fencing? Should it be long and narrow, short and wide, or somewhere in between? If we let x be the number of feet in the length, then $100 - x$ will represent the width, and we can write the area A as a function of x:

$$A(x) = x(100 - x) = 100x - x^2$$

(see Figure 12). Note that the domain of A is the interval consisting of all x

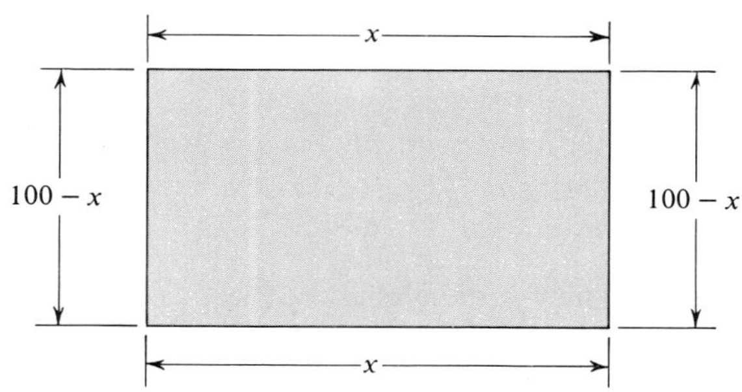

Figure 12

such that $0 < x < 100$. We want to find that value of x which will give a maximum value of A. Taking derivatives, we obtain $A'(x) = 100 - 2x$ and $A''(x) = -2$. Setting A' equal to zero, we find $x = 50$. Since $A'(50) = 0$ and $A''(50) = -2 < 0$, we know that 50 is that value of x which maximizes A. Thus the field, which is 50 feet long and $100 - 50 = 50$ feet wide, is the largest rectangular field which can be enclosed with 200 feet of fencing.

The problem of solving a maximum or minimum problem consists of setting up the function to be maximized or minimized and then taking derivatives. The theorems of this section tell how to proceed from there.

EXAMPLE 2. What are the dimensions of the cylindrical quart can which can be made from the least amount of tin? This problem is important to the manufacturer who produces tin cans and is more concerned with the amount of tin used than with the shape of the can. Should he make tall cans with a small radius or short cans with a large radius? We shall ignore seams and assume that tin cans are perfect cylinders. The volume of a cylinder is $\pi r^2 h$ and a quart contains $57\frac{3}{4}$ cubic inches. Thus if r is the radius of the can in inches and h is the height in inches, $\pi r^2 h = 57\frac{3}{4}$ or $\frac{231}{4}$ (see Figure 13).

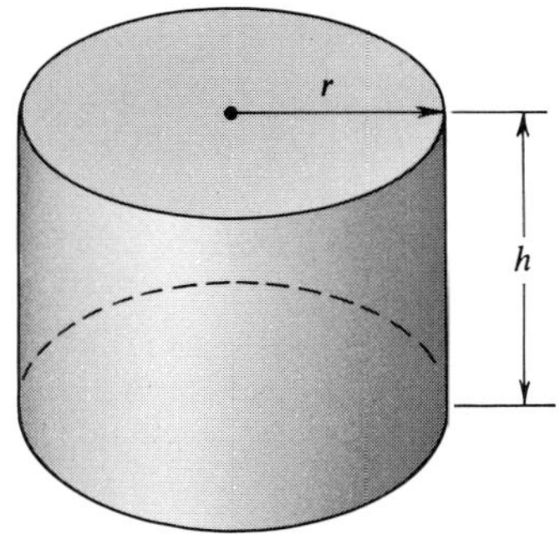

Figure 13

The area is the sum of the lateral surface area and the area of the bottom and the top: $A = 2\pi rh + 2\pi r^2$. The area depends on r and h, but we can use our volume equation to find h as a function of r: $h = \dfrac{231}{4\pi r^2}$, and then write

$$A(r) = 2\pi r\left(\frac{231}{4\pi r^2}\right) + 2\pi r^2 = \frac{231}{2r} + 2\pi r^2.$$

The domain of A is the set of positive real numbers. Taking derivatives, we have $A'(r) = -\dfrac{231}{2r^2} + 4\pi r$ and $A''(r) = \dfrac{231}{r^3} + 4\pi$. Setting A' equal to zero, we have $8\pi r^3 = 231$, or $r = \dfrac{1}{2}\sqrt[3]{\dfrac{231}{\pi}} = 2.10$ (approximately). Since $A'(2.10) = 0$ and $A''(2.10) > 0$, 2.10 gives a minimum value to A. $h = \dfrac{231}{4\pi(2.10)^2} = 4.20$ (approximately). The desired can will look square in profile, 2.10 inches in radius and 4.20 inches high. This problem could also have been solved by writing r as a function of $h : r = \sqrt{\dfrac{231}{4\pi h}}$ and then writing $A(h) = 2\pi\sqrt{\dfrac{231}{4\pi h}}\,h + 2\pi \cdot \dfrac{231}{4\pi h}$, although this area function is not as nice as $A(r)$. Another method involves thinking of h as a function of r, writing $A(r)$ containing both h and r, and differentiating implicity. Thus we write

$$\pi r^2 h = \frac{231}{4} \qquad \text{and} \qquad A(r) = 2\pi rh + 2\pi r^2.$$

Differentiating with respect to r, we obtain

$$\pi r^2 \frac{dh}{dr} + \pi h \cdot 2r = 0 \qquad \text{and} \qquad A'(r) = 2\pi r \frac{dh}{dr} + 2\pi h + 2\pi \cdot 2r.$$

From the first, $\dfrac{dh}{dr} = -\dfrac{2\pi rh}{\pi r^2} = -\dfrac{2h}{r}$. Substituting this in the second, we find $A'(r) = 2\pi r\left(-\dfrac{2h}{r}\right) + 2\pi h + 4\pi r = 4\pi r - 2\pi h$. Setting $A'(r) = 0$, we get

$$h = 2r.$$

Substitution in the volume equation yields $\pi r^2(2r) = \dfrac{231}{4}$, or $8\pi r^3 = 231$, as in the other solution. This same method could have been used with differentiation with respect to h.

In each of the preceding examples, the domain of the function was an interval of positive real numbers. However, if a problem involves the number of objects which will maximize or minimize a particular function, it will have a domain of positive integers. In this case, we may still do the problem as if it were one with the entire set of real numbers as the domain of the function, and then consider those positive integers which lie on either side of the x-coordinate of the critical point of this function.

EXAMPLE 3. A telephone company which serves a small community makes an annual profit of \$12 per subscriber if it has 725 subscribers or fewer. They decide to reduce the rate by a fixed sum for each subscriber over 725, thereby reducing the profit 1 cent per subscriber. Thus there will be a profit of \$11.99 on each of 726 subscribers, \$11.98 on each of 727, etc. What is the number of subscribers which will give them the greatest profit?

If we let x be the number of subscribers over 725, there will then be $725 + x$ subscribers and the profit per subscriber will be $1200 - x$ cents. The total profit will be $P(x) = (725 + x)(1200 - x) = 870{,}000 + 475x - x^2$. The domain of P is the set of positive integers, but let us treat the problem as if the domain were R. The derivatives are $P'(x) = 475 - 2x$ and $P''(x) = -2$. The value of x which makes P' zero is $237\frac{1}{2}$, and it also makes P'' negative, thereby ensuring a maximum for P. We find $P(237) = 926{,}406$ and $P(238) = 926{,}406$. Thus the profit is the same for $725 + 237 = 962$ subscribers and for $725 + 238 = 963$ subscribers. If we visualize the graph of P, we see a parabola concave downward with its maximum point at $(237\frac{1}{2}, 926{,}406\frac{1}{4})$. If we delete all points which do not have integral coefficients, then the points $(237, 926{,}406)$ and $(238, 926{,}406)$ are maximum points equally spaced on either side of the high point of the parabola and just lower than the high point. The profit of the telephone company will increase with each new subscriber until they have 962 subscribers. The addition of one more subscriber will not alter the profit, but it will then decrease with each new subscriber after the 963rd one.

EXAMPLE 4. Consider the function f defined by

$$f(x) = \frac{x^3}{3} - x^2 + \tfrac{2}{3}, \qquad -2 \leq x \leq 3.$$

The domain of f is the closed interval $[-2, 3]$. Find the maximum and minimum values. The derivatives are $f'(x) = x^2 - 2x = x(x - 2)$ and $f''(x) = 2x - 2 = 2(x - 1)$. Setting $f'(x) = 0$, we find solutions $x = 0$ and $x = 2$. By Theorem (2.3) we need evaluate $f(x)$ only where $x = 0$ and $x = 2$ [where $f'(x) = 0$] and at the endpoints of the interval, $x = -2$ and $x = 3$. These values are

$$\begin{aligned} f(-2) &= -6, \\ f(0) &= \tfrac{2}{3}, \\ f(2) &= -\tfrac{2}{3}, \\ f(3) &= \tfrac{2}{3}. \end{aligned}$$

Thus the maximum value of f is $\frac{2}{3}$, and the minimum value is -6. Note that the maximum value of the function occurs at two points, one of which is a local maximum and the other the right endpoint of the interval. The minimum value, on the other hand, does not occur at a local minimum point, but at the left endpoint. The graph of the function is shown in Figure 14.

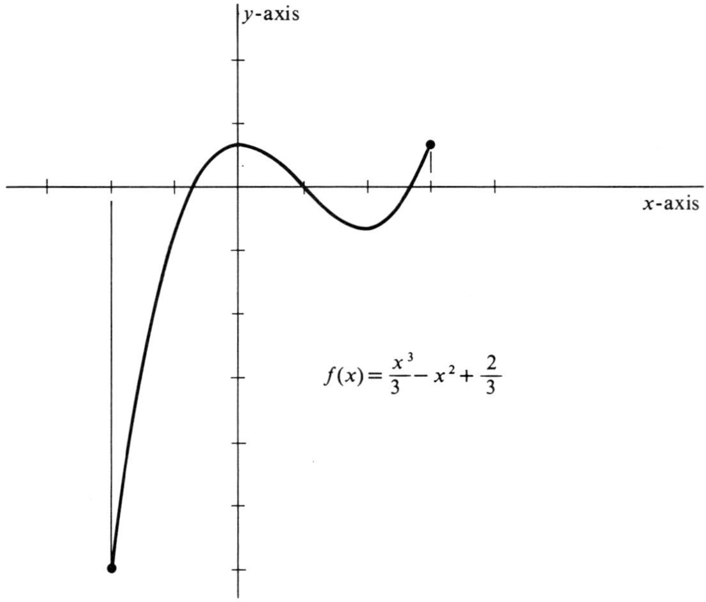

Figure 14

We summarize the techniques for solving maximum and minimum problems as follows:

1. Set up the function to be maximized or minimized. If it can easily be set up as a function of one variable, it should be. If it cannot easily be set up as a function of one variable, it should be accompanied by an equation relating the two variables.

2. Take first and second derivatives of the function. Set the first derivative equal to zero and solve the resulting equation. Evaluate the second derivative at those values of x for which the first derivative equals zero.

3. If the function is a "nice" function defined on an open interval, its maximum will occur where $f'(x) = 0$ and $f''(x) < 0$ and its minimum will occur where $f'(x) = 0$ and $f''(x) > 0$.

4. If the function has a closed interval for its domain, evaluate the function at both endpoints to see if maximum or minimum values occur there.

5. If the function has a domain restricted to integers, evaluate the function at integers near the values which give a maximum or a minimum value to the continuous function.

Sometimes the geometric or physical properties of the problem make it obvious that a critical point is of the desired type, i.e., a maximum or a minimum. If this happens, it is not necessary to compute the second derivative, although it may still be used as a check. The complete behavior of the function whose extreme points are desired can always be found by carefully plotting its graph.

PROBLEMS

1. Generalize on Example 1 to show that the largest rectangle with a fixed perimeter p is a square with side $\frac{p}{4}$.
2. A field is bounded on one side by a stone wall. A rectangular plot of ground is to be fenced off, using the stone wall as one boundary and 200 yards of fencing for the other three sides. What are the dimensions of the largest such plot?
3. Find the positive number which is such that the sum of the number and its reciprocal is a minimum.
4. List all local extreme points and all absolute extreme points for each of the following functions, noting carefully its domain of definition. Classify each extreme point by type.

   (a) $3x^5 - 5x^3 + 7$; domain: all real numbers.

   (b) $4x^3 + 3x^2 - 6x + 5$; domain: all real numbers.

   (c) $x + \dfrac{a^2}{x}$; domain: all nonzero real numbers.

   (d) $2x^3 - 21x^2 + 60x - 25$; domain: all nonnegative real numbers.

   (e) $\dfrac{x^2}{x - 1}$; domain: all real numbers except 1.

   (f) $3x^4 - 20x^3 - 36x^2 + 54$; domain: all nonpositive real numbers.

   (g) $(x - 1)^2(x + 1)^3$; domain: all nonnegative real numbers no greater than 2.

   (h) $2 - (x + 4)^{2/3}$; domain: all real numbers.

   (i) $(x - 1)^2(x - 4)$; domain: all nonnegative real numbers.

5. Generalize on Example 2 to show that the right circular cylinder with a fixed volume and the least total surface area has its diameter equal to its height.
6. Show that $f(x) = x^4$ has an extreme point where the second derivative is neither positive nor negative. What type of extreme point is it? Explain why this is not a contradiction of Theorem (2.2).
7. A line has positive intercepts on both axes and their sum is 8. Write an equation of the line if it cuts off in the first quadrant a triangle with area as large as possible.

8. Find two nonnegative numbers, x and y, such that $x + y = 6$ and x^2y is as large as possible.
9. Find all ordered pairs, (x, y), such that $xy = 9$ and $\sqrt{x^2 + y^2}$ is a minimum. Interpret your result geometrically.
10. (a) Graph the set of ordered pairs (x, y) such that $4x^2 + y^2 = 8$. The graph is called an ellipse.

    (b) Find all ordered pairs (x, y), such that $4x^2 + y^2 = 8$ and $4xy$ is a maximum.

    (c) Find the dimensions of the largest (in area) rectangle which has sides parallel to the x-axis and the y-axis and is inscribed in the ellipse of (a).
11. Find the dimensions of the largest rectangle which has its upper two vertices on the x-axis and the other two on the graph of $y = x^2 - 27$.
12. Find the dimensions of the rectangle which has its upper two vertices on the x-axis and the other two on the graph of $y = x^2 - 27$ and which has maximum perimeter.
13. A box without a top is to be made by cutting equal squares from the corners of a rectangular piece of tin 30 inches by 48 inches and bending up the sides. What size should the squares be if the volume of the box is to be a maximum? [*Hint:* If x is the side of a square, $V(x) = x(30 - 2x)(48 - 2x)$.]
14. (a) A box without a top is to be made by cutting equal squares from the corners of a square piece of tin, 18 inches on a side, and bending up the sides. How large should the squares be if the volume of the box is to be as large as possible?
    (b) Generalize (a) to the largest open-topped box which can be made from a square piece of tin, s inches on a side.
15. (a) Where should a wire 20 inches long be cut if one piece is to be bent into a circle, the other piece is to be bent into a square, and the two plane figures are to have areas the sum of which is a maximum?
    (b) Where should the cut be if the sum of areas is to be a minimum?
16. A man in a canoe is 6 miles from the nearest point of the shore of the lake. The shoreline is approximately a straight line and the man wants to reach a point on the shore 5 miles from the nearest point. If his rate of paddling is 4 miles per hour and he can run 5 miles per hour along the shore, where should he land to reach his destination in the shortest possible time?
17. Prove that the largest isosceles triangle which can be inscribed in a given circle is also equilateral.
18. Prove that the smallest isosceles triangle which can be circumscribed about a given circle is also equilateral.
19. What is the smallest positive number that can be written as the sum of two positive numbers x and y so that $\frac{1}{x} + \frac{2}{y} = 1$?
20. An excursion train is to be run for a lodge outing. The railroad company sets the rate at \$10 per person if less than 200 tickets are sold. They agree to lower the rate per person by 2 cents for each ticket sold above the 200 mark, but the train will only hold 450 people. What number of tickets will give the company the greatest income?

21. Consider the continuous real-valued function $f(x) = x$ with domain $0 < x < 1$. Does this function have an absolute maximum point or an absolute minimum point? Why is this function not a counterexample to Theorem (2.4)?
22. Let

$$f(x) = \begin{cases} \dfrac{1}{x}, & \text{for } \begin{cases} -1 \leq x < 0, \\ 0 < x \leq 1, \end{cases} \\ 0, & \text{for } \quad x = 0. \end{cases}$$

This real-valued function is defined on the closed interval $[-1, 1]$. Draw the graph of $f(x)$ and explain why this function has neither absolute maximum nor absolute minimum points.

3. Rates of Change with respect to Time. The values of many physical quantities depend on time and change with time. In a mathematical formulation such a quantity is usually denoted by a variable which is a function of time. In this section we are concerned with the instantaneous rates of change of time-dependent variables. Let u be a real-valued function of a real variable t, where we identify t with time. The rate of change of u with respect to time at a given instant $t = a$ can be determined by considering the derivative of u. We have already observed in Chapter 1 that the derivative $f'(a)$ of a function f at a particular number a is the rate of change of the value $f(x)$ of the function f with respect to x at a. It follows that the instantaneous rate of change of u with respect to time, when $t = a$, is equal to the derivative:

$$u'(a) = \frac{du}{dt}(a) = \lim_{d \to 0} \frac{u(a + d) - u(a)}{d}.$$

In a physical application the variable u might denote the number of gallons of water in a tank at time t, where t is measured in minutes. Then, $\dfrac{du}{dt}(a)$ is equal to the rate at which water is flowing in or out of the tank at time $t = a$ and is measured in gallons per minute. If $\dfrac{du}{dt}(a)$ is positive, then the quantity of water in the tank is increasing when $t = a$ and water is flowing into the tank. On the other hand, if $\dfrac{du}{dt}(a)$ is negative, then the amount of water is decreasing at that moment and water is draining out. Finally, if $\dfrac{du}{dt}(a) = 0$, then the amount is not changing at $t = a$.

An important example of rate of change with respect to time is velocity. For example, consider a car in motion on a straight road. To formulate the situation mathematically, we identify the road with a real number line, the car with a point on the line, and the location of the car at time t with the coordinate $s(t)$ of the point on the line. Thus, s is a real-valued function of the real variable t. The **average velocity** during the time interval from $t = a$ to $t = a + d$ is equal to the change in position divided by the change in time. Denoting this quantity by v_{av}, we have

$$v_{av} = \frac{s(a + d) - s(a)}{d}.$$

If we graph $s(t)$ on a time-position graph, as in Figure 15, we see that the average velocity is the slope of the line segment connecting the point $(a, s(a))$

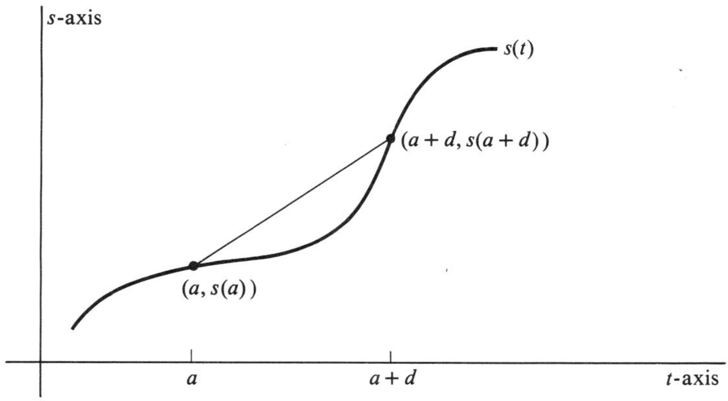

Figure 15

to the point $(a + d, s(a + d))$. If, keeping a fixed, we consider average velocities over successively shorter and shorter intervals of time, we obtain values nearer and nearer to the rate of change of s at $t = a$. We take this limit as d approaches zero as the definition of **velocity** at a and use the symbol $v(a)$ for it. Thus

$$v(a) = \lim_{d \to 0} v_{av} = \lim_{d \to 0} \frac{s(a + d) - s(a)}{d} = s'(a).$$

Hence velocity is the derivative of position with respect to time, and we write $v(t) = s'(t)$, or simply $v = s'$. Geometrically, the velocity at a is the slope of the tangent line to the graph of the function s at the point $(a, s(a))$.

For a particle moving on a real number line, a positive value of $v(t)$ means that the motion at time t is in the direction of increasing numerical values, which is called the positive direction (i.e., if the line is the x-axis, then the particle is moving to the right). If $v(t)$ is negative, then the particle is moving in the opposite direction. Finally, zero velocity indicates that the particle is at rest. The **speed** at time t is defined to be the absolute value $|v(t)|$ of the velocity. Obviously, the speed measures how fast the particle is moving without regard to its direction.

Velocity depends on time, and its rate of change with respect to time will tell us even more about the motion of a particle. This rate of change is called **acceleration** and is defined to be the derivative of velocity with respect to time. Denoted by a, it is the second derivative of position with respect to time:

$$a(t) = v'(t) = s''(t).$$

The acceleration is the rate of change of velocity with respect to time. It may be positive, indicating that the velocity is increasing; zero, telling us that the velocity is constant; or negative, indicating that the velocity is decreasing. For motion of a particle on a horizontal number line (or the x-axis) we have several possibilities. If the velocity is positive and the acceleration is positive, the motion is to the right and the speed of the particle is increasing. If the velocity is positive and the acceleration is negative, the motion is still to the right but the particle is slowing down. If the acceleration is zero, the velocity is momentarily not changing. If the velocity is negative, the particle is moving to the left and it is slowing down or speeding up, depending on whether the acceleration is, respectively, positive or negative.

EXAMPLE 1. A particle moves on the x-axis and its coordinate, as a function of time, is given by $x(t) = 2t^3 - 21t^2 + 60t - 14$, where t is measured in seconds. Describe its motion. We first take derivatives to find velocity and acceleration: $v(t) = 6t^2 - 42t + 60 = 6(t^2 - 7t + 10) = 6(t - 2)(t - 5)$ and $a(t) = 12t - 42 = 6(2t - 7)$. At zero time the particle is at $x = -14$, moving to the right with a velocity of 60 units per second. At that moment, acceleration is -42, and the particle is slowing down. At time $t = 2$, the particle is at rest ($v = 0$) at $x = 38$, and the acceleration is still negative: $a = -18$. For the next $1\frac{1}{2}$ seconds the particle moves to the left until, at $t = \frac{7}{2}$, it is at $x = 24\frac{1}{2}$, moving to the left with a speed of $13\frac{1}{2}$ units per second. At that moment, however, the acceleration is zero and, in the next instant, the velocity will begin to increase to the right and the particle begin to slow down. The particle continues to move to the left for the next $1\frac{1}{2}$ seconds, until $t = 5$. At that time, the particle is at rest at $x = 11$ and the acceleration is positive. From that time on, the particle will move to the right with ever-increasing velocity. Its motion is indicated in Figure 16.

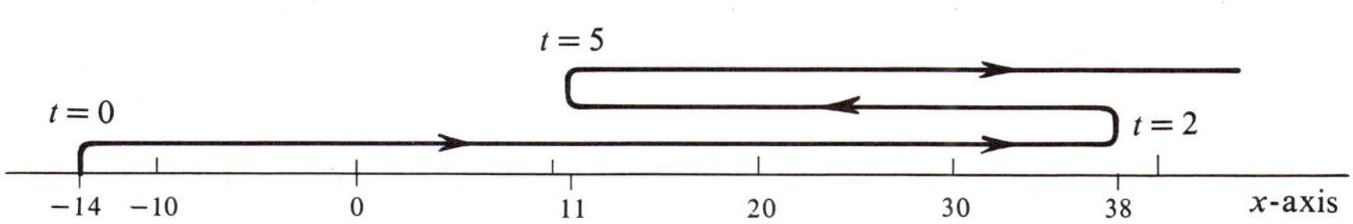

Figure 16

EXAMPLE 2. In a first course in physics, one encounters the formula for straight-line motion with constant acceleration, $s = s_0 + v_0t + \frac{1}{2}at^2$, where s is the distance from some fixed point, s_0 the initial distance, v_0 the initial velocity, and a the acceleration. Find v and a, thereby verifying another formula which usually accompanies the distance formula and also verifying that the acceleration is constant. Taking derivatives with respect to time, we obtain $v = s' = v_0 + \frac{1}{2}a(2t) = v_0 + at$ and $a = v' = a$. Thus we see that the derivative definitions do produce the familiar formulas.

If a particle is constrained to move in the xy-plane on a circle of radius 5, then the point where it is at any time has coordinates which satisfy the equation $x^2 + y^2 = 25$. Each of the coordinates, however, is a function of time and we may write $[x(t)]^2 + [y(t)]^2 = 25$. Here we have an equation which states that two functions of t, $[x(t)]^2 + [y(t)]^2$ and the constant function 25, are equal to each other. If the two functions are equal, they will change with respect to t at the same rate. Taking derivatives to find the common rate of change, we have

$$\frac{d}{dt}(x^2 + y^2) = \frac{d}{dt}25,$$

which implies

$$2x\frac{dx}{dt} + 2y\frac{dy}{dt} = 0.$$

We interpret $\frac{dx}{dt}$ as the rate of change of the abscissa of the particle with respect to time, or as the velocity in the horizontal direction. Similarly, we interpret $\frac{dy}{dt}$ as the velocity in the vertical direction. We use the symbols v_x for $\frac{dx}{dt}$ and v_y for $\frac{dy}{dt}$. Another interpretation of v_x and v_y is that they are horizontal and vertical components, respectively, of the velocity of the particle. Using this notation, we write $xv_x + yv_y = 0$ or $v_x = -\frac{y}{x}v_y$. These equations relate two rates of change, and problems of this type are called **related rate** problems.

EXAMPLE 3. A particle moves on the circle with equation $x^2 + y^2 = 10$. As it passes through the point $(-1, -3)$ the horizontal component of its velocity is 6 units per second. Find the vertical component. We first take derivatives with respect to time, $\frac{d}{dt}(x^2 + y^2) = \frac{d}{dt}10$, to get $2xv_x + 2yv_y = 0$.

We are given that $v_x = 6$ when $x = -1$ and $y = -3$. Substituting these values in the last equation, we have $2(-1)(6) + 2(-3)v_y = 0$. Hence $-12 - 6v_y = 0$, or $v_y = -2$. The vertical component of velocity is -2 units per second, indicating that the motion is, at that moment, downward and to the right, since v_x is given positive. It is, of course, obvious that a particle which is constrained to move on the circle must be moving downward if it is moving to the right in the third quadrant.

EXAMPLE 4. A spherical balloon is being blown up, and its volume is increasing at a rate of 4 cubic inches per second. At what rate is its radius increasing? The volume of a sphere is given by the equation, $V = \frac{4}{3}\pi r^3$. Since V and $\frac{4}{3}\pi r^3$ are both functions of t, their derivatives with respect to t are equal. Thus $\frac{dV}{dt} = \frac{4}{3}\pi \cdot 3r^2\frac{dr}{dt} = 4\pi r^2\frac{dr}{dt}$. Replacing $\frac{dV}{dt}$ by 4 and solving for $\frac{dr}{dt}$, we have $\frac{dr}{dt} = \frac{1}{\pi r^2}$. The rate at which the radius is increasing is not constant, but depends on the radius at a particular moment. When the radius is 2 inches, it is increasing $\frac{1}{4\pi}$ inches per second; when it is 5 inches, it is increasing $\frac{1}{25\pi}$ inches per second, etc.

Most related rate problems are solved by first finding an equation relating the variables. Then we may take derivatives to find an equation relating their rates of change with respect to time. Finally, we substitute those simultaneous values of the variables and rates which are given to us in the problem.

PROBLEMS

1. A ball is thrown upward with an initial velocity of 80 feet per second. Its distance above the ground t seconds later is given by $s = 80t - 16t^2$.

   (a) Show that it reaches its highest point when it has zero velocity.

   (b) Show that its acceleration is constant.

   (c) For how many seconds is it going up?

   (d) How high does it go?

   (e) Show that it strikes the ground with a speed of 80 feet per second.

2. A particle moves on the x-axis and its position t seconds after it starts is given by $x(t) = 4t^3 - 42t^2 + 135t - 100$. Describe its motion.

3. A particle moves on the y-axis and its position t seconds after it starts is given by $y(t) = 144t - 288 - 16t^2$. Describe its motion.
4. A particle moves on the x-axis and its position t seconds after it starts is given by $x(t) = t^3 - 3t^2 + 5$. Describe its motion.
5. A particle moves on the y-axis and its position t seconds after it starts is given by $y(t) = 3t^3 - 9t + 10$. Describe its motion.
6. A particle moves on the ellipse with equation $4x^2 + 9y^2 = 36$. When it is passing through the point $(-3, 0)$, what is the horizontal component of its velocity?
7. A particle moves on the circle with equation $x^2 + y^2 = 16$. Show that $\frac{v_y}{v_x}$ at (a, b) is equal to the slope of the tangent to the circle at (a, b).
8. Two ships leave the same dock at noon, one traveling due north at 15 knots (nautical miles per hour) and the other due east at 20 knots. At what rate is the distance between them increasing at 2 P.M.?
9. Water is pouring into a conical funnel and, although it is also running out of the bottom, the amount of water in the funnel is increasing at the rate of 3 cubic inches per minute. If the conical part of the funnel is 5 inches deep and the mouth of the funnel is 6 inches in diameter, how fast is the water rising when it is 2 inches deep?
10. (a) Show that, at any instant, the ratio of the rate at which the area of a circle is changing to the rate at which the radius is changing is equal to the circumference of the circle.
    (b) Show that, at any instant, the ratio of the rate at which the volume of a sphere is changing to the rate at which the radius is changing is equal to the surface area of the sphere.
11. A man 6 feet tall walks away from a lamppost 15 feet tall at a rate of 4 miles per hour. How fast is his shadow lengthening when he is 12 feet from the pole? How fast is the distance from the foot of the lamppost to the tip of his shadow lengthening?
12. At 3 P.M. a ship which is sailing due south at 12 knots is 5 miles west of a westbound ship which is making 16 knots.
    (a) At what rate is the distance between the ships changing at 3 P.M.?
    (b) At what time does the distance between the ships stop decreasing and start increasing?
    (c) What is the shortest distance between the ships?
13. A ladder 20 feet long leans against a vertical wall. The bottom of the ladder slides away from the wall at a constant rate of 1 foot per second. At what rate is the top coming down the wall when it is 12 feet from the ground?
14. A particle moves on the parabola with equation $y = x^2$. The horizontal component of the velocity at each point is equal to twice the abscissa of the point. Show that the vertical component of the velocity at each point is equal to four times the ordinate of the point.
15. Sand is being poured on the ground at a rate of 4 cubic feet per minute. At each moment, it forms a conical pile with the height of the cone $\frac{7}{3}$ of the radius of the base. How fast is the height of the pile rising when 21π cubic feet of sand is in the pile?

4. Approximate Values. If we can find the values of a function and its derivative at some particular number a, then there is a useful method for obtaining an approximation to the value of the function at any number near a. For example, knowing that $\sqrt{9} = 3$, we can easily obtain a good approximation to $\sqrt{9.1}$. Similarly, we can use this method to find simple approximations to such numbers as $\dfrac{1}{4.02}$, $\sqrt[3]{26.8}$, and $(32.1)^{1/5}$.

To obtain the approximation for $\sqrt{9.1}$, we consider the graph of the function f defined by $f(x) = \sqrt{x}$ and drawn in Figure 17. The tangent line

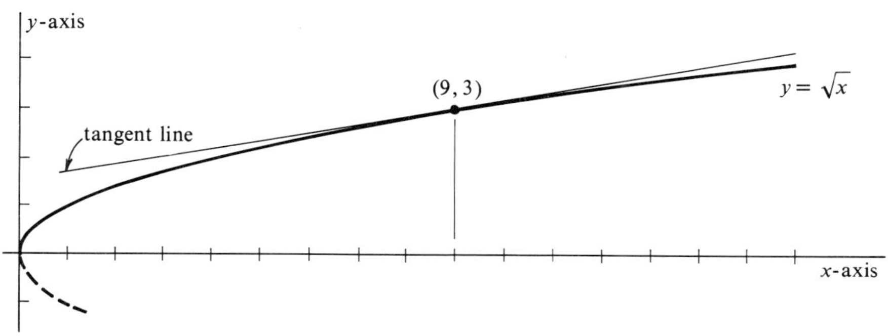

Figure 17

to the curve at the point (9, 3) touches the curve at that point and is not very far away from it for values of x near 9. Although it is tedious to find a decimal approximation for the ordinate of the point on the curve $y = \sqrt{x}$ with an abscissa of 9.1, it is relatively easy to find one for the point on the tangent line with that abscissa, and the two points are not very far apart. Since $\dfrac{d}{dx}\sqrt{x} = \dfrac{1}{2\sqrt{x}}$, the tangent line has a slope of $\dfrac{1}{2\sqrt{9}} = \dfrac{1}{6}$ and has an equation

$$y - 3 = \tfrac{1}{6}(x - 9).$$

If $x = 9.1$, $y = 3 + \frac{1}{6}(0.1) = 3\frac{1}{60}$, or 3.017. Thus 3.017 is our approximation for $\sqrt{9.1}$. That it is a good approximation may be seen by checking a table of square roots to find $\sqrt{9.1} = 3.016621$. The same tangent line may be used to approximate $\sqrt{10}$, but the accuracy will not be as good. If $x = 10$, $y = 3 + \frac{1}{6} = 3.167$. The tables give 3.162278 for $\sqrt{10}$.

The technique used in the problem above is the computation of an approximate value of $f(x)$ under the assumption that the difference $x - a$ is small in absolute value and that both $f(a)$ and $f'(a)$ are known, or can be easily evaluated. We write an equation of the tangent line to the graph of the function f at $(a, f(a))$ and take the ordinate of the point on the line with abscissa x as the approximation to $f(x)$. The tangent line has equation $y - f(a) = f'(a)(x - a)$, or, equivalently,

$$y = f(a) + f'(a)(x - a).$$

The function $f(a) + f'(a)(x - a)$ is a linear function of x, and is the linear function which best approximates $f(x)$ for values of x near a. The approximation consists of simply replacing the true value $f(x)$ by the corresponding value of the linear function. The result is summarized in the formula

$$\boxed{f(x) \approx f(a) + f'(a)(x - a),} \tag{1}$$

in which it is assumed that $|x - a|$ is small and the symbol $\approx$ indicates approximate equality.

EXAMPLE 1. Find an approximate value of $\dfrac{1}{4.02}$. If we define $f(x) = \dfrac{1}{x}$, then $f(4) = \frac{1}{4} = 0.25$ is easily evaluated. Moreover, $f'(4) = -\dfrac{1}{4^2} = -\dfrac{1}{16} = -0.0625$, and 0.02, the difference between 4.02 and 4, is small. Thus $\dfrac{1}{4.02}$ is approximately equal to $\frac{1}{4} - \frac{1}{16}(4.02 - 4) = 0.25 - (0.0625)(0.02) = 0.24875$.

EXAMPLE 2. Compute $\sqrt[3]{26.8}$ approximately. If we let $f(x) = x^{1/3}$, then $f(27) = 3$. Since $f'(x) = \frac{1}{3}x^{-2/3}$, we obtain $f'(27) = \frac{1}{3}(27)^{-2/3} = \frac{1}{27}$. The difference $|26.8 - 27| = 0.2$ is small. Thus we approximate $\sqrt[3]{26.8}$ by $3 + \frac{1}{27}(26.8 - 27) = 3 + \frac{1}{27}(-0.2) = 2.9926$. A table of cube roots gives a more exact value of 2.992574, but the linear approximation gives four-decimal accuracy.

An alternative point of view is obtained if we substitute $x = a + t$ in (1). The left side of the formula becomes $f(a + t)$ and the right side is then $f(a) + f'(a)\big((a + t) - a\big) = f(a) + tf'(a)$. Hence, we obtain the equivalent formula

$$\boxed{f(a + t) \approx f(a) + tf'(a),} \tag{2}$$

which gives an approximate value of $f(a + t)$ in terms of the known quantities $f(a)$, $f'(a)$, and t. The same result can also be obtained easily from the definition of the derivative of the function f at a,

$$f'(a) = \lim_{t \to 0} \frac{f(a + t) - f(a)}{t}.$$

It follows that if t is nonzero and small in absolute value, then $f'(a)$ is given approximately by

$$f'(a) \approx \frac{f(a + t) - f(a)}{t},$$

which immediately implies (2).

PROBLEMS

1. Give three-decimal approximations for each of the following numbers.

(a) $\sqrt{3.97}$ (d) $\dfrac{1}{(0.98)^3}$ (g) $\dfrac{1}{(4.02)^2}$

(b) $\sqrt[3]{64.2}$ (e) $\sqrt{16.6}$ (h) $(63.7)^{5/6}$

(c) $\sqrt[5]{31.85}$ (f) $\sqrt[4]{16.6}$ (i) $(0.95)^{3}$.

2. If $f(x) = \dfrac{1}{x + 5}$, find an approximation for $f(4.92)$.
3. If $f(x) = \sqrt{x - 2}$, find an approximation for $f(27.3)$.
4. If $f(x) = \sqrt{7x^2 - 3}$, find an approximation for $f(1.9)$.
5. (a) Find the volume of a sphere with a radius of 3 inches.
   (b) Find the approximate volume of a sphere with a radius of 3.1 inches.
6. (a) Find the volume of a cube 6 inches on an edge.
   (b) Find the approximate volume of a cube 5.9 inches on an edge.
7. Find an approximate value of $1.97\sqrt[3]{(1.97)^2 + 4}$.
8. Find an approximate value of the product $(63.2)^{1/3}(63.2)^{1/2}$.
9. (a) Find the area of an equilateral triangle 4 inches on a side.
   (b) Find an approximation for the area of an equilateral triangle 4.08 inches on a side.
10. The point $P = (2, 1)$ lies on the curve defined by the equation $x^3y + xy^3 = 10$. Find an approximation to the y-coordinate of the point on the curve near P with x-coordinate equal to 2.14.

5. Rolle's Theorem and Its Consequences. There are certain theoretical properties of differentiable functions which we are in a position to prove and which will aid us in our future work. Two of them express ideas which are geometrically obvious but, nevertheless, require proof. The first, due to the French mathematician Rolle and named for him, is illustrated in Figure 18. The assertion is that a differentiable function which has a graph crossing the x-axis at a and also at b must have on its graph at least one point between the crossing points where its tangent line is parallel to the x-axis.

(5.1) ROLLE'S THEOREM. *Assume that $a < b$ and that the function f is continuous on the closed interval $[a, b]$ and differentiable on the open interval*

(a, b). *If* $f(a) = f(b) = 0$, *then there exists a real number* c *such that* $a < c < b$ *and* $f'(c) = 0$.

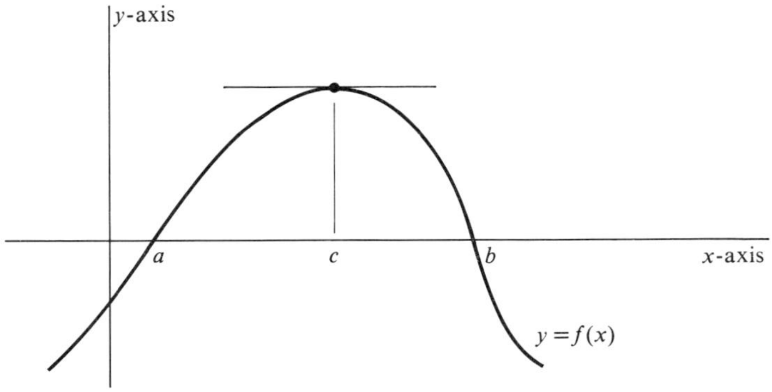

Figure 18

Proof. If $f(x) = 0$ for every x in $[a, b]$, there is nothing to prove because f is then a constant function and $f'(x) = 0$ for every x in the interval. So we assume that f is not constant on $[a, b]$. By Theorem (2.4) a function which is continuous at every point of a closed bounded interval has at least one absolute maximum point and at least one absolute minimum point. Since $f(x)$ does not equal zero for all x in the closed interval and since $f(a) = f(b) = 0$, the function f must have one of these extremes in the open interval. Let the abscissa of this point be c, and it follows by Theorem (2.3) that $f'(c) = 0$. This completes the proof.

The reader should try to construct functions which do not satisfy all the conditions of the theorem to see why the conclusion will not then hold, and hence why all the conditions are essential to the theorem. One such example was graphed in Figure 7.

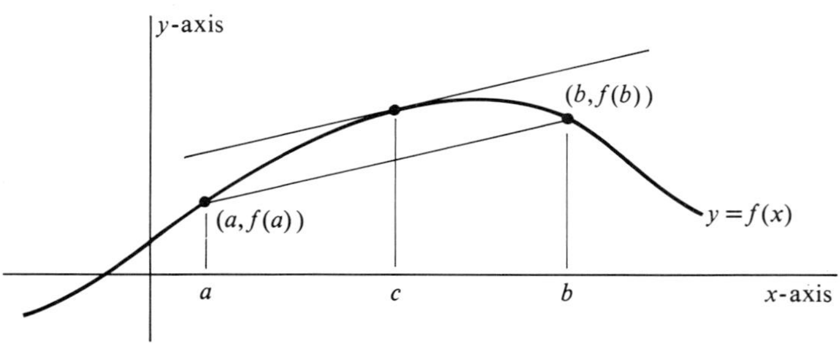

Figure 19

The second theorem tells us that a function which has a smooth graph between $(a, f(a))$ and $(b, f(b))$ has a point in between these two where the tangent to the graph is parallel to the line segment connecting these points (see Figure 19). Because the point lies between the other two and the tangent at this "in-between" point is parallel, this theorem is called the Mean Value Theorem.

(5.2) MEAN VALUE THEOREM. *Assume that $a < b$ and that the function f is continuous on the closed interval $[a, b]$ and differentiable on the open interval (a, b). Then there exists a real number c such that $a < c < b$ and*

$$f'(c) = \frac{f(b) - f(a)}{b - a},$$

or, equivalently, such that $f(b) - f(a) = f'(c)(b - a)$.

Proof. This theorem is proved as a corollary to Rolle's Theorem by constructing a function which satisfies the conditions of Rolle's Theorem and gives the same result as if we had tilted the graph of Figure 19 and dropped it down. Such a function is

$$F(x) = (x - a)f(b) + (a - b)f(x) + (b - x)f(a).$$

Since f is continuous on the closed interval and differentiable on the open interval, it follows that F is, too. We also observe that

$$F(a) = 0 + (a - b)f(a) + (b - a)f(a) = 0,$$

$$F(b) = (b - a)f(b) + (a - b)f(b) + 0 = 0.$$

Thus the function F satisfies all the conditions of Rolle's Theorem, and hence there is a real number c strictly between a and b such that $F'(c) = 0$. The derivative of F is given by

$$F'(x) = f(b) + (a - b)f'(x) - f(a).$$

It follows that

$$F'(c) = f(b) + (a - b)f'(c) - f(a) = 0,$$

which implies that $f(b) - f(a) = (b - a)f'(c)$. This completes the proof.

Here again it is a good idea to try various examples to see why all the hypotheses of the theorem are necessary. Note that the equation which forms the conclusion of the Mean Value Theorem is equivalent to the one obtained by interchanging a and b. Thus, if $b < a$, the Theorem remains true with $[a, b]$ and (a, b) replaced by $[b, a]$ and (b, a), respectively, and the inequalities $a < c < b$ replaced by $b < c < a$.

One of the most important consequences of the Mean Value Theorem is that a function which has a zero derivative on an interval must be a constant function on that interval.

(5.3) *If $f'(x) = 0$ for every x in an interval, then there exists a constant k such that $f(x) = k$ for every x in the interval.*

Proof. Pick an arbitrary point a in the interval, and set $k = f(a)$. Let b be any other point in the interval. We shall show that $f(b) = k$ also, and this will complete the proof. To be specific, let us assume that $a < b$; an exactly analogous argument can be made if $b < a$. By the Mean Value Theorem we know that $f(b) = f(a) + (b - a)f'(c)$ for some number c in the open interval (a, b), which is a subinterval of the larger interval referred to above. It follows from the hypothesis that $f'(c) = 0$. Since $f(a) = k$, we get $f(b) = k + (b - a) \cdot 0 = k$, and the result is proved.

The preceding theorem has an important corollary—that two functions with the same derivative over an interval differ by a constant.

(5.4) THEOREM. *If $f' = g'$ on an interval, then f and g differ by a constant function on the interval.*

Proof. Since $f' = g'$, we have $(f - g)'(x) = f'(x) - g'(x) = 0$ for every x in the interval. By (5.3) there exists a number k such that $k = (f - g)(x) = f(x) - g(x)$ for every x in the interval. This completes the proof.

The significance of Theorem (5.4), which will be fully exploited in the study of integration in Chapter 4, is that it gives a way of describing the set of all functions which have a given function as derivative. Specifically, let f be a function whose domain contains an interval I. Suppose that in one way or another we can find a function F with the property that $F'(x) = f(x)$, for every x in I. Then *the set $\mathfrak{F}$ of all functions whose derivatives equal f on I consists of all functions which on I differ from F by a constant.* To prove this assertion, we first observe that, for every real number c, the function defined by $F(x) + c$ has derivative equal to $F'(x) + 0 = f(x)$, for each x in I. Hence every function $F + c$ belongs to the set $\mathfrak{F}$. Conversely, if G is any function in the set $\mathfrak{F}$, then by definition $G' = f = F'$ on I. It follows by (5.4) that there exists a real number c (a constant) such that $G(x) - F(x) = c$, for every x in I. Thus on I the function G differs from F by a constant, and so the assertion is proved.

EXAMPLE 1. If f is the function defined by $f(x) = x^2 + 2x$, find the set of all functions with derivative equal to f. In this case the interval I is the set of all real numbers, and it is easy to see that one function in the set is $\frac{x^3}{3} + x^2$, since

$$\frac{d}{dx}\left(\frac{x^3}{3} + x^2\right) = x^2 + 2x = f(x).$$

Hence each function G in the set is defined by

$$G(x) = \frac{x^3}{3} + x^2 + c,$$

for some real number c. As c takes on all real number values, we get all members of the set. There are no other possibilities.

PROBLEMS

1. For each of the following functions find those values of x for which $f(x)$ or $f'(x)$ vanish and use them to verify Rolle's Theorem.

   (a) $f(x) = x^2 - 7x - 8$

   (b) $f(x) = 12x - x^3$

   (c) $f(x) = (x + 4)(x + 1)(x - 2)$

   (d) $f(x) = x^2(x^2 - 16)$

   (e) $f(x) = x - \frac{4}{x}$

   (f) $f(x) = (9 - x^2)^2$.

2. For each of the following functions and the specified values of a and b, find a number c such that $a < c < b$ and $f(b) = f(a) + (b - a)f'(c)$.

   (a) $f(x) = x^2 - 6x + 5$, $a = 1$, $b = 4$

   (b) $f(x) = x^3$, $a = 0$, $b = 1$

   (c) $f(x) = -\frac{1}{x}$, $a = 1$, $b = 3$

   (d) $f(x) = \frac{8}{x^2}$, $a = 1$, $b = 2$.

3. Consider the function $f(x) = 1 - |x|$ defined on the closed interval from -1 to 1. Which hypotheses of Rolle's Theorem does this function satisfy and which does it not satisfy? Does this function satisfy the conclusion of Rolle's Theorem?
4. Consider the function f defined on the closed interval $[4, 7]$ by

$$\begin{cases} f(x) = 0, & x = 4, \\ f(x) = 7 - x, & 4 < x \leq 7. \end{cases}$$

   Show where this function fails to satisfy the conditions of Rolle's Theorem, and that it does not satisfy the conclusion.
5. For each of the following functions f, find the set of all functions with derivative equal to f.

   (a) $f(x) = 4x$

   (b) $f(x) = 4x^3 + x^2 + 2$

   (c) $f(x) = \frac{1}{x^2}$

   (d) $f(x) = \frac{2x}{(x^2 + 1)^3}$.

6. Prove that, if on an automobile trip the average velocity was 45 miles per hour, then at some instant during the trip the speedometer registered precisely 45.

6. The Differential. If $y = f(x)$, we have denoted the derivative of f by f', or $\frac{df}{dx}$, or $\frac{dy}{dx}$. The value of the derivative at a number a is written $f'(a)$, or $\frac{df}{dx}(a)$, or $\frac{dy}{dx}(a)$. Up to this point, the expressions df, dy, and dx by themselves have had no meaning other than as parts of notations for the derivative. However, the cancellation suggested by the Chain Rule

$$\frac{dy}{dx} = \frac{dy}{du}\frac{du}{dx}$$

indicates that the derivative behaves like a ratio and suggests that it may be possible to sensibly regard it as such. In this section we shall define a mathematical object called the differential of a function, examples of which are df, dy, and dx. The ratio of df, or dy, to dx will be equal to the derivative.

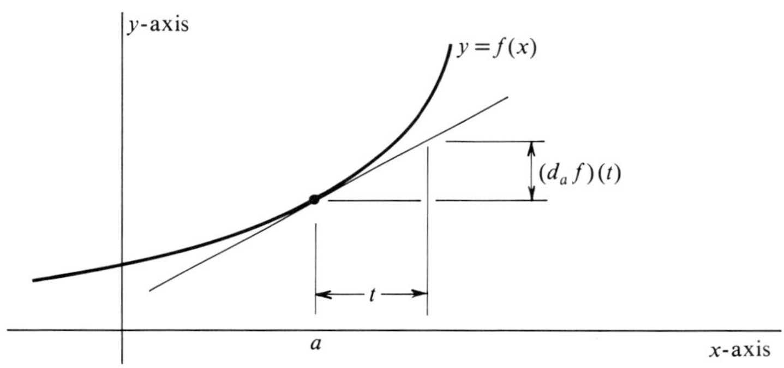

Figure 20

If f is a function having a derivative at a, we define its **differential at** a, denoted by d_af, to be the linear function whose value for any number t is

$$(d_af)(t) = f'(a)t.$$

For example, if $f(x) = x^2 - 2x$, then the differential d_af is the function of t defined by $f'(a)t = (2a - 2)t$. In particular,

$$(d_3f)(t) = [2 \cdot 3 - 2]t = 4t.$$

The value of the differential for a typical function f is illustrated in Figure 20.

By simply df we mean the rule (or function) that assigns the linear function d_xf to each number x in the domain of f'.

(6.1) *If f and u are differentiable functions, then*

$$\boxed{df(u) = f'(u)\, du.} \qquad (1)$$

Proof. This formula is an abbreviation of the equation

$$d_x f(u) = f'(u(x))\, d_x u. \tag{2}$$

The proof is an application of the Chain Rule. We first write down the two linear functions $d_x f(u)$ and $d_x u$. By the definition of the differential they are

$$(d_x u)(t) = u'(x)t,$$
$$(d_x f(u))(t) = [[f(u)]'(x)]t.$$

The Chain Rule says that $[f(u)]'(x) = f'(u(x))u'(x)$. Hence

$$\begin{aligned}(d_x f(u))(t) &= [f'(u(x))u'(x)]t\\ &= f'(u(x))(d_x u)(t).\end{aligned}$$

Thus (2) appears as an equality between linear functions, and the proof is complete.

If u is the independent variable x, then (6.1) reduces to the formula

$$\boxed{df(x) = f'(x)\, dx.} \tag{3}$$

EXAMPLE 1. Evaluate the following differentials:

$$\text{(a)}\quad d(x^2 + 2),$$
$$\text{(b)}\quad d\sqrt{x^2 + 3},$$
$$\text{(c)}\quad d(2x^2 - x)^7.$$

Using formula (3), we get immediately

$$\text{(a}'\text{)}\quad d(x^2 + 2) = 2x\, dx$$
$$\text{(b}'\text{)}\quad d\sqrt{x^2 + 3} = x(x^2 + 3)^{-1/2}\, dx$$
$$\text{(c}'\text{)}\quad d(2x^2 - x)^7 = 7(2x^2 - x)^6(4x - 1)\, dx.$$

It is worthwhile learning to use the stronger formula (1). In problem (b), let f be the function $f(u) = \sqrt{u}$. If we set $u = x^2 + 3$, then $du = 2x\, dx$ and

$$\begin{aligned}d\sqrt{x^2 + 3} &= df(u) = f'(u)\, du\\ &= \tfrac{1}{2}u^{-1/2}\, du\\ &= \tfrac{1}{2}(x^2 + 3)^{-1/2}2x\, dx.\end{aligned}$$

Let us also do problem (c) using (1), but without explicitly making the substitution $u = 2x^2 - x$. We get

$$\begin{aligned}d(2x^2 - x)^7 &= 7(2x^2 - x)^6\, d(2x^2 - x)\\ &= 7(2x^2 - x)^6\, (4x - 1)\, dx.\end{aligned}$$

Formula (3) establishes the fact that the ratio of df to dx is equal to the derivative f'. We can see this in greater detail by going back to the definitions:

$$(d_a f)(t) = f'(a)t,$$
$$(d_a x)(t) = x'(a)t.$$

Since x is the identity function, its derivative is the constant function 1. Hence $(d_a x)(t) = t$. The ratio of the two linear functions $d_a f$ and $d_a x$ is thus the constant function

$$\frac{d_a f}{d_a x} = \frac{f'(a)t}{t} = f'(a).$$

Having proved this formula for every a in the domain of f', we can write it simply as

$$\frac{df}{dx}(a) = f'(a) \qquad \text{or} \qquad \frac{df}{dx} = f'.$$

If f and g are differentiable functions, then it is easy to show that $d_a(f + g) = d_a f + d_a g$. The proof involves only the definition of the differential plus the fact that the derivative of a sum is the sum of the derivatives. In detail:

$$\begin{aligned}[d_a(f + g)](t) &= [(f + g)'(a)]\,t = [f'(a) + g'(a)]\,t \\ &= f'(a)t + g'(a)t = (d_a f)(t) + (d_a g)(t) \\ &= [d_a f + d_a g](t).\end{aligned}$$

The result is simply the equation

$$d(f + g) = df + dg.$$

An analogous argument using the Product Rule for differentiation shows that $d_a(fg) = f(a)d_a g + g(a)d_a f$, or, more simply,

$$d(fg) = f\,dg + g\,df.$$

For each one of the six differentiation rules proved in Section 7 of Chapter 1, there is an analogous rule in terms of differentials: Let u and v be differentiable functions, and c a constant. Then

(i) $d(u + v) = du + dv,$

(ii) $d(cu) = c\,du,$

(iii) $dc = 0,$

(iv) $d(uv) = u\,dv + v\,du,$

(v) $du^r = ru^{r-1}\,du,$ where r is any rational number,

(vi) $d\left(\frac{u}{v}\right) = \frac{v\,du - u\,dv}{v^2}.$

Note that we have replaced the analogue of (v) in the list in Section 7 of Chapter 1 by the formula corresponding to the more powerful theorem (8.2) of Chapter 1.

EXAMPLE 2. Find the differential $d(x^3 + \sqrt{x^2 + 2x})^7$. Applying the above formulas successively, we get

$$\begin{aligned}
&d(x^3 + \sqrt{x^2 + 2x})^7 \\
&\quad = 7(x^3 + \sqrt{x^2 + 2x})^6 \, d(x^3 + \sqrt{x^2 + 2x}) \qquad \text{by (v)} \\
&\quad = 7(x^3 + \sqrt{x^2 + 2x})^6 (dx^3 + d\sqrt{x^2 + 2x}) \qquad \text{by (i)} \\
&\quad = 7(x^3 + \sqrt{x^2 + 2x})^6 [3x^2 \, dx + \tfrac{1}{2}(x^2 + 2x)^{-1/2} \, d(x^2 + 2x)] \qquad \text{by (v)} \\
&\quad = 7(x^3 + \sqrt{x^2 + 2x})^6 [3x^2 \, dx + \tfrac{1}{2}(x^2 + 2x)^{-1/2}(2x \, dx + 2 \, dx)] \\
&\qquad\qquad \text{by (i) and (ii)} \\
&\quad = 7(x^3 + \sqrt{x^2 + 2x})^6 \left(3x^2 + \frac{x + 1}{\sqrt{x^2 + 2x}} \right) dx.
\end{aligned}$$

The derivative is therefore given by

$$\frac{d(x^3 + \sqrt{x^2 + 2x})^7}{dx} = 7(x^3 + \sqrt{x^2 + 2x})^6 \left(3x^2 + \frac{x + 1}{\sqrt{x^2 + 2x}} \right).$$

The task of computing the differential of a complicated function of x amounts to successively working the differential operator d through the given expression from left to right. At each stage one uses the correct one of formulas (i) through (vi), or formula (1), until one finally reaches dx, and the process stops. The derivative can then be obtained by dividing the resulting equation by dx. Note that an equation of the form $df(x) = \cdots$ will always contain the symbol d on the right side. Equations such as $dx^5 = 5x^4$ are not only false; they are nonsense. (Correct version: $dx^5 = 5x^4 \, dx$.)

EXAMPLE 3. Consider the functions

$$\text{(a)} \quad y = (4x^3 + 3x^2 + 1)^2,$$

$$\text{(b)} \quad y = \frac{x^2 - 1}{x^2 + 1},$$

$$\text{(c)} \quad z = 3y^{5/3}.$$

Find the differential of each:

$$\begin{aligned}
\text{(a}') \quad dy &= 2(4x^3 + 3x^2 + 1) \, d(4x^3 + 3x^2 + 1) \\
&= 2(4x^3 + 3x^2 + 1)(12x^2 \, dx + 6x \, dx) \\
&= 12x(4x^3 + 3x^2 + 1)(2x + 1) \, dx,
\end{aligned}$$

$$\text{(b}'\text{)}\quad dy = \frac{(x^2+1)\,d(x^2-1) - (x^2-1)\,d(x^2+1)}{(x^2+1)^2}$$

$$= \frac{(x^2+1)2x\,dx - (x^2-1)2x\,dx}{(x^2+1)^2} = \frac{4x\,dx}{(x^2+1)^2},$$

$$\text{(c}'\text{)}\quad dz = (3)(\tfrac{5}{3})y^{2/3}\,dy = 5y^{2/3}\,dy.$$

If we consider the composition of the function y in (b) with the function z in (c), we get for the differential of the composition

$$dz = 5y^{2/3}\frac{4x\,dx}{(x^2+1)^2} = 5\left(\frac{x^2-1}{x^2+1}\right)^{2/3}\frac{4x}{(x^2+1)^2}\,dx.$$

One traditional interpretation of the differential, which is especially useful in physics, is that of an "infinitesimal." If $y = f(x)$, we know that $dy = f'(x)\,dx$. Now dx is the function that assigns to every real number a the linear function defined by $(d_a x)(t) = x'(a)t = t$; i.e., it assigns the identity function. Hence we can interpret dx as simply another independent variable. Then dy is the variable whose value for a given x and dx is shown in Figure 21. (Compare this illustration with Figure 20.) The difference between the value of f at x and at $x + dx$ is denoted by Δy in the figure. If dx is chosen

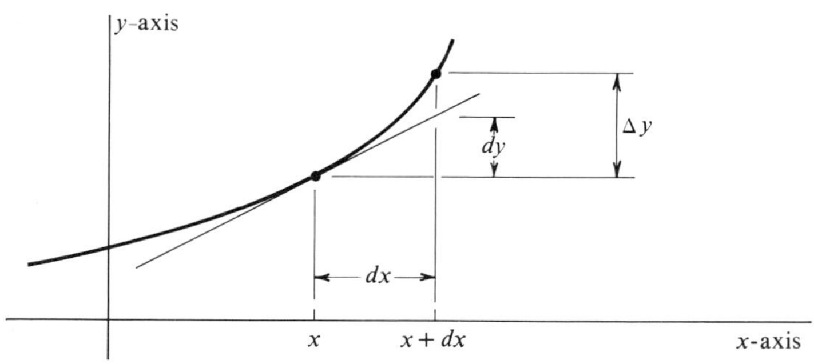

Figure 21

very small, then the difference between dy and Δy is relatively negligible. Hence dy measures the resulting change in the value of $y = f(x)$ corresponding to an infinitesimal change dx in the variable x.

EXAMPLE 4. The height h of a square pyramid is found to be 100 feet, and the length x of one edge of its base is measured to be 160 feet. The volume V of the pyramid is given by the formula $V = \frac{1}{3}hx^2$. What error in the computed volume will result from an error of 4 inches in the measurement of x? If we consider h as fixed, and V as a function of x, then

$$dV = \tfrac{1}{3}h\, dx^2 = \tfrac{2}{3}hx\, dx.$$

Since 4 inches is small compared with 160 feet, we set $dx = 4$ inches $= \frac{1}{3}$ foot. The resulting change in volume is then approximately

$$dV = \tfrac{2}{3}(100)(160)\tfrac{1}{3} = \tfrac{32}{9}\, 1000 = 3555 \text{ feet}^3.$$

The percentage error in volume is

$$\frac{dV}{V} = \frac{\frac{2}{3}hx\, dx}{\frac{1}{3}hx^2} = 2\frac{dx}{x}.$$

We compute $\dfrac{dx}{x} = \dfrac{\frac{1}{3}}{160} = 0.0021 = 0.21\%$, and so the percentage error in volume is only 0.42%.

PROBLEMS

1. Find the following differentials.

(a) $d(x^2 + x + 1) = \cdots$

(b) $d(7x + 2) = \cdots$

(c) $d(x^3 + 1)(5x - 1)^3 = \cdots$

(d) $d\left(\dfrac{x-1}{x+1}\right) = \cdots$

(e) $du^7 = \cdots$

(f) $d\left(\dfrac{u^2}{v^2}\right) = \cdots$

(g) $d(az^2 + bz + c) = \cdots$ (a, b, and c are constants)

(h) $d\sqrt{1 + \sqrt{1 + x}} = \cdots$

(i) $dx = \cdots$

(j) $d(u^2 + 2)(v^3 - 1) = \cdots.$

2. If $y = 7x^3 + 2x + 1$ and $w = \dfrac{1}{y}$, compute the differential of the composition of y with w. That is, compute dw in terms of x and dx.
3. If $x = 16t^2 + 2t$ and $y = \dfrac{1}{x}$ and $z = y^2 + 1$, compute dz in terms of t and dt.
4. Using Leibnitz's Rule (the Product Rule), prove that $d_a(fg) = f(a)d_a g + g(a)d_a f$, thereby establishing rule (iv) on page 118.
5. Using Theorem (8.2) of Chapter 1, prove that $d_a f^r = rf(a)^{r-1} d_a f$, thereby establishing rule (v) on page 118.
6. What is the approximate change in the volume of a sphere of radius 10 feet resulting from a change in the radius of 1 inch?

7. A metal cylinder is found by measurement to be 3 feet in diameter and 10 feet long. What will be the error in the computed volume of the cylinder resulting from an error of

   (a) 1 inch in the diameter?

   (b) 0.5 inch in the length?

   (c) both the errors in (a) and (b) combined?

8. If $f(x) = kx$, in what sense is f its own differential?

7. L'Hôpital's Rule. This section contains a number of theorems which provide an important technique for finding the limit of the quotient of two functions. These theorems are usually referred to collectively as L'Hôpital's Rule. Two examples of problems which are readily solved by this technique are the computations of the limits:

$$\lim_{x \to 2} \frac{\sqrt{x} - \sqrt{2}}{\sqrt[3]{x} - \sqrt[3]{2}} \qquad \text{and} \qquad \lim_{x \to \infty} \frac{\sqrt[3]{x+1}}{x+4}.$$

Properly speaking, this section is a continuation of Section 5, since all the results are corollaries of Rolle's Theorem and of the Mean Value Theorem.

The following proposition, known as the Generalized Mean Value Theorem, is the basic lemma used in proving the theorems which make up L'Hôpital's Rule. (A lemma is a theorem included as a reference in proving other theorems.)

(7.1) GENERALIZED MEAN VALUE THEOREM. *Assume that $a < b$, and let f and g be functions which are continuous on the closed interval $[a, b]$ and differentiable on the open interval (a, b). If $g'(x) \neq 0$ for every x in (a, b), then there exists a real number c in (a, b) such that*

$$\frac{f(b) - f(a)}{g(b) - g(a)} = \frac{f'(c)}{g'(c)}.$$

Note that $g(b) - g(a) \neq 0$. For otherwise the Mean Value Theorem would imply that $g'(x) = 0$ for some x in (a, b), which is contrary to hypothesis.

Proof. Let h be the function defined by

$$h(x) = f(x) - f(a) - \frac{f(b) - f(a)}{g(b) - g(a)}(g(x) - g(a)),$$

for every x in $[a, b]$. The function h is continuous on $[a, b]$, differentiable on (a, b), and, in addition, $h(a) = h(b) = 0$. Hence, by Rolle's Theorem, there exists a real number c in (a, b) such that $h'(c) = 0$. Since

$$h'(x) = f'(x) - \frac{f(b) - f(a)}{g(b) - g(a)} g'(x),$$

we obtain

$$0 = f'(c) - \frac{f(b) - f(a)}{g(b) - g(a)} g'(c),$$

and from this equation the conclusion of the theorem follows at once.

We first present L'Hôpital's Rule as a theorem about one-sided limits.

(7.2) L'HÔPITAL'S RULE I. *Let f and g be functions which are differentiable on a nonempty open interval (a, b) with $g'(x) \neq 0$ for every x in (a, b). If, in addition,*

(i) $\lim_{x \to a+} f(x) = \lim_{x \to a+} g(x) = 0,$

(ii) $\lim_{x \to a+} \frac{f'(x)}{g'(x)} = L,$

then

$$\lim_{x \to a+} \frac{f(x)}{g(x)} = L.$$

Proof. We may assume that $f(a) = g(a) = 0$. (If this is not the case to begin with, we simply define, or redefine, the values of f and g to be zero at a.) Thus we ensure that f and g are continuous on $[a, b)$. Let x be an arbitrary number in (a, b). Then f and g are continuous on $[a, x]$ (recall that differentiability at a point implies continuity) and are differentiable on (a, x). Moreover, the derivative g' does not take on the value zero in (a, x). Hence, by the Generalized Mean Value Theorem and the fact that $f(a) = g(a) = 0$, we obtain

$$\frac{f'(y)}{g'(y)} = \frac{f(x) - 0}{g(x) - 0} = \frac{f(x)}{g(x)},$$

for some number y in (a, x). As x approaches a from the right, so also does y, and hence $\lim_{x \to a+} \frac{f(x)}{g(x)} = \lim_{y \to a+} \frac{f'(y)}{g'(y)} = L$. This completes the proof.

EXAMPLE 1. Compute $\lim_{x \to 2+} \frac{\sqrt{x} - \sqrt{2}}{\sqrt{x - 2}}$. Let $f(x) = \sqrt{x} - \sqrt{2}$ and $g(x) = \sqrt{x - 2}$. Obviously, $\lim_{x \to 2+} f(x) = \lim_{x \to 2+} g(x) = 0$, and, since

$$f'(x) = \frac{1}{2\sqrt{x}} \quad \text{and} \quad g'(x) = \frac{1}{2\sqrt{x - 2}},$$

f and g are differentiable, and g' does not take on the value zero on any open interval with left endpoint equal to 2. We obtain

$$\lim_{x\to 2+} \frac{f'(x)}{g'(x)} = \lim_{x\to 2+} \frac{\dfrac{1}{2\sqrt{x}}}{\dfrac{1}{2\sqrt{x-2}}}$$

$$= \lim_{x\to 2+} \frac{\sqrt{x-2}}{\sqrt{x}} = \frac{0}{\sqrt{2}} = 0.$$

And it follows by L'Hôpital's Rule that $\lim_{x\to 2+} \dfrac{\sqrt{x}-\sqrt{2}}{\sqrt{x-2}} = 0.$

It is a simple matter to verify that Theorem (7.2) remains true if (a, b) is replaced throughout by (b, a), and $\lim_{x\to a+}$ is replaced throughout by $\lim_{x\to a-}$. This fact, significant in itself, also implies the following two-sided form of L'Hôpital's Rule.

(7.3) L'HÔPITAL'S RULE II. *Consider an open interval containing the number a, and let f and g be functions differentiable and with $g'(x) \neq 0$ at every point of the interval except possibly at a. If*

(i) $\lim_{x\to a} f(x) = \lim_{x\to a} g(x) = 0,$

(ii) $\lim_{x\to a} \dfrac{f'(x)}{g'(x)} = L,$

then

$$\lim_{x\to a} \frac{f(x)}{g(x)} = L.$$

The hypotheses of (7.3) have been taken as weak as possible. If, as frequently happens, the functions f and g are also continuous at a, then (i) can be replaced by the simpler condition $f(a) = g(a) = 0$.

EXAMPLE 2. Evaluate $\lim_{x\to a} \dfrac{x^{1/2} - a^{1/2}}{x^{1/3} - a^{1/3}}$, where $a > 0$. If $f(x) = x^{1/2} - a^{1/2}$ and if $g(x) = x^{1/3} - a^{1/3}$, then the derivatives are given by $f'(x) = \dfrac{1}{2x^{1/2}}$ and $g'(x) = \dfrac{1}{3x^{2/3}}$, and it is clear that f and g are differentiable (and hence continuous) and g' is not zero on an open interval containing a. Moreover,

$f(a) = g(a) = 0$. Hence, by L'Hôpital's Rule,

$$\lim_{x \to a} \frac{x^{1/2} - a^{1/2}}{x^{1/3} - a^{1/3}} = \lim_{x \to a} \frac{\dfrac{1}{2x^{1/2}}}{\dfrac{1}{3x^{2/3}}} = \frac{3a^{2/3}}{2a^{1/2}} = \frac{3}{2}a^{1/6}.$$

EXAMPLE 3. Compute $\lim_{x \to a} \dfrac{x - a}{x^2 - a^2}$ where $a \neq 0$. Doing this problem by L'Hôpital's Rule is somewhat akin to smashing a peanut with a sledgehammer. The fact that

$$\frac{x - a}{x^2 - a^2} = \frac{x - a}{(x - a)(x + a)} = \frac{1}{x + a} \qquad \text{if } x \neq a,$$

immediately implies that

$$\lim_{x \to a} \frac{x - a}{x^2 - a^2} = \lim_{x \to a} \frac{1}{x + a} = \frac{1}{2a}.$$

Of course, the same answer is obtained by L'Hôpital's Rule. If we let $f(x) = x - a$ and $g(x) = x^2 - a^2$, then $f(a) = g(a) = 0$ and $f'(x) = 1$ and $g'(x) = 2x$. Hence

$$\lim_{x \to a} \frac{x - a}{x^2 - a^2} = \lim_{x \to a} \frac{1}{2x} = \frac{1}{2a}.$$

It is important to realize that L'Hôpital's Rule II can be applied only if the function $\dfrac{f(x)}{g(x)}$ is undefined at $x = a$ and if $\lim_{x \to a} f(x) = \lim_{x \to a} g(x) = 0$. For example, if $f(x) = x^2 + 3x - 10$ and $g(x) = 3x$, then

$$\lim_{x \to 2} \frac{f(x)}{g(x)} = \lim_{x \to 2} \frac{x^2 + 3x - 10}{3x} = \frac{0}{6} = 0,$$

but

$$\lim_{x \to 2} \frac{f'(x)}{g'(x)} = \lim_{x \to 2} \frac{2x + 3}{3} = \frac{7}{3}.$$

If the hypotheses of Theorem (7.3) are satisfied for the functions f' and g', that is, for the derivatives of f and g, respectively, then we can conclude that

$$\lim_{x \to a} \frac{f'(x)}{g'(x)} = \lim_{x \to a} \frac{f''(x)}{g''(x)}.$$

This fact suggests the possibility of applying L'Hôpital's Rule more than once, and in some problems it is necessary to take second or higher derivatives to find the limit.

EXAMPLE 4. Evaluate $\lim_{x\to 1} \dfrac{3x^{1/3} - x - 2}{3x^2 - 6x + 3}$. Let $f(x) = 3x^{1/3} - x - 2$ and $g(x) = 3x^2 - 6x + 3$. Then $f(1) = g(1) = 0$, and the derivatives are given by $f'(x) = x^{-2/3} - 1$ and $g'(x) = 6x - 6$. However, the value of

$$\lim_{x\to 1} \frac{f'(x)}{g'(x)} = \lim_{x\to 1} \frac{x^{-2/3} - 1}{6x - 6}$$

is not obvious because $f'(1) = g'(1) = 0$. Taking derivatives again, we get $f''(x) = -\frac{2}{3}x^{-5/3}$ and $g''(x) = 6$, and it follows that

$$\lim_{x\to 1} \frac{f''(x)}{g''(x)} = \lim_{x\to 1} \frac{-\frac{2}{3}x^{-5/3}}{6} = -\frac{1}{9}.$$

Thus two applications of L'Hôpital's Rule yield

$$\lim_{x\to 1} \frac{f(x)}{g(x)} = \lim_{x\to 1} \frac{f'(x)}{g'(x)} = \lim_{x\to 1} \frac{f''(x)}{g''(x)} = -\frac{1}{9}.$$

A variation of (7.2), not difficult to prove, is the following:

(7.4) L'HÔPITAL'S RULE III. *Let f and g be differentiable on an open interval (a, ∞) with $g'(x) \neq 0$ for $x > a$. If*

(i) $\lim_{x\to\infty} f(x) = \lim_{x\to\infty} g(x) = 0$,

(ii) $\lim_{x\to\infty} \dfrac{f'(x)}{g'(x)} = L$,

then

$$\lim_{x\to\infty} \frac{f(x)}{g(x)} = L.$$

An analogous theorem holds if (a, ∞) is replaced by $(-\infty, a)$ and if $\lim_{x\to\infty}$ is replaced throughout by $\lim_{x\to -\infty}$.

Proof. The result is a corollary of (7.2) and the Chain Rule. Let $t = \dfrac{1}{x}$, and set $F(t) = f\left(\dfrac{1}{t}\right) = f(x)$ and $G(t) = g\left(\dfrac{1}{t}\right) = g(x)$. Since t approaches 0 from the right if and only if x increases without bound,

$$\lim_{t\to 0+} F(t) = \lim_{t\to 0+} f\left(\frac{1}{t}\right) = \lim_{x\to\infty} f(x) = 0,$$

$$\lim_{t\to 0+} G(t) = \lim_{t\to 0+} g\left(\frac{1}{t}\right) = \lim_{x\to\infty} g(x) = 0.$$

By the Chain Rule, $F'(t) = f'\left(\dfrac{1}{t}\right)\left(-\dfrac{1}{t^2}\right)$ and $G'(t) = g'\left(\dfrac{1}{t}\right)\left(-\dfrac{1}{t^2}\right)$.

Hence

$$\lim_{t\to 0+} \frac{F'(t)}{G'(t)} = \lim_{t\to 0+} \frac{f'\left(\frac{1}{t}\right)\left(-\frac{1}{t^2}\right)}{g'\left(\frac{1}{t}\right)\left(-\frac{1}{t^2}\right)} = \lim_{t\to 0+} \frac{f'\left(\frac{1}{t}\right)}{g'\left(\frac{1}{t}\right)}.$$

The last limit exists and is equal to L since

$$\lim_{t\to 0+} \frac{f'\left(\frac{1}{t}\right)}{g'\left(\frac{1}{t}\right)} = \lim_{x\to\infty} \frac{f'(x)}{g'(x)} = L.$$

By L'Hôpital's Rule I it follows that $\lim_{t\to 0+} \frac{F(t)}{G(t)} = L$. Hence

$$\lim_{x\to\infty} \frac{f(x)}{g(x)} = \lim_{t\to 0+} \frac{F(t)}{G(t)} = L,$$

and the proof is complete.

An important observation is that all the forms of L'Hôpital's Rule developed so far are valid whether L is finite or not. This fact requires no new proof and has really already been established. The reason is that the basic conclusion of Theorem (7.2) is the equation

$$\lim_{x\to a+} \frac{f(x)}{g(x)} = \lim_{x\to a+} \frac{f'(x)}{g'(x)},$$

and this holds good whether or not $\lim_{x\to a+} \frac{f'(x)}{g'(x)}$ is finite or infinite.

There is another significant variation of L'Hôpital's Rule, whose proof, although requiring only the Generalized Mean Value Theorem, cannot (as far as we know) be obtained from (7.2) by a simple substitution. It states that the several forms of condition (i), $\lim f(x) = \lim g(x) = 0$, can be replaced by $\lim |g(x)| = \infty$. The specific statement which we prove is the following:

(7.5) L'HÔPITAL'S RULE IV. *Let f and g be functions which are differentiable on a nonempty open interval (a, b) with $g'(x) \neq 0$ for every x in (a, b). If*

(i) $\lim_{x\to a+} |g(x)| = \infty$,

(ii) $\lim_{x\to a+} \frac{f'(x)}{g'(x)} = L$,

then

$$\lim_{x \to a+} \frac{f(x)}{g(x)} = L.$$

Proof. Let ϵ be an arbitrary positive number. By hypothesis (ii), there exists a real number c in (a, b) such that

$$\left| \frac{f'(x)}{g'(x)} - L \right| < \epsilon, \qquad \text{for every } x \text{ in } (a, c).$$

By hypothesis (i) there exists a real number d in (a, b), which we shall for convenience assume to be in (a, c), such that, for every x in (a, d), the following three inequalities hold:

$$g(x) \neq 0, \qquad \left| \frac{f(c)}{g(x)} \right| < \epsilon, \qquad \left| \frac{g(c)}{g(x)} \right| < \epsilon$$

(see Figure 22). It is a consequence of the last inequality that

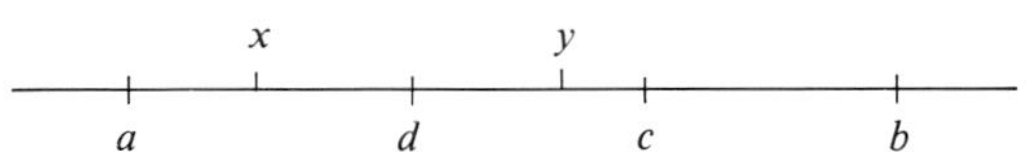

Figure 22

$$\left| 1 - \frac{g(c)}{g(x)} \right| < 1 + \epsilon, \qquad \text{for every } x \text{ in } (a, d).$$

Now let x be an arbitrary real number in (a, d). By the Generalized Mean Value Theorem, there exists a real number y in (x, c) such that

$$\frac{f(x) - f(c)}{g(x) - g(c)} = \frac{f'(y)}{g'(y)}.$$

Hence

$$f(x) = \frac{f'(y)}{g'(y)} (g(x) - g(c)) + f(c).$$

Dividing by $g(x)$, which cannot be zero, we get

$$\frac{f(x)}{g(x)} = \frac{f'(y)}{g'(y)} \left(1 - \frac{g(c)}{g(x)} \right) + \frac{f(c)}{g(x)}.$$

An equivalent equation is

$$\frac{f(x)}{g(x)} - L = \left(\frac{f'(y)}{g'(y)} - L \right) \left(1 - \frac{g(c)}{g(x)} \right) - L \frac{g(c)}{g(x)} + \frac{f(c)}{g(x)}.$$

From the general properties of the absolute value [see specifically (1.3) and (1.4), page 7], it follows that

$$\left|\frac{f(x)}{g(x)} - L\right| \leq \left|\frac{f'(y)}{g'(y)} - L\right| \left|1 - \frac{g(c)}{g(x)}\right| + |L| \left|\frac{g(c)}{g(x)}\right| + \left|\frac{f(c)}{g(x)}\right|.$$

Hence, the inequalities established in the first paragraph of the proof imply that

$$\left|\frac{f(x)}{g(x)} - L\right| \leq \epsilon(1 + \epsilon) + |L|\epsilon + \epsilon.$$

Since the right side of this inequality can be made arbitrarily small by taking ϵ sufficiently small, it follows that $\lim_{x\to a+} \frac{f(x)}{g(x)} = L$, and the proof is complete.

It is not difficult to derive variations of the preceding theorem analogous to the modified versions of (7.2) described above. Thus, with the obvious changes in the hypotheses, this last form of L'Hôpital's Rule also holds for two-sided limits and with a or L (or both) replaced by $\pm\infty$.

EXAMPLE 5. Compute $\lim_{x\to\infty} \frac{\sqrt[3]{x+1}}{x+4}$. Let f and g be the functions defined by $f(x) = \sqrt[3]{x+1}$ and $g(x) = x + 4$, respectively. Since $f'(x) = \frac{1}{3(x+1)^{2/3}}$ and $g'(x) = 1$, we see that f and g are differentiable on the interval $(1, \infty)$ and that $g'(x) \neq 0$. Moreover, $\lim_{x\to\infty} |g(x)| = \lim_{x\to\infty} |x + 4| = \infty$, and

$$\lim_{x\to\infty} \frac{f'(x)}{g'(x)} = \lim_{x\to\infty} \frac{\frac{1}{3(x+1)^{2/3}}}{1} = 0.$$

It follows by L'Hôpital's Rule that

$$\lim_{x\to\infty} \frac{\sqrt[3]{x+1}}{x+4} = \lim_{x\to\infty} \frac{f(x)}{g(x)} = 0.$$

The several forms of L'Hôpital's Rule which we have derived in this section fall naturally into two types, symbolically denoted as the $\frac{0}{0}$ type and the $\frac{*}{\infty}$ type. Theorems (7.2), (7.3), and (7.4) are all examples of the first type, whereas the harder Theorem (7.5) is the prototype of the second type. The full power of the $\frac{*}{\infty}$ forms will be realized later in the book in conjunction with the logarithmic, exponential, and trigonometric functions.

PROBLEMS

1. Evaluate each of the following limits *without* using L'Hôpital's Rule.

(a) $\lim_{x \to 3} \dfrac{x^2 - 9}{x^2 - 5x + 6}$

(b) $\lim_{x \to -2} \dfrac{x^3 + 8}{x^5 + 32}$

(c) $\lim_{x \to 2} \dfrac{x^3 - 6x + 4}{x^2 + 4}$

(d) $\lim_{x \to \infty} \dfrac{2x^2 + x - 1}{3x^2 - 2x + 1}$

(e) $\lim_{x \to 1} \dfrac{x^2 - 1}{x^3 - 1}$

(f) $\lim_{x \to 3} \dfrac{\sqrt{x} - \sqrt{3}}{x - 3}$

(g) $\lim_{t \to 2} \dfrac{t^2 + t + 6}{t^3 - 2t + 4}$

(h) $\lim_{t \to 0} \dfrac{t}{\sqrt{1 + t} - 1}$.

2. Evaluate each of the limits in Problem 1 using an appropriate form of L'Hôpital's Rule, if it is applicable.
3. Evaluate each of the following limits.

(a) $\lim_{x \to 4} \dfrac{x - 4}{x^n - 4^n}$, n a positive integer

(b) $\lim_{x \to 1+} \dfrac{x^{3/2} - 1}{\sqrt{x^3 - 1}}$

(c) $\lim_{x \to 2+} \dfrac{x^2 - 4x + 2}{\sqrt{x^2 - 4}}$

(d) $\lim_{x \to 1} \dfrac{x^{1/2} - x^{1/3}}{x - 1}$

(e) $\lim_{x \to 1} \dfrac{x^3 - x^2 - x + 1}{2x^3 - 3x^2 + 1}$

(f) $\lim_{t \to 0} \dfrac{3t^2}{3(1 + t)^{1/3} - t - 3}$.

4. Compute

(a) $\lim_{x \to \infty} \dfrac{\sqrt{x + 1}}{x + 2}$

(b) $\lim_{x \to \infty} \dfrac{(x^2 + 1)^{1/3}}{2x^2 - 3}$

(c) $\lim_{x \to \infty} \dfrac{x^{1/3} + 2}{x^{1/2} - 2}$

(d) $\lim_{x \to \infty} \dfrac{x^{1/3} + 2x + 1}{x^{1/2} + 3x - 2}$.

5. Compute each of the following limits directly using the $\dfrac{*}{\infty}$ form of L'Hôpital's Rule. Verify the result by writing the quotient in a different form and using either the $\dfrac{0}{0}$ form of the rule or some other method.

(a) $\lim_{x \to 0+} \dfrac{\sqrt{\dfrac{1}{x} + 1}}{\dfrac{1}{x} + 2}$

(b) $\lim_{x \to 0} \dfrac{\dfrac{1}{x^2} + 5}{\left(\dfrac{1}{x^2} - 1\right)^{1/3}}$.

6. Suppose that F is a function differentiable on the open interval $(0, \infty)$ and such that $F'(x) = \dfrac{1}{x}$, for every $x > 0$. Show that

(a) $\lim_{x \to \infty} \dfrac{F(x)}{x} = 0$

(b) $\lim_{x \to \infty} \dfrac{F(x)}{x^2} = 0$

(c) $\lim_{x \to \infty} \dfrac{F(x)}{x^n} = 0$, for every positive integer n.

CHAPTER 3

Conic Sections

We shall now consider a certain type of curve called a **conic section.** Each of these curves is the curve of intersection of a plane with a right circular cone and each is also the curve defined by a second-degree equation. It is also true that any second-degree equation in x and y defines one of these curves or a degenerate form of one of them. We encounter all of them—the circle, the parabola, the ellipse, and the hyperbola—frequently in mathematics and also in the physical world.

1. The Circle. We looked at a circle in Chapter 1 and have a definition from a first course in plane geometry. This is still the definition: A **circle** is the locus of points in a plane at a given distance from a fixed point. The given distance is called the **radius** and the fixed point is called the **center.**

If the center of the circle is at (h, k), the distance from the center to a variable point (x, y) is, by the distance formula, $\sqrt{(x-h)^2+(y-k)^2}$. If the radius is r, we have an equation of the circle given by

$$\sqrt{(x-h)^2+(y-k)^2} = r. \tag{1}$$

An equivalent equation which is more commonly used is

$$\boxed{(x-h)^2+(y-k)^2 = r^2.} \tag{2}$$

It is easy to see that, not only do all points at a distance r from (h, k) lie on the graph of (2), but also all points on the graph of (2) are at a distance r from (h, k).

EXAMPLE 1. (a) Write an equation of the circle with center at the origin and radius 3. (b) Write an equation of the circle with center at $(-1, 2)$ and radius 5.

(a) By the distance formula, the first circle has equation

$$\sqrt{(x-0)^2+(y-0)^2} = 3, \qquad \text{or} \qquad x^2+y^2 = 9.$$

(b) By the distance formula, an equation for the second circle is $\sqrt{[x - (-1)]^2 + (y - 2)^2} = 5$. Equivalent equations are

$$(x + 1)^2 + (y - 2)^2 = 25,$$
$$x^2 + y^2 + 2x - 4y = 20.$$

As the following examples will show, any equation of the form $ax^2 + ay^2 + bx + cy + d = 0$ is, loosely speaking, an equation of a circle. The words "loosely speaking" are inserted to cover possible degenerate cases. For example, if $a = 0$ and b and c are not both zero, the equation becomes an equation of a line. In another degenerate case "the circle" may be just a point (if its radius is zero), and in another there may be no locus at all.

EXAMPLE 2. Describe the graph of each of the following equations:

(a) $x^2 + y^2 - 6x + 8y - 75 = 0$,

(b) $x^2 + y^2 + 12x - 2y + 37 = 0$,

(c) $x^2 + y^2 - 4x - 5y + 12 = 0$,

(d) $3x^2 + 3y^2 - 9x + 10y - \frac{71}{12} = 0$.

The technique of completing the square is useful in problems of this type. In (a), we write equations equivalent to the given equation until we recognize the form.

$$x^2 - 6x + y^2 + 8y = 75,$$
$$x^2 - 6x + 9 + y^2 + 8y + 16 = 75 + 9 + 16,$$
$$(x - 3)^2 + (y + 4)^2 = 100.$$

The graph is a circle with center at $(3, -4)$ and radius 10.

Applying the same technique to (b), we have

$$x^2 + 12x + y^2 - 2y = -37,$$
$$x^2 + 12x + 36 + y^2 - 2y + 1 = -37 + 36 + 1,$$
$$(x + 6)^2 + (y - 1)^2 = 0.$$

The graph is a circle with center at $(-6, 1)$ and radius 0; i.e., it is just the point $(-6, 1)$. We may say that the graph consists of the single point, although we sometimes describe it as a point-circle.

The equation of (c) gives different results:

$$x^2 - 4x + y^2 - 5y = -12,$$

$$x^2 - 4x + 4 + y^2 - 5y + \tfrac{25}{4} = -12 + 4 + \tfrac{25}{4},$$

$$(x - 2)^2 + (y - \tfrac{5}{2})^2 = -\tfrac{7}{4}.$$

For any two real numbers x and y, the numbers $(x - 2)^2$ and $(y - \frac{5}{2})^2$ must both be nonnegative, while $-\frac{7}{4}$ is certainly negative. Hence there are no points in the plane satisfying this equation. However, the form of the last of the three equivalent equations is that of the equation of a circle, and we sometimes say that the graph is an imaginary circle with center at $(2, \frac{5}{2})$ and radius $\frac{1}{2}\sqrt{-7}$.

Equation (d) requires a bit more manipulation:

$$3x^2 - 9x + 3y^2 + 10y = \tfrac{71}{12},$$

$$x^2 - 3x + y^2 + \tfrac{10}{3}y = \tfrac{71}{36},$$

$$x^2 - 3x + \tfrac{9}{4} + y^2 + \tfrac{10}{3}y + \tfrac{25}{9} = \tfrac{71}{36} + \tfrac{9}{4} + \tfrac{25}{9},$$

$$(x - \tfrac{3}{2})^2 + (y + \tfrac{5}{3})^2 = 7.$$

The graph is a circle with center at $(\frac{3}{2}, -\frac{5}{3})$ and radius $\sqrt{7}$.

By completing the square, as in the above examples, one can show that any equation of the form

$$ax^2 + ay^2 + bx + cy + d = 0$$

is an equation of a circle of positive radius if and only if $a \neq 0$ and $b^2 + c^2 > 4ad$.

The circle is also the intersection of a right circular cone with a plane perpendicular to the axis of the cone. If the plane passes through the vertex of the cone, the intersection is a point.

We can use the techniques of the calculus, as well as our knowledge of Euclidean geometry, to write an equation for a circle or for a line tangent to a circle, from given geometric conditions.

EXAMPLE 3. Write an equation of the line which is tangent to the graph of $(x + 3)^2 + (y - 4)^2 = 25$ at $(1, 7)$. We may find the slope by use of the derivative, differentiating implicitly and remembering that one interpretation of the derivative is the slope of the tangent: $2(x + 3) + 2(y - 4)y' = 0$; hence $y' = -\dfrac{x + 3}{y - 4}$. The slope of the tangent is $-\dfrac{1 + 3}{7 - 4} = -\dfrac{4}{3}$. We may also find the slope of the tangent by noting first that the radius to (1, 7) has slope $\dfrac{7-4}{1-(-3)} = \dfrac{3}{4}$ and then by remembering that the tangent is perpendicular to the radius and hence has slope $-\frac{4}{3}$. Thus the tangent line has equation $y - 7 = -\frac{4}{3}(x - 1)$ or $4x + 3y = 25$.

PROBLEMS

1. Write an equation for each of the following.

   (a) A circle with center at the origin and radius 3.

   (b) A circle with center at the origin and radius $\frac{10}{3}$.

   (c) A circle with center at $(-2, 2)$ and radius 5.

   (d) A circle with center at $(3, 0)$ and radius 3.

   (e) A circle with center at $(0, -7)$ and radius 7.

   (f) A circle with center at $(-3, -3)$ and radius $3\sqrt{2}$.

   (g) A circle with center in the first quadrant, radius 4, and tangent to both axes.

   (h) A circle with center on the y-axis, radius $\frac{5}{2}$, and tangent to the x-axis (there are two such circles).

   (i) A circle with radius 2 and tangent to the x-axis and to the line defined by the equation $x = 5$ (there are four such circles).

2. For each of the following equations, describe the curve defined by it.

   (a) $x^2 + y^2 = 64$

   (b) $x^2 + y^2 = 32$

   (c) $x^2 + (y - 4)^2 = 9$

   (d) $(x + 2)^2 + y^2 = 16$

   (e) $(x - 2)^2 + (y + 7)^2 = 19$

   (f) $(2x - 3)^2 + (2y - 5)^2 = \frac{25}{4}$

   (g) $x^2 + y^2 - 8x - 12y + 27 = 0$

   (h) $x^2 + y^2 - 5x - 7y + \frac{5}{2} = 0$

   (i) $9x^2 + 9y^2 - 12x + 30y = 71$

   (j) $5x^2 + 5y^2 - 6x + 8y = 31$.

3. Show that, if $b^2 + c^2 > 4ad$ and $a \neq 0$, the equation $ax^2 + ay^2 + bx + cy + d = 0$ defines a circle with center at

$$\left(-\frac{b}{2a}, -\frac{c}{2a}\right) \text{ and radius } \sqrt{\frac{b^2 + c^2 - 4ad}{4a^2}}.$$

4. Show that the tangents from a point (x_1, y_1) outside the circle defined by $(x - h)^2 + (y - k)^2 = r^2$ to the circle are of length

$$\sqrt{(x_1 - h)^2 + (y_1 - k)^2 - r^2}.$$

5. Show that the line tangent to the circle defined by

$$ax^2 + ay^2 + bx + cy + d = 0 \text{ at } (x_1, y_1) \text{ has the equation}$$

$$ax_1x + ay_1y + \frac{b}{2}(x + x_1) + \frac{c}{2}(y + y_1) + d = 0.$$

6. Write an equation of the line which is tangent to

   (a) the circle defined by $x^2 + y^2 = 25$ at $(-3, 4)$.

   (b) the circle defined by $x^2 + y^2 = 9$ at $(0, 3)$.

   (c) the circle defined by $(x - 2)^2 + (y + 8)^2 = 169$ at $(7, 4)$.

   (d) the circle defined by $x^2 + y^2 - 10y = 33$ at $(7, 2)$.

7. Write an equation of the line containing the common chord of circles defined by $x^2 + y^2 - 8x - 12y = 48$ and $x^2 + y^2 - 4x + 6y = 23$.

8. Given a circle and a line tangent to it, the segment of the line between a given point and the point of tangency is commonly called the tangent from the point to the circle. Show that the locus of points from which the tangents to two unequal externally tangent circles have equal length is the common internal tangent line.

9. Write an equation for the circle which passes through

   (a) $(3, 4)$, $(-4, 3)$, and $(5, 0)$.

   (b) $(7, 1)$, $(6, 2)$, and $(-1, -5)$.

   (c) $(4, 16)$, $(-6, -8)$, and $(11, 9)$.

10. Use the results of Problems 3 and 5 to show that a line tangent to a circle is perpendicular to the radius drawn to the point of tangency.

2. The Parabola. A second conic section, which appears in a great many physical applications, is the parabola. In this section we shall derive its equation and study its properties.

By definition, a **parabola** is the locus of points in a plane equidistant from a given line and a given point not on the line. The line is called the **directrix,** and the point is called the **focus.** A simple equation for a parabola is found if a point on the x-axis is used for the focus and a line perpendicular to the x-axis and on the opposite side of the origin from the focus is used as the directrix. In Figure 1, the focus is at $(a, 0)$ and an equation of the directrix is $x = -a$.

The distance from an arbitrary point (x, y) to the focus $(a, 0)$ is, by the distance formula, $\sqrt{(x - a)^2 + (y - 0)^2}$. The perpendicular from (x, y) to the directrix intersects that line at $(-a, y)$, and so the distance from the point to the line is $\sqrt{(x + a)^2 + (y - y)^2} = |x + a|$. The point (x, y)

lies on the parabola if and only if the two distances are equal. Hence an equation of the parabola is

$$\sqrt{(x - a)^2 + y^2} = |x + a|.$$

An equivalent equation, which is simpler, is obtained by squaring both sides. We get

$$x^2 - 2ax + a^2 + y^2 = x^2 + 2ax + a^2,$$

or

$$\boxed{y^2 = 4ax.}$$

This equation has been derived in such a way that its graph contains all points and only those points equidistant from the focus and the directrix.

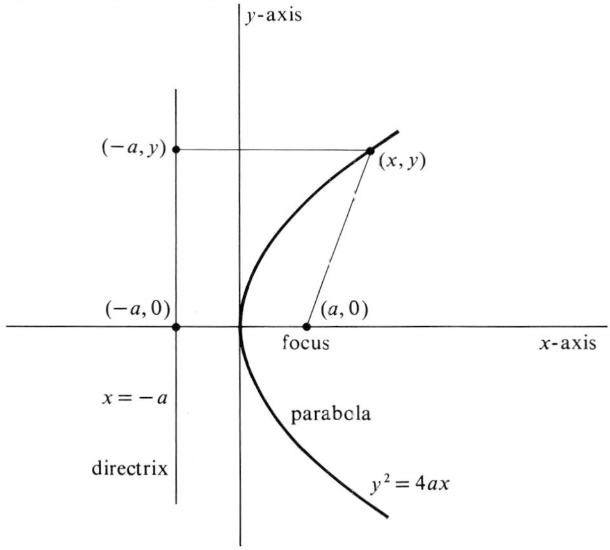

Figure 1

The point on a parabola nearest its directrix is called the **vertex.** For the parabola defined by $y^2 = 4ax$, the vertex is the origin, and there are no points on the curve to the left of the vertex. Since $|y|$ increases indefinitely as x increases, the graph is not a closed curve but opens to the right. We cannot, of course, draw the entire curve, but the incomplete graph in Figure 1 is sufficient to describe the entire curve for us. Note that the graph is symmetric with respect to the x-axis.

If we had chosen $(-a, 0)$ for the focus and the line $x = a$ for the directrix, we would have a parabola with equation $y^2 = -4ax$. This parabola would also have its vertex at the origin, also be symmetric with respect to the x-axis, but would open to the left.

With a horizontal directrix and the focus on the y-axis, the equation would be $x^2 = 4ay$ or $x^2 = -4ay$, opening upward or downward, respectively.

Thus the graphs of $y^2 = kx$ and $x^2 = ky$ are parabolas with their vertices at the origin. The line through the vertex and the focus is called the **axis** of the parabola. The graph of $y^2 = kx$ has the x-axis for its axis and opens to the right or left, depending on the sign of k. The graph of $x^2 = ky$ has the y-axis for its axis and opens upward or downward, depending on the sign of k. The absolute value of k determines the shape of the parabola. If $k = 0$, the two equations reduce to $y^2 = 0$ and $x^2 = 0$, whose graphs are the x-axis and the y-axis respectively. Thus a straight line may be regarded as a degenerate parabola.

EXAMPLE 1. Find the coordinates of the focus and an equation of the directrix of $x^2 = -7y$, and sketch its graph. An equivalent form of the equation is $x^2 = -4(\frac{7}{4})y$, from which it follows that the focus is $(0, -\frac{7}{4})$, that the directrix is $y = \frac{7}{4}$, and that the graph opens downward. It is shown in Figure 2.

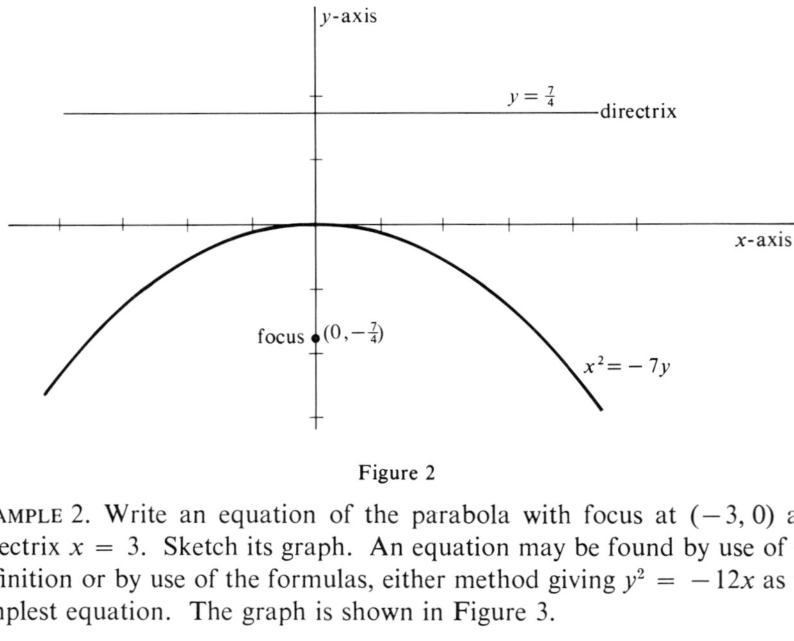

Figure 2

EXAMPLE 2. Write an equation of the parabola with focus at $(-3, 0)$ and directrix $x = 3$. Sketch its graph. An equation may be found by use of the definition or by use of the formulas, either method giving $y^2 = -12x$ as the simplest equation. The graph is shown in Figure 3.

In every case considered so far, the parabola has one of the coordinate axes for its axis and the focus and directrix are equally spaced on opposite sides of the origin. The equations are somewhat more involved if other vertical or horizontal lines are chosen as axes, and even more involved if the parabolas have axes which are not parallel to one of the coordinate axes.

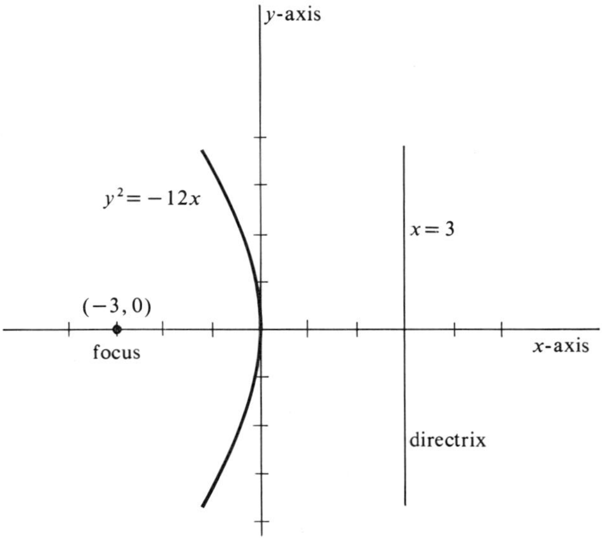

Figure 3

EXAMPLE 3. Write an equation of the parabola with focus at (2, 5) and directrix $x = -6$. Sketch the graph (see Figure 4). From the definition of

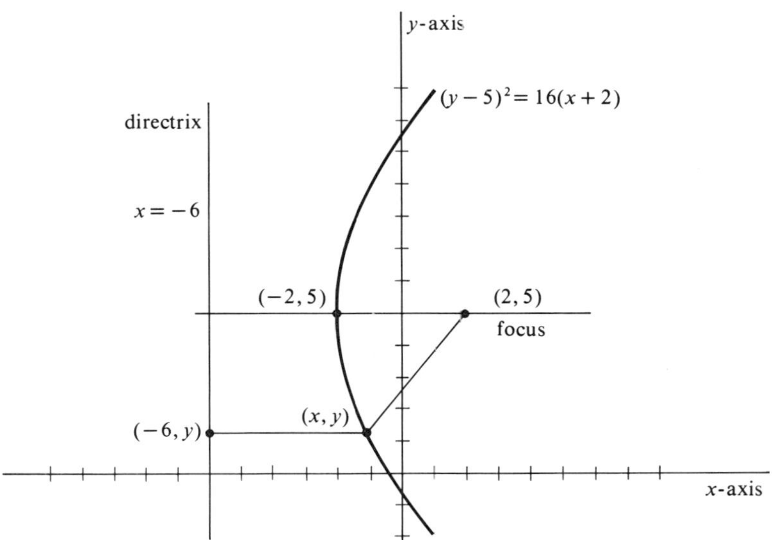

Figure 4

the parabola,

$$|x + 6| = \sqrt{(x - 2)^2 + (y - 5)^2},$$
$$x^2 + 12x + 36 = x^2 - 4x + 4 + (y - 5)^2,$$
$$16x + 32 = (y - 5)^2,$$
$$(y - 5)^2 = 16(x + 2).$$

Note the similarity of this form to the $y^2 = 4ax$ form. The x-coordinate of the vertex, which is located halfway along the perpendicular from the focus to the directrix, is equal to -2. That is, it is found by setting $x + 2$ equal to zero. Hence the y-coordinate of the vertex is found by setting $y - 5$ equal to zero. Thus the vertex is the point $(-2, 5)$.

Let us now write a general equation of an arbitrary parabola with a horizontal directrix (see Figure 5). Consider the perpendicular from the focus to the directrix, and denote the midpoint of this line segment by

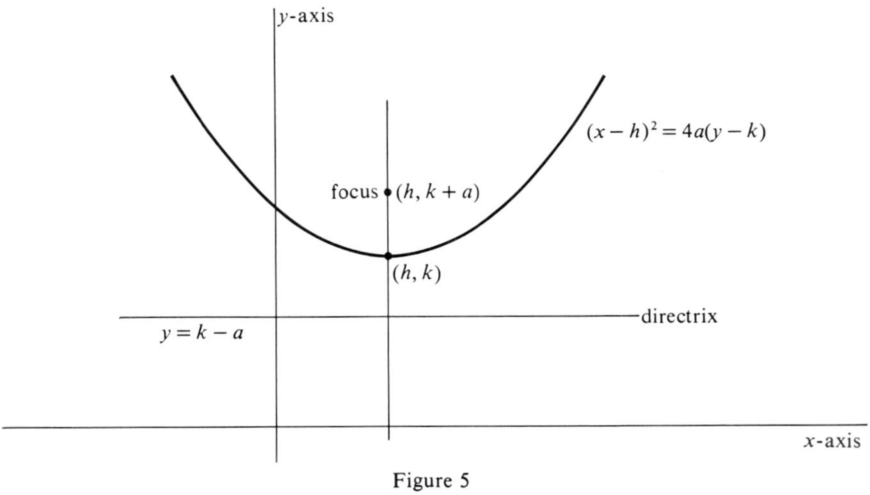

Figure 5

(h, k). This is the vertex, the point on the parabola nearest the directrix. Let the focus be the point $(h, k + a)$. Then the length of the segment is $|2a|$ and the directrix is the line $y = k - a$. By algebra similar to that in Example 3, we go from

$$|y - (k - a)| = \sqrt{(x - h)^2 + [y - (k + a)]^2}$$

to

$$(x - h)^2 = 4a(y - k). \qquad (1)$$

This parabola will have the same shape and the same orientation (opening upward or downward) as $x^2 = 4ay$, but will have its vertex at (h, k).

Similarly, $(y - k)^2 = 4a(x - h)$ will have the same shape and the same orientation as $y^2 = 4ax$, but will have its vertex at (h, k).

EXAMPLE 4. Write an equation of the parabola with focus at $(-3, 2)$ and directrix $y = 6$. The midpoint of the segment connecting focus and directrix and perpendicular to the directrix is at $(-3, 4)$ and the segment is of length 4. With directrix above the focus, the parabola opens downward and has the equation

$$(x + 3)^2 = -8(y - 4).$$

EXAMPLE 5. Describe the graph of $y^2 + 2x - 3y + 7 = 0$. We first write $y^2 - 3y = -2x - 7$ and, completing the square, add $\frac{9}{4}$ to each side of the equation:

$$y^2 - 3y + \tfrac{9}{4} = -2x - 7 + \tfrac{9}{4},$$
$$(y - \tfrac{3}{2})^2 = -2(x + \tfrac{19}{8}).$$

The graph is a parabola with vertex at $(-\frac{19}{8}, \frac{3}{2})$, focus at $(-\frac{23}{8}, \frac{3}{2})$, and directrix $x = -\frac{15}{8}$. It opens to the left and has the same shape and orientation as the graph of $y^2 = -2x$.

The graph of any function f defined by an equation $f(x) = ax^2 + bx + c$, where $a \neq 0$, is a parabola. The defining equation may be written

$$f(x) = a\left(x^2 + \frac{b}{a}x + \frac{b^2}{4a^2}\right) + c - \frac{b^2}{4a},$$

or

$$f(x) = a\left(x + \frac{b}{2a}\right)^2 + \frac{4ac - b^2}{4a}.$$

Hence the graph of the function, which is the graph of the equation $y = f(x)$, is the graph of $y = a\left(x + \dfrac{b}{2a}\right)^2 + \dfrac{4ac - b^2}{4a}$, or, equivalently, of

$$\left(x + \frac{b}{2a}\right)^2 = \frac{1}{a}\left(y - \frac{4ac - b^2}{4a}\right).$$

Comparing this equation with (1), we see that the graph is a parabola with a vertical axis and its vertex at $\left(-\dfrac{b}{2a}, \dfrac{4ac - b^2}{4a}\right)$.

PROBLEMS

1. Write an equation for each of the following.
   (a) The parabola with focus at $(-2, 0)$ and directrix $x = 2$.
   (b) The parabola with focus at $(0, 3)$ and directrix $y = -3$.

(c) The parabola with focus at $(0, -1)$ and directrix $y = 1$.

(d) The parabola with focus at $(4, 0)$ and vertex at the origin.

(e) The parabola with focus at $(0, -2)$ and vertex at $(0, 2)$.

(f) The parabola with vertex at $(-5, 0)$ and directrix $x + 1 = 0$.

2. Sketch the graph of each of the parabolas in Problem 1.
3. Sketch on the same set of axes the graph of each of the following equations. Compare and contrast the graphs, noting common features and differences.

(a) $y^2 = \frac{1}{2}x$ (c) $y^2 = 2x$

(b) $y^2 = x$ (d) $y^2 = 3x$.

4. Sketch on the same set of axes the graph of each of the following equations. Compare and contrast the graphs, noting common features and differences.

(a) $y^2 = 2x$ (c) $x^2 = 2y$

(b) $y^2 = -2x$ (d) $x^2 = -2y$.

5. Consider a point (x_1, y_1) on the graph of $y^2 = 4ax$.

(a) Find the slope of the tangent to the graph at (x_1, y_1).

(b) Write an equation of the tangent line in (a).

(c) Show that $yy_1 = 2a(x + x_1)$ is an equation of the tangent line.

6. Consider a point (x_1, y_1) on the graph of $x^2 = 4ay$. Show that $xx_1 = 2a(y + y_1)$ is an equation of the line tangent to the graph at (x_1, y_1).
7. If (x_1, y_1) lies on the graph of $y = ax^2 + bx + c$, show that $\frac{1}{2}(y + y_1) = axx_1 + \frac{1}{2}b(x + x_1) + c$ is an equation of the line tangent to the graph at (x_1, y_1).
8. Write an equation of a line which passes through the point $(8, 7)$ and is a tangent to the graph of $y^2 = 6x$.
9. (a) Find the point where the tangent to $y^2 = 4ax$ at the point (x_1, y_1) cuts the x-axis. Assume that $a \neq 0$.

(b) Show that the segment of the tangent line between (x_1, y_1) and the point found in (a) is bisected by the y-axis.

10. Write an equation for each of the following:

(a) The parabola with vertex $(1, 1)$ and directrix $x = -1$.

(b) The parabola with vertex $(1, 1)$ and directrix $y = 0$.

(c) The parabola with vertex $(4, 3)$ and directrix $x = -2$.

(d) The parabola with vertex $(-1, 2)$ and directrix $y = 4$.

11. Find the focus of each of the parabolas in Problem 10.
12. Do Problem 10 with each occurrence of the word "vertex" replaced by "focus."
13. Find the focus, vertex, and directrix of the parabola which is the graph of each of the following equations or functions. Sketch the graph.

(a) $y = x^2$ (d) $y^2 + 2y + 2x = 0$

(b) $y = x^2 - 2x$ (e) $g(y) = -y^2 - 4y + 4$

(c) $f(x) = x^2 - 6x + 1$ (f) $x^2 + x + y = 0$.

3. The Ellipse. A third conic section, quite fashionable these days as astronauts circle the earth in elliptical orbits, is the ellipse. In this section we shall derive its equation and study its properties.

By definition, an **ellipse** is the locus of points in a plane the sum of whose distances from two given points is a constant. The constant, of course, must be greater than the distance between the two given points. The two points are called the **foci** of the ellipse. A simple equation for an ellipse is found if the points $(-c, 0)$ and $(c, 0)$ are used for foci and $2a$ for the constant sum of distances. The foci and the arbitrary point (x, y) are shown in Figure 6. The

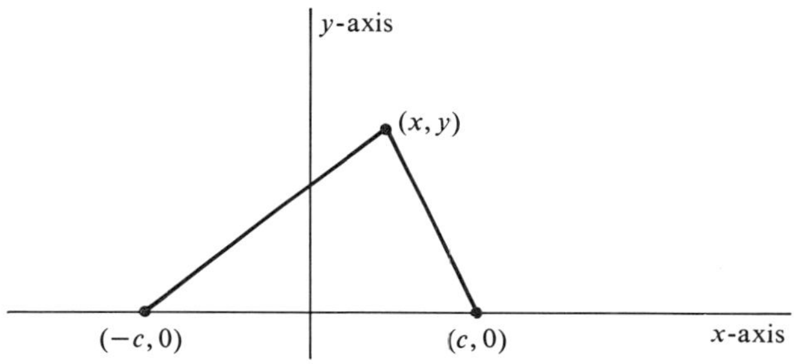

Figure 6

distance from (x, y) to $(-c, 0)$ is $\sqrt{(x + c)^2 + (y - 0)^2}$, and that from (x, y) to $(c, 0)$ is $\sqrt{(x - c)^2 + (y - 0)^2}$. The point (x, y) lies on the ellipse if and only if the sum of these two distances is equal to $2a$. Hence an equation of the ellipse is

$$\sqrt{(x + c)^2 + y^2} + \sqrt{(x - c)^2 + y^2} = 2a.$$

Simpler equivalent equations are found by subtracting $\sqrt{(x - c)^2 + y^2}$ from both sides of the equation

$$\sqrt{(x + c)^2 + y^2} = 2a - \sqrt{(x - c)^2 + y^2},$$

squaring both sides of the equation and simplifying

$$x^2 + 2cx + c^2 + y^2 = 4a^2 - 4a\sqrt{(x - c)^2 + y^2} + x^2 - 2cx + c^2 + y^2,$$

$$4a\sqrt{(x - c)^2 + y^2} = 4a^2 - 4cx,$$

dividing by 4 and squaring again

$$a^2(x^2 - 2cx + c^2 + y^2) = a^4 - 2a^2cx + c^2x^2,$$

and simplifying again,

$$x^2(a^2 - c^2) + a^2y^2 = a^2(a^2 - c^2).$$

Finally, dividing by $a^2(a^2 - c^2)$ we have

$$\frac{x^2}{a^2} + \frac{y^2}{a^2 - c^2} = 1.$$

Since $a > c$, it follows that $a^2 > c^2$, and we replace $a^2 - c^2$ by b^2. We then have the canonical form of an equation for the ellipse,

$$\boxed{\frac{x^2}{a^2} + \frac{y^2}{b^2} = 1.}$$

This equation is derived in such a way that its graph contains all points which satisfy the locus condition. One of the problems at the end of the section will be to show that the graph contains only those points.

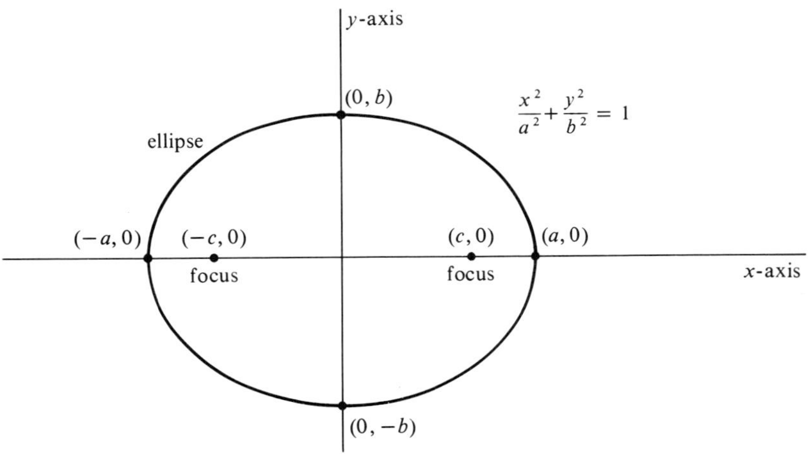

Figure 7

By placing y equal to 0, we obtain $x = \pm a$. The graph therefore cuts the x-axis at $(-a, 0)$ and at $(a, 0)$, and the numbers a and $-a$ are the x-intercepts of the graph. The line segment between $(-a, 0)$ and $(a, 0)$ is called the **major axis** of the ellipse. Similarly, the graph cuts the y-axis at $(0, -b)$ and at $(0, b)$, and the numbers b and $-b$ are the y-intercepts of the graph. The line segment between $(0, -b)$ and $(0, b)$ is called the **minor axis** of the ellipse. It is not difficult to see that there are no points on the graph for which $|x| > a$ or for $|y| > b$. Symmetry across both axes and across the origin can be seen by noting that $(-p, q)$, $(p, -q)$, and $(-p, -q)$ all lie on the graph if (p, q) does. The complete curve is shown in Figure 7.

There is a simple method of constructing an ellipse by placing thumbtacks at the foci and looping around them a string of length $2a + 2c$. If a pencil is placed in the loop so as to hold it taut and moved in a complete turn around the foci, the curve traced is an ellipse. This is readily seen from the definition.

EXAMPLE 1. Describe the graph of $4x^2 + 25y^2 = 100$ and draw it. An equivalent form of the equation is $\dfrac{x^2}{25} + \dfrac{y^2}{4} = 1$, from which it is apparent that the graph is an ellipse with x-intercepts of -5 and 5, major axis of length 10, y-intercepts of -2 and 2, and minor axis of length 4. The intersection of these axes, in this case the origin, is called the **center** of the ellipse. Using $a^2 = 25$, $a^2 - c^2 = 4$, we have $c^2 = 21$. Thus the foci are at $(-\sqrt{21}, 0)$ and $(\sqrt{21}, 0)$. The graph is drawn in Figure 8.

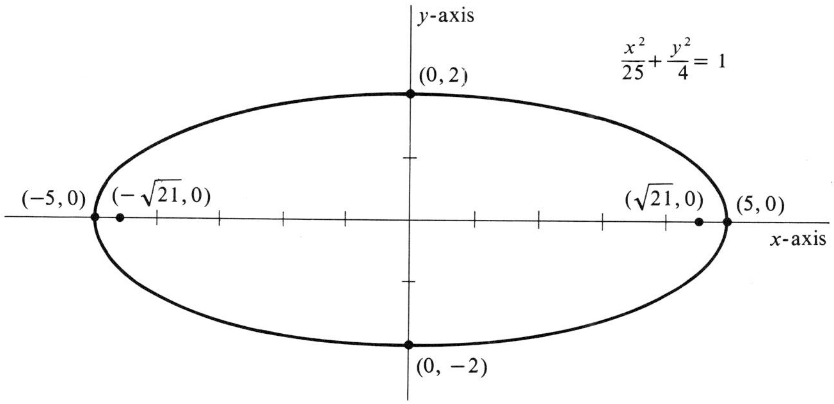

Figure 8

If the foci are located on the y-axis at $(0, c)$ and $(0, -c)$, an analogous derivation gives the equation

$$\frac{x^2}{a^2 - c^2} + \frac{y^2}{a^2} = 1$$

and its equivalent form

$$\frac{x^2}{b^2} + \frac{y^2}{a^2} = 1.$$

In this case, the major axis is the line segment between $(0, -a)$ and $(0, a)$, and the minor axis that between $(-b, 0)$ and $(b, 0)$.

EXAMPLE 2. Describe and draw the graph of $4x^2 + y^2 = 36$. An equivalent form of the equation is $\frac{x^2}{9} + \frac{y^2}{36} = 1$, from which we see that the positive y-intercept is larger than the positive x-intercept and therefore that the foci are on the y-axis. The x-intercepts are 3 and -3, the y-intercepts are 6 and -6, and the foci are at $(0, \sqrt{36 - 9})$ and $(0, -3\sqrt{3})$. The graph is drawn in Figure 9.

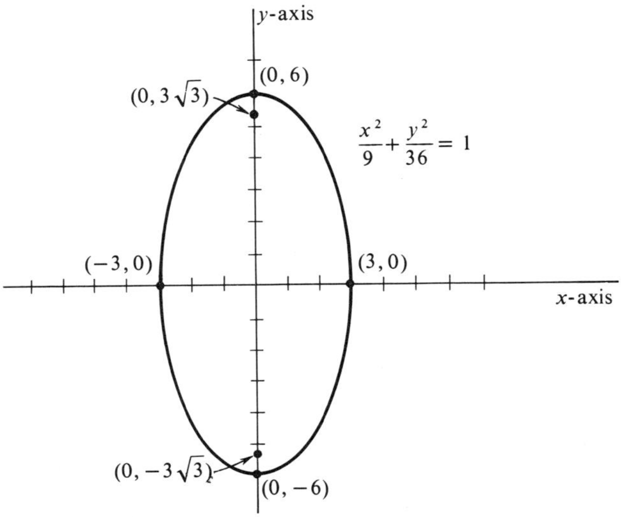

Figure 9

Examples 1 and 2 were each for an ellipse with its center at the origin. In a manner similar to that used in the last section, we can write an equation for an ellipse with center at (h, k) and foci at $(h - c, k)$ and $(h + c, k)$. The equation is then

$$\frac{(x - h)^2}{a^2} + \frac{(y - k)^2}{b^2} = 1.$$

If the center is at (h, k) and the foci at $(h, k - c)$ and $(h, k + c)$, an equation is

$$\frac{(y - k)^2}{a^2} + \frac{(x - h)^2}{b^2} = 1.$$

EXAMPLE 3. Describe and sketch the graph of $\frac{(x + 4)^2}{9} + \frac{(y - 7)^2}{25} = 1$. The graph is an ellipse with center at $(-4, 7)$. The foci are above and below the center, the distance being $\sqrt{25 - 9} = 4$. Thus the foci are at $(-4, 3)$ and $(-4, 11)$. The ends of the major axis are $(-4, 7 - 5)$ or $(-4, 2)$ and $(-4, 7 + 5)$ or $(-4, 12)$. The ends of the minor axis are $(-4 - 3, 7)$ or $(-7, 7)$ and $(-4 + 3, 7)$ or $(-1, 7)$. The graph is drawn in Figure 10.

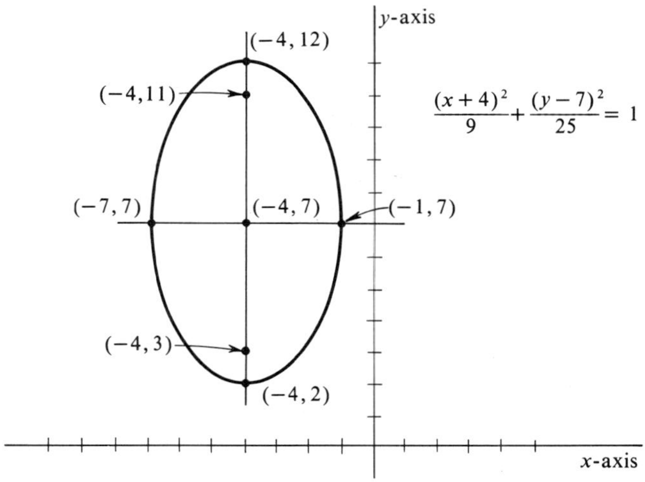

Figure 10

There is an alternative definition of the ellipse, one analogous to the definition of a parabola. It may be seen in considering the distance from a point (x, y) to the focus $(c, 0)$. This distance is $\sqrt{(x - c)^2 + y^2}$. If the point lies on the ellipse, then $\dfrac{x^2}{a^2} + \dfrac{y^2}{a^2 - c^2} = 1$, or $y^2 = a^2 - x^2 - c^2 + \dfrac{x^2c^2}{a^2}$. Thus the distance from the point to the focus is

$$\sqrt{x^2 - 2cx + c^2 + \left(a^2 - x^2 - c^2 + \frac{x^2c^2}{a^2}\right)},$$

or

$$\sqrt{\frac{x^2c^2}{a^2} - 2cx + a^2},$$

or

$$\left|\frac{xc}{a} - a\right|.$$

This latter expression may be written $\dfrac{c}{a}\left|x - \dfrac{a^2}{c}\right|$. But $\left|x - \dfrac{a^2}{c}\right|$ is the distance from the point (x, y) to the line $x = \dfrac{a^2}{c}$. Thus the distance from the point to the focus $(c, 0)$ is $\frac{c}{a}$ times the distance from the point to the line $x = \frac{a^2}{c}$. This result could also be stated by noting that the ratio of the two distances (that to the focus and that to the line) is a constant for all points on the ellipse. The line is called a **directrix** for the ellipse and the constant ratio, $\frac{c}{a}$, is called the **eccentricity** of the ellipse. The eccentricity is less than 1. For all points on a parabola, the ratio of the two distances (that to the focus and that to the directrix) is also a constant, namely 1. A parabola is said to have eccentricity 1. Even as the ellipse has two foci, it also has two directrices. The other is the line $x = -\frac{a^2}{c}$, and the ratio of the distance between a point on the ellipse and the focus $(-c, 0)$ to the distance from the same point to the directrix $x = -\frac{a^2}{c}$ is the same constant $\frac{c}{a}$.

Not all equations for ellipses are in canonical form, but equivalent equations in canonical form can be found by factoring and completing the square.

EXAMPLE 4. Describe and draw the graph of $16x^2 + 25y^2 - 64x + 150y - 111 = 0$. Equations equivalent to this one are

$$16(x^2 - 4x) + 25(y^2 + 6y) = 111,$$

$$16(x^2 - 4x + 4) + 25(y^2 + 6y + 9) = 111 + 16 \cdot 4 + 25 \cdot 9,$$

$$16(x - 2)^2 + 25(y + 3)^2 = 400,$$

$$\frac{(x - 2)^2}{25} + \frac{(y + 3)^2}{16} = 1.$$

From the last equation we can see that the graph is an ellipse with center at $(2, -3)$, horizontal major axis of length 10, vertical minor axis of length 8, and $c = \sqrt{25 - 16}$. Thus the foci are at $(-1, -3)$ and $(5, -3)$. The graph is shown in Figure 11.

As Example 4 illustrates, almost every equation of the type $ax^2 + by^2 + cx + dy + e = 0$ with $ab > 0$ has an ellipse for its graph. The reason for the "almost" can be seen as we write equivalent equations

$$a\left(x^2 + \frac{c}{a}x\right) + b\left(y^2 + \frac{d}{b}y\right) = -e,$$

$$a\left(x^2 + \frac{c}{a}x + \frac{c^2}{4a^2}\right) + b\left(y^2 + \frac{d}{b}y + \frac{d^2}{4b^2}\right) = \frac{c^2}{4a} + \frac{d^2}{4b} - e,$$

$$a\left(x + \frac{c}{2a}\right)^2 + b\left(y + \frac{d}{2b}\right)^2 = \frac{c^2}{4a} + \frac{d^2}{4b} - e.$$

If the expression on the right side of the last equation has the same sign as a (and hence b), then the graph is an ellipse. If the expression on the right side is equal to zero, the graph is the single point $\left(-\frac{c}{2a}, -\frac{d}{2b}\right)$. If the expression on the right side has sign opposite to that of a, there is no graph, although the equation is said to have an imaginary ellipse for its graph. If $a = b$, the foci coincide and the graph is a circle.

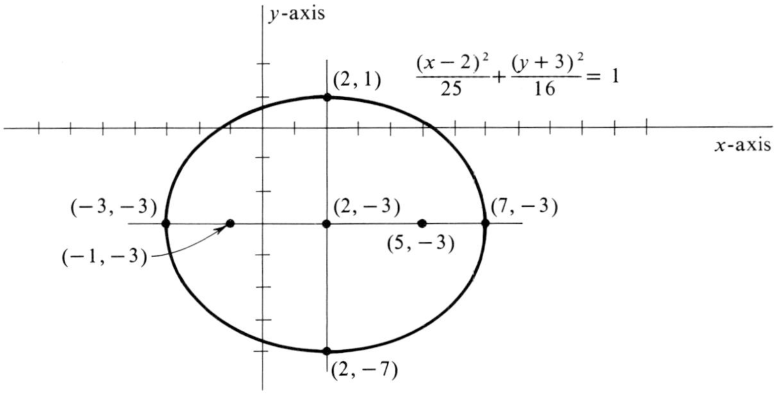

Figure 11

PROBLEMS

1. Describe and sketch the graph of each of the following equations. Label the foci and the endpoints of the major and minor axes.

(a) $\frac{x^2}{4} + \frac{y^2}{9} = 1$

(b) $\frac{x^2}{9} + \frac{y^2}{4} = 1$

(c) $\frac{x^2}{169} + \frac{y^2}{144} = 1$

(d) $\frac{x^2}{100} + \frac{y^2}{64} = 1$

(e) $\frac{x^2}{17} + \frac{y^2}{16} = 1.$

2. Write an equation for the ellipse satisfying the given conditions.

   (a) Foci at $(-5, 0)$ and $(5, 0)$. Minor axis of length 24.

   (b) Center at the origin. Major axis horizontal and of length 14, minor axis vertical and of length 8.

   (c) Center at the origin. Minor axis vertical and of length 4. Passing through $(3, 1)$.

   (d) Foci at $(-4, 0)$ and $(4, 0)$. Endpoints of major axis at $(-5, 0)$ and $(5, 0)$.

   (e) The locus of points the sum of whose distances from $(0, 2)$ and $(0, -2)$ is 7.

3. It has been shown that the distance between a point on the ellipse

$$\frac{x^2}{a^2} + \frac{y^2}{a^2 - c^2} = 1$$

and the focus $(c, 0)$ is $\left|\frac{xc}{a} - a\right|$.

   (a) Show that this distance is $a - \dfrac{xc}{a}$ for $|x| \leq a$.

   (b) Show that the distance between a point on the ellipse and the focus $(-c, 0)$ is $\left|\dfrac{xc}{a} + a\right|$ and that the distance is $a + \dfrac{xc}{a}$.

   (c) Show that the sum of the distances from a point on the ellipse to the foci is $2a$ and hence that the graph of $\dfrac{x^2}{a^2} + \dfrac{y^2}{a^2 - c^2} = 1$ contains only those points which satisfy the locus definition.

4. Describe and sketch the graph of each of the following equations.

   (a) $\dfrac{(x-2)^2}{25} + \dfrac{(y-4)^2}{9} = 1$

   (b) $\dfrac{(x+3)^2}{16} + \dfrac{(y-2)^2}{25} = 1$

   (c) $\dfrac{(x+5)^2}{169} + \dfrac{(y+2)^2}{144} = 4$

   (d) $25x^2 + 9(y+3)^2 = 225$

   (e) $9x^2 + 4y^2 + 36x - 24y + 36 = 0.$

5. The line segment which passes through a focus, is perpendicular to the major axis, and has its endpoints on the ellipse is called a **latus rectum.**

   (a) Find the length of a latus rectum of the ellipse $4x^2 + 9y^2 = 36$.

   (b) Find the length of a latus rectum of the ellipse $b^2x^2 + a^2y^2 = a^2b^2$. (Assume that $b < a$.)

   (c) Show that both latera recta of an ellipse are the same length.

6. Write equations of the directrices and find the eccentricity of each of the following ellipses.

   (a) $4x^2 + 9y^2 = 36$ (b) $9x^2 + 4y^2 = 144.$

7. Assume that $0 < c < a$.

   (a) Find the distance between (x, y) and $(-c, 0)$.

   (b) Find the distance between (x, y) and the line $x = -\dfrac{a^2}{c}\cdot$

   (c) Find the locus of points (x, y) such that the ratio between the distance in (a) and the distance in (b) is a constant $\dfrac{c}{a}\cdot$

8. Show that an ellipse becomes more nearly circular as its foci get closer and closer together.

9. Consider a point (x_1, y_1) on the graph of $b^2x^2 + a^2y^2 = a^2b^2$.

   (a) Find the slope of the tangent to the graph at (x_1, y_1).

   (b) Write an equation of the tangent line in (a).

   (c) Show that $b^2xx_1 + a^2yy_1 = a^2b^2$ is an equation of the tangent line.

10. Assume that the constants a, b, c, d, and e are such that $ax^2 + by^2 + cx + dy + e = 0$ is an equation of an ellipse. Consider a point (x_1, y_1) on this ellipse.

    (a) Find the slope of the tangent to the graph at (x_1, y_1).

    (b) Write an equation of the tangent line in (a).

    (c) Show that $axx_1 + byy_1 + \frac{1}{2}c(x + x_1) + \frac{1}{2}d(y + y_1) + e = 0$ is an equation of the tangent line.

11. Write an equation of the ellipse with horizontal and vertical axes satisfying the given data.

    (a) Foci at $(-3, 2)$ and $(5, 2)$. Eccentricity is $\frac{2}{3}$.

    (b) Center at $(2, -1)$. One focus at $(2, 2)$. Point $(2, 4)$ lies on ellipse.

    (c) Ends of major axis at $(2, 4)$ and $(12, 4)$. Ends of minor axis at $(7, 2)$ and $(7, 6)$.

12. In the definition of the ellipse on page 143, we asserted that the constant must be greater than the distance between the foci. What is the locus of points in the plane the sum of whose distances from $(-c, 0)$ and $(c, 0)$ is the constant $2c$?

13. What must be the relation between a and b for the major axis of the ellipse $\dfrac{x^2}{a^2} + \dfrac{y^2}{b^2} = 1$ to be horizontal? Vertical?

4. The Hyperbola. The fourth and last conic section is the hyperbola. By definition a **hyperbola** is the locus of points in a plane the absolute value of the difference of whose distances from two given points is a positive constant. The constant must be less than the distance between the two points, since the length of one side of a triangle must be greater than the absolute value of the difference between the lengths of the other two sides. The two given points are called the **foci** of the hyperbola.

If we select $(-c, 0)$ and $(c, 0)$ as foci and $2a$ as the difference of distances, the point (x, y) will lie on the hyperbola if and only if

$$|\sqrt{(x-c)^2+y^2} - \sqrt{(x+c)^2+y^2}| = 2a.$$

This equation is an abbreviation for the two equations

$$\sqrt{(x-c)^2+y^2} - \sqrt{(x+c)^2+y^2} = 2a,$$
$$\sqrt{(x+c)^2+y^2} - \sqrt{(x-c)^2+y^2} = 2a.$$

We shall simplify the first of these two equations and leave as Problem 1 at the end of the section the proof that a similar simplification of the second results in the same equation.

The steps are similar to those for the simplification of the defining equation of an ellipse:

$$\begin{aligned}
\sqrt{(x-c)^2+y^2} &= 2a + \sqrt{(x+c)^2+y^2},\\
x^2 - 2cx + c^2 + y^2 &= 4a^2 + 4a\sqrt{(x+c)^2+y^2} + x^2 + 2cx + c^2 + y^2,\\
-4cx - 4a^2 &= 4a\sqrt{(x+c)^2+y^2},\\
c^2x^2 + 2a^2cx + a^4 &= a^2(x^2 + 2cx + c^2 + y^2),\\
(c^2 - a^2)x^2 - a^2y^2 &= a^2(c^2 - a^2),
\end{aligned}$$

and finally

$$\frac{x^2}{a^2} - \frac{y^2}{c^2 - a^2} = 1.$$

This last equation is closely akin to the equation of an ellipse; however, the second term is $\dfrac{y^2}{c^2 - a^2}$ instead of $\dfrac{y^2}{a^2 - c^2}$. For the hyperbola, it is the case that $2a < 2c$ and so $a < c$, and therefore $c^2 - a^2 > 0$. Thus it is $c^2 - a^2$ which we replace by b^2 to obtain the canonical equation for the hyperbola

$$\boxed{\frac{x^2}{a^2} - \frac{y^2}{b^2} = 1.}$$

We know, from the derivation, that the graph of this equation contains all points (x, y) such that the distance between $(c, 0)$ and (x, y) is $2a$ more than the distance between $(-c, 0)$ and (x, y). In Problem 1 you will be asked to show that the graph also contains all points (x, y) such that the distance between $(-c, 0)$ and (x, y) is $2a$ more than the distance between $(c, 0)$ and (x, y). In Problem 3 you will be asked to show that the graph contains only those points.

By setting y equal to 0, we see that the graph cuts the x-axis at $(-a, 0)$ and $(a, 0)$. These points are called the **vertices** of the hyperbola and the line segment joining them the **transverse axis.** By writing the equivalent equation $y^2 = \frac{b^2}{a^2}(x^2 - a^2)$, we see that there are no points on the graph for $|x| < a$. Hence the graph cannot cut the y-axis. The curve is infinite in extent, since there are points on the graph for all x such that $|x| > a$, and $|y|$ increases indefinitely as $|x|$ increases. The curve is symmetric with respect to both axes and to the origin, since $(-p, q)$, $(p, -q)$, and $(-p, -q)$ all lie on the graph whenever (p, q) does. Although the graph is infinite in extent, the central portion of it is sketched in Figure 12. When we ask for the sketch of a hyperbola, it is the central part which is to be drawn.

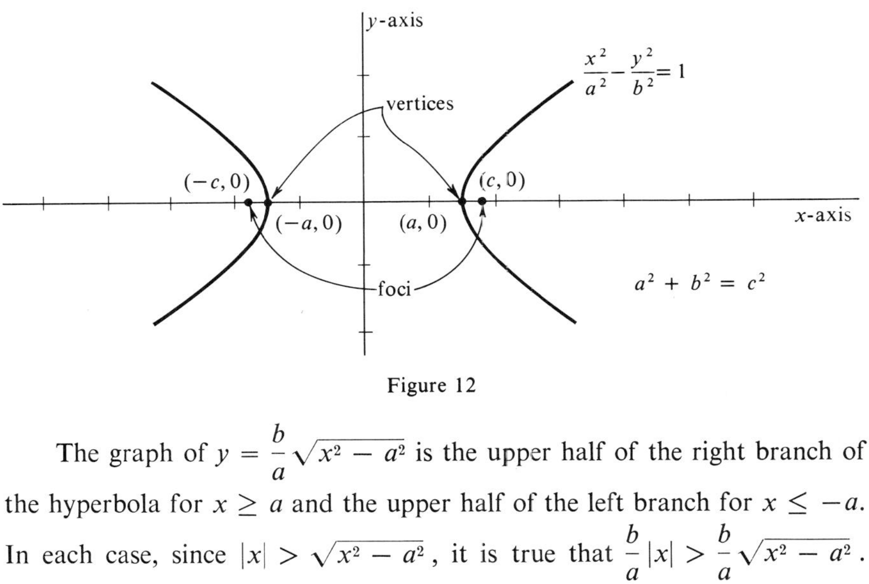

Figure 12

The graph of $y = \frac{b}{a}\sqrt{x^2 - a^2}$ is the upper half of the right branch of the hyperbola for $x \geq a$ and the upper half of the left branch for $x \leq -a$. In each case, since $|x| > \sqrt{x^2 - a^2}$, it is true that $\frac{b}{a}|x| > \frac{b}{a}\sqrt{x^2 - a^2}$. However, the difference between the two functions $\frac{b}{a}|x|$ and $\frac{b}{a}\sqrt{x^2 - a^2}$ gets less and less as $|x|$ increases. Thus the upper half of the hyperbola lies below the graph of $y = \frac{b}{a}|x|$ but gets closer to it as $|x|$ increases. Similarly, the graph of $y = -\frac{b}{a}\sqrt{x^2 - a^2}$ is the lower half of the hyperbola and it lies above the graph of $y = -\frac{b}{a}|x|$, but gets closer to it as $|x|$ increases. The union of the two graphs, those of $y = \frac{b}{a}|x|$ and $y = -\frac{b}{a}|x|$, is also the union of the graphs of the straight lines $y = \frac{b}{a}x$ and $y = -\frac{b}{a}x$. These lines, approached by the hyperbola, are called the **asymptotes** of the hyperbola. They are of use in sketching the graph of a hyperbola, giving guidelines approached by the hyperbola. They are easily drawn as the diagonals of the rectangle with vertical sides passing through the vertices of the hyperbola and horizontal sides passing through $(0, -b)$ and $(0, b)$.

EXAMPLE 1. Describe and sketch the graph of $\frac{x^2}{16} - \frac{y^2}{9} = 1$. Here we have $a = 4$, $b = 3$, and $c = 5$. The transverse axis is of length 8, the x-intercepts are -4 and 4, the foci are at $(-5, 0)$ and $(5, 0)$, and the asymptotes are the lines $y = \frac{3}{4}x$ and $y = -\frac{3}{4}x$. A very useful device for remembering equations of the asymptotes is obtained by replacing the "1" in the equation of the hyperbola by "0":

$$\frac{x^2}{16} - \frac{y^2}{9} = 0,$$

which is equivalent to

$$\left(\frac{x}{4} + \frac{y}{3}\right)\left(\frac{x}{4} - \frac{y}{3}\right) = 0.$$

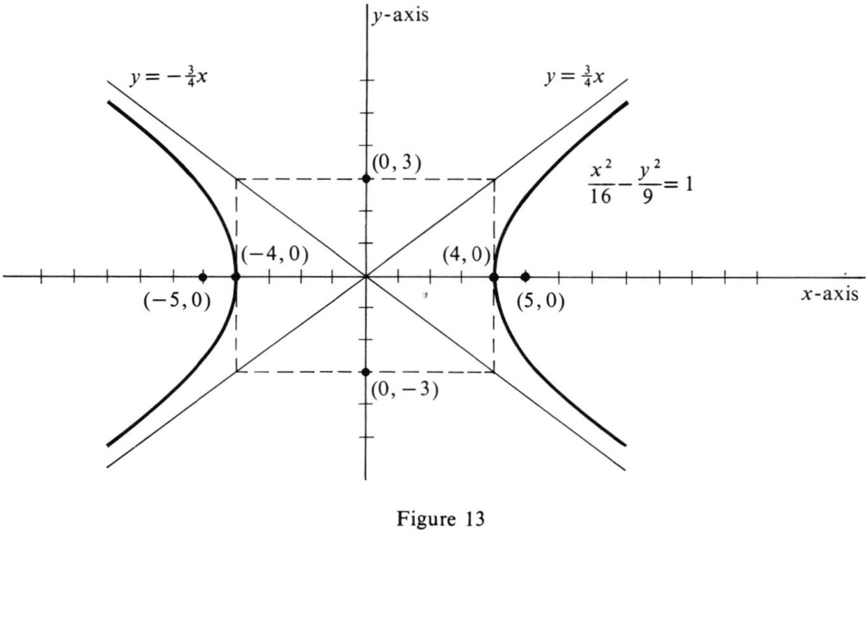

Figure 13

The graph of this equation is the union of the graphs of $\frac{x}{4} + \frac{y}{3} = 0$ and $\frac{x}{4} - \frac{y}{3} = 0$, equations of the asymptotes. The graph is sketched in Figure 13.

If the foci of the hyperbola are on the y-axis, at $(0, -c)$ and $(0, c)$ the equation will be $\frac{y^2}{a^2} - \frac{x^2}{b^2} = 1$. The graph will have branches opening upward and downward instead of to the right and left. The lines $y = \frac{a}{b}x$ and $y = -\frac{a}{b}x$ will be asymptotes.

EXAMPLE 2. Describe and sketch the graph of $4y^2 - 5x^2 = 20$. An equivalent equation is $\frac{y^2}{5} - \frac{x^2}{4} = 1$, from which we see that the vertices are at $(0, -\sqrt{5})$ and $(0, \sqrt{5})$. The transverse axis is vertical and of length $2\sqrt{5}$. The foci are at $(0, -3)$ and $(0, 3)$ and the lines $y = -\frac{1}{2}\sqrt{5}x$ and $y = \frac{1}{2}\sqrt{5}x$ are the asymptotes. The branches open upward and downward, as can be seen in Figure 14.

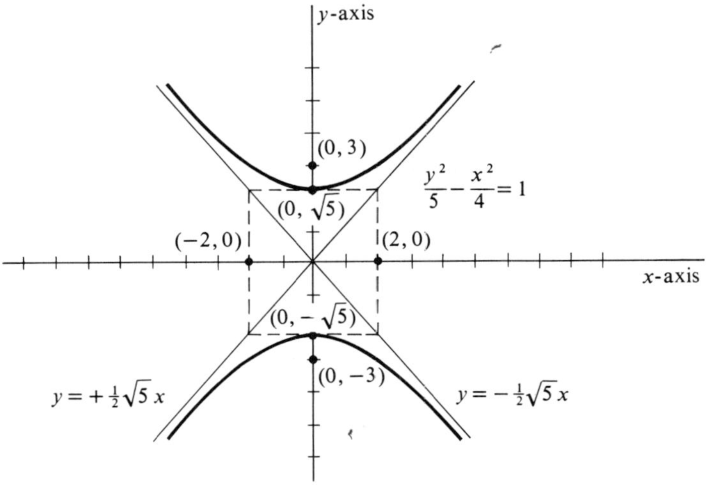

Figure 14

The equations $\frac{x^2}{a^2} - \frac{y^2}{b^2} = 1$ and $\frac{y^2}{b^2} - \frac{x^2}{a^2} = 1$ have the same asymptotes, the lines $y = \frac{b}{a}x$ and $y = -\frac{b}{a}x$, and they are called **conjugate hyperbolas.**

The first mentioned has its foci at $(-c, 0)$ and $(c, 0)$ and the latter has its foci at $(0, -c)$ and $(0, c)$, where $c = \sqrt{a^2 + b^2}$. The transverse axis of each is called the **conjugate axis** of the other.

Examples 1 and 2 have each been a hyperbola with its center at the origin. The foci, however, may be anywhere in the plane and the midpoint of the segment joining them will be the center. If they are on a line parallel to one of the axes, an equation analogous to that of the ellipse can be written. In particular, if the foci are at $(h - c, k)$ and $(h + c, k)$, an equation of the hyperbola is $\dfrac{(x-h)^2}{a^2} - \dfrac{(y-k)^2}{b^2} = 1$, and an equation for the asymptotes is $\dfrac{(x-h)^2}{a^2} - \dfrac{(y-k)^2}{b^2} = 0$. The latter equation is equivalent to the separate equations, $y - k = \dfrac{b}{a}(x - h)$ and $y - k = -\dfrac{b}{a}(x - h)$. If the foci are at $(h, k - c)$ and $(h, k + c)$, an equation for the hyperbola is

$$\frac{(y-k)^2}{a^2} - \frac{(x-h)^2}{b^2} = 1.$$

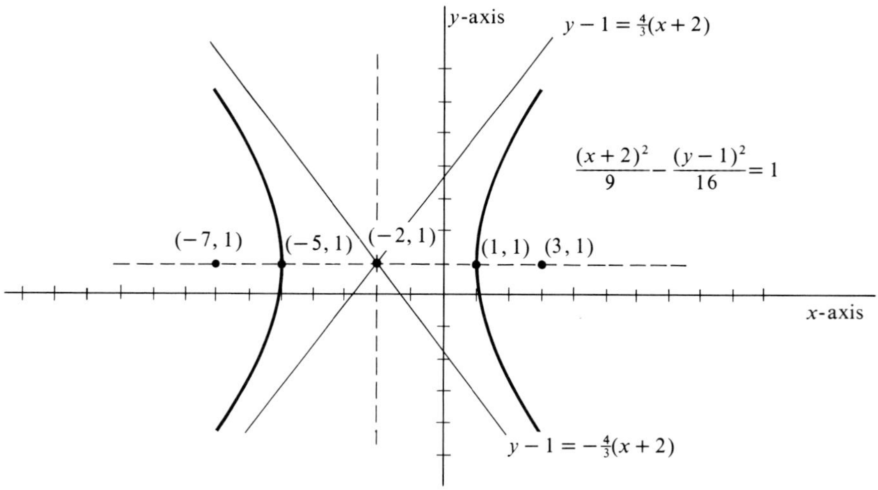

Figure 15

EXAMPLE 3. Describe and sketch the graph of $\dfrac{(x+2)^2}{9} - \dfrac{(y-1)^2}{16} = 1$. The center of the hyperbola is at $(-2, 1)$, the foci are at $(-7, 1)$ and $(3, 1)$, and the transverse axis is horizontal and of length 6. The asymptotes have equations $y - 1 = \frac{4}{3}(x + 2)$ and $y - 1 = -\frac{4}{3}(x + 2)$. The graph is sketched in Figure 15.

There is a focus-directrix definition of the hyperbola, analogous to those for the parabola and the ellipse. The distance from a point (x, y) to the focus $(c, 0)$ is $\sqrt{(x - c)^2 + y^2}$. The point lies on the hyperbola if and only if $\dfrac{x^2}{a^2} - \dfrac{y^2}{c^2 - a^2} = 1$, or $y^2 = a^2 - c^2 - x^2 + \dfrac{c^2}{a^2}x^2$. Thus the distance from a point on the hyperbola to the focus is

$$\sqrt{x^2 - 2cx + c^2 + \left(a^2 - c^2 - x^2 + \frac{c^2}{a^2}x^2\right)} = \sqrt{\frac{c^2}{a^2}x^2 - 2cx + a^2}$$
$$= \sqrt{\left(\frac{c}{a}x - a\right)^2}$$
$$= \left|\frac{c}{a}x - a\right|.$$

As with the ellipse, this distance is equal to $\dfrac{c}{a}\left|x - \dfrac{a^2}{c}\right|$, or $\dfrac{c}{a}$ times the distance to the line $x = \dfrac{a^2}{c}$. This line is again called the **directrix,** and the ratio $\dfrac{c}{a}$ the **eccentricity.** However, for the hyperbola the eccentricity is greater than 1. The hyperbola can therefore also be defined as the locus of points the ratio of whose distances to the focus and to the directrix is a constant greater than 1. Corresponding to the focus $(-c, 0)$ is a second directrix $x = -\dfrac{a^2}{c}$.

Not all hyperbolas with horizontal or vertical axes appear with equations in canonical form. But canonical equations can be found for them by factoring and completing the square.

EXAMPLE 4. Describe and sketch the graph of $16x^2 - 9y^2 - 32x - 54y - 641 = 0$. Equations equivalent to the given equations are

$$16(x^2 - 2x) - 9(y^2 + 6y) = 641,$$
$$16(x^2 - 2x + 1) - 9(y^2 + 6y + 9) = 641 + 16 \cdot 1 - 9 \cdot 9,$$
$$16(x - 1)^2 - 9(y + 3)^2 = 576,$$
$$\frac{(x - 1)^2}{36} - \frac{(y + 3)^2}{64} = 1.$$

The hyperbola, opening to the right and left, has its center at $(1, -3)$, its vertices at $(-5, -3)$ and $(7, -3)$, its foci at $(-9, -3)$ and $(11, -3)$, and the lines $y + 3 = \frac{4}{3}(x - 1)$ and $y + 3 = -\frac{4}{3}(x - 1)$ for asymptotes. The graph is sketched in Figure 16.

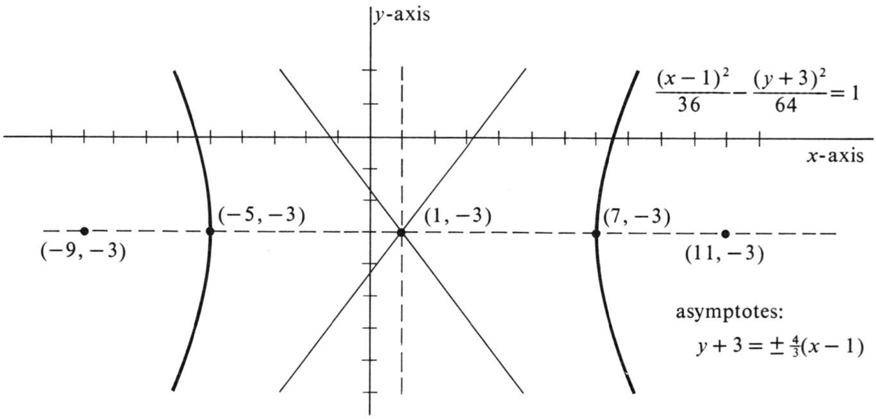

Figure 16

The method of Example 4 can be used to show that every equation of the type $ax^2 - by^2 + cx + dy + e = 0$ with $ab > 0$ has for its graph either a hyperbola or a pair of intersecting straight lines. Equivalent to the given equation are

$$a\left(x^2 + \frac{c}{a}x\right) - b\left(y^2 - \frac{d}{b}y\right) = -e,$$

$$a\left(x^2 + \frac{c}{a}x + \frac{c^2}{4a^2}\right) - b\left(y^2 - \frac{d}{b}y + \frac{d^2}{4b^2}\right) = \frac{c^2}{4a} - \frac{d^2}{4b} - e,$$

$$a\left(x + \frac{c}{2a}\right)^2 - b\left(y - \frac{d}{2b}\right)^2 = \frac{c^2}{4a} - \frac{d^2}{4b} - e.$$

If the expression on the right side of the equation has the same sign as a and b, the graph is a hyperbola centered at $\left(-\frac{c}{2a}, \frac{d}{2b}\right)$ opening to the right and to the left. If the expression has sign opposite to a and b, the graph is a hyperbola centered at $\left(-\frac{c}{2a}, \frac{d}{2b}\right)$ but opening upward and downward. If the expression is zero, the graph consists of the two straight lines

$$y - \frac{d}{2b} = \sqrt{\frac{a}{b}}\left(x + \frac{c}{2a}\right) \quad \text{and} \quad y - \frac{d}{2b} = -\sqrt{\frac{a}{b}}\left(x + \frac{c}{2a}\right).$$

One particular type of hyperbola with its axes neither horizontal nor vertical has a simple equation and appears frequently in mathematics. It is the hyperbola with foci at $(-a, -a)$ and (a, a) and the difference of distances equal to $2a$. By definition, a point lies on this hyperbola if and only if

$$|\sqrt{(x-a)^2 + (y-a)^2} - \sqrt{(x+a)^2 + (y+a)^2}| = 2a.$$

Squarings and simplifications yield the equation

$$xy = \tfrac{1}{2}a^2.$$

This hyperbola has the coordinate axes for its asymptotes and is drawn in Figure 17. Its vertices are at $\left(-\frac{a}{2}\sqrt{2}, -\frac{a}{2}\sqrt{2}\right)$ and $\left(\frac{a}{2}\sqrt{2}, \frac{a}{2}\sqrt{2}\right)$.

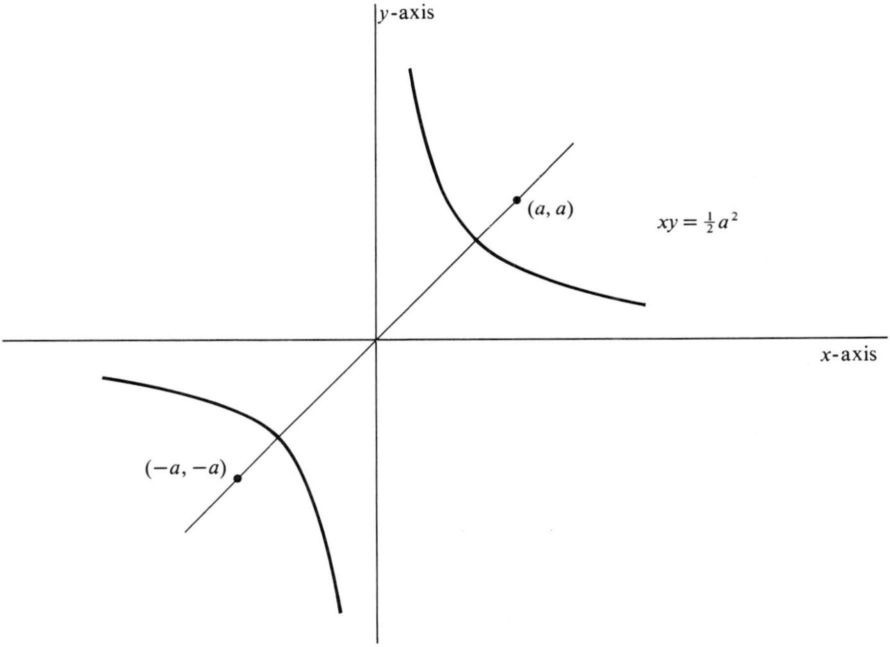

Figure 17

With foci at $(-a, a)$ and $(a, -a)$, the equation is $xy = -\frac{1}{2}a^2$ and the two branches lie in the second and fourth quadrants. This type of hyperbola is called **equilateral,** as are the hyperbolas $\frac{x^2}{a^2} - \frac{y^2}{b^2} = 1$ and $\frac{y^2}{a^2} - \frac{x^2}{b^2} = 1$ with $a^2 = b^2$.

PROBLEMS

1. Simplify $\sqrt{(x + c)^2 + y^2} - \sqrt{(x - c)^2 + y^2} = 2a$ and show that it results in the equation $\frac{x^2}{a^2} - \frac{y^2}{c^2 - a^2} = 1$.

2. Describe and sketch the graph of each of the following equations. Label the foci and vertices and sketch the asymptotes.

(a) $\dfrac{x^2}{9} - \dfrac{y^2}{16} = 1$

(b) $\dfrac{y^2}{18} - \dfrac{x^2}{7} = 1$

(c) $\dfrac{x^2}{25} - \dfrac{y^2}{144} = 1$

(d) $\dfrac{x^2}{225} - \dfrac{y^2}{64} = 1$

(e) $25y^2 - 9x^2 = 225$

(f) $xy = 8$

(g) $xy = -6$.

3. It has been shown that the distance between a point on the hyperbola $\dfrac{x^2}{a^2} - \dfrac{y^2}{c^2 - a^2} = 1$ and the focus $(c, 0)$ is $\left|\dfrac{c}{a}x - a\right|$. Call this distance d_1.

(a) Show that the distance d_2 between a point on the hyperbola and the focus $(-c, 0)$ is $\left|\dfrac{c}{a}x + a\right|$.

(b) Show that $x \geq a$ for a point on the right branch of the hyperbola and that for such a point $d_1 = \dfrac{c}{a}x - a$ and $d_2 = \dfrac{c}{a}x + a$.

(c) Show that $x \leq -a$ for a point on the left branch of the hyperbola and that for such a point $d_1 = -\dfrac{c}{a}x + a$ and $d_2 = -\dfrac{c}{a}x - a$.

(d) Hence show that the graph of $\dfrac{x^2}{a^2} - \dfrac{y^2}{c^2 - a^2} = 1$ contains only those points which satisfy the locus definition of hyperbola.

4. Write an equation for the hyperbola with horizontal and vertical axes satisfying the given conditions.

(a) Foci at $(-5, 0)$ and $(5, 0)$. Transverse axis of length 6.

(b) Center at the origin. Transverse axis horizontal and of length 4, conjugate axis vertical and of length 12.

(c) Center at the origin. Transverse axis vertical and of length 6. Passing through $(1, 2\sqrt{3})$.

(d) Center at the origin. Passing through $(1, 5)$ and $(2, 7)$.

5. (a) Show that the graph of $(x - 3)(y + 2) = 10$ is an equilateral hyperbola with center at $(3, -2)$ and asymptotes $x = 3$ and $y = -2$.

(b) Sketch the graph described in (a).

6. Describe and sketch the graph of each of the following equations. Label the foci and vertices and sketch the asymptotes.

(a) $\dfrac{(x-1)^2}{16} - \dfrac{(y+3)^2}{9} = 1$

(b) $\dfrac{(y+3)^2}{9} - \dfrac{(x-1)^2}{16} = 1$

(c) $(x-1)(y+3) = -12$

(d) $144(x+4)^2 - 25(y+2)^2 = 3600$

(e) $2xy - 4x - 4y - 17 = 0$

(f) $x^2 - y^2 - 2x + 2y - 2 = 0$

(g) $y^2 - 4x^2 + 4y - 16x - 28 = 0.$

7. Hyperbolas with the same pair of foci are said to be **confocal.** Show that the following equations have confocal hyperbolas for their graphs and sketch them all on the same set of axes.

(a) $x^2 - \dfrac{y^2}{24} = 1$

(b) $\dfrac{x^2}{9} - \dfrac{y^2}{16} = 1$

(c) $\dfrac{x^2}{16} - \dfrac{y^2}{9} = 1$

(d) $\dfrac{x^2}{24} - y^2 = 1.$

8. The line segment which passes through a focus, is perpendicular to the transverse axis extended, and has its endpoints on the hyperbola is called a **latus rectum.**

(a) Find the length of a latus rectum of the hyperbola $9x^2 - 16y^2 = 144$.

(b) Find the length of a latus rectum of the hyperbola $b^2x^2 - a^2y^2 = a^2b^2$.

(c) Show that both latera recta of a hyperbola are the same length.

9. Write equations of the directrices and find the eccentricity of each of the following hyperbolas.

(a) $9x^2 - 16y^2 = 144$

(b) $144y^2 - 25x^2 = 3600.$

10. Assume that $0 < a < c$.

(a) Find the distance between (x, y) and $(-c, 0)$.

(b) Find the distance between (x, y) and the line $x = -\dfrac{a^2}{c}\cdot$

(c) Find the locus of points (x, y) such that the ratio between the distance in (a) and the distance in (b) is a constant $\dfrac{c}{a}\cdot$

11. Consider points on the upper right branch of the hyperbola $x^2 - y^2 = 16$ and on its asymptote $x = y$. Find the y-coordinates of points on each for $x = 10$, 100, 1000, 10,000, and show that the vertical distance between corresponding points is decreasing as x increases.

12. Consider a point (x_1, y_1) on the graph of $b^2x^2 - a^2y^2 = a^2b^2$.

    (a) Find the slope of the tangent to the graph at (x_1, y_1).

    (b) Write an equation of the tangent line in (a).

    (c) Show that $b^2xx_1 - a^2yy_1 = a^2b^2$ is an equation of the tangent line.

13. Consider a point (x_1, y_1) on the hyperbola $ax^2 - by^2 + cx + dy + e = 0$ with $ab > 0$.

    (a) Find the slope of the tangent to the graph at (x_1, y_1).

    (b) Write an equation of the tangent line in (a).

    (c) Show that $axx_1 - byy_1 + \frac{1}{2}c(x + x_1) + \frac{1}{2}d(y + y_1) + e = 0$ is an equation of the tangent line.

14. Show that the product of the distances of a point on the hyperbola $xy = -12$ to its asymptotes is a constant.

15. If the difference of the distances of the point (x, y) from two foci is zero, show that the locus of (x, y) is the perpendicular bisector of the line segment joining the foci.

16. Describe and sketch the graph of each of the following equations. If the graph is a circle, give its center and radius. If the graph is a parabola, give its focus, directrix, vertex, and axis. If the graph is an ellipse, give its center, foci, directrices, eccentricity, and length of major and minor axes. If the graph is an hyperbola, give its center, foci, directrices, eccentricity, asymptotes, vertices, and length of transverse axis.

    (a) $x^2 + y^2 + 6x + 4y = 12$

    (b) $x^2 + 4y^2 + 6x + 4y + 6 = 0$

    (c) $x^2 + 6x + 4y + 2 = 0$

    (d) $x^2 - 4y^2 + 6x + 4y + 4 = 0$

    (e) $4y^2 + 6x + 4y + 13 = 0$

    (f) $xy + 6x + 4y = 3$

    (g) $3x^2 + 3y^2 + 6x - 18y = 162$

    (h) $4y^2 - x^2 + 6x + 4y = 11$

    (i) $y^2 = 9x^2 + 2y + 8$

    (j) $y^2 = 2y - 9x^2 + 8$

CHAPTER 4

Integration

There are two major topics in calculus, differentiation and integration. The theory of integration at first appears to have little connection with differentiation. However, we shall see that the two processes are indeed closely related.

Differentiation deals with derivatives of real-valued functions, integration with integrals of real-valued functions defined on closed intervals. In this chapter we shall define and study the definite integral from a to b of a real-valued function f, which is denoted by $\int_a^b f$. The integral has countless applications both in mathematics and the sciences. One of the most important enables us to find the area of a region more involved than those studied in plane geometry. If $f(x) \geq 0$ for every x in $[a, b]$, then we shall see that $\int_a^b f$ is equal to the area of the region lying above the x-axis, below the graph of f, and between the vertical lines $x = a$ and $x = b$ (see Figure 1). Thus the areas of many irregularly shaped regions can be found by integration.

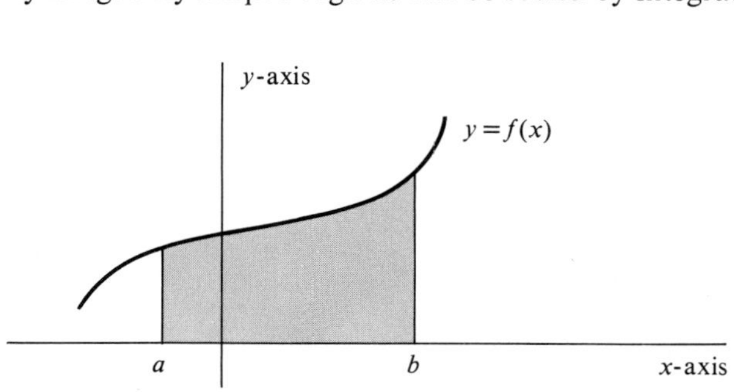

Figure 1

Another interesting application enables us to find the distance traveled by a particle which moves in a straight line. If the particle moves with a velocity equal to $v(t)$ at time t, then the distance traveled during the interval of time from $t = a$ to $t = b$ is given by the integral $\int_a^b |v|$.

It will be useful to recall some of the elementary ideas and notations of set theory. In particular, the **union** of two sets P and Q, denoted by $P \cup Q$, is the set of elements that belong either to P, or to Q, or to both. The **inter-**

section of P and Q, denoted by $P \cap Q$, is the set of elements that belong to both P and Q. If P and Q have no points in common, they are said to be **disjoint.** The set which contains no elements is called the **empty set** and will be denoted by $\emptyset$. Thus P and Q are disjoint if and only if $P \cap Q = \emptyset$. Finally, the **difference,** $P - Q$, is the set of all elements of P which do not belong to Q.

1. The Definite Integral. In defining the definite integral, we shall use the concepts of bounded sets discussed in Section 1 of Chapter 1. Recall that a number u is said to be an **upper bound** of a set S of real numbers if the inequality $x \leq u$ is satisfied for every number x in S. Thus the numbers 100, 5, and 1 are all upper bounds of the closed interval $[0, 1]$. The Least Upper Bound Property (see page 7) states that every nonempty set of real numbers which has an upper bound has a least upper bound. For example, the number 1 is obviously the least upper bound of the interval $[0, 1]$. Note that 1 is also the least upper bound of the open interval $(0, 1)$.

In the same way, a number l is called a **lower bound** of S provided $l \leq x$ for every x in S. The Greatest Lower Bound Property (see Problem 11, page 9) similarly asserts that if S is nonempty and has a lower bound, then it has a greatest lower bound. Finally, a set is simply said to be **bounded** if it has both an upper bound and a lower bound.

The notion of boundedness can be applied to functions. Specifically, a real-valued function f of a real variable is said to be **bounded on an interval** I if the following two conditions are satisfied:

(i) I is a subset of the domain of f.
(ii) There exists a real number k such that $|f(x)| \leq k$, for every x in I.

The reader should be able to supply the straightforward argument which shows that condition (ii) is equivalent to the assertion that the set S of all real numbers $f(x)$ for which x is in I is a bounded set. To illustrate the terminology, consider the two functions f and g defined by $f(x) = x^2$ and $g(x) = \frac{1}{x}$. The former is bounded on both the closed interval $[0, 1]$ and the open interval $(0, 1)$, whereas the latter is bounded on neither.

Let $[a, b]$ be a closed interval, and let $\sigma = \{x_0, \ldots, x_n\}$ be a finite subset of $[a, b]$ which contains the endpoints a and b. The set σ subdivides, or partitions, the interval into subintervals, and, for this reason, we shall call it a **partition** of $[a, b]$.

Let f be a function which is bounded on the closed interval $[a, b]$, and let $\sigma = \{x_0, \ldots, x_n\}$ be a partition of $[a, b]$ in which

$$a = x_0 \leq x_1 \leq \cdots \leq x_{n-1} \leq x_n = b.$$

Since f is bounded on the entire interval $[a, b]$, it is certainly bounded on each subinterval $[x_{i-1}, x_i]$, for $i = 1, \ldots, n$. Hence the set S consisting of all numbers $f(x)$ with x in $[x_{i-1}, x_i]$ has a least upper bound, which we denote by M_i. Similarly, of course, the set S has a greatest lower bound, which we denote by m_i. These numbers are illustrated for a typical subinterval in

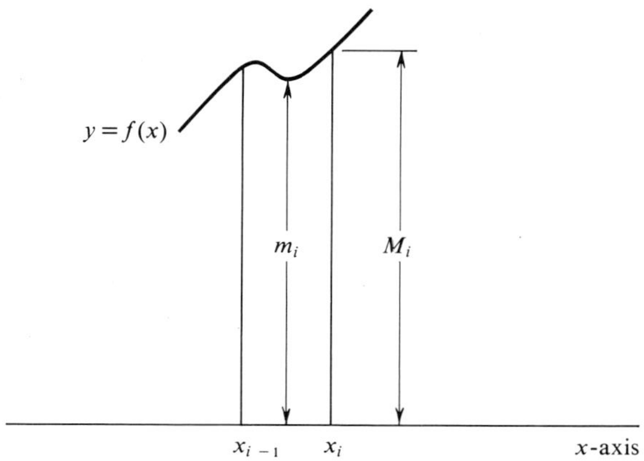

Figure 2

Figure 2. We now define two numbers U_σ and L_σ, called, respectively, the **upper sum** and the **lower sum** of f **relative to the partition** σ by the formulas

$$\boxed{\begin{aligned} U_\sigma &= M_1(x_1 - x_0) + M_2(x_2 - x_1) + \cdots + M_n(x_n - x_{n-1}), \\ L_\sigma &= m_1(x_1 - x_0) + m_2(x_2 - x_1) + \cdots + m_n(x_n - x_{n-1}). \end{aligned}} \tag{1}$$

Since $M_i \geq m_i$ and since $(x_i - x_{i-1}) \geq 0$, for each $i = 1, \ldots, n$, it follows that

$$U_\sigma - L_\sigma = (M_1 - m_1)(x_1 - x_0) + (M_2 - m_2)(x_2 - x_1) + \cdots + (M_n - m_n)(x_n - x_{n-1}) \geq 0.$$

Hence, we conclude that

$$L_\sigma \leq U_\sigma. \tag{2}$$

EXAMPLE 1. Let f be the function defined by $f(x) = \dfrac{1}{x}$, let $[a, b] = [1, 3]$, and consider the partition $\sigma = \{1, \frac{3}{2}, 2, \frac{5}{2}, 3\}$. There are four subintervals, $[1, \frac{3}{2}]$, $[\frac{3}{2}, 2]$, $[2, \frac{5}{2}]$, and $[\frac{5}{2}, 3]$, and each one is of length $\frac{1}{2}$. It is clear from

Figure 3 that the maximum value of f on each subinterval occurs at the left endpoint, and the minimum value occurs at the right endpoint. Hence

$$\begin{aligned} M_1 &= f(1) = 1, & m_1 &= f(\tfrac{3}{2}) = \tfrac{2}{3}, \\ M_2 &= f(\tfrac{3}{2}) = \tfrac{2}{3}, & m_2 &= f(2) = \tfrac{1}{2}, \\ M_3 &= f(2) = \tfrac{1}{2}, & m_3 &= f(\tfrac{5}{2}) = \tfrac{2}{5}, \\ M_4 &= f(\tfrac{5}{2}) = \tfrac{2}{5}, & m_4 &= f(3) = \tfrac{1}{3}. \end{aligned}$$

It follows that

$$\begin{aligned} U_\sigma &= 1 \cdot \tfrac{1}{2} + \tfrac{2}{3} \cdot \tfrac{1}{2} + \tfrac{1}{2} \cdot \tfrac{1}{2} + \tfrac{2}{5} \cdot \tfrac{1}{2}, \\ L_\sigma &= \tfrac{2}{3} \cdot \tfrac{1}{2} + \tfrac{1}{2} \cdot \tfrac{1}{2} + \tfrac{2}{5} \cdot \tfrac{1}{2} + \tfrac{1}{3} \cdot \tfrac{1}{2}. \end{aligned}$$

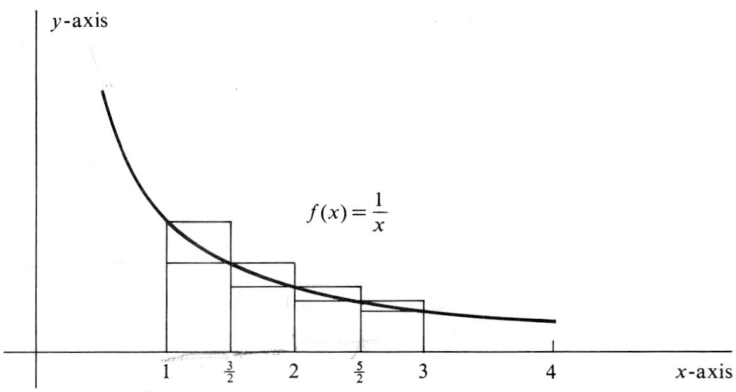

Figure 3

Continuing the computation, we obtain

$$U_\sigma = \frac{1}{2}\left(1 + \frac{2}{3} + \frac{1}{2} + \frac{2}{5}\right) = \frac{1}{2}\,\frac{30 + 20 + 15 + 12}{30} = \frac{77}{60},$$

$$L_\sigma = \frac{1}{2}\left(\frac{2}{3} + \frac{1}{2} + \frac{2}{5} + \frac{1}{3}\right) = \frac{1}{2}\,\frac{20 + 15 + 12 + 10}{30} = \frac{57}{60},$$

for the values of the upper and lower sums of f relative to σ.

In the paragraph preceding Example 1, it is proved that, for a given partition σ, the lower sum is less than or equal to the upper sum. We shall now prove the much stronger fact that all the lower sums are less than or equal to all the upper sums. More precisely,

(1.1) *Let f be bounded on $[a, b]$. If σ and τ are any two partitions of $[a, b]$, then $L_\sigma \leq U_\tau$.*

Proof. The argument will be divided into three parts.

(i) *If* τ *is obtained from* σ *by adjoining just one new number* y, *then* $L_\sigma \leq L_\tau \leq U_\tau \leq U_\sigma$.

Let $\sigma = \{x_0, \ldots, x_n\}$ and $a = x_0 \leq x_1 \leq \cdots \leq x_n = b$, and let the upper and lower sums U_σ and L_σ be defined as in (1), page 165. We shall assume that y lies in the kth subinterval $[x_{k-1}, x_k]$. Then

$$\tau = \{x_0, \ldots, x_{k-1}, y, x_k, \ldots, x_n\}$$

and

$$a = x_0 \leq \cdots \leq x_{k-1} \leq y \leq x_k \leq \cdots \leq x_n = b.$$

Abbreviating the expressions least upper bound and greatest lower bound by l.u.b. and g.l.b., respectively, let

M'_k = l.u.b. of the set of all numbers $f(x)$ with $x_{k-1} \leq x \leq y$,

M''_k = l.u.b. of the set of all numbers $f(x)$ with $y \leq x \leq x_k$,

m'_k = g.l.b. of the set of all numbers $f(x)$ with $x_{k-1} \leq x \leq y$,

m''_k = g.l.b. of the set of all numbers $f(x)$ with $y \leq x \leq x_k$.

These numbers are illustrated in Figure 4. The key idea in the entire proof is the fact that the least upper bound of f on any set S is greater than or equal to

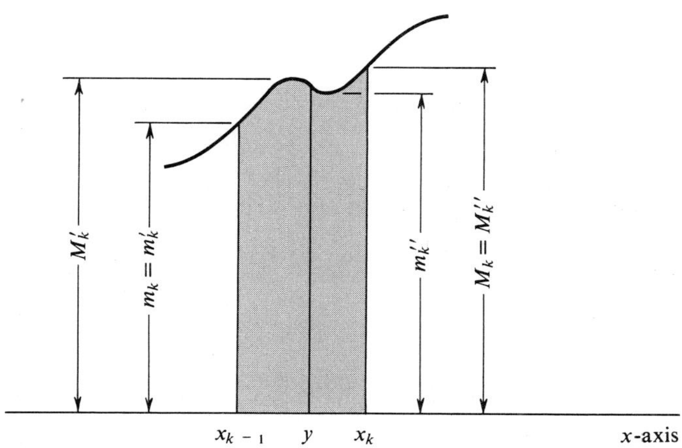

Figure 4

its least upper bound on any subset of S, and, similarly, the greatest lower bound of f on S is less than or equal to its greatest lower bound on a subset of S. This means that

$$M_k \geq M'_k \quad \text{and} \quad M_k \geq M''_k,$$

$$m_k \leq m'_k \quad \text{and} \quad m_k \leq m''_k,$$

as is borne out by Figure 4. Hence

$$\begin{aligned} M_k(x_k - x_{k-1}) &= M_k(x_k - y) + M_k(y - x_{k-1}) \\ &\geq M_k''(x_k - y) + M_k'(y - x_{k-1}), \\ m_k(x_k - x_{k-1}) &= m_k(x_k - y) + m_k(y - x_{k-1}) \\ &\leq m_k''(x_k - y) + m_k'(y - x_{k-1}). \end{aligned}$$

But the other terms in the upper and lower sums are the same for the two partitions. We conclude that $U_\sigma \geq U_\tau$ and that $L_\sigma \leq L_\tau$. Since it follows from (2) that $L_\tau \leq U_\tau$, we obtain

$$L_\sigma \leq L_\tau \leq U_\tau \leq U_\sigma. \tag{3}$$

(ii) *If σ is a subset of τ, then the preceding inequalities* (3) *are still satisfied.*

This follows by repeated applications of part (i), since the partition τ can be obtained from σ by adjoining one number at a time.

(iii) *If σ and τ are any two partitions, then $L_\sigma \leq U_\tau$.*

Both σ and τ are subsets of the partition $\sigma \cup \tau$ consisting of their union. Hence, one application of part (ii) gives $L_\sigma \leq L_{\sigma\cup\tau}$, and another application yields $U_{\sigma\cup\tau} \leq U_\tau$. Combining these inequalities with $L_{\sigma\cup\tau} \leq U_{\sigma\cup\tau}$, we obtain

$$L_\sigma \leq L_{\sigma\cup\tau} \leq U_{\sigma\cup\tau} \leq U_\tau,$$

and the proof of (1.1) is complete.

Theorem (1.1) states that, for a given function f bounded on an interval $[a, b]$, if we consider all partitions of $[a, b]$, then every lower sum is less than or equal to every upper sum. It is instructive to picture the relative positions of these numbers on the real line. If we indicate each lower sum by a right-hand parenthesis, ")", and each upper sum by a left-hand parenthesis, "(", the situation looks as shown in Figure 5 (except that in general there are

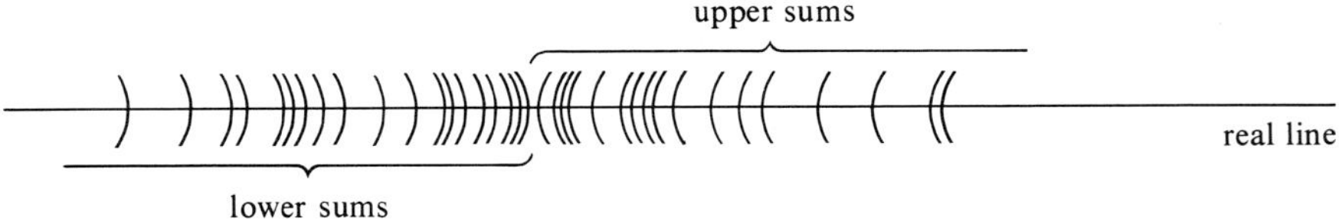

Figure 5

infinitely many sums of both kinds). The question naturally arises as to the existence of numbers in between the two sets, and this brings us to the definitions of integrability and of the definite integral: Let the function f be bounded on the closed interval $[a, b]$. Then f is said to be **integrable over** $[a, b]$ if there exists one and only one number J such that

$$L_\sigma \leq J \leq U_\tau, \tag{4}$$

for any two partitions σ and τ of $[a, b]$. If f is integrable over $[a, b]$, then the unique number J is called the **definite integral of f from a to b**, and is denoted by $\int_a^b f$. That is,

$$J = \int_a^b f.$$

Almost all the functions encountered in a first course in calculus are integrable over the closed intervals on which they are bounded. The reason is that for these functions the differences between the upper and lower sums can be made arbitrarily small by taking partitions which subdivide the interval into smaller and smaller subintervals. Many conditions which ensure that a function is integrable are known. Among these, we shall consider two [see Theorems (3.3) and (5.1)]. The second condition is continuity. We shall see that if f is continuous at every point of a closed interval $[a, b]$, then f is integrable over $[a, b]$.

If f is a function bounded on $[a, b]$, there are, according to the definition, two conditions which must be satisfied for f to be integrable over $[a, b]$. The first is that there must exist a number J such that the inequalities (4) hold for all partitions σ and τ of the interval. The second is that there must be only one such number. It is not hard to prove that the first condition is always satisfied (see Problem 9 at the end of this section). It is the second which may fail, as the following example illustrates. Let f be the function defined by

$$f(x) = \begin{cases} 0 & \text{if } x \text{ is rational,} \\ 1 & \text{if } x \text{ is irrational.} \end{cases}$$

This function is bounded on the interval $[0, 1]$. However, if σ is any partition whatever of $[0, 1]$, it is easy to see that $U_\sigma = 1$ and $L_\sigma = 0$. This means that every number J between, and including, 0 and 1 will satisfy (4). The function is therefore not integrable because J is not unique.

EXAMPLE 2. Assuming that the function f defined by $f(x) = \dfrac{1}{x}$ is integrable over the interval $[1, 3]$, prove that

$$\tfrac{57}{60} \le \int_1^3 f \le \tfrac{77}{60}.$$

This is the function and interval described in Example 1. For the partition $\sigma = \{1, \frac{3}{2}, 2, \frac{5}{2}, 3\}$, we saw that $L_\sigma = \frac{57}{60}$ and $U_\sigma = \frac{77}{60}$; hence the integral is bounded by these two numbers.

EXAMPLE 3. Consider the function f defined by $f(x) = x^2$. Assuming that f is integrable over the interval $[0, 1]$, show that

$$\tfrac{6}{25} \le \int_0^1 f \le \tfrac{11}{25}.$$

We use the partition $\sigma = \{0, \frac{1}{5}, \frac{2}{5}, \frac{3}{5}, \frac{4}{5}, 1\}$ and compute the upper and lower sums U_σ and L_σ. The points of the partition are given by $x_i = \dfrac{i}{5}$, for $i = 0, \ldots, 5$. Hence

$$x_i - x_{i-1} = \tfrac{1}{5}, \qquad \text{for } i = 1, \ldots, 5.$$

Since in this case the subintervals are all of length $\frac{1}{5}$, equations (1) can be simplified to read

$$U_\sigma = \tfrac{1}{5}(M_1 + \cdots + M_5),$$
$$L_\sigma = \tfrac{1}{5}(m_1 + \cdots + m_5).$$

It is clear from Figure 6 that the maximum value of f on each subinterval $[x_{i-1}, x_i]$ occurs at the right endpoint, and the minimum value occurs at the

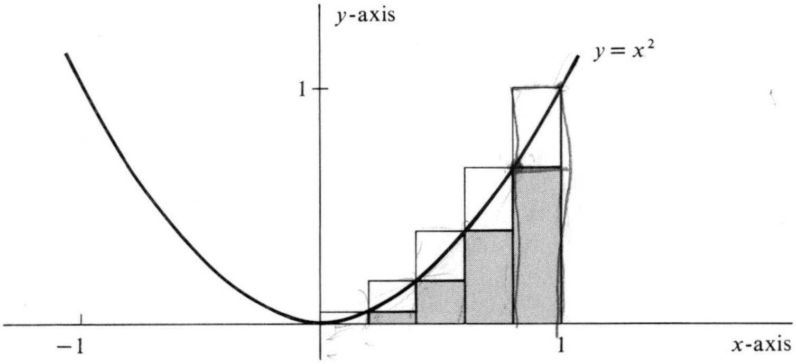

Figure 6

left endpoint. Hence $M_i = f(x_i)$ and $m_i = f(x_{i-1})$. Since $f(x) = x^2$, we have

$$\left.\begin{aligned} M_i &= x_i^2 = \frac{i^2}{25} \\ m_i &= x_{i-1}^2 = \frac{(i-1)^2}{25} \end{aligned}\right\} i = 1, \ldots, 5.$$

Thus

$$U_\sigma = \frac{1}{5}\left(\frac{1^2}{25} + \frac{2^2}{25} + \frac{3^2}{25} + \frac{4^2}{25} + \frac{5^2}{25}\right),$$

$$L_\sigma = \frac{1}{5}\left(\frac{0^2}{25} + \frac{1^2}{25} + \frac{2^2}{25} + \frac{3^2}{25} + \frac{4^2}{25}\right).$$

Since $1^2 + 2^2 + 3^2 + 4^2 + 5^2 = 55$ and $0^2 + 1^2 + 2^2 + 3^2 + 4^2 = 30$, it follows that

$$U_\sigma = \tfrac{1}{5}\tfrac{55}{25} = \tfrac{11}{25},$$
$$L_\sigma = \tfrac{1}{5}\tfrac{30}{25} = \tfrac{6}{25}.$$

This establishes the desired bounds, since $L_\sigma \leq \int_0^1 f \leq U_\sigma$.

An alternative notation for the integral, which we shall use interchangeably with $\int_a^b f$, is $\int_a^b f(x)\,dx$. This is the traditional way of writing the integral, and its usefulness will become increasingly apparent as we go on. In a later section we shall show how the dx which appears to the right of the integral sign may be interpreted as a differential. At present, however, it is important to realize that dx is only a part of the notation for the integral. The variable x which occurs in $\int_a^b f(x)\,dx$ is often called a **dummy variable.** This name serves as a reminder of the fact that the value of the integral depends only on the function f and the numbers a and b. Its value is not determined by giving a value of x. Thus

$$\int_a^b f = \int_a^b f(x)\,dx = \int_a^b f(y)\,dy = \int_a^b f(t)\,dt = \text{etc.},$$

and x, y, and t each occurs as a dummy variable. Thus the inequalities established in Examples 2 and 3 can alternatively be written

$$\frac{57}{60} \leq \int_1^3 \frac{1}{x}\,dx \leq \frac{77}{60},$$

$$\frac{6}{25} \leq \int_0^1 x^2\,dx \leq \frac{11}{25}.$$

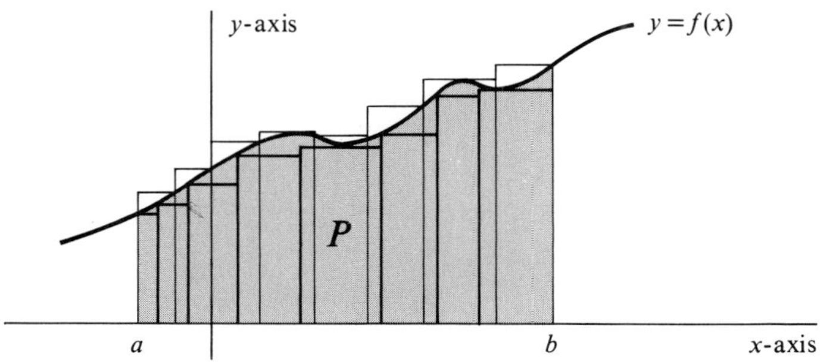

Figure 7

Although we shall not give a definition of area in this book, the basic properties of area can be used to establish its connection with the definite integral. Let the area of a set P be denoted by *area* (P). Two basic properties are:

(1.2) *The area of a set is never negative: area*$(P) \geq 0$.

(1.3) *If P is a subset of Q, then area*$(P) \leq$ *area*(Q).

In addition, we shall assume the elementary facts about the areas of rectangles.

Let f be a function which is integrable over the interval $[a, b]$, and which also satisfies the inequality $f(x) \geq 0$ for every x in $[a, b]$. Let P be the region under the curve. That is, P is the set of all points (x, y) in the plane such that $a \leq x \leq b$ and $0 \leq y \leq f(x)$ (see Figure 7). Next, consider two partitions σ and τ of $[a, b]$. The lower sum L_σ is the area of the union of rectangles contained in P; hence, by (1.3) we conclude that $L_\sigma \leq$ *area* (P). Conversely, P is a subset of the union of rectangles the sum of whose area is the upper sum U_τ. Hence *area* $(P) \leq U_\tau$, and we have shown that

$$L_\sigma \leq area(P) \leq U_\tau,$$

for any two partitions σ and τ. But the integrability of f asserts that the definite integral $\int_a^b f(x)\,dx$ is the only number with this property. Thus we have proved

$$\textbf{(1.4)} \qquad \boxed{\int_a^b f(x)\,dx = area(P), \qquad if\ f(x) \geq 0\ for\ every\ x\ in\ [a, b].}$$

EXAMPLE 4. Assuming the formula for the area of a circle and the integrability of the function $\sqrt{1 - x^2}$ over the interval $[0, 1]$, compute

$$\int_0^1 \sqrt{1 - x^2}\,dx.$$

The graph of the function $\sqrt{1 - x^2}$ is the upper half of the circle $x^2 + y^2 = 1$ shown in Figure 8. It follows from (1.4) that the integral in question is equal

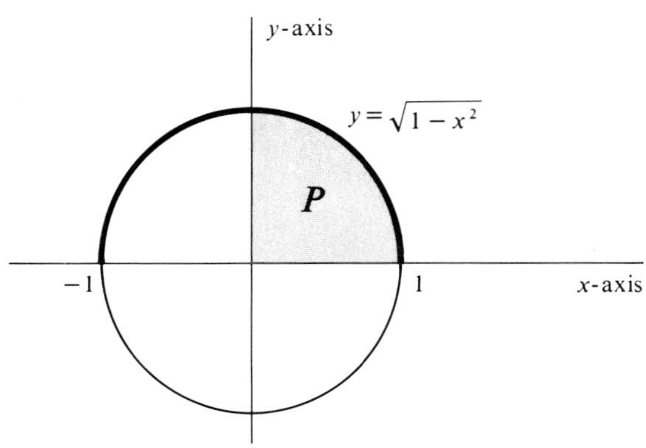

Figure 8

to one fourth of the area of the circle. Hence

$$\int_0^1 \sqrt{1 - x^2}\,dx = \frac{\pi}{4}\cdot$$

PROBLEMS

1. Draw the graph of the function f defined by $f(x) = \dfrac{1}{x}$, and answer the following questions.

   (a) Is f bounded on the closed interval $[2, 5]$?

   (b) Is f bounded on the open interval $(2, 5)$?

   (c) Does f have an upper bound on the interval $(0, 2)$? If so, give one.

   (d) Does f have a lower bound on the interval $(0, 2)$? If so, give one.

2. If the number M is the least upper bound of the set of all numbers $f(x)$ for x lying in an interval I, we say simply that M is the least upper bound of f on I. A similar remark holds for the greatest lower bound. Draw the graph of the function f defined by $f(x) = \dfrac{1}{x-1}$, and answer the following questions.

   (a) What is the least upper bound of f on the closed interval $[2, 3]$?

   (b) What is the greatest lower bound of f on $[2, 3]$?

   (c) What are the least upper bound and greatest lower bound of f on the open interval $(2, 3)$?

   (d) What is the greatest lower bound of f on the interval $(1, 2)$?

3. Compute the upper and lower sums U_σ and L_σ in each of the following examples.

   (a) $f(x) = \dfrac{1}{x}$, $[a, b] = [1, 4]$, and $\sigma = \{1, 2, 3, 4\}$.

   (b) $f(x) = \dfrac{x}{2}$, $[a, b] = [0, 2]$, and $\sigma = \{0, \frac{1}{3}, \frac{2}{3}, 1, \frac{4}{3}, \frac{5}{3}, 2\}$.

   (c) $g(x) = x^2 + 1$, $[a, b] = [0, 1]$, and $\sigma = \{x_0, x_1, x_2, x_3, x_4, x_5\}$, where $x_i = \dfrac{i}{5}$, $i = 0, \ldots, 5$.

   (d) $g(x) = x^3$, $[a, b] = [-1, 1]$, and $\sigma = \{-1, -\frac{1}{2}, 0, \frac{1}{2}, 1\}$.

4. True or false, and give your reason: If a function f is continuous at every x in a closed interval $[a, b]$, then f has both a least upper bound and a greatest lower bound on $[a, b]$.

5. Assume that the function f defined by $f(x) = x^2 + 1$ is integrable over the interval $[0, 1]$. Using the partition $\sigma = \{0, \frac{1}{5}, \frac{2}{5}, \frac{3}{5}, \frac{4}{5}, 1\}$, show that

$$\tfrac{31}{25} \le \int_0^1 f \le \tfrac{36}{25}.$$

6. Assuming that the function g defined by $g(x) = 2x$ is integrable over the interval $[0, 2]$, use the partition $\sigma = \{0, \frac{1}{2}, 1, \frac{3}{2}, 2\}$ to show that

$$3 \le \int_0^2 2x\,dx \le 5.$$

7. Assume that the function x^2 is integrable over the interval [0, 1]. Using the partition $\sigma = \{x_0, \ldots, x_n\}$, where $n = 10$ and $x_i = \dfrac{i}{10}$, for $0, \ldots, 10$, prove that

$$\tfrac{57}{200} \le \int_0^1 x^2\,dx \le \tfrac{77}{200}.$$

8. Compute the definite integral $\int_a^b f = \int_a^b f(x)\,dx = \int_a^b f(t)\,dt$ in each of the following examples. Assume that f is integrable, and use Theorem (1.4) and the standard formulas for the areas of simple plane figures. In each case, draw the graph of f and shade the region P.

(a) $\displaystyle\int_{-1}^{1} f$, where $f(x) = \sqrt{1 - x^2}$.

(b) $\displaystyle\int_1^2 f(t)\,dt$, where $f(t) = t - 1$.

(c) $\displaystyle\int_0^2 2x\,dx$

(d) $\displaystyle\int_0^1 (5 - 2x)\,dx$

(e) $\displaystyle\int_{-1}^{1} |x|\,dx$.

9. It is stated on page 169 that the first condition for integrability is always satisfied: If f is bounded on $[a, b]$, then there exists a real number J such that $L_\sigma \le J \le U_\tau$ for any two partitions σ and τ of $[a, b]$.

   (a) Show that one such number is the least upper bound of all the lower sums L_σ. (This number is called the **lower integral of *f* from *a* to *b*.**)

   (b) Show that another possibility is the greatest lower bound of all the upper sums U_τ. (This number is the **upper integral of *f* from *a* to *b*.**)

   (c) Show that f is integrable over $[a, b]$ if and only if the lower integral from a to b equals the upper integral, and that if the lower integral equals the upper then their common value is $\int_a^b f$.

2. Sequences and Summations. We shall return to the definite integral in Section 3. The purpose of the present digression is to develop some techniques, applicable not only to the study of the integral but also to many other parts of mathematics.

Most of the functions studied in this book have as domains intervals on the real line, or unions of intervals; e.g., the domain of the function $\dfrac{1}{x}$ is the union $(-\infty, 0) \cup (0, \infty)$. In this section, on the other hand, we are concerned with functions whose domains are sets of integers. An example is the function a defined by $a(n) = \sqrt{n - 2}$, for every integer greater than 1. If a is a function whose domain is a subset of the integers, it is common practice to denote its value at n by a_n. Thus

$$a_n = a(n).$$

A simple example in which the domain is a finite set of integers is a partition of an interval in which we have indexed the points of the partition as $x_0, \ldots, x_n$. In this case,

$$x_i = x(i), \qquad \text{for } i = 0, \ldots, n.$$

We come next to the definition of a sequence, which is a special case of a function defined on a set of integers. We shall accept the intuitive idea of a sequence to be that of a list (in mathematics, most likely, a list of numbers). With this in mind, we define a **sequence** to be a function whose domain D is a set of integers such that

(i) D is a set of consecutive integers; i.e., if i and j are in D, then every integer between i and j is also in D.
(ii) D contains a least element.

If s is a sequence and if l is the least, or smallest, integer in its domain, then $s(l) = s_l$ is the first member of the sequence, $s(l+1) = s_{l+1}$ is the second member, and so on. In the most common applications l is either 0 or 1, and so the values of the sequence appear as either $s_0, s_1, s_2, \ldots$ or as $s_1, s_2, \ldots$.

A sequence is **finite** or **infinite** according as its domain D is finite or infinite. Note that the range of an infinite sequence need not contain infinitely many numbers. The function a defined, for every positive integer n, by

$$a_n = a(n) = \begin{cases} 0, & \text{if } n \text{ is even,} \\ 1, & \text{if } n \text{ is odd,} \end{cases} \tag{1}$$

is the infinite sequence $1, 0, 1, 0, 1, 0, 1, \ldots$. An even simpler example of an infinite sequence is the constant function b defined by

$$b_n = 1, \qquad \text{for every positive integer } n.$$

A common notation for a sequence s, whether finite or infinite, is $\{s_n\}$. When a sequence is written in this way, the letter n is called an **index.** Like the variable of integration in a definite integral, it is a dummy symbol. Any letter can be used, although n, m, i, j, and k are the most common. Thus

$$s = \{s_n\} = \{s_m\} = \{s_i\} = \text{etc.}$$

Of course, a finite sequence can be described by simply enumerating its terms, e.g., $s_1, \ldots, s_n$, or $a_3, a_4, \ldots, a_{10}$.

We shall study two major topics in this section. The first is the limit of an infinite sequence. This is actually just an application of the idea of the limit of a function which we defined and studied in Chapter 1. As an example, let s be the infinite sequence defined by

$$s_n = \frac{2n^2 + n - 1}{3n^2 - 2n + 2}, \qquad \text{for every positive integer } n.$$

We ask for the limit of $\{s_n\}$ as n increases without bound, which we denote by $\lim_{n\to\infty} s_n$. Dividing both numerator and denominator of the above expression by n^2, we obtain

$$s_n = \frac{2 + \dfrac{1}{n} - \dfrac{1}{n^2}}{3 - \dfrac{2}{n} + \dfrac{2}{n^2}}.$$

If n is very large, it is clear that $2 + \dfrac{1}{n} - \dfrac{1}{n^2}$ is nearly equal to 2, and that $3 - \dfrac{2}{n} + \dfrac{2}{n^2}$ is nearly equal to 3. We conclude that the number which the values of the sequence are approaching, i.e., the limit, is $\frac{2}{3}$. Thus we write

$$\lim_{n\to\infty} s_n = \lim_{n\to\infty} \frac{2n^2 + n - 1}{3n^2 - 2n + 2} = \frac{2}{3}.$$

EXAMPLE 1. Let $\{s_n\}$ and $\{a_m\}$ be two infinite sequences defined, respectively, by

$$s_n = \frac{\sqrt{2n - 5}}{\sqrt{5n - 2}}, \qquad \text{for } n = 3, 4, 5, \ldots,$$

$$a_m = \frac{m^2 + 1}{m}, \qquad \text{for } m = 1, 2, 3, \ldots.$$

Find $\lim_{n\to\infty} s_n$ and $\lim_{m\to\infty} a_m$. For the sequence s, we divide numerator and denominator by $\sqrt{n}$, getting

$$s_n = \frac{\dfrac{1}{\sqrt{n}}\sqrt{2n - 5}}{\dfrac{1}{\sqrt{n}}\sqrt{5n - 2}} = \frac{\sqrt{2 - \dfrac{5}{n}}}{\sqrt{5 - \dfrac{2}{n}}}.$$

Both $\dfrac{5}{n}$ and $\dfrac{2}{n}$ obviously approach 0 as a limit as n increases without bound. We conclude that

$$\lim_{n\to\infty} s_n = \lim_{n\to\infty} \frac{\sqrt{2n - 5}}{\sqrt{5n - 2}} = \sqrt{\frac{2}{5}}.$$

For the sequence $\{a_m\}$ we have

$$a_m = \frac{m^2 + 1}{m} = m + \frac{1}{m}.$$

It is obvious that, as m increases without bound, so does $m + \dfrac{1}{m}$. Hence no limit exists. On the other hand, we can unambiguously express the fact that the values of the sequence are increasing without bound by writing

$$\lim_{m \to \infty} a_m = \lim_{m \to \infty} \frac{m^2 + 1}{m} = \infty.$$

As we have remarked, the definition of the limit of a sequence is included in the definition of the limit of a function. For emphasis, however, we shall give it in this special case. Let s be an infinite sequence of real numbers. Then **the limit as n increases without bound of s_n is equal to b**, written

$$\lim_{n \to \infty} s_n = b,$$

if, for any $\epsilon > 0$, there exists an integer m in the domain of s such that whenever $n > m$, then $|s_n - b| < \epsilon$. The definition can be phrased geometrically as follows: The limit of $\{s_n\}$ is b if, given an arbitrary open interval $(b - \epsilon, b + \epsilon)$, all the numbers s_n, from some integer on, lie in that interval. Thus for the oscillating sequence 1, 0, 1, 0, 1, 0, . . . defined in (1), no limit exists.

The second topic in the section is the study of a convenient notation for the sum of a finite number of consecutive terms of a sequence. Let a be a sequence (finite or infinite) of real numbers. If m and n are in the domain of the sequence, and if $m \leq n$, then the sum $a_m + a_{m+1} + \cdots + a_n$ is called a **series** and is abbreviated $\sum_{i=m}^{n} a_i$. Thus

$$\sum_{i=m}^{n} a_i = a_m + a_{m+1} + \cdots + a_n.$$

We call $\sum_{i=m}^{n} a_i$ the **summation of $\{a_i\}$ from m to n.**

EXAMPLE 2. Let $\{a_i\}$ be the sequence defined by $a_i = i^2$, for every positive integer i. Then

$$\sum_{i=1}^{5} a_i = \sum_{i=1}^{5} i^2 = 1^2 + 2^2 + 3^2 + 4^2 + 5^2 = 55.$$

Another series defined from the same sequence is

$$\sum_{i=3}^{6} a_i = \sum_{i=3}^{6} i^2 = 3^2 + 4^2 + 5^2 + 6^2 = 86.$$

On the other hand, we might be interested in the sum of the squares of the first n integers for an arbitrary positive integer n. This would be the series

$$\sum_{i=1}^{n} a_i = \sum_{i=1}^{n} i^2 = 1^2 + 2^2 + 3^2 + \cdots + n^2.$$

The symbol i which appears in the series $\sum_{i=m}^{n} a_i$ is called the **summation index.** It, too, is a dummy symbol, since the value of the series does not depend on i. Like the definite integral, $\sum_{i=m}^{n} a_i$ depends on three things: the sequence a (the function) and the two integers m and n. Thus

$$\sum_{i=m}^{n} a_i = \sum_{j=m}^{n} a_j = \sum_{k=m}^{n} a_k = a_m + a_{m+1} + \cdots + a_n.$$

EXAMPLE 3. Using the summation notation, write a series for the sum of all the odd integers from 11 to 101. An arbitrary odd integer can be written in the form $2i + 1$ for some integer i. It is not hard to see, therefore, that one answer to the problem is given by the series

$$\sum_{i=5}^{50} (2i + 1).$$

Another is the series

$$\sum_{i=6}^{51} (2i - 1).$$

It should be emphasized that the summation notation offers no new mathematical theory. It is merely a convenient shorthand for writing sums and manipulating them. The ability to manipulate comes from practice, but the techniques are based on the following properties:

(2.1) $$\sum_{i=m}^{n} (a_i + b_i) = \sum_{i=m}^{n} a_i + \sum_{i=m}^{n} b_i.$$

(2.2) $$\sum_{i=m}^{n} ca_i = c \sum_{i=m}^{n} a_i.$$

(2.3) $$\sum_{i=m}^{n} c = c(n - m + 1).$$

Proof. The proofs are very simple. For (2.1) we have

$$\begin{aligned}\sum_{i=m}^{n} (a_i + b_i) &= (a_m + b_m) + (a_{m+1} + b_{m+1}) + \cdots + (a_n + b_n) \\ &= (a_m + a_{m+1} + \cdots + a_n) + (b_m + b_{m+1} + \cdots + b_n) \\ &= \sum_{i=m}^{n} a_i + \sum_{i=m}^{n} b_i.\end{aligned}$$

For (2.2),

$$\begin{aligned}\sum_{i=m}^{n} ca_i &= ca_m + ca_{m+1} + \cdots + ca_n \\ &= c(a_m + a_{m+1} + \cdots + a_n) \\ &= c \sum_{i=m}^{n} a_i.\end{aligned}$$

To prove (2.3), one must understand that $\sum_{i=m}^{n} c$ means $\sum_{i=m}^{n} a_i$, where $\{a_i\}$ is the constant sequence defined by $a_i = c$. Hence

$$\sum_{i=m}^{n} c = \sum_{i=m}^{n} a_i = \overbrace{a_m + a_{m+1} + \cdots + a_n}^{n - m + 1 \text{ terms}}$$
$$= c + c + \cdots + c$$
$$= c(n - m + 1).$$

This completes the proof.

There are two other summation identities which are useful and which we shall include. They are the formulas for the sum of the first n positive integers and for the sum of the squares of the first n positive integers:

(2.4)
$$\boxed{\sum_{i=1}^{n} i = \frac{n(n+1)}{2}.}$$

(2.5)
$$\boxed{\sum_{i=1}^{n} i^2 = \frac{n(n+1)(2n+1)}{6}.}$$

Proof. There is a very clever proof of (2.4), which the great mathematician Carl Friedrich Gauss (1777–1855) is said to have figured out for himself in a few seconds in his first arithmetic class at the age of 10. Write the sum $S = \sum_{i=1}^{n} i$, once in natural order and, underneath it, the sum in reverse order as follows:

$$S = 1 + 2 + \cdots + (n - 1) + n,$$
$$S = n + (n - 1) + \cdots + 2 + 1.$$

If each number on the right side of the first equation is added to the number directly beneath it, the sum is $n + 1$. Hence the sum of the two right sides is a series consisting of n terms each equal to $n + 1$. It follows that

$$2S = n(n + 1),$$

from which (2.4) is an immediate corollary.

Formula (2.5) is probably most easily proved by induction on n. The proof is straightforward, and we omit it.

EXAMPLE 4. Evaluate

$$\text{(a)} \quad \sum_{i=1}^{n} (3i^2 + 5i - 2),$$

$$\text{(b)} \quad \sum_{i=1}^{n} \frac{(3i^2 + 5i - 2)}{n^3}.$$

Using the various properties of summation, we obtain for (a),

$$\begin{aligned}
\sum_{i=1}^{n} (3i^2 + 5i - 2) &= 3\sum_{i=1}^{n} i^2 + 5\sum_{i=1}^{n} i - 2\sum_{i=1}^{n} 1 \\
&= 3\,\frac{n(n+1)(2n+1)}{6} + 5\,\frac{n(n+1)}{2} - 2n \\
&= \frac{2n^3 + 3n^2 + n}{2} + \frac{5n^2 + 5n}{2} - \frac{4n}{2} \\
&= \frac{2n^3 + 8n^2 + 2n}{2} = n^3 + 4n^2 + n.
\end{aligned}$$

Part (b) is really a trivial modification of (a). The number n^3 which appears in the denominator is the same for each term in the sum, i.e., it is a constant, and can be factored out immediately. Thus

$$\sum_{i=1}^{n} \frac{3i^2 + 5i - 2}{n^3} = \frac{1}{n^3}\sum_{i=1}^{n} (3i^2 + 5i - 2).$$

Hence, using the answer from (a), we get

$$\begin{aligned}
\sum_{i=1}^{n} \frac{3i^2 + 5i - 2}{n^3} &= \frac{1}{n^3}(n^3 + 4n^2 + n) \\
&= 1 + \frac{4}{n} + \frac{1}{n^2}\cdot
\end{aligned}$$

We conclude the section with an example which combines the summation convention with the limit of an infinite sequence,

EXAMPLE 5. For every positive integer n, let S_n be defined by

$$S_n = \sum_{i=1}^{n} \frac{i^2 + 2}{n^3}\cdot$$

The numbers $S_1, S_2, S_3, \ldots$ form an infinite sequence, and the problem is to evaluate $\lim_{n\to\infty} S_n$. Using the properties of summation, we obtain

$$\begin{aligned}
S_n &= \frac{1}{n^3}\sum_{i=1}^{n} (i^2 + 2) \\
&= \frac{1}{n^3}\left(\sum_{i=1}^{n} i^2 + \sum_{i=1}^{n} 2\right) \\
&= \frac{1}{n^3}\left[\frac{n(n+1)(2n+1)}{6} + 2n\right] \\
&= \frac{1}{n^3}\left(\frac{2n^3 + 3n^2 + n}{6} + \frac{12n}{6}\right) \\
&= \frac{2n^3 + 3n^2 + 13n}{6n^3}\cdot
\end{aligned}$$

Hence

$$\lim_{n\to\infty} S_n = \lim_{n\to\infty} \frac{2n^3 + 3n^2 + 13n}{6n^3}$$

$$= \lim_{n\to\infty} \left(\frac{1}{3} + \frac{1}{2n} + \frac{13}{6n^2}\right) = \frac{1}{3},$$

which is the answer to the problem. Frequently the notations are compounded; i.e., we write

$$\lim_{n\to\infty} \sum_{i=1}^{n} \frac{i^2 + 2}{n^3} = \frac{1}{3}.$$

PROBLEMS

1. For each of the following sequences $\{s_n\}$, compute $\lim_{n\to\infty} s_n$ if the limit exists.

(a) $s_n = \dfrac{n+1}{n-1}$, for $n = 2, 3, \ldots$.

(b) $s_n = \dfrac{2n^2 - 3n + 1}{5n^2 + 7}$, for $n = 1, 2, 3, \ldots$.

(c) $s_n = 1 + \dfrac{1}{n}$, for every positive integer n.

(d) $s_n = \dfrac{n-2}{\sqrt{n}}$, for every positive integer n.

2. Let sequences $\{a_i\}$, $\{b_j\}$, and $\{s_n\}$ be defined by

$$\begin{cases} a_i = i^3, \\ b_j = j - 1, \\ s_n = \dfrac{1}{n+1}. \end{cases}$$

Evaluate

(a) $\sum_{i=1}^{4} a_i$

(b) $\sum_{j=-2}^{2} b_j$

(c) $\sum_{j=1}^{3} (2a_j + 5b_j)$

(d) $\sum_{i=1}^{4} \dfrac{a_i}{i+1}$

(e) $\sum_{i=1}^{3} s_i$

(f) $\sum_{j=0}^{3} a_j b_j$.

3. Compute

(a) $\sum_{i=1}^{5} (2i^2 - 3i + 4)$

(b) $\sum_{j=1}^{5} [(j+1)^2 - j^2]$

(c) $\sum_{k=0}^{3} x^k$

(d) $\sum_{j=1}^{4} \frac{x^j}{j}$

(e) $(1 - x)\sum_{k=0}^{3} x^k.$

4. For any sequence $a_0, \ldots, a_n$, show that $\sum_{k=1}^{n} (a_k - a_{k-1})$ depends only on the first and last terms.

5. Using the various properties of summation, evaluate

(a) $\sum_{i=1}^{n} (3i^2 + 2)$

(b) $\sum_{j=1}^{n} (j^2 - 2j + 1)$

(c) $\sum_{i=1}^{n} \frac{3i^2 + 2}{n^3}$

(d) $\sum_{i=1}^{n} \frac{2i - 2}{n^2}$

(e) $\sum_{i=0}^{n} (i^2 + i + 1).$

6. If $f(x) = x^2 - x + 1$, find

(a) $\sum_{i=1}^{4} f(i)$

(b) $\sum_{i=1}^{4} f\left(\frac{i}{4}\right)$

(c) $\sum_{i=1}^{n} f(i)$

(d) $\sum_{i=0}^{n} f\left(\frac{i}{n}\right).$

7. Compute $\lim_{n\to\infty} S_n$, if the limit exists, for each of the following sequences.

(a) $S_n = \sum_{i=1}^{n} \frac{3i^2 + 2}{n^3}$, for every positive integer n.

(b) $S_n = \sum_{i=1}^{n} \frac{2i - 2}{n^2}$, for every positive integer n.

(c) $S_n = \sum_{i=1}^{n} \frac{i + 1}{n^3}$, $n = 1, 2, \ldots.$

(d) $S_n = \sum_{j=1}^{n} \frac{j^2 + 1}{n}$, $n = 1, 2, \ldots.$

8. Evaluate

(a) $\lim_{n\to\infty} \sum_{i=1}^{n} \frac{i}{n^2}$ (b) $\lim_{n\to\infty} \sum_{i=1}^{n} \frac{6i^2 - 2i + 1}{n^3}$.

9. Prove (2.5) by induction on n.
10. Using the identity

$$\sum_{i=1}^{n} i^3 = \left(\frac{n(n+1)}{2}\right)^2,$$

prove that

$$\sum_{i=1}^{n} i^3 = \left(\sum_{i=1}^{n} i\right)^2.$$

Verify this result directly for $n = 1, 2, 3$.
11. (a) How many presents did I receive from my true love on the 12th day of Christmas, when she gave me "12 drummers drumming, 11 pipers piping, . . . , and a partridge in a pear tree"?
(b) How many presents did I receive during the entire 12 days of Christmas?
(Some familiarity with the words of the song is required.)

3. Integrability of Monotonic Functions. Let f be a given function bounded on a closed interval $[a, b]$. How do we know whether or not f is integrable over $[a, b]$, i.e., whether or not $\int_a^b f$ exists? In this section we shall give a partial answer, and also compute some integrals. Note that there is one situation where we know the answer immediately: If $a = b$, then all upper and lower sums are equal to zero. Hence f is integrable, and

(3.1)
$$\boxed{\int_a^a f = \int_a^a f(x)\,dx = 0.}$$

So we now assume that $a < b$. For every positive integer n, we shall denote by σ_n the partition which subdivides $[a, b]$ into n subintervals each of length $\frac{b-a}{n}$. Thus $\sigma_n = \{x_0, \ldots, x_n\}$, where

$$x_i = a + \left(\frac{b-a}{n}\right) i, \qquad i = 0, \ldots, n.$$

Moreover,

$$x_i - x_{i-1} = \frac{b-a}{n}, \qquad i = 1, \ldots, n.$$

The upper and lower sums of f relative to σ_n will be denoted simply U_n and L_n, respectively. That is, we abbreviate U_{σ_n} by U_n, and in the same way L_{σ_n} by L_n. One criterion for integrability is expressed in the following theorem.

(3.2) *If* $\lim_{n\to\infty} (U_n - L_n) = 0$, *then f is integrable over* $[a, b]$ *and*

$$\lim_{n\to\infty} L_n = \lim_{n\to\infty} U_n = \int_a^b f(x)\,dx.$$

Proof. We recall the basic theorem of Section 1—that the upper and lower sums of f relative to any two partitions σ and τ of the interval $[a, b]$ satisfy the inequality $L_\sigma \leq U_\tau$. This implies, in particular, that any upper sum U_τ is an upper bound of the set **L** of all lower sums L_σ. Hence, by the Least Upper Bound Property, the set **L** has a least upper bound which we denote by J. Since this number J is an upper bound of **L**, we know that $L_\sigma \leq J$ for every partition σ. Furthermore, since J is a *least* upper bound, we have $J \leq U_\tau$ for every partition τ. Thus

$$L_\sigma \leq J \leq U_\tau,$$

for all partitions σ and τ of $[a, b]$. As a special case of these inequalities, we have

$$L_n \leq J \leq U_n, \qquad \text{for every positive integer } n. \tag{1}$$

Since by hypothesis $\lim_{n\to\infty} (U_n - L_n) = 0$, it follows that this number J is the only number which can lie between all upper and lower sums. Hence, by the definition, f is integrable over $[a, b]$ and $J = \int_a^b f(x)\,dx$. From (1) we obtain the inequalities

$$0 \leq J - L_n \leq U_n - L_n, \qquad \text{for every positive integer } n.$$

Since the right side of the above inequalities approaches zero, the expression in the middle is caught in a squeeze and must also approach zero. Hence $\lim_{n\to\infty} (J - L_n) = 0$, or, equivalently,

$$\lim_{n\to\infty} L_n = J = \int_a^b f(x)\,dx.$$

Finally, consider the identity $U_n = J + (U_n - L_n) - (J - L_n)$. Since the two expressions in parentheses approach zero, it follows that

$$\lim_{n\to\infty} U_n = J = \int_a^b f(x)\,dx,$$

and the proof is complete.

An important class of functions to which the preceding theorem can be readily applied, and which we now define, is the class of monotonic functions.

To begin with, a real-valued function f is said to be **increasing on an interval** I if the domain of f contains I as a subset and if, for every x_1 and x_2 in I,

$$x_1 \leq x_2 \text{ implies } f(x_1) \leq f(x_2). \tag{2}$$

If (2) holds for every x_1 and x_2 in the entire domain of f, we say simply that f is an **increasing function.** Companion definitions are obtained by simultaneously replacing the second inequality in (2) by $f(x_1) \geq f(x_2)$ and the word **increasing** by the word **decreasing.** For example, the function f defined by $f(x) = x^2$ is increasing on the interval $[0, \infty)$ and decreasing on the interval $(-\infty, 0]$. The function g defined by $g(x) = -2x + 1$ is a decreasing function.

Note that, according to our definition, a constant function is both increasing and decreasing. Thus "increasing," as it is used here, literally means "nondecreasing," and in the same way "decreasing" means "nonincreasing."

A **monotonic function** is one which is either increasing or decreasing. Similarly, a function is **monotonic on an interval** if it is either increasing or decreasing on the interval. For such functions it is not difficult to prove the following integrability theorem.

(3.3) THEOREM. *If the function f is monotonic on the closed interval $[a, b]$, then f is integrable over $[a, b]$. Specifically,* $\lim_{n\to\infty} (U_n - L_n) = 0$.

Proof. For the sake of concreteness, we shall assume that f is increasing on $[a, b]$. An analogous argument works if f is decreasing. By far the best proof of this theorem is obtained from a picture, which provides a completely convincing

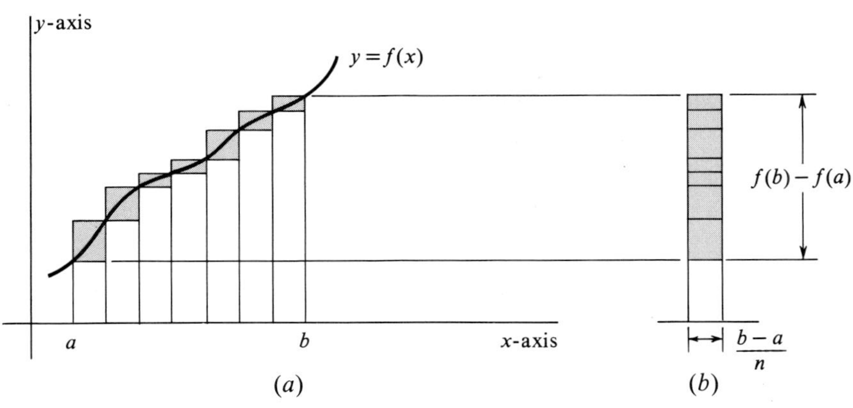

Figure 9

argument. A typical example of an increasing function together with a partition of the interval is shown in Figure 9(a). The difference $U_n - L_n$ is equal to the sum of the areas of the shaded rectangles. By sliding these rectangles under one another to form a single stack, we obtain the tall rectangle shown in Figure 9(b), whose area is also equal to $U_n - L_n$. This rectangle has base $\frac{b-a}{n}$ and altitude $f(b) - f(a)$. Its area is the product of these, and so

$$U_n - L_n = \left(\frac{b-a}{n}\right)\big(f(b) - f(a)\big).$$

This difference can be made arbitrarily small by taking n sufficiently large. It follows that $\lim_{n\to\infty} (U_n - L_n) = 0$, and we conclude from (3.2) that f is integrable over $[a, b]$. This completes the proof.

EXAMPLE 1. Evaluate $\int_0^2 x^2\,dx$. The function f to be integrated is defined by $f(x) = x^2$, and the interval of integration is $[0, 2]$. Since f is increasing on the interval, the integral certainly exists. The partition $\sigma_n = \{x_0, \ldots, x_n\}$ which subdivides $[0, 2]$ into n subintervals of equal length is given by

$$x_i = a + \frac{b-a}{n}i = 0 + \frac{2}{n}i = \frac{2i}{n},$$

for each $i = 0, \ldots, n$. Moreover,

$$x_i - x_{i-1} = \frac{b-a}{n} = \frac{2}{n}, \qquad i = 1, \ldots, n.$$

It follows from Theorems (3.2) and (3.3) that

$$\int_0^2 x^2\,dx = \lim_{n\to\infty} U_n = \lim_{n\to\infty} L_n.$$

That is, we may compute the integral using either the lower or the upper sums. Choosing the latter, we observe from Figure 10 that, on each subinterval $[x_{i-1}, x_i]$, the function f has its maximum value at the right endpoint, i.e., at x_i. Hence

$$M_i = f(x_i), \qquad i = 1, \ldots, n.$$

Since $f(x_i) = x_i^2$ and since $x_i = \frac{2i}{n}$, it follows that $M_i = x_i^2 = \frac{4i^2}{n^2}$. Substituting in the formula for the upper sum,

$$U_n = \sum_{i=1}^{n} M_i(x_i - x_{i-1}),$$

we obtain

$$U_n = \sum_{i=1}^{n}\left(\frac{4i^2}{n^2}\right)\left(\frac{2}{n}\right) = \sum_{i=1}^{n}\frac{8i^2}{n^3} = \frac{8}{n^3}\sum_{i=1}^{n} i^2.$$

From (2.5), we have

$$\sum_{i=1}^{n} i^2 = \frac{n(n+1)(2n+1)}{6} = \frac{2n^3 + 3n^2 + n}{6}.$$

Hence

$$U_n = \frac{8}{n^3}\,\frac{2n^3 + 3n^2 + n}{6} = \frac{4}{3}\left(2 + \frac{3}{n} + \frac{1}{n^2}\right),$$

and so

$$\lim_{n\to\infty} U_n = \lim_{n\to\infty} \frac{4}{3}\left(2 + \frac{3}{n} + \frac{1}{n^2}\right) = \frac{4}{3}\cdot 2 = \frac{8}{3}.$$

We conclude that

$$\int_0^2 x^2\,dx = \tfrac{8}{3}.$$

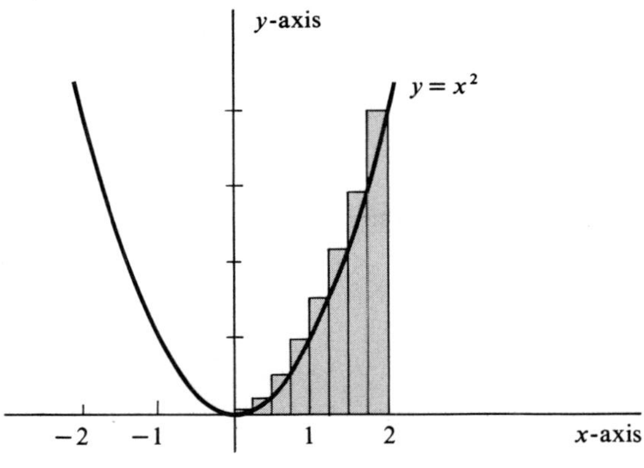

Figure 10

It was shown in Section 1 that the integral of a nonnegative function is equal to the area under its graph. It follows from the above example that the area of the region bounded by the parabola $y = x^2$, the x-axis, and the line $x = 2$ is equal to $\frac{8}{3}$.

EXAMPLE 2. Evaluate $\int_1^4 (5 - x)\,dx$. The function f, defined by $f(x) = 5 - x$, is linear and decreasing on the interval [1, 4]. Its graph is shown in Figure 11. The partition $\sigma_n = \{x_0, \ldots, x_n\}$ subdivides the interval [1, 4] into subintervals of length $\dfrac{4-1}{n} = \dfrac{3}{n}$, and the points are given by

$$x_i = 1 + \left(\frac{3}{n}\right) i, \qquad i = 0, \ldots, n.$$

In addition,

$$x_i - x_{i-1} = \frac{3}{n}, \qquad i = 1, \ldots, n.$$

We shall compute the integral as a limit of lower sums, and it follows from Theorems (3.2) and (3.3) that

$$\int_1^4 (5 - x)\, dx = \lim_{n\to\infty} L_n.$$

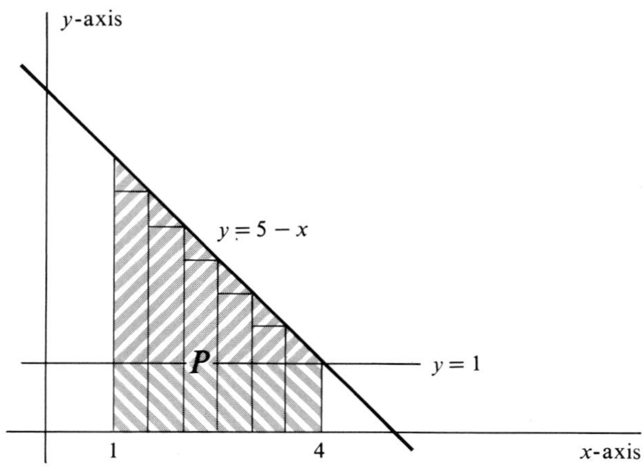

Figure 11

Since f is decreasing, its minimum value on each subinterval $[x_{i-1}, x_i]$ occurs at the right endpoint. Hence

$$m_i = f(x_i), \qquad i = 1, \ldots, n.$$

We have $x_i = 1 + \frac{3i}{n}$ and $f(x_i) = 5 - x_i$, and so

$$m_i = 5 - \left(1 + \frac{3i}{n}\right) = 4 - \frac{3i}{n}.$$

Since $x_i - x_{i-1} = \frac{3}{n}$, we get

$$L_n = \sum_{i=1}^{n} m_i(x_i - x_{i-1}) = \sum_{i=1}^{n} \left(4 - \frac{3i}{n}\right)\frac{3}{n}.$$

The rest of the problem uses the manipulative techniques of the summation convention.

$$\begin{aligned} L_n &= \sum_{i=1}^{n} \left(4 - \frac{3i}{n}\right)\frac{3}{n} = \sum_{i=1}^{n} \left(\frac{12}{n} - \frac{9i}{n^2}\right) \\ &= \sum_{i=1}^{n} \frac{12}{n} - \sum_{i=1}^{n} \frac{9i}{n^2} \\ &= \frac{12}{n}\sum_{i=1}^{n} 1 - \frac{9}{n^2}\sum_{i=1}^{n} i. \end{aligned}$$

Since $\sum_{i=1}^{n} 1 = n$ and since $\sum_{i=1}^{n} i = \dfrac{n(n+1)}{2}$, we get

$$\begin{aligned} L_n &= \frac{12}{n}n - \frac{9}{n^2}\frac{n(n+1)}{2} \\ &= 12 - \frac{9}{2}\left(1 + \frac{1}{n}\right). \end{aligned}$$

But it is easy to see that

$$\lim_{n\to\infty}\left[12 - \frac{9}{2}\left(1 + \frac{1}{n}\right)\right] = 12 - \frac{9}{2} = 7\frac{1}{2},$$

and we finally conclude that

$$\int_1^4 (5 - x)\,dx = \lim_{n\to\infty} L_n = 7\tfrac{1}{2}.$$

This answer can be checked by looking at Figure 11. The value of the integral is equal to the area of the shaded region P, which is divided by the horizontal line $y = 1$ into two pieces: a right triangle sitting on top of a rectangle. The area of the triangle is $\frac{1}{2}(3 \cdot 3) = \frac{9}{2}$, and that of the rectangle is $3 \cdot 1 = 3$. Hence

$$\int_1^4 (5 - x)\,dx = \mathit{area}(P) = \tfrac{9}{2} + 3 = 7\tfrac{1}{2}.$$

The excessive lengths of the computations in Examples 1 and 2 make it obvious that some powerful techniques are needed to streamline the process of evaluating definite integrals. The advent of modern high-speed computers is one answer to the problem, and occasionally, as in Example 2, a simple formula for area will do the trick. The classical solution to the problem, however, is the Fundamental Theorem of Calculus, which we shall study in detail in Section 5.

PROBLEMS

1. Evaluate the following definite integrals by finding the limits of the upper or lower sums.

(a) $\int_0^1 x^2\,dx$ (c) $\int_1^3 (x+1)\,dx$

(b) $\int_0^2 2x\,dx$ (d) $\int_0^1 (3x^2+1)\,dx.$

2. For each of the integrals in Problem 1, draw the region whose area is given by the integral.
3. Let f be the step function defined by $f(x) = i$, if $i - 1 < x \le i$, for every integer i. Draw the graph of f and compute the following integrals. (*Hint:* These problems are neither hard nor long. They require an understanding of the definition of integrability and possibly some ingenuity.)

(a) $\int_1^2 f$ (c) $\int_{-1}^3 f$

(b) $\int_0^3 f$ (d) $\int_{-2}^7 f.$

4. Every constant function is both increasing and decreasing. A stronger condition, which excludes constant functions, is obtained by defining f to be **strictly increasing** if

$$x < y \quad \text{implies} \quad f(x) < f(y),$$

for every x and y in the domain of f. The companion definitions of what it means for a function to be **strictly decreasing, strictly increasing on an interval,** etc., should be obvious. Using the Mean Value Theorem (page 113), prove that if a differentiable function f satisfies the inequality $f'(x) > 0$ for every x in an interval I, then f is strictly increasing on I.
5. Prove the converse of Theorem (3.2); i.e., if f is integrable over $[a, b]$, then $\lim_{n\to\infty} (U_n - L_n) = 0$. (This is a difficult problem.)

4. Properties of the Definite Integral. If a function f is integrable over an interval $[a, b]$, then in the definite integral

$$\int_a^b f = \int_a^b f(x)\,dx$$

the function f is called the **integrand,** and the numbers a and b the **limits of integration.**

The basic properties of the definite integral are contained in the following five theorems.

(4.1) *If $f(x) = k$ for every x in the interval $[a, b]$, then*

$$\int_a^b f(x)\,dx = \int_a^b k\,dx = k(b - a).$$

(4.2) *The function f is integrable over the intervals $[a, b]$ and $[b, c]$ if and only if it is integrable over their union $[a, c]$. Furthermore,*

$$\int_a^b f(x)\,dx + \int_b^c f(x)\,dx = \int_a^c f(x)\,dx.$$

(4.3) *If f and g are integrable over $[a, b]$ and if $f(x) \leq g(x)$ for every x in $[a, b]$, then*

$$\int_a^b f(x)\,dx \leq \int_a^b g(x)\,dx.$$

(4.4) *If f is integrable over $[a, b]$ and if k is any real number, then the product kf is integrable and*

$$\int_a^b kf(x)\,dx = k\int_a^b f(x)\,dx.$$

(4.5) *If f and g are integrable over $[a, b]$, then so is their sum and*

$$\int_a^b [f(x) + g(x)]\,dx = \int_a^b f(x)\,dx + \int_a^b g(x)\,dx.$$

None of the proofs of these theorems is deep in the sense of requiring great ingenuity or any techniques beyond the use of least upper bounds and greatest lower bounds. However, they vary considerably in the amount of detail required. The proof of (4.1) is a triviality. For if f has the constant value k on the interval $[a, b]$, then, for every partition σ of $[a, b]$, the upper sum U_σ of f relative to σ is equal to $k(b - a)$, and so is the lower sum. Thus

$$L_\sigma = k(b - a) = U_\sigma,$$

which proves both that f is integrable and that the value of the integral is $k(b - a)$.

The proof of (4.3) is slightly more difficult and probably most easily obtained by contradiction. Suppose the premise true and the conclusion false. That is, we assume that $\int_a^b f > \int_a^b g$. The definition of integrability asserts that if a function is integrable over an interval, then there exist upper and lower sums lying arbitrarily close to the definite integral. Therefore, since g is integrable and since $\int_a^b g < \int_a^b f$, there must exist an upper sum for g which is less than $\int_a^b f$. Specifically, there exists a partition σ of $[a, b]$ such that the upper sum of g relative to σ, which we shall denote by $U_\sigma(g)$, satisfies the inequality

$$\int_a^b g \leq U_\sigma(g) < \int_a^b f.$$

But since $f(x) \leq g(x)$ for every x in $[a, b]$, the corresponding upper sum of f, denoted $U_\sigma(f)$, is less than or equal to $U_\sigma(g)$. Thus we obtain the inequalities

$$U_\sigma(f) \leq U_\sigma(g) < \int_a^b f.$$

However, every upper sum of f is greater than or equal to the integral $\int_a^b f$. Hence we have arrived at a contradiction, and (4.3) is proved. The proofs of (4.2) and (4.5) are given in Appendix B, and that of (4.4) is assigned as a problem at the end of the section.

The additivity property of the integral stated in Theorem (4.2) obviously extends to any finite number of intervals. Thus if $\sigma = \{x_0, \ldots, x_n\}$ is a partition of $[a, b]$ with $a = x_0 \leq x_1 \leq \cdots \leq x_n = b$, and if f is integrable over each subinterval $[x_{i-1}, x_i]$, then by repeated application of (4.2) it follows that f is integrable over $[a, b]$ and that

$$\int_a^b f(x)\,dx = \sum_{i=1}^{n} \int_{x_{i-1}}^{x_i} f(x)\,dx. \tag{1}$$

In Section 3 it was proved that if a function is monotonic on a closed interval, then it is integrable over that interval. Theorem (4.2), as extended in equation (1), increases the scope of this result enormously. For although a function f may not be monotonic on a given interval $[a, b]$, it is frequently possible to partition $[a, b]$ into subintervals on each of which f is monotonic. It then follows that f is integrable over the entire interval; i.e., $\int_a^b f(x)\,dx$ exists.

EXAMPLE 1. For every nonnegative integer n and interval $[a, b]$, show that the definite integral

$$\int_a^b x^n\,dx$$

exists. To say that $\int_a^b x^n\,dx$ exists is just another way of saying that the function f defined by $f(x) = x^n$ is integrable over $[a, b]$. We now prove that this is so. For every nonnegative integer n, the function x^n is an increasing function on the interval $[0, \infty)$, and it is an increasing or a decreasing function on $(-\infty, 0]$ according as n is odd or even. Hence if $[a, b]$ is a subset of $[0, \infty)$ or a subset of $(-\infty, 0]$, then the function x^n is monotonic on $[a, b]$ and is therefore integrable over that interval. The remaining possibility is that

$a < 0 < b$. In this case, x^n is integrable over the intervals $[a, 0]$ and $[0, b]$ separately. It then follows that x^n is integrable over their union, which is $[a, b]$, and the proof is complete.

Just as Theorem (4.2) was generalized to more than two intervals, Theorem (4.5) can be extended to any finite number of functions. Thus if each one of the functions $f_1, \ldots, f_n$ is integrable over $[a, b]$, then by repeated applications of (4.5) it follows that the sum $f_1 + \cdots + f_n$ is integrable over $[a, b]$ and that

$$\int_a^b [f_1(x) + \cdots + f_n(x)]\, dx = \int_a^b f_1(x)\, dx + \cdots + \int_a^b f_n(x)\, dx. \qquad (2)$$

EXAMPLE 2. Consider an arbitrary polynomial

$$p(x) = a_n x^n + a_{n-1}x^{n-1} + \cdots + a_1 x + a_0$$

and a closed interval $[a, b]$. Then, for each $i = 0, \ldots, n$, we know from Example 1 that $\int_a^b x^i\, dx$ exists. It follows by (4.4) that each function $a_i x^i$ is integrable over $[a, b]$ and that $\int_a^b a_i x^i\, dx = a_i \int_a^b x^i\, dx$. We conclude from the preceding paragraph that the polynomial $p(x)$, which is the sum of the functions $a_i x^i$, is integrable and that

$$\int_a^b p(x)\, dx = \sum_{i=0}^{n} a_i \int_a^b x^i\, dx. \qquad (3)$$

As a concrete example of equation (3), consider the polynomial $7x^5 - 3x^3 + x^2 + 3$. We have immediately

$$\int_a^b (7x^5 - 3x^3 + x^2 + 3)\, dx$$
$$= 7\int_a^b x^5\, dx - 3\int_a^b x^3\, dx + \int_a^b x^2\, dx + 3\int_a^b 1\, dx.$$

Since we know from (4.1) that $\int_a^b 1\, dx = b - a$, the last term in the above equation can be replaced by $3(b - a)$.

Summarizing Examples 1 and 2, we conclude that all polynomials are integrable and that the problem of computing their integrals reduces to the problem of computing the integrals of the positive powers of x.

The interpretation of the definite integral as an area will now be generalized to include functions which may take on negative values. To begin with, suppose that a function f is integrable over $[a, b]$ and, in addition, that

$f(x) \leq 0$ for all x in $[a, b]$. The graphs of both f and $-f$ are drawn in Figure 12. As shown in the figure, we denote by P the region consisting of all

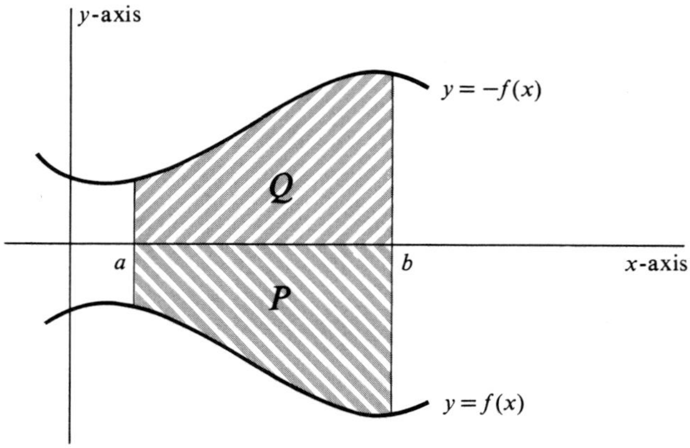

Figure 12

points (x, y) such that $a \leq x \leq b$ and $f(x) \leq y \leq 0$, and, similarly, by Q the region defined by $a \leq x \leq b$ and $0 \leq y \leq -f(x)$. It is obvious that

$$area(P) = area(Q).$$

It follows from Theorem (4.4), by taking $k = -1$, that the function $-f$ is integrable over $[a, b]$ and that

$$\int_a^b (-f(x))\, dx = -\int_a^b f(x)\, dx.$$

Since $-f(x) \geq 0$ for every x in $[a, b]$, we know that

$$\int_a^b (-f(x))\, dx = area(Q).$$

Combining the preceding three equations, we conclude that

$$\int_a^b f(x)\, dx = -area(P).$$

Next, we suppose that f is integrable over $[a, b]$ and takes on both positive and negative values. Specifically, let $[a, b]$ be partitioned by inequalities

$$a = x_0 \leq x_1 \leq \cdots \leq x_n = b$$

so that on each subinterval $[x_{i-1}, x_i]$ the function f is either nonnegative or nonpositive. We denote by P^+ the set of all points (x, y) such that $a \leq x \leq b$ and $0 \leq y \leq f(x)$, and by P^- the set of all points (x, y) such that $a \leq x \leq b$

and $f(x) \leq y \leq 0$ (see Figure 13). It follows from the conclusion of the preceding paragraph and from the additivity of the integral, as generalized in equation (1), that

(4.6)
$$\int_a^b f(x)\,dx = area(P^+) - area(P^-).$$

This is the principal geometric interpretation of the integral.

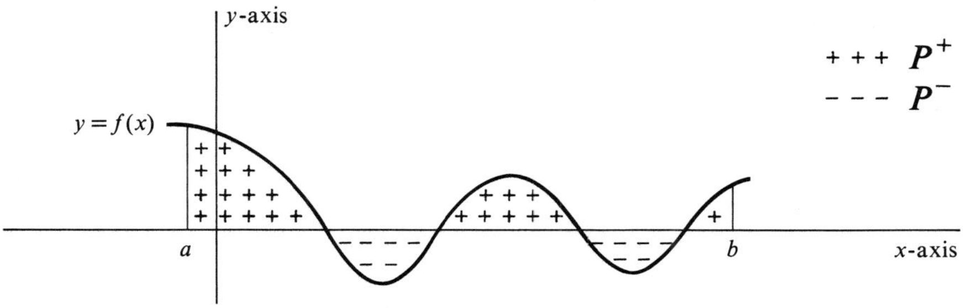

Figure 13

EXAMPLE 3. Evaluate $\int_{-2}^{2} (x^3 - 3x)\,dx$. The integrand, $f(x) = x^3 - 3x$, is an an odd function; i.e., the equation $f(-x) = -f(x)$ is satisfied for every x. Its graph, drawn in Figure 14, is therefore symmetric under reflection first about the x-axis and then about the y-axis. It follows that the region P^+ above the x-axis has the same areas as the region P^- below it. We conclude that

$$\int_{-2}^{2} (x^3 - 3x)\,dx = 0.$$

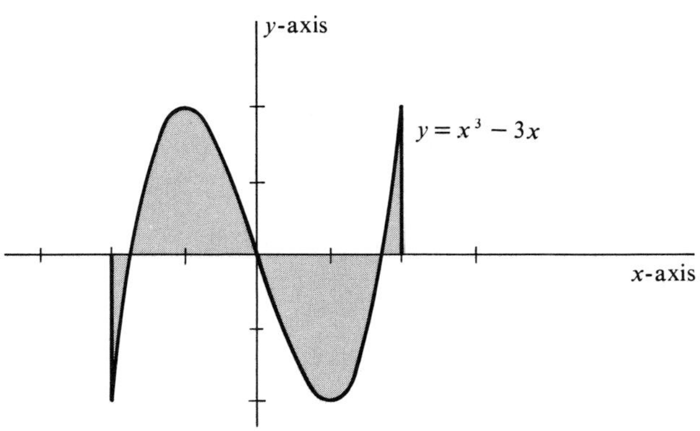

Figure 14

The final topic of this section is an extension of the definition of the integral. Up to this point, $\int_a^b f(x)\,dx$ has been defined only if $a \leq b$. It turns out to be algebraically more convenient to remove this restriction. We do so by decree: If f is integrable over the interval $[a, b]$, then we now define

$$\boxed{\int_b^a f(x)\,dx = -\int_a^b f(x)\,dx.} \tag{4}$$

It is a simple matter to verify that the equations which form the conclusions of Theorems (4.1), (4.4), and (4.5) remain true, in the light of the extended definition of the integral, if a and b are interchanged. Thus

$$\int_a^b k\,dx = k(b - a),$$

$$\int_a^b kf(x)\,dx = k\int_a^b f(x)\,dx,$$

$$\int_a^b [f(x) + g(x)]\,dx = \int_a^b f(x)\,dx + \int_a^b g(x)\,dx,$$

are valid equations regardless of whether $a \leq b$ or $b \leq a$.

On the other hand, if a and b are interchanged in the conclusion of Theorem (4.3), then the direction of the inequality must be reversed.

Less trivial to verify, but equally important, is the generalized form of (4.2):

(4.7) *If f is integrable over the smallest closed interval which contains the numbers a, b, and c, then*

$$\int_a^b f(x)\,dx + \int_b^c f(x)\,dx = \int_a^c f(x)\,dx.$$

The proof is obtained from (4.2) and the definition (4) by simply checking each of the six possible cases in turn:

(i) $a \leq b \leq c$.
(ii) $a \leq c \leq b$.
(iii) $b \leq a \leq c$.
(iv) $b \leq c \leq a$.
(v) $c \leq a \leq b$.
(vi) $c \leq b \leq a$.

The details are tedious, and we omit them.

PROBLEMS

1. Expand each of the following integrals. That is, write each one as a sum of constant multiples of the integrals of the powers of the variables.

(a) $\int_0^1 (x^2 + 5x)\, dx$

(b) $\int_2^3 (4x^5 - x - 2)\, dx$

(c) $\int_1^2 (3t^3 + 2t^2 + t)\, dt$

(d) $\int_5^3 (17y^{13} - 11y^7 + 4)\, dy$

(e) $\int_0^1 (x^2 + 2)^2\, dx.$

2. Given that $\int_0^1 x^n\, dx = \frac{1}{n+1}$, for every nonnegative integer n, evaluate

(a) $\int_0^1 (2x^2 + 3x)\, dx$

(b) $\int_0^1 (5x^3 - x^2 - 2)\, dx$

(c) $\int_0^1 (3t^2 - 1)\, dt$

(d) $\int_0^1 (x + 2)^2\, dx$

(e) $\int_0^1 (3y^2 - y + 1)\, dy.$

3. Use the result

$$\int_1^2 x^n\, dx = \frac{2^{n+1} - 1}{n+1}, \qquad n = 0, 1, 2, \ldots,$$

and the analogous result at the beginning of Problem 2 to evaluate

(a) $\int_1^2 (3x^2 - 2x + 1)\, dx$

(b) $\int_0^2 x^2\, dx$

(c) $\int_0^2 (4x^3 - 3x + 2)\, dx$

(d) $\int_0^2 (t^3 + t^2 + t)\, dt.$

4. Using the definition of integrability, prove Theorem (4.4). (*Suggestion:* Treat the cases $k \geq 0$ and $k \leq 0$ separately.)
5. Using (4.3), prove that if f and g are integrable over $[a, b]$ and if $f(x) = g(x)$, for every x in $[a, b]$, then

$$\int_a^b f(x)\, dx = \int_a^b g(x)\, dx.$$

6. Prove that if f is integrable over $[a, b]$ and if $f(x) \leq M$ for all x in $[a, b]$, then

$$\int_a^b f(x)\,dx \leq M(b - a).$$

7. Replace the symbol $*$ by either $\leq$ or $\geq$ so that the resulting expressions are correct. Give your reasons.

(a) $\displaystyle\int_0^1 x^2\,dx * \int_0^1 x^3\,dx$ (c) $\displaystyle\int_1^3 x^2\,dx * \int_1^3 x^3\,dx.$

(b) $\displaystyle\int_{-1}^1 x^2\,dx * \int_{-1}^1 x^3\,dx$

8. Plot the graph of the function $f(x) = 1 - x^2$, and indicate the region P^+ defined by the inequalities $0 \leq x \leq 2$ and $0 \leq y \leq f(x)$ and the region P^- defined by the inequality $0 \leq x \leq 2$ and $f(x) \leq y \leq 0$.

(a) Use the identities given in Problems 2 and 3 to evaluate the integrals $\int_0^1 f(x)\,dx$, $\int_1^2 f(x)\,dx$, and $\int_0^2 f(x)\,dx$.

(b) Find $area(P^+)$, $area(P^-)$, and $area(P^+ \cup P^-)$.

9. Draw the graph of the function $f(x) = x(x - 2)(x - 4) = x^3 - 6x^2 + 8x$, and indicate the region P^+ defined by the inequalities $0 \leq x \leq 3$ and $0 \leq y \leq f(x)$, and the region P^- defined by $0 \leq x \leq 3$ and $f(x) \leq y \leq 0$. Let $P = P^+ \cup P^-$, and suppose that $\int_0^2 f(x)\,dx = 4$ and $\int_0^3 f(x)\,dx = 2\frac{1}{4}$. Find $area(P^+)$, $area(P^-)$, and $area(P)$.

10. Prove case (iii) of Theorem (4.7).

11. Consider a function f which is integrable over $[a, b]$ and which, in addition, satisfies:

(i) f is continuous at every point of $[a, b]$.

(ii) $f(x) \geq 0$, for every x in $[a, b]$.

(iii) $f(c) > 0$ for at least one point c in $[a, b]$.

Prove that $\int_a^b f(x)\,dx > 0$.

5. The Fundamental Theorem of Calculus. In spite of the fact that we have thus far developed a significant amount of the theory of the definite integral, the actual evaluation of $\int_a^b f(x)\,dx$, for even a very simple function f, is generally an arduous task. For a wide class of functions, the problem of computation is solved by a theorem which relates the definite integral to the derivative and which has become known as the Fundamental Theorem of Calculus.

To understand clearly our presentation of this important result, it is necessary to be aware of the following integrability theorem.

(5.1) THEOREM. *If the function f is continuous at every x in the closed interval $[a, b]$, then f is integrable over $[a, b]$.*

We shall give only a brief outline of the proof, which shows that the result is plausible. Since f is continuous, its values do not vary widely over a small interval. Recall that the nth upper and lower sums are defined by

$$U_n = M_1(x_1 - x_0) + \cdots + M_n(x_n - x_{n-1}),$$

$$L_n = m_1(x_1 - x_0) + \cdots + m_n(x_n - x_{n-1}),$$

where M_i and m_i are, respectively, the maximum and minimum values of f on the ith subinterval. We assume that n is large and each subinterval small. By the continuity of f, therefore, the difference $M_i - m_i$ is small for each $i = 1, \ldots, n$. This in turn implies that $U_n - L_n$ is small. In fact, $U_n - L_n$ can be made arbitrarily small by taking n sufficiently large; i.e., $\lim_{n \to \infty} (U_n - L_n) = 0$. This limit is sufficient to prove that f is integrable over $[a, b]$, as shown in Theorem (3.2). To change this outline into a complete proof, it is necessary to introduce the concept of uniform continuity, which we shall not do in this book.

We shall now show how the definite integral can be used to define a new function. Suppose that f is a given function which is continuous at every x in some interval I. Let a be an arbitrary number in I. A new function F is defined, for every number t in I, by the equation

$$\boxed{F(t) = \int_a^t f(x)\,dx.}$$

The existence of $\int_a^t f(x)\,dx$ follows from the continuity of f and the integrability Theorem (5.1). Thus the function F is well defined by the above equation.

Using the interpretation of the integral as area, we can give geometric meaning to $F(t)$. Suppose that $a \leq t$ and that $f(x) \geq 0$ for every x in the interval $[a, t]$, as shown in Figure 15(a). The integral $\int_a^t f(x)\,dx$ is then equal to the area of the region P bounded by the graph of f, the x-axis, and the lines $x = a$ and $x = t$. Thus

$$F(t) = area(P).$$

On the other hand, if $t \leq a$ and $f(x) \geq 0$ for every x in $[t, a]$, which is the situation shown in Figure 15(b), then the area of P is equal to the integral

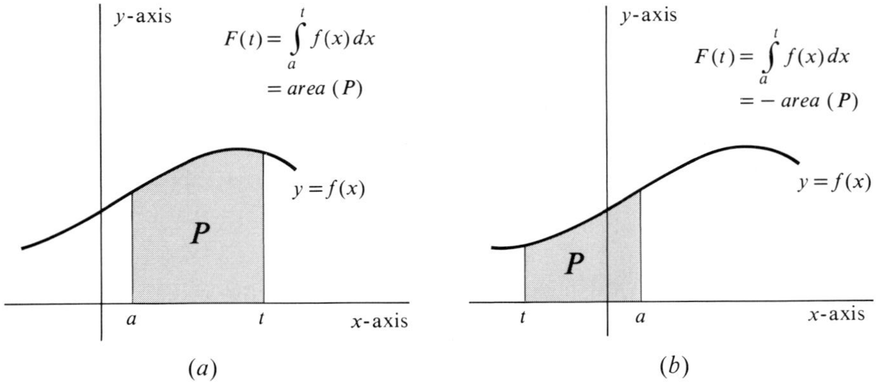

Figure 15

$\int_t^a f(x)\,dx$. Hence, in this case, we have

$$F(t) = \int_a^t f(x)\,dx = -\int_t^a f(x)\,dx = -area(P).$$

In the general case, of course, f may take on both positive and negative values. If the region bounded by the graph of f, the x-axis, and the lines $x = a$ and $x = t$ is expressed as the union of the part P^+ above the axis and the part P^- below the axis, then

$$F(t) = \pm[area(P^+) - area(P^-)],$$

where we take $+$ or $-$ according as $a \leq t$ or $t \leq a$.

We come now to the main result of the section.

(5.2) THE FUNDAMENTAL THEOREM OF CALCULUS. *Let f be continuous at every x in some interval I, and let a be a number in I. If the function F is defined by*

$$F(t) = \int_a^t f(x)\,dx, \qquad \textit{for every } t \textit{ in } I,$$

then F is a differentiable function and

$$F'(t) = f(t), \qquad \textit{for every } t \textit{ in } I.$$

Proof. According to the definition of the derivative, we must prove that

$$\lim_{h \to 0} \frac{F(t + h) - F(t)}{h} = f(t).$$

By the definition of the function F,

$$F(t+h) = \int_a^{t+h} f(x)\,dx.$$

Hence

$$F(t+h) - F(t) = \int_a^{t+h} f(x)\,dx - \int_a^t f(x)\,dx.$$

By Theorem (4.7) we have

$$\int_a^t f(x)\,dx + \int_t^{t+h} f(x)\,dx = \int_a^{t+h} f(x)\,dx,$$

and the preceding two equations therefore imply that

$$F(t+h) - F(t) = \int_t^{t+h} f(x)\,dx.$$

Consequently,

$$\frac{F(t+h) - F(t)}{h} = \frac{1}{h}\int_t^{t+h} f(x)\,dx. \tag{1}$$

These steps are illustrated geometrically in Figure 16.

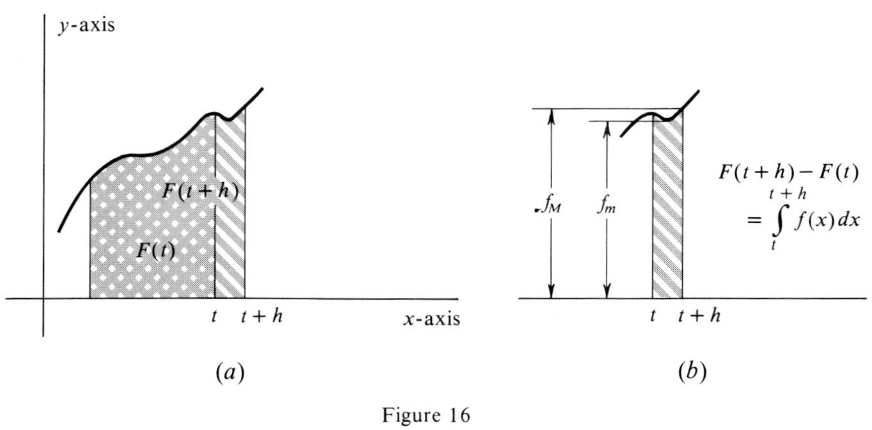

Figure 16

Let the maximum and minimum values of f between t and $t+h$ be denoted by f_M and f_m, respectively. Thus

$$f_m \le f(x) \le f_M,$$

for every x between t and $t + h$. If h is positive (as it is in Figure 16), then it follows by Theorem (4.3) that

$$\int_t^{t+h} f_m \, dx \leq \int_t^{t+h} f(x) \, dx \leq \int_t^{t+h} f_M \, dx.$$

Since f_m and f_M are constants, Theorem (4.1) implies that

$$\int_t^{t+h} f_m \, dx = f_m \cdot (t + h - t) = f_m \cdot h,$$

$$\int_t^{t+h} f_M \, dx = f_M \cdot (t + h - t) = f_M \cdot h.$$

Hence

$$f_m \cdot h \leq \int_t^{t+h} f(x) \, dx \leq f_M \cdot h,$$

or, equivalently,

$$f_m \leq \frac{1}{h} \int_t^{t+h} f(x) \, dx \leq f_M. \tag{2}$$

If, on the other hand, h is negative, it is a straightforward (and logically necessary) matter to verify that the same inequalities (2) follow. Combining (1) and (2), we therefore obtain

$$f_m \leq \frac{F(t + h) - F(t)}{h} \leq f_M. \tag{3}$$

Finally, since f is continuous at t, we know that

$$\lim_{h \to 0} f_m = f(t) = \lim_{h \to 0} f_M.$$

The fraction $\dfrac{F(t + h) - F(t)}{h}$ is thus seen in (3) to be caught between two quantities both of which approach $f(t)$ as h approaches zero. It, too, must therefore approach $f(t)$ as a limit. We conclude that

$$F'(t) = \lim_{h \to 0} \frac{F(t + h) - F(t)}{h} = f(t),$$

and the proof of the Fundamental Theorem is completed.

Before reaping the computational rewards of this theorem, we give a concrete example to emphasize precisely what the theorem says.

EXAMPLE 1. If $F(t) = \int_0^t \frac{1}{x^2 + 1}\,dx$, find $F'(1)$, $F'(2)$, and $F'(x)$. The integrand in this example is the continuous function f defined by $f(x) = \frac{1}{x^2 + 1}$. In this case, therefore, the interval I can be taken to be the whole real line. By the Fundamental Theorem,

$$F'(t) = f(t) = \frac{1}{t^2 + 1}.$$

In particular,

$$F'(1) = \frac{1}{1^2 + 1} = \frac{1}{2},$$

$$F'(2) = \frac{1}{2^2 + 1} = \frac{1}{5},$$

and, in general,

$$F'(x) = \frac{1}{x^2 + 1}, \qquad \text{for every real number } x.$$

As the preceding example illustrates, the conclusion of the Fundamental Theorem can equally well be written

$$F'(x) = f(x), \qquad \text{for every } x \text{ in } I.$$

We used the letter t in the statement of the theorem simply to avoid confusion with the dummy variable x which appears in the integral. We might just as well have written

$$\textit{If } F(x) = \int_a^x f(t)\,dt, \textit{ then } F'(x) = f(x),$$

or, perhaps better yet,

$$\textit{If } F(x) = \int_a^x f, \textit{ then } F'(x) = f(x).$$

The important thing to remember is that the derivative of F at any point in the given interval is equal to the value of the integrand f at that same point.

We now consider the implications of the Fundamental Theorem. By an **antiderivative** of a function f is meant any differentiable function F with the property that $F'(x) = f(x)$ for every x in the domain of f. Similarly, we shall say that a function F is an **antiderivative of f on an interval I** if $F'(x) = f(x)$ for every x in I. Thus the function $\frac{x^3}{3}$ is an antiderivative of x^2 because

$$\frac{d}{dx}\left(\frac{x^3}{3}\right) = \frac{3x^2}{3} = x^2.$$

Of course, $\frac{x^3}{3}$ is an antiderivative of x^2 on any interval we care to name. The Fundamental Theorem states that if f is continuous at every point of I, then the function F defined by

$$F(x) = \int_a^x f$$

is an antiderivative of f on I.

If a function f has one antiderivative F, then it has infinitely many because, for every constant c,

$$(F + c)' = F' + c' = F' + 0 = f.$$

Conversely, we have proved that any two functions which have the same derivative differ by a constant [see Theorem (5.4), page 114]. Hence, if $F' = f$, then the set of all antiderivatives of f is the set of all functions $F + c$ for every real number c. Combining these facts, we obtain

(5.3) COROLLARY OF THE FUNDAMENTAL THEOREM. *Let f be a function which is continuous at every x in some interval I. Then f has an antiderivative on I. Furthermore, if F is any antiderivative whatever of f on I, then, for any a and b in I,*

$$\int_a^b f(x)\,dx = F(b) - F(a).$$

This theorem is the computation tool which we have been seeking. Before giving the proof, which is easy, let us see how it works.

EXAMPLE 2. Evaluate the definite integrals

$$\text{(a)} \quad \int_0^2 x^2\,dx,$$

$$\text{(b)} \quad \int_1^4 (5 - x)\,dx.$$

Both integrands are obviously continuous functions. As already observed, the function F defined by $F(x) = \frac{x^3}{3}$ is an antiderivative of x^2. Hence, by Theorem (5.3)

$$\int_0^2 x^2\,dx = F(2) - F(0)$$

$$= \frac{2^3}{3} - \frac{0^3}{3} = \frac{8}{3}.$$

Similarly, we can see by inspection that the function G defined by $G(x) = 5x - \frac{x^2}{2}$ is an antiderivative of $5 - x$, since

$$\frac{d}{dx}\left(5x - \frac{x^2}{2}\right) = 5 - x.$$

It therefore follows by Theorem (5.3) that

$$\begin{aligned}\int_1^4 (5 - x)\,dx &= G(4) - G(1)\\ &= \left(5 \cdot 4 - \frac{4^2}{2}\right) - \left(5 \cdot 1 - \frac{1^2}{2}\right)\\ &= 12 - 4\tfrac{1}{2} = 7\tfrac{1}{2}.\end{aligned}$$

These are the two integrals whose values were computed in Section 3 by finding the limits of upper and lower sums. The difference in the magnitude of the computations there and here should render unnecessary any comments on the significance of the results of the present section.

Proof of Theorem (5.3). The assertion that f has an antiderivative on I is verified by the Fundamental Theorem. Let G be the antiderivative defined by

$$G(x) = \int_a^x f, \qquad \text{for every } x \text{ in } I. \tag{4}$$

Suppose now that F is an arbitrary antiderivative of f on I. Then

$$G'(x) = f(x) = F'(x), \qquad \text{for every } x \text{ in } I.$$

It follows by Theorem (5.4), page 114, that on the interval I the two functions G and F differ by a constant. That is, there exists a real number c such that

$$G(x) = F(x) + c, \qquad \text{for every } x \text{ in } I.$$

Substituting $x = b$ in equation (4), we obtain

$$G(b) = \int_a^b f.$$

Moreover, we know that $G(a) = \int_a^a f = 0$. Hence

$$\begin{aligned}\int_a^b f(x)\,dx = \int_a^b f &= G(b) - G(a)\\ &= [F(b) + c] - [F(a) + c]\\ &= F(b) - F(a),\end{aligned}$$

and the proof of (5.3) is complete.

The following is an extremely useful notational device. For any real-valued function F of one variable, we abbreviate $F(b) - F(a)$ by $F(x)|_a^b$. If F is an antiderivative of the continuous function f on some interval containing the numbers a and b, then we may write the value of the definite integral as

$$\int_a^b f(x)\,dx = F(x)\Big|_a^b.$$

The advantage of this notation is that the order of writing is the same as the logical order in which the problem is done. Thus one first writes the antiderivative, and then indicates the numbers at which it is to be evaluated. As a result, the whole problem can frequently be done in a single line. For example,

$$\int_0^2 x^2\,dx = \frac{x^3}{3}\Big|_0^2 = \frac{2^3}{3} - \frac{0^3}{3} = \frac{8}{3}.$$

EXAMPLE 3. Evaluate the definite integral $\int_{-1}^1 (y^5 - 3y^2 + 2)\,dy$.

Note that we get the same answer whether the dummy variable of integration is y, x, or anything else. The integral is the function f defined by $f(y) = y^5 - 3y^2 + 2$. An antiderivative of y^5 is easily seen to be $\frac{y^6}{6}$, an antiderivative of $3y^2$ is y^3, and an antiderivative of 2 is obviously $2y$. Hence $\frac{y^6}{6} - y^3 + 2y$ is an antiderivative of f. We conclude that

$$\begin{aligned}\int_{-1}^1 (y^5 - 3y^2 + 2)\,dy &= \left(\frac{y^6}{6} - y^3 + 2y\right)\Big|_{-1}^1 \\ &= \left(\frac{1^6}{6} - 1^3 + 2\cdot 1\right) - \left(\frac{(-1)^6}{6} - (-1)^3 + 2(-1)\right) \\ &= (\tfrac{1}{6} - 1 + 2) - (\tfrac{1}{6} + 1 - 2) \\ &= 2.\end{aligned}$$

PROBLEMS

1. Verify that $\frac{1}{r+1}x^{r+1}$ is an antiderivative of x^r, if r is any rational number except -1.

2. Find an antiderivative of each of the following functions.

(a) $f(x) = x^7$

(b) $f(x) = x^3 + \dfrac{1}{x^3}$

(c) $f(y) = 7y^{1/5}$

(d) $f(t) = 5t^4 + 3t^2 + 1$

(e) $g(x) = (3x + 1)^2$

(f) $f(x) = \dfrac{2x}{(x^2 + 1)^2}$.

3. Evaluate each of the following definite integrals by finding an antiderivative and using Theorem (5.3).

(a) $\displaystyle\int_0^1 3x^2\,dx$

(b) $\displaystyle\int_0^1 (4x^3 + 3x^2 + 2x + 1)\,dx$

(c) $\displaystyle\int_1^3 (5x - 1)\,dx$

(d) $\displaystyle\int_1^3 (5t - 1)\,dt$

(e) $\displaystyle\int_1^2 \left(x^2 + \frac{1}{x^2}\right) dx$

(f) $\displaystyle\int_3^2 x^{1/3}\,dx$

(g) $\displaystyle\int_{-2}^0 y^{1/5}\,dy$

(h) $\displaystyle\int_1^2 \left(\frac{2}{x^3} + \frac{1}{x^2} + 2\right) dx$

(i) $\displaystyle\int_{-1}^1 (y^2 - y + 1)\,dy$

(j) $\displaystyle\int_6^0 (x^3 - 9x^2 + 16x)\,dx$

(k) $\displaystyle\int_3^5 (2x - 1)^2\,dx$

(l) $\displaystyle\int_3^x (6t^2 - 4t + 2)\,dt$

(m) $\displaystyle\int_0^t (x^2 + 3x - 1)\,dx$

(n) $\displaystyle\int_0^{x^2} s^3\,ds$

(o) $\displaystyle\int_a^b dx$

(p) $\displaystyle\int_x^{3x} (4t - 1)\,dt$.

4. Let n be a positive integer.

(a) Evaluate $\int_a^b x^n\,dx$.

(b) Evaluate $\displaystyle\int_a^b \frac{1}{x^n}\,dx$ provided (i) $n \neq 1$, and (ii) a and b are either both positive or both negative.

(c) In (b), what is the reason for proviso (i)? For proviso (ii)?

5. Let F be an antiderivative of f and G an antiderivative of g.

(a) Prove that $F + G$ is an antiderivative of $f + g$.

(b) For any constant k, prove that kF is an antiderivative of kf.

6. What is the domain of the function F defined by $F(t) = \int_1^t \frac{1}{x}\,dx$?

7. Let $F(t) = \int_0^t (6x^2 - 4x + 1)\,dx$.

   (a) Using just the Fundamental Theorem and without evaluating F, find $F'(t)$, $F'(-1)$, $F'(2)$, and $F'(x)$.

   (b) Find $F(t)$ as a polynomial in t by finding a polynomial which is an antiderivative of $6x^2 - 4x + 1$.

   (c) Differentiate the answer in (b), and thereby check (a).

8. Let $G(x) = \int_1^x \left(t + \frac{1}{t^2}\right) dt$, for $x > 0$.

   (a) Using just the Fundamental Theorem, find $G'(x)$ and $G'(2)$.

   (b) Evaluate $G(x)$ as a rational function of x by finding an antiderivative of $t + \frac{1}{t^2}$.

   (c) Take the derivative of $G(x)$ as found in (b) and thereby check (a).

9. (a) Evaluate $F(t) = \int_0^{t^2} (3x^2 + 1)\,dx$.

   (b) Find $F'(t)$ and $F'(2)$ by taking the derivative of the answer to (a).

   (c) Find $F'(t)$ directly using just the Fundamental Theorem and the Chain Rule.

10. If f is continuous and g is differentiable and if $F(t) = \int_a^{g(t)} f(x)\,dx$, use the Fundamental Theorem and the Chain Rule to show that $F'(t) = f(g(t))g'(t)$.

11. In each of the following integrals evaluate $F'(t)$. Do not attempt to first find an antiderivative.

    (a) $F(t) = \int_0^t \sqrt{1 + x^3}\,dx$.

    (b) $F(t) = \int_t^1 \frac{1}{1 + x^2}\,dx$.

    (c) $F(t) = \int_0^{2t+1} \frac{1}{1 + x^2}\,dx$.

    (d) $F(t) = \int_t^{t^2} \frac{1}{x^2 + x + 1}\,dx$. $\left(\textit{Hint: } \int_t^{t^2} f = \int_t^1 f + \int_1^{t^2} f.\right)$

12. Is there anything wrong with the computation

$$\int_{-1}^1 \frac{1}{x^2}\,dx = -\frac{1}{x}\Big|_{-1}^1 = (-1) - (1) = -2?$$

    If so, what?

13. In each of the following, find the area of the subset P of the xy-plane bounded by the curve $y = f(x)$, the x-axis, and the lines $x = a$ and $x = b$. Sketch the curve and the subset P.

(a) $f(x) = x^2 + 1$, $a = -1$, and $b = 2$.

(b) $f(x) = x^2 + 2x$, $a = 0$, and $b = 2$.

(c) $f(x) = \frac{1}{2}x + 1$, $a = 2$, and $b = 4$.

(d) $f(x) = x^3$, $a = 0$, and $b = 2$.

(e) $f(x) = x^3 - 2x^2 + x$, $a = 0$, and $b = 2$.

14. Find the derivative of the function F defined by $F(x) = \displaystyle\int_0^x \frac{1}{t^2+1}\,dt$. Sketch the graph of F using the techniques of curve sketching discussed in Section 1 of Chapter 2. Label any maximum, minimum, or critical points and any points of inflection. What is the domain of F? $\left(\text{Do not attempt to find an explicit antiderivative of } \dfrac{1}{t^2+1}.\right)$

6. Indefinite Integrals. In this section we shall study the problem of finding a function given its derivative. The topic is a large one, and the present treatment is only an introduction. Many techniques for finding a function whose derivative is known have been developed, and some of these will be studied in Chapter 7.

Recall that an antiderivative of a function f is any differentiable function F with the property that $F'(x) = f(x)$ for every x in the domain of f. An antiderivative of f is also called an **indefinite integral** of f and is denoted by $\int f(x)\,dx$. If $F' = f$, we write

$$\int f(x)\,dx = F(x) + c.$$

Since the most we know about $\int f(x)\,dx$ and $F(x)$ is that they have the same derivative $f(x)$, they may very well differ by a nonzero constant. If the constant c is omitted, there is a very real possibility of making an error, since a particular indefinite integral may not be the one which is the solution to the problem at hand.

EXAMPLE 1. At every point $(x, f(x))$ on the graph of a given function f, there is a tangent line with slope equal to x^2. If the graph passes through the point (3, 2), find f. The solution is based on the fact that the slope of the tangent line is given by the derivative. Hence $f'(x) = x^2$. One function with this derivative is $\dfrac{x^3}{3}$, and so

$$f(x) = \frac{x^3}{3} + c, \tag{1}$$

for some constant c. We also write

$$\int x^2\,dx = \frac{x^3}{3} + c.$$

Since the point (3, 2) lies on the graph, we know that $f(3) = 2$. Thus

$$2 = f(3) = \frac{3^3}{3} + c = 9 + c,$$

whence $c = -7$, and we conclude that

$$f(x) = \frac{x^3}{3} - 7.$$

Omission of the c in equation (1) would have lead to the incorrect answer $f(x) = \frac{x^3}{3}$.

The reason for calling an antiderivative of a function f an indefinite integral and for denoting it by $\int f(x)\,dx$ is its close connection with the definite integral. Let f be continuous on an interval containing a and b. Since $\frac{d}{dx}\int f(x)\,dx = f(x)$, we obtain the formula

$$\boxed{\int_a^b f(x)\,dx = \int f(x)\,dx\Big|_a^b} \tag{2}$$

by applying Corollary (5.3) of the Fundamental Theorem of Calculus. The value of $\int f(x)\,dx|_a^b$ is the same for any two indefinite integrals of f and there is therefore no need to include the constant c in applications of equation (2). For example, even though

$$\int (2x + 1)\,dx = x^2 + x + c,$$

for an arbitrary real number c, we may write

$$\int_0^2 (2x + 1)\,dx = \int (2x + 1)\,dx\Big|_0^2 = (x^2 + x)\Big|_0^2 = 6.$$

The integration techniques that we shall consider are expressed in formulas for finding indefinite integrals. The first four of these, (6.1), (6.2), (6.3), and (6.4), have already been used in computing definite integrals. We write them down only to make them explicit. They are:

(6.1) $$\int [f(x) + g(x)]\, dx = \int f(x)\, dx + \int g(x)\, dx.$$

(6.2) $$\int kf(x)\, dx = k \int f(x)\, dx, \textit{ for every constant } k.$$

(6.3) $$\int dx = x + c.$$

(6.4) $$\int x^r\, dx = \frac{x^{r+1}}{r+1} + c, \textit{ where } r \textit{ is a rational number different from } -1.$$

Since an indefinite integral is determined only to within an additive constant, (6.1) and (6.2) are open to a possible (but unlikely) false interpretation. The precise statement of (6.1) is: *If F is an indefinite integral of f and if G is an indefinite integral of g, then $F + G$ is an indefinite integral of $f + g$.* The proof takes one line:

$$(F + G)' = F' + G' = f + g.$$

On the other hand, if F, G, and H are three arbitrary indefinite integrals of f, g, and $f + g$, respectively, we certainly cannot infer that $H = F + G$. All we know is that $H' = F' + G'$, whence $H = F + G + c$. Similarly, (6.2) should be read: *If F is an indefinite integral of f, then kF is an indefinite integral of kf.* The proof:

$$(kF)' = kF' = kf.$$

The proof of (6.4) is the equation

$$\frac{d}{dx}\left(\frac{x^{r+1}}{r+1} + c\right) = x^r,$$

and (6.3) is simply the special case of (6.4) obtained by setting $r = 0$. Note that each of these four integration formulas is the inverse of one of the basic rules for differentiation derived in Section 7 of Chapter 1.

EXAMPLE 2. Evaluate the following three integrals:

(i) $$\int \left(2x^2 + \frac{2}{x^2}\right) dx,$$

(ii) $$\int (y^3 + 2y^2 + 2y + 1)\, dy,$$

(iii) $$\int_1^5 (s^{2/3} + 1)\, ds.$$

Computation of the indefinite integrals follows directly from (6.1), (6.2), (6.3), and (6.4). Thus

$$\begin{aligned}\int\left(2x^2+\frac{2}{x^2}\right)dx &= 2\int x^2\,dx + 2\int x^{-2}\,dx\\ &= 2\frac{x^3}{3}+2\frac{x^{-1}}{-1}+c\\ &= \frac{2}{3}x^3-\frac{2}{x}+c.\end{aligned}$$

Since separately we would write $2\int x^2\,dx = \frac{2}{3}x^3 + c$ and also $2\int x^{-2}\,dx = -\frac{2}{x} + c$, one might think that two constants of integration should appear in the sum, i.e., that the answer should have been written

$$\int\left(2x^2+\frac{2}{x^2}\right)dx = \frac{2}{3}x^3 + c_1 - \frac{2}{x} + c_2.$$

Although this last equation is not false, it is unnecessarily complicated and also misleading. If F is one indefinite integral of a function, the specification of any other requires the specification of one additional number, not two. Remember that, for a given F, the set of all functions $F + c$ such that c is an arbitrary real number is identical to the set of all functions $F + c_1 + c_2$ such that c_1 and c_2 are arbitrary real numbers. The sum of two arbitrary constants is still an arbitrary constant.

To do (ii), one must realize that the sum rule (6.1) implies an analogous rule for integrating the sum of three functions, or four, or any finite number. We get

$$\begin{aligned}\int(y^3+2y^2+2y+1)\,dy &= \int y^3\,dy + 2\int y^2\,dy + 2\int y\,dy + \int dy\\ &= \frac{y^4}{4}+2\frac{y^3}{3}+2\frac{y^2}{2}+y+c\\ &= \tfrac{1}{4}y^4+\tfrac{2}{3}y^3+y^2+y+c.\end{aligned}$$

Finally, to evaluate (iii), we combine the above rules of integration with equation (2) to obtain

$$\begin{aligned}\int_1^5 (s^{2/3}+1)\,ds &= \left(\frac{s^{5/3}}{5/3}+s\right)\Bigg|_1^5\\ &= [\tfrac{3}{5}(5)^{5/3}+5]-[\tfrac{3}{5}(1)^{5/3}+1]\\ &= 3(5)^{2/3}+\tfrac{17}{5}.\end{aligned}$$

The Chain Rule provides an extremely useful technique for computing integrals. Suppose that F is an antiderivative of f and that g is a differentiable function. According to the Chain Rule,

$$[F(g)]' = F'(g)g'.$$

Since $F' = f$, we conclude that

$$[F(g)]' = F'(g)g' = f(g)g';$$

i.e., *the composition $F(g)$ is an antiderivative, or indefinite integral, of $f(g)g'$.* Thus we have proved:

(6.5) *If F is any indefinite integral of f, then*

$$\int f(g(x))g'(x)\,dx = F(g(x)) + c.$$

This formula tells us that we can integrate a function of the form $f(g(x))g'(x)$ provided we know how to integrate $f(x)$.

EXAMPLE 3. Compute $\int\sqrt{x^3 + x + 1}\,(3x^2 + 1)\,dx$. The integrand is the product of two functions. The first factor, $\sqrt{x^3 + x + 1}$, is the composition of $x^3 + x + 1$ with the square root, and we know how to integrate $\sqrt{x}$. The second factor is $3x^2 + 1$, which is the derivative of $x^3 + x + 1$. Hence (6.5) is applicable. We have

$$g(x) = x^3 + x + 1,$$
$$f(x) = \sqrt{x}.$$

Since

$$\int \sqrt{x}\,dx = \int x^{1/2}\,dx = \tfrac{2}{3}x^{3/2} + c,$$

we take $F(x) = \frac{2}{3}x^{3/2}$. According to (6.5), the answer to the problem is

$$F(g(x)) + c = \tfrac{2}{3}(x^3 + x + 1)^{3/2} + c.$$

That is,

$$\int \sqrt{x^3 + x + 1}\,(3x^2 + 1)\,dx = \tfrac{2}{3}(x^3 + x + 1)^{3/2} + c.$$

We can check this answer by taking its derivative. We obtain

$$\frac{d}{dx}[\tfrac{2}{3}(x^3 + x + 1)^{3/2} + c] = (x^3 + x + 1)^{1/2}(3x^2 + 1),$$

which is the original integrand.

EXAMPLE 4. Evaluate $\int(x^2 + 1)^5 x\,dx$. It is possible to do this problem by first expanding $(x^2 + 1)^5$ by the Binomial Theorem, but formula (6.5) makes this unnecessary. Again, the integrand is the product of two functions. The first is $(x^2 + 1)^5$, which is the composition $f(g(x))$ of the two functions $g(x) = x^2 + 1$ and $f(x) = x^5$. The latter we know how to integrate:

$$\int x^5\,dx = \frac{x^6}{6} + c,$$

so we take $F(x) = \dfrac{x^6}{6}$. The second factor in the integrand is x, which is not equal to $g'(x) = 2x$, but is a constant multiple of it. This is just as good because of the general rule $\int kf(x)\,dx = k\int f(x)\,dx$. In this case, we may write

$$\begin{aligned}\int (x^2 + 1)^5 x\,dx &= \tfrac{2}{2}\int (x^2 + 1)^5 x\,dx\\ &= \tfrac{1}{2}\int (x^2 + 1)^5 (2x)\,dx\\ &= \tfrac{1}{2}\int f(g(x))g'(x)\,dx\\ &= \tfrac{1}{2}F(g(x)) + c.\end{aligned}$$

Since $F(x) = \dfrac{x^6}{6}$, we have $F(g(x)) = \dfrac{(x^2 + 1)^6}{6}$ and so

$$\begin{aligned}\int (x^2 + 1)^5 x\,dx &= \left(\frac{1}{2}\right)\left[\frac{(x^2 + 1)^6}{6}\right] + c\\ &= \frac{(x^2 + 1)^6}{12} + c.\end{aligned}$$

The derivative of the indefinite integral should be the function which was integrated, i.e., the integrand. Checking, we get

$$\frac{d}{dx}\left[\frac{(x^2 + 1)^6}{12} + c\right] = \tfrac{6}{12}(x^2 + 1)^5 2x = (x^2 + 1)^5 x.$$

To summarize: Formula (6.5) is applicable if the integrand is a product of two functions one of which is a composition $f(g(x))$ and the other of which is $g'(x)$ or possibly a constant multiple of $g'(x)$. With a little practice the reader should be able to recognize immediately, for example, that of the three integrals

$$\int \sqrt{x^2 + 2}\, dx,$$

$$\int \sqrt{x^2 + 2}\, x\, dx,$$

$$\int \sqrt{x^2 + 2}\, x^2\, dx,$$

only the middle one can be successfully attacked by this method.

Formula (6.5) implies an analogous fact about definite integrals. Called the **Change of Variable Theorem for Definite Integrals,** it is the following:

(6.6) THEOREM. *If both integrands are continuous functions on their respective intervals of integration, then*

$$\int_a^b f(g(x))g'(x)\, dx = \int_{g(a)}^{g(b)} f(y)\, dy.$$

Proof. Since the integrands are continuous, both integrals exist. Let F be an indefinite integral of f. By the definition of the definite integral,

$$\int_{g(a)}^{g(b)} f(y)\, dy = F(y)\Big|_{g(a)}^{g(b)} = F(g(b)) - F(g(a)).$$

By (6.5),

$$\int_a^b f(g(x))g'(x)\, dx = F(g(x))\Big|_a^b = F(g(b)) - F(g(a)).$$

This completes the proof.

EXAMPLE 5. Compute $\displaystyle\int_{-2}^{2} \frac{x+1}{\sqrt{x^2+2x+2}}\, dx.$ We first check that the integrand is continuous on the interval $[-2, 2]$. Since the minimum value of $x^2 + 2x + 2$ is 1, which is positive, we know that the denominator is never zero, and so the integral is defined. Set $g(x) = x^2 + 2x + 2$ and $f(y) = y^{-1/2}$. Then $g'(x) = 2x + 2$, $g(-2) = 2$, and $g(2) = 10$. Hence

$$\begin{aligned}
\int_{-2}^{2} \frac{x+1}{\sqrt{x^2+2x+2}}\, dx &= \frac{1}{2}\int_{-2}^{2} \frac{2x+2}{\sqrt{x^2+2x+2}}\, dx \\
&= \tfrac{1}{2}\int_{-2}^{2} f(g(x))g'(x)\, dx \\
&= \tfrac{1}{2}\int_{2}^{10} y^{-1/2}\, dy.
\end{aligned}$$

Since $\int y^{-1/2}\,dy = 2y^{1/2} + c$, we obtain

$$\tfrac{1}{2}\int_2^{10} y^{-1/2}\,dy = y^{1/2}\Big|_2^{10} = \sqrt{10} - \sqrt{2}.$$

We conclude that

$$\int_{-2}^{2} \frac{x+1}{\sqrt{x^2+2x+2}}\,dx = \sqrt{10} - \sqrt{2}.$$

The differential of a function F was defined in Section 6 of Chapter 2, and was shown to satisfy the basic equation $dF(x) = F'(x)\,dx$. If $F' = f$, we therefore obtain

$$dF(x) = F'(x)\,dx = f(x)\,dx. \tag{3}$$

In this section we have expressed the fact that F is an antiderivative of f by writing

$$\int f(x)\,dx = F(x) + c. \tag{4}$$

Equations (3) and (4) suggest that we interpret the symbol dx that appears to the right of the integral sign not merely as a piece of notation but as a differential. With this interpretation, the symbol $\int$ becomes a notation for the operation which is the inverse of taking differentials. Thus, for any differentiable function F, we define

$$\boxed{\int dF(x) = F(x) + c.} \tag{5}$$

If $f(x)$ is given and we find $F(x)$ such that $dF(x) = f(x)\,dx$, then

$$\int f(x)\,dx = \int dF(x) = F(x) + c.$$

If a function is denoted by a variable u, the definition (5) has the simple form $\int du = u + c$. Moreover, in terms of differentials, Theorem (6.5) also has the following simple form:

(6.7) *If $F' = f$ and if u is a differentiable function, then*

$$\int f(u)\,du = F(u) + c.$$

Proof. If $u = g(x)$, then $du = g'(x)\,dx$. The above equation therefore becomes

$$\int f(g(x))g'(x)\,dx = F(g(x)) + c,$$

and this is just (6.5). Another way of proving (6.7) is to start from the equation

$$dF(u) = F'(u)\,du, \qquad (6)$$

[see Theorem (6.1), Chapter 2]. From this follows $dF(u) = f(u)\,du$, and so

$$F(u) + c = \int dF(u) = \int f(u)\,du.$$

It is worth noting that Theorem (6.5) is simply an inverse statement of the Chain Rule. The Chain Rule was also the *raison d'être* behind equation (6). The differential is a handy device solely because this important theorem is true.

EXAMPLE 6. Evaluate the integrals

$$\text{(i)} \quad \int \sqrt{5x+2}\,dx,$$

$$\text{(ii)} \quad \int s(s^2-1)^{125}\,ds.$$

To do (i), set $u = 5x + 2$. Then $du = 5\,dx$, and so $dx = \frac{1}{5}\,du$. Hence

$$\int \sqrt{5x+2}\,dx = \frac{1}{5}\int \sqrt{u}\,du = \frac{1}{5}\,\frac{2}{3}u^{3/2} + c$$
$$= \frac{2}{15}(5x+2)^{3/2} + c.$$

Similarly, in (ii), let $u = s^2 - 1$. We get $du = 2s\,ds$ and

$$\int s(s^2-1)^{125}\,ds = \frac{1}{2}\int u^{125}\,du = \frac{1}{2}\,\frac{1}{126}u^{126} + c$$
$$= \frac{1}{252}(s^2-1)^{126} + c.$$

In each of these examples the reader should verify that the derivative of the answer gives back the original integrand.

Each of the integral formulas (6.1), (6.2), (6.3), and (6.4) can be written as a fact about the integral of certain differentials. Let u and v be differentiable functions and c an arbitrary constant. Then

(6.1′) $\int (du + dv) = \int du + \int dv.$

(6.2′) $\int k\,du = k \int du$, *for every constant k.*

(6.3′) $\int du = u + c.$

(6.4′) $\int u^r\,du = \frac{u^{r+1}}{r+1} + c$, *where r is a rational number and* $r \neq -1$.

PROBLEMS

1. Evaluate the following indefinite integrals.

(a) $\int (x^2 + x + 1)\,dx$

(b) $\int \left(3x^3 - \frac{1}{3x^3}\right) dx$

(c) $\int (6t^2 - 2t + 5)\,dt$

(d) $\int (2y + 1)(y - 3)\,dy$

(e) $\int (2x - 1)^{3/2}\,dx$

(f) $\int (3x^3 + 2)^5 x^2\,dx$

(g) $\int x\sqrt{a^2 - x^2}\,dx$

(h) $\int \frac{t + 2}{\sqrt{t^2 + 4t + 5}}\,dt$

(i) $\int s(s^3 + 3s^2 + 5)(s + 2)\,ds$

(j) $\int |x|\,dx.$

2. Among the following integrals identify those that can be evaluated using the techniques in this section. Evaluate them.

(a) $\int \left(x + \frac{1}{x}\right) dx$

(b) $\int \left(\sqrt{x} + \frac{1}{\sqrt{x}}\right) dx$

(c) $\int y^2(y^3 + 7)^4\,dy$

(d) $\int y(y^3 + 7)^4\,dy$

(e) $\int t\sqrt{t^3 - 1}\,dt$

(f) $\int \frac{x + 1}{x - 1}\,dx$

(g) $\int (3x^2 - 1)(x + 2)\,dx$

(h) $\int (s + 1)(s^2 + 2s - 3)^4\,ds$

(i) $\int \frac{x^2 - 1}{x + 1}\,dx$

(j) $\int \frac{y + 2}{y^2 + 1}\,dy$

(k) $\int \frac{x - 1}{(x + 1)^3}\,dx.$

3. The curve defined by $y = f(x)$ passes through the point $(1, 4)$. In addition, at each point $(x, f(x))$, the slope of the curve is $8x^3 + 2x$. Find $f(x)$.
4. The line tangent to the graph of the differentiable function f at each point $(x, f(x))$ has slope $3x^2 + 1$, and the graph passes through the point $(2, 9)$. Find $f(x)$.
5. If $f''(x) = 12x^2 + 2$ and the graph of $y = f(x)$ passes through $(0, -2)$ with a slope of 5, find $f(x)$.
6. Evaluate the following definite integrals.

(a) $\int_0^1 (3x^2 + 4x + 1)\,dx$ (f) $\int_0^2 \frac{1}{(x+1)^2}\,dx$

(b) $\int_{-1}^1 (2t^3 + t)\,dt$ (g) $\int_{-2}^2 \sqrt{4 - x^2}\,x\,dx$

(c) $\int_{-1}^1 (x^3 + 1)^{17}x^2\,dx$ (h) $\int_{-2}^2 (2|x| + 1)\,dx$

(d) $\int_{-1}^2 \frac{s+1}{\sqrt{s^2 + 2s + 3}}\,ds$ (i) $\int_0^1 t(t^3 + 3t^2 - 1)^3(t + 2)\,dt$

(e) $\int_1^3 \left(x^2 + \frac{1}{x}\right)^3\left(2x - \frac{1}{x^2}\right)dx$ (j) $\int_1^2 \frac{x^4 + 2x^3 - 2}{x^2}\,dx.$

7. If $f''(x) = 18x + 10$ and $f'(0) = 2$, find $f'(x)$. If, in addition, $f(0) = 1$, find $f(x)$.
8. (a) If $g''(x) = \sqrt{x}$ and $g'(1) = 0$ and $g(0) = \sqrt{2}$, find $g(x)$.

   (b) If $f'''(t) = 6$ and $f''(1) = 8$ and $f'(0) = 1$ and $f(1) = 4$, find $f(t)$.
9. If the slope of the curve $y = f(x)$ is equal to 6 at the point $(1, 4)$ and, more generally, equals $6x$ at $(x, f(x))$, what is the area bounded by the curve $y = f(x)$, the x-axis, and the lines $x = 1$ and $x = 3$?
10. Sketch the region bounded by the curves $y = \frac{1}{\sqrt{x+1}}$, $x = 0$, and $y = \frac{1}{2}$.

    Find its area.

7. Area between Curves. Let f be a function which is integrable over the interval $[a, b]$ and whose graph may cross the x-axis a finite number of times in the interval. It was shown in Section 4 that the relation between area and the definite integral is given by the formula

$$\int_a^b f(x)\,dx = area(P^+) - area(P^-), \tag{1}$$

where P^+ is the set of all points (x, y) such that $a \le x \le b$ and $0 \le y \le f(x)$, and P^- is the set of all points (x, y) such that $a \le x \le b$

and $f(x) \leq y \leq 0$. If we exclude their boundaries, then either or both of the regions P^+ and P^- may consist of several pieces. This possibility is illustrated in Figure 17, which shows the graph of a continuous function f

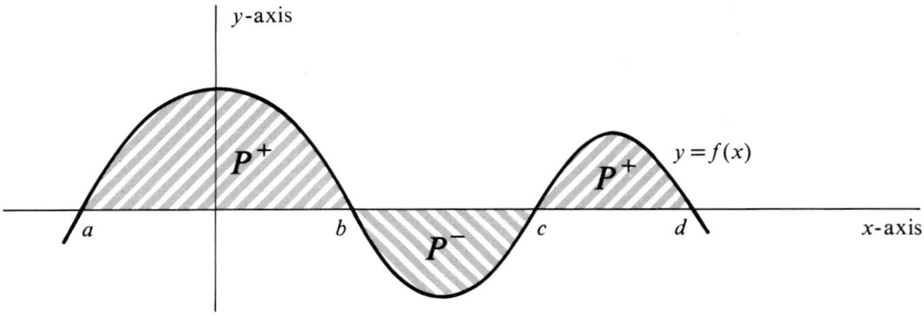

Figure 17

which crosses the x-axis at the four points a, b, c, and d. For the regions shown in the figure, we have

$$area(P^+) = \int_a^b f(x)\,dx + \int_c^d f(x)\,dx,$$

$$area(P^-) = -\int_b^c f(x)\,dx.$$

Since $area(P^+ \cup P^-) = area(P^+) + area(P^-)$, it follows that

$$area(P^+ \cup P^-) = \int_a^b f(x)\,dx - \int_b^c f(x)\,dx + \int_c^d f(x)\,dx.$$

Suppose in this example that F is an antiderivative of f on the interval $[a, d]$; i.e., we have $F'(x) = f(x)$ for every x in $[a, d]$. Then

$$\int_a^b f(x)\,dx = F(b) - F(a),$$

$$\int_b^c f(x)\,dx = F(c) - F(b),$$

$$\int_c^d f(x)\,dx = F(d) - F(c).$$

Hence the total area is given by

$$\begin{aligned} area(P^+ \cup P^-) &= [F(b) - F(a)] - [F(c) - F(b)] + [F(d) - F(c)] \\ &= -F(a) + 2F(b) - 2F(c) + F(d). \end{aligned}$$

In this section we shall extend (1) to compute the areas of regions with more complicated boundaries. The principal result here is the following:

(7.1) *Let the functions f and g be integrable over a closed interval $[a, b]$, and suppose that $g(x) \leq f(x)$ for every x in $[a, b]$. If R is the set of all points (x, y) such that $a \leq x \leq b$ and $g(x) \leq y \leq f(x)$, then*

$$area(R) = \int_a^b [f(x) - g(x)]\, dx.$$

Proof. The region R is illustrated in Figure 18(a). The only complicating feature of the proof is the fact that g or f, or both, may take on negative values in the interval $[a, b]$. If this happens [as it does in Figure 18(a)], we shall simply

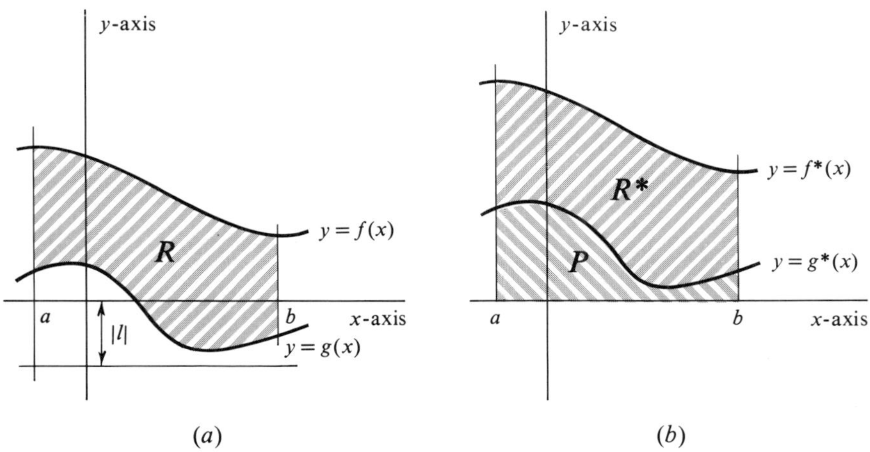

Figure 18

slide, or translate, the graphs of g and f upward so that the region between them lies above the x-axis and its area is unchanged. The sliding, or translation, is shown geometrically in Figure 18(b), and can be accomplished analytically as follows. Let l be a lower bound for the values of the function g on the interval $[a, b]$; i.e., we choose a number l such that $l \leq g(x)$ for every x in $[a, b]$. Such a number certainly exists since g is assumed to be integrable over $[a, b]$, and is therefore bounded on $[a, b]$. In addition, if it does happen that $g(x) \geq 0$ for every x in $[a, b]$, then we take $l = 0$. We now define functions f^* and g^* by

$$f^*(x) = f(x) + |l|,$$
$$g^*(x) = g(x) + |l|.$$

Their graphs and the region between them, which is denoted by R^*, is shown in Figure 18(b). It is obvious that

$$area(R) = area(R^*).$$

Let P be the set of points bounded by the graph of g^*, the x-axis, and the lines $x = a$ and $x = b$. By the integral formula for area, we know that

$$area(P \cup R^*) = \int_a^b f^*(x)\,dx,$$

$$area(P) = \int_a^b g^*(x)\,dx.$$

Since $area(R^*) = area(P \cup R^*) - area(P)$, we obtain

$$\begin{aligned} area(R) = area(R^*) &= \int_a^b f^*(x)\,dx - \int_a^b g^*(x)\,dx \\ &= \int_a^b [f^*(x) - g^*(x)]\,dx. \end{aligned}$$

However,

$$\begin{aligned} f^*(x) - g^*(x) &= [f(x) + |l|] - [g(x) + |l|] \\ &= f(x) - g(x). \end{aligned}$$

Hence

$$area(R) = \int_a^b [f(x) - g(x)]\,dx,$$

and the proof is complete.

It should be remarked that, rigorously speaking, the above proof omits some logical details. These occur in our use of the function *area*. To begin with, we have not in this book attempted to give a mathematical definition of the area of a set, although we have shown that if *area* does exist and satisfies certain simple properties, then the integral formula (1) is valid. Moreover, in the preceding paragraph we have tacitly assumed some of the properties of area which are obvious to the intuition, but logically would require proof. For example, we assumed that area is invariant under translation when we asserted that $area(R) = area(R^*)$. In the same way, our statement that $area(R^*) = area(P \cup R^*) - area(P)$ [see Figure 18(b)] was based solely on geometric intuition. There is nothing wrong with omitting these details, but it is important that a careful reader realize that the omissions are there.

EXAMPLE 1. Compute the area of the region R bounded by the graphs of the functions $f(x) = 2x - 1$ and $g(x) = -(2x - 1)^2$ and the lines $x = 1$ and $x = 2$. Since

$$g(x) \leq 0, \qquad \text{for every } x,$$

$$f(x) \geq 0, \qquad \text{if } 1 \leq x \leq 2,$$

it follows that $g(x) \leq f(x)$ on the interval [1, 2]. Both functions are continuous and hence integrable, and so (7.1) is applicable. We obtain

$$area(R) = \int_1^2 [f(x) - g(x)]\, dx.$$

Since $f(x) - g(x) = (2x - 1) + (2x - 1)^2 = 4x^2 - 2x$, the answer is

$$area(R) = \int_1^2 (4x^2 - 2x)\, dx = (\tfrac{4}{3}x^3 - x^2)\Big|_1^2$$
$$= (\tfrac{32}{3} - 4) - (\tfrac{4}{3} - 1) = \tfrac{19}{3}.$$

EXAMPLE 2. Find the area of the region R lying between the lines $x = -1$ and $x = 2$ and between the curves $y = x^2$ and $y = x^3$. We first sketch the

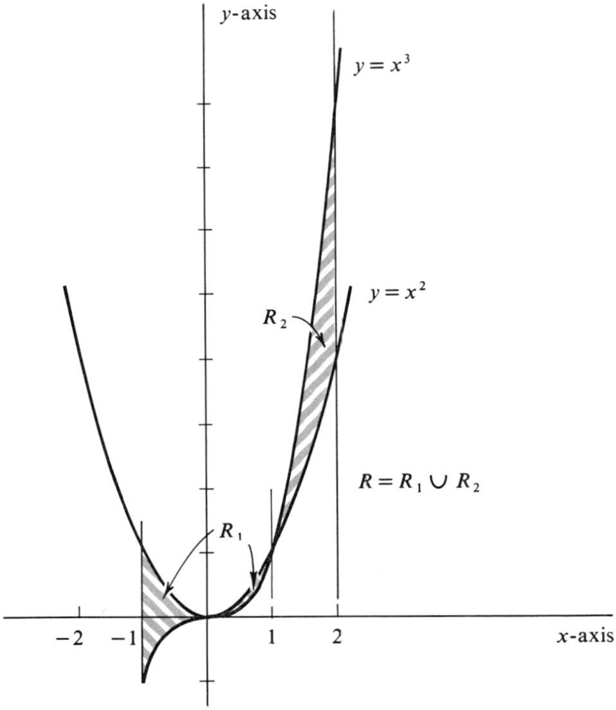

Figure 19

lines and curves in question and indicate the region R by shading (see Figure 19). Observe that

$$x^3 \leq x^2, \qquad \text{if } -1 \leq x \leq 1,$$
$$x^2 \leq x^3, \qquad \text{if } 1 \leq x \leq 2.$$

Writing R as the union of regions R_1 and R_2, as is done in the figure, we have, by (7.1),

$$area(R_1) = \int_{-1}^{1} (x^2 - x^3)\, dx,$$

$$area(R_2) = \int_{1}^{2} (x^3 - x^2)\, dx.$$

Hence

$$\begin{aligned} area(R) &= area(R_1) + area(R_2) \\ &= \int_{-1}^{1} (x^2 - x^3)\, dx + \int_{1}^{2} (x^3 - x^2)\, dx. \end{aligned}$$

The function $F(x) = \dfrac{x^3}{3} - \dfrac{x^4}{4}$ is clearly an indefinite integral of $x^2 - x^3$. Thus

$$\int_{-1}^{1} (x^2 - x^3)\, dx = F(x)\Big|_{-1}^{1} = F(1) - F(-1),$$

$$\begin{aligned} \int_{1}^{2} (x^3 - x^2)\, dx &= -\int_{1}^{2} (x^2 - x^3)\, dx \\ &= -F(x)\Big|_{1}^{2} = -F(2) + F(1). \end{aligned}$$

Finally, therefore,

$$\begin{aligned} area(R) &= F(1) - F(-1) - F(2) + F(1) \\ &= \tfrac{1}{12} + \tfrac{7}{12} + \tfrac{4}{3} + \tfrac{1}{12} = \tfrac{25}{12}. \end{aligned}$$

Because of symmetry, each of the integral formulas for area which we have developed has an obvious counterpart for functions of y. For example, if f is the function of y whose graph is drawn in Figure 20, then

$$\int_{c}^{d} f(y)\, dy = area(P^+) - area(P^-),$$

where P^+ and P^- are the regions indicated in the figure.

Sometimes a curve in the xy-plane can be defined as the graph of a function of y and not as a function of x. An example is the parabola defined by the equation $(y - 1)^2 = x - 1$ and illustrated in Figure 21. Although this equation cannot be solved uniquely for y in terms of x, it is easy to do the opposite. We get

$$x = (y - 1)^2 + 1 = y^2 - 2y + 2,$$

and so the curve is the graph of the function f defined by $f(y) = y^2 - 2y + 2$. The area of the region P bounded by the parabola, the two coordinate axes, and the horizontal line $y = 2$ is given by

$$\begin{aligned} area(P) &= \int_0^2 f(y)\,dy = \int_0^2 (y^2 - 2y + 2)\,dy \\ &= \left(\frac{y^3}{3} - y^2 + 2y\right)\Big|_0^2 = \frac{8}{3}. \end{aligned}$$

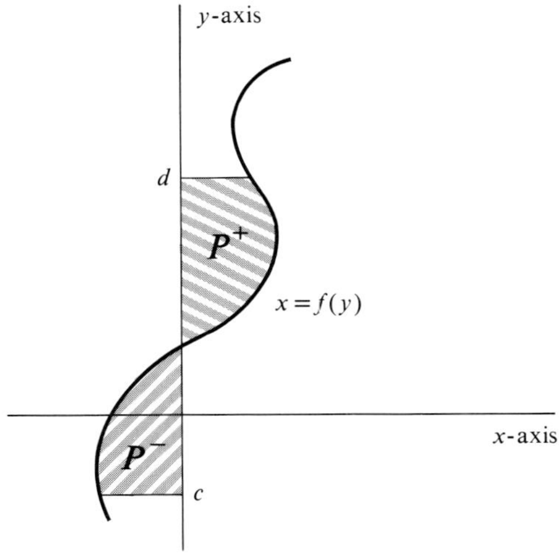

Figure 20

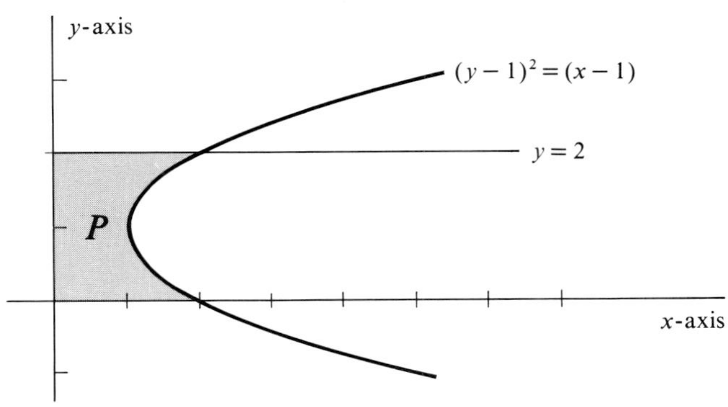

Figure 21

EXAMPLE 3. Express the area of the shaded region Q in Figure 22(a) as a sum of definite integrals. There are many different ways to do this. We shall begin by subdividing Q into four subregions Q_1, Q_2, Q_3, and Q_4 using the two vertical lines $x = a_1$ and $x = a_2$ as shown in Figure 22(b).

Consider for a moment the region Q_1 alone. Every horizontal line L which intersects Q_1 cuts its boundary in at most two points. Suppose that L crosses the y-axis at the point $(0, y)$. Moving along L from left to right, we denote the x-coordinate of the first point encountered on the boundary of Q_1 by $g_1(y)$. In this way a function g_1 is defined whose graph forms the "western boundary" of Q_1. In fact, Q_1 is the region between $y = b_1$ and

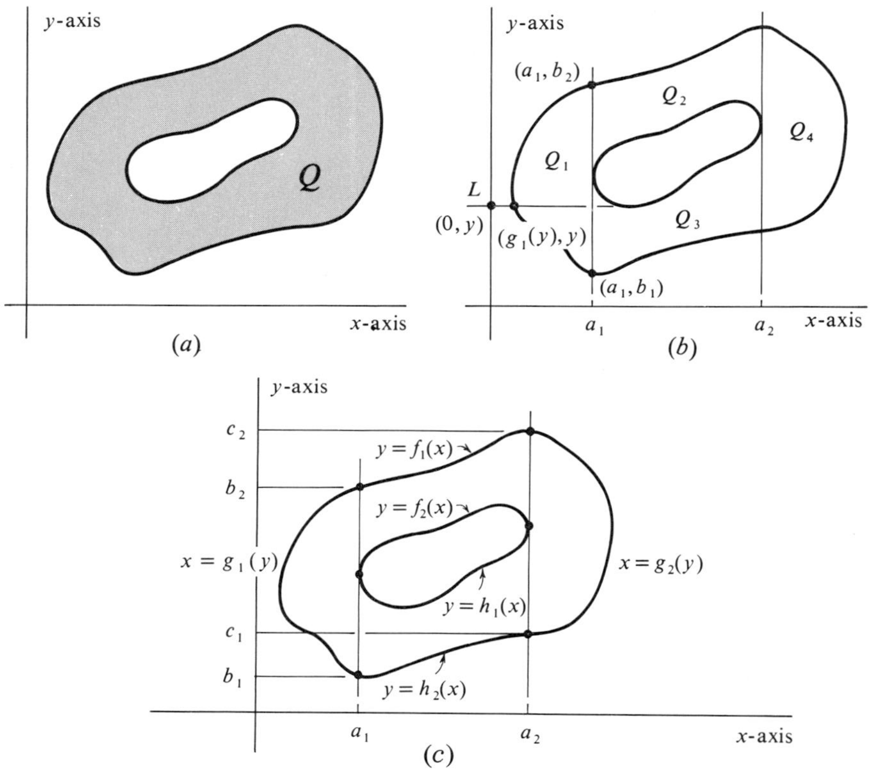

Figure 22

$y = b_2$ [see Figure 22(b)] and also between the curves $x = g_1(y)$ and $x = a_1$. By the counterpart of (7.1) for functions of y, we therefore obtain

$$area(Q_1) = \int_{b_1}^{b_2} [a_1 - g_1(y)]\, dy.$$

Turning next to the region Q_2, we see that the "northern boundary" is the graph of a function f_1, and the "southern boundary" is the graph of a function f_2. These functions are indicated in Figure 22(c). Hence

$$area(Q_2) = \int_{a_1}^{a_2} [f_1(x) - f_2(x)]\,dx.$$

Similarly, we have

$$area(Q_3) = \int_{a_1}^{a_2} [h_1(x) - h_2(x)]\,dx,$$

and also

$$area(Q_4) = \int_{c_1}^{c_2} [g_2(y) - a_2]\,dy.$$

Since $area(Q) = area(Q_1) + area(Q_2) + area(Q_3) + area(Q_4)$, we therefore obtain finally

$$\begin{aligned} area(Q) = &\int_{b_1}^{b_2} [a_1 - g_1(y)]\,dy + \int_{a_1}^{a_2} [f_1(x) - f_2(x)]\,dx \\ &+ \int_{a_1}^{a_2} [h_1(x) - h_2(x)]\,dx + \int_{c_1}^{c_2} [g_2(y) - a_2]\,dy. \end{aligned}$$

PROBLEMS

1. For each of the following functions and intervals: Compute $\int_a^b f(x)\,dx$; draw the graph of f; label the region P^+ on or above the x-axis which is bounded by the curve $y = f(x)$, the x-axis, and the lines $x = a$ and $x = b$; label the analogous region P^- on or below the x-axis; evaluate $area(P^+)$ and $area(P^-)$; and check formula (1) at the beginning of the section.

   (a) $f(x) = x - 1$, $a = 0$, and $b = 4$.

   (b) $f(x) = -x^2 + x + 2$, $a = 0$, and $b = 3$.

   (c) $f(x) = -x^2 + x + 2$, $a = -2$, and $b = 3$.

   (d) $f(x) = (x - 1)^3$, $a = 0$, and $b = 2$.

2. In each of the following find the area of the subset $P^+ \cup P^-$ of the xy-plane bounded by the curve $y = f(x)$, the x-axis, and the lines $x = a$ and $x = b$.

   (a) $f(x) = x^5$, $a = -1$, and $b = 1$.

   (b) $f(x) = x^2 - 3x + 2$, $a = 0$, and $b = 2$.

   (c) $f(x) = (x + 1)(x - 1)(x - 3)$, $a = 0$, and $b = 2$.

   (d) $f(x) = |x^2 - 1|$, $a = -2$, and $b = 2$.

3. Let f be a continuous function. Using areas, show that

   (a) If f is an odd function, then $\int_{-a}^{a} f(x)\,dx = 0$.

   (b) If f is an even function, then $\int_{-a}^{a} f(x)\,dx = 2\int_0^a f(x)\,dx$.

4. Prove (a) and (b) of Problem 3 analytically using the Fundamental Theorem of Calculus. [More specifically, use Theorems (5.2) and (5.3).]
5. Draw the region R bounded by the lines $x = 0$ and $x = 2$ and lying between the graphs of the functions $f(x) = x + 2$ and $g(x) = (x - 1)^2$. Find the area of R.
6. Draw the region Q lying to the right of the y-axis and bounded by the curves $x = 0$, $3y - x + 3 = 0$, and $3y + 3x^2 - 8x = 3$. Compute $area(Q)$.
7. Find the area of the subset R of the xy-plane lying between the lines $x = \frac{1}{2}$ and $x = 2$, and between the graphs of the functions $f(x) = \dfrac{1}{x^2}$ and $g(x) = x^2$. Draw the relevant lines and curves and indicate the region R.
8. Find the area of the region bounded by the two parabolas $y = -x^2 + x + 2$ and $y = x^2 - 2x$.
9. Draw the graphs of the equations $y = x^2$ and $y = 4$, and label the region R bounded by them.

   (a) Express the area of R as an integral with respect to x using (7.1). Evaluate the integral.

   (b) Similarly, express the area of R as an integral with respect to y using the counterpart of (7.1) for functions of y. Evaluate the integral and check the answer to (a).

10. (a) If $f(y) = -y^2 + y + 2$, sketch the region bounded by the curve $x = f(y)$, the y-axis, and the lines $y = 0$ and $y = 1$. Find its area.

   (b) Find the area bounded by the curve $x = -y^2 + y + 2$ and the y-axis.

   (c) The equation $x + y^2 = 4$ can be solved for x as a function of y, or for y as plus or minus a function of x. Sketch the region in the first quadrant bounded by the curve $x + y^2 = 4$, and find its area first by integrating a function of y and then by integrating a function of x.

11. The function f is continuous at every point of the interval $[a, b]$ and may cross the x-axis at a finite number of points in the interval. Let P^+ and P^- have their usual meaning [as in formula (1)].

   (a) Is $|f(x)|$ continuous at every point of $[a, b]$?

   (b) Show that

$$area(P^+ \cup P^-) = \int_a^b |f(x)|\, dx.$$

12. Find the area of the region bounded by the parabola $y = x^2$, the x-axis, and the line tangent to the parabola at the point $(2, 4)$. Do the problem

   (a) using x as the variable of integration.

   (b) using y as the variable of integration.

13. Do Problem 12 for the line tangent to the parabola at the general point (a, a^2).

14. Express the area of the ellipse $\frac{x^2}{a^2} + \frac{y^2}{b^2} = 1$ as a definite integral of a function of x, and as a definite integral of a function of y. (The resulting indefinite integrals cannot be evaluated with the theory so far developed.)
15. Find the area of the shaded region in Figure 23. The curves are parabolas. The inscribed square has area 4, and the circumscribed square has area 16.

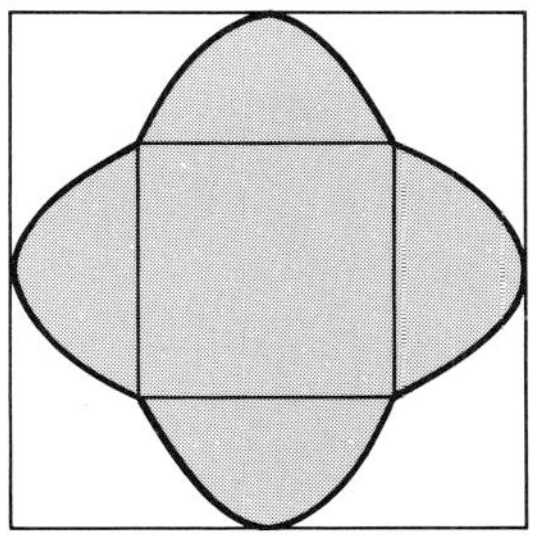

Figure 23

8. Integrals of Velocity and Acceleration. In this section we shall develop some of the integral formulas associated with velocity and acceleration. Among these is the formula for the distance traveled by an object, or particle, which moves with velocity $v(t)$ during the time interval from $t = a$ to $t = b$.

Consider a particle which, during some interval of time, moves along a straight line. We take the straight line to be a coordinate axis of real numbers, and denote the position of the particle on the line at time t by $s(t)$. Thus s is a real-valued function of a real variable with value $s(t)$ for every t in some interval. We shall assume that s is differentiable. In Section 3 of Chapter 2 the velocity v of the particle is defined to be the derivative of s. That is,

$$v(t) = s'(t).$$

Equivalent to this equation is the statement that s is an antiderivative of v. It follows that

$$\boxed{s(t) = \int v(t)\,dt + c.} \tag{1}$$

If we consider the motion of the particle from $t = a$ to $t = b$, and add the assumption that v is continuous on $[a, b]$, then, by Corollary (5.3) of the Fundamental Theorem of Calculus (see page 204), we obtain the formula

$$\boxed{s(b) - s(a) = \int_a^b v(t)\,dt.} \tag{2}$$

EXAMPLE 1. An object dropped from a cliff at time $t = 0$ falls with a velocity given by $v(t) = kt$. If we take the direction of increasing distance to be downward, and measure distance in feet and time in seconds, then $k = 32$ and $v(t)$ is in units of feet per second. How high is the cliff if the object hits the bottom 3 seconds after being dropped? The height equals the difference $s(3) - s(0)$. Hence

$$\begin{aligned} s(3) - s(0) &= \int_0^3 v(t)\,dt \\ &= \int_0^3 32t\,dt = 16t^2\Big|_0^3 \\ &= 144 \text{ feet.} \end{aligned}$$

Acceleration is defined in Section 3 of Chapter 2 to be the derivative of velocity. Thus

$$a(t) = v'(t),$$

and, as before, an equivalent statement is that velocity is an antiderivative of acceleration. We therefore have the formula

$$\boxed{v(t) = \int a(t)\,dt + c.} \tag{3}$$

EXAMPLE 2. A body in free fall under the earth's gravitational pull falls with a constant acceleration g, equal in magnitude to 32 feet per second per second. Suppose that at time $t = 0$ a ball is projected straight up from the ground with an initial velocity $v_0 = 256$ feet per second. Write formulas for the subsequent velocity $v(t)$ and distance from the ground $s(t)$. What is the maximum height the ball attains? In this example we shall choose the direction of increasing distance to be upward. As a result, the gravitational acceleration is negative, and the starting point of our calculations is the equation $a(t) = -32$. We have

$$\begin{aligned} v(t) &= \int a(t)\,dt + c = \int (-32)\,dt + c \\ &= -32t + c, \end{aligned}$$

whence

$$v(0) = -32 \cdot 0 + c = c.$$

The initial velocity is given as $v_0 = v(0) = 256$ feet per second. Hence $c = 256$ and

$$v(t) = -32t + 256,$$

which is one of the formulas asked for. The second integration yields

$$\begin{aligned} s(t) &= \int v(t)\,dt + c \\ &= \int (-32t + 256)\,dt + c \\ &= -16t^2 + 256t + c. \end{aligned}$$

The constant of integration c in the preceding equations has, of course, nothing to do with the one obtained from integrating $a(t)$. Here $s(0) = -16 \cdot 0^2 + 256 \cdot 0 + c = c$. Since the ball is at ground level when $t = 0$, we conclude that $0 = s(0) = c$, and so

$$s(t) = -16t^2 + 256t,$$

which is the second formula required. To find the maximum value of the function s, we compute its derivative and set it equal to zero:

$$s'(t) = v(t) = -32t + 256 = 0,$$

whence it follows that

$$t = \tfrac{256}{32} = 8 \text{ seconds.}$$

Since $s''(t) = v'(t) = a(t) = -32$, which is negative, we know that s has a local maximum when $t = 8$, and it is easy to see that this local maximum is an absolute maximum. Thus the maximum height attained by the ball is equal to

$$\begin{aligned} s(8) &= -16 \cdot 8^2 + 256 \cdot 8 \\ &= -2^{10} + 2^{11} \\ &= 1024 \text{ feet.} \end{aligned}$$

It is important to realize that the quantity $s(b) - s(a)$ in (2) does not necessarily equal the distance traveled by the object, or particle, during the time interval from $t = a$ to $t = b$. This is because the number $s(t)$ simply gives the position of the particle on the line at time t. Thus in the preceding example of the ball we showed that

$$s(t) = -16t^2 + 256t.$$

Substituting $t = 0$ and $t = 16$, respectively, we get

$$\begin{aligned} s(0) &= -16 \cdot 0^2 + 256 \cdot 0 = 0, \\ s(16) &= -16 \cdot 16^2 + 256 \cdot 16 = 0. \end{aligned}$$

The interpretation of these equations is clear: The ball left the ground at time $t = 0$, and 16 seconds later it had fallen back. However, an insect who accompanied the ball on its flight would probably not report to his admiring friends and relatives that the total distance traveled was

$$s(16) - s(0) = 0 - 0 = 0 \text{ feet.}$$

A similar situation is an automobile trip 10 miles down a road and back again. The distance traveled is presumably 20 miles, and not zero.

It is not hard to guess the proper modification of formula (2) to obtain a true distance formula. If we denote the distance traveled during the time interval from $t = a$ to $t = b$, where $a \leq b$, by $distance|_a^b$, then

$$\boxed{distance\Big|_a^b = \int_a^b |v(t)|\, dt.} \tag{4}$$

Logically, however, there is no way to prove (4), since we have not given a mathematical definition of $distance|_a^b$. As a result, we shall take (4) as a definition after checking that it corresponds to our intuition. First of all, if $v(t)$ does not change sign from $t = a$ to $t = b$, then

$$\int_a^b |v(t)|\, dt = \left| \int_a^b v(t)\, dt \right| \cdot$$

(See Problem 4 at the end of this section.) Hence, by formula (2),

$$\int_a^b |v(t)|\, dt = \left| \int_a^b v(t)\, dt \right| = |s(b) - s(a)|.$$

The assumption that $v(t)$ does not change sign means that the direction of the motion does not change. In this case, we would certainly expect $|s(b) - s(a)|$, which equals the distance on the real line between the initial position $s(a)$ and the final position $s(b)$, to be the total distance traveled.

Further motivation for the definition in formula (4) is obtained by going back to the definition of the definite integral. We assume that the function $|v(t)|$ is integrable over the interval $[a, b]$. Consider an arbitrary partition $\sigma = \{t_0, \ldots, t_n\}$ of $[a, b]$ such that

$$a = t_0 \leq t_1 \leq \cdots \leq t_{n-1} \leq t_n = b.$$

For each $i = 1, \ldots, n$, we denote by M_i the least upper bound of the set of all numbers $|v(t)|$, where t is in the subinterval $[t_{i-1}, t_i]$. Similarly, let m_i be the greatest lower bound. Thus the number M_i is the maximum speed of the particle in the subinterval $[t_{i-1}, t_i]$, and m_i is the minimum speed. Intuitively, therefore, the distance traveled by the particle during the subinterval of time $[t_{i-1}, t_i]$ must be less than or equal to $M_i(t_i - t_{i-1})$ and greater than or equal to $m_i(t_i - t_{i-1})$. Consequently, the upper and lower sums,

$$U_\sigma = \sum_{i=1}^{n} M_i(t_i - t_{i-1}),$$

$$L_\sigma = \sum_{i=1}^{n} m_i(t_i - t_{i-1}),$$

are upper and lower bounds, respectively, of the total distance traveled. But the function $|v(t)|$ has been assumed to be integrable over $[a, b]$. It follows that there exists one and only one number, $\int_a^b |v(t)|\,dt$, such that

$$L_\sigma \leq \int_a^b |v(t)|\,dt \leq U_\sigma,$$

for every partition σ of $[a, b]$. This justifies the adoption of formula (4) as the definition of $distance\big|_a^b$.

EXAMPLE 3. A particle moves on the x-axis, and its position at time t is given by $x(t) = 2t^3 - 21t^2 + 60t - 14$. Its velocity is the function v defined by

$$\begin{aligned} v(t) = x'(t) &= 6t^2 - 42t + 60 \\ &= 6(t^2 - 7t + 10) = 6(t - 2)(t - 5). \end{aligned}$$

Find the total distance traveled by the particle during the time interval $t = 0$ to $t = 6$. Clearly, $v(t) = 0$ when $t = 2$ and $t = 5$, and the sign of $v(t)$ is as indicated by

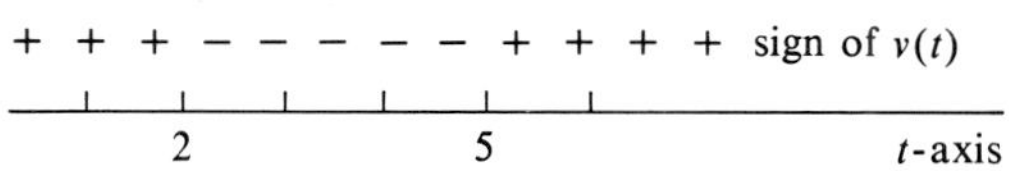

Hence

$$|v(t)| = \begin{cases} v(t) & \text{if } -\infty < t \leq 2, \\ -v(t) & \text{if } 2 \leq t \leq 5, \\ v(t) & \text{if } 5 \leq t < \infty. \end{cases}$$

Consequently,

$$\begin{aligned} distance\Big|_0^6 &= \int_0^6 |v(t)|\,dt \\ &= \int_0^2 v(t)\,dt - \int_2^5 v(t)\,dt + \int_5^6 v(t)\,dt. \end{aligned}$$

Since $x(t)$ is an antiderivative of $v(t)$,

$$\int_0^2 v(t)\,dt = x(2) - x(0),$$

$$\int_2^5 v(t)\,dt = x(5) - x(2),$$

$$\int_5^6 v(t)\,dt = x(6) - x(5),$$

and so

$$\int_0^6 |v(t)|\,dt = -x(0) + 2 \cdot x(2) - 2 \cdot x(5) + x(6).$$

Substitution in the equation for $x(t)$ yields $x(0) = -14$, $x(2) = 38$, $x(5) = 11$, and $x(6) = 22$. Hence

$$\begin{aligned} distance\Big|_0^6 &= \int_0^6 |v(t)|\,dt \\ &= 14 + 2 \cdot 38 - 2 \cdot 11 + 22 \\ &= 90. \end{aligned}$$

The motion described in this example is the same as in Example 1, page 105. By looking at Figure 16, Chapter 2, one can see that the distance traveled by the particle during the time interval [0, 6] agrees with the value just obtained.

The integral formulas derived in this section all presuppose motion along a straight line. The reason for this restriction is that the definition of velocity, and consequently of acceleration, has been based on the possibility of representing the position of the particle at time t by a real number $s(t)$ along a coordinate axis. A coordinate system on a line in turn is defined in terms of the distance between two points, and thus far the only measure of distance between points which we have is straight-line distance. In Chapter 10 we shall introduce the notion of arc length along a curve and shall study the notions of velocity and acceleration for curvilinear motion. At this point, however, it is worth noting that if we think of obtaining a curve by bending a coordinate axis without stretching it, and if $s(t)$ measures position on the curve in the obvious way, then formulas (1), (2), (3), and (4) still hold. Thus if the speedometer reading on a car is given by some nonnegative and integrable function f of time, then the distance traveled during the time interval $[a, b]$ is equal to $\int_a^b f(t)\,dt$ whether the road is straight or not.

We conclude this section with another type of problem illustrating the integration of rates of change.

EXAMPLE 4. Air is escaping from a spherical balloon so that its radius r is decreasing at the rate of 2 inches per minute. Find the rate of change of the volume V as a function of time if we are given that $r = 10$ inches when $t = 0$. What is the volume of the balloon when $t = 2$ minutes, and at what time will $V = 0$? The volume of a sphere is given by $V = \frac{4}{3}\pi r^3$, and we are given that

$$\frac{dr}{dt} = -2.$$

Hence

$$\begin{aligned} r &= \int \frac{dr}{dt}\,dt + c = \int (-2)\,dt + c \\ &= -2t + c. \end{aligned}$$

Setting $t = 0$, we see that the constant of integration c is the value of the radius at $t = 0$, namely, 10 inches. Hence

$$r = -2t + 10.$$

Since

$$\frac{dV}{dt} = \frac{d}{dt}\left(\frac{4}{3}\pi r^3\right) = \frac{4}{3}\pi \cdot 3r^2 \cdot \frac{dr}{dt},$$

we obtain for the answer to the first part of the problem

$$\frac{dV}{dt} = 4\pi(-2t + 10)^2(-2) = -32\pi(-t + 5)^2,$$

where t is measured in minutes and $\frac{dV}{dt}$ in cubic inches per minute. Clearly,

$$V(t) = \tfrac{4}{3}\pi[r(t)]^3 = \tfrac{4}{3}\pi(-2t + 10)^3.$$

The volume of the balloon when $t = 2$ minutes is therefore

$$V(2) = \tfrac{4}{3}\pi(-2 \cdot 2 + 10)^3 = \tfrac{4}{3}\pi 6^3 = 288\pi \text{ cubic inches,}$$

and the volume will be zero when

$$V(t) = \tfrac{4}{3}\pi(-2t + 10)^3 = 0,$$

i.e., when $t = 5$ minutes.

PROBLEMS

1. A straight highway connects towns A and B. A car starts at $t = 0$ from A and goes toward B with a velocity given by $v(t) = 60t - 12t^2$, measured in miles per hour. When the car arrives at B, it is slowing down and its speed is 48 miles per hour.
   (a) How far apart are the two towns?
   (b) What are the maximum and minimum speeds obtained during the trip? When are they reached, and at what distances?

2. A straight highway connects towns A and B. A car, initially stopped, starts at $t = 0$ from A and accelerates at 240 miles per hour per hour until reaching a speed of 60 miles per hour.

   (a) How long does this take, both in time and distance?

   Assume that the car travels at the constant speed of 60 miles per hour once it has reached that speed, and that it slows down to a stop at town B in the same way that it left A.

   (b) How far apart are A and B if the whole trip takes 5 hours?

3. A projectile is fired straight up with an initial velocity of 640 feet per second (see Example 2).

   (a) Find the velocity $v(t)$.

   (b) How far does the projectile travel during the first 10 seconds of its flight?

   (c) How far does the projectile go, and how many seconds after takeoff is this maximum height reached?

   (d) What is the total distance traveled by the projectile during the first 30 seconds of its flight?

   (e) What is its velocity when the projectile returns to the ground?

4. Let the function f be integrable over the interval $[a, b]$, and suppose that $f(x)$ does not change sign on the interval. Prove that

$$\int_a^b |f(x)|\, dx = \left| \int_a^b f(x)\, dx \right| \cdot$$

   (This is an easy problem. Consider separately the two cases: First, $f(x) \geq 0$ for every x in $[a, b]$, and second, $f(x) \leq 0$ for every x in $[a, b]$.)

5. A particle moves on the x-axis with velocity given by $v(t) = -4t + 20$.

   (a) In which direction is the particle moving at time $t = 0$?

   (b) Find $s(t)$, the position of the particle at time t, if its coordinate is -30 when $t = 1$.

   (c) Find the distance traveled by the particle during the time interval from $t = 0$ to $t = 4$.

   (d) Find the distance traveled by the particle during the time interval from $t = 0$ to $t = 8$.

   (e) When is $s(t) = 0$?

6. A particle moves on the y-axis with acceleration given by $a(t) = 6t - 2$. Denote its velocity and position at time t by $v(t)$ and $y(t)$, respectively. At time $t = 1$, the particle is at rest at the zero position.

   (a) Find $v(t)$ and $y(t)$.

   (b) How far does the particle move during the time interval from $t = 1$ to $t = 3$?

   (c) What is the distance traveled by the particle from $t = -1$ to $t = 2$?

7. A road borders a rectangular forest, and a car is driven around it. The car starts from rest at one corner and accelerates at 120 miles per hour per hour until it reaches the next corner 15 minutes later. The second side is 20 miles long and the car is driven along it at constant velocity equal to the final velocity reached on the first side. The car continues at this same speed on the third side. On the fourth side, however, the car slows down with a constant acceleration and comes to a stop at its original starting place. Find

   (a) the dimensions of the rectangle.

   (b) the acceleration on the fourth side.

   (c) the time taken for the whole trip.

8. Let the function f be integrable over the interval $[a, b]$. From the definition of integrability in Section 1, prove that $\int_a^b f$ is the only number such that

$$L_\sigma \leq \int_a^b f \leq U_\sigma,$$

for every partition σ of $[a, b]$.

9. A conical funnel of height 36 inches and base with radius 12 inches is initially filled with sand. At $t = 0$, the sand starts running out the bottom (apex of the cone) so that the volume V of sand remaining in the funnel is decreasing at the constant rate of 10 cubic inches per minute.

   (a) Find V as a function of time t, and determine how long it takes for all the sand to run out.

   (b) Assuming that the sand retains its original conical shape during the process, find the radius r of the base of the cone of sand as a function of t.

10. A particle moves on the parabola $y = x^2$, and its horizontal component of velocity is given by $x'(t) = \dfrac{1}{(t+1)^2}$, $t \geq 0$. At time $t = 0$ the particle is at the origin.

   (a) What are the x and y coordinates of the particle when $t = 1$? When $t = 3$?

   (b) As t increases without bound what happens to the particle?

11. At time $t = 0$, an object is dropped from an airplane which is moving horizontally with velocity v_0. As the object falls, its horizontal component of velocity $x'(t)$ is constant and equal to v_0. Its downward acceleration $y''(t)$ is constant and equal to $-g$. Measure the positive direction of x in the direction of motion of the airplane and the positive direction of y upward. Also assume that $x(0) = y(0) = 0$.

   (a) Find $x(t)$ and $y(t)$.

   (b) By eliminating t from the equations in (a), find the equation in terms of x and y in which the object falls. What is the name of the curve?

CHAPTER 5

Logarithms and Exponential Functions

1. The Natural Logarithm. If f is any real-valued function of a real variable which is continuous on some interval, we have seen that there exists an antiderivative, or indefinite integral, F such that $F'(x) = f(x)$ for every x in the interval. For some functions we have been able to write their antiderivatives explicitly. For example, if $f(x) = x^r$ and r is a rational number different from -1, then the general antiderivative is

$$\int x^r\,dx = \frac{x^{r+1}}{r+1} + c,$$

where c is an arbitrary constant. This formula is not applicable if $r = -1$. Nevertheless, $\frac{1}{x}$ is certainly continuous on the interval $(0, \infty)$, and therefore some function has it for a derivative. For every positive number a, we can define such a function by writing

$$F(x) = \int_a^x \frac{dt}{t}, \qquad \text{for every } x > 0.$$

For then by the Fundamental Theorem of Calculus [see Theorem (5.2), page 200],

$$F'(x) = \frac{1}{x}\cdot$$

To be specific, we choose $a = 1$ and select $\int_1^x \frac{dt}{t}$ for a particular antiderivative of $\frac{1}{x}$. As we shall see, this function is interesting enough to have a special name and a special notation. The notation we shall use is $\ln(x)$, or, more briefly, $\ln x$, and we shall save the name and reason for its choice until we have investigated its properties. For now, we define

$$\boxed{\ln x = \int_1^x \frac{dt}{t}, \qquad \text{for every } x > 0.}$$

The reason that x must be positive in this definition is that the function $\frac{1}{t}$ has a discontinuity at $t = 0$. If x is negative, the integral does not exist and $\ln x$ is not defined.

Geometrically, $\ln x$ is an area or the negative of an area. If $x > 1$, then $\ln x$ is the area of the region bounded by the hyperbola $y = \frac{1}{t}$, the t-axis, and the lines $t = 1$ and $t = x$ [see Figure 1(a)]. On the other hand, if $0 < x < 1$, we have

$$\ln x = \int_1^x \frac{dt}{t} = -\int_x^1 \frac{dt}{t},$$

and $\ln x$ is the negative of the area shown in Figure 1(b). Thus we have the following properties of the function ln.

$$\textbf{(1.1)} \qquad \begin{cases} \ln 1 = 0, & \\ \ln x > 0, & \text{if} \quad x > 1, \\ \ln x < 0, & \text{if} \quad 0 < x < 1. \end{cases}$$

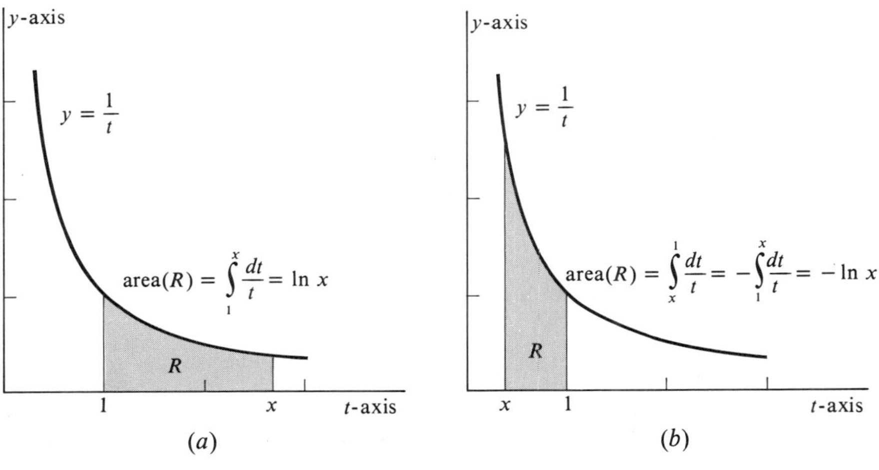

Figure 1

Another interesting property is obtained by taking the derivative $\frac{d}{dx} \ln kx$, where k is an arbitrary positive constant. By the Chain Rule,

$$\frac{d}{dx} \ln kx = \frac{1}{kx} \frac{d}{dx} kx = \frac{1}{kx} \cdot k = \frac{1}{x}.$$

Hence the functions $\ln kx$ and $\ln x$ have the same derivative. We know that two functions which have the same derivative over an interval, in this case all positive real numbers, differ by a constant. Hence

$$\ln kx - \ln x = c,$$

for some real number c and all positive x. We evaluate c by substituting a particular value for x. Since the value of ln 1 is known, we let $x = 1$, getting

$$\ln k - \ln 1 = c.$$

Since $\ln 1 = 0$, we know that $c = \ln k$. Hence $\ln kx - \ln x = \ln k$, or, equivalently,

$$\ln kx = \ln k + \ln x.$$

In deriving this equation we have used the fact that k is a constant and x a variable. Once we have derived it, and know that it is valid for every positive k and x, we can forget the distinction and write

(1.2) $\ln ab = \ln a + \ln b,$ *for all positive real numbers a and b.*

If a is positive, then $\frac{1}{a}$ is also positive and a substitution of $\frac{1}{a}$ for b in (1.2) gives the equation $\ln 1 = \ln a + \ln \frac{1}{a}$. Since $\ln 1 = 0$, we have

(1.3) $$\ln \frac{1}{a} = -\ln a, \qquad a > 0.$$

Applying (1.2) to the product of the two numbers a and $\frac{1}{b}$, and using (1.3), we obtain $\ln \frac{a}{b} = \ln\left(a \cdot \frac{1}{b}\right) = \ln a + \ln \frac{1}{b} = \ln a - \ln b$. That is,

(1.4) $$\ln \frac{a}{b} = \ln a - \ln b, \qquad a > 0 \textit{ and } b > 0.$$

In summary: The function ln applied to a product is equal to the sum of the values obtained when the function is applied to the factors. Applied to the quotient, the value of ln is the difference between the values of the function applied to the numerator and to the denominator.

Let r be any rational number, and consider the derivative $\frac{d}{dx} \ln x^r$. By the Chain Rule again,

$$\frac{d}{dx} \ln x^r = \frac{1}{x^r} \frac{d}{dx} x^r = \frac{1}{x^r} r x^{r-1} = r \frac{1}{x}.$$

Moreover,

$$\frac{d}{dx} r \ln x = r \frac{d}{dx} \ln x = r \frac{1}{x}.$$

Thus $\ln x^r$ and $r \ln x$ have the same derivative and so must differ by a constant:

$$\ln x^r - r \ln x = c.$$

Substitution of 1 for x tells us that $c = 0$, and therefore $\ln x^r = r \ln x$ for every positive number x. For uniformity in appearance with properties (1.2), (1.3), and (1.4), we set $x = a$ and obtain

(1.5) $\ln a^r = r \ln a$, *for every rational number r and every positive real number a.*

The properties we have derived for the function ln should be recognized by anyone who has been exposed to logarithms. However, ln is not the same as the function $\log_{10}$ or the function $\log_2$, which the student may have seen earlier. It has many of the same properties though and is called the **natural logarithm,** hence the abbreviation ln.

We next draw the graph of the function ln. Since $\frac{d}{dx} \ln x = \frac{1}{x}$ and since $\frac{1}{x}$ is positive for all x for which $\ln x$ is defined, the slope of the tangent line to the graph is always positive. From this we can see that $\ln x$ always increases as x increases; that is, *the function* ln *is strictly increasing.* To prove this analytically, we use the Mean Value Theorem, page 113. Suppose that $0 < a < b$. By the Mean Value Theorem there exists a number c such that $a < c < b$ and

$$\ln b - \ln a = (b - a)\left(\frac{d}{dx} \ln x\right)(c) = (b - a)\frac{1}{c} > 0.$$

Hence $\ln a < \ln b$, and the monotonicity of the natural logarithm is proved. Since $\frac{d^2}{dx^2} \ln x = -\frac{1}{x^2}$, the graph of $\ln x$ is concave downward for all x.

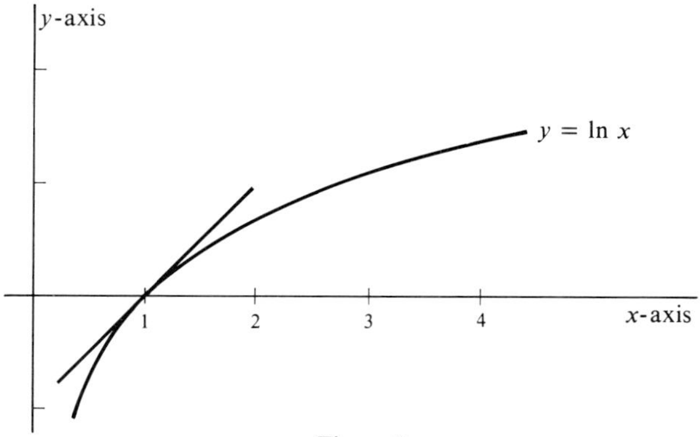

Figure 2

With this information and the fact that $\ln 1 = 0$, we can make a reasonable sketch of the graph. This appears in Figure 2. Note that the graph is steep and the values of the function negative for very small values of x. The values increase and the curve goes up to the right, passing through (1, 0) with a slope of 1. Since $\frac{d}{dx} \ln x = \frac{1}{x}$, the slope decreases as x increases. This raises the question of whether $\ln x$ becomes arbitrarily large as x increases without bound, or possibly tends toward some limiting value. However, it is obvious that the inequality $\frac{1}{t} \geq \frac{1}{2}$ is true for every real number in the closed interval [1, 2]. It follows from one of the fundamental properties of the integral [specifically, from (4.3) on page 191] that

$$\int_1^2 \frac{1}{t}\,dt \geq \int_1^2 \frac{1}{2}\,dt.$$

The left side of this inequality is equal to ln 2, and the right side to $\frac{1}{2}$. Thus we have proved that

$$\ln 2 \geq \tfrac{1}{2},$$

[a fact which can also be obtained geometrically by looking at Figure 1(a) and considering the area under the curve $y = \frac{1}{t}$ between $t = 1$ and $t = 2$]. Hence for every rational number r, we have $\ln 2^r = r \ln 2 \geq \frac{r}{2}$. By taking r large enough, we can make $\frac{r}{2}$ and, consequently $\ln 2^r$, as large as we like. We conclude that

$$\lim_{x\to\infty} \ln x = \infty.$$

EXAMPLE 1. If $\ln 2 = a$ and $\ln 3 = b$, evaluate each of the following in terms of a and b.

(a) $\ln 4$, (c) $\ln \frac{1}{6}$, (e) $\ln \frac{2}{9}$,

(b) $\ln 6$, (d) $\ln 24$, (f) $\ln 2^m 3^n$.

Using the various properties of logarithms developed above, we find

(a) $\ln 4 = \ln 2^2 = 2 \ln 2 = 2a$,

(b) $\ln 6 = \ln(2 \cdot 3) = \ln 2 + \ln 3 = a + b$,

(c) $\ln \frac{1}{6} = -\ln 6 = -(a + b) = -a - b$,

(d) $\ln 24 = \ln(8 \cdot 3) = \ln 8 + \ln 3 = \ln 2^3 + \ln 3 = 3 \ln 2 + \ln 3 = 3a + b$,

(e) $\ln \frac{2}{9} = \ln 2 - \ln 9 = \ln 2 - \ln 3^2 = \ln 2 - 2 \ln 3 = a - 2b$,

(f) $\ln 2^m 3^n = \ln 2^m + \ln 3^n = m \ln 2 + n \ln 3 = ma + nb$.

The derivative of the natural logarithm of a differentiable function is found by means of the Chain Rule. Thus *if* $F(x) = \ln(f(x))$, *then* $F'(x) = \ln'(f(x))f'(x) = \dfrac{1}{f(x)} \cdot f'(x) = \dfrac{f'(x)}{f(x)}$. If the variable u is used to denote a differentiable function of x, then the same result can be written

$$\frac{d}{dx} \ln u = \frac{1}{u} \cdot \frac{du}{dx}.$$

EXAMPLE 2. Find the derivatives of (a) $\ln(x^2 - 3)$, (b) $\ln\sqrt{4x + 7}$. Using the Chain Rule in (a), we obtain

$$\frac{d}{dx} \ln(x^2 - 3) = \frac{1}{x^2 - 3} \cdot 2x = \frac{2x}{x^2 - 3}.$$

Note that the original function, $\ln(x^2 - 3)$, is defined only for $|x| > \sqrt{3}$, although the function which is its derivative can be defined for all x except $\pm\sqrt{3}$. We can do (b) either by use of the Chain Rule directly, as

$$\frac{d}{dx} \ln \sqrt{4x + 7} = \frac{1}{\sqrt{4x + 7}} \cdot \tfrac{1}{2}(4x + 7)^{-1/2} \cdot 4 = \frac{2}{4x + 7},$$

or more simply by noting that $\ln \sqrt{4x + 7} = \frac{1}{2} \ln(4x + 7)$. Then

$$\frac{d}{dx} \ln\sqrt{4x + 7} = \frac{1}{2} \frac{d}{dx} \ln(4x + 7) = \frac{1}{2} \frac{1}{4x + 7} \cdot 4 = \frac{2}{4x + 7}.$$

The latter method would be much shorter for finding the derivative of $\ln(x^2 + 2)(x - 3)(x + 5)^3$. The expanded form $\ln(x^2 + 2) + \ln(x - 3) + 3 \ln(x + 5)$ is certainly simpler to differentiate.

Since the natural logarithm is defined only for positive numbers, it follows that the function $\ln x$ is an antiderivative of $\dfrac{1}{x}$ only for $x > 0$. It is natural to ask whether or not $\dfrac{1}{x}$, which is defined and continuous for all x except 0, has an antiderivative for $x < 0$. The answer is yes; $\ln(-x)$ is an antiderivative. Of course, if x is negative, then $-x$ is positive and so $\ln(-x)$ is defined. By the Chain Rule,

$$\frac{d}{dx} \ln(-x) = \frac{1}{(-x)} \frac{d}{dx}(-x) = \frac{1}{(-x)} \cdot (-1) = \frac{1}{x}.$$

This equation may be combined with the equation $\frac{d}{dx}\ln x = \frac{1}{x}$, which holds for all positive x, into the single equation

$$\frac{d}{dx}\ln|x| = \frac{1}{x}, \qquad \text{for all } x \text{ except } 0.$$

The corresponding formula for the indefinite integral is

(1.6)
$$\int \frac{dx}{x} = \ln|x| + c.$$

If f is a differentiable function, the Chain Rule implies that

$$\frac{d}{dx}\ln|f(x)| = \frac{1}{f(x)} \cdot f'(x) = \frac{f'(x)}{f(x)}.$$

Hence, we have the integration formula

(1.7)
$$\int \frac{f'(x)}{f(x)}\,dx = \ln|f(x)| + c.$$

EXAMPLE 3. Integrate (a) $\int \frac{dx}{3x}$, (b) $\int \frac{x+7}{x^2+14x+5}\,dx$. To do (a), we use (1.6) and the fact that the integral of a constant times a function is the constant times the integral of the function.

$$\int \frac{dx}{3x} = \frac{1}{3}\int \frac{dx}{x} = \frac{1}{3}\ln|x| + c.$$

For (b), formula (1.7) is applicable because the numerator is $\frac{1}{2}$ times the derivative of the denominator.

$$\int \frac{x+7}{x^2+14x+5}\,dx = \frac{1}{2}\int \frac{2x+14}{x^2+14x+5}\,dx = \tfrac{1}{2}\ln|x^2+14x+5| + c.$$

Alternative ways of writing the integration formula (1.7) are obtained by letting $u = f(x)$. We then get

$$\int \frac{1}{u}\frac{du}{dx}\,dx = \ln|u| + c.$$

Using the theory of differentials, we have $du = \frac{du}{dx}\,dx$, and the formula becomes

$$\int \frac{du}{u} = \ln|u| + c.$$

EXAMPLE 4. Compute the definite integral $\int_0^2 \frac{x^2dx}{x^3 - 17}$. Let $u = x^3 - 17$. Then $du = 3x^2dx$, or, equivalently, $x^2dx = \frac{du}{3}$. Hence

$$\frac{x^2\,dx}{x^3 - 17} = \frac{1}{u}\frac{du}{3} = \frac{1}{3}\frac{du}{u},$$

and so

$$\int \frac{x^2\,dx}{x^3 - 17} = \frac{1}{3}\int \frac{du}{u} = \frac{1}{3}\ln|u| + c$$
$$= \frac{1}{3}\ln|x^3 - 17| + c.$$

Finally, therefore,

$$\int_0^2 \frac{x^2\,dx}{x^3 - 17} = \frac{1}{3}\ln|x^3 - 17|\Big|_0^2 = \frac{1}{3}\ln|8 - 17| - \frac{1}{3}\ln|-17|$$
$$= \frac{1}{3}\ln 9 - \frac{1}{3}\ln 17 = \frac{1}{3}\ln\frac{9}{17}.$$

Note that, if we had neglected the absolute value, we would have encountered the undefined quantities $\ln(-9)$ and $\ln(-17)$.

Since $\log_{10}10 = 1$ and $\log_2 2 = 1$, it is reasonable to ask for that value of x for which $\ln x = 1$ and to call this number the base for natural logarithms. We know that such a number exists, since $\ln x$ increases as x increases and since the graph of $y = \ln x$ crosses the line $y = 1$. An accurate graph would show that the abscissa of the point where they cross is between 2.7 and 2.8, more accurately between 2.71 and 2.72, and more accurately still between 2.718 and 2.719. Although this number is irrational, we can find better and better decimal approximations to it. They start out 2.71828 . . . . This number, denoted by e, is therefore defined by the equation $\ln e = 1$.

The sequence of rational numbers $1 + 1, 1 + \frac{1}{2}, 1 + \frac{1}{3}, 1 + \frac{1}{4}, \ldots, 1 + \frac{1}{n}, \ldots$ approaches 1 as a limit. The chord connecting points $(1, 0)$ and $\left(1 + \frac{1}{n}, \ln\left(1 + \frac{1}{n}\right)\right)$ on the graph of $y = \ln x$ has a slope equal to

$$\frac{\ln\left(1 + \frac{1}{n}\right) - 0}{\left(1 + \frac{1}{n}\right) - 1} = n\ln\left(1 + \frac{1}{n}\right).$$

Hence, by (1.5), the slope is $\ln\left(1 + \frac{1}{n}\right)^n$. As $n \to \infty$, we have $\frac{1}{n} \to 0$, and the chord approaches the tangent to the curve at (1, 0), which has slope equal to 1. Hence

$$\lim_{n\to\infty} \ln\left(1 + \frac{1}{n}\right)^n = 1.$$

Furthermore, as can be seen from Figure 3, the slope of the chord increases

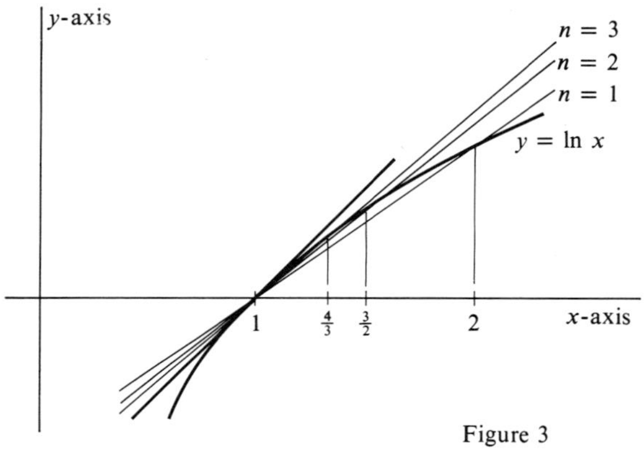

Figure 3

TABLE 1

n	$\left(1+\frac{1}{n}\right)^n$
1	2.
2	2.25
10	2.594
50	2.692
1000	2.717
e	2.71828 . . .

as n increases. Hence $\ln\left(1 + \frac{1}{n}\right)^n$ increases, and because ln is an increasing function, it follows that the numbers $\left(1 + \frac{1}{n}\right)^n$ also increase. Since

$$\ln\left(1 + \frac{1}{n}\right)^n < 1 = \ln e,$$

we know that

$$\left(1 + \frac{1}{n}\right)^n < e.$$

It follows immediately from the least upper bound property of the real numbers that a bounded sequence of increasing real numbers must have a limit. Hence

$$\lim_{n\to\infty} \left(1 + \frac{1}{n}\right)^n \text{ exists.}$$

Since ln is a continuous function, the limit of a natural logarithm approaches the natural logarithm of the limit, and so

$$1 = \lim_{n\to\infty} \ln\left(1 + \frac{1}{n}\right)^n = \ln\left[\lim_{n\to\infty}\left(1 + \frac{1}{n}\right)^n\right].$$

Since $1 = \ln e$ and since there is only one number whose logarithm is 1, we conclude that

$$\boxed{\lim_{n\to\infty}\left(1+\frac{1}{n}\right)^n = e.}$$

In Table 1 values of $\left(1+\frac{1}{n}\right)^n$ for several values of n are compared with the limiting value e.

PROBLEMS

1. Find the derivative with respect to x of each of the following functions.

(a) $\ln x^2$
(b) $\ln(7x+2)$
(c) $\ln\sqrt{(x-3)(x+4)}$
(d) $\ln(x^2-9x+3)$
(e) $\ln\dfrac{2}{x}$
(f) $\ln\sqrt[7]{7x^3}$
(g) $\displaystyle\int_1^{x^2+3}\frac{dt}{t}$
(h) $(\ln x)^3$
(i) $\ln(\ln x)$
(j) $\ln x\sqrt{x-1}$
(k) $\ln\dfrac{x-3}{x+1}$
(l) $\ln\dfrac{x^2-2x+4}{x^2+1}$
(m) $\ln\dfrac{x}{2-x^2}$.

2. Show that $|x^2+2x+3| = x^2+2x+3$ and hence that $\ln(x^2+2x+3)$ is defined for all x.
3. If $\ln 2 = p$, $\ln 3 = q$, and $\ln 5 = r$, write each of the following as a function of p, q, and r.

(a) $\ln 10$
(b) $\ln 0.25$
(c) $\ln 6000$
(d) $\ln 0.625$
(e) $\ln 0.03$
(f) $\ln 1728$.

4. Integrate each of the following.

(a) $\displaystyle\int\frac{2\,dx}{x}$
(b) $\displaystyle\int\frac{x\,dx}{x^2+1}$
(c) $\displaystyle\int\frac{(x-3)\,dx}{x^2-6x+2}$
(d) $\displaystyle\int x\sqrt{x^2+3}\,dx$

(e) $\displaystyle\int \frac{13x^2}{x^3 - 6}\,dx$

(f) $\displaystyle\int \left(\frac{2}{x+1} + \frac{3}{2x-1} - \frac{4}{3x+5}\right) dx$

(g) $\displaystyle\int \frac{x\,dx}{x-1}$

(h) $\displaystyle\int \frac{2\,dx}{(2x-1)^2}$

(i) $\displaystyle\int \frac{1}{x} \ln x\,dx.$

5. Sketch the graph of each of the following functions. Label all extreme points and points of inflection, and give the values of x at which these occur. Classify each extreme point as a local maximum or minimum.

(a) $f(x) = \ln(x + 4)$

(b) $f(x) = \ln x^2$

(c) $f(x) = \ln(1 + x^2)$

(d) $f(x) = x^2 - 4 \ln x^2$

(e) $f(x) = x^2 + 4 \ln x^2.$

6. Compute the following definite integrals.

(a) $\displaystyle\int_0^1 \frac{dt}{t+1}$

(b) $\displaystyle\int_1^3 \frac{dx}{5x-3}$

(c) $\displaystyle\int_0^1 \frac{s-2}{s^2 - 4s + 4}\,ds$

(d) $\displaystyle\int_2^x \frac{t\,dt}{t^2 - 3}$

(e) $\displaystyle\int_1^{x^2} \frac{dt}{t^2}$

(f) $\displaystyle\int_{-5}^{-1} \frac{dx}{x}.$

7. In each of the following examples, find the area of the region bounded by the graph of $y = f(x)$, the x-axis, and the two vertical lines whose equations are given.

(a) $f(x) = \dfrac{1}{x}$, $x = 3$ and $x = 7$.

(b) $f(x) = \dfrac{\ln x}{x}$, $x = 1$ and $x = 4$.

(c) $f(x) = \dfrac{1}{x^2}$, $x = 9$ and $x = 11$.

8. If $F(x) = \displaystyle\int_2^x \frac{t\,dt}{t^2 - 3}$, what is the domain of the function F?

9. (a) For what values of x is $\displaystyle\int_{-2}^x \frac{dt}{t-1}$ defined?

(b) What, if anything, is wrong with the computation

$$\int_{-2}^{2} \frac{dx}{x-1} = \ln|x - 1| \Big|_{-2}^{2} = \ln 1 - \ln 3 = -\ln 3?$$

10. Use the appropriate form of L'Hôpital's Rule to evaluate each of the following limits.

(a) $\lim_{x\to\infty} \frac{\ln x}{x}$

(b) $\lim_{x\to\infty} \frac{\ln x}{x^n}$, n an arbitrary positive integer

(c) $\lim_{x\to 0+} x \ln x$

(d) $\lim_{x\to 1} \frac{\ln x}{1 - x}$

(e) $\lim_{x\to 0+} x^n \ln x$, n an arbitrary positive integer.

2. The Exponential Function. According to the definition in Chapter 1, page 13, a function f is a set of ordered pairs with the property that the first member x of any pair (x, y) in the set determines the second member y, which we call $f(x)$. For example, the set $\{(1, 3), (2, 5), (4, 3)\}$ is a function. Suppose, for a given function f, we consider the set of all pairs (y, x) such that (x, y) is in f. This set may or may not be a function. In our example it is the set $\{(3, 1), (5, 2), (3, 4)\}$, which is not a function because it contains both $(3, 1)$ and $(3, 4)$. Hence the first member does not determine the second uniquely. However, if the new set is a function, it is called the **inverse function** of f and is denoted by f^{-1}.

If f is the function defined by $f(x) = 7x$, for every real number x, then the inverse function f^{-1} exists and is defined by $f^{-1}(x) = \frac{1}{7}x$. On the other hand, the function f defined by

$$f(x) = x^2, \qquad -\infty < x < \infty,$$

does not have an inverse. The reason is that if $f(x) = x^2 = 4$, for example, we have no way of knowing whether $x = 2$ or $x = -2$. If we restrict the domain of f to the set of nonnegative real numbers, i.e.,

$$f(x) = x^2, \qquad 0 \leq x,$$

then f^{-1} exists and is the function defined by

$$f^{-1}(x) = \sqrt{x}, \qquad 0 \leq x.$$

The following three elementary properties of functions and their inverses should be noted:

(2.1) $$(f^{-1})^{-1} = f.$$

(2.2) $y = f(x)$ *if and only if* $x = f^{-1}(y)$.

(2.3) *Two functions f and g are inverses of each other if and only if*

$$f(g(x)) = x, \qquad \textit{for every } x \textit{ in the domain of } g,$$

$$g(f(x)) = x, \qquad \textit{for every } x \textit{ in the domain of } f.$$

The first of these, $(f^{-1})^{-1} = f$, follows at once from the definition of f^{-1}.

To prove (2.2), suppose first that $y = f(x)$. This means that the ordered pair (x, y) belongs to the set f. Hence the pair (y, x) belongs to f^{-1}. But this says that $x = f^{-1}(y)$. To prove the converse, suppose that $x = f^{-1}(y)$. Hence, by what we have just proved, we know that $y = (f^{-1})^{-1}(x)$. Since $(f^{-1})^{-1} = f$, we obtain $y = f(x)$.

The importance of (2.3) is that it provides a simple criterion for deciding when two functions are inverses of each other. For example, if $f(x) = 3x + 5$ and $g(x) = \dfrac{x - 5}{3}$, then

$$f(g(x)) = f\left(\frac{x-5}{3}\right) = 3\left(\frac{x-5}{3}\right) + 5 = x,$$

$$g(f(x)) = g(3x + 5) = \frac{(3x + 5) - 5}{3} = x,$$

and we may therefore conclude that $g = f^{-1}$ and that $g^{-1} = f$. The proof is completely straightforward (and rather tedious), and we omit it.

The only way a function f can fail to have an inverse is if there exist at least two elements a and b in its domain for which $a \neq b$ and $f(a) = f(b)$. Suppose that f is a strictly increasing function and that a and b are two distinct numbers in its domain. Then either $a < b$ or $b < a$. If $a < b$, then $f(a) < f(b)$; and if $b < a$, then $f(b) < f(a)$. Thus it is impossible that $f(a) = f(b)$, and we conclude that f^{-1} exists. A similar argument applies to any function that is strictly decreasing, and so

(2.4) *If f is strictly increasing or strictly decreasing, then f has an inverse.*

The natural logarithm has been shown to be a strictly increasing function, and hence must have an inverse. Let us call this function $\exp(x)$ and justify the name after we have looked at its properties. Thus, by (2.2),

$$\boxed{y = \exp(x) \qquad \textit{if and only if } x = \ln y.}$$

Since $\ln y$ is defined only for positive y, we see immediately that $\exp(x)$ is always positive. Furthermore, for any real number x, there exists a number y such that $x = \ln y$ because the graph of the equation $x = \ln y$ crosses every vertical line. Hence $\exp(x)$ is defined for every real number x. Finally, since $0 = \ln 1$, we obtain $1 = \exp(0)$. Summarizing, we have

(2.5)
$$\begin{cases} \exp(x) > 0, & -\infty < x < \infty, \\ \exp(0) = 1. \end{cases}$$

To develop the algebraic properties of the function exp, let $p = \exp(a)$ and $q = \exp(b)$. Then $a = \ln p$ and $b = \ln q$. Hence

$$a + b = \ln p + \ln q = \ln pq,$$

and therefore $pq = \exp(a + b)$. Replacing p and q in this last equation, we obtain the important fact that

$$\exp(a) \cdot \exp(b) = \exp(a + b). \tag{2.6}$$

Similarly, $-a = -\ln p = \ln \frac{1}{p}$, and therefore $\frac{1}{p} = \exp(-a)$. Replacing p in this equation, we find that

$$\frac{1}{\exp(a)} = \exp(-a). \tag{2.7}$$

If we apply (2.6) to the sum of the two numbers a and $-b$, we get

$$\exp(a - b) = \exp[a + (-b)] = \exp(a) \cdot \exp(-b).$$

Since

$$\exp(-b) = \frac{1}{\exp(b)},$$

it follows that

$$\frac{\exp(a)}{\exp(b)} = \exp(a - b). \tag{2.8}$$

If a function f has an inverse, then the ordered pairs which comprise it become the ordered pairs of f^{-1} when we interchange the order of each pair. Hence the graph of f^{-1} may be obtained from the graph of f by interchanging

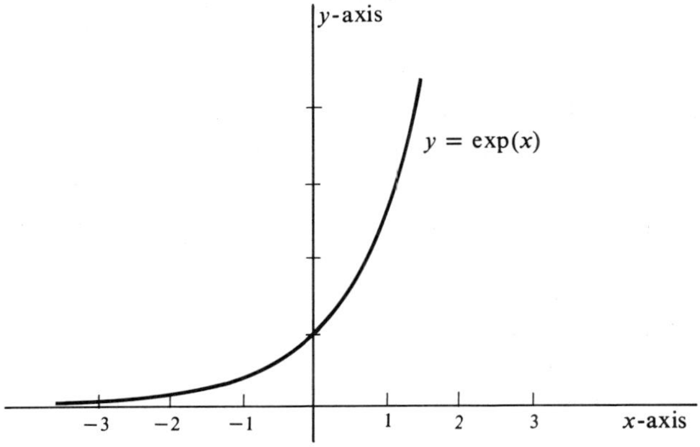

Figure 4

x and y. This is equivalent to a reflection across the line $y = x$. The graph of $y = \exp(x)$ is thus the reflection of the graph of $y = \ln x$ across this line, and it is shown in Figure 4. The curve passes through (0, 1) and gets closer and closer to the x-axis as x decreases without bound. As x increases without bound, so also does $\exp(x)$.

The graph of $y = \exp(x)$ is a smooth curve, and it is obvious geometrically that there is a tangent line at every point. We conclude that exp is a differentiable function. [For an analytic proof of this fact, see Theorem (3.4) in the next section.] We may compute the derivative by implicit differentiation. Consider the equation $y = \exp(x)$ and its equivalent equation $\ln y = x$. The latter implicitly defines $\exp(x)$ since $\ln[\exp(x)] = x$. Hence from $\ln y = x$ we obtain

$$\frac{d}{dx}\ln y = \frac{d}{dx}x,$$

$$\frac{1}{y}\frac{dy}{dx} = 1,$$

$$\frac{dy}{dx} = y.$$

Replacing y by $\exp(x)$, we get

(2.9) $$\boxed{\frac{d}{dx}\exp(x) = \exp(x).}$$

Alternatively we may write: *If* $f(x) = \exp(x)$, *then* $f'(x) = \exp(x)$ *for every real number* x.

Thus $\exp(x)$ is a most remarkable function, one which is equal to its own derivative. It is quite easy to show that this function and constant multiples of it are the only functions with this property. [See Problem 8 at the end of this section.]

Another property of exp arises as a consequence of the logarithmic equation, $\ln p^r = r \ln p$, for positive p and rational r. If we again let $a = \ln p$, then $p = \exp(a)$ and $ar = r \ln p = \ln p^r$. Equivalent to $ar = \ln p^r$ is $p^r = \exp(ar)$. Since p stood for $\exp(a)$, we conclude that

(2.10) $$[\exp(a)]^r = \exp(ar), \qquad \textit{for all } a \textit{ and rational } r.$$

We know that $\exp(0) = 1$ and may wonder what $\exp(1)$ is. If we set $y = \exp(1)$, then $1 = \ln y$. But e is the only number with a natural logarithm equal to 1. Hence, $\exp(1) = e$.

As an application of (2.10), we see that if x is a rational number, then

$$e^x = [\exp(1)]^x = \exp(1 \cdot x) = \exp(x).$$

What about e^x if x is real but not rational? Note that if a is any positive number, we have previously encountered a raised only to rational powers. For example, at this point we have no idea what $3^{\sqrt{2}}$ even means. For the number e, however, there is a very natural way to define e^x for all real numbers x. We have just shown that if x is rational, then $e^x = \exp(x)$.

Since the function exp has every real number in its domain, we shall *define* e^x to be $\exp(x)$ if x is not rational. Hence

$$e^x = \exp(x), \qquad \textit{for every real number } x.$$

We shall define a^x, for an arbitrary positive number a and real x, in Section 4, and then $3^{\sqrt{2}}$ will make sense.

The reason for the term "exp," which is an abbreviation for "exponential," should now be clear. The function agrees with our previous idea of an exponential for rational values, and it has the following properties translated from those derived earlier in this section:

$$\begin{cases} e^0 = 1, \\ e^a \cdot e^b = e^{a+b}, \\ e^{-a} = \dfrac{1}{e^a}, \\ \dfrac{e^a}{e^b} = e^{a-b}. \end{cases} \tag{1}$$

Thus e^x obeys the familiar laws of exponents. In addition,

$$\boxed{\frac{d}{dx} e^x = e^x.} \tag{2}$$

If u is a differentiable function of x, then the Chain Rule implies that $\dfrac{d}{dx} e^u = \left(\dfrac{d}{du} e^u\right) \dfrac{du}{dx}$. Since $\dfrac{d}{du} e^u = e^u$, we have

$$\frac{d}{dx} e^u = e^u \frac{du}{dx}. \tag{3}$$

EXAMPLE 1. Find the derivative of each of the composite functions

$$\text{(a)}\ e^{2x+7}, \qquad \text{(b)}\ e^{x^2} \qquad \text{(c)}\ \frac{1}{e^{2x^3}}.$$

For the first, we have $\dfrac{d}{dx} e^{2x+7} = e^{2x+7} \dfrac{d}{dx}(2x + 7) = 2e^{2x+7}$. For (b), $\dfrac{d}{dx} e^{x^2} = e^{x^2} \dfrac{d}{dx} x^2 = 2xe^{x^2}$. For (c), we write $\dfrac{1}{e^{2x^3}}$ as e^{-2x^3} and differentiate to get $e^{-2x^3}(-6x^2)$ or $-\dfrac{6x^2}{e^{2x^3}}$.

Since e^x is its own derivative, it is also its own indefinite integral. Hence

(2.11) $$\int e^x\,dx = e^x + c.$$

More generally, from (3) [or, equivalently, from (6.5), page 213] we obtain the integral formula

(2.12) $$\int e^u \frac{du}{dx}\,dx = e^u + c.$$

EXAMPLE 2. Compute the following integrals:

$$\text{(a)}\ \int e^{5x}\,dx, \qquad \text{(b)}\ \int x^2 e^{x^3+7}\,dx, \qquad \text{(c)}\ \int \frac{e^x\,dx}{4e^x - 3}.$$

To solve (a), we let $u = 5x$. Then $\dfrac{du}{dx} = 5$, and multiplying by $\frac{5}{5}$, we have

$$\int e^{5x}\,dx = \frac{1}{5}\int e^u 5\,dx = \frac{1}{5}\int e^u \frac{du}{dx}\,dx$$
$$= \tfrac{1}{5}e^u + c = \tfrac{1}{5}e^{5x} + c.$$

In the same manner we solve (b) by letting $u = x^3 + 7$. Then $\dfrac{du}{dx} = 3x^2$. Omitting the explicit substitution of the variable u, we write

$$\int x^2 e^{x^3+7}\,dx = \tfrac{1}{3}\int e^{x^3+7} 3x^2\,dx = \tfrac{1}{3}e^{x^3+7} + c.$$

Part (c) combines logarithms and exponentials. Since e^x lacks only a factor of 4 to be the derivative of $4e^x - 3$, we can supply the 4, and the integral is then of the form $\displaystyle\int \frac{1}{u}\frac{du}{dx}\,dx$, which is equal to $\ln|u| + c$. Hence

$$\int \frac{e^x\,dx}{4e^x - 3} = \frac{1}{4}\int \frac{1}{4e^x - 3}4e^x\,dx = \tfrac{1}{4}\ln|4e^x - 3| + c.$$

Each of these answers can be checked by differentiating to see if we get back the original integrand. For example, in (b) we get

$$\frac{d}{dx}(\tfrac{1}{3}e^{x^3+7} + c) = \tfrac{1}{3}e^{x^3+7}\frac{d}{dx}(x^3 + 7)$$
$$= \tfrac{1}{3}e^{x^3+7}\,3x^2 = x^2 e^{x^3+7}.$$

In (c),

$$\frac{d}{dx}\left(\tfrac{1}{4}\ln|4e^x - 3| + c\right) = \frac{1}{4}\,\frac{1}{4e^x - 3}\,\frac{d}{dx}(4e^x - 3)$$

$$= \frac{e^x}{4e^x - 3}.$$

Since the natural logarithm and the exponential functions are inverses of each other, an application of (2.3) gives the two useful formulas

$$\boxed{\begin{aligned} \ln e^x &= x, \qquad -\infty < x < \infty, \\ e^{\ln x} &= x, \qquad 0 < x < \infty. \end{aligned}}$$

EXAMPLE 3. Simplify (a) $e^{2\ln x}$ and (b) $e^{3+5\ln x}$. For the first, since $2\ln x = \ln x^2$, we have

$$e^{2\ln x} = e^{\ln x^2} = x^2.$$

For part (b),

$$e^{3+5\ln x} = e^3 e^{5\ln x} = e^3 e^{\ln x^5} = e^3 x^5.$$

PROBLEMS

1. One must be careful to distinguish between the inverse of a function and the reciprocal of the function. If $f(x) = 3x + 4$, write

   (a) $f^{-1}(x)$, the value of the inverse function f^{-1} at x.

   (b) $[f(x)]^{-1}$, the reciprocal of $f(x)$.

2. If f has an inverse f^{-1}, what relations hold between the domains and ranges of f and f^{-1}?
3. Find the derivative with respect to x of each of the following functions.

   (a) e^{7x}

   (b) $\frac{1}{3}e^{3x+2}$

   (c) xe^x

   (d) e^{-x^2}

   (e) $e^x \ln x$

   (f) $\dfrac{e^x + e^{-x}}{2}$

   (g) $\dfrac{e^x - e^{-x}}{2}$

   (h) $\dfrac{3}{2e^x}$

   (i) $\dfrac{e^x}{x}$

   (j) $\dfrac{5\ln x}{x}$

   (k) e^{3x^2-4x+5}

   (l) e^{ax+b}.

4. Solve each of the following integrals.

(a) $\int e^{3x}\,dx$

(b) $\int 2xe^{x^2}\,dx$

(c) $\int \frac{1}{x}e^{\ln x}\,dx$

(d) $\int \frac{e^x + e^{-x}}{e^x - e^{-x}}\,dx$

(e) $\int \frac{xe^{x^2}\,dx}{4e^{x^2} + 5}$

(f) $\int \frac{3\,dx}{2e^{4x}}$

(g) $\int \frac{x^2\,dx}{e^{x^3-2}}$

(h) $\int (x + 1)e^{x^2+2x}\,dx$

(i) $\int e^{\pi}\,dx$

(j) $\int e^{ax+b}\,dx.$

5. Evaluate each of the following integrals.

(a) $\int_1^2 \frac{dx}{x}$

(b) $\int_0^3 e^{2x}\,dx$

(c) $\int_x^{x^2} e^t\,dt$

(d) $\int_4^8 \frac{dx}{e^x}.$

6. Sketch the graph of each of the following equations. Label all extreme points and points of inflection, and give the values of x at which these occur. Classify each extreme point as a local maximum or minimum.

(a) $y = e^{3x}$

(b) $y = x \ln x$

(c) $y = xe^{-x}$

(d) $y = x^2e^{-x}$

(e) $y = e^{-x^2}.$

7. In each of the following examples, find the area of the region above the x-axis, below the graph of the function f, and between the two vertical lines whose equations are given.

(a) $f(x) = 2e^{4x}$, $x = 0$ and $x = 2$.

(b) $f(x) = xe^{x^2}$, $x = 2$ and $x = 4$.

(c) $f(x) = \frac{1}{e^x}$, $x = 5$ and $x = 7$.

8. Suppose f is a function which has the property that it is equal to its own derivative; i.e., $f' = f$.

(a) Compute the derivative of the quotient $\frac{f(x)}{e^x}$.

(b) Using the result of (a), prove that $f(x) = ke^x$ for some constant k.

9. Let f be the function with domain $[0, 1]$ and defined by

$$f(x) = \sqrt{2x - x^2}, \qquad 0 \leq x \leq 1.$$

Draw the graph of f and the graph of the inverse function f^{-1}.

10. Compute each of the following limits using L'Hôpital's Rule or some other method if you prefer.

(a) $\lim\limits_{x \to 0} \dfrac{e^x - 1}{x}$

(b) $\lim\limits_{x \to 0} \dfrac{e^x - 1 - x}{x^2}$

(c) $\lim\limits_{x \to 1} \dfrac{e^x - e}{x - 1}$

(d) $\lim\limits_{x \to 0} \dfrac{x^2 e^x}{1 - e^{x^2}}$

(e) $\lim\limits_{x \to \infty} \dfrac{x}{e^x}$

(f) $\lim\limits_{x \to \infty} \dfrac{e^x}{x^6}$

(g) $\lim\limits_{x \to 0+} \dfrac{\ln x}{e^{1/x}}$

(h) $\lim\limits_{x \to 0} \dfrac{e^x - e^{-x}}{x}$.

11. Show by means of an example that half of (2.3) is not enough. That is, define two functions f and g such that $f(g(x)) = x$, for every x in the domain of g, but $g \neq f^{-1}$.

12. Prove that $\lim\limits_{x \to 0} \dfrac{e^x - 1}{x} = 1$, using $\dfrac{d}{dx} e^x = e^x$ and the definition of the derivative at 0.

13. A function of x is a solution of a differential equation if it and its derivatives make the equation true. For what value (or values) of m is $y = e^{mx}$ a solution of $\dfrac{d^2y}{dx^2} - 3\dfrac{dy}{dx} + 2y = 0$?

14. Find each of the following limits.

(a) $\lim\limits_{n \to \infty} \left(1 + \dfrac{1}{n}\right)^{2n}$, n an integer.

(b) $\lim\limits_{x \to \infty} \left(1 + \dfrac{1}{x}\right)^{-3x}$, x an integer.

15. Let f be a function differentiable on some unbounded interval (a, ∞). Prove that if $\lim\limits_{x \to \infty} [f(x) + f'(x)] = L$, then $\lim\limits_{x \to \infty} f(x) = L$. [*Hint:* Consider the quotient $\dfrac{e^x f(x)}{e^x}$.]

3. Inverse Function Theorems. In this section we shall prove some basic theorems about continuous functions and the inverses of monotonic functions, which we have already used in studying the exponential function. These theorems are all geometrically obvious. We shall show that they also follow logically from the definitions of continuity and monotonicity using the least upper bound property of the real numbers. This shows, as much as anything, that these definitions say what we want them to say. To put it facetiously, if these theorems could not be proved, we would change the definitions until they could be.

(3.1) INTERMEDIATE VALUE THEOREM. *If f is continuous on the closed interval $[a, b]$ and w is any real number such that $f(a) < w < f(b)$, then there exists at least one real number c such that $a < c < b$ and $f(c) = w$.*

A similar theorem is obtained if the inequalities $f(a) < w < f(b)$ are replaced by $f(a) > w > f(b)$. The proof, in all essentials, is the same. To see that (3.1) is geometrically obvious, look at Figure 5. The horizontal line $y = w$ must certainly cut the curve $y = f(x)$ at least once.

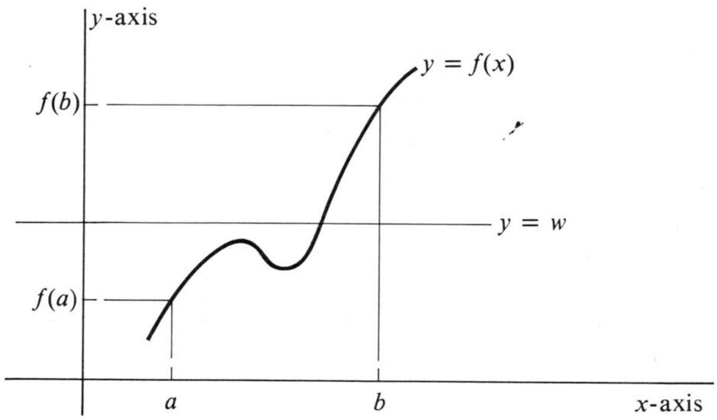

Figure 5

Proof. Consider the subset L of the interval $[a, b]$ that consists of all numbers x in $[a, b]$ such that $f(x) < w$. The set L is not empty because in particular it contains a. Since every number in L is less than b, the number b is an upper bound for L. It follows by the Least Upper Bound Property of the real numbers (page 7) that L has a least upper bound, which we denote by c. Moreover, c lies in $[a, b]$. There are three possibilities:

(i) $f(c) < w$,

(ii) $f(c) > w$,

(iii) $f(c) = w$.

We shall show that (i) and (ii) are, in fact, not possible. Suppose that (i) holds. Since $f(b) > w$, it follows that $c < b$. Set $\epsilon = w - f(c)$, which is positive. Since f is continuous at c, there exists a positive number δ such that whenever $|x - c| < \delta$ and x is in $[a, b]$, then $|f(x) - f(c)| < \epsilon$. Hence, there exist numbers x in $[a, b]$ larger than c for which $|f(x) - f(c)| < \epsilon$. For any such x,

$$f(x) - f(c) \leq |f(x) - f(c)| < \epsilon = w - f(c),$$

and so $f(x) < w$, which implies that x belongs to L. Thus there are numbers in L which are larger than c, which contradicts the fact that c is an upper bound.

Next suppose that (ii) holds. Since $f(a) < w$, it follows that $a < c$. This time set $\epsilon = f(c) - w$, which is positive. The continuity of f at c implies the existence of a positive number δ such that if $|x - c| < \delta$ and x is in $[a, b]$, then $|f(x) - f(c)| < \epsilon$. Pick x_0 in $[a, b]$ such that $|x_0 - c| < \delta$ and $x_0 < c$. We contend that x_0 is also an upper bound for L. For if x is any number such that $x_0 \leq x \leq c$, then $|f(x) - f(c)| < \epsilon$. Consequently,

$$-f(x) + f(c) \leq |f(x) - f(c)| < \epsilon = f(c) - w,$$

and so $-f(x) < -w$, or, equivalently, $f(x) > w$, which implies that x is not in L. This proves the contention that x_0 is an upper bound for L, which contradicts the fact that c is the least upper bound.

The only remaining possibility is (iii). Hence $f(c) = w$, and the proof is complete.

This theorem, (3.1), was used in Section 2, where it was asserted that, for any real number x, there exists a number y such that $x = \ln y$. We have previously shown that the natural logarithm takes on arbitrarily large positive and negative values. Hence we can "surround" a given number x with values of ln. That is, there exist numbers a and b for which $\ln a < x < \ln b$. The existence of a number y such that $x = \ln y$ now follows immediately from (3.1).

An interval was defined on page 4 to be any subset I of the set of all real numbers with the property that, if a and c belong to I and $a \leq b \leq c$, then b also belongs to I. The following proposition is therefore fully equivalent to Theorem (3.1). [More precisely, it is equivalent to the conjunction of (3.1) and its companion theorem with the inequality $f(a) > w > f(b)$.]

(3.2) *If the domain of a continuous real-valued function is an interval, then so is its range.*

The reader should convince himself that (3.1) and (3.2) are equivalent.

We have already proved that every strictly monotonic function has an inverse [see (2.4), page 250]. However, more is needed than simply existence:

(3.3) *If f is a strictly increasing continuous function whose domain is an interval, then the same is true of the inverse function f^{-1}.*

A companion theorem is obtained if "increasing" is replaced by "decreasing."

Proof. There are three things to be proved: (i) f^{-1} is strictly increasing, (ii) the domain of f^{-1} is an interval, and (iii) f^{-1} is a continuous function. The first is completely straightforward, and we leave it as a problem. The second follows at once from (3.2) and the observation that the domain of f^{-1} is equal to the range of f. In proving (iii), we shall assume that the interval which is the domain of f is neither empty nor consists of a single point. This is reasonable, because in these two cases the assumption that f is continuous is not particularly meaningful. Let b be a number in the domain of f^{-1}, let $a = f^{-1}(b)$, and let ϵ be an arbitrary positive number. If a is an endpoint of the interval which is the domain of f, the following argument must be modified slightly. We shall assume that a is not an endpoint and also that ϵ is sufficiently small that both $a + \epsilon$ and $a - \epsilon$ are in the domain of f (see Figure 6). Set δ

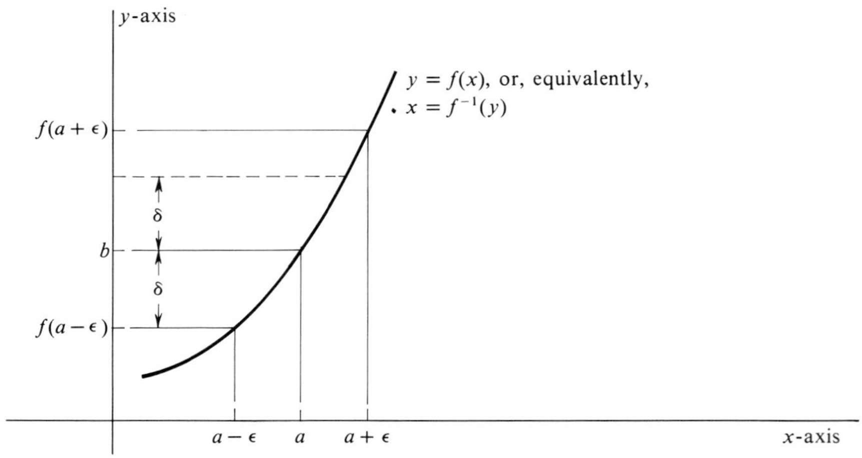

Figure 6

equal to the smaller of the two numbers $f(a + \epsilon) - b$ and $b - f(a - \epsilon)$. Then, if y is any number in the domain of f^{-1} such that $|y - b| < \delta$, we know that

$$f(a - \epsilon) < y < f(a + \epsilon).$$

Since f^{-1} is strictly increasing, we have

$$a - \epsilon = f^{-1}\big(f(a - \epsilon)\big) < f^{-1}(y) < f^{-1}\big(f(a + \epsilon)\big) = a + \epsilon,$$

which implies that $-\epsilon < f^{-1}(y) - a < \epsilon$. Since $a = f^{-1}(b)$, the latter inequalities are equivalent to

$$|f^{-1}(y) - f^{-1}(b)| < \epsilon,$$

which proves that f^{-1} is continuous. This completes the proof.

Our final theorem concerns the differentiability of an inverse function. It was used in Section 2, where we asserted that the exponential function exp is differentiable.

(3.4) *Let f be a strictly monotonic differentiable function whose domain is an interval. If $b = f(a)$ and if $f'(a) \neq 0$, then f^{-1} is differentiable at b. Moreover,*

$$(f^{-1})'(b) = \frac{1}{f'(a)}.$$

Proof. According to the definition of the derivative, we must show that

$$\lim_{y \to b} \frac{f^{-1}(y) - f^{-1}(b)}{y - b} = \frac{1}{f'(a)}.$$

Let $x = f^{-1}(y)$. Then $y = f(x)$ and, of course, $a = f^{-1}(b)$. Hence

$$\frac{f^{-1}(y) - f^{-1}(b)}{y - b} = \frac{x - a}{f(x) - f(a)} = \frac{1}{\dfrac{f(x) - f(a)}{x - a}}.$$

Since the reciprocal of a limit is the limit of the reciprocal, we know that

$$\frac{1}{f'(a)} = \frac{1}{\lim\limits_{x \to a} \dfrac{f(x) - f(a)}{x - a}} = \lim_{x \to a} \frac{1}{\dfrac{f(x) - f(a)}{x - a}}.$$

It is a simple matter to finish the proof provided we know that x approaches a as y approaches b. But this is just the assertion that f^{-1} is continuous at b, which we know to be true as a result of Theorem (3.3). Hence

$$\frac{1}{f'(a)} = \lim_{y \to b} \frac{1}{\dfrac{f(x) - f(a)}{x - a}}$$

$$= \lim_{y \to b} \frac{f^{-1}(y) - f^{-1}(b)}{y - b} = (f^{-1})'(b),$$

and the proof is complete.

We have also used Theorem (3.4) before to establish the differentiability of the function g defined by $g(x) = x^{1/n}$, where n is a positive integer and x is any positive real number (see page 72). The inverse function f, defined by $f(x) = x^n$, for every positive real number x, is strictly increasing and has a positive derivative at every point in the interval $(0, \infty)$. Theorem (3.4) tells us at once that g is a differentiable function.

PROBLEMS

1. Prove that if f is a strictly increasing function, then the inverse function f^{-1} is also strictly increasing.
2. Does the assertion in Problem 1 remain true if "increasing" is replaced by "decreasing"?
3. Show by giving an example that a strictly increasing function is not necessarily continuous.
4. Give an example of a differentiable strictly increasing function defined for all values of x whose inverse is not differentiable everywhere.
5. Show that Theorem (3.4) is geometrically obvious. [*Hint:* The derivative is the slope of the tangent line, and the graph of $y = f(x)$ is the same as that of $x = f^{-1}(y)$.]
6. Supply the details which prove that Theorem (3.2) is equivalent to (3.1) [i.e., to the conjunction of (3.1) and its companion].

4. Other Exponential and Logarithm Functions. If m and n are integers and $n > 0$, then

$$2^{m/n} = \sqrt[n]{2^m}.$$

Hence, for any rational number r, the number 2^r is defined. But what is 2^x if x is not rational? More generally, how should a^x be defined for an arbitrary real number x and a positive number a?

If x is a rational number and a is positive, we have shown that $\ln a^x = x \ln a$, and therefore $a^x = e^{x \ln a}$. However, $e^{x \ln a}$ is defined for every real number x. We shall take advantage of this fact, and, if x is real but not rational, we *define* a^x to be $e^{x \ln a}$. Consequently, for every real number x, we have

$$\boxed{a^x = e^{x \ln a}, \qquad a > 0.}$$

This function, so defined, has all the familiar properties of an exponential function:

$$(4.1) \qquad \begin{cases} a^x > 0, & -\infty < x < \infty, \\ a^0 = 1, & \\ a^1 = a, & \\ a^x \cdot a^y = a^{x+y}, & \\ a^{-x} = \dfrac{1}{a^x}, & \\ \dfrac{a^x}{a^y} = a^{x-y}. & \end{cases}$$

The proofs follow readily from the properties of the functions ln and exp. For example,

$$a^1 = e^{1 \ln a} = e^{\ln a} = a,$$

$$a^x \cdot a^y = e^{x \ln a} \cdot e^{y \ln a} = e^{x \ln a + y \ln a} = e^{(x+y)\ln a} = a^{x+y}.$$

The derivative of a^x is easily computed from its defining equation. Since

$$\frac{d}{dx} a^x = \frac{d}{dx} e^{x \ln a} = e^{x \ln a} \frac{d}{dx}(x \ln a) = a^x \ln a,$$

we have the formula

(4.2)

$$\boxed{\frac{d}{dx} a^x = a^x \ln a.}$$

More generally, if u is a differentiable function of x, the Chain Rule implies that

$$\frac{d}{dx} a^u = a^u \ln a \frac{du}{dx}. \qquad (1)$$

EXAMPLE 1. Compute the derivative of each of the following functions:

(a) 2^x, (b) $2^{(x^2)}$, (c) $2^{(2^x)}$.

For (a) we get

$$\frac{d}{dx} 2^x = 2^x \ln 2;$$

for (b),

$$\frac{d}{dx} 2^{(x^2)} = 2^{(x^2)} \ln 2 \frac{d}{dx} x^2 = 2^{(x^2)}(\ln 2)(2x)$$
$$= x 2^{x^2+1} \ln 2;$$

and for (c),

$$\frac{d}{dx} 2^{(2^x)} = 2^{(2^x)} \ln 2 \frac{d}{dx} 2^x = 2^{(2^x)}(\ln 2) 2^x \ln 2$$
$$= 2^{2^x + x}(\ln 2)^2.$$

If $a = 1$, then $a^x = e^{x \ln 1} = e^0 = 1$ for every real number x. Hence 1^x is the constant function 1.

If $a > 1$, then the graph of the function a^x resembles the graph of e^x. The slope of the tangent line to the graph is always positive, for if $a > 1$, then $\ln a > 0$, and, since $a^x > 0$, we see that

$$\frac{d}{dx} a^x = a^x \ln a > 0.$$

This means that a^x is a strictly increasing function (see Problem 10 at the end of this section). The second derivative is also always positive, since

$$\frac{d^2}{dx^2} a^x = \frac{d}{dx}(a^x \ln a) = a^x(\ln a)^2 > 0.$$

Hence the graph is concave upward for all x. Moreover, there are no extreme points, critical points, or points of inflection. The graph is drawn in Figure 7. It is relatively flat on the left, passes through $\left(-1, \frac{1}{a}\right)$, $(0, 1)$,

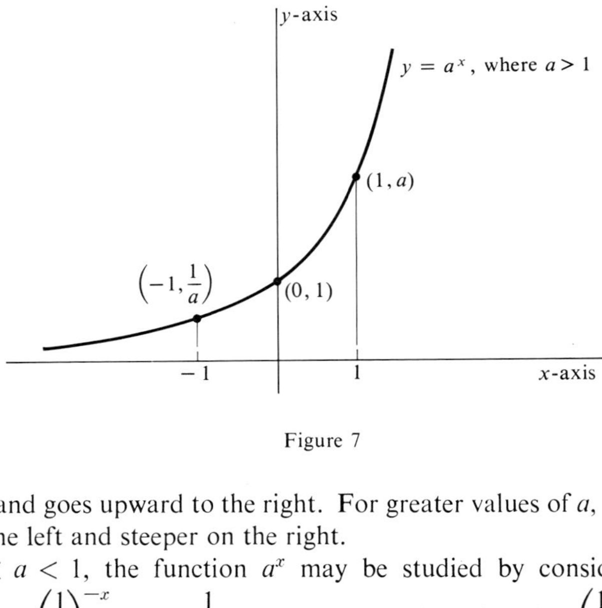

Figure 7

and $(1, a)$, and goes upward to the right. For greater values of a, the graph is flatter on the left and steeper on the right.

If $0 < a < 1$, the function a^x may be studied by considering it in another form, $\left(\frac{1}{a}\right)^{-x}$. Since $\frac{1}{a} > 1$, the graph of the function $\left(\frac{1}{a}\right)^x$ is of the type described in the preceding paragraph, and the graph of a^x, which is equal to $\left(\frac{1}{a}\right)^{-x}$, is the same curve reflected across the y-axis. It is steep on the left, passes through $\left(-1, \frac{1}{a}\right)$, $(0, 1)$, and $(1, a)$, and flattens out as it goes to to the right. It is drawn in Figure 8.

Every derivative formula has a corresponding integral formula. Since

$$\frac{d}{dx}\left(\frac{a^x}{\ln a}\right) = \frac{1}{\ln a}\frac{d}{dx}a^x = a^x,$$

the integral formula corresponding to (4.2) is

(4.3) $$\boxed{\int a^x\,dx = \frac{a^x}{\ln a} + c.}$$

As always, the Chain Rule provides a generalization. If u is a differentiable function of x, then

$$\int a^u \frac{du}{dx}\,dx = \frac{a^u}{\ln a} + c. \tag{2}$$

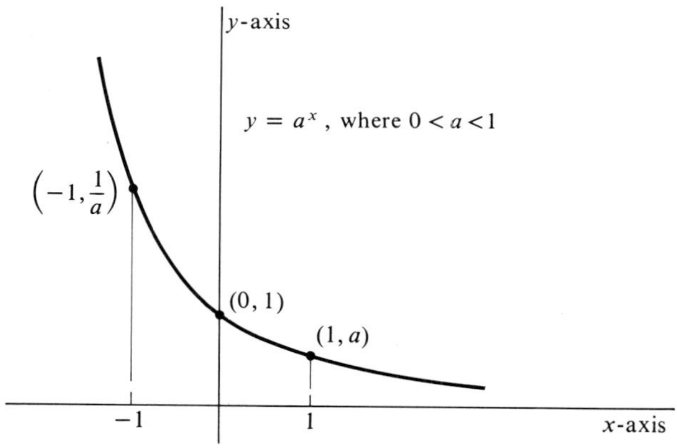

Figure 8

EXAMPLE 2. Compute each of the following indefinite integrals:

$$\text{(a)}\ \int 3^y\,dy, \qquad \text{(b)}\ \int x10^{x^2-7}\,dx, \qquad \text{(c)}\ \int \frac{1}{x}(2.31)^{\ln x}\,dx.$$

A direct use of (4.3) gives for (a)

$$\int 3^y\,dy = \frac{3^y}{\ln 3} + c.$$

Since $\frac{d}{dx}(x^2 - 7) = 2x$, integral (b) can be written $\frac{1}{2}\int 10^{x^2-7}\cdot 2x\cdot dx$, which by (2) is equal to $\frac{1}{2}\frac{10^{x^2-7}}{\ln 10} + c$. Hence

$$\int x10^{x^2-7}\,dx = \frac{10^{x^2-7}}{2\ln 10} + c.$$

For part (c) we note that $\frac{d}{dx}\ln x = \frac{1}{x}$, and therefore that the integral is of the form in (2). Thus

$$\int (2.31)^{\ln x}\frac{1}{x}\,dx = \frac{(2.31)^{\ln x}}{\ln 2.31} + c.$$

It was proved on page 241 that $\ln a^r = r \ln a$, for every rational number r and every positive real number a. We are now in a position to remove the restriction that r be rational. Let x be an arbitrary real number. Then $a^x = e^{x \ln a}$, and so $\ln a^x = \ln e^{x \ln a} = \ln \exp(x \ln a)$. Since ln and exp are inverse functions of each other it follows that $\ln \exp(x \ln a) = x \ln a$. We have therefore proved that

(4.4) $\ln a^x = x \ln a$, *for every real number x and every positive real number a.*

Another of the well-known laws of exponents now follows easily:

(4.5) $(a^x)^y = a^{xy}$, *for all real numbers x and y and every positive real number a.*

Proof. If we let $a^x = b$, then $(a^x)^y = b^y = e^{y \ln b}$. Replacing b in the last expression, we have

$$(a^x)^y = e^{y \ln a^x},$$

and, using (4.4),

$$e^{y \ln a^x} = e^{y(x \ln a)} = e^{xy \ln a}.$$

Since $e^{xy \ln a} = a^{xy}$, it follows that $(a^x)^y = a^{xy}$, and the proof is complete.

In particular, $(e^x)^y = e^{xy}$ for all real numbers x and y.

Let a be any real number, and consider the function f defined for every positive real number x by

$$f(x) = x^a.$$

Hitherto in this section we have considered the function a^x. Now we reverse the roles of constant and variable. One of the basic rules of differentiation proved in Chapter 1 states that, if a is a rational number, then

$$f'(x) = \frac{d}{dx}x^a = ax^{a-1}.$$

We now remove the restriction that a be rational. Observe first that x^a is certainly a differentiable function, since it is the composition of differentiable functions:

$$x^a = e^{a \ln x} = \exp(a \ln x).$$

Knowing this, we use implicit differentiation to compute its derivative. Let $y = x^a$. Then $\ln y = \ln x^a = a \ln x$, and so

$$\frac{d}{dx} \ln y = \frac{d}{dx} a \ln x,$$

$$\frac{1}{y} \frac{dy}{dx} = a \frac{1}{x},$$

$$\frac{dy}{dx} = \frac{ay}{x}.$$

Since $y = x^a$, it follows that $\dfrac{ay}{x} = \dfrac{ax^a}{x} = ax^{a-1}$. Thus we have proved that

(4.6) $$\frac{d}{dx} x^a = ax^{a-1}, \textit{for any real number } a.$$

The technique of taking logarithms and differentiating implicitly, which was used in proving (4.6), can also be used to compute the derivative of a positive differentiable function which is raised to a power which is itself a differentiable function. For example, to compute $\dfrac{d}{dx} x^x$, we let $y = x^x$. Then

$$\ln y = \ln x^x = x \ln x,$$

and it follows that

$$\frac{1}{y} \frac{dy}{dx} = \frac{d}{dx}(x \ln x) = x \frac{1}{x} + \ln x = 1 + \ln x,$$

$$\frac{dy}{dx} = y(1 + \ln x) = x^x(1 + \ln x), \qquad (x > 0).$$

This technique is known as **logarithmic differentiation** and is a basic tool for finding derivatives. We can use it to derive a formula for $\dfrac{d}{dx} u^v$, where u is a positive differentiable function of x and v is any differentiable function of x. Let $y = u^v$, and then $\ln y = v \ln u$. Hence

$$\frac{1}{y} \frac{dy}{dx} = \frac{d}{dx}(v \ln u) = v \frac{1}{u} \frac{du}{dx} + \ln u \frac{dv}{dx},$$

$$\frac{dy}{dx} = y \left(\frac{v}{u} \frac{du}{dx} + \ln u \frac{dv}{dx} \right),$$

$$\frac{dy}{dx} = u^v \left(\frac{v}{u} \frac{du}{dx} + \ln u \frac{dv}{dx} \right),$$

and finally, therefore,

$$\frac{d}{dx} u^v = vu^{v-1}\frac{du}{dx} + u^v \ln u \frac{dv}{dx}. \tag{3}$$

We do not suggest that the reader memorize this formula. It is more important to be able to use the method of logarithmic differentiation.

EXAMPLE 3. Find $\frac{d}{dx}(x^2 + 1)^{e^x}$. Letting $y = (x^2 + 1)^{e^x}$ and taking natural logarithms, we have

$$\ln y = \ln (x^2 + 1)^{e^x} = e^x \ln (x^2 + 1).$$

Differentiating, we obtain

$$\frac{1}{y}\frac{dy}{dx} = e^x \frac{1}{x^2 + 1} 2x + e^x \ln (x^2 + 1),$$

$$\frac{dy}{dx} = ye^x\left[\frac{2x}{x^2 + 1} + \ln (x^2 + 1)\right].$$

Hence

$$\frac{d}{dx}(x^2 + 1)^{e^x} = e^x(x^2 + 1)^{e^x}\left[\frac{2x}{x^2 + 1} + \ln (x^2 + 1)\right].$$

The function a^x is strictly monotonic if a is positive and not equal to 1, increasing if $a > 1$ and decreasing if $0 < a < 1$. Moreover, it has a nonzero derivative at every x. It follows by Theorem (3.4), page 261, that a^x has a differentiable inverse function. Even as the inverse function of e^x is the natural logarithm, we call the inverse function of a^x the **logarithm to the base *a*.** Hence

$$\boxed{y = \log_a x \qquad \textit{if and only if } x = a^y.}$$

We emphasize that a must be a positive number different from 1 and that $\log_a x$ is defined only for positive values of x. The so-called **common logarithm,** usually denoted by simply log and encountered in the usual tables of logarithms, is the logarithm to the base 10. Thus $\log 100 = \log_{10} 100 = 2$, since $10^2 = 100$. The logarithm to the base a has the same algebraic properties as the natural logarithm:

(4.7)
$$\begin{cases} \log_a 1 = 0, \\ \log_a a = 1, \\ \log_a pq = \log_a p + \log_a q, \\ \log_a \dfrac{p}{q} = \log_a p - \log_a q, \\ \log_a p^b = b \log_a p. \end{cases}$$

The above properties hold for every positive real number a different from 1, for all positive real numbers p and q, and for every real number b. Each one may be proved by considering the corresponding exponential function. Note that since a^x and $\log_a x$ are inverse functions of each other,

$$\begin{cases} \log_a a^x = x, & \textit{for all real } x, \\ a^{\log_a x} = x, & \textit{for all positive real } x. \end{cases}$$

For example, if we let $x = \log_a p$ and $y = \log_a q$, then we have $p = a^x$ and $q = a^y$, and so

$$\begin{aligned} \log_a pq &= \log_a(a^x a^y) = \log_a a^{x+y} = x + y \\ &= \log_a p + \log_a q. \end{aligned}$$

The other properties are proved in the same way.

To compute the derivative of $\log_a x$, we let $y = \log_a x$. The equivalent exponential equation is $x = a^y$, from which it follows that $\ln x = \ln a^y = y \ln a$. By implicit differentiation, therefore,

$$\frac{d}{dx}(y \ln a) = \frac{d}{dx} \ln x,$$

$$\ln a \frac{dy}{dx} = \frac{1}{x}.$$

Solving for $\frac{dy}{dx}$, which equals $\frac{d}{dx} \log_a x$, we obtain

$$\boxed{\frac{d}{dx} \log_a x = \frac{1}{\ln a} \frac{1}{x}.} \tag{4.8}$$

PROBLEMS

1. Find the derivative with respect to x of each of the following functions.

(a) 4^{x+1}
(b) $\log_{10}(x^2 + 1)$
(c) $\log_{10} 4^{x+1}$
(d) e^{x^2+x+2}
(e) xa^x
(f) $2^x x^2$
(g) xe^{-x}
(h) $\log_4(x^2 - 4^x)$
(i) x^{x-1}
(j) $x^{(x^2)}$
(k) $(x^x)^2$.

2. If a and b are positive numbers not equal to 1, prove that $\log_a b = \frac{1}{\log_b a}$.

3. Prove that

(a) $\ln x = (\ln a)(\log_a x)$ (b) $\ln x = \dfrac{\log_a x}{\log_a e}$.

4. Integrate each of the following.

(a) $\int 7^x\,dx$

(b) $\int x^2 2^{3x^3+4}\,dx$

(c) $\int \frac{1}{x+2}\ln|x+2|\,dx$

(d) $\int \frac{1}{x}\ln\left|\frac{1}{x}\right|\,dx$

(e) $\int \log_2 e^{7x-5}\,dx$

(f) $\int \frac{1}{x+3}3^{\ln|x+3|}\,dx$

(g) $\int e^x 5^{(e^x)}\,dx$.

5. (a) Differentiate logarithmically $y = \sqrt{\dfrac{(x-1)(x+3)}{(x+2)(x-4))}}$.

(b) For what values of x is the differentiation in (a) valid?

6. If u is a positive function of x and a is positive but not equal to 1, show that $\log_a u = \dfrac{\ln u}{\ln a}$.

7. Differentiate each of the following functions with respect to x.

(a) $x^{\ln x}$

(b) $(\ln x)^x$

(c) $(e^x)^{x^2+1}$

(d) $(\ln x)^{\ln x}$.

8. (a) Given only that $\log_a 1 = 0$ and that $\log_a pq = \log_a p + \log_a q$, prove that $\log_a\left(\dfrac{1}{p}\right) = -\log_a p$.

(b) Then prove that $\log_a \dfrac{p}{q} = \log_a p - \log_a q$.

9. Prove all the properties listed in (4.7) (see Problem 8).

10. Using the Mean Value Theorem, page 113, prove that if $f'(x) > 0$ for all x, then f is a strictly increasing function.

11. Assume that $a > 1$.

(a) Using the definition of a^x, show that

$$\lim_{x\to\infty} a^x = \infty.$$

(b) Using the result of (a), prove that

$$\lim_{x\to-\infty} a^x = 0.$$

(c) Using (a), show that

$$\lim_{x\to\infty} \frac{d}{dx} a^x = \infty.$$

(d) Using (b), show that

$$\lim_{x\to-\infty} \frac{d}{dx} a^x = 0.$$

(e) What do (a), (b), (c), and (d) say geometrically about the graph of the function a^x?

12. Assume that $a_1 > a_2 > 1$.

(a) Using the definition of a^x, show that, if $x > 0$, then $a_1^x > a_2^x$.

(b) Using (a), show that, if $x < 0$, then $a_1^x < a_2^x$.

13. Evaluation of a limit of the form $\lim_{x\to a} f(x)^{g(x)}$ is not obvious if any one of the following three possibilities occurs.

(i) $\lim_{x\to a} f(x) = \lim_{x\to a} g(x) = 0$.

(ii) $\lim_{x\to a} f(x) = 1$ and $\lim_{x\to a} g(x) = \infty$.

(iii) $\lim_{x\to a} f(x) = \infty$ and $\lim_{x\to a} g(x) = 0$.

These three types are usually referred to, respectively, as the **indeterminate forms** 0^0, 1^∞, and ∞^0. The standard attack, akin to logarithmic differentiation, is the following: Let

$$h(x) = \ln f(x)^{g(x)} = g(x) \ln f(x) = \frac{\ln f(x)}{\dfrac{1}{g(x)}}.$$

One then applies L'Hôpital's Rule to the quotient, thereby hopefully discovering that $\lim_{x\to a} h(x)$ exists and what its value is. If it does exist, it follows by the continuity of the exponential function that

$$e^{\left[\lim_{x\to a} h(x)\right]} = \lim_{x\to a} e^{h(x)}.$$

But, since

$$e^{h(x)} = e^{\ln f(x)^{g(x)}} = f(x)^{g(x)},$$

we therefore conclude that

$$\lim_{x\to a} f(x)^{g(x)} = e^{\left[\lim_{x\to a} h(x)\right]},$$

and the problem is solved.

Apply this method to evaluate the following limits.

(a) $\lim_{x\to 0+} x^x$ (b) $\lim_{x\to\infty} x^{1/x}$ (c) $\lim_{x\to 0} (1 + 2x)^{1/x}$.

5. Introduction to Differential Equations. For a given differentiable function, we are frequently interested in an equation which contains the derivative of the function and which is true for every number in the domain of the function. These equations arise naturally in physics and in many applied branches of mathematics. An example of such an equation is obtained if y is the function of x defined by $y = 2e^{3x}$. Since $\frac{dy}{dx} = 3(2e^{3x}) = 3y$, the equation

$$\frac{dy}{dx} - 3y = 0$$

holds for this particular function y and all real values of x. For another example, let y be the function defined by $y = x^3 - x^2$. It is easy to verify by differentiation and substitution that the equation

$$x\frac{dy}{dx} - 3y = x^2$$

is true for this function and all real values of x.

The two equations in the preceding paragraph are examples of differential equations. They are called first-order differential equations because they involve the first derivative of the function but no higher derivatives. In each example above we started with a function and then found an equation containing its derivative. More commonly we encounter the differential equation and then set out to find the function. For example, for what function y is the equation

$$\frac{dy}{dx} = \frac{x}{y}$$

true for all values of x in the domain of y? If such a function exists, it is called a solution to the differential equation. Generally speaking, if a differential equation has one solution, it has infinitely many. We may be required to find one solution to a given differential equation, or possibly all solutions.

Let us try to fix the ideas in the above examples by giving a general definition. Consider an equation in three variables x, y, and z, which we write

$$F(x, y, z) = 0.$$

Not all the variables need occur in the equation, but at least z must. Substituting $\frac{dy}{dx}$ for z, we obtain the equation

$$F\left(x, y, \frac{dy}{dx}\right) = 0, \tag{1}$$

which is a **first-order differential equation.** This equation, however, is merely a formal statement of equality containing the symbols x, y, and $\frac{dy}{dx}$. As such, it is neither true nor false. By a **solution** to (1) we mean any differentiable function f such that the equation

$$F(x, f(x), f'(x)) = 0$$

is true for every x in the domain of f.

The reader should realize, of course, that there is nothing sacred about the letters x and y which we have thus far used to denote the independent and dependent variable, respectively. For example, the differential equation

$$t\frac{dx}{dt} + x = e^t$$

has for a solution the function of t defined by $x = \frac{e^t}{t}$.

In this section we shall consider some simple types of first-order differential equations and the techniques for solving them. Other first-order differential equations and differential equations of higher order will be studied in Chapters 6 and 11.

The first type to be studied has already been solved in this book. Let f be a given continuous function, and consider the differential equation

$$\boxed{\frac{dy}{dx} = f(x).} \tag{2}$$

A solution is any function y with the property that its derivative is the function f. That is, a function is a solution if and only if it is an antiderivative, or indefinite integral, of f. Hence, if $F'(x) = f(x)$, we have

$$\boxed{y = \int f(x)\,dx = F(x) + c,} \tag{3}$$

where c is an arbitrary constant. Thus solving the differential equation is the same thing as finding the indefinite integral. As c ranges over all real numbers, we get all antiderivatives and therefore all solutions to the differential equation (2). For this reason, (3) is called the **general solution** to the differential equation.

EXAMPLE 1. Find the general solution of each of the following differential equations:

(a) $\frac{dy}{dx} = 3x^2 + 2x - 1,$

(b) $\frac{dx}{dt} = e^t - 1,$

(c) $x\frac{dy}{dx} = (\ln x)^2.$

Solving (a), we obtain

$$y = \int (3x^2 + 2x - 1)\,dx = x^3 + x^2 - x + c.$$

Similarly, for (b),

$$x = \int (e^t - 1)\,dt = e^t - t + c.$$

As it stands, (c) is not in the form of (2). However, an equivalent equation is $\dfrac{dy}{dx} = \dfrac{1}{x}(\ln x)^2$, and so

$$y = \int (\ln x)^2 \frac{1}{x}\,dx.$$

The integral is of the form $\displaystyle\int u^2 \frac{du}{dx}\,dx$, where $u = \ln x$; hence

$$y = \frac{(\ln x)^3}{3} + c.$$

The second type of differential equation which we consider in this section arises when we are given two continuous functions f and g and form the differential equation

$$\boxed{\frac{dy}{dx} = \frac{f(x)}{g(y)}.} \tag{4}$$

A differential equation of this form is called **separable.** An equivalent equation is $g(y)\dfrac{dy}{dx} = f(x)$, where the variables have been "separated" in the sense that on the right we have a function of x and on the left a function of y and the derivative of y. The differential equation can be readily solved provided we can find antiderivatives of f and g. From the latter equation we obtain

$$\int g(y)\frac{dy}{dx}\,dx = \int f(x)\,dx. \tag{5}$$

Suppose that $F'(x) = f(x)$ and that $G'(y) = g(y)$. That is,

$$\int f(x)\,dx = F(x) + c,$$

$$\int g(y)\,dy = G(y) + k.$$

Then it is also true that

$$\int g(y)\frac{dy}{dx}\,dx = G(y) + k,$$

because, by the Chain Rule,

$$\frac{d}{dx}[G(y) + k] = G'(y)\frac{dy}{dx} = g(y)\frac{dy}{dx}\cdot$$

It follows from (5) that $G(y) + k = F(x) + c$. This tells us that $G(y)$ and $F(x)$ differ by the constant $c - k$, which for convenience we rename simply c. Therefore, we finally obtain the equation

$$\boxed{G(y) = F(x) + c,} \tag{6}$$

which implicitly defines any solution y of the original differential equation.

Conversely, if we differentiate (6) with respect to x, we get (4) again:

$$\frac{d}{dx}G(y) = \frac{d}{dx}[F(x) + c],$$

$$G'(y)\frac{dy}{dx} = f(x),$$

$$g(y)\frac{dy}{dx} = f(x),$$

$$\frac{dy}{dx} = \frac{f(x)}{g(y}\cdot$$

Thus, for any value of the constant c, every differentiable function y defined implicitly by (6) is a solution. Hence (6) defines the **general solution** to the separable differential equation (4).

EXAMPLE 2. (a) Find the general solution to the differential equation $\dfrac{dy}{dx} = \dfrac{x}{y}\cdot$ (b) Find the particular solution whose graph passes through the point (2, 1). This is a separable differential equation, and "separating variables" we replace it by the equivalent form $y\dfrac{dy}{dx} = x$, $y \neq 0$. It follows that

$$\int y\frac{dy}{dx}\,dx = \int x\,dx,$$

and, integrating both sides, we obtain

$$\frac{y^2}{2} = \frac{x^2}{2} + c.$$

If we multiply by 2, we get $y^2 = x^2 + 2c$. But twice an arbitrary constant is still an arbitrary constant, so we replace $2c$ by simply c. Hence the general solution to $\frac{dy}{dx} = \frac{x}{y}$ is implicitly defined by the equation

$$y^2 = x^2 + c. \tag{7}$$

Solving for y explicitly, we obtain

$$y = \pm\sqrt{x^2 + c}$$

as the answer to part (a). To find the particular solution that passes through (2, 1), we substitute $x = 2$ and $y = 1$ in (7) to get $1 = 4 + c$, whence $c = -3$. Since y takes on the value 1, which is positive, we choose the positive square root, and the answer to part (b) is therefore the function defined by

$$y = \sqrt{x^2 - 3}.$$

A first-order differential equation $F\left(x, y, \frac{dy}{dx}\right) = 0$ is called **linear** if the corresponding function $F(x, y, z)$ is a polynomial of first degree in y and z, i.e., if $F(x, y, z) = f(x)y + g(x)z + h(x)$. Thus among the differential equations

$$\frac{dy}{dx} - 3y = 0,$$

$$x\frac{dy}{dx} - 3y = x^2,$$

$$\frac{dy}{dx} = \frac{x}{y},$$

the first two are linear and the third is not. The last type of differential equation which we study in this section is the simplest linear type,

$$\boxed{\frac{dy}{dx} + ky = 0,} \tag{8}$$

where k is an arbitrary constant.

Actually every such differential equation is also separable, since it can be written in the form $\frac{dy}{dx} = -\frac{k}{1/y}$. We treat it as a third type because it is linear and because it has many interesting applications. Solving it as a separable differential equation, however, we first replace it by the equivalent equation $\frac{1}{y}\frac{dy}{dx} = -k$. Then

$$\int \frac{1}{y}\frac{dy}{dx}\,dx = -\int k\,dx.$$

Integrating, we obtain

$$\ln|y| = -kx + c,$$

which defines the general solution implicitly. Since the natural logarithm and the exponential are inverse functions, we can solve for $|y|$, getting

$$|y| = e^{-kx+c} = e^c e^{-kx},$$

$$y = (\pm e^c)e^{-kx}. \tag{9}$$

As c takes on all real values, the quantity e^c takes on all positive real values. Thus e^c is simply an arbitrary positive constant, and $\pm e^c$ is therefore an arbitrary nonzero constant. The original differential equation $\frac{dy}{dx} + ky = 0$ certainly also has the constant function $y = 0$ as a solution. From this fact and (9) we conclude that the general solution to the linear differential equation (8) is

$$\boxed{y = ce^{-kx},} \tag{10}$$

where c is an arbitrary constant.

EXAMPLE 3. Let x be the amount of radium present in a pile at time t. Thus x is a function of t. It is known that the rate of radioactive decay of the pile of radium is proportional to the amount x that remains in the pile.

(a) Show that the length of time T for an amount x to diminish by radioactive decay to an amount $\frac{x}{2}$ is independent of x. The number T is called the **half-life** of radium. It is equal to approximately 1550 years.

(b) If 0.01 grams of radium is present at $t = 0$, how much is present after 500 years?

The rate of change of the amount of radium present with respect to time is given by the derivative $\frac{dx}{dt}$. This is positive for growth and negative for decay. Since the rate of decay is proportional to the amount present, i.e., to x, we know that

$$\frac{dx}{dt} = -kx,$$

where k is some positive constant of proportionality. This is the differential equation governing the physical process. We have shown that the general solution is

$$x = ce^{-kt}, \tag{11}$$

where c is an arbitrary constant. If after an interval of time equal to T the amount has dwindled to $\frac{x}{2}$, we have

$$\frac{x}{2} = ce^{-k(t+T)}.$$

Solving this equation for x and using (11), we get $x = 2ce^{-k(t+T)} = ce^{-kt}$, from which it follows that $2e^{-kT} = 1$, or $2 = e^{kT}$. Hence, $\ln 2 = \ln(e^{kT}) = kT$, and we conclude that

$$T = \frac{\ln 2}{k},$$

which is independent of x.

To do part (b), let us denote by x_0 the amount of radium present at time $t = 0$. Hence from (11) we get

$$x_0 = ce^{-k\cdot 0} = c,$$

and so $x = x_0e^{-kt}$. Since $k = \dfrac{\ln 2}{T}$, we obtain the formula

$$x = x_0e^{-(\ln 2/T)t},$$

which expresses the amount of radium present at time t in terms of the original amount at time $t = 0$ and the half-life of radium. In our problem $x_0 = 0.01$ grams, $t = 500$ years, and $T = 1550$ years. Hence, the answer is x grams, where

$$x = (0.01)e^{-(500/1550)\ln 2} = 0.008 \text{ (approximately)}.$$

PROBLEMS

1. Find the general solution of each of the following differential equations.

(a) $\dfrac{dy}{dx} - 3y = 0$

(b) $\dfrac{dy}{dt} = t(t^2 + 1)$

(c) $x\dfrac{dy}{dx} + y = 0$

(d) $(x + 1)y\dfrac{dy}{dx} = (y^2 + 1)$

(e) $\dfrac{dy}{dt} = e^{t-y}$

(f) $xy\dfrac{dy}{dx} = y^2 - 2.$

2. For each of the following differential equations, find the particular solution whose graph passes through the point indicated.

   (a) $\frac{dy}{dx} = -\frac{y}{x}$, passing through (1, 1).

   (b) $2\frac{dy}{dx} = 3y$, passing through (0, 5).

   (c) $y\frac{dy}{dx} = 18x^3$, passing through (2, −9).

   (d) $\frac{dy}{dx} + \frac{x}{y} = 0$, passing through (5, 0).

3. A curve defined by $y = f(x)$ has slope m at every point (x, y) given by $m = 2y$. If the curve passes through the point $(0, -1)$, find $f(x)$.

4. Find all solutions to the differential equation $\frac{dy}{dx} = -\frac{x}{y}$. Sketch the graphs of the different solutions.

5. Find all solutions to the differential equation $\frac{dy}{dx} = \frac{x}{y}$. Sketch the graphs of the different solutions.

6. Classify each of the following differential equations as separable, linear, both, or neither.

   (a) $\ln y \frac{dy}{dx} = \frac{y}{x}$

   (b) $x^2 \frac{dy}{dx} + y = e^x$

   (c) $y\frac{dy}{dx} + x + y = 0$

   (d) $\frac{dx}{dt} - 7t = 0$

   (e) $\frac{dx}{dt} - 7x = 0$

   (f) $\sqrt{y^2 + 1}\frac{dy}{dx} + x^2 y = 0$

   (g) $\frac{1}{x}\frac{dx}{dt} = 3$

   (h) $\frac{dy}{dx} + \frac{x}{y} = x^2$

   (i) $\left(\frac{dy}{dx}\right)^2 + 3y = 7x.$

7. An alternative approach to solving the linear differential equation $\frac{dy}{dx} + ky = 0$ is to write it as $\frac{dy}{dx} = -ky$. The latter equation is similar to $\frac{dy}{dx} = y$, which has e^x for a solution. With this similarity in mind, it is not hard to guess, and then verify, that $y = e^{-kx}$ is a solution to the original equation. The problem is now to show that *every* solution is a constant multiple of e^{-kx}. Prove this fact by assuming that $y = f(x)$ is an arbitrary solution of $\frac{dy}{dx} + ky = 0$ and then showing that the derivative of the quotient $\frac{f(x)}{e^{-kx}}$ is zero. (See Problem 8, page 256.)

8. A radioactive substance has a half-life of 10 hours. What fraction of an amount of this substance decays in 15 hours?
9. If a certain population increases at a rate proportional to the number in the population and it doubles in 45 years, in how many years is it multiplied by a factor of 3?
10. Find the constant of proportionality relating a radioactive substance and its rate of decay if the substance has a half-life of 16 hours.
11. The number y of bacteria in a culture grows at a rate $\frac{dy}{dt}$ proportional to the number present. If the number doubles in 3 days and there are 10^7 bacteria present at the beginning of the experiment, how many are there after 24 hours?
12. A toy block lying on the floor is given a kick. The resulting acceleration v' (which is negative) is equal to $-5v$. If the kick gives it an initial velocity of 6 feet per second, how many seconds later is the velocity equal to 2 feet per second?
13. A car sliding along a track slows down at a rate proportional to its velocity. If it has one-half its initial velocity after 30 seconds, at what fraction of its initial velocity is it traveling after 1 minute?
14. Let a and b be constants with $a \neq 0$. Show that the differential equation

$$\frac{dy}{dx} + ay = b \tag{12}$$

reduces to $\frac{dz}{dx} + az = 0$ if we let $z = y - \frac{b}{a}$. As a result, find the general solution of (12).
15. Use the substitution described in Problem 14 to find the particular solution of the differential equation $\frac{dy}{dx} - 2y = 6$ which passes through the point $(0, 4)$.

CHAPTER 6

Trigonometric Functions

Although the reader is probably familiar with the geometry of sines and cosines, etc., in terms of angles, our definitions will emphasize them as functions of a real variable. Thus the stage is set for the development of the differential and integral calculus of these important functions. Later in the chapter we introduce complex numbers where exponential and trigonometric functions are blended in the famous equation $e^{ix} = \cos x + i \sin x$. We conclude with an application to linear differential equations.

1. Sine and Cosine. In this section we shall define and study the elementary properties of two real-valued functions, the sine and the cosine, abbreviated sin and cos, respectively. Both functions have as domain the entire set of all real numbers, and, as we shall see in Section 2, both are differentiable functions.

The definitions will be made in terms of what is called arc length, by which is meant the distance from one point on a curve to another measured along the curve. This is a new concept, for although we have defined the straight-line distance between two points on page 11, we have not yet treated distance along a curve. Actually we shall postpone the discussion of arc length in general to Section 2 of Chapter 10 because here we need it only for distance along a circle, in fact, only along the particular circle C which is the graph of the equation $x^2 + y^2 = 1$ in the xy-plane. However, we shall assume that the idea of distance along this curve is understood. For example, we assume the familiar fact that the arc length of the whole circle C, i.e., its circumference, is equal to 2π.

Let t be an arbitrary positive real number. We denote by $P(t)$ the point on the circle C whose distance from the fixed point $(1, 0)$ along C in the counterclockwise direction is equal to t. Intuitively, we take a piece of string of length t, fasten one end at $(1, 0)$, and wrap the string counterclockwise around C. Then $P(t)$ is the point on the circle to which the other end of the string reaches. Next, for every negative number t, let $P(t)$ be the point on C whose distance from the same fixed point $(1, 0)$ along the curve in the clockwise direction is equal to $-t$. That is, this time we wrap the string in the opposite direction. Finally, for $t = 0$, we set $P(0) = (1, 0)$. The definition is illustrated in Figure 1. Thus, to every real number t, we have assigned a point $P(t)$ which is an ordered pair of real numbers. Note that, because the circumference of the circle C is 2π, it follows that

$$P(t + 2\pi) = P(t), \tag{1}$$

for every real number t.

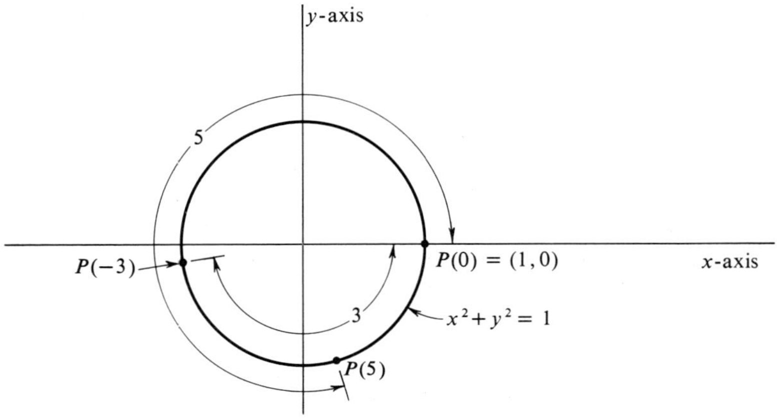

Figure 1

The cosine and sine are now defined as follows: cos(t) is the x-coordinate of $P(t)$, and sin(t) is the y-coordinate. More briefly, we write $\cos t$ and $\sin t$. Thus

$$\boxed{P(t) = (\cos t, \sin t).}$$

For example, from Figure 1, in which $P(5)$ is seen to be in the fourth quadrant, we may conclude that cos 5 is positive and that sin 5 is negative. It is clear geometrically that, if the difference between two real numbers t and a is small, then the point $P(t)$ is close to the point $P(a)$ and hence the differences between their corresponding coordinates are small. More precisely, both $|\cos t - \cos a|$ and $|\sin t - \sin a|$ can be made arbitrarily small by choosing $|t - a|$ sufficiently small. It follows that the cosine and the sine are continuous functions. Their common domain is the set of all real numbers.

For certain values of t which are simple fractions of the total circumference 2π, it is easy to locate the point $P(t)$ on the circle and then to read off the coordinates $\cos t$ and $\sin t$. For example (see Figure 2),

$$P(0) = (1, 0),$$

$$P\left(\frac{\pi}{4}\right) = (\tfrac{1}{2}\sqrt{2}, \tfrac{1}{2}\sqrt{2}),$$

$$P\left(\frac{\pi}{2}\right) = (0, 1),$$
$$P(\pi) = (-1, 0),$$
$$P\left(\frac{3\pi}{2}\right) = (0, -1),$$

from which it follows at once that

$$\cos 0 = 1 \text{ and } \sin 0 = 0,$$
$$\cos\frac{\pi}{4} = \sin\frac{\pi}{4} = \tfrac{1}{2}\sqrt{2},$$
$$\cos\frac{\pi}{2} = 0 \text{ and } \sin\frac{\pi}{2} = 1,$$
$$\cos\pi = -1 \text{ and } \sin\pi = 0,$$
$$\cos\frac{3\pi}{2} = 0 \text{ and } \sin\frac{3\pi}{2} = -1.$$

The reader should be thoroughly familiar with all these values—not by sheer memory, but from an understanding of $P(t)$ and its location on the circle C.

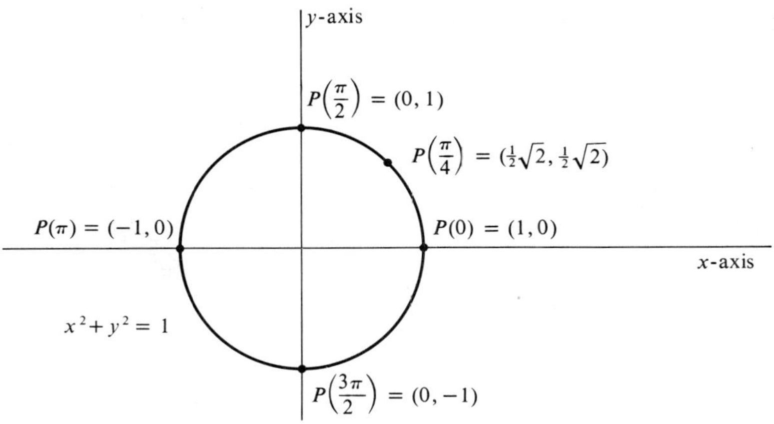

Figure 2

Most of the important properties of the functions cos and sin can be expressed in a few equations called trigonometric identities. Some of these are obvious from the definition of $P(t)$. To begin with, it follows from (1) that

$$\bigl(\cos(t + 2\pi), \sin(t + 2\pi)\bigr) = P(t + 2\pi) = P(t) = (\cos t, \sin t),$$

and so

(1.1) $$\left.\begin{aligned}\cos(t + 2\pi) &= \cos t\\ \sin(t + 2\pi) &= \sin t\end{aligned}\right\}\ \textit{for every real number } t.$$

The property of cos and sin stated in these two equations is expressed in words by saying that cos and sin are **periodic functions** with **period** 2π. That is, each time the value of the variable is increased by 2π, the value of each function is repeated.

Next, since $P(t)$ lies on the circle defined by $x^2 + y^2 = 1$, the coordinates of $P(t)$ must satisfy this equation. Hence

(1.2) $$(\cos t)^2 + (\sin t)^2 = 1, \quad \textit{for every real number } t.$$

There is a strong tradition for abbreviating $(\cos t)^n$ and $(\sin t)^n$ by $\cos^n t$ and $\sin^n t$, respectively, provided n is a positive integer. [However, one *never* writes $\sin^{-1} t$ for $(\sin t)^{-1}$.] As a result, (1.2) is usually written

$$\cos^2 t + \sin^2 t = 1.$$

The third basic property of cos and sin comes from the relation between $P(t) = (\cos t, \sin t)$ and $P(-t) = \big(\cos(-t), \sin(-t)\big)$. It is not hard to see that the difference between measuring the arc length $|t|$ in the counterclockwise direction from $(1, 0)$ and measuring the same distance in the clockwise direction will be only a difference of sign in the y-coordinate of $P(t)$. That is,

$$P(-t) = \big(\cos(-t), \sin(-t)\big) = (\cos t, -\sin t).$$

It follows that

(1.3) $$\left.\begin{aligned} \cos(-t) &= \cos t \\ \sin(-t) &= -\sin t \end{aligned}\right\} \quad \textit{for every real number } t.$$

Thus the cosine is an even function, and the sine is an odd function (see pages 90 and 92).

If a and b are any two real numbers, what can we say about the relative positions of the points $P(a)$, $P(b)$ and $P(a + b)$ on the circle C defined by $x^2 + y^2 = 1$? It follows from the definition of $P(t)$ that the point $P(a + b)$ is obtained by moving from $P(0) = (1, 0)$ a distance $|a + b|$ along C counterclockwise or clockwise according as $a + b$ is positive or negative. However, it is important to realize that $P(a + b)$ can be reached in two steps another way: First, move from $P(0)$ a distance $|a|$ along C, counterclockwise or clockwise according as a is positive or negative. This move will take us to $P(a)$. Second, move from $P(a)$ a distance $|b|$ along C, counterclockwise if b is positive and clockwise if b is negative. By examining the different cases—$a > 0$ and $b > 0$, then $a < 0$ and $b > 0$, etc.—one can verify that the final point reached in these two steps is $P(a + b)$. Note, however, that in the second step we move from $P(a)$ to $P(a + b)$ in exactly the same way that we would move from $P(0)$ to $P(b)$ according to the definition of $P(b)$—by moving along C the same distance and in the same direction. Thus the distance moved along the circle from $P(a)$ to $P(a + b)$ is equal to the distance moved from $P(0)$ to $P(b)$. It follows that the straight-line distances are the same, too. That is, *the straight-line distance between $P(a + b)$ and $P(a)$ is equal to the straight-line distance between $P(b)$ and $P(0)$.* This important fact is illustrated in Figure 3.

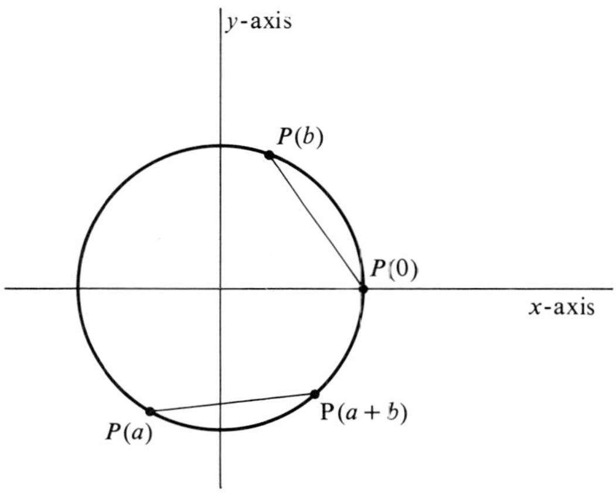

Figure 3

We can now derive a formula for the cosine of the difference of two numbers, $\cos(c - d)$, in terms of the cosines and sines of c and d. Let $a = d$ and $b = c - d$. Then $a + b = c$, and it follows directly from the conclusion of the preceding paragraph that the straight-line distance between $P(c)$ and $P(d)$ is equal to the straight-line distance between $P(c - d)$ and $(1, 0)$. But

$$\begin{aligned} P(c) &= (\cos c, \sin c), \\ P(d) &= (\cos d, \sin d), \\ P(c - d) &= \big(\cos(c - d), \sin(c - d)\big). \end{aligned}$$

Hence by the formula for the distance between two points,

$$\sqrt{(\cos c - \cos d)^2 + (\sin c - \sin d)^2} \\ = \sqrt{\big(\cos (c - d) - 1\big)^2 + \big(\sin (c - d) - 0\big)^2}.$$

Squaring both sides and multiplying out, we get

$$\cos^2 c - 2 \cos c \cos d + \cos^2 d + \sin^2 c - 2 \sin c \sin d + \sin^2 d \\ = \cos^2(c - d) - 2 \cos(c - d) + 1 + \sin^2(c - d).$$

This equation can be greatly simplified by use of the formula $\cos^2 t + \sin^2 t = 1$ three times. The result is

$$2 - 2\cos c \cos d - 2\sin c \sin d = 2 - 2\cos(c - d),$$

from which follows the identity

(1.4) $\cos(c - d) = \cos c \cos d + \sin c \sin d,$ *for all real numbers c and d.*

A similar formula for $\cos(c + d)$ can be obtained easily from (1.3) and (1.4). We have

$$\cos(c + d) = \cos(c - (-d)) = \cos c \cos(-d) + \sin c \sin(-d).$$

Since $\cos(-d) = \cos d$, and $\sin(-d) = -\sin d$, we get

(1.5) $\cos(c + d) = \cos c \cos d - \sin c \sin d,$ *for all real numbers c and d.*

Taking $c = \frac{\pi}{2}$ in (1.4), we get $\cos\left(\frac{\pi}{2} - d\right) = \cos\frac{\pi}{2}\cos d + \sin\frac{\pi}{2}\sin d$.

Since $\cos\frac{\pi}{2} = 0$ and $\sin\frac{\pi}{2} = 1$, the result is the useful equation $\cos\left(\frac{\pi}{2} - d\right) = \sin d$. This equation implies its mate. If we write it letting $d = \frac{\pi}{2} - a$, then $\frac{\pi}{2} - d = a$ and we obtain $\cos a = \sin\left(\frac{\pi}{2} - a\right)$. Thus we have proved the symmetric pair of identities

(1.6)
$$\left.\begin{aligned}\cos\left(\frac{\pi}{2} - a\right) &= \sin a\\ \sin\left(\frac{\pi}{2} - a\right) &= \cos a\end{aligned}\right\}\ \textit{for every real number } a.$$

The remaining two identities are the formulas for the sine of the sum and difference of two numbers. The first follows easily from (1.4) and (1.6):

$$\begin{aligned}\sin(a + b) &= \cos\left(\frac{\pi}{2} - (a + b)\right) = \cos\left(\left(\frac{\pi}{2} - a\right) - b\right)\\ &= \cos\left(\frac{\pi}{2} - a\right)\cos b + \sin\left(\frac{\pi}{2} - a\right)\sin b\\ &= \sin a \cos b + \cos a \sin b.\end{aligned}$$

Thence, by (1.3),

$$\begin{aligned}\sin(a - b) &= \sin\big(a + (-b)\big) \\ &= \sin a \cos(-b) + \cos a \sin(-b) \\ &= \sin a \cos b - \cos a \sin b.\end{aligned}$$

We write these together in the formula

(1.7) $\sin(a \pm b) = \sin a \cos b \pm \cos a \sin b,$ *for all real numbers a and b.*

An alternative approach to the trigonometric functions is made with a domain of angles instead of real numbers. We shall show that the two approaches are in no way contradictory.

It is assumed that the reader knows what an angle is and what its initial side, its terminal side, and its vertex are. An angle α is said to be in standard position on a Cartesian grid if it has its vertex at the origin and its initial side lies along the positive x-axis. If any point, excluding the vertex, on the terminal side is chosen, it has an abscissa x and an ordinate y and lies at a distance d from the origin. Although each point on the terminal side has its x, its y, and its d, it is easy to see that the ratios $\frac{x}{d}$ and $\frac{y}{d}$ are independent of the choice of the point, i.e., in Figure 4 we have $\frac{x}{d} = \frac{x_1}{d_1}$ and $\frac{y}{d} = \frac{y_1}{d_1}$.

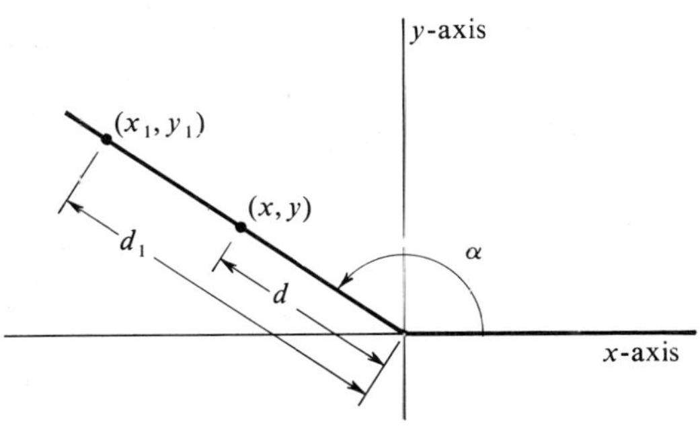

Figure 4

Since the location of the terminal side depends on the angle, each of these ratios is a function of the angle and we define the cosine of α to be $\frac{x}{d}$ and the sine of α to be $\frac{y}{d}$. Since we may choose any point (not the origin) on the terminal side, we could simplify the process by choosing the point where

$d = 1$. This, of course, is the point where the terminal side cuts the circle with equation $x^2 + y^2 = 1$. Then $\cos \alpha$ is the abscissa of that point and $\sin \alpha$ its ordinate. Hence in many ways the two approaches are the same.

We have not yet mentioned units for measuring angles. If we pick a unit to agree with the arc length, we would have exact agreement: The cosine of an angle of u such units is equal to the cosine of the real number u and the sine of an angle of u such units is equal to the sine of the real number u. This unit is called the **radian** and it should be obvious that there are 2π radians in an angle of one revolution. Another unit, probably more familiar to the reader, is the **degree,** which is $\frac{1}{360}$ of a revolution and $\dfrac{\pi}{180}$ of a radian. Thus the cosine of an angle of d degrees is equal to the cosine of the real number $\dfrac{\pi d}{180}$.

PROBLEMS

1. Find the values of

(a) $\sin(-\pi)$ and $\cos(-\pi)$ (d) $\sin\left(\dfrac{\pi}{6}\right)$ and $\cos\left(\dfrac{\pi}{6}\right)$

(b) $\sin(\frac{3}{4}\pi)$ and $\cos(\frac{3}{4}\pi)$ (e) $\sin\left(\dfrac{\pi}{3}\right)$ and $\cos\left(\dfrac{\pi}{3}\right)$

(c) $\sin\left(-\dfrac{\pi}{2}\right)$ and $\cos\left(-\dfrac{\pi}{2}\right)$ (f) $\sin\left(\dfrac{5\pi}{4}\right)$ and $\cos\left(\dfrac{5\pi}{4}\right)$.

2. Make a table like the one below showing the sign of $\cos t$ and $\sin t$ in each of the four quadrants. Put + or − in each entry of the table.

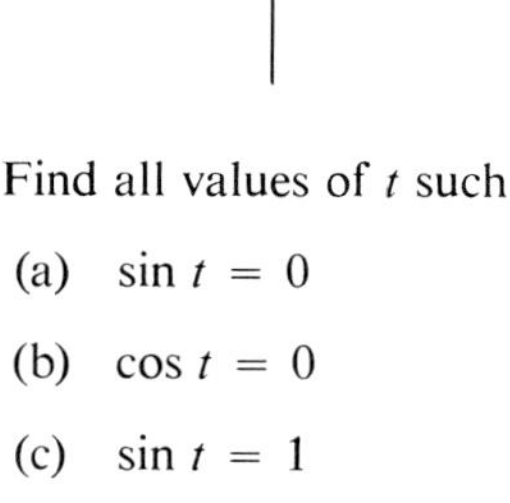

Quadrant	$\cos t$	$\sin t$
I		
II		
III		
IV		

3. Find all values of t such that

(a) $\sin t = 0$ (e) $\sin t = -1$

(b) $\cos t = 0$ (f) $\cos t = -1$

(c) $\sin t = 1$ (g) $\sin t = 2$

(d) $\cos t = 1$ (h) $\cos t = 2$.

4. What is the domain and range of each of the functions cos and sin?

5. If k is an arbitrary integer, find

(a) $\cos k\pi$

(b) $\sin k\pi$

(c) $\cos\left(\frac{\pi}{2} + k\pi\right)$

(d) $\sin\left(\frac{\pi}{2} + k\pi\right)$.

6. Remembering that $\frac{\pi}{4} + \frac{\pi}{6} = \frac{5\pi}{12}$ and $\frac{\pi}{4} - \frac{\pi}{6} = \frac{\pi}{12}$, find

(a) $\sin\frac{5\pi}{12}$

(b) $\cos\frac{5\pi}{12}$

(c) $\sin\frac{\pi}{12}$

(d) $\cos\frac{\pi}{12}$.

7. If f is a function with the property that $f(t + 2\pi) = f(t)$, for every real number t, show from this that

(a) $f(t - 2\pi) = f(t)$, for every real number t.

(b) $f(t + 2\pi n) = f(t)$, for every real number t and every integer n. (Use induction.)

8. (a) Use (1.5) to write a formula for $\cos 2a$ in terms of $\cos a$ and $\sin a$.

(b) Similarly, use (1.7) to write a formula for $\sin 2a$.

(c) Write a formula for $\cos a$ and another for $\sin a$ in terms of $\cos\frac{a}{2}$ and $\sin\frac{a}{2}$.

9. Use the formula for $\cos 2a$ [Problem 8(a)] and the identity $1 = \cos^2 a + \sin^2 a$ to derive a formula for

(a) $\cos^2 a$ in terms of $\cos 2a$

(b) $\sin^2 a$ in terms of $\cos 2a$.

10. Let f be a function which is periodic with period 2π, i.e., $f(t + 2\pi) = f(t)$, and suppose that the graph of f for $0 \leq t \leq 2\pi$ is as shown in Figure 5. Draw the graph of f for $-2\pi \leq t \leq 6\pi$.

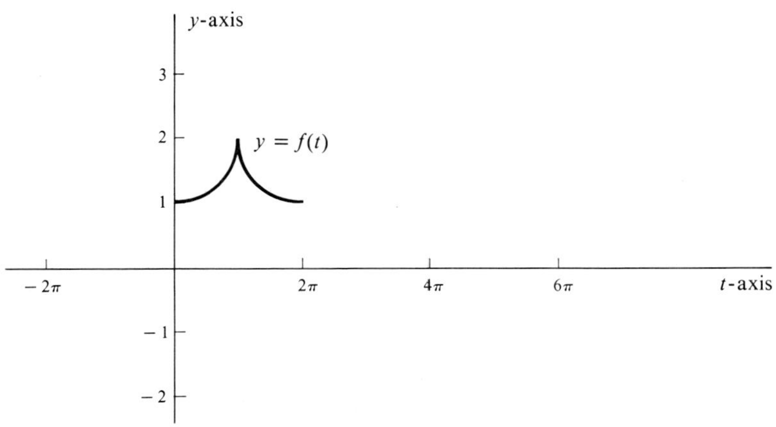

Figure 5

2. Calculus of Sine and Cosine. The formulas for the derivative and integral of the functions sin and cos follow in a straightforward way from one fundamental limit theorem. It is

(2.1)
$$\lim_{t \to 0} \frac{\sin t}{t} = 1.$$

Proof. It is convenient first to impose the restriction that $t > 0$ and prove that the limit from the right equals 1; i.e.,

$$\lim_{t \to 0+} \frac{\sin t}{t} = 1. \tag{1}$$

Since, in proving (1), we are concerned only with small values of t, we may assume that $t < \frac{\pi}{2}$. Thus we have $0 < t < \frac{\pi}{2}$ and, as a consequence, $\sin t > 0$ and $\cos t > 0$. Let S be the region in the plane bounded by the circle $x^2 + y^2 = 1$, the positive x-axis, and the line segment which joins the origin to the point $(\cos t, \sin t)$; i.e., S is the shaded sector in Figure 6. Since

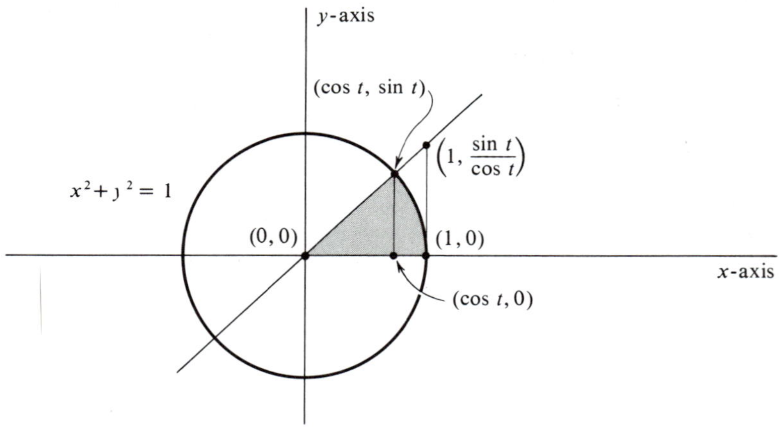

Figure 6

the area of the circle is π and the circumference is 2π, the area of S is equal to $\frac{t}{2\pi} \cdot \pi = \frac{t}{2}$. Next, consider the right triangle T_1 with vertices $(0, 0)$, $(\cos t, \sin t)$, and $(\cos t, 0)$. Since the area of any triangle is one half the base times the altitude, it follows that $area(T_1) = \frac{1}{2} \cos t \sin t$. The line which passes through $(0, 0)$ and $(\cos t, \sin t)$ has slope $\frac{\sin t}{\cos t}$ and equation $y = \frac{\sin t}{\cos t} x$. Setting $x = 1$, we see that it passes through the point

$\left(1, \frac{\sin t}{\cos t}\right)$, as shown in Figure 6. Hence if T_2 is the right triangle with vertices $(0, 0)$, $\left(1, \frac{\sin t}{\cos t}\right)$, and $(1, 0)$, then

$$area(T_2) = \frac{1}{2} \cdot 1 \cdot \frac{\sin t}{\cos t} = \frac{1}{2} \frac{\sin t}{\cos t}.$$

Since T_1 is a subset of S and since S is a subset of T_2, it follows by a fundamental property of area [see (1.3), page 171] that

$$area(T_1) \leq area(S) \leq area(T_2).$$

Hence

$$\frac{1}{2} \cos t \sin t \leq \frac{t}{2} \leq \frac{1}{2} \frac{\sin t}{\cos t}.$$

If we multiply through by $\frac{2}{\sin t}$, we get

$$\cos t \leq \frac{t}{\sin t} \leq \frac{1}{\cos t}.$$

Taking reciprocals and reversing the direction of the inequalities, we obtain finally

$$\frac{1}{\cos t} \geq \frac{\sin t}{t} \geq \cos t. \tag{2}$$

With these inequalities, the proof of (1) is essentially finished. Since the function cos is continuous, we have $\lim_{t \to 0+} \cos t = \cos 0 = 1$. Moreover, the limit of a quotient is the quotient of the limits, and so $\lim_{t \to 0+} \frac{1}{\cos t} = \frac{1}{1} = 1$. Thus $\frac{\sin t}{t}$ lies between two quantities both of which approach 1 as t approaches zero from the right. It follows that

$$\lim_{t \to 0+} \frac{\sin t}{t} = 1.$$

It is now a simple matter to remove the restriction $t > 0$. Since $\frac{\sin t}{t} = \frac{-\sin t}{-t} = \frac{\sin(-t)}{-t}$, we know that

$$\frac{\sin t}{t} = \frac{\sin|t|}{|t|}. \tag{3}$$

As t approaches zero, so does $|t|$; and as $|t|$ approaches zero, we have just proved that the right side of (3) approaches 1. The left side, therefore, also approaches 1, and so the proof is complete.

It is interesting to compare actual numerical values of t and $\sin t$. Table 1 illustrates the limit theorem (2.1) quite effectively.

TABLE 1

t	$\sin t$
0.50	0.4794
0.40	0.3894
0.30	0.2955
0.20	0.1987
0.10	0.0998
0.08	0.0799
0.06	0.0600
0.04	0.0400
0.02	0.0200

A useful corollary of (2.1) is

$$\lim_{t \to 0} \frac{1 - \cos t}{t} = 0. \tag{2.2}$$

Proof. Using trigonometric identities, we write $\frac{1 - \cos t}{t}$ in such a form that (2.1) is applicable.

$$1 = \cos^2 \frac{t}{2} + \sin^2 \frac{t}{2},$$

$$\cos t = \cos\left(\frac{t}{2} + \frac{t}{2}\right) = \cos^2 \frac{t}{2} - \sin^2 \frac{t}{2}.$$

Hence $1 - \cos t = 2 \sin^2 \frac{t}{2}$, and

$$\frac{1 - \cos t}{t} = \frac{2}{t} \sin^2 \frac{t}{2} = \left(\frac{\sin \frac{t}{2}}{\frac{t}{2}}\right) \sin \frac{t}{2}.$$

As t approaches zero, $\frac{t}{2}$ also approaches zero, so, by (2.1), the quantity

$$\frac{\sin \frac{t}{2}}{\frac{t}{2}}$$

approaches 1. Moreover, sin is a continuous function, and therefore $\sin \frac{t}{2}$ approaches $\sin 0 = 0$. The product therefore approaches $1 \cdot 0 = 0$, and the proof is complete.

In writing values of the functions sin and cos, we have thus far avoided the letter x and have not written $\sin x$ and $\cos x$ simply because the point on the circle $x^2 + y^2 = 1$ whose coordinates define the value of cos and sin has nothing to do with, and generally does not lie on, the x-axis. However, when we study sin and cos as two real-valued functions of a real variable, it is natural to use x as the independent variable. We shall not hesitate to do so from now on.

EXAMPLE 1. Evaluate the limits

$$\text{(a)}\ \lim_{x\to 0}\frac{\sin 3x}{\sin 7x}, \qquad \text{(b)}\ \lim_{x\to 0}\frac{1-\cos^2 x}{x}, \qquad \text{(c)}\ \lim_{x\to 0}\frac{\cos x}{\sin x}.$$

We evaluate the first two limits by writing the quotients in such a form that the fundamental trigonometric limit theorem, $\lim_{x\to 0}\dfrac{\sin x}{x} = 1$, is applicable. For (a),

$$\frac{\sin 3x}{\sin 7x} = \frac{\sin 3x}{3x}\,\frac{7x}{\sin 7x}\,\frac{3}{7}.$$

As x approaches zero, so does $3x$ and so does $7x$. Hence $\dfrac{\sin 3x}{3x}$ approaches 1, and $\dfrac{7x}{\sin 7x} = \left(\dfrac{\sin 7x}{7x}\right)^{-1}$ approaches $1^{-1} = 1$. We conclude that

$$\lim_{x\to 0}\frac{\sin 3x}{\sin 7x} = 1\cdot 1\cdot \tfrac{3}{7} = \tfrac{3}{7}.$$

To do (b), we use the identity $\cos^2 x + \sin^2 x = 1$. Thus

$$\frac{1-\cos^2 x}{x} = \frac{\sin^2 x}{x} = \sin x\,\frac{\sin x}{x}.$$

As x approaches zero, $\sin x$ approaches $\sin 0 = 0$, and $\dfrac{\sin x}{x}$ approaches 1. Hence

$$\lim_{x\to 0}\frac{1-\cos^2 x}{x} = 0\cdot 1 = 0.$$

For (c), no limit exists. The numerator approaches 1, and the denominator approaches zero. Note that we cannot even write the limit as $+\infty$ or $-\infty$ because $\sin x$ may be either positive or negative. As a result, $\dfrac{\cos x}{\sin x}$ takes on both arbitrarily large positive values and arbitrarily large negative values as x approaches zero.

We are now ready to find $\frac{d}{dx}\sin x$. The value of the derivative at an arbitrary number a is by definition

$$\left(\frac{d}{dx}\sin x\right)(a) = \lim_{t\to 0}\frac{\sin(a+t)-\sin a}{t}.$$

As always, the game is to manipulate the quotient into a form in which we can see what the limit is. Since $\sin(a+t) = \sin a\cos t + \cos a\sin t$, we have

$$\begin{aligned}\frac{\sin(a+t)-\sin a}{t} &= \frac{\sin a\cos t + \cos a\sin t - \sin a}{t}\\ &= \cos a\frac{\sin t}{t} - \sin a\frac{1-\cos t}{t}.\end{aligned}$$

As t approaches 0, the quantities $\cos a$ and $\sin a$ stay fixed. Moreover, $\frac{\sin t}{t}$ approaches 1, and $\frac{1-\cos t}{t}$ approaches 0. Hence, the right side of the above equation approaches $(\cos a)\cdot 1 - (\sin a)\cdot 0 = \cos a$. We conclude that

$$\left(\frac{d}{dx}\sin x\right)(a) = \cos a, \qquad \textit{for every real number } a.$$

Writing this result as an equality between functions, we get the simpler form

(2.3)
$$\boxed{\frac{d}{dx}\sin x = \cos x.}$$

The derivative of the cosine may be found from the derivative of the sine using the Chain Rule and the twin identities $\cos x = \sin\left(\frac{\pi}{2}-x\right)$ and $\sin x = \cos\left(\frac{\pi}{2}-x\right)$ [see (1.6), page 286].

$$\begin{aligned}\frac{d}{dx}\cos x = \frac{d}{dx}\sin\left(\frac{\pi}{2}-x\right) &= \cos\left(\frac{\pi}{2}-x\right)\frac{d}{dx}\left(\frac{\pi}{2}-x\right)\\ &= \cos\left(\frac{\pi}{2}-x\right)(-1) = -\sin x.\end{aligned}$$

Writing this result in a single equation, we have

(2.4)
$$\boxed{\frac{d}{dx}\cos x = -\sin x.}$$

EXAMPLE 2. Find the following derivatives.

(a) $\dfrac{d}{dx}\sin(x^2+1)$, (c) $\dfrac{d}{dt}\sin e^t$,

(b) $\dfrac{d}{dx}\cos 7x$, (d) $\dfrac{d}{dx}\ln(\cos x)^2$.

These are routine exercises which combine the basic derivatives with the Chain Rule. For (a) we have

$$\frac{d}{dx}\sin(x^2+1) = \cos(x^2+1)\frac{d}{dx}(x^2+1) = 2x\cos(x^2+1).$$

The solution to (b) is

$$\frac{d}{dx}\cos 7x = -\sin 7x\frac{d}{dx}7x = -7\sin 7x.$$

For (c),

$$\frac{d}{dt}\sin e^t = \cos e^t\frac{d}{dt}e^t = e^t\cos e^t,$$

and for (d),

$$\begin{aligned}\frac{d}{dx}\ln(\cos x)^2 &= \frac{1}{(\cos x)^2}\frac{d}{dx}(\cos x)^2\\ &= \frac{1}{(\cos x)^2}2\cos x\frac{d}{dx}\cos x\\ &= \frac{-2\cos x\sin x}{(\cos x)^2} = -\frac{2\sin x}{\cos x}.\end{aligned}$$

Every derivative formula has its corresponding integral formula. For the trigonometric functions sin and cos, they are

(2.5)
$$\boxed{\begin{aligned}\int\sin x\,dx &= -\cos x + c,\\ \int\cos x\,dx &= \sin x + c.\end{aligned}}$$

The proofs consist of simply verifying that the derivative of the proposed integral is the integrand. For example,

$$\frac{d}{dx}(-\cos x + c) = -\frac{d}{dx}\cos x = \sin x.$$

EXAMPLE 3. Find the following integrals.

$$\text{(a)} \int \sin 8x\, dx, \qquad \text{(b)} \int x \cos(x^2)\, dx, \qquad \text{(c)} \int \cos^5 x \sin x\, dx.$$

The solutions use only the basic integral formulas and the fact that if $F' = f$, then $\int f(u)\frac{du}{dx}\, dx = F(u) + c$. Integral (a) is simple enough to write down at a glance:

$$\int \sin 8x\, dx = -\tfrac{1}{8}\cos 8x + c.$$

To do (b), let $u = x^2$. Then $\frac{du}{dx} = 2x$, and

$$\begin{aligned} \int x\cos(x^2)\, dx &= \tfrac{1}{2}\int (\cos(x^2))2x\, dx \\ &= \tfrac{1}{2}\int (\cos u)\frac{du}{dx}\, dx \\ &= \tfrac{1}{2}\sin u + c \\ &= \tfrac{1}{2}\sin(x^2) + c. \end{aligned}$$

For (c), we let $u = \cos x$. Then $\frac{du}{dx} = -\sin x$. Hence

$$\begin{aligned} \int \cos^5 x \sin x\, dx &= -\int \cos^5 x(-\sin x)\, dx \\ &= -\int u^5 \frac{du}{dx}\, dx \\ &= -\tfrac{1}{6}u^6 + c \\ &= -\tfrac{1}{6}\cos^6 x + c. \end{aligned}$$

The graphs of the functions sin and cos are extremely interesting and important curves. To begin with, let us consider the graph of $\sin x$ only for $0 \le x \le \frac{\pi}{2}$. A few isolated points can be plotted immediately (see Table 2).

TABLE 2

x	$y = \sin x$
0	0
$\frac{\pi}{6}$	$\frac{1}{2}$
$\frac{\pi}{4}$	$\frac{1}{2}\sqrt{2} = 0.71$ (approximately)
$\frac{\pi}{3}$	$\frac{1}{2}\sqrt{3} = 0.87$ (approximately)
$\frac{\pi}{2}$	1

The slope of the graph is given by the derivative, $\frac{d}{dx}\sin x = \cos x$. At the origin it is $\cos 0 = 1$, and, where $x = \frac{\pi}{2}$ the slope is $\cos\frac{\pi}{2} = 0$. Since

$$\frac{d}{dx}\sin x = \cos x > 0 \qquad \text{if } 0 < x < \frac{\pi}{2},$$

we know that $\sin x$ is a strictly increasing function on the open interval $\left(0, \frac{\pi}{2}\right)$. In addition, there are no points of inflection on the open interval and the curve is concave downward there because

$$\frac{d^2}{dx^2}\sin x = \frac{d}{dx}\cos x = -\sin x < 0 \qquad \text{if } 0 < x < \frac{\pi}{2}.$$

On the other hand, the second derivative changes sign at $x = 0$, and so there is a point of inflection at the origin. With all these facts we can draw quite an accurate graph. It is shown in Figure 7.

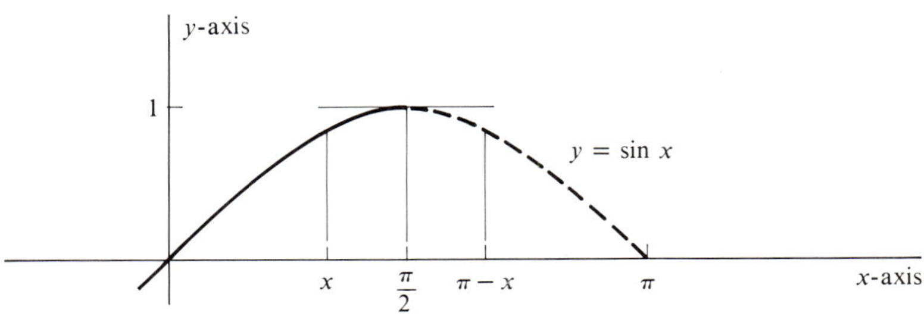

Figure 7

It is now a simple matter to fill in as much of the rest of the graph of $\sin x$ as we like. For every real number x, the points x and $\pi - x$ on the real number line are symmetrically located about the point $\frac{\pi}{2}$. The midpoint between x and $\pi - x$ is given by $\frac{x + (\pi - x)}{2} = \frac{\pi}{2}$. As x increases from 0 to $\frac{\pi}{2}$, the number $\pi - x$ decreases from π to $\frac{\pi}{2}$. Moreover,

$$\begin{aligned} \sin(\pi - x) &= \sin \pi \cos x - \cos \pi \sin x \\ &= 0 \cdot \cos x - (-1) \cdot \sin x \\ &= \sin x. \end{aligned}$$

It follows that the graph of $\sin x$ on the interval $\left[\frac{\pi}{2}, \pi\right]$ is the mirror image of the graph on $\left[0, \frac{\pi}{2}\right]$ reflected across the line $x = \frac{\pi}{2}$. This is the dashed curve in Figure 7. Now, because $\sin x$ is an odd function, its graph for $x \leq 0$ is obtained by reflecting the graph for $x \geq 0$ about the origin (i.e., reflecting first about one coordinate axis and then the other). This gives us the graph for $-\pi \leq x \leq \pi$. Finally, since $\sin x$ is a periodic function with period 2π, its values repeat after intervals of length 2π. It follows that the entire graph of $\sin x$ is the infinite wave, part of which is shown in Figure 8.

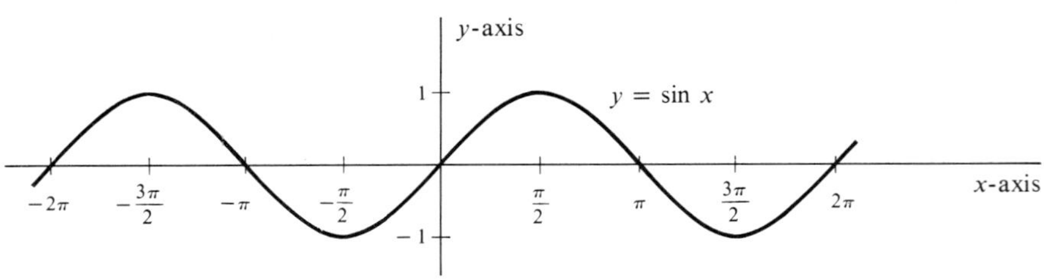

Figure 8

The graph of $\cos x$ is obtained by translating (sliding) the graph of $\sin x$ to the left a distance $\frac{\pi}{2}$. This geometric assertion is equivalent to the algebraic equation $\cos x = \sin\left(x + \frac{\pi}{2}\right)$. But this follows from the trigonometric identity

$$\begin{aligned} \sin\left(x + \frac{\pi}{2}\right) &= \sin x \cos \frac{\pi}{2} + \cos x \sin \frac{\pi}{2} \\ &= (\sin x) \cdot 0 + (\cos x) \cdot 1 \\ &= \cos x. \end{aligned}$$

The graphs of $\cos x$ and $\sin x$ are shown together in Figure 9.

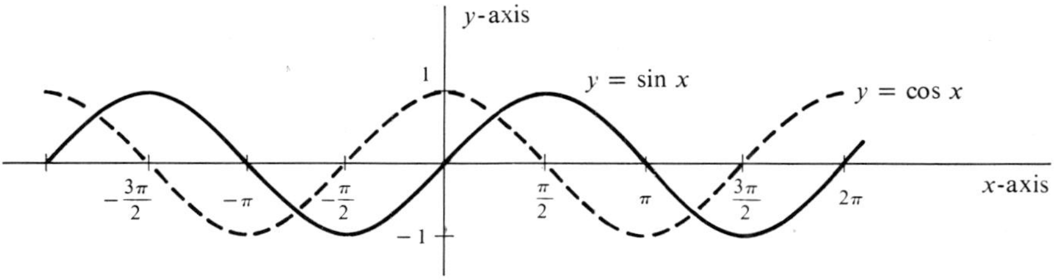

Figure 9

PROBLEMS

1. Evaluate the following limits.

(a) $\lim_{t\to 0} \frac{\sin^2 t}{t^2}$

(b) $\lim_{t\to 0} \frac{\sin t \cos t}{t}$

(c) $\lim_{x\to 0} \frac{1 - \cos^2 x}{x^2}$

(d) $\lim_{x\to 0} \frac{\sin 2x}{x}$

(e) $\lim_{x\to \pi} \frac{\sin(\pi - x)}{x(\pi - x)}$

(f) $\lim_{t\to 0} \frac{\sin 2t}{\sin 3t}$

(g) $\lim_{x\to \pi/2} \frac{\cos x}{\left(\frac{\pi}{2} - x\right)}$

(h) $\lim_{x\to 0} \frac{\cos x}{x}$

(i) $\lim_{x\to 0} \frac{1 - \cos 2x}{x^2}$

(j) $\lim_{x\to 0} \frac{1 - \cos x}{x \sin x}$.

2. Find the derivatives of the following functions.

(a) $\sin(x^2 + 3)$

(b) $\cos e^x$

(c) $\cos t \sin t$

(d) $\cos^2 x + \sin^2 x$

(e) $\cos(\sin x)$

(f) $\ln \sin x$

(g) $\sin^5 x^5$

(h) $\frac{\sin x}{\cos x}$

(i) $\frac{\cos x}{\sin x}$

(j) $e^{-x} \sin x$.

3. Evaluate the following integrals.

(a) $\int \cos 7x\, dx$

(b) $\int (\cos 2x + \sin 3x)\, dx$

(c) $\int e^x \cos e^x\, dx$

(d) $\int \sin(x + a)\, dx$

(e) $\int (\cos x)e^{-\sin x}\, dx$

(f) $\int (\cos t)\cos(\sin t)\, dt$

(g) $\int \dfrac{\sin x}{\cos x}\, dx$

(h) $\int \cos^6 x \sin x\, dx$

(i) $\int \sin^6 x \cos x\, dx$

(j) $\int (\cos^2 x + \sin^2 x)\, dx$

4. Find the integrals

(a) $\int \cos^n x \sin x\, dx$

(b) $\int \sin^n x \cos x\, dx.$

The next two integrals can be reduced to sums of integrals of the forms (a) and (b) by using the identity $\sin^2 x + \cos^2 x = 1$.

(c) $\int \sin^3 x\, dx$

(d) $\int \cos^4 x \sin^3 x\, dx.$

5. Express $\cos^2 x$ in terms of $\cos 2x$, and thence evaluate $\int \cos^2 x\, dx$.
6. Express $\sin^2 x$ in terms of $\cos 2x$, and thence evaluate $\int \sin^2 x\, dx$.
7. Solve the differential equations

(a) $\dfrac{dy}{dx} = e^x \sin e^x$ (b) $\dfrac{dy}{dx} = \dfrac{\sin x}{\cos y}$ (c) $y\dfrac{dy}{dx} = \dfrac{\cos x}{\sin(y^2)}$.

8. Evaluate the following limits using L'Hôpital's Rule.

(a) $\lim_{x \to 0} \dfrac{\sin x}{x}$

(b) $\lim_{x \to 0} \dfrac{1 - \cos x}{x}$

(c) $\lim_{x \to 0} \dfrac{\sin^2 x}{x^2}$

(d) $\lim_{t \to 0} \dfrac{e^t - 1}{t}$

(e) $\lim_{x \to 0} \dfrac{e^x - 1 - x}{x^2}$

(f) $\lim_{x \to 0} \dfrac{1 - \cos x}{x \sin x}$.

9. Draw the graphs of the equations

(a) $y = 3 \sin x$

(b) $y = \cos\left(\dfrac{\pi}{2} x\right)$

(c) $y = \sin(2\pi x)$

(d) $y = 2 \sin\left(x + \dfrac{\pi}{6}\right)$

3. Other Trigonometric Functions. The other trigonometric functions are the tangent, cotangent, secant, and cosecant. They are abbreviated tan, cot (or ctn), sec, and csc, respectively, and the definitions are

$$\tan x = \frac{\sin x}{\cos x} \qquad \sec x = \frac{1}{\cos x}$$

$$\cot x = \frac{\cos x}{\sin x} \qquad \csc x = \frac{1}{\sin x}\,.$$

Unlike sin and cos, these functions are not defined for all real values of x since the denominators in the defining expressions are zero for some values of x. The set of all solutions to the equation $\cos x = 0$ is the set consisting of all odd multiples of $\frac{\pi}{2}$. Hence $\tan x$ and $\sec x$ are defined if and only if x is not an odd multiple of $\frac{\pi}{2}$. Similarly, $\cot x$ and $\csc x$ are defined for all real numbers x except integer multiples of π.

Although sine and cosine were first defined with a domain of real numbers, we have shown that they also can be considered as functions with a domain of angles. Since the other four functions are defined in terms of sine and cosine, they may also be regarded as functions with a domain of angles. Thus it makes sense to speak of the tangent of the angle α, written $\tan \alpha$, and of the cosecant of an angle of 30°, written csc 30°. The former is defined if and only if the radian measure of α is not an odd multiple of $\frac{\pi}{2}$, or, alternatively, if the degree measure of α is not an odd multiple of 90. The latter is the reciprocal of sin 30° and is equal to $\frac{1}{\frac{1}{2}} = 2$.

Two useful trigonometric identities, which are simply alternative statements of the basic equation $\cos^2 x + \sin^2 x = 1$, are derived as follows: Dividing first by $\cos^2 x$, we have

$$\begin{array}{ccccc} \dfrac{\cos^2 x}{\cos^2 x} & + & \dfrac{\sin^2 x}{\cos^2 x} & = & \dfrac{1}{\cos^2 x} \\ \| & & \| & & \| \\ 1 & & \tan^2 x & & \sec^2 x, \end{array}$$

hence $1 + \tan^2 x = \sec^2 x$. On the other hand, if we divide by $\sin^2 x$, we have

$$\begin{array}{ccccc} \dfrac{\cos^2 x}{\sin^2 x} & + & \dfrac{\sin^2 x}{\sin^2 x} & = & \dfrac{1}{\sin^2 x} \\ \| & & \| & & \| \\ \cot^2 x & & 1 & & \csc^2 x, \end{array}$$

and so $\cot^2 x + 1 = \csc^2 x$. Summarizing, we write

(3.1)
$$\boxed{\begin{aligned} 1 + \tan^2 x &= \sec^2 x, \\ \cot^2 x + 1 &= \csc^2 x. \end{aligned}}$$

Another formula which we shall find useful is that for the tangent of the difference of two numbers, $a - b$, in terms of $\tan a$ and $\tan b$. First,

$$\tan(a - b) = \frac{\sin(a - b)}{\cos(a - b)} = \frac{\sin a \cos b - \cos a \sin b}{\cos a \cos b + \sin a \sin b}.$$

Dividing both numerator and denominator by $\cos a \cos b$, we get

$$\tan(a - b) = \frac{\dfrac{\sin a \cos b}{\cos a \cos b} - \dfrac{\cos a \sin b}{\cos a \cos b}}{\dfrac{\cos a \cos b}{\cos a \cos b} + \dfrac{\sin a \sin b}{\cos a \cos b}}$$

$$= \frac{\dfrac{\sin a}{\cos a} - \dfrac{\sin b}{\cos b}}{1 + \dfrac{\sin a}{\cos a}\,\dfrac{\sin b}{\cos b}}.$$

Hence

(3.2)
$$\boxed{\tan(a - b) = \frac{\tan a - \tan b}{1 + \tan a \tan b}.}$$

The trigonometric identities developed in this section are handy tools, and we shall not hesitate to use them. In themselves, however, they are of secondary importance. Any one of them can be derived quickly and in a completely routine way from the basic identities in sin and cos derived in Section 1.

An important application of the tangent function is in connection with the slope of a straight line. We define the angle of inclination α of a straight line L as follows: If L is horizontal, then $\alpha = 0$. If L is not horizontal, then it intersects an arbitrary horizontal line H in a single point P. Let α be the angle with vertex P, initial side the part of H to the right of P, terminal side the part of L above P, and whose measure in radians satisfies the inequality $0 < \alpha < \pi$ (see Figure 10). We contend that

(3.3) *The slope of a line is equal to the tangent of its angle of inclination.*

Proof. We refer again to Figure 10. The given line is L, its inclination is α, and its slope is m. If L' is drawn through the origin parallel to L, then L' also has inclination α and slope m. Furthermore, the point $(\cos\alpha, \sin\alpha)$ lies on L'. Since $(0, 0)$ also lies on L', the definition of slope yields

$$m = \frac{\sin\alpha - 0}{\cos\alpha - 0} = \frac{\sin\alpha}{\cos\alpha} = \tan\alpha,$$

which completes the proof. If L is vertical, its slope is not defined. But then the angle of inclination is $\frac{\pi}{2}$ and $\tan\frac{\pi}{2}$ is not defined either.

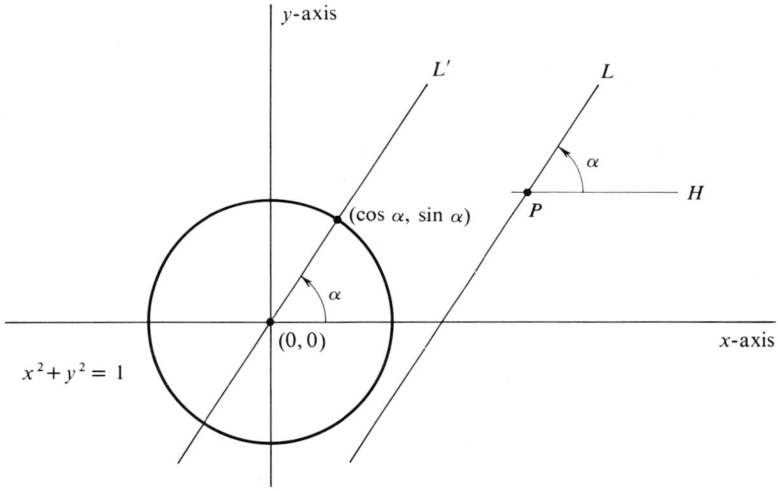

Figure 10

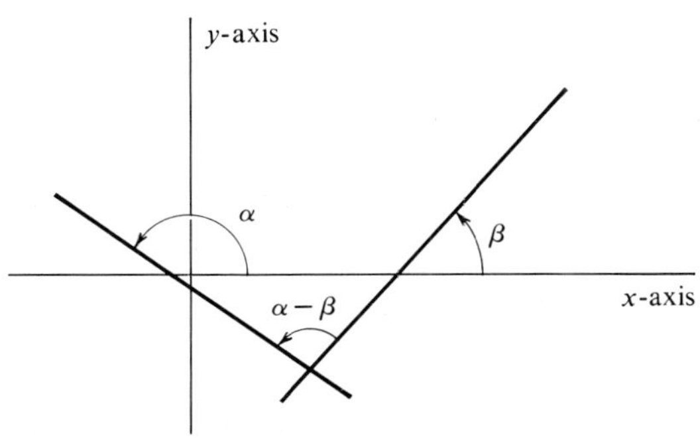

Figure 11

The importance of (3.2) is apparent when we try to determine the angle between two nonvertical intersecting lines. Let α be the angle of inclination of one line and β the angle of inclination of the other. For convenience, we assume that $\alpha > \beta$. It follows that $0 < \alpha - \beta < \pi$ and, from Figure 11, that the difference $\alpha - \beta$ is an angle between the two lines. We denote the slope of the first line by m_1, and that of the second by m_2. That is, we have $m_1 = \tan\alpha$ and $m_2 = \tan\beta$. By (3.2),

$$\tan(\alpha - \beta) = \frac{\tan\alpha - \tan\beta}{1 + \tan\alpha\tan\beta} = \frac{m_1 - m_2}{1 + m_1 m_2}.$$

If this number is positive, then $\alpha - \beta$ is the acute angle between the lines. If this number is negative, then $\alpha - \beta$ is the obtuse angle between the lines. If this number is undefined, then $\alpha - \beta = \frac{\pi}{2}$ and the lines are perpendicular. The number $\frac{m_1 - m_2}{1 + m_1 m_2}$ is undefined if and only if $1 + m_1 m_2 = 0$. Since this equation is equivalent to $m_1 m_2 = -1$, it follows that we have proved the statement made on page 44 that two nonvertical lines with slopes m_1 and m_2 are perpendicular if and only if $m_1 m_2 = -1$.

The formulas for the derivatives of the remaining four trigonometric functions are found using the derivatives of sin and cos together with the usual rules of differentiation. They are

(3.3)

$$\frac{d}{dx}\tan x = \sec^2 x,$$

$$\frac{d}{dx}\cot x = -\csc^2 x,$$

$$\frac{d}{dx}\sec x = \sec x \tan x,$$

$$\frac{d}{dx}\csc x = -\csc x \cot x.$$

Proving the first of these, we have

$$\frac{d}{dx}\tan x = \frac{d}{dx}\frac{\sin x}{\cos x} = \frac{\cos x \frac{d}{dx}\sin x - \sin x \frac{d}{dx}\cos x}{\cos^2 x}$$

$$= \frac{\cos^2 x + \sin^2 x}{\cos^2 x} = \frac{1}{\cos^2 x} = \sec^2 x.$$

The others are left as exercises. We observe the following mnemonic device. From any one of the six formulas for differentiating trigonometric functions another one is obtained by adding the prefix "co" to every function which does not have one, removing the prefix "co" from each function which has it already, and changing the sign. For example, this procedure transforms the equation $\frac{d}{dx}\sec x = \sec x \tan x$ into $\frac{d}{dx}\csc x = -\csc x \cot x$ and transforms $\frac{d}{dx}\cos x = -\sin x$ into $\frac{d}{dx}\sin x = \cos x$. Hence the number of derivative formulas which need to be memorized can be cut in half.

The integrals corresponding to the above derivatives are

(3.4)
$$\begin{aligned}
\int \sec^2 x\,dx &= \tan x + c,\\
\int \csc^2 x\,dx &= -\cot x + c,\\
\int \sec x \tan x\,dx &= \sec x + c,\\
\int \csc x \cot x\,dx &= -\csc x + c.
\end{aligned}$$

EXAMPLE 1. Find the following integrals:

(a) $\int x^2 \sec^2(x^3 + 1)\,dx$,

(b) $\int \tan x\,dx$,

(c) $\int \csc^2 x \cot^5 x\,dx$.

In (a) we observe that $\frac{d}{dx}(x^3 + 1) = 3x^2$, or, equivalently,

$$x^2 = \frac{1}{3}\frac{d}{dx}(x^3 + 1).$$

Hence

$$\begin{aligned}
\int x^2 \sec^2(x^3 + 1)\,dx &= \tfrac{1}{3}\int [\sec^2(x^3 + 1)]\frac{d}{dx}(x^3 + 1)\,dx\\
&= \tfrac{1}{3}\tan(x^3 + 1) + c.
\end{aligned}$$

For (b), we have $\int \tan x\,dx = \int \frac{\sin x}{\cos x}\,dx$. If $u = \cos x$, then $\frac{du}{dx} = -\sin x$, and so

$$\begin{aligned}
\int \tan x\,dx = \int \frac{\sin x}{\cos x}\,dx &= -\int \frac{1}{u}\frac{du}{dx}\,dx\\
&= -\ln|u| + c = -\ln|\cos x| + c.
\end{aligned}$$

Finally, to do (c), we see that since $\frac{d}{dx}\cot x = -\csc^2 x$, the integral is, except for a minus sign, of the form $\int u^5 \frac{du}{dx}\,dx$. Thus

$$\int \csc^2 x \cot^5 x\,dx = -\int (\cot^5 x)\frac{d}{dx}\cot x\,dx$$
$$= -\tfrac{1}{6}\cot^6 x + c.$$

Each of these integrals can be checked by differentiation.

The graph of tan x is an interesting curve, which we now describe. Note, first of all, that tan is an odd function,

$$\tan(-x) = \frac{\sin(-x)}{\cos(-x)} = \frac{-\sin x}{\cos x} = -\tan x,$$

and the graph is therefore symmetric about the origin. Moreover,

$$\tan(x+\pi) = \frac{\sin(x+\pi)}{\cos(x+\pi)} = \frac{\sin x\cos\pi + \cos x\sin\pi}{\cos x\cos\pi - \sin x\sin\pi}$$
$$= \frac{-\sin x}{-\cos x} = \tan x.$$

Thus tan is a periodic function with period π. The slope of the graph is given by the derivative,

$$\frac{d}{dx}\tan x = \sec^2 x,$$

which is positive for every value of x for which tan x is defined. Hence tan x is a strictly increasing function in the interval $-\frac{\pi}{2} < x < \frac{\pi}{2}$. From the definition, $\tan x = \frac{\sin x}{\cos x}$, we see that tan x is positive when both functions are positive, as they are for $0 < x < \frac{\pi}{2}$; is zero when $\sin x = 0$, as it is for $x = 0$; and is negative when the two functions have opposite sign, as they do for $-\frac{\pi}{2} < x < 0$. We also see that tan x takes on arbitrarily large positive values as x approaches $\frac{\pi}{2}$ from the left, since sin x approaches 1 and cos x approaches 0. Thus

$$\lim_{x\to\pi/2-} \tan x = \infty.$$

The second derivative is given by

$$\frac{d^2}{dx^2}\tan x = 2\sec^2 x \tan x \begin{cases} < 0 & \text{if } -\dfrac{\pi}{2} < x < 0, \\ = 0 & \text{if } x = 0, \\ > 0 & \text{if } 0 < x < \dfrac{\pi}{2}, \end{cases}$$

from which it follows that the graph is concave downward for $-\frac{\pi}{2} < x < 0$, concave upward for $0 < x < \frac{\pi}{2}$, and has a point of inflection at the origin. Combining all these facts with the few isolated values shown in Table 3, we obtain the graph shown in Figure 12.

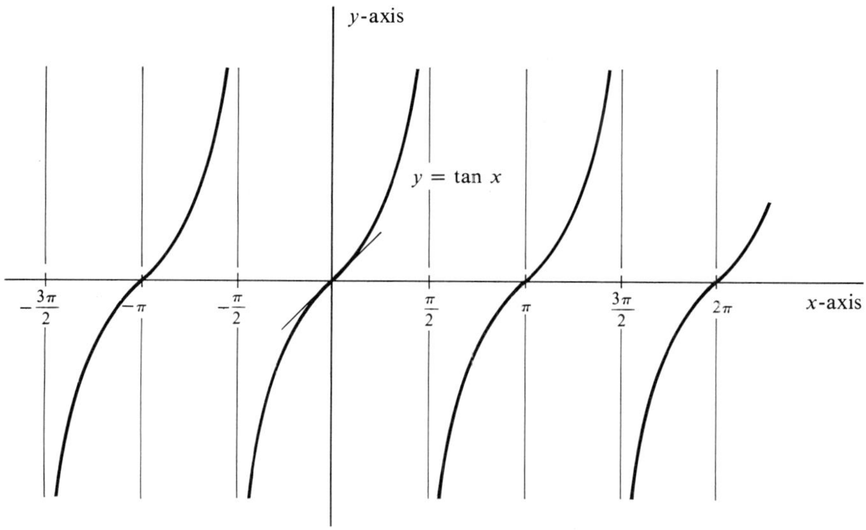

Figure 12

TABLE 3

x	$y = \tan x$	$\frac{dy}{dx} = \sec^2 x$
0	0	1
$\frac{\pi}{6}$	$\frac{1}{\sqrt{3}} = 0.58$ (approx.)	$\frac{4}{3} = 1.33$ (approx.)
$\frac{\pi}{4}$	1	2
$\frac{\pi}{3}$	$\sqrt{3} = 1.73$ (approx.)	4

The graph of cot x can be obtained in the same way in which we worked out the graph of tan x. However, there is a quicker way based on an identity. Since

$$\tan\left(x + \frac{\pi}{2}\right) = \frac{\sin\left(x + \frac{\pi}{2}\right)}{\cos\left(x + \frac{\pi}{2}\right)} = \frac{\sin x \cos \frac{\pi}{2} + \cos x \sin \frac{\pi}{2}}{\cos x \cos \frac{\pi}{2} - \sin x \sin \frac{\pi}{2}}$$

$$= \frac{\cos x}{-\sin x} = -\cot x,$$

we know that

$$\cot x = -\tan\left(x + \frac{\pi}{2}\right) \cdot$$

The geometric significance of this identity is that the graph of cot x is obtained by translating (sliding) the graph of tan x to the left a distance $\frac{\pi}{2}$ and then reflecting about the x-axis.

From the definition, $\sec x = \dfrac{1}{\cos x}$, it is apparent that

$$\sec n\pi = \frac{1}{\cos n\pi} = \begin{cases} 1 & \text{if } n \text{ is even,} \\ -1 & \text{if } n \text{ is odd.} \end{cases}$$

If a is any odd multiple of $\frac{\pi}{2}$, then $\cos a = 0$, and so

$$\lim_{x \to a} |\sec x| = \lim_{x \to a} \frac{1}{|\cos x|} = \infty.$$

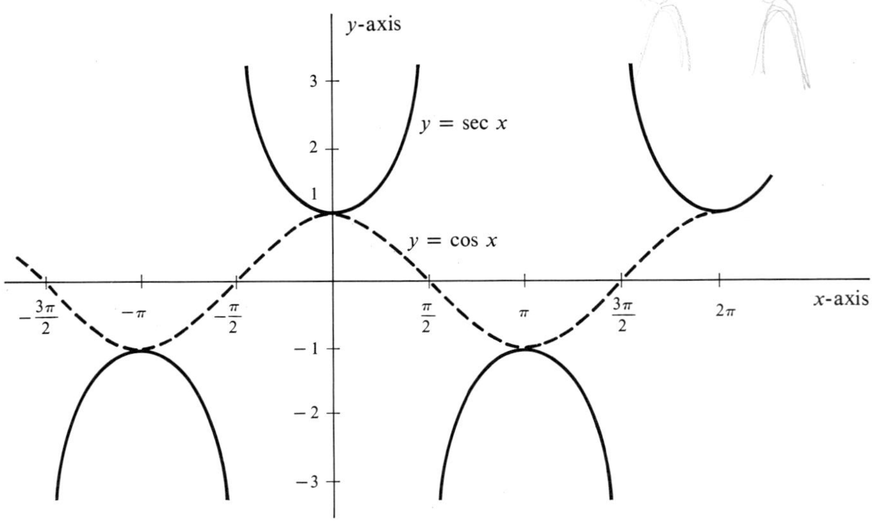

Figure 13

Moreover, on an interval where one function is increasing, its reciprocal function is decreasing, and vice versa. It follows that the over-all shape of the graph of sec x can be ascertained quite easily from the graph of its reciprocal function cos x. The graph of sec x is shown in Figure 13.

The graph of csc x is related to that of sec x in the same way as the graph of sin x is related to the graph of cos x.

PROBLEMS

1. Which of the six trigonometric functions are odd functions and which are even functions?
2. Derive the formulas for the derivatives of the functions cot, sec, and csc.
3. Find the following derivatives.

(a) $\dfrac{d}{dx}\sec^2 x$

(b) $\dfrac{d}{dx}\tan(2x^2 - 1)$

(c) $\dfrac{d}{dx}\ln|\sec x|$

(d) $\dfrac{d}{dy}\cos y \tan y$

(e) $\dfrac{d}{dt}(\sec^2 t - 1)$

(f) $\dfrac{d}{dx}\csc(x^3 - 1)$

(g) $\dfrac{d}{dx}\tan x \cot x$

(h) $\dfrac{d}{dt}\ln|\cot t|$.

4. Prove each of the following identities from the basic identities in sine and cosine developed in Section 1.

(a) $\tan(x + y) = \dfrac{\tan x + \tan y}{1 - \tan x \tan y}$

(b) $\csc x = \sec\left(x - \dfrac{\pi}{2}\right)$

(c) $\cot(a + b) = \dfrac{\cot a \cot b - 1}{\cot a + \cot b}$

(d) $\cot(x + \pi) = \cot x$.

5. Find the following integrals.

(a) $\displaystyle\int \tan 5x\, dx$

(b) $\displaystyle\int \cot x\, dx$

(c) $\displaystyle\int e^x \sec^2 e^x\, dx$

(d) $\displaystyle\int \tan^2 x\, dx$ [*Hint:* Use $\tan^2 x + 1 = \sec^2 x$.]

(e) $\displaystyle\int \tan^4 x \sec^2 x\, dx$

(f) $\int \sec^4 x \tan x \, dx$

(g) $\int \frac{1}{\sqrt{x}} \csc\sqrt{x} \cot\sqrt{x} \, dx$

(h) $\int \csc^4 x \, dx.$

6. Find $\frac{d}{dx}(\sec x + \tan x)$ and use the result to evaluate the integral

$$\int \sec x \, dx = \int \sec x \frac{\sec x + \tan x}{\sec x + \tan x} dx.$$

7. Find $\int \tan x \, dx = \int \frac{\tan x \sec x}{\sec x} dx$ by substituting $u = \sec x$ and $\frac{du}{dx}$ in the right side. Compare the answer obtained with Example 1(b).
8. Draw the graph of (a) $\cot x$, (b) $\csc x$.
9. Evaluate the following limits.

(a) $\lim_{x\to 0} \frac{\tan x}{x}$ (b) $\lim_{x\to 0} x \cot 2x.$

10. Find the tangent of the angle between

(a) the straight lines $y - 2x = 1$ and $2y - x = 4$.

(b) the straight lines $y + 2x = 2$ and $2y - x = 2$.

(c) the tangent lines to the curves $y = x^2$ and $x^2 + y^2 = 1$ at their point of intersection in the first quadrant.

11. What is the domain and the range of each one of the six trigonometric functions?
12. Evaluate each of the following indeterminate forms (see Problem 13, page 271):

(a) $\lim_{x\to 0+} (\sin x)^{\tan x}$ (b) $\lim_{x\to 0+} x^{(1-\cos x)}$.

13. If $\lim_{x\to a} f(x) = \lim_{x\to a} g(x) = \pm\infty$, it is not immediately apparent whether or not $\lim_{x\to a} (f(x) - g(x))$ exists. Such limits are commonly called **indeterminate forms** of the type $\infty - \infty$. The usual method of evaluation is to express the difference $f(x) - g(x)$ as a quotient and then to try to find its limit. For example, we write

$$\frac{e^x}{e^x - 1} - \frac{1}{x} = \frac{xe^x - (e^x - 1)}{x(e^x - 1)},$$

and, as x approaches zero, the limit of the right side can be obtained by two applications of L'Hôpital's Rule. Evaluate

(a) $\lim_{x\to 0} \left(\frac{e^x}{e^x - 1} - \frac{1}{x}\right)$ (b) $\lim_{x\to 0} \left[\frac{(x^2 + 8)^{1/3}}{2x^2} - \frac{1}{x^2}\right]$

(c) $\lim_{x \to 0} \left(\frac{x^2 + 3x + 5}{\sin x} - \frac{5}{x} \right)$ (e) $\lim_{x \to 0+} \left(\frac{1}{x} + \ln x \right)$

(d) $\lim_{t \to 0} \left(\cot t - \frac{1 - 2t}{t} \right)$ (f) $\lim_{x \to \pi/2} (\sec x - \tan x)$.

4. Inverse Trigonometric Functions. The function sin does not have an inverse function. The reason is that it is perfectly possible to have $a \neq b$ and $\sin a = \sin b$. Another way to reach the same conclusion is to consider the equation $x = \sin y$. Its graph is the curve in Figure 14. It does not

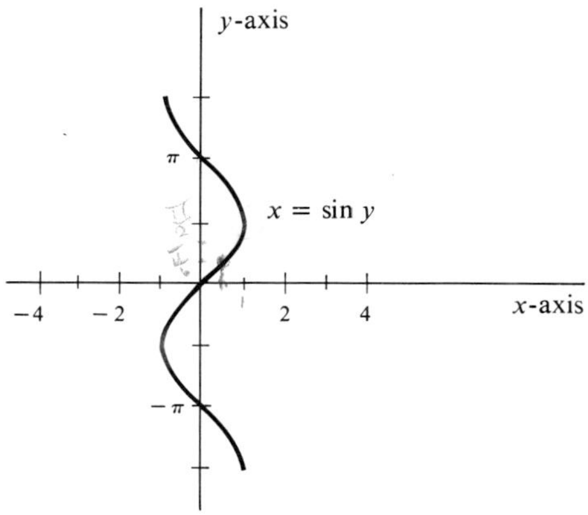

Figure 14

define a function of x because it does not satisfy the condition in the definition of function (see page 14) which asserts that every vertical line intersects the graph of a function in at most one point.

However, on the interval $\left[-\frac{\pi}{2}, \frac{\pi}{2}\right]$ the function sin is a strictly increasing function. Hence, although sin does not have an inverse, it follows by Theorem (2.4), page 250, that the function sin *with domain restricted to* $\left[-\frac{\pi}{2}, \frac{\pi}{2}\right]$ does have an inverse. This inverse function is denoted by either $\sin^{-1}$ or arcsin, and in this book we shall use the latter notation. Thus

$$y = \arcsin x \quad \textit{if and only if} \quad x = \sin y \quad \textit{and} \; -\frac{\pi}{2} \le y \le \frac{\pi}{2}.$$

The graph of the function arcsin is shown in Figure 15(b). It is that part of the graph of the equation $x = \sin y$ for which y satisfies the inequality $-\frac{\pi}{2} \leq y \leq \frac{\pi}{2}$. Note that the graph of arcsin is obtained from the graph of the restricted function sin by reflection across the diagonal line $y = x$. It follows both from the definition of arcsin and also from the illustration that

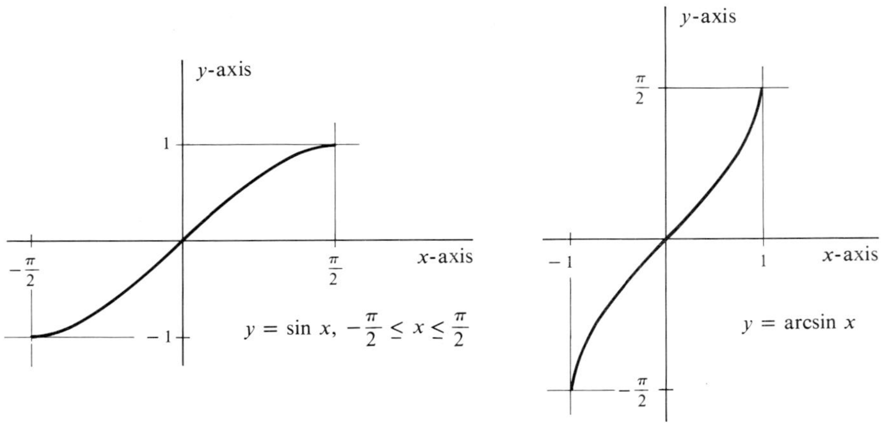

(a) (b)

Figure 15

the domain of arcsin is the closed interval $[-1, 1]$ *and the range is the closed interval* $\left[-\frac{\pi}{2}, \frac{\pi}{2}\right]$.

It is a consequence of Theorem (3.4), page 261, that the function arcsin is differentiable at every point of its domain except at -1 and $+1$. [In applying (3.4), let f be the function sin restricted to $\left[-\frac{\pi}{2}, \frac{\pi}{2}\right]$, and then $f^{-1} = \arcsin$.] We may compute the formula for the derivative either directly from (3.4) or by implicit differentiation. Choosing the latter method, we begin with $y = \arcsin x$ and seek to find $\frac{dy}{dx}$. If $y = \arcsin x$, then $x = \sin y$, and so

$$\frac{d}{dx}x = \frac{d}{dx}\sin y \qquad \text{or} \qquad 1 = \cos y\frac{dy}{dx}.$$

Hence

$$\frac{dy}{dx} = \frac{1}{\cos y}.$$

To express $\cos y$ in terms of x, we use the identity $\cos^2 y + \sin^2 y = 1$ and the fact that $x = \sin y$. Hence $\cos^2 y + x^2 = 1$, and therefore

$$\cos y = \pm\sqrt{1 - x^2}.$$

However, y is restricted by the inequality $-\frac{\pi}{2} \leq y \leq \frac{\pi}{2}$, and in this interval $\cos y$ is never negative. Hence the positive square root is the correct one, and we conclude that $\frac{dy}{dx} = \frac{1}{\sqrt{1 - x^2}}$. Thus

$$\boxed{\frac{d}{dx} \arcsin x = \frac{1}{\sqrt{1 - x^2}}.} \tag{4.1}$$

EXAMPLE 1. Find the domain, range, and derivative of each of the composite functions

$$\text{(a)} \quad \arcsin \frac{1}{1 + x^2}, \qquad \text{(b)} \quad \arcsin(\ln x).$$

The quantity $\frac{1}{1 + x^2}$ is defined for every real number x and also satisfies the inequalities $0 < \frac{1}{1 + x^2} \leq 1$. Hence, $\arcsin \frac{1}{1 + x^2}$ is defined for every x; i.e., its domain is the set of all real numbers. The range of the function $\frac{1}{1 + x^2}$, however, is the half-open interval $(0, 1]$. It can be seen from Figure 15(b) that the function arcsin maps the interval $(0, 1]$ on the x-axis onto the interval $\left(0, \frac{\pi}{2}\right]$ on the y-axis. It follows that the range of the composition $\arcsin \frac{1}{1 + x^2}$ is the half-open interval $\left(0, \frac{\pi}{2}\right]$. The derivative is found using (4.1) and the Chain Rule:

$$\begin{aligned}
\frac{d}{dx} \arcsin \frac{1}{1 + x^2} &= \frac{1}{\sqrt{1 - \left(\frac{1}{1 + x^2}\right)^2}} \frac{d}{dx}\left(\frac{1}{1 + x^2}\right) \\
&= \frac{1}{\sqrt{\frac{(1 + x^2)^2 - 1}{(1 + x^2)^2}}} \frac{-2x}{(1 + x^2)^2} \\
&= \frac{-2x}{(1 + x^2)\sqrt{2x^2 + x^4}} = \frac{-2x}{(1 + x^2)|x|\sqrt{2 + x^2}}.
\end{aligned}$$

For any particular value of x, the quantity $\arcsin(\ln x)$ is defined if and only if $\ln x$ is defined and also lies in the domain of the function arcsin, which is the interval $[-1, 1]$. Thus x must be positive, and $\ln x$ must satisfy the inequalities $-1 \leq \ln x \leq 1$. Hence x must satisfy $\frac{1}{e} \leq x \leq e$. The domain of $\arcsin(\ln x)$ is therefore the interval $\left[\frac{1}{e}, e\right]$, and the range is the same as that of arcsin, i.e., the interval $\left[-\frac{\pi}{2}, \frac{\pi}{2}\right]$. The derivative is given by

$$\frac{d}{dx}\arcsin(\ln x) = \frac{1}{\sqrt{1 - (\ln x)^2}}\frac{d}{dx}\ln x = \frac{1}{x\sqrt{1 - (\ln x)^2}}.$$

The integral formula corresponding to (4.1) is

$$\textbf{(4.2)} \qquad \int \frac{dx}{\sqrt{1 - x^2}} = \arcsin x + c.$$

EXAMPLE 2. Find $\int \frac{x\,dx}{\sqrt{4 - x^4}}$. The first thing we do is to write the denominator, as closely as possible, in the form $\sqrt{1 - u^2}$.

$$\frac{1}{\sqrt{4 - x^4}} = \frac{1}{\sqrt{4\left(1 - \frac{x^4}{4}\right)}} = \frac{1}{2\sqrt{1 - \left(\frac{x^2}{2}\right)^2}}.$$

Hence, letting $u = \frac{x^2}{2}$, we have $\frac{du}{dx} = x$ and

$$\int \frac{x\,dx}{\sqrt{4 - x^4}} = \frac{1}{2}\int \frac{x\,dx}{\sqrt{1 - \left(\frac{x^2}{2}\right)^2}} = \frac{1}{2}\int \frac{1}{\sqrt{1 - u^2}}\frac{du}{dx}\,dx.$$

By (4.2),

$$\frac{1}{2}\int \frac{1}{\sqrt{1 - u^2}}\frac{du}{dx}\,dx = \tfrac{1}{2}\arcsin u + c.$$

Finally, substituting $\frac{x^2}{2}$ for u, we obtain

$$\int \frac{x\,dx}{\sqrt{4 - x^4}} = \frac{1}{2}\arcsin\frac{x^2}{2} + c.$$

The function cos does not have an inverse for the same reason that sin does not. However, a partial inverse can be obtained, just as before, by restricting the domain to an interval on which cos is either increasing or decreasing. Any such interval can be chosen. With the function sin it was natural to choose the largest possible interval containing the number 0—the closed interval $\left[-\frac{\pi}{2}, \frac{\pi}{2}\right]$. With cos the choice is less obvious. However, we shall select the interval $[0, \pi]$, on which cos is strictly decreasing [see Figure 16(a)]. The function cos with domain restricted to $[0, \pi]$ then has an

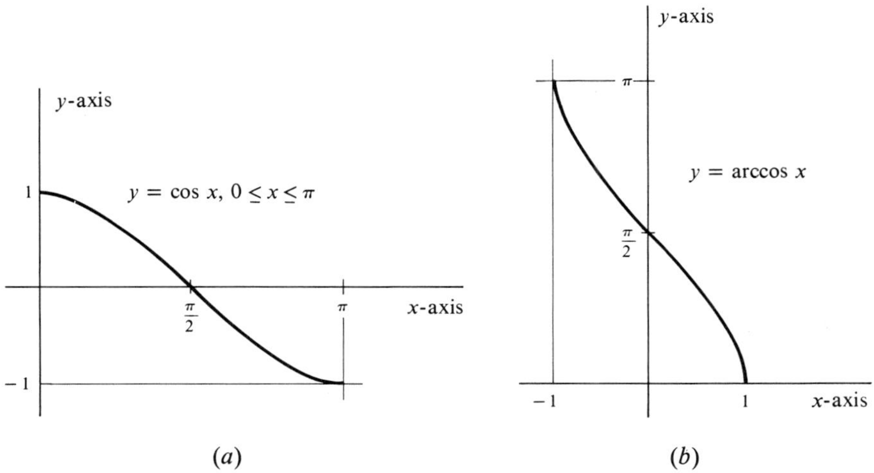

Figure 16

inverse, which is denoted $\cos^{-1}$ or arccos. As before, we shall use the second notation. Thus

$$y = \arccos x \quad \textit{if and only if} \quad x = \cos y \text{ and } 0 \le y \le \pi.$$

The graph of arccos is shown in Figure 16(b).

It should come as no surprise that the two functions arcsin and arccos are closely related. In fact,

(4.3) $$\arcsin x = \frac{\pi}{2} - \arccos x.$$

Proof. Let $y = \frac{\pi}{2} - \arccos x$. Then $\arccos x = \frac{\pi}{2} - y$, and so $x = \cos\left(\frac{\pi}{2} - y\right)$.

Hence

$$x = \cos\left(\frac{\pi}{2} - y\right) = \cos\frac{\pi}{2}\cos y + \sin\frac{\pi}{2}\sin y = \sin y.$$

Since $0 \leq \arccos x \leq \pi$, it follows that $-\pi \leq -\arccos x \leq 0$ and hence $-\frac{\pi}{2} \leq \frac{\pi}{2} - \arccos x \leq \frac{\pi}{2}$, or, equivalently, $-\frac{\pi}{2} \leq y \leq \frac{\pi}{2}$. This, together with $x = \sin y$, implies that $y = \arcsin x$, and the proof is complete.

Note that the validity of (4.3) depends on our having chosen arccos so that its range is the interval $[0, \pi]$.

It follows from (4.3) that the derivative of arccos is the negative of the derivative of arcsin. Thus

(4.4)
$$\frac{d}{dx}\arccos x = -\frac{1}{\sqrt{1 - x^2}}.$$

From (4.4) we see that another indefinite integral of $\frac{1}{\sqrt{1 - x^2}}$ is $-\arccos x$.

Obviously, not one of the six trigonometric functions with unrestricted domain has an inverse. The function tan with its domain restricted to the open interval $\left(-\frac{\pi}{2}, \frac{\pi}{2}\right)$ is strictly increasing and so has an inverse function, which we denote $\tan^{-1}$ or arctan.

$y = \arctan x$ *if and only if* $x = \tan y$ *and* $-\frac{\pi}{2} < y < \frac{\pi}{2}$.

The graph of arctan is obtained by reflecting the graph of tan with domain restricted to $\left(-\frac{\pi}{2}, \frac{\pi}{2}\right)$ across the diagonal line $y = x$. It is shown in Figure 17(b). As can be seen from Figure 17(a), the function tan maps the open interval $\left(-\frac{\pi}{2}, \frac{\pi}{2}\right)$ onto the entire set of real numbers. That is, for every real number y, there exists a real number x in $\left(-\frac{\pi}{2}, \frac{\pi}{2}\right)$ such that $y = \tan x$. Hence *the domain of arctan is the whole real line* $(-\infty, \infty)$. *The range is the interval* $\left(-\frac{\pi}{2}, \frac{\pi}{2}\right)$.

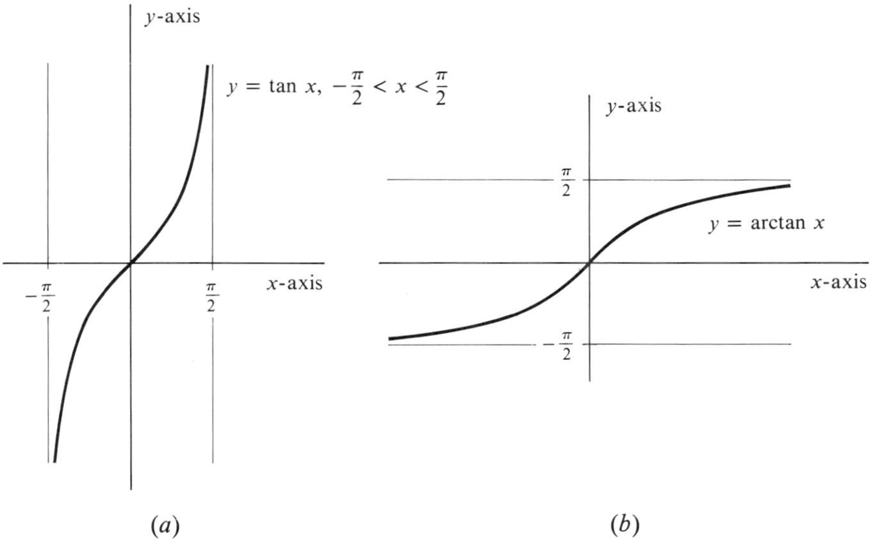

Figure 17

It is a corollary of Theorem (3.4), page 261, that arctan is a differentiable function. We compute the derivative by implicit differentiation. Let $y = \arctan x$. Then $x = \tan y$, and

$$\frac{d}{dx}x = \frac{d}{dx}\tan y \qquad \text{or} \qquad 1 = \sec^2 y\,\frac{dy}{dx}.$$

Hence

$$\frac{dy}{dx} = \frac{1}{\sec^2 y}.$$

From the identity $\sec^2 y = 1 + \tan^2 y$, we get $\sec^2 y = 1 + x^2$. It follows that $\dfrac{dy}{dx} = \dfrac{1}{1 + x^2}$. Thus

(4.5)
$$\frac{d}{dx}\arctan x = \frac{1}{1 + x^2}.$$

The corresponding integral formula is

(4.6)
$$\int \frac{dx}{1 + x^2} = \arctan x + c.$$

EXAMPLE 3. Compute the definite integral $\int_0^1 \frac{dx}{1+x^2}$. We get immediately

$$\int_0^1 \frac{dx}{1+x^2} = \arctan x \Big|_0^1 = \arctan 1 - \arctan 0.$$

Since $\tan 0 = 0$ and $\tan \frac{\pi}{4} = 1$, we know that $0 = \arctan 0$ and $\frac{\pi}{4} = \arctan 1$.

Hence

$$\int_0^1 \frac{dx}{1+x^2} = \frac{\pi}{4}.$$

This is a fascinating result: The number π is equal to four times the area bounded by the curve $y = \frac{1}{1+x^2}$, the x-axis, and the lines $x = 0$ and $x = 1$.

The function cot is strictly decreasing on the open interval $(0, \pi)$. With its domain restricted to this interval, cot therefore has an inverse function, which we denote $\cot^{-1}$ or arccot. The relation between the two functions arccot and arctan is the same as that between arccos and arcsin,

$$\textbf{(4.7)} \qquad \operatorname{arccot} x = \frac{\pi}{2} - \arctan x.$$

The proof is analogous to the proof of (4.3) and is left to the reader as an exercise. It is a corollary that the derivative of arccot is the negative of the derivative of arctan. Hence

$$\textbf{(4.8)} \qquad \boxed{\frac{d}{dx} \operatorname{arccot} x = -\frac{1}{1+x^2}.}$$

Here again, we see another indefinite integral of $\frac{1}{1+x^2}$, the function $-\operatorname{arccot} x$.

The union of the two half-open intervals $\left[0, \frac{\pi}{2}\right)$ and $\left(\frac{\pi}{2}, \pi\right]$ consists of all real numbers x such that $0 \leq x \leq \pi$ and $x \neq \frac{\pi}{2}$. It can be seen from the graph of the equation $y = \sec x$ in Figure 13, page 308, that if a and b are two numbers in the union $\left[0, \frac{\pi}{2}\right) \cup \left(\frac{\pi}{2}, \pi\right]$ and if $a \neq b$, then

$\sec a \neq \sec b$. We omit $\frac{\pi}{2}$ because the secant of that number is not defined. It follows that the function sec with domain restricted to $\left[0, \frac{\pi}{2}\right) \cup \left(\frac{\pi}{2}, \pi\right]$ has an inverse, which is denoted $\sec^{-1}$ or arcsec. Thus

$$y = \operatorname{arcsec} x \quad \textit{if and only if} \quad x = \sec y \text{ and } y \text{ is in } \left[0, \frac{\pi}{2}\right) \cup \left(\frac{\pi}{2}, \pi\right].$$

The graph of the function arcsec is shown in Figure 18(b). *Its range is the union* $\left[0, \frac{\pi}{2}\right) \cup \left(\frac{\pi}{2}, \pi\right]$. As can be seen from Figure 18(a), the function sec maps the set $\left[0, \frac{\pi}{2}\right) \cup \left(\frac{\pi}{2}, \pi\right]$ onto the set of all real numbers with absolute

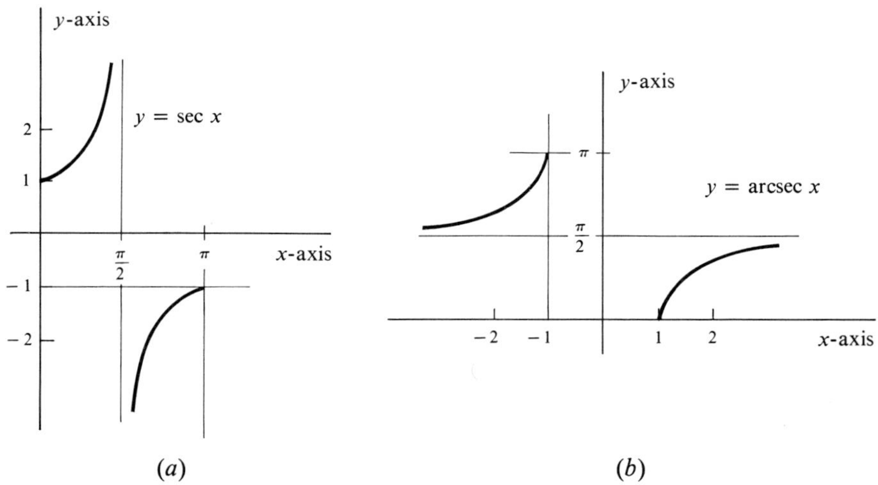

Figure 18

value greater than or equal to 1. Hence *the domain of* arcsec *is the set of all real numbers x such that* $|x| \geq 1$.

The derivative can again be found by implicit differentiation. Let $y = \operatorname{arcsec} x$. Then $x = \sec y$ and

$$\frac{d}{dx} x = \frac{d}{dx} \sec y, \qquad \text{which implies} \qquad 1 = \sec y \tan y \frac{dy}{dx}.$$

Thus

$$\frac{dy}{dx} = \frac{1}{\sec y \tan y}.$$

Using the identity $\sec^2 y = 1 + \tan^2 y$ and the equation $x = \sec y$, we obtain

$$\sec y \tan y = \pm x\sqrt{x^2 - 1}.$$

If $x \geq 1$, then $0 \leq y < \frac{\pi}{2}$, and so $\sec y \tan y$ is nonnegative. Hence

$$\sec y \tan y = x\sqrt{x^2 - 1} \qquad \text{if } x \geq 1.$$

On the other hand, if $x \leq -1$, then $\frac{\pi}{2} < y \leq \pi$, and in this case both $\sec y$ and $\tan y$ are nonpositive. Their product is therefore again nonnegative; i.e.,

$$\sec y \tan y = -x\sqrt{x^2 - 1} \qquad \text{if } x \leq -1.$$

It follows that both cases are covered by the single equation $\sec y \tan y = |x|\sqrt{x^2 - 1}$. Hence $\frac{dy}{dx} = \frac{1}{|x|\sqrt{x^2 - 1}}$, and we have derived the formula

$$\boxed{\frac{d}{dx} \operatorname{arcsec} x = \frac{1}{|x|\sqrt{x^2 - 1}}.} \tag{4.9}$$

The final inverse trigonometric function is the inverse of the cosecant with domain restricted to the union $\left[-\frac{\pi}{2}, 0\right) \cup \left(0, \frac{\pi}{2}\right]$. This function, denoted $\csc^{-1}$ or arccsc, has range equal to $\left[-\frac{\pi}{2}, 0\right) \cup \left(0, \frac{\pi}{2}\right]$ and domain equal to the set of all real numbers x such that $|x| \geq 1$. The analogue of (4.3) and (4.7) is valid. That is,

$$\operatorname{arcsec} x = \frac{\pi}{2} - \operatorname{arccsc} x. \tag{4.10}$$

Proof. The proof mimics that of (4.3). Let $y = \frac{\pi}{2} - \operatorname{arccsc} x$. Then

$\operatorname{arccsc} x = \frac{\pi}{2} - y$, and so $x = \csc\left(\frac{\pi}{2} - y\right)$. Thus

$$x = \csc\left(\frac{\pi}{2} - y\right) = \frac{1}{\sin\left(\frac{\pi}{2} - y\right)} = \frac{1}{\cos y} = \sec y.$$

Since $-\frac{\pi}{2} \leq \operatorname{arccsc} x \leq \frac{\pi}{2}$, it follows that $0 \leq \frac{\pi}{2} - \operatorname{arccsc} x \leq \pi$, or, equivalently, that $0 \leq y \leq \pi$. This, together with $x = \sec y$, implies that $y = \operatorname{arcsec} x$, and the proof is complete.

From this it follows at once that

$$\frac{d}{dx} \operatorname{arccsc} x = -\frac{1}{|x|\sqrt{x^2 - 1}}. \tag{4.11}$$

PROBLEMS

1. Evaluate the following.

(a) $\arcsin 1$

(b) $\arcsin \frac{1}{2}$

(c) $\arccos \frac{1}{2}$

(d) $\arcsin\left(-\frac{\sqrt{3}}{2}\right)$

(e) $\arctan \sqrt{3}$

(f) $\operatorname{arccot} \sqrt{3}$

(g) $\operatorname{arcsec}\left(\frac{2}{\sqrt{3}}\right)$

(h) $\operatorname{arccsc} 2$

(i) $\arcsin(\sin a)$

(j) $\arctan\left(\tan \frac{\pi}{7}\right)$

(k) $\arctan\left(\cot \frac{\pi}{7}\right)$

(l) $\arcsin(\cos a)$

(m) $\tan[\arctan(-1)]$

(n) $\arcsin(2 \sin x \cos x)$

(o) $\arcsin\left(\sin \frac{3\pi}{4}\right)$

(p) $\arctan\left(\cot \frac{\pi}{6}\right)$.

2. Find $\frac{dy}{dx}$.

(a) $y = \arcsin x^2$

(b) $y = \arctan \sqrt{x}$

(c) $y = \arcsin \frac{x - 1}{x + 1}$

(d) $y = \arccos(\cos x)$

(e) $y = \arccos(\sin x)$

(f) $y = \operatorname{arcsec}(1 + x^2)$

(g) $y = \arcsin(x + 1) + \arccos(x + 1)$

(h) $y = \arctan x^3 - \operatorname{arccot} x^3$

(i) $y = \arctan(\ln x)$

(j) $y = \arccos\left(\frac{1}{x}\right) - \operatorname{arcsec} x$.

3. What is the domain and range of each of the functions y of x in Problem 2?
4. Find the following integrals.

(a) $\displaystyle\int \frac{dx}{x^2 + 2}$

(b) $\displaystyle\int \frac{dx}{\sqrt{2 - x^2}}$

(c) $\displaystyle\int \frac{y\,dy}{1 + y^4}$

(d) $\displaystyle\int \frac{dx}{x^2 + 2x + 2}$

(e) $\displaystyle\int \frac{dy}{\sqrt{2y - y^2}}$

(f) $\displaystyle\int \frac{(x + 1)\,dx}{\sqrt{1 - (x + 1)^4}}$

(g) $\displaystyle\int \frac{dx}{\sqrt{x^4 - x^2}}$

(h) $\displaystyle\int \frac{x\,dx}{x^4 + 2x^2 + 2}$.

5. Prove the identity (4.7).
6. Find $\dfrac{d}{dx} \arccos x$ by implicit differentiation.
7. Evaluate the following definite integrals.

(a) $\displaystyle\int_0^{1/2} \frac{dx}{\sqrt{1 - x^2}}$

(b) $\displaystyle\int_{1/2}^{1} \frac{dx}{\sqrt{2x - x^2}}$

(c) $\displaystyle\int_0^{1} \frac{dt}{1 + 3t^2}$

(d) $\displaystyle\int_0^{\sqrt{3}} \frac{dt}{\sqrt{4 - t^2}}$

(e) $\displaystyle\int_0^{x} \frac{dt}{1 + t^2}$

(f) $\displaystyle\int_1^{\sqrt{2}} \frac{dx}{x\sqrt{x^2 - 1}}$.

8. (a) Draw the graph of the function arccot.

   (b) What is the domain and range of arccot?

   (c) Find $\dfrac{d}{dx} \cot x$ by implicit differentiation.

9. Identify the function F defined in Example 1, page 203.
10. Prove the identities

(a) $\operatorname{arcsec} x = \arccos\left(\dfrac{1}{x}\right)$ (b) $\operatorname{arccsc} x = \arcsin\left(\dfrac{1}{x}\right)$.

11. Draw the graph of the function arccsc, and specify its domain and range.

12. Prove or disprove

(a) $\arcsin(-x) = -\arcsin x$ (c) $\arctan(-x) = -\arctan x$

(b) $\arccos(-x) = \arccos x$ (d) $\arccos(-x) = \pi - \arccos x$.

5. Algebraic and Transcendental Functions. We recall that a real-valued function f of one variable is a polynomial if there exist real numbers a_0, $a_1, \ldots, a_n$ such that, for every real number x,

$$f(x) = a_0 + a_1 x + \cdots + a_n x^n = \sum_{k=0}^{n} a_k x^k.$$

Thus, among the different functions f defined respectively by

(a) $f(x) = x^3 + 17x - 2$, (f) $f(x) = \dfrac{1}{x}$,

(b) $f(x) = \pi x^2$, (g) $f(x) = \dfrac{1}{1 + x^2}$,

(c) $f(x) = e^x$, (h) $f(x) = 5$,

(d) $f(x) = x^e$, (i) $f(x) = \sqrt{x + 1}$,

(e) $f(x) = \sin x$, (j) $f(x) = e(x - 1)$,

only those defined in (a), (b), (h), and (j) are polynomials, and the rest are not. In asserting, for example, that the trigonometric function sin is not a polynomial, it is important to realize that we are stating more than just the obvious fact that $\sin x$ does not look like a finite sum of terms of the form $a_k x^k$. We are asserting that it is impossible to write $\sin x$ in this form. The easiest way to prove this is to find some one property which every polynomial has and which sin does not have. For example, if f is a polynomial, then the derivative f' is a polynomial of degree one less. Hence the jth derivative $f^{(j)}$ is the constant function zero, if j is chosen big enough. On the other hand, the jth derivative $\dfrac{d^j}{dx^j} \sin x$ is equal to $\pm\cos x$ or $\pm\sin x$, and is therefore never a constant. This proves that the function sin is not a polynomial.

Similarly, a real-valued function F of two variables will be defined to be a **polynomial** if there exist real numbers a_{ij}, $i = 0, \ldots, m$ and $j = 0, \ldots, n$, such that, for every pair of real numbers x and y,

$$F(x, y) = \sum_{i=0}^{m} \sum_{j=0}^{n} a_{ij} x^i y^j.$$

An alternative formulation, which avoids writing the double sum, is to define a polynomial in two variables to be a function which is the sum of functions each one of which is defined by an expression ax^iy^j, where a is a constant, and i and j are nonnegative integers. Examples of polynomials in two variables are those functions defined by

(a) $F(x, y) = x^3 + 2x^2y + xy^3$,
(b) $G(x, y) = (x + y)yx$,
(c) $f(x, y) = 17xy$,
(d) $H(x, y) = 7x + 2$,
(e) $h(x, y) = 3$.

We come now to the two principal definitions of this section. A function f of one variable is said to be an **algebraic function** if there exists a polynomial F in two variables such that $F(x, f(x)) = 0$, for every x in the domain of f. A **transcendental function** is one which is not algebraic.

EXAMPLE 1. The two functions

(a) $f(x) = \dfrac{x^2 + 1}{2x^3 - 1}$, (b) $g(x) = \sqrt{x^3 + 2}$,

are both algebraic. To show that f is an algebraic function, let $y = \dfrac{x^2 + 1}{2x^3 - 1}$. Then $y(2x^3 - 1) = x^2 + 1$, or, equivalently, $2x^3y - y - x^2 - 1 = 0$. Hence if we let F be the polynomial defined by

$$F(x, y) = 2x^3y - y - x^2 - 1,$$

it will be true that $F(x, f(x)) = 0$. This is not surprising, since the polynomial $F(x, y)$ was invented precisely to make the last equation true. Checking, we get

$$\begin{aligned} F(x, f(x)) &= 2x^3 \frac{x^2 + 1}{2x^3 - 1} - \frac{x^2 + 1}{2x^3 - 1} - x^2 - 1 \\ &= (2x^3 - 1)\frac{x^2 + 1}{2x^3 - 1} - x^2 - 1 \\ &= x^2 + 1 - x^2 - 1 = 0. \end{aligned}$$

The function g can be shown to be algebraic by letting $y = \sqrt{x^3 + 2}$. Squaring both sides, we obtain $y^2 = x^3 + 2$, which is equivalent to $y^2 - x^3 - 2 = 0$. Hence if we define the polynomial F by the equation

$$F(x, y) = y^2 - x^3 - 2,$$

then $F(x, g(x)) = 0$. Checking, we have

$$\begin{aligned} F(x, g(x)) &= (\sqrt{x^3 + 2})^2 - x^3 - 2 \\ &= x^3 + 2 - x^3 - 2 = 0. \end{aligned}$$

A rational function is one which can be expressed as the ratio of two polynomials. That is, a function f of one variable is rational if there exist polynomials p and q of one variable such that $f(x) = \dfrac{p(x)}{q(x)}$. The technique used in Example 1 to show that the function $\dfrac{x^2 + 1}{2x^3 - 1}$ is algebraic can be applied to any rational function of one variable. Thus we have the theorem

(5.1) *Any rational function f of one variable is algebraic.*

Proof. Since f is rational, there exist polynomials p and q such that $f(x) = \dfrac{p(x)}{q(x)}$. Letting $y = \dfrac{p(x)}{q(x)}$, we obtain $yq(x) = p(x)$, which is equivalent to $yq(x) - p(x) = 0$. The function F defined by

$$F(x, y) = yq(x) - p(x)$$

is a polynomial in x and y. Substituting $f(x)$ for y, and then $\dfrac{p(x)}{q(x)}$ for $f(x)$, we obtain

$$\begin{aligned} F(x, f(x)) &= f(x)q(x) - p(x) \\ &= \frac{p(x)}{q(x)}q(x) - p(x) \\ &= p(x) - p(x) = 0, \end{aligned}$$

which completes the proof.

The function g defined by $g(x) = \sqrt{x^3 + 2}$ is an example of an algebraic function which is not rational. (A simple proof of this fact is suggested in Problem 2.) Thus the set of all rational functions of one variable is a proper subset of the larger set of all algebraic functions of one variable.

It is by no means obvious that transcendental functions exist. However, we have actually encountered quite a few such functions already. Although a proof of the next theorem is too advanced to give in this book, it is important to know that it is true.

(5.2) *The following functions are transcendental:*

(i) $\ln x$,
(ii) e^x,
(iii) a^x, for any $a > 0$, $a \neq 1$,
(iv) $\log_a x$, for any $a > 0$, $a \neq 1$,
(v) $\sin x$, $\cos x$, $\tan x$, $\cot x$, $\sec x$, $\csc x$,
(vi) $\arcsin x$, $\arccos x$, $\arctan x$, $\text{arccot}\, x$, $\text{arcsec}\, x$, $\text{arccsc}\, x$.

Another theorem [not so deep as (5.2), but still beyond the scope of this book] states that if f is an algebraic function, then the derivative f' is also algebraic. However, the converse is false. In particular, we know that

$$\frac{d}{dx} \ln x = \frac{1}{x},$$

and $\frac{1}{x}$ is not only algebraic, but also rational. In addition, the formulas in Section 4 for the derivatives of the inverse trigonometric functions show that every one of these six transcendental functions has a derivative which is algebraic.

PROBLEMS

1. In each of the following examples identify the function f as a polynomial or not. If it is not a polynomial, give a reason. (Consider such things as the vanishing of higher-order derivatives, or the behavior of f, or of some derivative $f^{(j)}$, near a point of discontinuity.)

(a) $f(x) = \frac{1}{x}$ (e) $f(x) = \sqrt{x^2 - 2}$

(b) $f(x) = \frac{x-1}{x+1}$ (f) $f(x) = x(x^2 - 7)$

(c) $f(x) = \pi x^2 + ex + 2$ (g) $f(x) = e^x$

(d) $f(x) = x^{2/3} + x^{1/3}$ (h) $f(x) = \tan x$.

2. Prove that the algebraic function g defined by $g(x) = \sqrt{x^3 + 2}$ is not rational. [*Hint:* Suppose it is rational. Then there exist polynomials p and q such that $\sqrt{x^3 + 2} = \frac{p(x)}{q(x)}$, for every $x \geq -\sqrt[3]{2}$. But then

$$x^3 + 2 = \left[\frac{p(x)}{q(x)}\right]^2,$$

or, equivalently,

$$(x^3 + 2)[q(x)]^2 - [p(x)]^2 = 0, \qquad \text{for all } x \geq -\sqrt[3]{2}.$$

The left side of this equation is a polynomial which is not identically zero. (Why?) How many roots can such a polynomial have?]

3. Prove that each of the following function f is algebraic by exhibiting a polynomial $F(x, y)$ and showing that $F(x, f(x)) = 0$.

(a) $f(x) = \sqrt{\frac{x+1}{x-1}}$ (d) $f(x) = \frac{d}{dx} \operatorname{arcsec} x$

(b) $f(x) = \frac{d}{dx} \arctan x$ (e) $f(x) = \ln 5^x$

(c) $f(x) = \frac{d}{dx} \arcsin x$ (f) $f(x) = 2x + \sqrt{4x^2 - 1}$.

6. Complex Numbers. Since the square of a real number is never negative, the equation $x^2 = -1$ has no solution in the set R of all real numbers. However, we shall show that R can be considered as a subset of a larger set C which has the following properties: (i) The sum and product of any two elements in C are defined, and addition and multiplication obey the ordinary laws of algebra. (ii) There is an element i in C such that $i^2 = -1$. (iii) Every element in C can be written in the form $x + iy$, where x and y are real numbers. The elements of the set C are called **complex numbers.** Let us assume, for the moment, that the existence of C, obeying the three properties, has already been demonstrated. Then the sum of two complex numbers $x_1 + iy_1$ and $x_2 + iy_2$ is given by

$$(x_1 + iy_1) + (x_2 + iy_2) = (x_1 + x_2) + i(y_1 + y_2). \tag{1}$$

For the product, we have

$$\begin{aligned}(x_1 + iy_1)(x_2 + iy_2) &= x_1x_2 + iy_1x_2 + iy_2x_1 + i^2y_1y_2 \\ &= x_1x_2 + i(x_1y_2 + x_2y_1) + i^2y_1y_2.\end{aligned}$$

However, since $i^2 = -1$, we get

$$(x_1 + iy_1)(x_2 + iy_2) = (x_1x_2 - y_1y_2) + i(x_1y_2 + x_2y_1). \tag{2}$$

For example,

$$\begin{aligned}(2 + i) + (5 - i3) &= 7 - i2, \\ (2 + i)(5 - i3) &= 10 + i5 - i6 - i^23 \\ &= 10 - i - (-1)3 \\ &= 13 - i.\end{aligned}$$

We turn now to the task of showing that there is a set C having the properties listed in (i), (ii), and (iii). We shall take for C the set R^2 of all ordered pairs of real numbers, i.e., the xy-plane. Thus a complex number is by definition an ordered pair (x, y) of real numbers. Up to this point we have not ascribed any algebraic structure to the xy-plane, and so we must define what we mean by addition and multiplication of ordered pairs of real numbers. Later in this section we shall show how to express the ordered pair (x, y) in the traditional form $x + iy$. Anticipating this, however, we use equations (1) and (2) to motivate the definitions of the sum and product of ordered pairs. We define

$$(x_1, y_1) + (x_2, y_2) = (x_1 + x_2, y_1 + y_2), \tag{3}$$

$$(x_1, y_1)(x_2, y_2) = (x_1x_2 - y_1y_2, x_1y_2 + x_2y_1). \tag{4}$$

It was stated that addition and multiplication of complex numbers are to obey the ordinary laws of algebra. By this we mean that the following six propositions are true. The basic algebraic properties which they describe are the same as those for the real numbers, and these six statements should be compared with the corresponding list on page 2. Abbreviating (x_1, y_1), (x_2, y_2), and (x_3, y_3) by z_1, z_2, and z_3, respectively, we have

(6.1) ASSOCIATIVE LAWS.

$$z_1 + (z_2 + z_3) = (z_1 + z_2) + z_3, \qquad z_1(z_2z_3) = (z_1z_2)z_3.$$

(6.2) COMMUTATIVE LAWS.

$$z_1 + z_2 = z_2 + z_1, \qquad z_1z_2 = z_2z_1.$$

(6.3) DISTRIBUTIVE LAW.

$$(z_1 + z_2)z_3 = z_1z_3 + z_2z_3.$$

(6.4) EXISTENCE OF IDENTITIES. *The two complex numbers* $0' = (0, 0)$ *and* $1' = (1, 0)$ *have the properties that* $0' + z = z$ *and* $1'z = z$ *for every* z in C.

(6.5) EXISTENCE OF SUBTRACTION. *For every complex number* $z = (x, y)$, *the complex number* $(-x, -y)$ *is denoted by* $-z$ *and has the property that* $z + (-z) = 0'$. [The expression $z_1 - z_2$ is an abbreviation for $z_1 + (-z_2)$.]

(6.6) EXISTENCE OF DIVISION. *For every complex number* $z = (x, y)$ *different from* $0'$, *the complex number* $\left(\frac{x}{x^2 + y^2}, \frac{-y}{x^2 + y^2}\right)$ *is denoted by* z^{-1} *or* $\frac{1}{z}$ *and has the property that* $zz^{-1} = 1'$. $\Big($The expression $\frac{z_1}{z_2}$ is an abbreviation for $z_1z_2^{-1}$.$\Big)$

Proof. The proofs are simple exercises using the definitions and the algebraic properties of real numbers. We give the proofs of (6.4) and (6.6) and leave the others for the reader to supply. It is asserted in (6.4) that the complex number $(0, 0)$, which is abbreviated $0'$ (and later simply as 0), is an additive identity. Letting $z = (x, y)$, we have

$$0' + z = (0, 0) + (x, y) = (0 + x, 0 + y) = (x, y) = z,$$

which proves the assertion. Similarly, for the multiplicative identity $1' = (1, 0)$ (later to be abbreviated simply by 1), we obtain

$$1'z = (1, 0)(x, y) = (1x - 0y,\ 1y + x0) = (x, y) = z,$$

and the proof of (6.4) is complete.

To prove (6.6), let $z = (x, y)$ be a complex number different from $0' = (0, 0)$. It follows that $x^2 + y^2$ is positive. Hence the ordered pair $\left(\frac{x}{x^2 + y^2}, \frac{-y}{x^2 + y^2}\right)$ is defined, and (in anticipation of the proof) is denoted z^{-1}. Multiplying, we get

$$\begin{aligned} zz^{-1} &= (x, y)\left(\frac{x}{x^2 + y^2}, \frac{-y}{x^2 + y^2}\right) \\ &= \left(\frac{x^2}{x^2 + y^2} - \frac{-y^2}{x^2 + y^2}, \frac{-xy}{x^2 + y^2} + \frac{xy}{x^2 + y^2}\right) \\ &= \left(\frac{x^2 + y^2}{x^2 + y^2}, 0\right) = (1, 0) \\ &= 1', \end{aligned}$$

which completes the proof of (6.6).

Assuming that the remaining four propositions have been proved, we have now satisfied requirement (i) in the first paragraph of the section for the set C of complex numbers: Addition and multiplication are defined and obey the ordinary laws of algebra. But what about the prior assumption that the set R of all real numbers can be considered a subset of C? Of course, it is not actually a subset, since no real number is also an ordered pair of real numbers. However, there is a subset of C which has all the properties of R. This subset is the x-axis, the set of all complex numbers whose second coordinate is zero. Speaking informally, we shall identify R with the x-axis in C by identifying an arbitrary real number x with the complex number $(x, 0)$. Proceeding formally, we define a function whose value for each real number x is denoted by x' and defined by $x' = (x, 0)$. This function sets up a one-to-one correspondence between the set R and the x-axis in C. Essential, however, is the fact that this correspondence preserves the algebraic operations of addition and multiplication. To show that this is so, let x_1 and x_2 be any two real numbers. Then, for addition,

$$x_1' + x_2' = (x_1, 0) + (x_2, 0) = (x_1 + x_2, 0) = (x_1 + x_2)',$$

and for multiplication,

$$\begin{aligned} x_1'x_2' = (x_1, 0)(x_2, 0) &= (x_1x_2 - 0 \cdot 0, x_1 \cdot 0 + x_2 \cdot 0) \\ &= (x_1x_2, 0) = (x_1x_2)'. \end{aligned}$$

It follows that the algebraic properties of the set R of all real numbers are identical with those of the x-axis in C. It is therefore legitimate to make the identification, and henceforth we shall denote x' simply by x. Note that in so doing, the additive identity $0'$ and the multiplicative identity $1'$, referred to in (6.4), become simply 0 and 1, respectively.

We now define i to be the complex number (0, 1). Requirement (ii) in the first paragraph of this section is easily seen to be satisfied:

$$\begin{aligned} i^2 = (0, 1)(0, 1) &= (0 \cdot 0 - 1 \cdot 1, 0 \cdot 1 + 0 \cdot 1) \\ &= (-1, 0) = -(1, 0). \end{aligned}$$

Since we have agreed to write $(1, 0) = 1' = 1$, we therefore obtain the famous equation

(6.7) $$\boxed{i^2 = -1.}$$

Requirement (iii) is

(6.8) *If* $z = (x, y)$ *is an arbitrary complex number, then* $z = x + iy$.

Proof. The expression $x + iy$ is an abbreviation for the more formal $x' + iy'$. Hence

$$\begin{aligned} x + iy &= (x, 0) + (0, 1)(y, 0) \\ &= (x, 0) + (0 \cdot y - 1 \cdot 0,\ 0 \cdot 0 + y \cdot 1) \\ &= (x, 0) + (0, y) = (x, y) \\ &= z, \end{aligned}$$

completing the proof, and also our construction of the set C of complex numbers.

EXAMPLE 1. If $z_1 = 4 + i3$, $z_2 = 4 - i3$, and $z_3 = 7 - i2$, then find $z_1 + z_2$, z_1z_2, $3z_1 - 2z_3$, and z_1z_3. These are simply routine exercises involving the addition, subtraction, and multiplication of complex numbers.

$$\begin{aligned} z_1 + z_2 &= (4 + i3) + (4 - i3) = 8, \\ z_1z_2 &= (4 + i3)(4 - i3) = 16 + i12 - i12 - i^29 \\ &= 16 - (-1)9 = 25, \\ 3z_1 - 2z_3 &= 3(4 + i3) - 2(7 - i2) \\ &= 12 + i9 - 14 + i4 = -2 + i13, \\ z_1z_3 &= (4 + i3)(7 - i2) = 28 + i21 - i8 - i^26 \\ &= 28 - (-1)6 + i13 = 34 + i13. \end{aligned}$$

EXAMPLE 2. If $z_1 = 2 + i3$ and $z_2 = 3 - i$, plot z_1, z_2, and $z_1 + z_2$ on the complex plane. We have

$$z_1 + z_2 = (2 + i3) + (3 - i) = 5 + i2,$$

and the three points are shown in Figure 19.

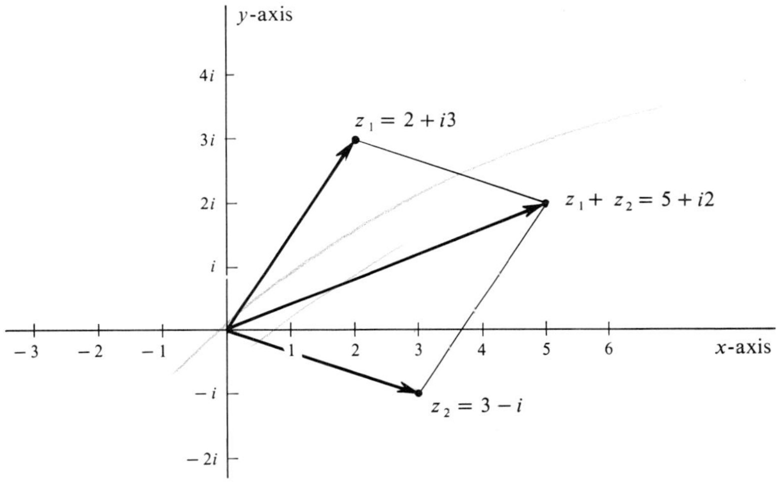

Figure 19

A complex number is a point in the xy-plane and can be indicated in a picture by a dot. Another useful geometric representation of z is an arrow with its tail at the origin and its head at the point z. We have drawn these arrows in Figure 19. Note that *if P is the parallelogram whose adjacent sides are the arrows representing z_1 and z_2, then the diagonal of P which has the origin as an endpoint is the arrow representing the sum $z_1 + z_2$.* The definition of addition in C implies that this parallelogram principle is valid for every pair of complex numbers. It provides a good method for adding complex numbers geometrically.

When the complex number i was first introduced in mathematics, it was regarded as highly mysterious and was called an **imaginary number,** and this terminology has survived. If $z = x + iy$ is an arbitrary complex number, then by definition x is the **real part** of z, and y is the **imaginary part** of z. Note that the imaginary part of a complex number is a real number. A complex number whose real part is zero, i.e., one that lies on the y-axis, is called **pure imaginary.** It is important to remember that

(6.9) *Two complex numbers are equal if and only if their real parts are equal and their imaginary parts are equal. That is,* $x_1 + iy_1 = x_2 + iy_2$ *if and only if* $x_1 = x_2$ *and* $y_1 = y_2$.

Proof. We know that $x_1 + iy_1 = (x_1, y_1)$ and $x_2 + iy_2 = (x_2, y_2)$. But two ordered pairs are equal if and only if their first coordinates are equal and their second coordinates are equal, and so (6.9) is proved. Another proof, which uses only the algebraic properties of complex numbers rather than their explicit construction as ordered pairs, is the following. Suppose that $x_1 + iy_1 = x_2 + iy_2$. Then $x_1 - x_2 = i(y_2 - y_1)$, and squaring both sides we obtain

$$(x_1 - x_2)^2 = i^2(y_2 - y_1)^2 = -(y_2 - y_1)^2.$$

The left side is a nonnegative real number, and the right side is a nonpositive real number. They can be equal only if both are zero. Hence $x_1 = x_2$ and $y_2 = y_1$. The converse proposition is, of course, trivial.

The **absolute value,** or **modulus,** of a complex number $z = x + iy$ is a nonnegative real number denoted by $|z|$ and defined by

$$|z| = \sqrt{x^2 + y^2}.$$

The geometric significance of $|z|$ is that it is the distance between the point z and the origin in the complex plane. If z is represented by an arrow, then $|z|$ is the length of the arrow. Note that if z is real, i.e., if its imaginary part is equal to zero, then the absolute value of z is simply its absolute value as a real number. Thus, if $z = x + iy$ and $y = 0$, then

$$|z| = \sqrt{x^2 + 0^2} = \sqrt{x^2} = |x|.$$

An important property of the absolute value is the following.

(6.10) *The absolute value of the product of two complex numbers is the product of their absolute values;* i.e., $|z_1z_2| = |z_1|\,|z_2|$.

Proof. Let $z_1 = x_1 + iy_1$ and $z_2 = x_2 + iy_2$. Then we have $z_1z_2 = x_1x_2 - y_1y_2 + i(x_1y_2 + x_2y_1)$. Hence, by the definition of absolute value,

$$|z_1z_2|^2 = (x_1x_2 - y_1y_2)^2 + (x_1y_2 + x_2y_1)^2.$$

Simplifying, we get

$$\begin{aligned}|z_1z_2|^2 &= x_1^2x_2^2 - 2x_1x_2y_1y_2 + y_1^2y_2^2 + x_1^2y_2^2 + 2x_1x_2y_1y_2 + x_2^2y_1^2\\ &= x_1^2(x_2^2 + y_2^2) + y_1^2(x_2^2 + y_2^2)\\ &= (x_1^2 + y_1^2)(x_2^2 + y_2^2)\\ &= |z_1|^2\,|z_2|^2.\end{aligned}$$

Thus $|z_1z_2|^2 = |z_1|^2\,|z_2|^2$, and the proof is completed by taking the positive square root of each side of the equation.

As an illustration of Theorem (6.10), consider the complex numbers z_1 and z_2 shown in Figure 19. The product z_1z_2 is equal to

$$\begin{aligned} z_1z_2 &= (2 + i3)(3 - i) = 6 + i9 - i2 - i^23 \\ &= 9 + i7. \end{aligned}$$

The absolute values are

$$\begin{aligned} |z_1| &= \sqrt{2^2 + 3^2} = \sqrt{13}, \\ |z_2| &= \sqrt{3^2 + (-1)^2} = \sqrt{10}, \\ |z_1z_2| &= \sqrt{9^2 + 7^2} = \sqrt{130}, \end{aligned}$$

which is in agreement with (6.10).

If $z = x + iy$, then the **complex conjugate** of z, denoted by $\bar{z}$, is defined to be the complex number

$$\bar{z} = x - iy.$$

The product of a complex number and its complex conjugate is always a nonnegative real number, since

$$z\bar{z} = (x + iy)(x - iy) = x^2 + y^2.$$

Since $x^2 + y^2 = |z|^2$, we obtain the formula

$$\boxed{z\bar{z} = |z|^2.}$$

The complex conjugate is a useful tool for computing the real and imaginary parts of the quotient of two complex numbers. If z_1 and z_2 are given and if $z_2 \neq 0$, then

$$\frac{z_1}{z_2} = \frac{z_1}{z_2}\frac{\bar{z}_2}{\bar{z}_2} = \frac{z_1\bar{z}_2}{|z_2|^2},$$

and the denominator of the right side is a real number.

EXAMPLE 3. Compute the real and imaginary parts of the complex number $\dfrac{5 + i3}{7 - i2}$. The complex conjugate of $7 - i2$ is the number $7 + i2$. Hence

$$\frac{5 + i3}{7 - i2} = \frac{5 + i3}{7 - i2}\,\frac{7 + i2}{7 + i2} = \frac{(5 + i3)(7 + i2)}{7^2 + 2^2}.$$

Since $(5 + i3)(7 + i2) = 35 - 6 + i21 + i10 = 29 + i31$, we obtain

$$\frac{5 + i3}{7 - i2} = \frac{29 + i31}{53} = \frac{29}{53} + i\frac{31}{53}.$$

Thus the real part is $\frac{29}{53}$, and the imaginary part is $\frac{31}{53}$.

PROBLEMS

1. Prove Propositions (6.1), (6.2), (6.3), and (6.5).
2. Perform each of the indicated operations and write the answer in the form $x + iy$.

(a) $(3 + i4) + (7 - i3)$

(b) $(-2 + i\sqrt{2}) + (-2 - i\sqrt{2})$

(c) $(-2 + i\sqrt{2})(-2 - i\sqrt{2})$

(d) $\dfrac{3 - i7}{2 + i}$

(e) $(a + ib)(2a - i2b)$

(f) $2(4 - i3) + 7(-2 + i5)$

(g) $\dfrac{-3 + i4}{4 - i3}$

(h) $\dfrac{7 + i}{-2 - i5}$

(i) $\dfrac{a + ib}{3a - i3b}$

(j) $\dfrac{-2 - i}{2 - i}$

(k) $\dfrac{25}{3 - i4}$

(l) $(2 + i7)(2 - i5)$.

3. Let $z_1 = 2 + i3$, $z_2 = -1 - i$, and $z_3 = i$. Plot each of the following complex numbers in the complex plane.

(a) z_1

(b) z_2

(c) $z_1 + z_2$

(d) z_1z_3

(e) $z_1 - 2z_2$

(f) $\dfrac{z_1}{z_2}$.

4. Find the complex conjugate of each of the following complex numbers.

(a) $2 - i3$

(b) $5 + i4$

(c) $(2 - i3) + (5 + i4)$

(d) $(2 - i3)(5 + i4)$

(e) $4(2 - i3)$

(f) -7

(g) $2i$.

5. (a) In Problem 4, compute the sum of conjugates formed in (a) and (b), and compare with the conjugate of the sum in (c).

   (b) In Problem 4, compute the product of the conjugates found in (a) and (b), and compare with the conjugate of the product found in (d).

   (c) In Problem 4, multiply the conjugate found in (a) by 4, and compare with the answer found in (e).

6. For any complex numbers z_1 and z_2, prove that

   (a) $\overline{z_1 + z_2} = \bar{z}_1 + \bar{z}_2$ (b) $\overline{z_1z_2} = \bar{z}_1\bar{z}_2$ (c) $\overline{kz_1} = k\bar{z}_1$, k real.

7. If a is real and positive and z complex, prove that $|az| = a|z|$.

8. (a) Prove that the sum and product of two complex numbers which are conjugates of each other are real.

   (b) Prove that the difference of two complex numbers which are conjugates of each other is pure imaginary.

9. Graph all points z satisfying

   (a) $|z| = 2$
   (b) $|z| < 3$
   (c) $|z| > 1$
   (d) $|z| \leq 2$
   (e) $2 < |z| < 4$
   (f) $|z - 2| = 2$
   (g) $|z - z_0| = 3$, for a fixed z_0
   (h) $1 \leq |z - 3| \leq 2$.

10. Given two complex numbers z_1 and z_2, plot them and give a geometric interpretation of $|z_1 - z_2|$.

11. (a) Show that -1 has two square roots in C.

    (b) Show that every nonzero real number has two square roots.

7. The Complex Exponential Function e^z. Consider the function φ defined by

$$\varphi(x) = \cos x + i \sin x,$$

for every real number x. This is a complex-valued function of a real variable. The domain of φ is the set R of all real numbers. For every real number x, we have

$$|\varphi(x)| = \sqrt{\cos^2 x + \sin^2 x} = \sqrt{1} = 1.$$

It follows that $\varphi(x)$ is a point on the unit circle in the complex plane, i.e., the circle with center at the origin and radius 1. Conversely, every point on the unit circle is equal to $(\cos x, \sin x)$, for some real number x, and we know that $(\cos x, \sin x) = \cos x + i \sin x$. It follows that the range of φ is the unit circle.

The function φ has the following properties:

(7.1) $$\varphi(0) = 1.$$

(7.2) $$\varphi(a)\varphi(b) = \varphi(a + b).$$

(7.3) $$\frac{\varphi(a)}{\varphi(b)} = \varphi(a - b).$$

(7.4) $$\varphi(-a) = \frac{1}{\varphi(a)}\cdot$$

Proof. The proofs are completely straightforward. Thus (7.1) follows from the equations

$$\varphi(0) = \cos 0 + i \sin 0 = \cos 0 = 1.$$

To prove (7.2), we write

$$\begin{aligned}\varphi(a)\varphi(b) &= (\cos a + i \sin a)(\cos b + i \sin b) \\ &= \cos a \cos b - \sin a \sin b + i(\sin a \cos b + \cos a \sin b).\end{aligned}$$

The trigonometric identities for the cosine and sine of the sum of two numbers then imply that

$$\varphi(a)\varphi(b) = \cos(a + b) + i\sin(a + b),$$

and the right side is by definition equal to $\varphi(a + b)$. Thus (7.2) is proved. As a special case of (7.2), we have

$$\varphi(a - b)\varphi(b) = \varphi(a - b + b) = \varphi(a).$$

On dividing by $\varphi(b)$, which is never zero, we get (7.3). The last result, (7.4), is obtained by taking $a = 0$ in (7.3) and then substituting 1 for $\varphi(0)$ in accordance with (7.1). Thus

$$\frac{\varphi(0)}{\varphi(b)} = \varphi(0 - b),$$

$$\frac{1}{\varphi(b)} = \varphi(-b).$$

The above four properties of φ are also shared by the real-valued exponential function exp [we recall that $\exp(x) = e^x$]. This fact suggests the possibility of extending the domain and range of exp into the complex plane. That is, it suggests that the functions φ and exp can be combined to give a complex-valued exponential function of a complex variable which will have the property that when its domain is restricted to the real numbers, it is simply exp. We define such a function now. For every complex number $z = x + iy$, let Exp be the function defined by

$$\text{Exp}(z) = \exp(x)\varphi(y).$$

Thus

$$\text{Exp}(z) = e^x(\cos y + i\sin y).$$

If $z = x + i0$, then $z = x$ and $\text{Exp}(z) = \exp(x)\varphi(0) = \exp(x)$. Hence *the function* Exp *is an extension of the function* exp.

It is a routine matter to show that the function Exp has the exponential properties listed above for φ. Following the practice for the real-valued exponential, we shall write $\text{Exp}(z)$ as e^z. In this notation therefore, if $z = x + iy$, the definition reads

$$\boxed{e^z = e^x(\cos y + i\sin y).}$$

The exponential properties are

(7.1′) $$e^0 = 1.$$

(7.2′) $$e^{z_1}e^{z_2} = e^{z_1+z_2}.$$

(7.3′) $$\frac{e^{z_1}}{e^{z_2}} = e^{z_1-z_2}.$$

(7.4′) $$\frac{1}{e^z} = e^{-z}.$$

Proof. The proofs simply use the fact that the functions exp and φ separately have these properties. Thus

$$e^0 = e^{0+i0} = \exp(0)\varphi(0) = 1 \cdot 1 = 1.$$

Letting $z_1 = x_1 + iy_1$ and $z_2 = x_2 + iy_2$, we have

$$\begin{aligned} e^{z_1}e^{z_2} &= \exp(x_1)\varphi(y_1)\exp(x_2)\varphi(y_2) \\ &= \exp(x_1 + x_2)\varphi(y_1 + y_2). \end{aligned}$$

Since $z_1 + z_2 = (x_1 + x_2) + i(y_1 + y_2)$, the right side is by definition equal to $\text{Exp}(z_1 + z_2)$, which is $e^{z_1+z_2}$. The last two propositions, (7.3′) and (7.4′), are corollaries of (7.1′) and (7.2′) in exactly the same way that (7.3) and (7.4) follow from (7.1) and (7.2).

If x is an arbitrary real number, then

$$e^{ix} = e^{0+ix} = e^0(\cos x + i \sin x).$$

Thus we have the equation

(7.5) $$e^{ix} = \cos x + i \sin x, \quad \textit{for every real number } x.$$

Thus if x is any real number, the complex number e^{ix} is the ordered pair $(\cos x, \sin x)$. Hence e^{ix} is the point on the unit circle obtained by starting

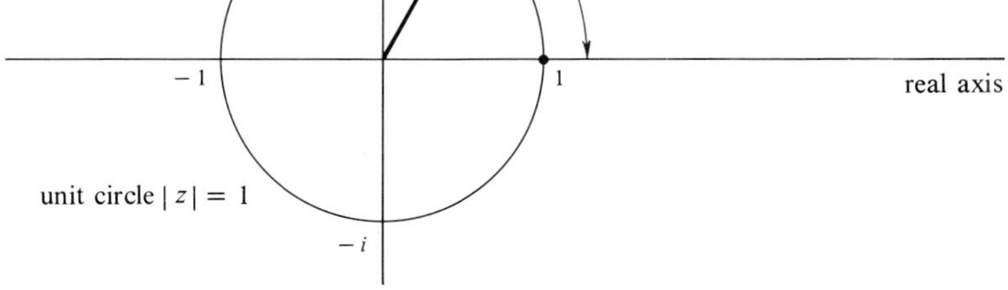

Figure 20

at the complex number 1 and measuring along the circle at a distance equal to the absolute value of x, measuring in the counterclockwise direction if x is positive and in the clockwise direction if it is negative (see Figure 20). In terms of angle, x is the radian measure of the angle whose initial side is the positive half of the real axis and whose terminal side contains the arrow representing z.

Letting $x = \pi$ in (7.5), we get $e^{i\pi} = \cos\pi + i\sin\pi$. Since $\cos\pi = -1$ and $\sin\pi = 0$, it follows that $e^{i\pi} = -1$, which is equivalent to the equation

$$\boxed{e^{i\pi} + 1 = 0.}$$

This equation is most famous since it combines in a simple formula the three special numbers π, e, and i with the additive and multiplicative identities 0 and 1.

One of the most important features of the complex exponential function is that it provides an alternative way of writing complex numbers. We have

(7.6) *Every complex number z can be written in the form $z = |z|e^{it}$, for some real number t. Furthermore, if $z = x + iy$ and $z \neq 0$, then $z = |z|e^{it}$ if and only if* $\cos t = \dfrac{x}{|z|}$ *and* $\sin t = \dfrac{y}{|z|}$.

Proof. If $z = 0$, then $|z| = 0$, and so $0 = z = |z|e^{it}$, for every real number t. Next we suppose that $z \neq 0$. Then $|z| \neq 0$, and $\dfrac{z}{|z|}$ is defined and lies on the unit circle because

$$\left|\frac{z}{|z|}\right| = \frac{|z|}{|z|} = 1.$$

Hence there exists a real number t such that $\dfrac{z}{|z|} = e^{it}$, and this proves the first statement in the theorem. Suppose that $z = x + iy$ and that $z \neq 0$. If $z = |z|e^{it}$, then

$$x + iy = |z|e^{it} = |z|(\cos t + i\sin t).$$

Two complex numbers are equal if and only if their real parts are equal and their imaginary parts are equal. Hence $x = |z|\cos t$ and $y = |z|\sin t$. Since $|z| \neq 0$, we conclude that $\cos t = \dfrac{x}{|z|}$ and $\sin t = \dfrac{y}{|z|}$. Conversely, if we start from the last two equations, it follows that

$$x + iy = |z|(\cos t + i\sin t).$$

The left side is equal to z, and the right side to $|z|e^{it}$. This completes the proof of the theorem.

A complex number written as $z = |z|e^{it}$ is said to be in **exponential form.** The number $|z|$ is, of course, the absolute value of z, and the number t

is called the **angle,** or **argument,** of z. The latter is not uniquely determined by z. Since the trigonometric functions sin and cos have period 2π, it follows that

$$z = |z|e^{it} = |z|e^{i(t+2\pi n)},$$

for every integer n.

Consider two complex numbers written in exponential form:

$$z_1 = |z_1|e^{it_1} \quad \text{and} \quad z_2 = |z_2|e^{it_2}.$$

The product and ratio are given by

$$z_1z_2 = |z_1|\,|z_2|e^{it_1}e^{it_2}, \qquad \frac{z_1}{z_2} = \frac{|z_1|}{|z_2|}\frac{e^{it_1}}{e^{it_2}}.$$

Hence by formulas (7.2′) and (7.3′) for the product and ratio of exponentials, we have

$$z_1z_2 = |z_1|\,|z_2|e^{i(t_1+t_2)}, \qquad \frac{z_1}{z_2} = \frac{|z_1|}{|z_2|}e^{i(t_1-t_2)}.$$

That is, two complex numbers are multiplied by multiplying their absolute values and adding their angles. They are divided by dividing their absolute values and subtracting their angles.

EXAMPLE 1. Let $z_1 = 3 + i4$ and $z_2 = -2i$. Express z_1, z_2, z_1z_2, and $\frac{z_1}{z_2}$ in the exponential form $|z|e^{it}$, and plot the resulting arrows in the complex plane. To begin with,

$$|z_1| = \sqrt{3^2 + 4^2} = 5,$$

$$|z_2| = \sqrt{0^2 + (-2)^2} = 2.$$

We next seek a real number t_1 such that $\cos t_1 = \frac{3}{5}$ and $\sin t_1 = \frac{4}{5}$, and also a number t_2 such that $\cos t_2 = 0$ and $\sin t_2 = -1$. These are given by

$$t_1 = \arccos \tfrac{3}{5} = 0.93 \text{ (approximately)},$$

$$t_2 = \arcsin(-1) = -\frac{\pi}{2}.$$

Then

$$z_1 = |z_1|e^{it_1} = 5e^{i(0.93)},$$

$$z_2 = |z_2|e^{it_2} = 2e^{-i(\pi/2)}.$$

Since $t_1 + t_2 = 0.93 - \frac{\pi}{2} = -0.64$ (approximately) and $t_1 - t_2 = 0.93 - \left(-\frac{\pi}{2}\right) = 2.50$ (approximately), we obtain

$$z_1z_2 = |z_1|\,|z_2|e^{i(t_1+t_2)} = 10e^{-i(0.64)},$$

$$\frac{z_1}{z_2} = \frac{|z_1|}{|z_2|}\,e^{i(t_1-t_2)} = \frac{5}{2}\,e^{i(2.50)}.$$

The arrows representing z_1, z_2, z_1z_2, and $\frac{z_1}{z_2}$ are shown in Figure 21. To

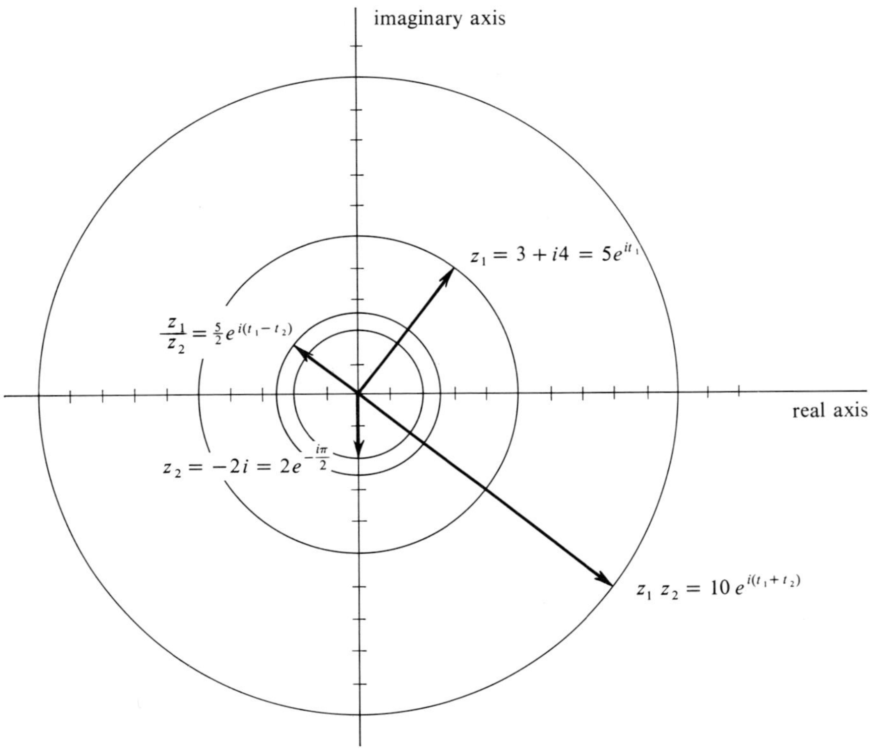

Figure 21

locate these numbers geometrically using a ruler and protractor marked off in degrees, we would compute

$$\begin{aligned} t_1 &= 0.93 \text{ radian} = 53 \text{ degrees}, \\ t_2 &= -\frac{\pi}{2} \text{ radians} = -90 \text{ degrees}, \\ t_1 + t_2 &= -0.64 \text{ radian} = -37 \text{ degrees}, \\ t_1 - t_2 &= 2.50 \text{ radians} = 143 \text{ degrees}. \end{aligned}$$

Of course, we can find the real and imaginary parts of z_1z_2 and $\frac{z_1}{z_2}$ by the computations

$$z_1z_2 = (3 + i4)(-2i) = 8 - i6,$$

$$\frac{z_1}{z_2} = \frac{3 + i4}{-2i} = \frac{3 + i4}{-2i}\frac{2i}{2i} = \frac{-8 + i6}{4} = -2 + i\tfrac{3}{2}.$$

If z is a complex number, then z^n can be defined inductively, for every nonnegative integer n, by

$$z^0 = 1, \tag{1}$$

$$z^n = z(z^{n-1}), \qquad \text{for } n > 0. \tag{2}$$

Another useful property of the complex exponential function is

(7.7) $(e^z)^n = e^{nz}$, *for every nonnegative integer n.*

Proof. By induction. If $n = 0$, then $(e^z)^n = (e^z)^0$. Since e^z is a complex number, $(e^z)^0 = 1$, by equation (1). Moreover, in this case, $e^{nz} = e^{0z} = 1$, by (7.1′). Next suppose that $n > 0$. By equation (2), we have $(e^z)^n = e^z(e^z)^{n-1}$, and by hypothesis of induction $(e^z)^{n-1} = e^{(n-1)z}$. Hence, by (7.2′),

$$(e^z)^n = e^z e^{(n-1)z} = e^{z+(n-1)z},$$

and, since $z + (n - 1)z = nz$, the proof is finished.

Let z be a complex number and n a positive integer. A complex number w is said to be an ***n*th root** of z if $w^n = z$. We shall now show that

(7.8) *If $z \neq 0$, then there exist n distinct* nth *roots of z.*

Proof. Let us write z in exponential form: $z = |z|e^{it}$. By $|z|^{1/n}$ we mean the positive nth root of the real number $|z|$ (which we assume exists and is unique). Consider the complex number

$$w_0 = |z|^{1/n}e^{i(t/n)}.$$

It is easy to see that w_0 is an nth root of z, since

$$\begin{aligned} w_0{}^n &= (|z|^{1/n})^n(e^{i(t/n)})^n \\ &= |z|e^{it} \\ &= z. \end{aligned}$$

However, w_0 is not the only nth root. We have already observed that

$$z = |z|e^{it} = |z|e^{i(t+2\pi k)},$$

for every integer k. If we set

$$w_k = |z|^{1/n}e^{i\frac{t+2\pi k}{n}},$$

then all these numbers are seen to be nth roots of z, since each one satisfies the equation $w_k{}^n = z$. However, they are not all different. Note that w_{k+1} is equal to the product $w_k e^{i(2\pi/n)}$. The angle of $e^{i(2\pi/n)}$ is $\frac{2\pi}{n}$ radians, and $\frac{2\pi}{n}$ is one nth the entire circumference of the unit circle. Thus w_{k+1} is obtained from w_k by adding an angle of $\frac{2\pi}{n}$ radians, or, equivalently, by rotating w_k exactly $\frac{1}{n}$ of an entire rotation. If we begin with w_k and form $w_{k+1}, w_{k+2}, \ldots$ by successive rotations, when we get to w_{k+n} we will be back at w_k, where we started. Thus there are only n distinct complex numbers among all the w's. In particular,

$$w_k = |z|^{1/n} e^{i\frac{t+2\pi k}{n}}, \qquad k = 0, \ldots, n-1,$$

are n distinct nth roots of z. This completes the proof.

An nth root of z is a solution of the complex polynomial equation $w^n - z = 0$. It is a well-known theorem of algebra that a polynomial equation of degree n cannot have more than n roots. Hence we can strengthen the statement of (7.8) to read that every nonzero complex number z has precisely n distinct nth roots.

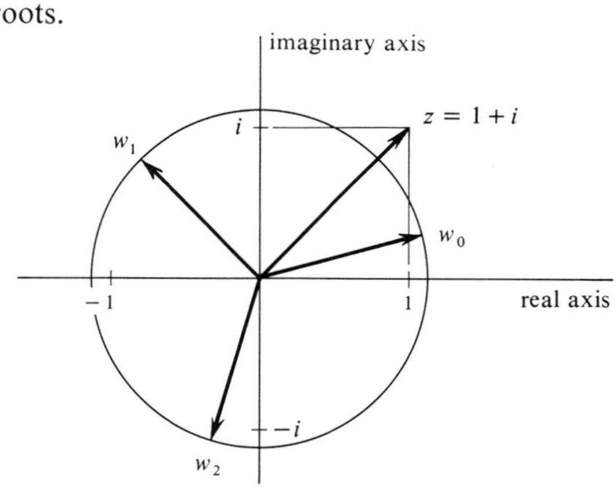

Figure 22

EXAMPLE 2. Find the three cube roots of the complex number $z = 1 + i$, and plot them in the complex plane. Writing z in exponential form, we have

$$z = \sqrt{2}\, e^{i(\pi/4)}$$

(see Figure 22). Hence the three cube roots are

$$w_k = (\sqrt{2})^{1/3} e^{i\frac{\pi/4+2\pi k}{3}}, \qquad k = 0, 1, 2.$$

Since $(\sqrt{2})^{1/3} = 2^{1/6} = 1.12$ (approximately) and $\frac{1}{3}\left(\frac{\pi}{4}\right)$ radius $= 15$ degrees, we see that w_0 is the complex number lying on the circle of radius 1.12 about the origin and making an angle of 15 degrees with the positive x-axis. The other two roots lie on the same circle and have angles of $15 + 120$ degrees and $15 + 240$ degrees, respectively. The three roots are thus $\sqrt[6]{2}\, e^{i(\pi/12)}$, $\sqrt[6]{2}\, e^{i(3\pi/4)}$, and $\sqrt[6]{2}\, e^{i(17\pi/12)}$.

PROBLEMS

1. Write each of the following complex numbers in the exponential form $|z|e^{it}$.

   (a) $1 + i$
   (b) $1 - i$
   (c) $1 + i\sqrt{3}$
   (d) $-2\sqrt{3} + 2i$
   (e) -5
   (f) e^z, where $z = 2 + i\frac{\pi}{4}$
   (g) 5
   (h) i.

2. Let $z_1 = \sqrt{3} + i$ and $z_2 = 1 - i$. Write each of the following complex numbers in the exponential form $|z|e^{it}$ and plot it in the complex plane.

   (a) z_1
   (b) z_2
   (c) $z_1 z_2$
   (d) $\frac{z_1}{z_2}$
   (e) $2z_1$
   (f) $(z_1)^6$.

3. Find the real and imaginary parts of each of the following complex numbers.

   (a) $e^{i\pi}$
   (b) $2e^{i(\pi/4)}$
   (c) $e^{i\pi} 2e^{i(\pi/4)}$
   (d) $e^{i\pi} + 2e^{i(\pi/4)}$
   (e) $\sqrt{34}\, e^{it}$, where $t = \arcsin \frac{5}{\sqrt{34}}$
   (f) $\sqrt{13}\, e^{it}$, where $\sin t = -\frac{2}{\sqrt{13}}$.

4. If $z_1 = |z_1|e^{it}$, what is the exponential form of its complex conjugate $\bar{z}_1$?
5. Derive (7.3′) and (7.4′) using (7.1′) and (7.2′).
6. Let n be a positive integer, and let z^n be defined as in the text. If $z \neq 0$, define $z^{-n} = \frac{1}{z^n}$, and thence show that $(e^z)^{-n} = e^{-nz}$. As a result, we know that Theorem (7.7) holds for all integers.
7. Find and plot the nth roots of z in each of the following cases.

   (a) $n = 3$ and $z = 8i$
   (b) $n = 2$ and $z = i$
   (c) $n = 3$ and $z = 2$
   (d) $n = 4$ and $z = 1$
   (e) $n = 5$ and $z = 2i$
   (f) $n = 3$ and $z = 1 + i\sqrt{3}$.

8. How would you define the function 5^z?
9. (a) Using the equation $e^{ix} = \cos x + i \sin x$ and the fact that $(e^{ix})^n = e^{inx}$, prove that

$$\cos nx + i \sin nx = (\cos x + i \sin x)^n.$$

This is known as *de Moivre's Formula.*

(b) Using de Moivre's Formula and the Binomial Theorem, find trigonometric identities for $\cos 3x$ and $\sin 3x$ in terms of $\cos x$ and $\sin x$.

10. Every complex-valued function f of a real variable determines two real-valued functions f_1 and f_2 of a real variable defined by

$$f_1(x) = \text{real part of } f(x),$$

$$f_2(x) = \text{imaginary part of } f(x).$$

Thus $f(x) = f_1(x) + if_2(x)$ for every x in the domain of f. We define the **derivative** f' by the formula

$$f'(x) = f_1'(x) + i f_2'(x).$$

Applying this definition to the function $f(x) = e^{ix}$, show that

$$\frac{d}{dx} e^{ix} = ie^{ix}.$$

8. Differential Equations. In this section we shall show how to obtain the general solution of any differential equation of the form

$$\boxed{\frac{d^2y}{dx^2} + a\frac{dy}{dx} + by = 0,} \tag{1}$$

where a and b are real constants. Differential equations of this type occur frequently in mechanics and also in the theory of electric circuits. Equation (1) is a **second-order differential equation,** since it contains the second derivative $\frac{d^2y}{dx^2}$ but no higher derivative. It is called **linear** because each one of y, $\frac{dy}{dx}$, and $\frac{d^2y}{dx^2}$ occurs, if at all, to the first power. That is, if we set $\frac{dy}{dx} = z$ and $\frac{d^2y}{dx^2} = w$, then (1) becomes $w + az + by = 0$, and the left side is a linear polynomial, or polynomial of first degree, in w, z, and y. A second-order linear differential equation more general than (1) is

$$\frac{d^2y}{dx^2} + f(x)\frac{dy}{dx} + g(x)y = h(x),$$

where f, g, and h are given functions of x. Equation (1) is a special case, called **homogeneous,** because h is the zero function, and said to have **constant coefficients,** since f and g are constant functions. Thus the topic of this section becomes: the study of second-order, linear, homogeneous differential equations with constant coefficients.

An important and easily proved property of differential equations of this kind is the following:

(8.1) *If y_1 and y_2 are any two solutions of the differential equation* (1), *and if c_1 and c_2 are any two real numbers, then $c_1y_1 + c_2y_2$ is also a solution.*

Proof. The proof uses only the elementary properties of the derivative. We know that

$$\frac{d}{dx}(c_1y_1 + c_2y_2) = c_1\frac{dy_1}{dx} + c_2\frac{dy_2}{dx}.$$

Hence

$$\begin{aligned}\frac{d^2}{dx^2}(c_1y_1 + c_2y_2) &= \frac{d}{dx}\left(c_1\frac{dy_1}{dx} + c_2\frac{dy_2}{dx}\right)\\ &= c_1\frac{d^2y_1}{dx^2} + c_2\frac{d^2y_2}{dx^2}.\end{aligned}$$

To test whether or not $c_1y_1 + c_2y_2$ is a solution, we substitute it for y in the differential equation:

$$\begin{aligned}&\frac{d^2}{dx^2}(c_1y_1 + c_2y_2) + a\frac{d}{dx}(c_1y_1 + c_2y_2) + b(c_1y_1 + c_2y_2)\\ &\qquad = c_1\frac{d^2y_1}{dx^2} + c_2\frac{d^2y_2}{dx^2} + ac_1\frac{dy_1}{dx} + ac_2\frac{dy_2}{dx} + bc_1y_1 + bc_2y_2\\ &\qquad = c_1\left(\frac{d^2y_1}{dx^2} + a\frac{dy_1}{dx} + by_1\right) + c_2\left(\frac{d^2y_2}{dx^2} + a\frac{dy_2}{dx} + by_2\right).\end{aligned}$$

The expressions in parentheses in the last line are both zero because y_1 and y_2 are by assumption solutions of the differential equation. Hence the top line is also zero, and so $c_1y_1 + c_2y_2$ is a solution. This completes the proof.

It follows in particular that the sum and difference of any two solutions of (1) is a solution, and also that any constant multiple of a solution is again a solution. Finally, note that the constant function 0 is a solution of (1) for any constants a and b.

In Section 5 of Chapter 5 we found that the general solution of the differential equation $\frac{dy}{dx} + ky = 0$ is the function $y = ce^{-kx}$, where c is an arbitrary real number. This differential equation is first-order, linear, homogeneous, and with constant coefficients. Let us see whether by any chance an exponential function might also be a solution of the second-order differential equation $\frac{d^2y}{dx^2} + b\frac{dy}{dx} + cy = 0$. Let $y = e^{rx}$, where r is any real number. Then

$$\frac{dy}{dx} = re^{rx}, \qquad \frac{d^2y}{dx^2} = r^2e^{rx}.$$

Hence

$$\begin{aligned}\frac{d^2y}{dx^2} + a\frac{dy}{dx} + by &= r^2e^{rx} + are^{rx} + be^{rx} \\ &= (r^2 + ar + b)e^{rx}.\end{aligned}$$

Since e^{rx} is never zero, the right side is zero if and only if $r^2 + ar + b = 0$. That is, we have shown that

(8.2) *The function* e^{rx} *is a solution of* $\frac{d^2y}{dx^2} + a\frac{dy}{dx} + by = 0$ *if and only if the real number* r *is a solution of* $t^2 + at + b = 0$.

The latter equation is called the **characteristic equation** of the differential equation.

EXAMPLE 1. Consider the differential equation $\frac{d^2y}{dx^2} - \frac{dy}{dx} - 6y = 0$. Its characteristic equation is $t^2 - t - 6 = 0$. Since $t^2 - t - 6 = (t - 3)(t + 2)$, the two solutions, or roots, are 3 and -2. Hence, by (8.2), both functions e^{3x} and e^{-2x} are solutions of the differential equation. It follows by (8.1) that the function

$$y = c_1e^{3x} + c_2e^{-2x}$$

is a solution for any two real numbers c_1 and c_2.

The form of the general solution of the differential equation

$$\frac{d^2y}{dx^2} + a\frac{dy}{dx} + by = 0$$

depends on the roots of the characteristic equation $t^2 + at + b = 0$. There are three different cases to be considered.

Case 1. The characteristic equation has distinct real roots. This is the simplest case. We have

$$t^2 + at + b = (t - r_1)(t - r_2),$$

where r_1 and r_2 are real numbers and $r_1 \neq r_2$. Both functions $e^{r_1 x}$ and $e^{r_2 x}$ are solutions of the differential equation, and so is any linear combination $c_1 e^{r_1 x} + c_2 e^{r_2 x}$. Moreover it can be shown, although we defer the proof until Chapter 11, that if y is any solution of the differential equation, then

$$\boxed{y = c_1 e^{r_1 x} + c_2 e^{r_2 x},} \tag{2}$$

for some two real numbers c_1 and c_2. Hence we say that (2) is the general solution. In Example 1 the function $c_1 e^{3x} + c_2 e^{-2x}$ is therefore the general solution of the differential equation $\dfrac{d^2y}{dx^2} - \dfrac{dy}{dx} - 6y = 0$.

Case 2. The characteristic equation has complex roots. The roots of $t^2 + at + b = 0$ are given by the quadratic formula

$$r_1, r_2 = \frac{-a \pm \sqrt{a^2 - 4b}}{2}.$$

Since a and b are real, r_1 and r_2 are complex if and only if $a^2 - 4b < 0$, which we now assume. Setting $\alpha = -\dfrac{a}{2}$ and $\beta = \dfrac{\sqrt{4b - a^2}}{2}$, we have

$$r_1 = \alpha + i\beta, \qquad r_2 = \alpha - i\beta.$$

Note that r_1 and r_2 are complex conjugates of each other.

Motivated by the situation in Case 1, in which r_1 and r_2 were real, we consider the complex-valued function $c_1 e^{r_1 x} + c_2 e^{r_2 x}$, where we now allow c_1 and c_2 to be complex numbers. We shall show that

(8.3) *If c_1 and c_2 are any two complex conjugates of each other and if r_1 and r_2 are complex solutions of the characteristic equation, then the function*

$$y = c_1 e^{r_1 x} + c_2 e^{r_2 x}$$

is real-valued. Moreover, it is a solution of the differential equation (1).

Proof. Let $c_1 = \gamma + i\delta$ and $c_2 = \gamma - i\delta$. Since $r_1 = \alpha + i\beta$ and $r_2 = \alpha - i\beta$, we have

$$\begin{aligned} c_1 e^{r_1 x} + c_2 e^{r_2 x} &= (\gamma + i\delta)e^{(\alpha + i\beta)x} + (\gamma - i\delta)e^{(\alpha - i\beta)x} \\ &= e^{\alpha x}[(\gamma + i\delta)e^{i\beta x} + (\gamma - i\delta)e^{-i\beta x}]. \end{aligned}$$

Recall that $e^{i\beta x} = \cos \beta x + i \sin \beta x$ and $e^{-i\beta x} = \cos(-\beta x) + i \sin(-\beta x) = \cos \beta x - i \sin \beta x$. Substituting, we get

$$\begin{aligned} c_1 e^{r_1 x} + c_2 e^{r_2 x} &= e^{\alpha x}[(\gamma + i\delta)(\cos \beta x + i \sin \beta x) \\ &\qquad + (\gamma - i\delta)(\cos \beta x - i \sin \beta x)] \\ &= e^{\alpha x}(2\gamma \cos \beta x - 2\delta \sin \beta x). \end{aligned}$$

The right side is certainly real-valued, and this proves the first statement of the theorem. Since γ and δ are arbitrary real numbers, so are 2γ and -2δ. We may therefore replace 2γ by k_1 and -2δ by k_2. We prove the second statement of the theorem by showing that the function

$$y = e^{\alpha x}(k_1 \cos \beta x + k_2 \sin \beta x) \tag{3}$$

is a solution of the differential equation. Let $y_1 = e^{\alpha x}\cos \beta x$ and $y_2 = e^{\alpha x}\sin \beta x$. Since $y = k_1 y_1 + k_2 y_2$, it follows by (8.1) that it is enough to show that y_1 and y_2 separately are solutions of the differential equation. We give the proof for y_1 and leave it to the reader to check it for y_2. By the product rule,

$$\begin{aligned} \frac{dy_1}{dx} &= \frac{d}{dx}(e^{\alpha x}\cos \beta x) = \alpha e^{\alpha x}\cos \beta x - \beta e^{\alpha x}\sin \beta x \\ &= e^{\alpha x}(\alpha \cos \beta x - \beta \sin \beta x). \end{aligned}$$

Hence

$$\begin{aligned} \frac{d^2 y_1}{dx^2} &= \alpha e^{\alpha x}(\alpha \cos \beta x - \beta \sin \beta x) + e^{\alpha x}(-\alpha\beta \sin \beta x - \beta^2 \cos \beta x) \\ &= e^{\alpha x}[(\alpha^2 - \beta^2) \cos \beta x - 2\alpha\beta \sin \beta x]. \end{aligned}$$

Thus

$$\begin{aligned} \frac{d^2 y_1}{dx^2} + a\frac{dy_1}{dx} + by_1 &= e^{\alpha x}[(\alpha^2 - \beta^2) \cos \beta x - 2\alpha\beta \sin \beta x] \\ &\quad + ae^{\alpha x}(\alpha \cos \beta x - \beta \sin \beta x) + be^{\alpha x}\cos \beta x \\ &= e^{\alpha x}\big([(\alpha^2 - \beta^2) + a\alpha + b] \cos \beta x - \beta(2\alpha + a) \sin \beta x\big). \end{aligned}$$

But, remembering that $r_1 = \alpha + i\beta$ and $r_2 = \alpha - i\beta$ and that these are the roots of the characteristic equation, we read from the quadratic formula that

$$\alpha = -\frac{a}{2}, \qquad \beta = \frac{\sqrt{4b - a^2}}{2}.$$

Hence

$$\begin{aligned} (\alpha^2 - \beta^2) + a\alpha + b &= \left(-\frac{a}{2}\right)^2 - \left(\frac{\sqrt{4b - a^2}}{2}\right)^2 + a\left(-\frac{a}{2}\right) + b \\ &= \frac{a^2}{4} - b + \frac{a^2}{4} - \frac{a^2}{2} + b = 0, \end{aligned}$$

and also

$$2\alpha + a = 2\left(-\frac{a}{2}\right) + a = -a + a = 0,$$

whence we get

$$\frac{d^2y_1}{dx^2} + a\frac{dy_1}{dx} + by_1 = e^{\alpha x}(0 \cdot \cos \beta x - \beta \cdot 0 \cdot \sin \beta x)$$
$$= 0,$$

and so y_1 is a solution. Assuming the analogous proof for y_2, it follows that y, as defined by (3), is also a solution and the proof is complete.

It can be shown, although again we defer the proof, that if y is any real solution to the differential equation (1), and if the roots r_1 and r_2 of the characteristic equation are complex, then

$$\boxed{y = c_1e^{r_1x} + c_2e^{r_2x},} \tag{4}$$

for some complex number c_1 and its complex conjugate c_2. Hence, if the roots are complex, the general solution of the differential equation can be written either as (4), or in the equivalent form,

$$\boxed{y = e^{\alpha x}(k_1 \cos \beta x + k_2 \sin \beta x),} \tag{5}$$

where $r_1 = \alpha + i\beta$ and $r_2 = \alpha - i\beta$, and k_1 and k_2 are arbitrary real numbers. Note that solutions (2) and (4) look the same, even though they involve different kinds of r's and different kinds of c's.

EXAMPLE 2. Find the general solution of the differential equation

$$\frac{d^2y}{dx^2} + 4\frac{dy}{dx} + 13y = 0.$$

The characteristic equation is $t^2 + 4t + 13 = 0$. Using the quadratic formula, we find the roots

$$r_1, r_2 = \frac{-4 \pm \sqrt{16 - 4 \cdot 13}}{2} = \frac{-4 \pm \sqrt{-36}}{2}$$
$$= -2 \pm 3i.$$

Hence, by (4), the general solution can be written

$$y = c_1e^{(-2+3i)x} + c_2e^{(-2-3i)x},$$

where c_1 and c_2 are complex conjugates of each other. Unless otherwise stated, however, the solution should appear as an obviously real-valued function. That is, it should be written without the use of complex numbers as in (5). Hence the preferred form of the general solution is

$$y = e^{-2x}(k_1 \cos 3x + k_2 \sin 3x).$$

We now consider the remaining possibility.

Case 3. The characteristic equation $t^2 + at + b = 0$ has only one root r. In this case, we have $t^2 + at + b = (t - r)(t - r)$, and the quadratic formula yields $r = -\frac{a}{2}$ and $\sqrt{a^2 - 4b} = 0$.

Theorem (8.2) is still valid, of course, and so one solution of the differential equation $\frac{d^2y}{dx^2} + a\frac{dy}{dx} + by = 0$ is obtained by taking $y = e^{rx}$. We shall show that, in the case of only one root, xe^{rx} is also a solution. Setting $y = xe^{rx}$, we obtain

$$\begin{aligned}\frac{dy}{dx} &= e^{rx} + xre^{rx} = e^{rx}(1 + rx),\\ \frac{d^2y}{dx^2} &= re^{rx}(1 + rx) + e^{rx} \cdot r\\ &= re^{rx}(2 + rx).\end{aligned}$$

Hence

$$\begin{aligned}\frac{d^2y}{dx^2} + a\frac{dy}{dx} + by &= re^{rx}(2 + rx) + ae^{rx}(1 + rx) + bxe^{rx}\\ &= e^{rx}(2r + r^2x + a + arx + bx)\\ &= e^{rx}[x(r^2 + ar + b) + (a + 2r)].\end{aligned}$$

Since r is a root of $t^2 + at + b$, we know that $r^2 + ar + b = 0$. Moreover, we have seen that $r = -\frac{a}{2}$, and so $a + 2r = 0$. It follows that the last expression in the above equations is equal to zero, which shows that the function xe^{rx} is a solution of the differential equation.

Thus e^{rx} is one solution, and xe^{rx} is another. It follows by (8.1) that, for any two real numbers c_1 and c_2, a solution is given by

$$y = c_1xe^{rx} + c_2e^{rx} = (c_1x + c_2)e^{rx}.$$

Conversely, it can be shown that if y is any solution of the differential equation (1), and if the characteristic equation has only one root r, then

$$\boxed{y = (c_1x + c_2)e^{rx},} \qquad (6)$$

for some pair of real numbers c_1 and c_2. The general solution in the case of a single root is therefore given by (6).

EXAMPLE 3. Find the general solution of the differential equation $9y'' - 6y' + y = 0$. Here we have used the common notation y' and y'' for the first and second derivatives of the unknown function y. Dividing the equation by 9 to obtain a leading coefficient of 1, we get $y'' - \frac{2}{3}y' + \frac{1}{9}y = 0$, for which the characteristic equation is $t^2 - \frac{2}{3}t + \frac{1}{9} = 0$. Since $t^2 - \frac{2}{3}t + \frac{1}{9} = (t - \frac{1}{3})(t - \frac{1}{3})$, there is only one root, $r = \frac{1}{3}$. Hence

$$y = (c_1x + c_2)e^{x/3}$$

is the general solution.

The solution of a differential equation can be checked just as simply as an indefinite integral, by differentiation and substitution.

PROBLEMS

1. Find the general solution of each of the following differential equations. If the characteristic equation has complex roots, write your solution in trigonometric form.

(a) $2\frac{dy}{dx} + 3y = 0$

(b) $y' = 5y$

(c) $\frac{d^2y}{dx^2} + 4\frac{dy}{dx} - 5y = 0$

(d) $\frac{d^2y}{dx^2} - 2\frac{dy}{dx} + 5y = 0$

(e) $y'' + 8y' + 16y = 0$

(f) $4\frac{d^2y}{dx^2} - 4\frac{dy}{dx} + y = 0$

(g) $y'' - 7y' = 0$

(h) $\frac{d^2y}{dx^2} + 4y = 0$

(i) $\frac{d^2y}{dx^2} - 9y = 0$

(j) $\frac{dy}{dx} + 13y = 0$

(k) $\frac{d^2y}{dx^2} + 13\frac{dy}{dx} = 0$

(l) $\frac{d^2y}{dx^2} + \frac{dy}{dx} + y = 0$

(m) $y'' + 14y' + 50y = 0$

(n) $y'' + 14y' + 49y = 0$.

2. (a) Find the general solution of the differential equation

$$\frac{d^2y}{dx^2} - 8\frac{dy}{dx} + 16y = 0.$$

(b) Find the particular solution y of the equation in part (a) with the property that $y = 2$ and $\frac{dy}{dx} = 9$ when $x = 0$. (*Hint*: Use these two conditions to evaluate the arbitrary constants which appear in the general solution.)

3. Find the particular solution y of each of the following differential equations such that y and $\frac{dy}{dx}$ have the prescribed values when $x = 0$.

(a) $\frac{d^2y}{dx^2} + 4\frac{dy}{dx} + 3y = 0, \qquad y = 3$ and $\frac{dy}{dx} = -5$ when $x = 0$.

(b) $25\frac{d^2y}{dx^2} + 10\frac{dy}{dx} + y = 0, \qquad y = 1$ and $\frac{dy}{dx} = \frac{14}{5}$ when $x = 0$.

(c) $\frac{d^2y}{dx^2} + 3\frac{dy}{dx} = 0, \qquad y = 3$ and $\frac{dy}{dx} = 6$ when $x = 0$.

(d) $\frac{d^2y}{dx^2} + 2\frac{dy}{dx} + 2y = 0, \qquad y = 5$ and $\frac{dy}{dx} = -5$ when $x = 0$.

(e) $y'' + 3y' + 5y = 0, \qquad y(0) = 2$ and $y'(0) = 6$.

4. Show by differentiation and substitution that $e^{\alpha x}\sin\beta x$ is a solution of $\frac{d^2y}{dx^2} + a\frac{dy}{dx} + by = 0$ if $\alpha + i\beta$ is a root of $t^2 + at + b = 0$.

5. (a) Show that $\overline{e^z} = e^{\bar{z}}$.

(b) Use (a) and Problem 6, page 334, to show that $ce^z + \bar{c}e^{\bar{z}}$ is real for any complex numbers c and z.

6. (a) Multiply $a\cos x + b\sin x$ by $\dfrac{\sqrt{a^2+b^2}}{\sqrt{a^2+b^2}}$ and hence show that it can be written in the form $c\sin(x + k)$, where $c = \sqrt{a^2+b^2}$, $\sin k = \dfrac{a}{\sqrt{a^2+b^2}}$, and $\cos k = \dfrac{b}{\sqrt{a^2+b^2}}$.

(b) Multiply $a\cos x + b\sin x$ by $\dfrac{\sqrt{a^2+b^2}}{\sqrt{a^2+b^2}}$ and hence show that it can be written in the form $c\cos(x + k)$, where

$$c = \sqrt{a^2+b^2}, \sin k = \frac{-b}{\sqrt{a^2+b^2}}, \text{ and } \cos k = \frac{a}{\sqrt{a^2+b^2}}.$$

7. (a) Show that $y = c_1 e^{\alpha x}\sin(\beta x + c_2)$ is a solution of the differential equation $y'' - 2\alpha y' + (\alpha^2 + \beta^2)y = 0$.

(b) Show that $y = c_1 e^{\alpha x}\cos(\beta x + c_2)$ is also a solution of the differential equation in (a).

8. For what value or values of r is e^{rx} a solution of the third-order linear differential equation $y''' - 6y'' + 5y' = 0$?

9. (a) Solve the homogeneous differential equation

$$\frac{d^2y}{dx^2} - 8\frac{dy}{dx} + 12y = 0.$$

(b) Substitute the linear polynomial $Ax + B$ for y in the nonhomogeneous differential equation

$$\frac{d^2y}{dx^2} - 8\frac{dy}{dx} + 12y = 24x + 12.$$

Hence find values of A and B for which this polynomial is a particular solution of the differential equation.

(c) Show that the function which is the sum of the solutions found in (a) and (b) is also a solution to the differential equation in (b).

10. If y_h is a solution of $\frac{d^2y}{dx^2} + a\frac{dy}{dx} + by = 0$ and y_p is a solution of $\frac{d^2y}{dx^2} + a\frac{dy}{dx} + by = h(x)$, show that $y_h + y_p$ is also a solution of

$$\frac{d^2y}{dx^2} + a\frac{dy}{dx} + by = h(x).$$

CHAPTER 7

Techniques of Integration

We have found indefinite integrals for many functions. Nevertheless, there are many seemingly simple functions for which we have as yet no method of integration. Among these, for example, are $\ln x$ and $\sqrt{a^2 - x^2}$, since we have no way of finding $\int \ln x\, dx$ and $\int \sqrt{a^2 - x^2}\, dx$. In this chapter we shall develop a number of techniques, each of which will enlarge the set of functions which we can integrate.

1. Integration by Parts. One of the most powerful methods of integration comes from the formula for finding the derivative of the product of two functions. If u and v are both differentiable functions of x, we recall that $\frac{d}{dx} uv = u\frac{dv}{dx} + v\frac{du}{dx}$. As a consequence, of course, it is true that $u\frac{dv}{dx} = \frac{d}{dx} uv - v\frac{du}{dx}$. From this last equation it follows that

$$\int u\frac{dv}{dx}\,dx = \int \left(\frac{d}{dx} uv - v\frac{du}{dx}\right) dx.$$

Since the integral of a difference is equal to the difference of the integrals, the last equation is equivalent to

$$\int u\frac{dv}{dx}\,dx = \int \frac{d}{dx}(uv)\,dx - \int v\frac{du}{dx}\,dx.$$

But $\int \frac{d}{dx}(uv)\,dx = uv + c$. Hence, leaving the constant of integration as a by-product of the last integral, we obtain

$$\boxed{\int u\frac{dv}{dx}\,dx = uv - \int v\frac{du}{dx}\,dx.} \tag{1.1}$$

This is the formula for the technique of integration by parts. To use it, the function to be integrated must be factored into a product of two functions, one to be labeled u and the other $\dfrac{dv}{dx}$. If we can recover v from $\dfrac{dv}{dx}$, then we hope that $\int v \dfrac{du}{dx}\, dx$ is easier to find than $\int u \dfrac{dv}{dx}\, dx$. As we shall see, the trick is to find the right factorization. Sometimes it is obvious, and sometimes it is not.

EXAMPLE 1. Integrate

$$\text{(a)} \int \ln x\, dx, \qquad\qquad \text{(b)} \int x \sin x\, dx.$$

For (a), consider the factorization $\ln x \cdot 1$, and let $u = \ln x$ and $\dfrac{dv}{dx} = 1$. Then $\dfrac{du}{dx} = \dfrac{1}{x}$ and $v = x$ (we normally do not concern ourselves with a constant of integration at this point—we look for *some* v, not the most general v). Integrating by parts, we have

$$\int \underset{\substack{\updownarrow \\ u}}{\ln x} \cdot \underset{\substack{\updownarrow \\ \frac{dv}{dx}}}{1} \cdot dx = (\underset{\substack{\updownarrow \\ u}}{\ln x})(\underset{\substack{\updownarrow \\ v}}{x}) - \int \underset{\substack{\updownarrow \\ v}}{x} \cdot \underset{\substack{\updownarrow \\ \frac{du}{dx}}}{\frac{1}{x}} \cdot dx,$$

or

$$\begin{aligned} \int \ln x\, dx &= x \ln x - \int dx \\ &= x \ln x - x + c. \end{aligned}$$

If this is the correct answer, its derivative will be the integrand. Checking, we get

$$\begin{aligned} \frac{d}{dx}(x \ln x - x + c) &= 1 \cdot \ln x + x \cdot \frac{1}{x} - 1 + 0 \\ &= \ln x + 1 - 1 = \ln x. \end{aligned}$$

For (b) we have choices. We can try letting $u = x$ and $\dfrac{dv}{dx} = \sin x$, or we can try $u = \sin x$ and $\dfrac{dv}{dx} = x$, or even $u = x \sin x$ and $\dfrac{dv}{dx} = 1$. Trial and error shows that the first suggestion works and the others do not. If $u = x$ and $\dfrac{dv}{dx} = \sin x$, then we have $\dfrac{du}{dx} = 1$ and $v = -\cos x$. The integration-by-parts formula implies that

$$\int x \sin x\, dx = (x)(-\cos x) - \int (-\cos x)(1)\, dx,$$

or

$$\int x \sin x \, dx = -x \cos x + \int \cos x \, dx$$
$$= -x \cos x + \sin x + c.$$

Although it is not true that every function can be written as a product of functions which will lead to a simplification of integrals—this technique is of little use if $\int v \frac{du}{dx} dx$ is not easier to integrate than $\int u \frac{dv}{dx} dx$—it is true that many can and that they can be integrated as were the two integrals in Example 1. The problem is to find the best function to call u. Sometimes the technique of integration by parts must be used more than once in a problem.

EXAMPLE 2. Integrate $\int (3x^2 - 4x + 7)e^{2x} dx$. If we let $u = 3x^2 - 4x + 7$ and $\frac{dv}{dx} = e^{2x}$, then $\frac{du}{dx} = 6x - 4$ and $v = \frac{1}{2}e^{2x}$.

$$\int (3x^2 - 4x + 7)e^{2x} dx = (3x^2 - 4x + 7)(\tfrac{1}{2}e^{2x}) - \int (\tfrac{1}{2}e^{2x})(6x - 4)\, dx$$
$$= \tfrac{1}{2}(3x^2 - 4x + 7)e^{2x} - \int (3x - 2)e^{2x} dx.$$

This last integral again involves a product of a polynomial and e^{2x}, so we apply the technique again. Let $u_1 = 3x - 2$ and $\frac{dv_1}{dx} = e^{2x}$. Then $\frac{du_1}{dx} = 3$ and $v_1 = \frac{1}{2}e^{2x}$. Thus

$$\int (3x - 2)e^{2x} dx = (3x - 2)(\tfrac{1}{2}e^{2x}) - \int (\tfrac{1}{2}e^{2x})(3)\, dx$$
$$= \tfrac{1}{2}(3x - 2)e^{2x} - \tfrac{3}{2}\int e^{2x} dx$$
$$= \tfrac{1}{2}(3x - 2)e^{2x} - \tfrac{3}{4}e^{2x} + c_1.$$

Substituting, we have

$$\int (3x^2 - 4x + 7)e^{2x} dx$$
$$= \tfrac{1}{2}(3x^2 - 4x + 7)e^{2x} - [\tfrac{1}{2}(3x - 2)e^{2x} - \tfrac{3}{4}e^{2x} + c_1]$$
$$= [(\tfrac{3}{2}x^2 - 2x + \tfrac{7}{2} - \tfrac{3}{2}x + 1 + \tfrac{3}{4})e^{2x}] + c$$
$$= (\tfrac{3}{2}x^2 - \tfrac{7}{2}x + \tfrac{21}{4})e^{2x} + c.$$

Generally, faced with the product of a polynomial and a trigonometric or exponential function, it is best to let the polynomial be u and the transcendental function be $\dfrac{dv}{dx}$. In this way, the degree of the polynomial is reduced by one each time the product is integrated by parts. However, faced with a product of transcendental functions, the choice may not be quite so obvious.

EXAMPLE 3. Integrate $\int e^{2x}\cos 3x\,dx$. Let us select e^{2x} as u and $\cos 3x$ as $\dfrac{dv}{dx}$. Then $\dfrac{du}{dx} = 2e^{2x}$ and $v = \frac{1}{3}\sin 3x$. Thus

$$\int e^{2x}\cos 3x\,dx = (e^{2x})(\tfrac{1}{3}\sin 3x) - \int(\tfrac{1}{3}\sin 3x)(2e^{2x})\,dx$$

$$= \tfrac{1}{3}e^{2x}\sin 3x - \tfrac{2}{3}\int e^{2x}\sin 3x\,dx.$$

At this point a second integration by parts is necessary. The reader should check that the choice of $\sin 3x$ for u_1 will lead back to an identity. We choose e^{2x} for u_1 and $\sin 3x$ for $\dfrac{dv_1}{dx}$. Then $\dfrac{du_1}{dx} = 2e^{2x}$ and $v_1 = -\frac{1}{3}\cos 3x$. Hence

$$\int e^{2x}\sin 3x\,dx = (e^{2x})(-\tfrac{1}{3}\cos 3x) - \int(-\tfrac{1}{3}\cos 3x)(2e^{2x})\,dx$$

$$= -\tfrac{1}{3}e^{2x}\cos 3x + \tfrac{2}{3}\int e^{2x}\cos 3x\,dx.$$

Substituting, we have

$$\int e^{2x}\cos 3x\,dx = \tfrac{1}{3}e^{2x}\sin 3x - \tfrac{2}{3}\left(-\tfrac{1}{3}e^{2x}\cos 3x + \tfrac{2}{3}\int e^{2x}\cos 3x\,dx\right)$$

or

$$\int e^{2x}\cos 3x\,dx = \tfrac{1}{3}e^{2x}\sin 3x + \tfrac{2}{9}e^{2x}\cos 3x - \tfrac{4}{9}\int e^{2x}\cos 3x\,dx.$$

This does not look much simpler, since we have found an integral in terms of itself. But, if we add $\frac{4}{9}\int e^{2x}\cos 3x\,dx$ to each side of the equation (supplying the constant of integration at the same time), we have

$$\tfrac{13}{9}\int e^{2x}\cos 3x\,dx = \frac{e^{2x}}{9}(3\sin 3x + 2\cos 3x) + c_1.$$

Finally, therefore,

$$\int e^{2x}\cos 3x\,dx = \frac{e^{2x}}{13}(3\sin 3x + 2\cos 3x) + c.$$

If we had chosen $\cos 3x$ as u in the first integration by parts and $\sin 3x$ as u_1 in the second integration by parts, the integral could be found in the same way that we just found it.

The differential of a function was introduced in Section 6 of Chapter 2 and was shown to satisfy the equation $dF(x) = F'(x)\,dx$. As a result, the symbol dx which occurs in an indefinite integral $\int f(x)\,dx$ may be legitimately regarded as a differential since, if $F'(x) = f(x)$, then $dF(x) = f(x)\,dx$. Moreover, if u is a differentiable function of x, then

$$dF(u) = F'(u)\,du = f(u)\,du,$$

so we write

$$F(u) + c = \int f(u)\,du.$$

[see (6.7) on page 216]. Using differentials, we obtain a very compact form for the formula for integration by parts. Since $du = \frac{du}{dx}\,dx$ and $dv = \frac{dv}{dx}\,dx$, substitution in (1.1) yields

(1.1′)
$$\boxed{\int u\,dv = uv - \int v\,du.}$$

We have less to write when we use this form of the formula, but the result is the same.

EXAMPLE 4. Integrate $\int x \ln x\,dx$. If we use (1.1′), we set $u = \ln x$ and $dv = x\,dx$. Then $du = \frac{1}{x}\,dx$ and $v = \frac{1}{2}x^2$. Hence

$$\begin{aligned}\int x \ln x\,dx &= (\ln x)(\tfrac{1}{2}x^2) - \int (\tfrac{1}{2}x^2)\left(\frac{1}{x}\,dx\right)\\ &= \tfrac{1}{2}x^2 \ln x - \tfrac{1}{2}\int x\,dx\\ &= \tfrac{1}{2}x^2 \ln x - \tfrac{1}{4}x^2 + c.\end{aligned}$$

In the remainder of this chapter we shall take full advantage of the streamlined notation offered by the differential and shall use it freely when making substitutions in indefinite integrals.

EXAMPLE 5. Find $\int x \ln(x+1)\,dx$. This example illustrates the fact that a judicious choice of a constant of integration can sometimes simplify the computation. Set $u = \ln(x+1)$ and $dv = x\,dx$. Then $du = \dfrac{1}{x+1}\,dx$ and we may take $v = \dfrac{x^2}{2}$. If we do this, we have

$$\begin{aligned}\int x\ln(x+1)\,dx &= \ln(x+1)\frac{x^2}{2} - \int \frac{x^2}{2}\,\frac{1}{x+1}\,dx \\ &= \tfrac{1}{2}x^2\ln(x+1) - \frac{1}{2}\int \frac{x^2}{x+1}\,dx.\end{aligned}$$

Dividing, we find that $\dfrac{x^2}{x+1} = x - 1 + \dfrac{1}{x+1}$, and hence

$$\begin{aligned}\int x\ln(x+1)\,dx &= \tfrac{1}{2}x^2\ln(x+1) - \frac{1}{2}\int\left(x - 1 + \frac{1}{x+1}\right)dx \\ &= \tfrac{1}{2}x^2\ln(x+1) - \tfrac{1}{2}(\tfrac{1}{2}x^2 - x + \ln|x+1|) + c \\ &= \tfrac{1}{2}(x^2-1)\ln(x+1) - \tfrac{1}{4}x^2 + \tfrac{1}{2}x + c.\end{aligned}$$

[Since $\ln(x+1)$ makes sense only if $x + 1 > 0$, we have replaced $\ln|x+1|$ by $\ln(x+1)$.] The same result is reached more quickly if we take $v = \dfrac{x^2}{2} + c = \dfrac{x^2+k}{2}$. Then integration by parts gives

$$\int x\ln(x+1)\,dx = \ln(x+1)\frac{x^2+k}{2} - \int \frac{x^2+k}{2}\,\frac{1}{x+1}\,dx.$$

Since we have a free choice for k, we shall choose $k = -1$ and then $\dfrac{x^2+k}{2}\,\dfrac{1}{x+1} = \dfrac{x^2-1}{2}\,\dfrac{1}{x+1} = \dfrac{x-1}{2}$. The problem becomes

$$\begin{aligned}\int x\ln(x+1)\,dx &= \ln(x+1)\frac{x^2-1}{2} - \int\frac{x-1}{2}\,dx \\ &= \tfrac{1}{2}(x^2-1)\ln(x+1) - \tfrac{1}{4}x^2 + \tfrac{1}{2}x + c.\end{aligned}$$

The method of integration by parts can frequently be used to obtain a recursion formula for one integral in terms of a simpler one. As an example, we derive the useful identity

(1.2) *For every integer* $n \geq 2$,

$$\boxed{\int \cos^n x\,dx = \frac{\cos^{n-1}x\sin x}{n} + \frac{n-1}{n}\int\cos^{n-2}x\,dx.}$$

Proof. We write $\cos^n x$ as the product $\cos^{n-1}x \cos x$, and let $u = \cos^{n-1}x$ and $dv = \cos x\, dx$. Then $v = \sin x$ and $du = (n-1)\cos^{n-2}x\,(-\sin x\, dx) = -(n-1)\cos^{n-2}x \sin x\, dx$. Hence

$$\int \cos^n x\, dx = \cos^{n-1}x \sin x - \int \sin x[-(n-1)\cos^{n-2}x \sin x\, dx]$$

$$= \cos^{n-1}x \sin x + (n-1)\int \cos^{n-2}x \sin^2 x\, dx.$$

Replacing $\sin^2 x$ by $1 - \cos^2 x$, the equation becomes

$$\int \cos^n x\, dx = \cos^{n-1}x \sin x + (n-1)\int \cos^{n-2}x\,(1 - \cos^2 x)\, dx$$

$$= \cos^{n-1}x \sin x + (n-1)\int \cos^{n-2}x\, dx - (n-1)\int \cos^n x\, dx.$$

Adding $(n-1)\int \cos^n x\, dx$ to both sides of the equation, we obtain

$$n\int \cos^n x\, dx = \cos^{n-1}x \sin x + (n-1)\int \cos^{n-2}x\, dx,$$

whence (1.2) follows at once upon division by n.

Thus by repeated applications of the recursion formula (1.2), the integral $\int \cos^n x\, dx$ can be reduced eventually to a polynomial in $\sin x$ and $\cos x$. If n is even, the final integral is

$$\int \cos^0 x\, dx = \int dx = x + c,$$

and, if n is odd, it is

$$\int \cos x\, dx = \sin x + c.$$

EXAMPLE 6. Use the recursion formula (1.2) to find $\int \cos^5 2x\, dx$. We first write $\int \cos^5 2x\, dx = \frac{1}{2}\int \cos^5 2x\, d(2x)$ and then

$$\int \cos^5 2x\, d(2x) = \frac{\cos^4 2x \sin 2x}{5} + \frac{4}{5}\int \cos^3 2x\, d(2x).$$

A second application of the formula yields

$$\int \cos^3 2x\, d(2x) = \frac{\cos^2 2x \sin 2x}{3} + \frac{2}{3}\int \cos 2x\, d(2x),$$

and, of course,

$$\int \cos 2x\, d(2x) = \sin 2x + c_1.$$

Combining, we have

$$\int \cos^5 2x\,dx = \frac{1}{2}\left[\frac{\cos^4 2x \sin 2x}{5} + \frac{4}{5}\left\{\frac{\cos^2 2x \sin 2x}{3} + \frac{2}{3}(\sin 2x + c_1)\right\}\right]$$

$$= \tfrac{1}{10}\cos^4 2x \sin 2x + \tfrac{2}{15}\cos^2 2x \sin 2x + \tfrac{4}{15}\sin 2x + c.$$

PROBLEMS

1. Integrate each of the following.

(a) $\int x \cos x\,dx$ (f) $\int (x^3 - 7x + 2)\sin x\,dx$

(b) $\int (x + 2)e^{2x}dx$ (g) $\int e^{3x}\sin 2x\,dx$

(c) $\int \arctan x\,dx$ (h) $\int \sec^3 x\,dx$

(d) $\int x^2 \sin 7x\,dx$ (i) $\int x^2 \ln(x + 1)\,dx$

(e) $\int x^3 \ln x\,dx$ (j) $\int \ln(x^2 + 2)\,dx.$

2. Do Example 3 again, first letting $u = \cos 3x$ and then $u_1 = \sin 3x$.
3. Find formulas for

(a) $\int e^{ax}\cos bx\,dx$ (b) $\int e^{ax}\sin bx\,dx.$

4. Derive the recursion formula analogous to (1.2): *For every integer* $n \geq 2$,

$$\int \sin^n x\,dx = -\frac{\sin^{n-1}x\cos x}{n} + \frac{n-1}{n}\int \sin^{n-2}x\,dx.$$

5. Evaluate

(a) $\int \cos^2 x\,dx$ (b) $\int \sin^2 x\,dx$

by the recursion formulas [see (1.2) and Problem 4], and also using the trigonometric identities

$$\cos^2 x = \tfrac{1}{2}(1 + \cos 2x),$$
$$\sin^2 x = \tfrac{1}{2}(1 - \cos 2x).$$

Show that the results obtained are the same. (The preceding identities can be read off at once from two more basic ones:

$$1 = \cos^2 x + \sin^2 x$$
$$\cos 2x = \cos^2 x - \sin^2 x.)$$

6. Use integration by parts to find recursion formulas, expressing the given integral in terms of an integral with a lower power:

(a) Show that $\int x^n e^x dx = x^n e^x - n\int x^{n-1}e^x dx.$

(b) Show that $\int \sec^n x\, dx = \dfrac{\sec^{n-2}x \tan x}{n-1} + \dfrac{n-2}{n-1}\int \sec^{n-2}x\, dx.$

(c) Find a reduction formula, expressing $\int (\ln|ax+b|)^n dx$ in terms of $\int (\ln|ax+b|)^{n-1}dx$.

7. Use the formulas derived in (1.2) and in Problems 4 and 6 to find

(a) $\int x^5 e^x dx$

(b) $\int \sin^4 x\, dx$

(c) $\int \cos^3 5x\, dx$

(d) $\int (\ln|3x+7|)^6 dx$

(e) $\int_0^{\pi/2} \sin^3 x\, dx$

(f) $\int_0^1 x^3 e^x dx.$

2. Integrals of Trigonometric Functions. Products of trigonometric functions, powers of trigonometric functions, and products of their powers are all functions which we need to integrate at various times. In this section techniques will be developed for finding antiderivatives of the commonly encountered functions of these types.

The first and simplest occur with the integrals

$$\int \cos ax \cos bx\, dx,$$
$$\int \sin ax \sin bx\, dx, \qquad (1)$$
$$\int \sin ax \cos bx\, dx, \qquad \text{in which } a \neq b.$$

None of these can be integrated directly, but each of the three integrands is a term in the expansions of $\cos(ax + bx)$ and $\cos(ax - bx)$ or in the expansions of $\sin(ax + bx)$ and $\sin(ax - bx)$. We can use these addition formulas to change products to sums or differences, and the latter can be integrated easily.

EXAMPLE 1. Integrate:

(a) $\int \sin 8x \sin 3x\, dx,$

(b) $\int \sin 7x \cos 2x\, dx.$

The integrand $\sin 8x \sin 3x$ in (a) is one term in the expansion of $\cos(8x + 3x)$ and also in the expansion of $\cos(8x - 3x)$. That is, we have

$$\cos(8x + 3x) = \cos 8x \cos 3x - \sin 8x \sin 3x,$$

$$\cos(8x - 3x) = \cos 8x \cos 3x + \sin 8x \sin 3x.$$

Subtracting the first from the second, we get

$$\cos(8x - 3x) - \cos(8x + 3x) = 2 \sin 8x \sin 3x.$$

Hence, since $8x - 3x = 5x$ and $8x + 3x = 11x$, we obtain

$$\sin 8x \sin 3x = \tfrac{1}{2}(\cos 5x - \cos 11x),$$

and so

$$\begin{aligned}\int \sin 8x \sin 3x \, dx &= \tfrac{1}{2}\int (\cos 5x - \cos 11x)\, dx \\ &= \tfrac{1}{10}\sin 5x - \tfrac{1}{22}\sin 11x + c.\end{aligned}$$

For the integral in (b), we use the formulas for the sine of the sum and difference of two numbers:

$$\sin(7x + 2x) = \sin 7x \cos 2x + \cos 7x \sin 2x,$$

$$\sin(7x - 2x) = \sin 7x \cos 2x - \cos 7x \sin 2x.$$

Adding, we have

$$\sin(7x + 2x) + \sin(7x - 2x) = 2 \sin 7x \cos 2x.$$

Hence

$$\sin 7x \cos 2x = \tfrac{1}{2}(\sin 9x + \sin 5x),$$

and

$$\begin{aligned}\int \sin 7x \cos 2x \, dx &= \tfrac{1}{2}\int (\sin 9x + \sin 5x)\, dx \\ &= -\tfrac{1}{18}\cos 9x - \tfrac{1}{10}\cos 5x + c.\end{aligned}$$

It should be clear that, using the formulas for the cosine and sine of the sum and difference of two numbers as in Example 1, we can readily evaluate any integral of the type given in equations (1).

We next consider integrals of the type

$$\int \cos^m x \sin^n x \, dx, \tag{2}$$

in which at least one of the exponents m and n is an odd positive integer (the other exponent need only be a real number). Suppose that $m = 2k + 1$, where k is a nonnegative integer. Then

$$\begin{aligned}\cos^m x \sin^n x &= \cos^{2k+1} x \sin^n x \\ &= (\cos^2 x)^k \sin^n x \cos x.\end{aligned}$$

Using the identity $\cos^2 x = 1 - \sin^2 x$, we obtain

$$\int \cos^m x \sin^n x \, dx = \int (1 - \sin^2 x)^k \sin^n x \cos x \, dx.$$

The factor $(1 - \sin^2 x)^k$ can be expanded by the Binomial Theorem, and the result is that $\int \cos^m x \sin^n x \, dx$ can be written as a sum of constant multiples of integrals of the form $\int \sin^q x \cos x \, dx$. Since

$$\int \sin^q x \cos x \, dx = \begin{cases} \dfrac{1}{q+1} \sin^{q+1} x + c & \text{if } q \neq -1, \\ \ln|\sin x| + c & \text{if } q = -1, \end{cases}$$

it follows that $\int \cos^m x \sin^n x \, dx$ can be readily evaluated. An entirely analogous argument follows if the exponent n is an odd positive integer.

EXAMPLE 2. Integrate

$$\text{(a)} \int \cos^3 4x \, dx, \qquad \text{(b)} \int \sin^5 x \cos^4 x \, dx.$$

The integral in (a) illustrates that the method just described is applicable to odd positive integer powers of the sine or cosine (i.e., either m or n may be zero). We obtain

$$\begin{aligned}\int \cos^3 4x \, dx &= \int \cos^2 4x \cos 4x \, dx \\ &= \int (1 - \sin^2 4x) \cos 4x \, dx \\ &= \int \cos 4x \, dx - \int \sin^2 4x \cos 4x \, dx \\ &= \tfrac{1}{4} \sin 4x - \tfrac{1}{12} \sin^3 4x + c.\end{aligned}$$

In (b) it is the exponent of the sine which is an odd positive integer.

Hence

$$\begin{aligned}\int \sin^5 x \cos^4 x\,dx &= \int (\sin^2 x)^2 \cos^4 x \sin x\,dx \\ &= \int (1 - \cos^2 x)^2 \cos^4 x \sin x\,dx \\ &= \int (1 - 2\cos^2 x + \cos^4 x)\cos^4 x \sin x\,dx \\ &= \int \cos^4 x \sin x\,dx - 2\int \cos^6 x \sin x\,dx + \int \cos^8 x \sin x\,dx \\ &= -\tfrac{1}{5}\cos^5 x + \tfrac{2}{7}\cos^7 x - \tfrac{1}{9}\cos^9 x + c.\end{aligned}$$

The third type of integral we consider consists of those of the form

$$\int \cos^m x \sin^n x\,dx, \tag{3}$$

in which both m and n are even nonnegative integers. These functions are not so simple to integrate as those containing an odd power. We first consider the special case in which either $m = 0$ or $n = 0$. The simplest nontrivial examples are the two integrals $\int \cos^2 x\,dx$ and $\int \sin^2 x\,dx$, which can be integrated by means of the identities

$$\begin{aligned}\cos^2 x &= \tfrac{1}{2}(1 + \cos 2x), \\ \sin^2 x &= \tfrac{1}{2}(1 - \cos 2x).\end{aligned}$$

These are useful enough to be worth memorizing, but they can also be derived quickly by addition and subtraction from the two more primitive identities

$$\begin{aligned}1 &= \cos^2 x + \sin^2 x, \\ \cos 2x &= \cos^2 x - \sin^2 x.\end{aligned}$$

Evaluation of the two integrals is now a simple matter. We get

$$\begin{aligned}\int \cos^2 x\,dx &= \tfrac{1}{2}\int (1 + \cos 2x)\,dx = \frac{x}{2} + \frac{1}{4}\sin 2x + c, \\ \int \sin^2 x\,dx &= \tfrac{1}{2}\int (1 - \cos 2x)\,dx = \frac{x}{2} - \frac{1}{4}\sin 2x + c.\end{aligned}$$

Going on to the higher powers, consider the integral $\int \cos^{2i} x\,dx$, where i is an arbitrary positive integer. We write

$$\begin{aligned}\cos^{2i} x = (\cos^2 x)^i &= [\tfrac{1}{2}(1 + \cos 2x)]^i \\ &= \frac{1}{2^i}(1 + \cos 2x)^i.\end{aligned}$$

The factor $(1 + \cos 2x)^i$ can be expanded by the Binomial Theorem. The result is that $\cos^{2i}x$ can be written as a sum of constant multiples of functions of the form $\cos^j 2x$, and in each of these $j < 2i$. The terms in this sum for which j is odd are all of the type already shown to be integrable. The terms for which j is even are of the type now under consideration. However, the exponents j are all smaller than the original power $2i$. For each function $\cos^j 2x$ with j even and nonzero, we repeat the process just described. Again, the resulting even powers of the cosine will be reduced. By repetition of these expansions, the even powers of the cosine can eventually all be reduced to zero. It follows that, although the process may be a tedious one, the integral $\int \cos^{2i}x\, dx$ can always be evaluated. The argument for $\int \sin^{2i}x\, dx$ is entirely analogous.

EXAMPLE 3. Integrate $\int \sin^6 2x\, dx$. We write

$$\begin{aligned} \sin^6 2x &= (\sin^2 2x)^3 = [\tfrac{1}{2}(1 - \cos 4x)]^3 \\ &= \tfrac{1}{8}(1 - 3\cos 4x + 3\cos^2 4x - \cos^3 4x) \\ &= \tfrac{1}{8}[1 - 3\cos 4x + \tfrac{3}{2}(1 + \cos 8x) - \cos^3 4x] \\ &= \tfrac{5}{16} - \tfrac{3}{8}\cos 4x + \tfrac{3}{16}\cos 8x - \tfrac{1}{8}\cos^3 4x. \end{aligned}$$

Hence

$$\int \sin^6 2x\, dx = \frac{5x}{16} - \frac{3}{32}\sin 4x + \frac{3}{128}\sin 8x - \frac{1}{8}\int \cos^3 4x\, dx.$$

In Example 2 we have shown that

$$\int \cos^3 4x\, dx = \tfrac{1}{4}\sin 4x - \tfrac{1}{12}\sin^3 4x + c.$$

We conclude that

$$\begin{aligned} \int \sin^6 2x\, dx &= \frac{5x}{16} - \frac{3}{32}\sin 4x + \frac{3}{128}\sin 8x - \frac{1}{32}\sin 4x + \frac{1}{96}\sin^3 4x + c \\ &= \frac{5x}{16} - \frac{1}{8}\sin 4x + \frac{3}{128}\sin 8x + \frac{1}{96}\sin^3 4x + c. \end{aligned}$$

Returning to the general case, we can now integrate $\int \cos^m x \sin^n x\, dx$, where m and n are arbitrary nonnegative even integers. For, setting $m = 2i$ and $n = 2j$, we can write

$$\begin{aligned} \cos^m x \sin^n x &= \cos^{2i}x\,(\sin^2 x)^j \\ &= \cos^{2i}x\,(1 - \cos^2 x)^j. \end{aligned}$$

When expanded, the right side is a sum of constant multiples of even powers of $\cos x$, and we have shown that each of these can be integrated. This completes the argument. Actually, if neither m nor n is zero, we can save time by using the identity

$$\sin x \cos x = \tfrac{1}{2} \sin 2x,$$

as illustrated in the following example.

EXAMPLE 4. Integrate $\int \cos^4 x \sin^2 x \, dx$. Since $\sin^2 x$ is the factor with the smaller exponent, we write

$$\begin{aligned} \cos^4 x \sin^2 x &= \cos^2 x \, (\cos^2 x \sin^2 x) \\ &= \cos^2 x \, (\sin x \cos x)^2 \\ &= [\tfrac{1}{2}(1 + \cos 2x)](\tfrac{1}{2} \sin 2x)^2. \end{aligned}$$

Expanding, we get

$$\begin{aligned} \cos^4 x \sin^2 x &= \tfrac{1}{8}(1 + \cos 2x) \sin^2 2x \\ &= \tfrac{1}{8} \sin^2 2x + \tfrac{1}{8} \sin^2 2x \cos 2x \\ &= \tfrac{1}{16}(1 - \cos 4x) + \tfrac{1}{8} \sin^2 2x \cos 2x. \end{aligned}$$

Hence

$$\begin{aligned} \int \cos^4 x \sin^2 x \, dx &= \tfrac{1}{16} \int dx - \tfrac{1}{16} \int \cos 4x \, dx + \tfrac{1}{8} \int \sin^2 2x \cos 2x \, dx \\ &= \frac{x}{16} - \frac{1}{64} \sin 4x + \frac{1}{48} \sin^3 2x + c. \end{aligned}$$

An important alternative method for integrating positive integer powers of the sine and cosine is by means of recursion (or reduction) formulas. In Section 1 [see (1.2), page 359], such a formula was developed, expressing $\int \cos^n x \, dx$ in terms of $\int \cos^{n-2} x \, dx$. Following the derivation, $\int \cos^5 2x \, dx$ is evaluated with two applications of the formula. A similar reduction formula for $\int \sin^n x \, dx$ was given in Problem 4, page 361. Certainly no one should memorize these formulas, but, if they are available, they undoubtedly provide the most automatic way of performing the integration.

We next turn to the problem of evaluating

$$\int \tan^n x \, dx, \tag{4}$$

where n is an arbitrary positive integer. For $n = 1$, the integral is an elementary one:

$$\int \tan x\, dx = \int \frac{\sin x}{\cos x}\, dx = \begin{cases} -\ln|\cos x| + c, \\ \\ \ln|\sec x| + c. \end{cases}$$

For $n \geq 2$, there is a reduction formula, which is easily derived as follows. Using the identity $\sec^2 x - 1 = \tan^2 x$, we have

$$\begin{aligned} \int \tan^n x\, dx &= \int \tan^{n-2} x \tan^2 x\, dx \\ &= \int \tan^{n-2} x (\sec^2 x - 1)\, dx \\ &= \int \tan^{n-2} x \sec^2 x\, dx - \int \tan^{n-2} x\, dx. \end{aligned}$$

Since $\dfrac{d}{dx} \tan x = \sec^2 x$, the first integral on the right is equal to

$$\int \tan^{n-2} x\, d \tan x.$$

Hence we obtain

$$(2.1) \qquad \boxed{\int \tan^n x\, dx = \frac{1}{n-1} \tan^{n-1} x - \int \tan^{n-2} x\, dx.}$$

However, we generally perform such integrations without explicit use of the reduction formula (2.1). We simply carry out this technique of replacing $\tan^2 x$ by $\sec^2 x - 1$ as often as necessary.

EXAMPLE 5. Integrate $\int \tan^5 x\, dx$. Factoring and substituting, we get

$$\begin{aligned} \int \tan^5 x\, dx &= \int \tan^3 x \tan^2 x\, dx \\ &= \int \tan^3 x (\sec^2 x - 1)\, dx \\ &= \int \tan^3 x \sec^2 x\, dx - \int \tan^3 x\, dx \\ &= \tfrac{1}{4} \tan^4 x - \int \tan x (\sec^2 x - 1)\, dx \\ &= \tfrac{1}{4} \tan^4 x - \int \tan x \sec^2 x\, dx + \int \tan x\, dx \\ &= \tfrac{1}{4} \tan^4 x - \tfrac{1}{2} \tan^2 x + \ln|\sec x| + c. \end{aligned}$$

The difficulty in evaluating the integral

$$\int \sec^n x \, dx, \tag{5}$$

where n is a positive integer, depends on whether n is even or odd. If $n = 2i$, for some positive integer i, then

$$\sec^n x = (\sec^2 x)^{i-1} \sec^2 x = (1 + \tan^2 x)^{i-1} \sec^2 x.$$

Hence, if n is even, $\sec^n x$ can be expanded into a sum of multiples of integrals of the form

$$\int \tan^j x \sec^2 x \, dx = \frac{1}{j+1} \tan^{j+1} x + c.$$

If n is odd, the problem is not so simple. We shall use the reduction formula

$$\text{(2.2)} \qquad \boxed{\int \sec^n x \, dx = \frac{\sec^{n-2} x \tan x}{n-1} + \frac{n-2}{n-1} \int \sec^{n-2} x \, dx.}$$

This formula is derived by integration by parts [see Problem 6(b), page 362] and is applicable for any integer $n \geq 2$, whether even or odd. With a finite number of applications, $\int \sec^n x \, dx$ can therefore be reduced to an expression in which the only remaining integral is $\int dx$ or $\int \sec x \, dx$, according as n is even or odd. Hence, if n is odd, we need to know $\int \sec x \, dx$. An ingenious method of integration is to consider the pair of functions $\sec x$ and $\tan x$ and to observe that the derivative of each one is equal to $\sec x$ times the other. Writing this fact in terms of differentials, we have

$$d \sec x = \sec x \tan x \, dx,$$

$$d \tan x = \sec^2 x \, dx = \sec x \sec x \, dx.$$

Adding and factoring, we obtain

$$d(\sec x + \tan x) = \sec x (\tan x + \sec x) \, dx.$$

Hence

$$\sec x \, dx = \frac{d(\sec x + \tan x)}{\sec x + \tan x},$$

from which follows the useful formula

$$\text{(2.3)} \qquad \boxed{\int \sec x \, dx = \ln|\sec x + \tan x| + c.}$$

EXAMPLE 6. Integrate $\int \sec^5 x\,dx$. Using the reduction formula (2.2) twice, we have

$$\int \sec^5 x\,dx = \frac{\sec^3 x \tan x}{4} + \frac{3}{4}\int \sec^3 x\,dx$$
$$= \frac{\sec^3 x \tan x}{4} + \frac{3}{4}\left(\frac{\sec x \tan x}{2} + \frac{1}{2}\int \sec x\,dx\right)\cdot$$

From this and (2.3), we conclude that

$$\int \sec^5 x\,dx = \frac{\sec^3 x \tan x}{4} + \frac{3 \sec x \tan x}{8} + \frac{3}{8}\ln|\sec x + \tan x| + c.$$

Of course, the integration of $\int \cot^n x\,dx$ parallels the technique for integrating $\int \tan^n x\,dx$, and the integration of $\int \csc^n x\,dx$ parallels that for $\int \sec^n x\,dx$. The reduction formula corresponding to (2.2) is

$$(2.4)\qquad \boxed{\int \csc^n x\,dx = -\frac{\csc^{n-2} x \cot x}{n-1} + \frac{n-2}{n-1}\int \csc^{n-2} x\,dx.}$$

The last type of integral to be discussed consists of those of the form

$$\int \sec^m x \tan^n x\,dx, \qquad (6)$$

where m and n are positive integers. There are a number of variations, depending on whether each of m and n is even or odd. We shall consider three cases:

Case 1. m is even. Then $m = 2k$, for some positive integer k. Hence

$$\int \sec^m x \tan^n x\,dx = \int \sec^{2k} x \tan^n x\,dx$$
$$= \int \sec^{2k-2} x \tan^n x \sec^2 x\,dx$$
$$= \int (\sec^2 x)^{k-1} \tan^n x \sec^2 x\,dx$$
$$= \int (1 + \tan^2 x)^{k-1} \tan^n x \sec^2 x\,dx.$$

We can now expand $(1 + \tan^2 x)^{k-1}$, and the result is that the original integral can be written as a sum of constant multiples of integrals of the form $\int \tan^j x \sec^2 x\,dx$. As we have seen, each of these is equal to $\int u^j du$, with $u = \tan x$, and is easily integrated.

Case 2. n is odd. Then $n = 2k + 1$, for some nonnegative integer k. We write

$$\begin{aligned}\int \sec^m x \tan^n x\, dx &= \int \sec^m x \tan^{2k+1} x\, dx \\ &= \int \sec^{m-1} x\, (\tan^2 x)^k \sec x \tan x\, dx \\ &= \int \sec^{m-1} x\, (\sec^2 x - 1)^k \sec x \tan x\, dx.\end{aligned}$$

Again we expand by use of the Binomial Theorem. In this case, the original integral becomes a sum of constant multiples of integrals of the form $\int \sec^j x \sec x \tan x\, dx$, each of which can be integrated, since

$$\begin{aligned}\int \sec^j x \sec x \tan x\, dx &= \int \sec^j x\, d(\sec x) \\ &= \frac{1}{j+1} \sec^{j+1} x + c.\end{aligned}$$

Case 3. n is even. Then $n = 2k$, for some positive integer k. In this case, we have

$$\begin{aligned}\int \sec^m x \tan^n x\, dx &= \int \sec^m x \tan^{2k} x\, dx \\ &= \int \sec^m x\, (\tan^2 x)^k dx \\ &= \int \sec^m x\, (\sec^2 x - 1)^k dx.\end{aligned}$$

This time, if we expand the integrand, we get a sum of constant multiples of integrals of the type $\int \sec^j x\, dx$, and we can use the reduction formula (2.2) on each of them.

The three cases discussed are not mutually exclusive, and one may have a choice. For example, if m is even and n odd, the integral may be found by the techniques of Case 1 or that of Case 2. If m and n are both even, either the techniques described in Case 1 or Case 3 may be used.

EXAMPLE 7. Evaluate the integrals

(a) $\int \sec^4 x \tan^6 x\, dx$, (b) $\int \sec^3 x \tan^5 x\, dx$.

For (a) we write

$$\begin{aligned}\int \sec^4 x \tan^6 x\, dx &= \int \sec^2 x \tan^6 x \sec^2 x\, dx \\ &= \int (1 + \tan^2 x) \tan^6 x \sec^2 x\, dx \\ &= \int \tan^6 x \sec^2 x\, dx + \int \tan^8 x \sec^2 x\, dx \\ &= \tfrac{1}{7} \tan^7 x + \tfrac{1}{9} \tan^9 x + c.\end{aligned}$$

It is also possible to evaluate this integral by the technique described in Case 3. However, the resulting computation would be so much longer that it would be foolish to do so.

For (b) we use the method of Case 2. Factoring, we get

$$\begin{aligned}\int \sec^3 x \tan^5 x\, dx &= \int \sec^2 x \tan^4 x \sec x \tan x\, dx \\ &= \int \sec^2 x (\sec^2 x - 1)^2 \sec x \tan x\, dx \\ &= \int \sec^2 x (\sec^4 x - 2 \sec^2 x + 1) \sec x \tan x\, dx \\ &= \int \sec^6 x \sec x \tan x\, dx - 2 \int \sec^4 x \sec x \tan x\, dx \\ &\quad + \int \sec^2 x \sec x \tan x\, dx \\ &= \tfrac{1}{7} \sec^7 x - \tfrac{2}{5} \sec^5 x + \tfrac{1}{3} \sec^3 x + c.\end{aligned}$$

We conclude with the remark that techniques for integrating $\int \csc^m x \cot^n x$ are analogous to those for $\int \sec^m x \tan^n x\, dx$.

PROBLEMS

1. Integrate each of the following.

(a) $\int \cos 5x \sin 2x\, dx$ (e) $\int \cos 3x \sin \pi x\, dx$

(b) $\int \cos 3x \cos x\, dx$ (f) $\int \cos 2\pi y \sin \pi y\, dy$

(c) $\int \sin 3x \sin x\, dx$ (g) $\int \sin x \sin 6x\, dx$

(d) $\int \cos 4z \cos 7z\, dz$ (h) $\int \cos 7w \sin 17w\, dw$.

2. (a) Integrate $\int \sin^3\theta \, d\theta$ by using the fact that the exponent of $\sin\theta$ is an odd positive integer.
   (b) Integrate $\int \sin^3\theta \, d\theta$ by making use of the identity $\sin 3\theta = 3\sin\theta - 4\sin^3\theta$.
   (c) Show that the answers obtained in (a) and (b) differ by a constant.
3. (a) Integrate $\int \sin^5 2x \, dx$ by using the fact that the exponent of the sine is an odd positive integer.
   (b) Integrate $\int \sin^5 2x \, dx$ using the recursion formula given in Problem 4, page 361.
   (c) Show that the answers obtained in (a) and (b) differ by a constant.
4. Integrate each of the following.

(a) $\displaystyle\int \cos^3 2x \, dx$ (f) $\displaystyle\int \sqrt{\sin x}\cos^5 x \, dx$

(b) $\displaystyle\int \cos^4 x \, dx$ (g) $\displaystyle\int \cos^6 3x \, dx$

(c) $\displaystyle\int \sin^4 3x \, dx$ (h) $\displaystyle\int \sin^4 y \cos^5 y \, dy$

(d) $\displaystyle\int \sin^3 x \cos^{44} x \, dx$ (i) $\displaystyle\int \cos^2 y \sin^4 y \, dy$

(e) $\displaystyle\int \sin^2\theta \cos^2\theta \, d\theta$ (j) $\displaystyle\int \sin^3 x \, (\cos x)^{5/2} dx.$

5. (a) Integrate $\int \cos^2\theta \, d\theta$ using the identity $\cos^2\theta = \frac{1}{2}(1 + \cos 2\theta)$.
   (b) Integrate $\int \cos^2\theta \, d\theta$ by parts.
   (c) Show that the answers obtained in (a) and (b) differ by a constant.
6. Evaluate $\int \sec^2 x \tan x \, dx$ in two different ways: first using the fact that the secant has an even exponent, and then using the fact that the tangent has an odd exponent. Show that the two solutions differ by a constant.
7. Integrate each of the following.

(a) $\displaystyle\int \tan^4 x \, dx$ (f) $\displaystyle\int \sec^3 x \tan^3 x \, dx$

(b) $\displaystyle\int \tan^3 4y \, dy$ (g) $\displaystyle\int \sec^3 x \tan^2 x \, dx$

(c) $\displaystyle\int \sec^4\theta \, d\theta$ (h) $\displaystyle\int \sec^4 x \tan^5 x \, dx$

(d) $\displaystyle\int \sec^3 2x \, dx$ (i) $\displaystyle\int \sec^6 x \sqrt{\tan x} \, dx$

(e) $\displaystyle\int \sec^4 x \tan^4 x \, dx$ (j) $\displaystyle\int \frac{dx}{\sec x \tan x}.$

8. (a) Let $n \geq 2$ be an integer, and derive a reduction formula for $\int \cot^n x\, dx$ analogous to (2.1).

   (b) Use the formula derived in (a) to integrate $\int \cot^5 3\theta\, d\theta$.

9. By a method analogous to that used previously to find $\int \sec x\, dx$, prove that

$$\int \csc x\, dx = -\ln|\csc x + \cot x| + c.$$

10. (a) Use integration by parts to derive the reduction formula (2.4) for $\int \csc^n x\, dx$.

    (b) Use this formula to integrate $\int \csc^6 y\, dy$.

11. Integrate each of the following.

(a) $\displaystyle\int \csc^5\theta\, d\theta$ (d) $\displaystyle\int \csc^4 x \cot^2 x\, dx$

(b) $\displaystyle\int \sin 3x \cot 3x\, dx$ (e) $\displaystyle\int \csc^3 2y \cot^3 2y\, dy$

(c) $\displaystyle\int \cot^4 y\, dy$ (f) $\displaystyle\int \csc^3\varphi \cot^2\varphi\, d\varphi.$

3. Trigonometric Substitutions. In this section we shall study a technique of integration which is particularly useful for finding integrals of functions of $\sqrt{a^2 - x^2}$, $\sqrt{a^2 + x^2}$, and $\sqrt{x^2 - a^2}$. The technique is that of trigonometric substitutions and is based on some of the elementary trigonometric identities developed in Chapter 6. We shall develop the method by doing specific examples.

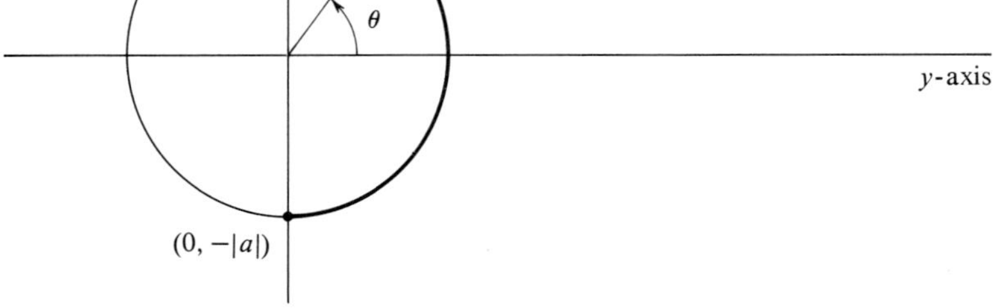

Figure 1

Consider the problem of evaluating $\int \sqrt{a^2 - x^2}\, dx$. The domain of the function $\sqrt{a^2 - x^2}$ is the closed interval $[-|a|, |a|]$. In what follows it will be convenient to view this interval as lying on the vertical axis, and, for this reason, the traditional positions of the x-axis and the y-axis will be interchanged. For every real number x in the domain $[-|a|, |a|]$, the point $(\sqrt{a^2 - x^2}, x)$ lies on the semicircle in the right half-plane with radius $|a|$ and center at the origin (see Figure 1). It follows from the definition of the trigonometric functions sine and cosine on page 282 that this point $(\sqrt{a^2 - x^2}, x)$ is equal to $(|a| \cos \theta, |a| \sin \theta)$, where θ is the radian measure of the angle denoted by the same letter in Figure 1. Hence

$$\begin{cases} \sqrt{a^2 - x^2} = |a| \cos \theta, \\ x = |a| \sin \theta. \end{cases} \tag{1}$$

We shall restrict θ to the interval $\left[-\frac{\pi}{2}, \frac{\pi}{2}\right]$. Then θ is uniquely defined by equations (1), and, as θ takes on all values in this interval, x assumes all values in the domain of the function $\sqrt{a^2 - x^2}$. Using equations (1), we obtain $dx = |a| \cos \theta\, d\theta$ and

$$\int \sqrt{a^2 - x^2}\, dx = \int |a| \cos \theta \cdot |a| \cos \theta\, d\theta = a^2 \int \cos^2\theta\, d\theta.$$

Since $\cos^2\theta = \frac{1}{2}(\cos 2\theta + 1)$,

$$\begin{aligned} \int \sqrt{a^2 - x^2}\, dx &= \frac{a^2}{2} \int (\cos 2\theta + 1)\, d\theta \\ &= \frac{a^2}{2} \left(\tfrac{1}{2} \sin 2\theta + \theta\right) + c. \end{aligned}$$

But $\theta = \arcsin \dfrac{x}{|a|}$ and

$$\tfrac{1}{2} \sin 2\theta = \sin \theta \cos \theta = \frac{x}{|a|} \frac{\sqrt{a^2 - x^2}}{|a|} = \frac{x\sqrt{a^2 - x^2}}{a^2}.$$

Substituting back, we get

$$\begin{aligned} \int \sqrt{a^2 - x^2}\, dx &= \frac{a^2}{2} \left(\frac{x\sqrt{a^2 - x^2}}{a^2} + \arcsin \frac{x}{|a|}\right) + c \\ &= \tfrac{1}{2} \left(x\sqrt{a^2 - x^2} + a^2 \arcsin \frac{x}{|a|}\right) + c. \end{aligned}$$

In general, with any integral involving $\sqrt{a^2 - x^2}$, we make the trigonometric substitutions based on equations (1); i.e., we replace x by $|a| \sin \theta$, $\sqrt{a^2 - x^2}$ by $|a| \cos \theta$, and dx by $|a| \cos \theta\, d\theta$. Note that there is an equivalent alternative procedure: We may set $x = |a| \cos \theta$ and restrict θ to the interval $[0, \pi]$. Then x can take on all values in the interval $[-|a|, |a|]$ as before, and, in addition, $\sin\theta$ will be nonnegative. Since $\cos^2\theta + \sin^2\theta = 1$, we will have

$$\sqrt{a^2 - x^2} = \sqrt{a^2 - a^2\cos^2\theta} = \sqrt{a^2\sin^2\theta} = |a|\sin\theta.$$

Thus the integral may be evaluated equally well using the equations

$$\begin{cases} x = |a|\cos\theta, \\ \sqrt{a^2 - x^2} = |a|\sin\theta. \end{cases} \tag{2}$$

Of course, if these substitutions are used, then $dx = -|a|\sin\theta\, d\theta$. Geometrically, equations (2) are obtained by starting from the point $(x, \sqrt{a^2 - x^2})$, which lies on the semicircle in the *upper* half-plane instead of the *right* half-plane.

A definite integral may be simpler to evaluate than an indefinite integral, since we may use the Change of Variable Theorem for Definite Integrals (see page 215) and thereby avoid the substitution back to the original variable.

EXAMPLE 1. Evaluate the definite integral

$$\int_2^{2\sqrt{3}} \frac{x^2 dx}{\sqrt{16 - x^2}}.$$

Using equations (1), we define θ by setting $x = 4\sin\theta$ and $\sqrt{a^2 - x^2} = 4\cos\theta$. It follows that $dx = 4\cos\theta\, d\theta$. If $x = 2$, then $\sin\theta = \frac{1}{2}$ and so $\theta = \frac{\pi}{6}$. If $x = 2\sqrt{3}$, then $\sin\theta = \frac{\sqrt{3}}{2}$ and $\theta = \frac{\pi}{3}$. Hence

$$\int_2^{2\sqrt{3}} \frac{x^2 dx}{\sqrt{16 - x^2}} = \int_{\pi/6}^{\pi/3} \frac{16\sin^2\theta \cdot 4\cos\theta\, d\theta}{4\cos\theta}$$

$$= 16\int_{\pi/6}^{\pi/3} \sin^2\theta\, d\theta.$$

Since $\sin^2\theta = \frac{1}{2}(1 - \cos 2\theta)$, we obtain

$$\int_2^{2\sqrt{3}} \frac{x^2 dx}{\sqrt{16 - x^2}} = 8\int_{\pi/6}^{\pi/3} (1 - \cos 2\theta)\, d\theta$$

$$= (8\theta - 4\sin 2\theta)\Big|_{\pi/6}^{\pi/3}$$

$$= \left(8\cdot\frac{\pi}{3} - 4\sin\frac{2\pi}{3}\right) - \left(8\cdot\frac{\pi}{6} - 4\sin\frac{2\pi}{6}\right)$$

$$= \frac{8\pi}{3} - 4\cdot\frac{\sqrt{3}}{2} - \frac{8\pi}{6} + 4\cdot\frac{\sqrt{3}}{2} = \frac{4\pi}{3}.$$

Next we consider integrals involving $\sqrt{a^2 + x^2}$. The domain of the function $\sqrt{a^2 + x^2}$ is the set of all real numbers, i.e., the unbounded interval $(-\infty, \infty)$. Geometrically, we shall again find it convenient to place this domain on the vertical axis and to interchange the usual x-axis and y-axis in the picture. For every real number x, consider the point $(|a|, x)$, and let

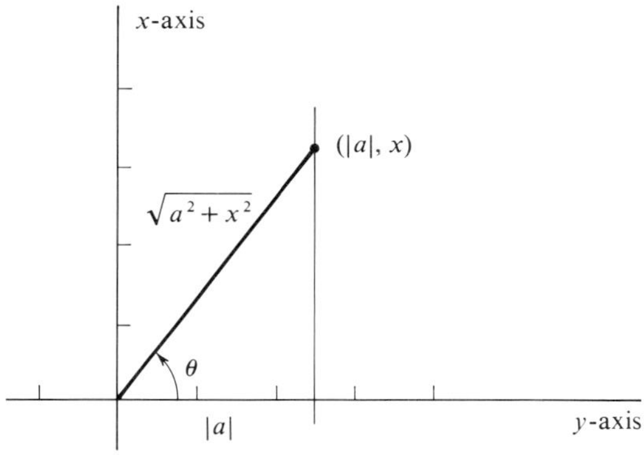

Figure 2

θ be the radian measure of the angle shown in Figure 2. It follows that

$$\begin{cases} x = |a| \tan \theta, \\ \sqrt{a^2 + x^2} = |a| \sec \theta. \end{cases} \tag{3}$$

We restrict θ to the open interval $\left(-\frac{\pi}{2}, \frac{\pi}{2}\right)$. In this way, θ is uniquely determined by x according to equations (3), and the interval $(-\infty, \infty)$ of all possible values of x corresponds in a one-to-one fashion to the interval $\left(-\frac{\pi}{2}, \frac{\pi}{2}\right)$ of all values of θ. It follows from the equation $x = |a| \tan \theta$ that $dx = |a| \sec^2\theta \, d\theta$. Hence, if we substitute $|a| \tan \theta$ for x, we substitute $|a| \sec \theta$ for $\sqrt{a^2 + x^2}$ and $|a| \sec^2\theta \, d\theta$ for dx.

Algebraically, the substitutions given by equations (3) arise from the trigonometric identity $1 + \tan^2\theta = \sec^2\theta$. If we set $x = |a| \tan \theta$, then x will assume all real values as θ takes on all values in the open interval $\left(-\frac{\pi}{2}, \frac{\pi}{2}\right)$. Moreover, $\sec \theta$ is positive in this interval. It follows that

$$\sqrt{a^2 + x^2} = \sqrt{a^2(1 + \tan^2\theta)} = \sqrt{a^2\sec^2\theta} = |a| \sec \theta.$$

EXAMPLE 2. Integrate $\int \frac{dx}{\sqrt{a^2 + x^2}}$. Letting $x = |a| \tan \theta$, we obtain $\sqrt{a^2 + x^2} = |a| \sec \theta$ and $dx = |a| \sec^2\theta \, d\theta$.

Hence

$$\int \frac{dx}{\sqrt{a^2 + x^2}} = \int \frac{|a| \sec^2\theta \, d\theta}{|a| \sec \theta} = \int \sec \theta \, d\theta.$$

It was shown in Section 2 that

$$\int \sec \theta \, d\theta = \ln|\sec \theta + \tan \theta| + c.$$

Consequently,

$$\begin{aligned} \int \frac{dx}{\sqrt{a^2 + x^2}} &= \ln|\sec \theta + \tan \theta| + c \\ &= \ln \left| \frac{\sqrt{a^2 + x^2}}{|a|} + \frac{x}{|a|} \right| + c = \ln \left| \frac{\sqrt{a^2 + x^2} + x}{|a|} \right| + c. \end{aligned}$$

Using the properties of the logarithm, we may write

$$\ln \left| \frac{\sqrt{a^2 + x^2} + x}{|a|} \right| = \ln|\sqrt{a^2 + x^2} + x| - \ln|a|.$$

Since $-\ln|a| + c$ is no more or less arbitrary as a constant than c itself, we conclude that

$$\int \frac{dx}{\sqrt{a^2 + x^2}} = \ln|\sqrt{a^2 + x^2} + x| + c.$$

By trigonometric substitutions, functions of $\sqrt{x^2 - a^2}$ may frequently be put in a form so that integration is possible. Since $\sqrt{x^2 - a^2}$ is defined if and only if $|x| \geq |a|$, the domain of the function $\sqrt{x^2 - a^2}$, unlike the others, is the union of two intervals: $(-\infty, -|a|]$ and $[|a|, \infty)$. In this case we shall set $x = |a| \sec \theta$. Using the identity $1 + \tan^2\theta = \sec^2\theta$, we obtain

$$\sqrt{x^2 - a^2} = \sqrt{a^2(\sec^2\theta - 1)} = \sqrt{a^2\tan^2\theta} = |a \tan \theta|.$$

If θ is restricted to the interval $\left[0, \frac{\pi}{2}\right)$, then $\tan \theta$ is nonnegative, and, as θ takes on all values in this interval, x assumes all values in $[|a|, \infty)$. Similarly, if θ is restricted to the interval $\left[-\pi, -\frac{\pi}{2}\right)$, then $\tan \theta$ is again nonnegative, and, as θ runs through this interval, x correspondingly traverses $(-\infty, -|a|]$ (in the opposite direction). Thus we have defined a new variable θ by the equations

$$\begin{cases} x = |a| \sec \theta, \\ \sqrt{x^2 - a^2} = |a| \tan \theta. \end{cases} \tag{4}$$

These equations can also be obtained geometrically. Figure 3 illustrates the situation for $x \geq |a|$. For every such x, consider the point $(|a|, \sqrt{x^2 - a^2})$ in the plane, and let θ be the radian measure of the angle shown. Since x appears only as the hypotenuse of a right triangle, we have used letters other than x and y in labeling the horizontal and vertical axes.

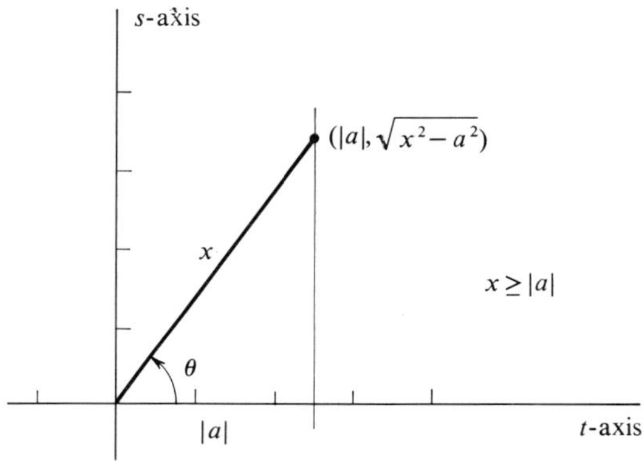

Figure 3

It follows from $x = |a| \sec\theta$ that $dx = |a| \sec\theta \tan\theta \, d\theta$. Hence if we make the trigonometric substitutions based on equations (4), we substitute $|a| \sec\theta \tan\theta \, d\theta$ for dx.

EXAMPLE 3. Find the indefinite integral

$$\text{(a)} \quad \int \frac{dx}{\sqrt{x^2 - a^2}},$$

and evaluate the definite integral

$$\text{(b)} \quad \int_{-6}^{-3\sqrt{2}} x^3\sqrt{x^2 - 9}\, dx.$$

For part (a), we let $x = |a| \sec\theta$ and $\sqrt{x^2 - a^2} = |a| \tan\theta$. Then $dx = |a| \sec\theta \tan\theta \, d\theta$ and

$$\int \frac{dx}{\sqrt{x^2 - a^2}} = \int \frac{|a| \sec\theta \tan\theta \, d\theta}{|a| \tan\theta} = \int \sec\theta \, d\theta.$$

It was shown in Section 2 that

$$\int \sec\theta \, d\theta = \ln|\sec\theta + \tan\theta| + c.$$

Hence

$$\int \frac{dx}{\sqrt{x^2 - a^2}} = \ln|\sec\theta + \tan\theta| + c.$$

But $\sec\theta = \dfrac{x}{|a|}$ and $\tan\theta = \dfrac{\sqrt{x^2 - a^2}}{|a|}$ and, substituting back, we therefore obtain

$$\int \frac{dx}{\sqrt{x^2 - a^2}} = \ln\left|\frac{x}{|a|} + \frac{\sqrt{x^2 - a^2}}{|a|}\right| + c.$$

Since $\ln\left|\dfrac{x}{|a|} + \dfrac{\sqrt{x^2 - a^2}}{|a|}\right| = \ln|x + \sqrt{a^2 - x^2}| - \ln|a|$ and since c is an arbitrary constant, we may incorporate the term $-\ln|a|$ into the constant of integration and conclude that

$$\int \frac{dx}{\sqrt{x^2 - a^2}} = \ln|x + \sqrt{x^2 - a^2}\,| + c.$$

For the definite integral in part (b), we introduce θ as the variable of integration by letting $x = 3\sec\theta$. Then $\sqrt{x^2 - 9} = 3\tan\theta$ and $dx = 3\sec\theta\tan\theta\,d\theta$. Restricting θ to the interval $\left[-\pi, -\dfrac{\pi}{2}\right)$, we see that, when $x = -6$, $\sec\theta = -2$ and so $\theta = -\dfrac{2\pi}{3}$. Similarly, when $x = -3\sqrt{2}$, $\sec\theta = -\sqrt{2}$ and $\theta = -\dfrac{3\pi}{4}$. With these substitutions and the Change of Variable Theorem for Definite Integrals, we obtain

$$\begin{aligned}\int_{-6}^{-3\sqrt{2}} x^3\sqrt{x^2 - 9}\,dx &= \int_{-2\pi/3}^{-3\pi/4} 27\sec^3\theta \cdot 3\tan\theta \cdot 3\sec\theta\tan\theta\,d\theta \\ &= 243\int_{-2\pi/3}^{-3\pi/4} \sec^4\theta\tan^2\theta\,d\theta.\end{aligned}$$

To integrate this, we replace a factor of $\sec^2\theta$ in the integrand by $1 + \tan^2\theta$. The integral then becomes

$$\begin{aligned}\int_{-6}^{-3\sqrt{2}} x^3\sqrt{x^2 - 9}\,dx \\ &= 243\int_{-2\pi/3}^{-3\pi/4} (1 + \tan^2\theta)\tan^2\theta\sec^2\theta\,d\theta \\ &= 243\int_{-2\pi/3}^{-3\pi/4} \tan^2\theta\sec^2\theta\,d\theta + 243\int_{-2\pi/3}^{-3\pi/4} \tan^4\theta\sec^2\theta\,d\theta.\end{aligned}$$

Since $d \tan \theta = \sec^2\theta\, d\theta$, it is easy to find antiderivatives of $\tan^2\theta \sec^2\theta$ and $\tan^4\theta \sec^2\theta$. Hence

$$\begin{aligned}\int_{-6}^{-3\sqrt{2}} x^3\sqrt{x^2 - 9}\, dx &= 243(\tfrac{1}{3}\tan^3\theta + \tfrac{1}{5}\tan^5\theta)\Big|_{-2\pi/3}^{-3\pi/4} \\ &= 243[(\tfrac{1}{3}\cdot 1 + \tfrac{1}{5}\cdot 1) - (\tfrac{1}{3}\cdot 3\sqrt{3} + \tfrac{1}{5}\cdot 9\sqrt{3})] \\ &= 243\left(\frac{8}{15} - \frac{14\sqrt{3}}{5}\right) \\ &= \tfrac{81}{5}(8 - 42\sqrt{3}),\end{aligned}$$

and the example is finished.

Although the trigonometric substitutions developed in this section have been primarily directed at integrands containing certain square roots, we can equally well apply them to other functions of $a^2 - x^2$, $a^2 + x^2$, and $x^2 - a^2$. For example, if we let $x = |a| \tan \theta$, then $a^2 + x^2 = a^2(1 + \tan^2\theta) = a^2 \sec^2\theta$ and $dx = |a| \sec^2\theta\, d\theta$. We then obtain

$$\begin{aligned}\int \frac{dx}{a^2 + x^2} = \int \frac{|a| \sec^2\theta\, d\theta}{a^2\sec^2\theta} = \frac{1}{|a|}\int d\theta &= \frac{1}{|a|}\theta + c \\ &= \frac{1}{|a|}\arctan\frac{x}{|a|} + c.\end{aligned}$$

Given a choice, one would probably not evaluate $\int \frac{dx}{a^2 + x^2}$ by this method. It is more likely that one would do the problem directly, remembering the formula $\int \frac{dx}{1 + x^2} = \arctan x + c$.

We conclude this section with consideration of the integral

$$\int \frac{dx}{(ax^2 + bx + c)^n}, \tag{5}$$

where n is any positive integer and the polynomial $ax^2 + bx + c$ is irreducible over the real numbers. To say that a quadratic polynomial is **irreducible** means that it cannot be written as the product of two linear factors. It follows from the familiar quadratic formula that $ax^2 + bx + c$ is irreducible over the real numbers if and only if $b^2 - 4ac < 0$. We shall show, by means of the trigonometric substitution used in the preceding paragraph, that the integral (5) can be changed to

$$K\int \cos^{2n-2}\theta\, d\theta, \tag{6}$$

where K is a constant. This latter integral, as we saw in Section 2, can always be integrated, and this means that it is always possible to integrate (5). This fact, which is of interest in itself, will play a part in a more general theory to be developed in Section 4.

By first factoring and then completing the square, we obtain

$$\begin{aligned} ax^2 + bx + c &= a\left(x^2 + \frac{b}{a}x + \frac{c}{a}\right) \\ &= a\left(x^2 + \frac{b}{a}x + \frac{b^2}{4a^2} + \frac{c}{a} - \frac{b^2}{4a^2}\right) \\ &= a\left[\left(x + \frac{b}{2a}\right)^2 + \left(\frac{\sqrt{4ac - b^2}}{2a}\right)^2\right]. \end{aligned}$$

Note that, since $b^2 - 4ac < 0$, we know both that $a \neq 0$ and that $\sqrt{4ac - b^2}$ is real. For convenience, we shall let $y = x + \frac{b}{2a}$ and $k = \frac{\sqrt{4ac - b^2}}{2a}$. Then

$$ax^2 + bx + c = a(y^2 + k^2),$$

and $dy = dx$. Making the trigonometric substitution $y = |k| \tan \theta$, we have

$$ax^2 + bx + c = a(y^2 + k^2) = ak^2(\tan^2\theta + 1) = ak^2\sec^2\theta,$$

and $dx = dy = |k| \sec^2\theta \, d\theta$. It therefore follows that

$$\begin{aligned} \int \frac{dx}{(ax^2 + bx + c)^n} &= \int \frac{|k|\sec^2\theta \, d\theta}{(ak^2\sec^2\theta)^n} = \frac{|k|}{a^n k^{2n}} \int \frac{1}{(\sec\theta)^{2n-2}} d\theta \\ &= K \int \cos^{2n-2}\theta \, d\theta, \end{aligned}$$

where $K = \frac{|k|}{a^n k^{2n}}$. Thus, every integral (5) can be integrated by first changing it into the integral of a power of a cosine by trigonometric substitutions and then by reducing the power of the cosine with the reduction formula on page 359.

PROBLEMS

1. Evaluate $\int \sqrt{a^2 - x^2}\, dx$ using equations (2), page 376, and compare your answer with that found using equations (1).
2. (a) Write a set of equations for integrating functions of $\sqrt{a^2 + x^2}$ which are analogous to equations (3), page 377, but are based on the identity $1 + \cot^2\theta = \csc^2\theta$.

   (b) Select an interval to which θ can be restricted so that it is uniquely determined by the equations in part (a) and so that x can take on all real number values.
3. Evaluate $\int \frac{dx}{\sqrt{a^2 + x^2}}$ using the substitution described in Problem 2.

4. What is the set to which θ should be restricted if the substitution of $|a| \csc \theta$ for x makes $\sqrt{x^2 - a^2}$ equal to $|a| \cot \theta$, defines θ unambiguously, and also lets x take on all real values such that $|x| \geq |a|$?
5. Evaluate the following integrals.

(a) $\displaystyle\int \frac{\sqrt{x^2 - 9}}{x}\, dx$ (f) $\displaystyle\int \sqrt{(x^2 + 4)^3}\, dx$

(b) $\displaystyle\int \sqrt{(x^2 - 1)^3}\, dx$ (g) $\displaystyle\int \frac{x\, dx}{\sqrt{(a^2 + x^2)^3}}$

(c) $\displaystyle\int x\sqrt{16 - x^2}\, dx$ (h) $\displaystyle\int \frac{x^3 dx}{\sqrt{4 - x^2}}$

(d) $\displaystyle\int x^3\sqrt{x^2 - 4}\, dx$ (i) $\displaystyle\int \sqrt{x^2 - a^2}\, dx$

(e) $\displaystyle\int \frac{dx}{x^2 - 9}$ (j) $\displaystyle\int \frac{dx}{\sqrt{(x^2 - 25)^3}}$.

6. Evaluate $2\int_{a-h}^{a} \sqrt{a^2 - x^2}\, dx$, and hence find the area of a segment of height h in a circle of radius a.
7. Evaluate the following definite integrals.

(a) $\displaystyle\int_0^4 \frac{dx}{\sqrt{9 + x^2}}$ (d) $\displaystyle\int_{4/\sqrt{3}}^4 \frac{dx}{\sqrt{(x^2 - 4)^3}}$

(b) $\displaystyle\int_0^{\sqrt{5}} \frac{x\, dx}{\sqrt{4 + x^2}}$ (e) $\displaystyle\int_3^4 x\sqrt{25 - x^2}\, dx$

(c) $\displaystyle\int_0^{1/4} \frac{dx}{\sqrt{1 - 4x^2}}$ (f) $\displaystyle\int_{5/2}^{7/2} \frac{dx}{\sqrt{4x^2 - 9}}$.

8. Integrate

(a) $\displaystyle\int \frac{dx}{5 + x^2}$ (e) $\displaystyle\int \frac{dx}{(2x^2 + 6)^2}$

(b) $\displaystyle\int \frac{dx}{2x^2 + 8}$ (f) $\displaystyle\int \frac{dx}{(x^2 - 4x + 8)^2}$

(c) $\displaystyle\int \frac{dx}{x^2 + 2x + 5}$ (g) $\displaystyle\int \frac{dx}{(x^2 + 9)^3}$

(d) $\displaystyle\int \frac{dx}{2x^2 + 12x + 20}$ (h) $\displaystyle\int \frac{dx}{(x^2 + 2x + 2)^3}$.

9. By the substitutions used to change equation (5) to (6) and by the reduction formula on page 359, verify the following reduction formula (where $b^2 - 4ac < 0$):

$$\int \frac{dx}{(ax^2 + bx + c)^n} = \frac{2ax + b}{(n - 1)(4ac - b^2)(ax^2 + bx + c)^{n-1}} + \frac{2a(2n - 3)}{(n - 1)(4ac - b^2)} \int \frac{dx}{(ax^2 + bx + c)^{n-1}}.$$

4. Partial Fractions. A rational function is by definition one which can be expressed as the ratio of two polynomials. A simple example is the function f, defined by

$$f(x) = \frac{1}{(x^2 + 1)(x - 2)} = \frac{1}{x^3 - 2x^2 + x - 2},$$

for every real value of x except 2. At present, we have no way of integrating this function. However, in this section we shall develop a method of integration which is applicable to any rational function. It is called the method of partial fractions.

To illustrate the method, consider the equation

$$\frac{1}{x - 2} - \frac{x + 2}{x^2 + 1} = \frac{5}{(x^2 + 1)(x - 2)},$$

which is easily seen to be true for all real values of x except 2. It follows that

$$\frac{1}{(x^2 + 1)(x - 2)} = \frac{1}{5}\frac{1}{x - 2} - \frac{1}{5}\frac{x + 2}{x^2 + 1}$$
$$= \frac{1}{5}\frac{1}{x - 2} - \frac{1}{5}\frac{x}{x^2 + 1} - \frac{2}{5}\frac{1}{x^2 + 1}.$$

Hence

$$\int \frac{dx}{(x^2 + 1)(x - 2)} = \frac{1}{5}\int \frac{dx}{x - 2} - \frac{1}{5}\int \frac{x\,dx}{x^2 + 1} - \frac{2}{5}\int \frac{dx}{x^2 + 1}$$
$$= \tfrac{1}{5}\ln|x - 2| - \tfrac{1}{10}\ln(x^2 + 1) - \tfrac{2}{5}\arctan x + c.$$

Thus $\dfrac{1}{(x^2 + 1)(x - 2)}$ can be integrated, since it can be written as a sum of simpler rational functions, each of which can be integrated separately. The method of integration by partial fractions is based on the fact that such a decomposition exists for every rational function. In this example, we have given no indication of how the decomposition is to be found. However, the general method, which we now describe, consists of just such a prescription.

We begin with the result from algebra that it is always possible by means of division to express any given rational function as the sum of a polynomial and a rational function in which the degree of the numerator is less than the degree of the denominator. Stated formally, this theorem says that, if $N(x)$ and $D(x)$ are any two polynomials and if $D(x)$ is not the zero function, then there exist uniquely determined polynomials $Q(x)$ and $R(x)$ such that

$$\boxed{\frac{N(x)}{D(x)} = Q(x) + \frac{R(x)}{D(x)},} \tag{1}$$

and such that the degree of $R(x)$ is less than the degree of $D(x)$. (The letters N, D, Q, and R have been used to suggest, respectively, the words "numerator," "denominator," "quotient," and "remainder.") The first step in the method of integration by partial fractions is to write the given rational function $\frac{N(x)}{D(x)}$ in the form of equation (1). Since we can obviously integrate the polynomial $Q(x)$, we need next consider only rational functions in which the degree of the numerator is less than the degree of the denominator. If we start with such a function, then no division is necessary.

EXAMPLE 1. Write the function $\dfrac{x^4 - 4x^3 + 8x^2 - 7x + 3}{x^3 - 2x^2 + 3x - 4}$ as the sum of a polynomial and a rational function in which the degree of the numerator is less than the degree of the denominator. Dividing, we have

$$\begin{array}{r|l}
 & x - 2 \\
x^3 - 2x^2 + 3x - 4 & x^4 - 4x^3 + 8x^2 - 7x + 3 \\
 & x^4 - 2x^3 + 3x^2 - 4x \\
\hline
 & -2x^3 + 5x^2 - 3x + 3 \\
 & -2x^3 + 4x^2 - 6x + 8 \\
\hline
 & x^2 + 3x - 5.
\end{array}$$

It follows that

$$\frac{x^4 - 4x^3 + 8x^2 - 7x + 3}{x^3 - 2x^2 + 3x - 4} = (x - 2) + \frac{x^2 + 3x - 5}{x^3 - 2x^2 + 3x - 4},$$

which gives the required sum.

Another algebraic fact about polynomials, which we shall not prove, but shall assume and use, is that any nonconstant polynomial (i.e., of degree at least 1) with real coefficients can be written as a product of linear and quadratic factors, each with real coefficients. By a linear factor we mean a polynomial $L(x)$ of degree 1; that is, $L(x) = ax + b$ and $a \neq 0$. Similarly, a quadratic factor is a polynomial $Q(x)$ of degree 2; thus $Q(x) = cx^2 + dx + e$ and $c \neq 0$. The theorem states that, for any polynomial

$$f(x) = a_n x^n + \cdots + a_1 x + a_0$$

with real coefficients a_i and with $n \geq 1$ and $a_n \neq 0$, there exist linear factors $L_1(x), \ldots, L_p(x)$ and quadratic factors $Q_1(x), \ldots, Q_q(x)$ with real coefficients such that

$$f(x) = L_1(x) \cdots L_p(x) Q_1(x) \cdots Q_q(x).$$

Note that either p or q may be zero. In actual practice, such a factorization of $f(x)$ may be very difficult to find, but the theorem assures us that it exists.

A polynomial is said to be **irreducible** if it cannot be written as the product of two polynomials each of degree greater than or equal to 1. The degree of the product of two polynomials is equal to the sum of the degrees of the factors, and it therefore follows that every linear polynomial is irreducible. It was pointed out in Section 3 that a quadratic polynomial $cx^2 + dx + e$ is irreducible over the reals if and only if its discriminant $d^2 - 4ce$ is negative. For example, the polynomials $x^2 + 1$ and $x^2 + x + 1$ are irreducible, whereas $x^2 + 2x + 1$ and $x^2 + x - 1$ are not. If a quadratic polynomial is not irreducible, it can be factored and written as the product of two linear polynomials. Hence the factorization of an arbitrary nonconstant polynomial into linear and quadratic factors, as described in the preceding paragraph, can always be done so that all the factors are irreducible.

Returning specifically to the method of integration by partial fractions, we consider a rational function $\frac{N(x)}{D(x)}$, with the degree of $D(x)$ greater than the degree of $N(x)$. The second step is to write the denominator $D(x)$ as a product of irreducible factors. Having done so, we have

$$\boxed{D(x) = L_1(x) \cdots L_p(x) Q_1(x) \cdots Q_q(x),} \tag{2}$$

where, for each $i = 1, \ldots, p$,

$$L_i(x) = a_i x + b_i, \qquad a_i \neq 0,$$

and, for each $j = 1, \ldots, q$,

$$Q_j(x) = c_j x^2 + d_j x + e_j, \qquad d_j^2 - 4c_j e_j < 0.$$

There is no reason to suppose that the factors which appear in equation (2) will all be distinct, and it may very well happen that $L_1(x) = L_2(x)$, etc.

However, the theory is simpler if no repetitions occur, and we shall consider that case first.

Case 1. The irreducible factors of $D(x)$ are all distinct. The algebraic theory of partial fractions, which we shall assume, tells us that we can write $\frac{N(x)}{D(x)}$ as a sum of rational functions each of which has one of the factors of $D(x)$ as its denominator and such that in each term the degree of the numerator is less than the degree of the denominator. Moreover, given the factorization of $D(x)$ into irreducibles, this decomposition is unique except for the order in which the terms are written. Thus

$$\begin{aligned}\frac{N(x)}{D(x)} = \frac{A_1}{a_1x + b_1} + \frac{A_2}{a_2x + b_2} + \cdots + \frac{A_p}{a_px + b_p} \\ + \frac{B_1x + C_1}{c_1x^2 + d_1x + e_1} + \frac{B_2x + C_2}{c_2x^2 + d_2x + e_2} \\ + \cdots + \frac{B_qx + C_q}{c_qx^2 + d_qx + e_q},\end{aligned}$$

where each of the letters A_i, B_j, and C_k represents a uniquely determined real constant. The rational functions which appear on the right side are called the **partial fractions** of the decomposition of $\frac{N(x)}{D(x)}$.

We shall show by means of examples how the constants in the partial fractions decomposition are determined. Consider the rational function $\frac{1}{(x^2 + 1)(x - 2)}$ discussed at the beginning of the section. The degree of the numerator, zero, is already less than 3, the degree of the denominator. Moreover, the denominator is already factored into irreducibles. Hence, we seek constants A, B, and C such that

$$\frac{1}{(x^2 + 1)(x - 2)} = \frac{A}{x - 2} + \frac{Bx + C}{x^2 + 1}.$$

Adding the two fractions on the right side, we have

$$\frac{1}{(x^2 + 1)(x - 2)} = \frac{A(x^2 + 1) + (Bx + C)(x - 2)}{(x^2 + 1)(x - 2)}.$$

The fact that a nonzero polynomial of degree n has at most n distinct roots implies that two rational functions with the same denominator are equal if and only if their numerators are equal (see Problem 8 at the end of this section). Hence the equation

$$1 = A(x^2 + 1) + (Bx + C)(x - 2) \tag{3}$$

is true for all real values of x. There are two common ways to find the constants A, B, and C. One is to multiply and regroup the terms in (3) to obtain the equation

$$0 = (A + B)x^2 + (C - 2B)x + (A - 2C - 1),$$

which also holds for all real values of x. The only polynomial with infinitely many roots is the zero polynomial, i.e., the polynomial with no nonzero coefficients. Hence the three coefficients on the right side of the preceding equation are all equal to zero, and we can therefore find A, B, and C by solving the system of equations

$$\begin{aligned} A + B &= 0, \\ -2B + C &= 0, \\ A - 2C &= 1. \end{aligned}$$

Usually simpler is the technique in which we take advantage of the fact that (3) must be true for all values of x, and we choose values cleverly to help evaluate the constants. For example, letting $x = 2$ in (3), we have

$$\begin{aligned} 1 &= A(2^2 + 1) + (B \cdot 2 + C)(2 - 2) \\ &= 5A. \end{aligned}$$

Hence $A = \frac{1}{5}$. If we then let $x = 0$, we have

$$\begin{aligned} 1 &= \tfrac{1}{5}(0^2 + 1) + (B \cdot 0 + C)(0 - 2) \\ &= \tfrac{1}{5} - 2C, \end{aligned}$$

or, equivalently, $2C = \frac{1}{5} - 1$, and so $C = -\frac{2}{5}$. Finally, choosing $x = 1$, we get

$$\begin{aligned} 1 &= \tfrac{1}{5}(1^2 + 1) + (B \cdot 1 - \tfrac{2}{5})(1 - 2) \\ &= \tfrac{2}{5} - B + \tfrac{2}{5}, \end{aligned}$$

from which we conclude that $B = -\frac{1}{5}$. Whichever method we use, we have $A = \frac{1}{5}$, $B = -\frac{1}{5}$, and $C = -\frac{2}{5}$, from which it follows that

$$\frac{1}{(x^2 + 1)(x - 2)} = \frac{1}{5}\,\frac{1}{x - 2} - \frac{1}{5}\,\frac{x + 2}{x^2 + 1},$$

the form which we integrated at the beginning of the section.

EXAMPLE 2. Integrate $\displaystyle\int \frac{dx}{a^2 - x^2}$. Since $a^2 - x^2 = (a - x)(a + x)$, we decompose $\dfrac{1}{a^2 - x^2}$ into partial fractions $\dfrac{A}{a + x}$ and $\dfrac{B}{a - x}$. Thus

$$\frac{1}{a^2 - x^2} = \frac{A}{a + x} + \frac{B}{a - x} = \frac{A(a - x) + B(a + x)}{a^2 - x^2}.$$

Equating numerators on the left and right, we get

$$1 = A(a - x) + B(a + x).$$

Letting $x = a$, we obtain the equation $1 = A \cdot 0 + B \cdot 2a$, and so $B = \frac{1}{2a}$. Similarly, setting $x = -a$, we get $1 = A \cdot 2a + B \cdot 0$, from which it follows that $A = \frac{1}{2a}$. Thus

$$\frac{1}{a^2 - x^2} = \frac{1}{2a}\frac{1}{a + x} + \frac{1}{2a}\frac{1}{a - x},$$

and, therefore,

$$\begin{aligned}\int \frac{dx}{a^2 - x^2} &= \frac{1}{2a}\int \frac{dx}{a + x} + \frac{1}{2a}\int \frac{dx}{a - x}\\ &= \frac{1}{2a}\ln|a + x| - \frac{1}{2a}\ln|a - x| + c\\ &= \frac{1}{2a}\ln\left|\frac{a + x}{a - x}\right| + c.\end{aligned}$$

Thus the third step in applying this method of integration is the decomposition into partial fractions, and the fourth and final step is the integration of the partial fractions. We shall show later in the section that it is always possible to carry out the last step, but, as the next example shows, doing so can be tedious.

EXAMPLE 3. Integrate $\int \frac{3x^3 + x^2 - 14x + 46}{(x^2 + x + 1)(x^2 - 5x - 14)}\,dx$. The degree of the numerator is 3 and that of the denominator is 4, so we proceed to the factorization of the denominator into irreducibles. It is already written as the product of two quadratics, of which $x^2 + x + 1$ is irreducible but $x^2 - 5x - 14$ is not, since $x^2 - 5x - 14 = (x - 7)(x + 2)$. Hence the form of the partial fractions decomposition is

$$\frac{3x^3 + x^2 - 14x + 46}{(x^2 + x + 1)(x - 7)(x + 2)} = \frac{Ax + B}{x^2 + x + 1} + \frac{C}{x - 7} + \frac{D}{x + 2}.$$

The sum of the three fractions on the right side is

$$\frac{(Ax + B)(x - 7)(x + 2) + C(x^2 + x + 1)(x + 2) + D(x^2 + x + 1)(x - 7)}{(x^2 + x + 1)(x - 7)(x + 2)}.$$

Equating numerators, we have

$$(Ax + B)(x - 7)(x + 2) + C(x^2 + x + 1)(x + 2) + D(x^2 + x + 1)(x - 7) = 3x^3 + x^2 - 14x + 46.$$

If we set $x = 7$ in this equation, then

$$C \cdot 513 = 1026 \quad \text{or} \quad C = 2.$$

Letting $x = -2$, we obtain

$$D \cdot (-27) = 54 \quad \text{or} \quad D = -2.$$

If we let $x = 0$, then

$$B \cdot (-14) + C \cdot 2 + D \cdot (-7) = 46$$

or

$$-14B = 46 - 2C + 7D = 46 - 4 - 14 = 28 \quad \text{or} \quad B = -2.$$

Finally, letting $x = -1$, we have

$$(-A + B) \cdot (-8) + C + D \cdot (-8) = 58$$

or

$$8A = 58 + 8B - C + 8D = 58 - 16 - 2 - 16 = 24 \quad \text{or} \quad A = 3.$$

The partial fractions decomposition is, therefore,

$$\frac{3x^3 + x^2 - 14x + 46}{(x^2 + x + 1)(x^2 - 5x - 14)} = \frac{3x - 2}{x^2 + x + 1} + \frac{2}{x - 7} - \frac{2}{x + 2}. \quad (4)$$

Except for the first, the terms on the right side are easily integrated. The first term can be integrated by writing it as the sum of two fractions. We use the identity

$$\frac{Bx + C}{cx^2 + dx + e} = \frac{B}{2c}\frac{2cx + d}{cx^2 + dx + e} + \left(C - \frac{dB}{2c}\right)\frac{1}{cx^2 + dx + e}.$$

Note that the numerator $2cx + d$ is the derivative of $cx^2 + dx + e$. Both of these fractions can be integrated. In this example, we have

$$\frac{3x - 2}{x^2 + x + 1} = \frac{3}{2}\frac{2x + 1}{x^2 + x + 1} - \frac{7}{2}\frac{1}{x^2 + x + 1}.$$

Then

$$\int \frac{2x + 1}{x^2 + x + 1}\,dx = \ln(x^2 + x + 1) + c.$$

For the second fraction, we complete the square in the denominator. The result is

$$\frac{1}{x^2 + x + 1} = \frac{1}{\left(x + \frac{1}{2}\right)^2 + \left(\frac{\sqrt{3}}{2}\right)^2},$$

and, since $\displaystyle\int \frac{dy}{y^2 + a^2} = \frac{1}{|a|} \arctan \frac{y}{|a|}$, it follows that

$$\int \frac{dx}{x^2 + x + 1} = \int \frac{dx}{\left(x + \frac{1}{2}\right)^2 + \left(\frac{\sqrt{3}}{2}\right)^2} = \frac{2}{\sqrt{3}} \arctan \left(\frac{x + \frac{1}{2}}{\frac{\sqrt{3}}{2}}\right) + c$$

$$= \frac{2}{\sqrt{3}} \arctan \left(\frac{2x + 1}{\sqrt{3}}\right) + c.$$

Hence

$$\int \frac{3x - 2}{x^2 + x + 1}\, dx = \frac{3}{2} \ln(x^2 + x + 1) - \frac{7}{\sqrt{3}} \arctan \left(\frac{2x + 1}{\sqrt{3}}\right) + c.$$

Returning to equation (4), we therefore get the final integral

$$\int \frac{3x^3 + x^2 - 14x + 46}{(x^2 + x + 1)(x^2 - 5x - 14)}\, dx$$

$$= \frac{3}{2} \ln(x^2 + x + 1) - \frac{7}{\sqrt{3}} \arctan \left(\frac{2x + 1}{\sqrt{3}}\right)$$

$$+ 2 \ln|x - 7| - 2 \ln|x + 2| + c,$$

and this completes the example.

We consider next the situation in which the factorization of the denominator of $\dfrac{N(x)}{D(x)}$, as shown in equation (2), contains repeated factors.

Case 2. The irreducible factors of $D(x)$ are not all distinct. We assume, as in Case 1, that the degree of $N(x)$ is less than the degree of $D(x)$. There is still a unique decomposition of $\dfrac{N(x)}{D(x)}$ into the sum of partial fractions, but now it is more complicated. By regrouping, we may write the factorization of $D(x)$ into irreducibles as

$$\boxed{D(x) = [L_1(x)]^{m_1} \cdots [L_r(x)]^{m_r} [Q_1(x)]^{n_1} \cdots [Q_s(x)]^{n_s},} \qquad (5)$$

where $m_1, \ldots, m_r$ and $n_1, \ldots, n_s$ are positive integers, the factors $L_i(x) = a_ix + b_i$ are all distinct, and the factors $Q_j(x) = c_jx^2 + d_jx + e_j$ are all distinct. In this case, $\dfrac{N(x)}{D(x)}$ is the total sum of the following individual sums of partial fractions: For each $i = 1, \ldots, r$, there is the sum

$$\boxed{\frac{A_{i1}}{a_ix + b_i} + \frac{A_{i2}}{(a_ix + b_i)^2} + \cdots + \frac{A_{im_i}}{(a_ix + b_i)^{m_i}},}$$

in which the A_{ik} are uniquely determined real constants. Similarly, for each $j = 1, \ldots, s$, there is the sum

$$\boxed{\frac{B_{j1}x + C_{j1}}{c_jx^2 + d_jx + e_j} + \frac{B_{j2}x + C_{j2}}{(c_jx^2 + d_jx + e_j)^2} + \cdots + \frac{B_{jn_j}x + C_{jn_j}}{(c_jx^2 + d_jx + e_j)^{n_j}},}$$

in which the B_{jk} and C_{jk} are uniquely determined real constants.

EXAMPLE 4. Integrate the rational function $\dfrac{2x^2 + x + 2}{x(x-1)^3}$ by the method of partial fractions. The degree of the numerator, 2, is less than that of the denominator, 4. So we turn at once to the decomposition into partial fractions. Since the irreducible factor $x - 1$ is repeated three times, the decomposition is of the form

$$\frac{2x^2 + x + 2}{x(x-1)^3} = \frac{A}{x-1} + \frac{B}{(x-1)^2} + \frac{C}{(x-1)^3} + \frac{D}{x}.$$

The sum on the right side is equal to

$$\frac{Ax(x-1)^2 + Bx(x-1) + Cx + D(x-1)^3}{x(x-1)^3}.$$

Equating numerators, we obtain the equation

$$Ax(x-1)^2 + Bx(x-1) + Cx + D(x-1)^3 = 2x^2 + x + 2, \quad (6)$$

which holds for all real values of x. Setting $x = 1$, we obtain

$$C \cdot 1 = 2 \cdot 1^2 + 1 + 2, \qquad \text{whence } C = 5.$$

If we let $x = 0$, then

$$D \cdot (-1)^3 = 2, \qquad \text{whence } D = -2.$$

Thus, equation (6) has become

$$Ax(x-1)^2 + Bx(x-1) + 5x - 2(x-1)^3 = 2x^2 + x + 2.$$

In this equation we let $x = 2$, getting

$$A \cdot 2 + B \cdot 2 + 10 - 2 = 8 + 2 + 2$$

or

$$2A + 2B = 4. \tag{7}$$

Next, setting $x = -1$, we have

$$A \cdot (-1)(-2)^2 + B \cdot (-1)(-2) - 5 + 16 = 2 - 1 + 2$$

or

$$-4A + 2B = -8. \tag{8}$$

Subtracting (8) from (7), we get $6A = 12$ and so $A = 2$. It follows that $B = 0$, and we have therefore found the partial fractions decomposition to be

$$\frac{2x^2 + x + 2}{x(x-1)^3} = \frac{2}{x-1} + \frac{5}{(x-1)^3} - \frac{2}{x}.$$

Hence

$$\int \frac{2x^2 + x + 2}{x(x-1)^3}\,dx = 2\int \frac{dx}{x-1} + 5\int \frac{dx}{(x-1)^3} - 2\int \frac{dx}{x}$$
$$= 2\ln|x-1| - \frac{5}{2}\frac{1}{(x-1)^2} - 2\ln|x| + c,$$

and this completes the example.

Since any rational function can be written as the sum of a polynomial and a series of partial fractions, the general problem of integrating a rational function reduces to three integration problems: (1) integration of a polynomial; (2) integration of functions of the form $\dfrac{A}{(ax+b)^m}$, where m is a positive integer and $a \neq 0$; and (3) integration of functions of the form $\dfrac{Bx + C}{(cx^2 + dx + e)^n}$, where n is a positive integer and $d^2 - 4ce < 0$. The first, of course, offers no difficulties whatever. The second is also simple, since $\dfrac{A}{(ax+b)^m} = \dfrac{A}{a}\dfrac{a}{(ax+b)^m}$, and so

$$\int \frac{A}{(ax+b)^m}\,dx = \frac{A}{a}\int \frac{a}{(ax+b)^m}\,dx$$
$$= \begin{cases} \dfrac{A}{a(1-m)}\dfrac{1}{(ax+b)^{m-1}} + c & \text{if } m \neq 1, \\[2ex] \dfrac{A}{a}\ln|ax+b| + c & \text{if } m = 1. \end{cases}$$

The third problem can be solved, but we have seen in Example 3 that it is complicated even with $n = 1$. It is attacked by writing

$$\frac{Bx + C}{(cx^2 + dx + e)^n} = \frac{B}{2c}\,\frac{2cx + d}{(cx^2 + dx + e)^n} + \left(C - \frac{dB}{2c}\right)\frac{1}{(cx^2 + dx + e)^n}.$$

The integral $\int \frac{2cx + d}{(cx^2 + dx + e)^n}\,dx$ is easily found, since it is of the form $\int \frac{du}{u^n}$ with $u = cx^2 + dx + e$. The problem therefore reduces to finding $\int \frac{dx}{(cx^2 + dx + e)^n}$. However, it was demonstrated at the end of Section 3 that this integral can be evaluated by trigonometric substitution, which reduces it to an integral of the form $\int \cos^{2n-2}\theta\,d\theta$. An alternative method is to use the reduction formula given in Problem 9, page 384.

Thus all possible partial fractions resulting from the decomposition of a rational function can be integrated. It follows that every rational function can be integrated. The factoring of the denominator into irreducibles may be difficult, and the decomposition into partial fractions and the resulting integrations may be tedious, but the following important result has been established.

(4.1) THEOREM. *Every rational function can be integrated by the method of partial fractions.*

A table of integrals will show how to integrate many of the functions which are the partial fractions of a rational fraction. For this reason, there is no need to memorize the formulas for integration. However, it is necessary to know the technique of separating a rational function into its partial fractions in order to replace an apparently nonintegrable function by a sum of obviously integrable functions.

PROBLEMS

1. Separate each of the following into the sum of a polynomial and a sum of partial fractions.

(a) $\dfrac{5}{(x - 2)(x + 3)}$

(b) $\dfrac{x + 2}{(2x + 1)(x + 1)}$

(c) $\dfrac{2x^3 + 3x^2 - 2}{2x^2 + 3x + 1}$

(d) $\dfrac{4x^2 - 5x + 10}{(x - 4)(x^2 + 2)}$

(e) $\dfrac{3x^3 + 5x^2 - 27x + 8}{x^2 + 4x}$

(f) $\dfrac{x^2 + 1}{x^2 + x + 1}$.

2. Integrate each of the following.

(a) $\displaystyle\int \frac{5}{(x-2)(x+3)}\,dx$ (c) $\displaystyle\int \frac{2x^3+3x^2-2}{2x^2+3x+1}\,dx$

(b) $\displaystyle\int \frac{x+2}{(2x+1)(x+1)}\,dx$ (d) $\displaystyle\int \frac{4x^2-5x+10}{(x-4)(x^2+2)}\,dx.$

3. Find the partial fractions decomposition of each of the following rational functions.

(a) $\dfrac{x-8}{x^2-x-6}$ (e) $\dfrac{4}{x^2(x+2)}$

(b) $\dfrac{18}{x^2+8x+7}$ (f) $\dfrac{6x^2-x+13}{(x+1)(x^2+4)}$

(c) $\dfrac{x+1}{(x-1)^2}$ (g) $\dfrac{(x+2)^2}{(x+3)^3}$

(d) $\dfrac{8x+25}{x^2+5x}$ (h) $\dfrac{x^2+2x+5}{(2x-1)(x^2+1)^2}.$

4. Evaluate each of the following integrals.

(a) $\displaystyle\int \frac{x-8}{x^2-x-6}\,dx$ (d) $\displaystyle\int \frac{8x+25}{x^2+5x}\,dx$

(b) $\displaystyle\int \frac{18}{x^2+8x+7}\,dx$ (e) $\displaystyle\int \frac{6x^2-x+13}{(x+1)(x^2+4)}\,dx$

(c) $\displaystyle\int \frac{x+1}{(x-1)^2}\,dx$ (f) $\displaystyle\int \frac{(x+2)^2}{(x+3)^3}\,dx.$

5. (a) Show directly that $\dfrac{2x-3}{(x-2)^2}$ can be written in the form $\dfrac{A}{x-2}+\dfrac{B}{(x-2)^2}$ by first writing $\dfrac{2x-3}{(x-2)^2}=\dfrac{2(x-2)+1}{(x-2)^2}.$

(b) Following the method in (a), show that $\dfrac{ax+b}{(x-k)^2}$ can always be written $\dfrac{A}{x-k}+\dfrac{B}{(x-k)^2}$, where A and B are constants.

(c) Extend the result in (b) by factoring, completing the square, and dividing to show directly that

$$\frac{ax^2+bx+c}{(x-k)^3} \text{ can be written } \frac{A}{x-k}+\frac{B}{(x-k)^2}+\frac{C}{(x-k)^3}$$

where A, B, and C are constants.

[*Note:* Without knowledge of the algebraic theory of partial fractions, it would not be unreasonable to assume that a decomposition of a rational function $\frac{N(x)}{P(x)(x-k)^3}$ would necessarily contain fractions $\frac{A}{x-k}$, $\frac{Bx+C}{(x-k)^2}$, and $\frac{Dx^2+Ex+F}{(x-k)^3}$. This problem shows, however, that in the complete decomposition $B = D = E = 0$.]

6. Why can there not be an irreducible cubic polynomial with real coefficients?
7. Integrate each of the following.

(a) $\displaystyle\int \frac{(3x+1)\,dx}{x^3+2x^2+x}$

(b) $\displaystyle\int \frac{(x^2+1)\,dx}{x^2-3x+2}$

(c) $\displaystyle\int \frac{(x-2)\,dx}{(2x+1)(x^2+1)}$

(d) $\displaystyle\int \frac{x^2-3x-2}{(x-2)^2(x-3)}\,dx$

(e) $\displaystyle\int \frac{dx}{x^2+2x+2}$

(f) $\displaystyle\int \frac{(2x+1)\,dx}{x^2+2x+2}$

(g) $\displaystyle\int \frac{\sec^2 x\,dx}{\tan^2 x-4\tan x+3}$

(h) $\displaystyle\int \frac{\sec y\tan y\,dy}{2\sec^2 y+5\sec y+2}$

(i) $\displaystyle\int \frac{y^2+1}{y^2+y+1}\,dy$

(j) $\displaystyle\int \frac{10+5z-z^2}{(z+4)(z^2+z+1)}\,dz$

(k) $\displaystyle\int \frac{(6x+3)\,dx}{(x-1)(x+2)(x^2+x+1)}$

(l) $\displaystyle\int \frac{(x^3+4)\,dx}{(x+1)(x+2)^2}$

(m) $\displaystyle\int \frac{x^2+2x+5}{(2x-1)(x^2+1)^2}\,dx$

(n) $\displaystyle\int \frac{dx}{(x^2+x+5)^3}$.

8. Prove the statement in the text, page 387, that, since a nonzero polynomial of degree n has at most n distinct roots, two rational functions with the same denominator are equal if and only if their numerators are equal. [*Hint:* Suppose that $\frac{N_1(x)}{D(x)} = \frac{N_2(x)}{D(x)}$, where the polynomial $D(x)$ is not the zero function. Then $\frac{N_1(x)-N_2(x)}{D(x)} = 0$, and so the polynomial equation $N_1(x) - N_2(x) = 0$ holds for every real number x for which $D(x) \neq 0$.]

5. Other Substitutions. We have seen in Section 4 that any rational function $\frac{N(x)}{D(x)}$ can be integrated by the method of partial fractions. This result can be extended to show that any rational function of the six trigonometric functions can also be integrated. Such a function is defined as the result of replacing each occurrence of x in $\dfrac{N(x)}{D(x)}$ by any one of the six possibilities: $\sin x$, $\cos x$, $\tan x$, $\cot x$, $\sec x$, or $\csc x$. An example is the function F defined by

$$F(x) = \frac{\sin^2 x \cos x + 2 \tan^2 x + \sec x}{\cos^2 x + 3 \cot x + 1},$$

which is obtained in the manner just described from the rational function $\dfrac{x^3 + 2x^2 + x}{x^2 + 3x + 1}$. Since each one of the four functions $\tan x$, $\cot x$, $\sec x$, and $\csc x$ is a simple rational function $\sin x$ and $\cos x$,

$$\tan x = \frac{\sin x}{\cos x}, \qquad \sec x = \frac{1}{\cos x},$$

$$\cot x = \frac{\cos x}{\sin x}, \qquad \csc x = \frac{1}{\sin x},$$

it follows that every rational function of the six trigonometric functions is equal to a rational function of $\sin x$ and $\cos x$. Thus, in the above example, we have

$$F(x) = \frac{\sin^2 x \cos x + 2\dfrac{\sin^2 x}{\cos^2 x} + \dfrac{1}{\cos x}}{\cos^2 x + 3\dfrac{\cos x}{\sin x} + 1}$$

$$= \frac{\sin^3 x \cos^3 x + 2 \sin^3 x + \sin x \cos x}{\sin x \cos^4 x + 3 \cos^3 x + \sin x \cos^2 x}.$$

It is therefore sufficient to show that every rational function of $\sin x$ and $\cos x$ can be integrated.

Surprisingly enough, a simple substitution will transform any rational function of $\sin x$ and $\cos x$ into a rational function of a single variable. The substitution consists of defining y, a new variable of integration, by the equation

$$\boxed{y = \tan \frac{x}{2}.} \tag{1}$$

We can express $\cos x$ in terms of y by first writing

$$\cos x = \cos 2 \cdot \frac{x}{2} = \cos^2 \frac{x}{2} - \sin^2 \frac{x}{2}.$$

Since $\cos^2 \dfrac{x}{2} + \sin^2 \dfrac{x}{2} = 1$, we have

$$\cos x = \frac{\cos^2 \dfrac{x}{2} - \sin^2 \dfrac{x}{2}}{1} = \frac{\cos^2 \dfrac{x}{2} - \sin^2 \dfrac{x}{2}}{\cos^2 \dfrac{x}{2} + \sin^2 \dfrac{x}{2}}.$$

Dividing numerator and denominator by $\cos^2 \frac{x}{2}$, we get

$$\cos x = \frac{1 - \tan^2 \frac{x}{2}}{1 + \tan^2 \frac{x}{2}} = \frac{1 - y^2}{1 + y^2}.$$

Thus we have obtained the equation

$$\boxed{\cos x = \frac{1 - y^2}{1 + y^2}.} \tag{2}$$

In a similar fashion,

$$\begin{aligned} \sin x &= \sin 2 \cdot \frac{x}{2} = 2 \sin \frac{x}{2} \cos \frac{x}{2} \\ &= 2 \frac{\sin \frac{x}{2}}{\cos \frac{x}{2}} \cos^2 \frac{x}{2} = 2 \tan \frac{x}{2} [\tfrac{1}{2}(1 + \cos x)] \\ &= \tan \frac{x}{2} (1 + \cos x) \\ &= y \left(1 + \frac{1 - y^2}{1 + y^2}\right) = \frac{2y}{1 + y^2}. \end{aligned}$$

Hence

$$\boxed{\sin x = \frac{2y}{1 + y^2}.} \tag{3}$$

Finally, since $\frac{x}{2} = \arctan y$, or, equivalently, $x = 2 \arctan y$, we have

$$\boxed{dx = \frac{2\, dy}{1 + y^2}.} \tag{4}$$

By means of the substitutions given in formulas (2), (3), and (4), any integral of a rational function of $\sin x$ and $\cos x$ can be transformed into an integral of a rational function of y. Since the latter can be integrated by partial fractions, we have proved that *every rational function of* $\sin x$ *and* $\cos x$ *can be integrated.*

EXAMPLE 1. Integrate $\int \frac{\cos x\,dx}{1 + \cos x}$. If we let $y = \tan \frac{x}{2}$, then, as we have seen, we may replace $\cos x$ by $\frac{1 - y^2}{1 + y^2}$, and dx by $\frac{2\,dy}{1 + y^2}$. The integral then becomes

$$\int \frac{\cos x\,dx}{1 + \cos x} = \int \frac{\frac{1 - y^2}{1 + y^2}\,\frac{2\,dy}{1 + y^2}}{1 + \frac{1 - y^2}{1 + y^2}}$$

$$= \int \frac{2(1 - y^2)\,dy}{(1 + y^2)^2 + (1 + y^2)(1 - y^2)}$$

$$= \int \frac{2(1 - y^2)}{2(1 + y^2)}\,dy = \int \frac{1 - y^2}{1 + y^2}\,dy.$$

By division one finds that

$$\frac{1 - y^2}{1 + y^2} = -1 + \frac{2}{1 + y^2}.$$

Hence

$$\int \frac{\cos x\,dx}{1 + \cos x} = \int \left(-1 + \frac{2}{1 + y^2}\right) dy$$

$$= -y + 2 \arctan y + c.$$

But $y = \tan \frac{x}{2}$ and $x = 2 \arctan y$, and we therefore conclude that

$$\int \frac{\cos x\,dx}{1 + \cos x} = -\tan \frac{x}{2} + x + c.$$

We do not recommend that the above substitution formulas be memorized. However, one should remember the simple fact that any rational function of the six trigonometric functions is equal to a rational function of the sine and cosine, and one should also remember that a routine substitution procedure exists by which the integral of a function of the latter type can be reduced to the integral of a rational function. For the details, one will probably want to refer directly to formulas (1), (2), (3), and (4).

There are other substitutions which simplify integrals, but none of them is as standard and automatic as the one just described. For example, $\int \frac{\sqrt{x}\,dx}{1 + \sqrt{x}}$ is not readily integrated. However, if we define a new variable of integration y by the equation $y = \sqrt{x}$, the substitution yields a simple integral.

EXAMPLE 2. Evaluate the indefinite integral $\int \frac{\sqrt{x}\,dx}{1+\sqrt{x}}$. Let $y = \sqrt{x}$. Then $y^2 = x$ and $2y\,dy = dx$. Substituting for $\sqrt{x}$ and dx, we obtain

$$\int \frac{\sqrt{x}\,dx}{1+\sqrt{x}} = \int \frac{y \cdot 2y\,dy}{1+y} = \int \frac{2y^2}{1+y}\,dy.$$

Division yields the identity

$$\frac{2y^2}{1+y} = 2y - 2 + \frac{2}{1+y}\cdot$$

Hence

$$\int \frac{\sqrt{x}\,dx}{1+\sqrt{x}} = \int \left(2y - 2 + \frac{2}{1+y}\right) dy$$
$$= y^2 - 2y + 2\ln|1+y| + c.$$

Since $\sqrt{x}$ is nonnegative, we have $|1+y| = |1+\sqrt{x}| = 1+\sqrt{x}$. Thus

$$\int \frac{\sqrt{x}\,dx}{1+\sqrt{x}} = x - 2\sqrt{x} + 2\ln(1+\sqrt{x}) + c.$$

The same integral can be evaluated by a different substitution. Let us define the variable z by the equation $z = 1 + \sqrt{x}$. Then $\sqrt{x} = z - 1$ and $x = (z-1)^2$ and, as a result, $dx = 2(z-1)\,dz$. After substitution, the integral becomes

$$\int \frac{\sqrt{x}\,dx}{1+\sqrt{x}} = \int \frac{(z-1)\cdot 2(z-1)\,dz}{z}$$
$$= \int \frac{2(z^2 - 2z + 1)\,dz}{z}$$
$$= \int \left(2z - 4 + \frac{2}{z}\right) dz$$
$$= z^2 - 4z + 2\ln|z| + c.$$

Again, since $\sqrt{x}$ is nonnegative, we have $|z| = |1+\sqrt{x}| = 1+\sqrt{x}$. Hence, after substituting back, we get

$$\int \frac{\sqrt{x}\,dx}{1+\sqrt{x}} = (1+\sqrt{x})^2 - 4(1+\sqrt{x}) + 2\ln(1+\sqrt{x}) + c$$
$$= 1 + 2\sqrt{x} + x - 4 - 4\sqrt{x} + 2\ln(1+\sqrt{x}) + c$$
$$= x - 2\sqrt{x} + 2\ln(1+\sqrt{x}) - 3 + c.$$

The two solutions in Example 2 differ by a constant, in accordance with Theorem (5.4), page 114. The two substitutions differ in the initial goal: In the first, we decided that the integral would be simpler if we replaced the radical by a new variable, and in the second we decided to replace the denominator. There is little to choose between the two methods.

EXAMPLE 3. Integrate $\int \frac{x^2 - 3}{(2x + 5)^{1/3}}\, dx$. If we define the variable y by the equation $y = (2x + 5)^{1/3}$, then $y^3 = 2x + 5$ and $3y^2 dy = 2\, dx$. Hence $x = \frac{y^3 - 5}{2}$ and $dx = \frac{3y^2 dy}{2}$. Substituting, we get

$$\begin{aligned}
\int \frac{x^2 - 3}{(2x + 5)^{1/3}}\, dx &= \int \frac{\left[\left(\frac{y^3 - 5}{2}\right)^2 - 3\right]}{y} \frac{3y^2 dy}{2} \\
&= \tfrac{3}{2}\int y\left(\frac{y^6 - 10y^3 + 25}{4} - 3\right) dy \\
&= \tfrac{3}{8}\int (y^7 - 10y^4 + 13y)\, dy \\
&= \tfrac{3}{8}(\tfrac{1}{8}y^8 - \tfrac{10}{5}y^5 + \tfrac{13}{2}y^2) + c \\
&= \tfrac{3}{64}(2x + 5)^{8/3} - \tfrac{3}{4}(2x + 5)^{5/3} + \tfrac{39}{16}(2x + 5)^{2/3} + c.
\end{aligned}$$

There are no universal rules for integration by substitution. In most cases we are interested in replacing an involved function forming part of the integrand by a simpler one, frequently by a single new variable.

In this chapter we have developed a number of techniques for finding indefinite integrals, or antiderivatives. However, it is by no means the case that these techniques will yield an antiderivative for every integrable function. For example, it is impossible to integrate $\int e^{-x^2} dx$ in the sense that the word "integrate" has been used in this chapter. (Since e^{-x^2} is everywhere continuous, an antiderivative certainly exists. In particular, the function F defined by

$$F(t) = \int_0^t e^{-x^2} dx, \qquad \text{for every real number } t,$$

is an antiderivative as a result of the Fundamental Theorem of Calculus, page 200. However, it can be proved that no antiderivative of e^{-x^2} can be expressed algebraically in terms of functions defined by compositions of rational functions, trigonometric functions, and exponential and logarithmic functions.) Nevertheless, the methods discussed in this and the preceding section have significantly increased the set of functions whose indefinite integrals we can find.

The reader should be aware of the fact that there are in existence excellent tables of integrals in which frequently encountered integrals are tabulated. No such table contains all tractable integrals, but some are quite complete, and they are of immense practical value for those people whose work often leads them to problems requiring integration.

PROBLEMS

1. (a) Integrate $\int \sec x\, dx$ by the technique for integrating rational functions of trigonometric functions.

   (b) We have already shown (see page 369) that
$$\int \sec x\, dx = \ln|\sec x + \tan x| + c.$$
   Show that this solution agrees with the one found in (a) for an appropriate choice of the constant c.

2. (a) Integrate $\int \csc x\, dx$ by the technique for integrating rational functions of trigonometric functions.

   (b) The formula $\int \csc x\, dx = -\ln|\csc x + \cot x| + c$ is given on page 374. Show that this integral agrees with the one obtained in (a) for an appropriate choice of c.

3. Integrate each of the following.

(a) $\displaystyle\int \frac{x\, dx}{\sqrt[3]{1 + x}}$

(b) $\displaystyle\int \frac{x\, dx}{1 + \sqrt[3]{x}}$

(c) $\displaystyle\int \frac{\sin x\, dx}{1 + \sin x}$

(d) $\displaystyle\int \frac{3\, dx}{\sin x + \cos x}$

(e) $\displaystyle\int \frac{dt}{2 + \cos t}$

(f) $\displaystyle\int \frac{(1 + x)^{1/5}}{(1 + x)^{1/3}}\, dx$

(g) $\displaystyle\int \frac{dy}{\sqrt{y} + \sqrt[3]{y}}$

(h) $\displaystyle\int \frac{\tan x\, dx}{1 + \tan^2 x}$

(i) $\displaystyle\int \frac{x^2}{\sqrt{5x + 3}}\, dx$

(j) $\displaystyle\int \frac{dx}{(1 + \sqrt{x})^5}$

(k) $\displaystyle\int \frac{e^x\, dx}{\sqrt{1 + e^x}}$

(l) $\displaystyle\int \frac{dx}{\sqrt{1 + e^x}}.$

4. Evaluate each of the following definite integrals.

(a) $\displaystyle\int_{\pi/6}^{\pi/3} \frac{dx}{\cos^2 x \sin x}$

(b) $\displaystyle\int_0^3 \frac{x\,dx}{\sqrt{1+x}}$

(c) $\displaystyle\int_0^{\pi/2} \sin^5 x \cos^2 x\,dx$

(d) $\displaystyle\int_1^{64} \frac{dx}{\sqrt[3]{x} + 2\sqrt{x}}$.

SUPPLEMENTARY PROBLEMS

Integrate each of the following.

1. $\int x^2 \cos 2x\,dx$
2. $\int \frac{dy}{16+y^2}$
3. $\int \frac{dy}{16-y^2}$
4. $\int \frac{x^2 dx}{1-x^2}$
5. $\int 2x^2 e^{x^3} dx$
6. $\int \frac{x\,dx}{\sqrt{x+1}}$
7. $\int (e^x + x + \cos x)\,dx$
8. $\int (t^2 + t) \ln|t|\,dt$
9. $\int \sqrt{4-x^2}\,dx$
10. $\int \sqrt{4+x^2}\,dx$
11. $\int \tan^5 x\,dx$
12. $\int \sin^5 2x \cos^{10} 2x\,dx$
13. $\int \sqrt{4x^2+1}\,dx$
14. $\int \frac{1}{2}(e^z - e^{-z})\,dz$
15. $\int \frac{dx}{\sqrt{x} + \sqrt[4]{x}}$
16. $\int \frac{dy}{y^3 - 4y}$
17. $\int e^{7t} \cos t\,dt$
18. $\int \frac{dx}{1+e^x}$
19. $\int \frac{dx}{1+\cos(2x+1)}$
20. $\int \frac{dx}{\sqrt{4+x^2}}$
21. $\int \frac{dx}{x(2+\ln x)}$
22. $\int \frac{x\,dx}{x^4+4}$
23. $\int \frac{e^{2x} - e^{-2x}}{e^{-2x} + e^{2x}}\,dx$
24. $\int (\ln y)^2 dy$

25. $\int \sin^5 x \, dx$

26. $\int (x^2 + 1) \ln x \, dx$

27. $\int \frac{\sec t \tan t}{7 + 4 \sec t} dt$

28. $\int \frac{3 \, dx}{4e^{5x} + 2}$

29. $\int x^n \ln x \, dx, \quad n \neq -1$

30. $\int \frac{\ln x}{x} dx$

31. $\int \sec^5 y \, dy$

32. $\int \ln(x^2 - 1) \, dx$

33. $\int \cos 4y \cos 5y \, dy$

34. $\int \cos 3x \sin 5x \, dx$

35. $\int \frac{dx}{x^2 + 4x - 5}$

36. $\int \frac{dx}{x^2 + 4x + 5}$

37. $\int x2^x dx$

38. $\int x^5 e^x dx$

39. $\int \sqrt{2x - x^2} \, dx$

40. $\int \frac{e^x dx}{(3 + 2e^x)^4}$

41. $\int \frac{\sqrt{t} \, dt}{t^3 + 9}$

42. $\int \frac{x^4 dx}{x^2 + 4}$

43. $\int \frac{2x - 4}{5 + 3x} dx$

44. $\int e^{\sec x} \sec x \tan x \, dx$

45. $\int 3x^2 5^{x^3} dx$

46. $\int \frac{\cos x \, dx}{1 + 4 \sin x}$

47. $\int x^2 \arccos x \, dx$

48. $\int \frac{(5t - 11) \, dt}{t^2 - 3t - 10}$

49. $\int \frac{3x \, dx}{4 + \sqrt[5]{x}}$

50. $\int \frac{x^6 - 1}{x - 1} dx.$

CHAPTER 8

The Definite Integral (Continued)

1. Average Value of a Function. Let f be a real-valued function of a real variable which is bounded on the closed interval $[a, b]$. Furthermore, let f be integrable over $[a, b]$. Then the **mean,** or **average value,** of f on the interval $[a, b]$ will be denoted by $M_a^b(f)$ and is defined by

$$M_a^b(f) = \begin{cases} \dfrac{1}{b-a}\displaystyle\int_a^b f, & \text{if } a < b, \\ f(a), & \text{if } a = b. \end{cases}$$

If $a < b$, then it follows at once from the definition that

$$(b - a)M_a^b(f) = \int_a^b f.$$

This equation is also true if $a = b$, for then both sides are equal to zero. We conclude that

$$\text{(1.1)} \qquad \boxed{\int_a^b f = (b - a)M_a^b(f).}$$

If f is nonnegative on $[a, b]$, i.e., if $f(x) \geq 0$ for every x such that $a \leq x \leq b$, then (1.1) yields a good geometric interpretation of the mean $M_a^b(f)$. Let P be the set of all points (x, y) such that $a \leq x \leq b$ and $0 \leq y \leq f(x)$, as shown in Figure 1. Then

$$\textit{area}(P) = \int_a^b f = (b - a)M_a^b(f).$$

It follows that $M_a^b(f)$ is equal to the height of a rectangle with the same base and the same area as P.

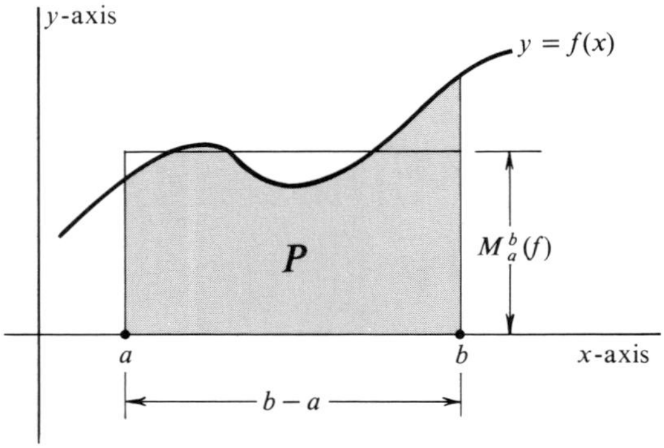

Figure 1

EXAMPLE 1. Let f be the function defined by $f(x) = x^2 - x + 1$. Find the average value of f on the interval [0, 2], draw the graph of f, and show on it the rectangle with base [0, 2] and area equal to the area under the curve. The graph is shown in Figure 2. The mean, or average value, of f is given by

$$\begin{aligned} M_0^2(f) &= \frac{1}{2-0}\int_0^2 f(x)\,dx \\ &= \tfrac{1}{2}\int_0^2 (x^2 - x + 1)\,dx \\ &= \frac{1}{2}\left(\frac{x^3}{3} - \frac{x^2}{2} + x\right)\Big|_0^2 = \frac{4}{3}. \end{aligned}$$

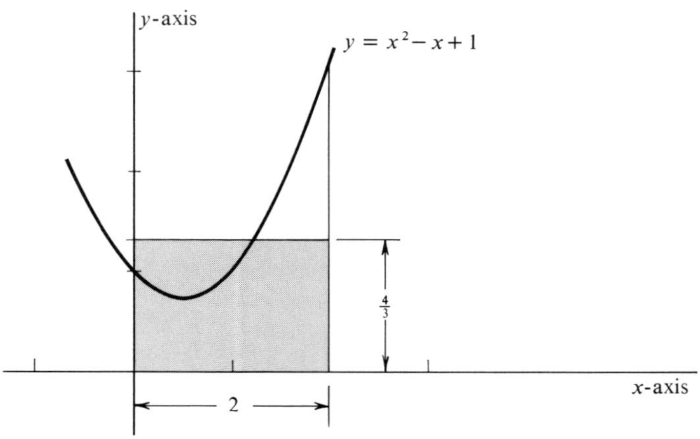

Figure 2

The words "mean" and "average value" are common in our vocabularies and have intuitive meaning for most of us. To use them as names for $M_a^b(f)$ is a sensible thing to do only if this quantity, as we have defined it, has the properties we associate with these words. We shall now show that it does.

First, let us verify that the average value of a velocity function agrees with our earlier definition of average velocity. We consider a particle moving along a straight line, which we take to be a coordinate axis. The position and instantaneous velocity of the particle at time t are denoted by $s(t)$ and $v(t)$, respectively, and we know that $s'(t) = v(t)$. Suppose that the interval of motion is from time $t = a$ to time $t = b$ and that $a < b$. Assuming that v is a continuous function, we have

$$\int_a^b v(t)\,dt = s(b) - s(a).$$

According to the definition on page 104, the average velocity v_{av} is equal to

$$v_{av} = \frac{s(b) - s(a)}{b - a}.$$

The mean, or average value, of the function v on the interval $[a, b]$ is given by

$$\begin{aligned} M_a^b(v) &= \frac{1}{b-a}\int_a^b v(t)\,dt \\ &= \frac{s(b) - s(a)}{b - a} = v_{av}. \end{aligned}$$

Hence the two definitions agree.

The basic properties of the average value of a function correspond closely to the basic properties of the definite integral as they are enumerated at the beginning of Section 4 of Chapter 4. To begin with, we would expect a function which is constant on an interval to have, on that interval, an average value equal to the constant value of the function. The following proposition states that this is so.

(1.2) *If* $f(x) = k$ *for every* x *in the interval* $[a, b]$, *then* $M_a^b(f) = k$.

The proof is an immediate corollary of the definition of the mean $M_a^b(f)$ and of Theorem (4.1), page 191. The reader should supply the details.

If one function is less than or equal to another function on some interval, then the lesser one should have the smaller average value. Thus we expect the theorem:

(1.3) *If* f *and* g *are integrable over* $[a, b]$ *and if* $f(x) \leq g(x)$ *for every* x *in* $[a, b]$, *then* $M_a^b(f) \leq M_a^b(g)$.

The proof follows easily from Theorem (4.3), page 191.

We introduce the third property of the average value of a function by means of an example. Suppose that on a 5-hour automobile trip the average velocity is 45 miles per hour during the first 3 hours and 30 during the last 2 hours. What is the average velocity for the whole trip? To get the answer, we observe that the total distance traveled is

$$45 \cdot 3 + 30 \cdot 2 = 195 \text{ miles.}$$

The average velocity over 5 hours is, therefore,

$$\tfrac{195}{5} = 39 \text{ mph.}$$

If we denote the instantaneous velocity of the automobile by $v(t)$, and assume that the trip began at time $t = 0$, then we can express the fact that the average velocity over the first 3 hours was 45 miles per hour by the equation $M_0^3(v) = 45$. Similarly, we are given $M_3^5(v) = 30$ and have shown that $M_0^5(v) = 39$. Since $3 \cdot 45 + 2 \cdot 30 = 5 \cdot 39$, we can write

$$(3 - 0)M_0^3(v) + (5 - 3)M_3^5(v) = (5 - 0)M_0^5(v).$$

Abstracting from this example, we conclude that the average value of a function should have the property expressed in the proposition:

(1.4) *If f is integrable over $[a, b]$ and $[b, c]$, then*

$$(b - a)M_a^b(f) + (c - b)M_b^c(f) = (c - a)M_a^c(f).$$

Proof. Since $(b - a)M_a^b(f) = \int_a^b f$, the conclusion of (1.4) is equivalent to the equation

$$\int_a^b f + \int_b^c f = \int_a^c f.$$

But this is one of the basic properties of the definite integral [see Theorem (4.2), page 191], so the proof is complete.

The next theorem states the properties of the mean corresponding to Theorems (4.4) and (4.5), page 191.

(1.5) *If f and g are integrable over $[a, b]$ and if k is any real number, then*

(i) $M_a^b(kf) = kM_a^b(f)$,

(ii) $M_a^b(f + g) = M_a^b(f) + M_a^b(g)$.

The proofs are left as exercises.

EXAMPLE 2. Let us see whether the definition of average value of a function agrees with our intuition in a simple example. Let f be the linear function defined by

$$f(x) = \tfrac{1}{2}x + 1,$$

whose graph is shown in Figure 3. What is the average value of f between 2 and 6?

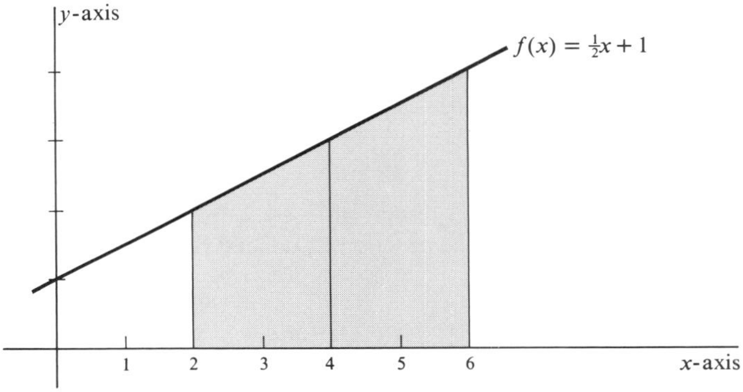

Figure 3

We have $f(2) = \frac{1}{2} \cdot 2 + 1 = 2$ and $f(6) = \frac{1}{2} \cdot 6 + 1 = 4$. Since the graph of f is a straight line, the region below the curve is a trapezoid. It would seem natural for the average value of f on the interval to be the length of the median, which is given by

$$\frac{f(2) + f(6)}{2} = \frac{2 + 4}{2} = 3.$$

Computation of $M_2^6(f)$ yields

$$\begin{aligned} M_2^6(f) &= \frac{1}{6 - 2} \int_2^6 (\tfrac{1}{2}x + 1)\, dx \\ &= \frac{1}{4}\left(\frac{x^2}{4} + x\right)\Bigg|_2^6 \\ &= \tfrac{1}{4}[(\tfrac{36}{4} + 6) - (\tfrac{4}{4} + 2)] \\ &= \tfrac{1}{4}(15 - 3) = 3. \end{aligned}$$

In motivating the definition of the mean, or average value, of a function, we have seen its very close connection with the definite integral. Since a beginning student of calculus probably has a greater feeling for the idea of average than for that of an integral, it is fruitful to reverse our point of view.

That is, if we were to ask the question "What really is the definite integral of a function?", one answer is that it is a weighted average. Specifically, as stated in (1.1), the integral $\int_a^b f$ is equal to the product of $b - a$ and the average value of f on the interval $[a, b]$.

We conclude this section with a theorem which is sometimes called the integral form of the Mean Value Theorem. It asserts that if f is continuous, the number $M_a^b(f)$, which we have called an average value, is quite literally the value of the function f for some number between a and b.

(1.6) INTEGRAL FORM OF MEAN VALUE THEOREM. *If $a < b$ and if f is continuous on the interval $[a, b]$, then there exists a number c such that $a < c < b$ and $M_a^b(f) = f(c)$.*

Proof. Since f is continuous at every point of $[a, b]$, it follows by the Fundamental Theorem of Calculus that the function F, defined by

$$F(x) = \int_a^x f(t)\,dt, \qquad \text{for every } x \text{ in } [a, b],$$

is differentiable. Furthermore,

$$F'(x) = f(x), \qquad \text{for every } x \text{ in } [a, b].$$

A differentiable function is necessarily continuous [see (6.1), page 55], and so F more than satisfies the hypotheses of the Mean Value Theorem, (5.2), page 113. That theorem therefore implies that there exists a number c such that $a < c < b$ and

$$F(b) - F(a) = (b - a)F'(c).$$

But $F'(c) = f(c)$, and

$$F(b) - F(a) = \int_a^b f(x)\,dx.$$

Hence

$$\int_a^b f(x)\,dx = (b - a)f(c),$$

and so

$$f(c) = \frac{1}{b - a}\int_a^b f(x)\,dx = M_a^b(f).$$

This completes the proof.

PROBLEMS

1. Find the average value $M_a^b(f)$ of f on the interval $[a, b]$, where

(a) $f(x) = x^2 - 2x + 1$ and $[a, b] = [0, 2]$.

(b) $f(x) = 2x^3$ and $[a, b] = [-1, 1]$.

(c) $f(x) = \frac{1}{x}$ and $[a, b] = [1, 2]$.

(d) $f(x) = \frac{1}{x}$ and $[a, b] = [1, n]$, where n is a positive integer.

(e) $f(x) = \sin x$ and $[a, b] = [0, \pi]$.

(f) $f(x) = \ln x$ and $[a, b] = [1, 5]$.

2. In each of the following find $M_a^b(f)$, draw the graph of f, and superimpose on the graph a rectangle with base $[a, b]$ and area equal to the area under the curve $y = f(x)$ between a and b.

(a) $f(x) = x^2$, $a = -1$, and $b = 1$.

(b) $f(x) = x^3$, $a = 0$, and $b = 1$.

(c) $f(x) = 4 - (x - 1)^2$, $a = 0$, and $b = 3$.

(d) $f(x) = e^x$, $a = 0$, and $b = 2$.

(e) $f(x) = \cos x$, $a = 0$, and $b = \frac{\pi}{2}$.

3. Each of the propositions (1.2), (1.3), (1.4), and (1.5) corresponds to one of the basic properties of the definite integral as they are enumerated in Theorems (4.1) through (4.5), page 191. In general, the proof of each is obtained by checking the special case $a = b$ separately and then using the formula

$$M_a^b(f) = \frac{1}{b - a} \int_a^b f(x)\,dx, \qquad \text{for } a < b,$$

together with the appropriate property of the integral.

(a) Prove (1.2) (b) Prove (1.3) (c) Prove (1.5).

4. A stone dropped from a cliff 400 feet high falls to the bottom with a constant acceleration equal to 32 feet per second per second. That is,

$$a(t) = v'(t) = s''(t) = 32,$$

where the direction of increasing s is downward. If the stone is dropped at time $t = 0$, find the time it takes to reach the bottom of the cliff, and the mean velocity during the fall.

5. A typist's speed over an interval from $t = 0$ to $t = 4$ hours increases as she warms up and then decreases as she gets tired. Measured in words per minute,

suppose that her speed is given by $v(t) = 6[4^2 - (t - 1)^2]$. Find her speed at the beginning, at the end, her maximum speed, and her average speed over the 4-hour interval. How many words did she type during the 4 hours?

6. A particle moves during the interval of time from $t = 1$ second to $t = 3$ seconds with a velocity given by $v(t) = t^2 + 2t + 1$ feet per second. Find the total distance that the particle has moved and also the average velocity.
7. For each of the functions and intervals in Problem 2, find a number c such that $a < c < b$ and $M_a^b(f) = f(c)$.
8. An arbitrary linear function f is defined by $f(x) = Ax + B$ for some constants A and B. Show that

$$M_a^b(f) = \frac{f(a) + f(b)}{2}.$$

9. Let $x(t)$ be the number of bacteria in a culture at time t, and let $x_0 = x(0)$. The number grows at a rate proportional to the number present, and doubles in a time interval of length T. Find an expression for $x(t)$ in terms of x_0 and T, and find the average number of bacteria present over the time interval $[0, T]$.

2. Riemann Sums and the Trapezoid Rule. This section is divided into two parts. The first is devoted to an alternative approach to the definite integral, which is useful for many purposes. The second is an application of the first part to the problem of computing definite integrals by numerical approximations. We begin by reviewing briefly the definitions in Section 1 of Chapter 4. If the function f is bounded on the closed interval $[a, b]$, then, for every partition σ of $[a, b]$, there are defined an upper sum U_σ and a lower sum L_σ, which approximate the definite integral from above and below, respectively. The function f is defined to be integrable over $[a, b]$ if there exists one and only one number, denoted by $\int_a^b f$, with the property that

$$L_\sigma \leq \int_a^b f \leq U_\tau,$$

for every pair of partitions σ and τ of $[a, b]$.

In the alternative description of the integral, the details are similar, but not the same. As above, let $\sigma = \{x_0, \ldots, x_n\}$ be a partition of $[a, b]$, which satisfies the inequalities

$$a = x_0 \leq x_1 \leq \cdots \leq x_n = b.$$

In each subinterval $[x_{i-1}, x_i]$ we select an arbitrary number, which we shall denote by x_i^*. Then the sum

$$\boxed{R_\sigma = \sum_{i=1}^{n} f(x_i^*)(x_i - x_{i-1})}$$

is called a **Riemann sum** for f relative to the partition σ. (The name commemorates the great mathematician Bernhard Riemann, 1826–1866.) It is important to realize that since x_i^* may be any number which satisfies $x_{i-1} \leq x_i^* \leq x_i$, there are in general infinitely many Riemann sums R_σ for a given f and partition σ. However, every R_σ lies between the corresponding upper and lower sums U_σ and L_σ. For, if the least upper bound of the values of f on $[x_{i-1}, x_i]$ is denoted by M_i and the greatest lower bound by m_i, then $f(x_i^*)$ is an intermediate value and

$$m_i \leq f(x_i^*) \leq M_i,$$

as shown in Figure 4. It follows that

$$\sum_{i=1}^{n} m_i(x_i - x_{i-1}) \leq \sum_{i=1}^{n} f(x_i^*)(x_i - x_{i-1}) \leq \sum_{i=1}^{n} M_i(x_i - x_{i-1}),$$

which states that

$$L_\sigma \leq R_\sigma \leq U_\sigma,$$

for every Riemann sum R_σ for f relative to σ.

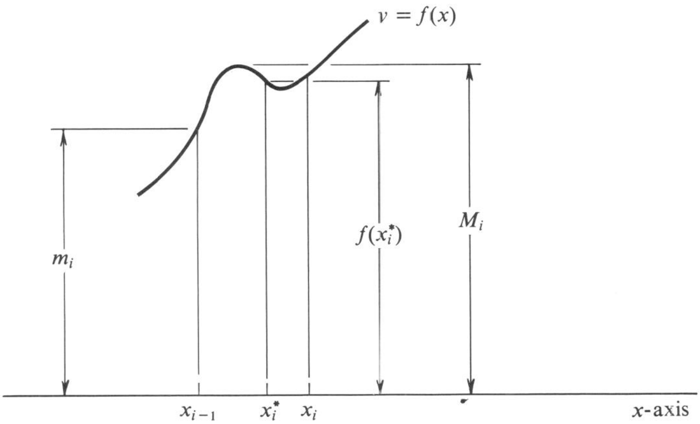

Figure 4

Every Riemann sum is an approximation to the definite integral. By taking partitions which subdivide the interval of integration into smaller and smaller subintervals, we should expect to get better and better approximations to $\int_a^b f$. One number which measures the fineness of a given partition σ is the length of the largest subinterval into which σ subdivides $[a, b]$. This number is called the **mesh** of the partition and is denoted by $\|\sigma\|$. Thus if $\sigma = \{x_0, \ldots, x_n\}$ is a partition of $[a, b]$ with

$$a = x_0 \leq x_1 \leq \cdots \leq x_n = b,$$

then

$$\|\sigma\| = \underset{1 \le i \le n}{\text{maximum}} \{(x_i - x_{i-1})\}.$$

The following definition states precisely what we mean when we say that the Riemann sums approach a limit as the mesh tends to zero. Let the function f be bounded on the interval $[a, b]$. We shall write

$$\lim_{\|\sigma\| \to 0} R_\sigma = L,$$

where L is a real number, if the difference between the number L and any Riemann sum R_σ for f is arbitrarily small provided the mesh $\|\sigma\|$ is sufficiently small. Stated formally, the limit exists if: For any positive real number ϵ, there exists a positive real number δ such that, if R_σ is any Riemann sum for f relative to a partition σ of $[a, b]$ and if $\|\sigma\| < \delta$, then $|R_\sigma - L| < \epsilon$.

The fundamental fact that integrability is equivalent to the existence of the limit of Riemann sums is expressed in the following theorem.

(2.1) THEOREM. *Let f be bounded on $[a, b]$. Then f is integrable over $[a, b]$ if and only if $\lim_{\|\sigma\| \to 0} R_\sigma$ exists. If this limit exists, then it is equal to $\int_a^b f$.*

The proof is given in Appendix C.

EXAMPLE 1. Using the fact that the definite integral is the limit of Riemann sums, evaluate

$$\lim_{n \to \infty} \frac{\sqrt{1} + \sqrt{2} + \cdots + \sqrt{n}}{n^{3/2}}.$$

The numerator of this fraction suggests trying the function f defined by $f(x) = \sqrt{x}$. If $\sigma = \{x_0, \ldots, x_n\}$ is the partition of the interval $[0, 1]$ into subintervals of length $\frac{1}{n}$, then

$$x_0 = 0, \quad x_1 = \frac{1}{n}, \quad x_2 = \frac{2}{n}, \ldots, x_i = \frac{i}{n},$$

$$x_i - x_{i-1} = \frac{1}{n}, \qquad \text{for } i = 1, \ldots, n.$$

In each subinterval $[x_{i-1}, x_i]$ we take $x_i^* = x_i$. Then

$$f(x_i^*) = f(x_i) = \sqrt{\frac{i}{n}} = \frac{\sqrt{i}}{\sqrt{n}},$$

and the resulting Riemann sum, denoted by R_n, is given by

$$\begin{aligned} R_n &= \sum_{i=1}^{n} f(x_i^*)(x_i - x_{i-1}) = \sum_{i=1}^{n} \frac{\sqrt{i}}{\sqrt{n}} \frac{1}{n} \\ &= \frac{1}{n^{3/2}} \sum_{i=1}^{n} \sqrt{i} \\ &= \frac{\sqrt{1} + \sqrt{2} + \cdots + \sqrt{n}}{n^{3/2}}. \end{aligned}$$

The function $\sqrt{x}$ is continuous and hence integrable over the interval $[0, 1]$. Since $\|\sigma\| \to 0$ as $n \to \infty$, it follows from Theorem (2.1) that

$$\int_a^b f = \int_0^1 \sqrt{x}\, dx = \lim_{n\to\infty} R_n.$$

Hence

$$\begin{aligned} \lim_{n\to\infty} \frac{\sqrt{1} + \sqrt{2} + \cdots + \sqrt{n}}{n^{3/2}} &= \int_0^1 \sqrt{x}\, dx \\ &= \tfrac{2}{3} x^{3/2} \Big|_0^1 = \tfrac{2}{3}, \end{aligned}$$

and the problem is solved.

Theorem (2.1) shows that it is immaterial whether we define the definite integral in terms of upper and lower sums (as is done in this book) or as the limit of Riemann sums. Hence there is no logical necessity for introducing the latter at all. However, a striking illustration of the practical use of Riemann sums arises in studying the problem of evaluating definite integrals by numerical methods.

In spite of the variety of techniques which exist for finding antiderivatives and the existence of tables of indefinite integrals,there are still many functions for which we cannot find an antiderivative. More often than not, the only way of computing $\int_a^b f(x)\, dx$ is by numerical approximation. However, the increasing availability of high-speed computers has placed these methods in an entirely new light. Evaluating $\int_a^b f(x)\, dx$ by numerical approximation is no longer to be regarded as a last resort to be used only if all else fails. It is an interesting, instructive, and simple task to write a machine program to do the job, and the hundreds, thousands, or even millions of arithmetic operations which may be needed to obtain the answer to a desired accuracy can be performed by a machine in a matter of seconds or minutes.

One of the simplest and best of the techniques of numerical integration is the Trapezoid Rule, which we now describe. Suppose that the function f is integrable over the interval $[a, b]$. For every positive integer n, let σ_n be the partition which subdivides $[a, b]$ into n subintervals of equal length h. Thus $\sigma_n = \{x_0, \ldots, x_n\}$ and

$$h = \frac{b-a}{n} = x_i - x_{i-1}, \qquad i = 1, \ldots, n.$$

It is convenient to set

$$y_i = f(x_i), \qquad i = 0, 1, \ldots, n.$$

Then the Riemann sum obtained by choosing x_i^* to be the left endpoint of each subinterval $[x_{i-1}, x_i]$, i.e., by choosing $x_i^* = x_{i-1}$, is given by

$$\sum_{i=1}^{n} f(x_i^*)(x_i - x_{i-1}) = \sum_{i=1}^{n} f(x_{i-1})h = h\sum_{i=1}^{n} y_{i-1}.$$

Similarly, the Riemann sum obtained by choosing x_i^* to be the right endpoint, i.e., by taking $x_i^* = x_i$, is

$$\sum_{i=1}^{n} f(x_i^*)(x_i - x_{i-1}) = \sum_{i=1}^{n} f(x_i)h = h\sum_{i=1}^{n} y_i.$$

The approximation to $\int_a^b f$ prescribed by the Trapezoid Rule, which we denote by T_n, is by definition the average of these two Riemann sums. Thus

$$\begin{aligned} T_n &= \frac{h\sum_{i=1}^{n} y_{i-1} + h\sum_{i=1}^{n} y_i}{2} \\ &= \frac{h}{2}\left(\sum_{i=1}^{n} y_{i-1} + \sum_{i=1}^{n} y_i\right). \end{aligned} \tag{1}$$

The last expression above can be simplified by observing that

$$\sum_{i=1}^{n} y_{i-1} + \sum_{i=1}^{n} y_i = y_0 + 2\sum_{i=1}^{n-1} y_i + y_n.$$

Hence

$$T_n = h(\tfrac{1}{2}y_0 + \sum_{i=1}^{n-1} y_i + \tfrac{1}{2}y_n). \tag{2}$$

We shall express the fact that T_n is an approximation to the integral $\int_a^b f$ by writing $\int_a^b f \approx T_n$. The **Trapezoid Rule** then appears as the formula

$$\int_a^b f \approx T_n = h(\tfrac{1}{2}y_0 + y_1 + \cdots + y_{n-1} + \tfrac{1}{2}y_n). \tag{2.2}$$

Why is this formula called the Trapezoid Rule? Suppose that $f(x) \geq 0$ for every x in $[a, b]$, and observe that

$$\int_a^b f = \int_{x_0}^{x_1} f + \int_{x_1}^{x_2} f + \cdots + \int_{x_{n-1}}^{x_n} f.$$

The area of the shaded trapezoid shown in Figure 5 is an approximation to $\int_{x_{i-1}}^{x_i} f$. This trapezoid has bases $y_{i-1} = f(x_{i-1})$ and $y_i = f(x_i)$ and altitude h. By a well-known formula, its area is therefore equal to $\dfrac{h}{2}(y_{i-1} + y_i)$.

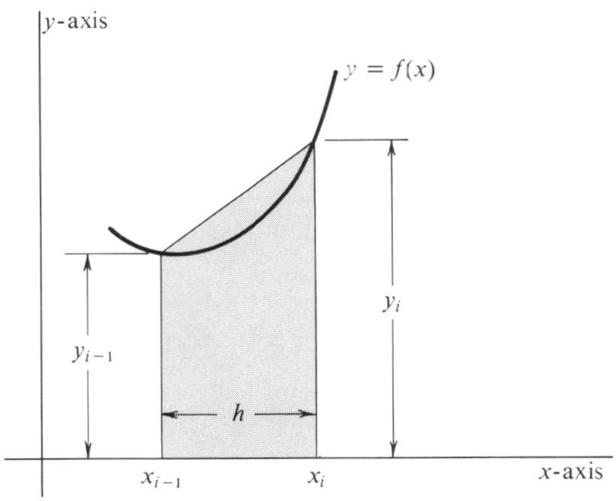

Figure 5

The integral $\int_a^b f$ is therefore approximated by the sum of the areas of the trapezoids, which is equal to

$$\sum_{i=1}^{n} \frac{h}{2}(y_{i-1} + y_i) = \frac{h}{2}\left(\sum_{i=1}^{n} y_{i-1} + \sum_{i=1}^{n} y_i\right).$$

Equation (1) shows that this number is equal to T_n.

EXAMPLE 2. Using the Trapezoid Rule, find an approximate value for $\int_{-1}^{1} (x^3 + x^2)\,dx$. We shall subdivide the interval $[-1, 1]$ into $n = 8$ subintervals each of length $h = \frac{1}{4}$. The relevant numbers are compiled in Table 1.

TABLE 1

i	x_i	$y_i = x_i^2 + x_i^3$
0	-1	$1 - 1 = 0$
1	$-\frac{3}{4}$	$\frac{9}{16} - \frac{27}{64} = \frac{9}{64}$
2	$-\frac{2}{4}$	$\frac{4}{16} - \frac{8}{64} = \frac{8}{64}$
3	$-\frac{1}{4}$	$\frac{1}{16} - \frac{1}{64} = \frac{3}{64}$
4	0	$0 + 0 = 0$
5	$\frac{1}{4}$	$\frac{1}{16} + \frac{1}{64} = \frac{5}{64}$
6	$\frac{2}{4}$	$\frac{4}{16} + \frac{8}{64} = \frac{24}{64}$
7	$\frac{3}{4}$	$\frac{9}{16} + \frac{27}{64} = \frac{63}{64}$
8	1	$1 + 1 = 2$

Hence

$$\begin{aligned} T_8 &= \tfrac{1}{4}(\tfrac{1}{2}y_0 + y_1 + \cdots + y_7 + \tfrac{1}{2}y_8) \\ &= \frac{1}{4}\left(0 + \frac{9 + 8 + 3 + 0 + 5 + 24 + 63}{64} + 1\right) \\ &= \tfrac{1}{4}\,\tfrac{176}{64} = \tfrac{11}{16}. \end{aligned}$$

With the Fundamental Theorem of Calculus it is easy in this case to compute the integral exactly. We get

$$\int_{-1}^{1} (x^3 + x^2)\,dx = \left(\frac{x^4}{4} + \frac{x^3}{3}\right)\Bigg|_{-1}^{1} = \tfrac{2}{3}.$$

The error obtained using the Trapezoid Rule is, therefore,

$$\tfrac{11}{16} - \tfrac{2}{3} = \tfrac{33}{48} - \tfrac{32}{48} = \tfrac{1}{48}.$$

In Example 2, the error can be reduced by taking a larger value of n, or, equivalently, a smaller value of $h = \dfrac{b-a}{n}$. Indeed, it is easy to show that in any application of the Trapezoid Rule the error approaches zero as n increases without bound (or as h approaches zero). That is, we have the theorem

(2.3)
$$\lim_{n\to\infty} T_n = \int_a^b f(x)\,dx.$$

Proof. It follows from equation (1) that

$$T_n = \frac{1}{2}\left(h\sum_{i=1}^{n} y_{i-1} + h\sum_{i=1}^{n} y_i\right).$$

Both $h\sum_{i=1}^{n} y_{i-1}$ and $h\sum_{i=1}^{n} y_i$ are Riemann sums for f. Hence, by Theorem (2.1), both approach the integral as $n \to \infty$. Thus

$$\lim_{n\to\infty} h\sum_{i=1}^{n} y_{i-1} = \int_a^b f(x)\,dx,$$

$$\lim_{n\to\infty} h\sum_{i=1}^{n} y_i = \int_a^b f(x)\,dx,$$

and so

$$\lim_{n\to\infty} T_n = \frac{1}{2}\left[\int_a^b f(x)\,dx + \int_a^b f(x)\,dx\right] = \int_a^b f(x)\,dx,$$

and the proof is complete.

As a practical aid to computation, Theorem (2.3) is actually of little value. What is needed instead is a method of estimating the error, which is equal to

$$\left|\int_a^b f - T_n\right|,$$

for a particular choice of n used in a particular application of the Trapezoid Rule. For this purpose the following theorem is useful.

(2.4) *If the second derivative f'' is continuous at every point of* $[a, b]$, *then there exists a number c such that $a < c < b$ and*

$$\int_a^b f = T_n - \frac{b-a}{12} f''(c)h^2.$$

An outline of a proof of this theorem can be found in J. M. H. Olmsted, *Advanced Calculus*, Appleton-Century-Crofts, 1961, pages 118 and 119.

To see how this theorem can be used, consider Example 2, in which $f(x) = x^3 + x^2$ and in which the interval of integration is $[-1, 1]$. In this case $f''(x) = 6x + 2$, from which it is easy to see that

$$f''(x) \leq 8, \qquad \text{for every } x \text{ in } [-1, 1].$$

Applying (2.4), we have

$$\int_{-1}^{1} (x^3 + x^2)\,dx - T_n = -\frac{1-(-1)}{12} f''(c)h^2.$$

Hence the error satisfies

$$\begin{aligned}\left|\int_{-1}^{1} (x^3 + x^2)\,dx - T_n\right| &= \tfrac{1}{6}|f''(c)|h^2 \\ &\leq \tfrac{1}{6} \cdot 8 \cdot h^2 = \tfrac{4}{3}h^2.\end{aligned}$$

It follows that by halving h, we could have quartered the error. If we were to replace h by $\frac{h}{10}$, which would be no harder with a machine, we would divide the error by 100.

PROBLEMS

1. For each of the following limits, find a function $f(x)$ such that the limit is equal to $\int_0^1 f(x)\,dx$. Evaluate the limit.

(a) $\displaystyle\lim_{n\to\infty} \frac{1 + 2^2 + 3^2 + \cdots + n^2}{n^3}$.

(b) $\displaystyle\lim_{n\to\infty} \frac{(1 + n^2) + (2^2 + n^2) + (3^2 + n^2) + \cdots + (n^2 + n^2)}{n^3}$.

(c) $\displaystyle\lim_{n\to\infty} \frac{\sqrt{1+n} + \sqrt{2+n} + \sqrt{3+n} + \cdots + \sqrt{n+n}}{n^{3/2}}$

(d) $\displaystyle\lim_{n\to\infty} \frac{1}{\sqrt{n}}\left(\frac{1}{\sqrt{1+n}} + \frac{1}{\sqrt{2+n}} + \frac{1}{\sqrt{3+n}} + \cdots + \frac{1}{\sqrt{n+n}}\right)$.

2. Prove that

(a) $\displaystyle\ln 2 = \lim_{n\to\infty}\left(\frac{1}{n+1} + \frac{1}{n+2} + \cdots + \frac{1}{n+n}\right)$.

(b) $\displaystyle\pi = \lim_{n\to\infty} \frac{4}{n^2}\left(\sqrt{n^2 - 1} + \sqrt{n^2 - 2^2} + \cdots + \sqrt{n^2 - n^2}\right)$.

(c) $\displaystyle\int_1^3 (x^2 + 1)\,dx = \lim_{n\to\infty} \frac{4}{n^3} \sum_{i=1}^{n} (n^2 + 2in + 2i^2)$.

(d) $\displaystyle\frac{\pi}{6} = \lim_{n\to\infty}\left(\frac{1}{\sqrt{4n^2 - 1}} + \frac{1}{\sqrt{4n^2 - 2^2}} + \cdots + \frac{1}{\sqrt{4n^2 - n^2}}\right)$.

3. Use the Trapezoid Rule with $n = 4$ to compute approximations to the following integrals. In (a), (b), (c), and (d), compare the approximation obtained with the true value.

(a) $\int_0^1 (x^2 + 1)\,dx$

(b) $\int_0^2 (x^2 + 1)\,dx$

(c) $\int_{-1}^3 (4x - 1)\,dx$

(d) $\int_1^3 \frac{1}{x^2}\,dx$

(e) $\int_0^1 \frac{1}{1 + x}\,dx$

(f) $\int_0^1 \frac{dx}{1 + x^2}$

(g) $\int_0^1 e^{-x^2}\,dx$

(h) $\int_0^1 \frac{1}{x^2 + x + 1}\,dx$

(i) $\int_0^1 \frac{x^2 - 1}{x^2 + 1}\,dx$

(j) $\int_0^\pi \frac{\sin x}{x}\,dx.$

4. Show geometrically, without appealing to Theorem (2.4), that the approximation T_n obtained with the Trapezoid Rule has the following properties.

(a) If $f''(x) \geq 0$ for every x in $[a, b]$, then $T_n \geq \int_a^b f$.

(b) If $f''(x) \leq 0$ for every x in $[a, b]$, then $T_n \leq \int_a^b f$.

5. For each of the following integrals, use Theorem (2.4) as the basis for finding the smallest integer n such that the error $|\int_a^b f - T_n|$ in applying the Trapezoid Rule is less than (i) $\frac{1}{100}$, (ii) $\frac{1}{10{,}000}$, and (iii) 10^{-8}.

(a) $\int_1^4 \frac{1}{6x^2}\,dx$

(b) $\int_0^1 (8x^3 - 5x + 3)\,dx$

(c) $\int_{-1}^2 (3x + 1)\,dx$

(d) $\int_1^2 \frac{1}{x}\,dx$

(e) $\int_0^{12} \frac{1}{16x + 2}\,dx$

(f) $\int_0^1 e^{-x^2}\,dx.$

6. (For those who have access to a high-speed digital computer and know how to use it.) Compute the Trapezoid approximation T_n to each of the following integrals.

(a) $\int_0^1 \frac{1}{1 + x^3}\,dx$, for $n = 10$, 100, and 1000.

(b) $\int_0^1 \frac{1}{1+x^2}\,dx$, for successive values of $n = 10, 100, 1000, \ldots$, until the error is less than 10^{-6}.

(c) $\int_0^1 \sqrt{1+x^3}\,dx$, for $n = 5, 50$, and 500.

(d) $\int_0^\pi \frac{\sin x}{x}\,dx$, for $n = 2, 4, 8, 16$, and 100.

(e) $\int_0^1 e^{-x^2}\,dx$, for $n = 10, 100$, and 1000.

3. Numerical Approximations (Continued). Two additional methods of integration by numerical approximations, which we shall describe in this section, are the Midpoint Rule and Simpson's Rule.

In the Midpoint Rule the approximation to the integral $\int_a^b f$ is a Riemann sum $\sum_{i=1}^n f(x_i^*)(x_i - x_{i-1})$ in which each x_i^* is taken to be the midpoint of the subinterval $[x_{i-1}, x_i]$. In more detail: Let f be a function which is integrable over $[a, b]$. For every positive integer n, let $h = \dfrac{b-a}{n}$, and let $\sigma_n = \{x_0, \ldots, x_n\}$ be the partition defined by

$$x_i = a + ih, \qquad i = 0, \ldots, n.$$

As a result, it follows that

$$x_i - x_{i-1} = h, \qquad i = 1, \ldots, n.$$

If we take x_i^* to be the midpoint of the subinterval $[x_{i-1}, x_i]$, then

$$x_i^* = \frac{x_{i-1} + x_i}{2}, \qquad i = 1, \ldots, n.$$

The Riemann sum used as the approximation to the integral in the Midpoint Rule will be denoted by M_n. It is given by

$$M_n = \sum_{i=1}^n f(x_i^*)(x_i - x_{i-1}) = h\sum_{i=1}^n f\left(\frac{x_{i-1} + x_i}{2}\right).$$

In studying the Trapezoid Rule, we found it convenient to use the abbreviation $y_i = f(x_i)$, for $i = 0, \ldots, n$. By analogy, we shall here let

$$y_{i-1/2} = f\left(\frac{x_{i-1} + x_i}{2}\right), \qquad i = 1, \ldots, n.$$

Then

$$M_n = h\sum_{i=1}^n y_{i-1/2} = h(y_{1/2} + y_{3/2} + \cdots + y_{n-1/2}),$$

and we express the **Midpoint Rule** for numerical integration by the formula

$$\int_a^b f \approx M_n = h(y_{1/2} + y_{3/2} + \cdots + y_{n-1/2}). \tag{3.1}$$

If $f(x) \geq 0$ for every x in $[a, b]$, the Midpoint Rule approximates the integral $\int_a^b f$, which is the area under the curve, by a sum of areas of rectangles, as illustrated in Figure 6.

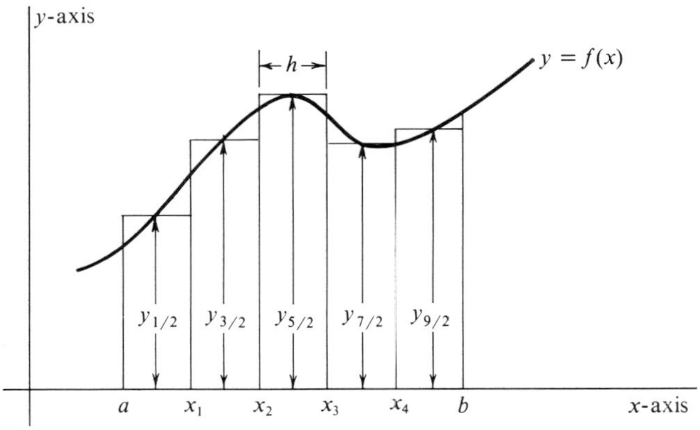

Figure 6

An alternative to approximating the integral by a Riemann sum is to use straight-line segments which are tangent to the curve $y = f(x)$ at the midpoints of the subintervals. An example is shown in Figure 7, in which

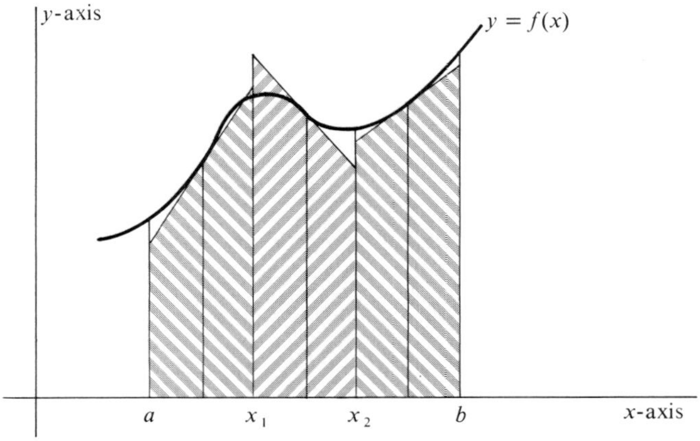

Figure 7

$\int_a^b f$ is approximated by the sum of the areas of the three shaded trapezoids. This method yields the so-called **Tangent Formula.** It turns out, however, that the Tangent Formula is the same as the Midpoint Rule. The reason can be seen by looking at Figure 8, in which the area of the shaded trapezoid with one side tangent to the curve is the ith term in the approximating sum

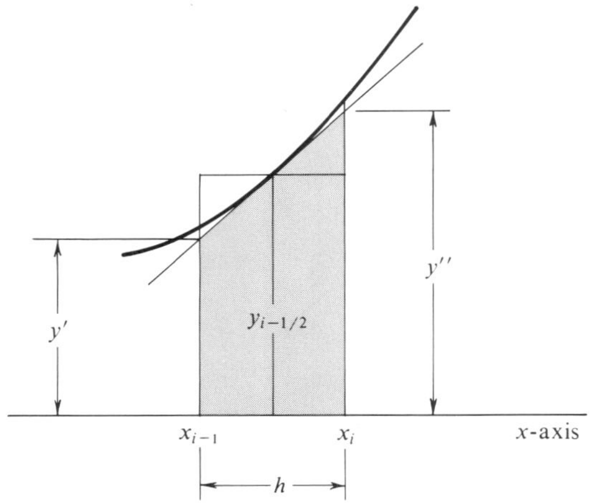

Figure 8

used in the Tangent Formula. The area of this trapezoid is equal to $\frac{h}{2}(y' + y'')$, where y' and y'' are the lengths of the bases. However, by elementary geometry the trapezoid can be seen to have the same area as the rectangle with base $[x_{i-1}, x_i]$ and altitude $y_{i-1/2}$. The area of the rectangle is $h \cdot y_{i-1/2}$, and so

$$\frac{h}{2}(y' + y'') = h \cdot y_{i-1/2}.$$

(Incidentally, note that this equation is true regardless of whether y', y'', and $y_{i-1/2}$ are positive, negative, or zero.) Since the product $h \cdot y_{i-1/2}$ is the ith term in the midpoint approximation M_n, we conclude that the Tangent Formula and the Midpoint Rule, although differently motivated, are in fact the same.

EXAMPLE 1. Approximate $\int_{-1}^{1} (x^2 + x^3)\,dx$ using the Midpoint Rule. This is the same integral which we evaluated in Section 2 by the Trapezoid Rule. To compare the two methods, we shall again take $n = 8$ and

$$h = \frac{1 - (-1)}{8} = \frac{1}{4}.$$

Since, for an arbitrary interval $[a, b]$, we have

$$x_i = a + ih, \qquad i = 0, \ldots, n,$$

it follows that

$$x_i^* = \frac{x_{i-1} + x_i}{2} = \frac{[a + (i - 1)h] + (a + ih)}{2}$$
$$= a + (i - \tfrac{1}{2})h.$$

In addition, since $y_{i-1/2} = f(x_i^*)$, we have a pair of useful formulas:

$$\left.\begin{aligned} x_i^* &= a + (i - \tfrac{1}{2})h \\ y_{i-1/2} &= f(a + (i - \tfrac{1}{2})h) \end{aligned}\right\} i = 1, \ldots, n.$$

In the present example, $a = -1$, $h = \frac{1}{4}$, and $f(x) = x^2 + x^3$. Hence

$$x_i^* = -1 + (i - \tfrac{1}{2})\tfrac{1}{4} = \frac{2i - 9}{8},$$

$$y_{i-1/2} = \left(\frac{2i - 9}{8}\right)^2 + \left(\frac{2i - 9}{8}\right)^3 = \frac{8(2i - 9)^2 + (2i - 9)^3}{8^3}$$
$$= \frac{(2i - 9)^2(2i - 1)}{8^3}.$$

Table 2 contains the numbers for the computation of M_8.

TABLE 2

i	$y_{i-1/2} = \dfrac{(2i - 9)^2(2i - 1)}{8^3}$
1	$y_{1/2} = \dfrac{49}{8^3}$
2	$y_{3/2} = \dfrac{25 \cdot 3}{8^3} = \dfrac{75}{8^3}$
3	$y_{5/2} = \dfrac{9 \cdot 5}{8^3} = \dfrac{45}{8^3}$
4	$y_{7/2} = \dfrac{7}{8^3}$
5	$y_{9/2} = \dfrac{9}{8^3}$
6	$y_{11/2} = \dfrac{9 \cdot 11}{8^3} = \dfrac{99}{8^3}$
7	$y_{13/2} = \dfrac{25 \cdot 13}{8^3} = \dfrac{325}{8^3}$
8	$y_{15/2} = \dfrac{49 \cdot 15}{8^3} = \dfrac{735}{8^3}$

Hence we obtain

$$
\begin{aligned}
M_8 &= \tfrac{1}{4}(y_{1/2} + \cdots + y_{15/2}) \\
&= \frac{1}{4 \cdot 8^3}(49 + 75 + 45 + 7 + 9 + 99 + 325 + 735) \\
&= \frac{1344}{4 \cdot 8^3} = \frac{21}{32}
\end{aligned}
$$

as an approximation to the integral $\int_{-1}^{1}(x^2 + x^3)\,dx$. The value obtained earlier with the Trapezoid Rule was $T_8 = \frac{11}{16}$. Since the true value is given by

$$\int_{-1}^{1}(x^2 + x^3)\,dx = \tfrac{2}{3},$$

it follows that the error using the Midpoint Rule is equal to

$$\left|\tfrac{2}{3} - \tfrac{21}{32}\right| = \tfrac{1}{96}.$$

This is one half the error obtained using the Trapezoid Rule with the same value of h.

In any application of the Midpoint Rule, the error $\left|\int_a^b f - M_n\right|$ can be made arbitrarily small by taking n sufficiently large. That is, we assert that

(3.2)
$$\boxed{\lim_{n\to\infty} M_n = \int_a^b f.}$$

This theorem is easier to prove than the corresponding one for the Trapezoid Rule because every approximation M_n is, as it stands, a Riemann sum of f relative to the partition σ_n of $[a, b]$. Since $\|\sigma_n\| \to 0$ as $n \to \infty$, it is a direct corollary of the fundamental theorem on the limit of Riemann sums [(2.1), page 414] that $\lim\limits_{n\to\infty} M_n = \int_a^b f$.

A means of estimating the error $\left|\int_a^b f - M_n\right|$ in a particular application of the Midpoint Rule is provided by the following theorem:

(3.3) *If the second derivative f'' is continuous at every point of $[a, b]$, then there exists a number c such that $a < c < b$ and*

$$\int_a^b f = M_n + \frac{b-a}{24} f''(c)h^2.$$

As with the analogous theorem for the Trapezoid Rule [(2.4), page 419], this theorem can be proved by first reducing it to the case $n = 1$. A discussion of the error term can be found in R. Courant and F. John, *Introduction to Calculus and Analysis*, Volume I, Interscience Publishers (Wiley), 1965, pages 486 and 487.

Theorem (3.3) can be used to obtain an upper bound on the error $|\int_a^b f - M_n|$ provided the second derivative f'' is bounded on the interval $[a, b]$. That is, if there exists a real number K such that

$$|f''(x)| \leq K, \qquad \text{for every } x \text{ in } [a, b],$$

then, in particular $|f''(c)| \leq K$, and so

$$\begin{aligned}\left|\int_a^b f - M_n\right| &= \frac{(b-a)h^2}{24}|f''(c)| \\ &\leq \frac{(b-a)K}{24}h^2.\end{aligned}$$

For the one integral we computed by both methods, the Midpoint Rule gave a better approximation than the Trapezoid Rule by a factor of 2. Comparison of Theorem (3.3) with (2.4) shows that in general this ratio can be expected.

Geometrically, the different methods of numerical integration described thus far in this and the preceding section are all based on approximating the area under a curve by a sum of areas of rectangles or trapezoids. Analytically, in each method the approximation to $\int_a^b f$ has been obtained by replacing f over every subinterval by a linear function. In Simpson's Rule, however, we shall replace f over each subinterval by a quadratic polynomial $Ax^2 + Bx + C$. The corresponding area problem is the simple one of finding the area under a parabola (or a straight line, if $A = 0$). For most integrals, Simpson's Rule gives much greater accuracy for the same value of h.

Let f be a function which is integrable over the interval $[a, b]$. The procedure differs from the others in that we consider only partitions of $[a, b]$ into an even number of subintervals. Thus for an arbitrary *even* integer $n > 0$, we set $h = \frac{b-a}{n}$, and let

$$\begin{aligned}x_i &= a + ih, \\ y_i &= f(x_i), \qquad \text{for } i = 0, \ldots, n.\end{aligned}$$

Since n is even, there is an integral number of "double" intervals $[x_{2i-2}, x_{2i}]$, $i = 1, \ldots, \frac{n}{2}$, as illustrated in Figure 9, and

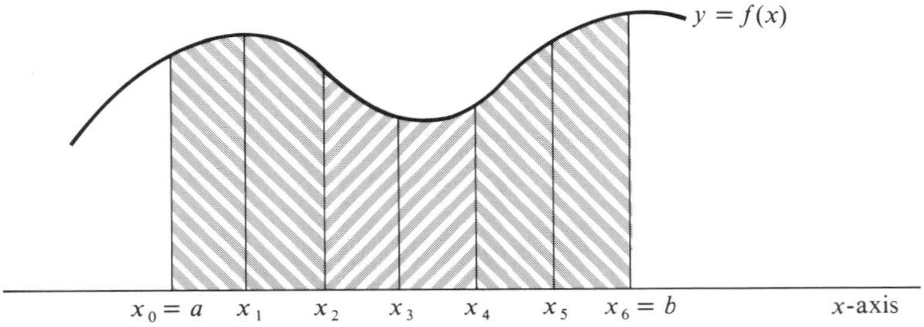

Figure 9

$$\int_a^b f(x)\,dx = \sum_{i=1}^{n/2} \int_{x_{2i-2}}^{x_{2i}} f(x)\,dx.$$

There exists one and only one polynomial $q_i(x) = A_i x^2 + B_i x + C_i$ of degree less than three whose graph passes through the three points (x_{2i-2}, y_{2i-2}), (x_{2i-1}, y_{2i-1}), and (x_{2i}, y_{2i}) (see Figure 10). Over each double interval $[x_{2i-2}, x_{2i}]$ we shall approximate the integral of f by the integral of q_i. Let us assume for the moment, and later prove, the fact that

$$\int_{x_{2i-2}}^{x_{2i}} q_i(x)\,dx = \frac{h}{3}(y_{2i-2} + 4y_{2i-1} + y_{2i}). \tag{1}$$

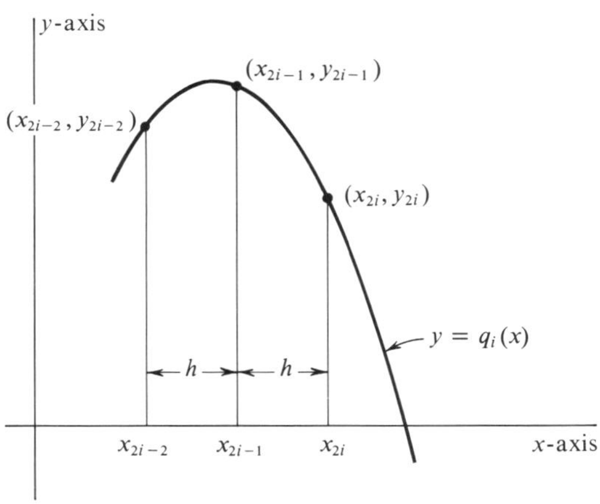

Figure 10

The sum of these integrals, which we shall denote S_n, is the approximation to $\int_a^b f$ prescribed by Simpson's Rule. Hence

$$S_n = \frac{h}{3}\sum_{i=1}^{n/2}(y_{2i-2} + 4y_{2i-1} + y_{2i}).$$

If this sum is expanded, note the pattern of the coefficient of the y_i's: If i is odd, the coefficient of y_i is 4. If i is even, the coefficient is 2, with the exception of y_0 and y_n, each of which has coefficient 1. Thus **Simpson's Rule** is expressed in the formula

$$\int_a^b f \approx S_n = \frac{h}{3}(y_0 + 4y_1 + 2y_2 + 4y_3 + 2y_4 + \cdots + 2y_{n-2} + 4y_{n-1} + y_n). \tag{3.4}$$

We now prove equation (1). The algebra is significantly simpler if we write $q_i(x)$ in terms of $x - x_{2i-1}$ instead of x (see Figure 10). Hence we let

$$q_i(x) = \alpha_i(x - x_{2i-1})^2 + \beta_i(x - x_{2i-1}) + \gamma_i. \tag{2}$$

The integral $\int_{x_{2i-2}}^{x_{2i}} q_i(x)\,dx$ may be computed by substituting $u = x - x_{2i-1}$ and using the Change of Variable Theorem for Definite Integrals (see page 215). Since $x_{2i-2} - x_{2i-1} = -h$ and $x_{2i} - x_{2i-1} = h$, the result is

$$\begin{aligned}\int_{x_{2i-2}}^{x_{2i}} q_i(x)\,dx &= \int_{-h}^{h}(\alpha_i u^2 + \beta_i u + \gamma_i)\,du\\ &= \left(\frac{\alpha_i u^3}{3} + \frac{\beta_i u^2}{2} + \gamma_i u\right)\Bigg|_{-h}^{h}\\ &= \frac{h}{3}(2\alpha_i h^2 + 6\gamma_i).\end{aligned} \tag{3}$$

Setting first $x = x_{2i-1}$ in equation (2), we obtain

$$y_{2i-1} = q_i(x_{2i-1}) = \alpha_i \cdot 0^2 + \beta_i \cdot 0 + \gamma_i = \gamma_i.$$

Next we let $x = x_{2i-2}$ and $x = x_{2i}$ to get

$$y_{2i-2} = q_i(x_{2i-2}) = \alpha_i h^2 - \beta_i h + \gamma_i,$$

and

$$y_{2i} = q_i(x_{2i}) = \alpha_i h^2 + \beta_i h + \gamma_i.$$

Adding, we obtain

$$y_{2i-2} + y_{2i} = 2\alpha_i h^2 + 2\gamma_i.$$

Since $\gamma_i = y_{2i-1}$, it follows that

$$2\alpha_i h^2 + 6\gamma_i = y_{2i-2} + 4y_{2i-1} + y_{2i}.$$

Substituting this result in (3) yields the desired result (1), and the derivation of Simpson's Rule is complete.

EXAMPLE 2. Using $n = 4$, find an approximate value of $\int_0^1 \frac{1}{1+x^2}\,dx$ by Simpson's Rule. We have $h = \frac{1-0}{4} = \frac{1}{4}$,

$$x_i = \frac{i}{4}, \qquad i = 0, 1, \ldots, 4,$$

$$y_i = \frac{1}{1+x_i^2} = \frac{1}{1+\frac{i^2}{16}} = \frac{16}{16+i^2}, \qquad i = 0, 1, \ldots, 4.$$

Table 3 contains the numbers necessary for the computation.

TABLE 3

i	0	1	2	3	4
$y_i = \frac{16}{16+i^2}$	1	$\frac{16}{17}$	$\frac{16}{20}$	$\frac{16}{25}$	$\frac{16}{32}$

Hence

$$\begin{aligned} S_4 &= \frac{h}{3}(y_0 + 4y_1 + 2y_2 + 4y_3 + y_4) \\ &= \tfrac{1}{12}(1 + \tfrac{64}{17} + \tfrac{32}{20} + \tfrac{64}{25} + \tfrac{16}{32}) \\ &= 0.785392\ldots. \end{aligned}$$

We know that

$$\int_0^1 \frac{1}{1+x^2}\,dx = \arctan x\Big|_0^1 = \frac{\pi}{4} = 0.785398\ldots.$$

Hence the error $|\int_a^b f - S_4|$ is approximately 0.000006.

Just as with the other two methods of integration by numerical approximation, the error $|\int_a^b f - S_n|$ can be made arbitrarily small by taking n sufficiently large. That is, we have the following theorem, which we state without proof.

(3.5)
$$\lim_{n\to\infty} S_n = \int_a^b f.$$

In addition, the next theorem can be used to estimate the error in a particular application of Simpson's Rule.

(3.6) *If the fourth derivative $f^{(4)}$ is continuous at every point of $[a, b]$, then there exists a number c such that $a < c < b$ and*

$$\int_a^b f = S_n - \frac{b-a}{180} f^{(4)}(c)h^4.$$

For an outline of a proof, see J. M. H. Olmsted, *Advanced Calculus,* Appleton-Century-Crofts, 1961, page 119.

The fourth derivative of every cubic polynomial is identically zero, for if

$$f(x) = a_3x^3 + a_2x^2 + a_1x + a_0,$$

then $f^{(4)}(x) = 0$. It is therefore a rather surprising corollary of Theorem(3.6) that Simpson's Rule will always give the exact value of the integral when applied to any polynomial of degree less than 4.

PROBLEMS

1. Use the Midpoint Rule with $n = 4$ to compute approximations to the following integrals. In (a), (b), (c), (d), and (e) compare the result obtained with the true value.

(a) $\int_0^1 (x^2 + 1)\, dx$

(b) $\int_{-1}^3 (6x - 5)\, dx$

(c) $\int_1^3 \frac{1}{x^2}\, dx$

(d) $\int_0^3 \frac{1}{1+x}\, dx$

(e) $\int_0^2 \sqrt{1+x}\, dx$

(f) $\int_0^{2\pi} \sin^2 x\, dx$

(g) $\int_0^1 e^{-x^2} dx$

(h) $\int_0^1 \sqrt{1+x^3}\, dx.$

2. (a) through (h). Compare M_4, the Midpoint Approximation computed in Problem 1, to T_4, the corresponding Trapezoid Approximation.
3. (a) through (h). Use Simpson's Rule with $n = 4$ to compute an approximation to the corresponding definite integral in Problem 1.
4. Show geometrically that, if the graph of f is concave up at every point of the interval $[a, b]$, then the Midpoint Approximation is too small and the Trapezoid Approximation is too big; i.e.,

$$M_n < \int_a^b f < T_n.$$

5. Do Problem 4 analytically by using the remainder formulas (2.4) and (3.3).
6. Show that Simpson's Approximation is the weighted average of the Trapezoid Approximation and the Midpoint Approximation. Specifically, for any even positive integer $n = 2m$, show that

$$S_n = \tfrac{1}{3}T_m + \tfrac{2}{3}M_m.$$

7. Prove Theorem (3.5), i.e., if f is integrable over $[a, b]$, then

$$\lim_{n\to\infty} S_n = \int_a^b f,$$

by showing that it is a direct corollary of the result of Problem 6 and the two corresponding theorems, $\lim_{n\to\infty} T_n = \int_a^b f$ and $\lim_{n\to\infty} M_n = \int_a^b f$.

8. For each of the following integrals and each of the three methods of numerical integration (Trapezoid Rule, Midpoint Rule, and Simpson's Rule), find the smallest integer n such that the error obtained is less than 10^{-4}. As the basis for finding n, use Theorems (2.4) (3.3), and (3.6).

(a) $\displaystyle\int_1^4 \left(\frac{1}{2x^2} + \frac{x^2}{2}\right) dx$ (b) $\displaystyle\int_0^2 \frac{1}{2x+1}\,dx.$

9. This problem is analogous to Problem 6. Show that, for any positive integer n,

$$T_{2n} = \tfrac{1}{2}T_n + \tfrac{1}{2}M_n.$$

10. Suppose that the graph of f is concave up at every point of the interval $[a, b]$.
    (a) Using the results of Problems 4 and 9, show that

$$T_{2n} - (T_n - T_{2n}) < \int_a^b f < T_{2n},$$

    for every positive integer n.

    (b) Hence show that the error $\left|\int_a^b f - T_{2n}\right|$ in the Trapezoid Approximation satisfies

$$\left|\int_a^b f - T_{2n}\right| < |T_n - T_{2n}|.$$

    (c) Show that (b) also holds if the graph of f is concave down at every point of $[a, b]$.

4. Volume. In this section we shall use Riemann sums to derive several integral formulas for the volume of solids. The concept of volume is a familiar one, and we shall not give a mathematical definition of it. As a result, the formulas will be derived from properties of volume which seem intuitively natural and which we shall assume. By a solid we mean a subset of three-dimensional space.

A subset Q of three-dimensional space is said to be **bounded** if there exists a real number k such that, for any two points p and q in Q, the straight-line distance in space between p and q is less than k. Alternatively, Q is bounded if and only if there exists a sphere in space which contains Q in its interior. In finding volumes of solids, we shall restrict ourselves to bounded subsets of three-dimensional space.

Let Q be a bounded solid, such as the one shown in Figure 11. We choose an arbitrary straight line in space for a coordinate axis. Any point

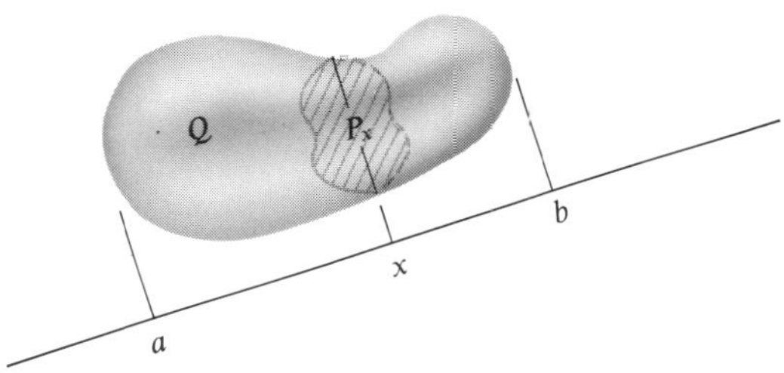

Figure 11

on the line may be chosen for the origin and either direction as the direction of increasing numbers. The scale on the line must agree with the scale of distance in space. That is, if two points on the line correspond to the numbers x and y, then $|x - y|$ must equal the straight-line distance in space between the two points.

For every number x, let P_x be the plane figure which is the intersection of the solid Q and the plane perpendicular to the coordinate axis at x (see Figure 11). The second assumption we make about Q is that every set P_x has a well-defined area. We then define a **cross-sectional area function** A by setting

$$A(x) = area(P_x).$$

One consequence of the fact that Q is bounded is that there exists an interval $[a, b]$ such that, for every x outside of $[a, b]$, the set P_x is empty and so $A(x) = 0$. Another consequence is that the function A is bounded on $[a, b]$.

Let us now consider a partition $\sigma = \{x_0, \ldots, x_n\}$ of the interval $[a, b]$ which satisfies the inequalities

$$a = x_0 \leq x_1 \leq \cdots \leq x_n = b.$$

In each subinterval $[x_{i-1}, x_i]$, we choose an arbitrary number x_i^*. The product $A(x_i^*)(x_i - x_{i-1})$ is equal to the volume of a right cylindrical slab with constant cross-sectional area $A(x_i^*)$ and thickness $x_i - x_{i-1}$. If $x_i - x_{i-1}$ is small, then we should expect $A(x_i^*)(x_i - x_{i-1})$ to be a good approximation to the volume of the slice of Q which lies between the parallel planes at

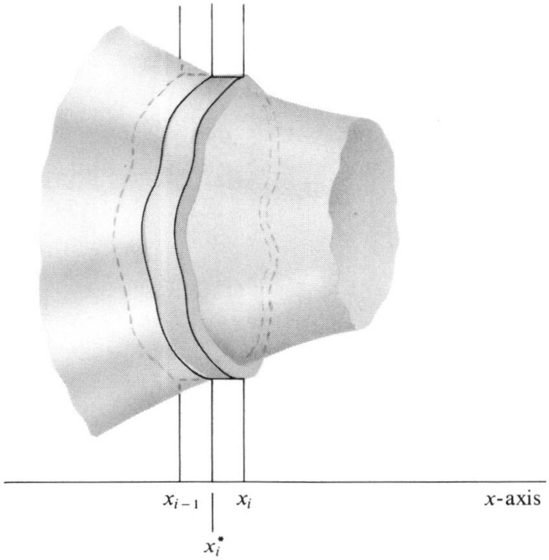

Figure 12

x_{i-1} and x_i (see Figure 12). Hence if the mesh of the partition σ is small, then the Riemann sum

$$\sum_{i=1}^{n} A(x_i^*)(x_i - x_{i-1})$$

will be a good approximation to the volume of Q. Moreover, the smaller the mesh, the better the approximation ought to be. Therefore, we shall assume that if the volume of Q is defined, then it is given by

$$vol(Q) = \lim_{\|\sigma\| \to 0} \sum_{i=1}^{n} A(x_i^*)(x_i - x_{i-1}).$$

It follows immediately from the integrability criterion, Theorem (2.1), page 414, that

(4.1)
$$vol(Q) = \int_a^b A(x)\, dx.$$

This is a basic integral formula for volume.

EXAMPLE 1. Find the volume V of a pyramid of height h with a square base of length a on a side. The pyramid is shown in Figure 13(a). We choose a vertical x-axis with origin at the apex and which cuts the center of the base at $x = h$. The cross-sectional area $A(x)$ of the intersection of the pyramid

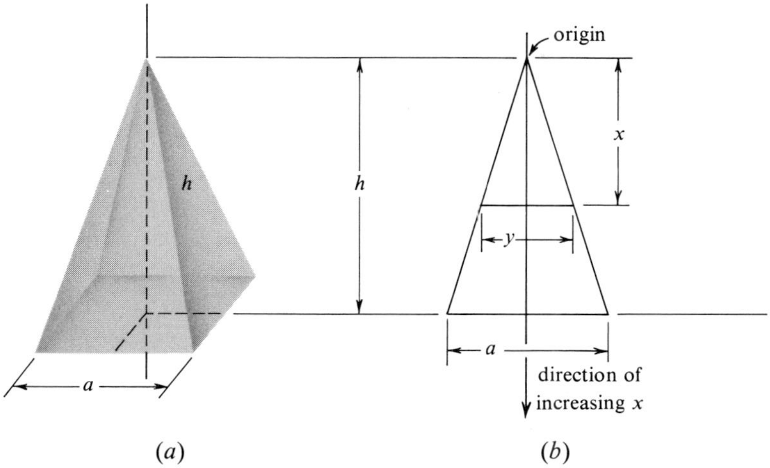

Figure 13

and the plane perpendicular to the coordinate axis at x is easily seen from the side view [Figure 13(b)]. Since corresponding parts of similar triangles are proportional, it follows that

$$\frac{y}{x} = \frac{a}{h}.$$

Hence

$$A(x) = y^2 = \frac{a^2x^2}{h^2},$$

and so

$$\begin{aligned} V &= \int_0^h A(x)\,dx = \frac{a^2}{h^2}\int_0^h x^2 dx \\ &= \frac{a^2}{h^2}\frac{h^3}{3} = \frac{1}{3}a^2h. \end{aligned}$$

This is the well-known result that the volume of a pyramid is equal to one third the product of the height and the area of the base.

EXAMPLE 2. The axes of two right circular cylinders P and Q of equal radii a intersect at right angles as shown in Figure 14(a). Find the volume of the intersection $P \cap Q$. We choose an x-axis perpendicular to the axes of both cylinders and which passes through their point of intersection. This point is chosen for the origin, and the direction of increasing x is upward.

Figure 14(b) helps one to see that the cross sections of $P \cap Q$ perpendicular to the x-axis are squares. From the end-on view in Figure 14(c), it is ap-

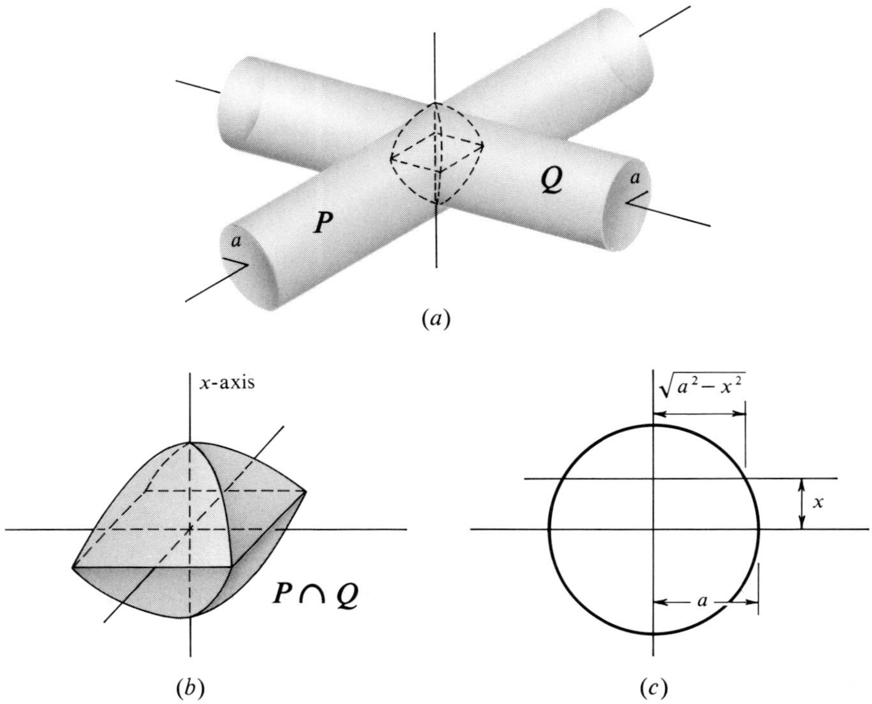

Figure 14

parent that the cross section at x is a square with an edge of length $2\sqrt{a^2 - x^2}$. Hence

$$A(x) = (2\sqrt{a^2 - x^2})^2 = 4(a^2 - x^2), \qquad -a \leq x \leq a.$$

By the integral formula for volume, therefore,

$$vol(P \cap Q) = \int_{-a}^{a} A(x)\, dx = 4\int_{-a}^{a} (a^2 - x^2)\, dx.$$

Since $a^2 - x^2$ is an even function, the integral from $-a$ to a is twice the integral from 0 to a. Thus

$$\begin{aligned} vol(P \cap Q) &= 8\int_{0}^{a} (a^2 - x^2)\, dx \\ &= 8\left(a^2x - \frac{x^3}{3}\right)\Big|_0^a = 8\left(a^3 - \frac{a^3}{3}\right) \\ &= \frac{16a^3}{3}. \end{aligned}$$

Consider a plane in three-dimensional space containing a region R and a line L. The solid generated by rotating R in space about L is called a **solid of revolution.** Among the solids of revolution are those described by rotating a portion of the graph of a function in the xy-plane about the x-axis (or the y-axis). We shall give an integral formula for the volumes of these solids, which is a special case of (4.1). Let f be a function integrable over the closed interval $[a, b]$, and consider the solid of revolution Q swept out by rotating about the x-axis the region bounded by the graph of f, the x-axis, and the

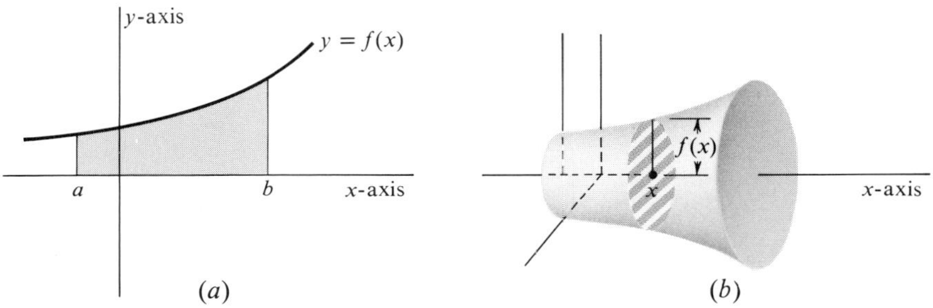

Figure 15

lines $x = a$ and $x = b$. Such a region is illustrated in Figure 15(a), and the corresponding solid in Figure 15(b). For every x in $[a, b]$, the cross section of Q perpendicular to the x-axis at x is a circular disk, of radius $|f(x)|$. Hence

$$A(x) = \pi|f(x)|^2 = \pi[f(x)]^2.$$

By formula (4.1), therefore, the volume of the solid of revolution Q is given by

(4.2)
$$vol(Q) = \pi \int_a^b [f(x)]^2 dx.$$

EXAMPLE 3. Let R be the region in the upper half-plane bounded by the parabola $y^2 = 4x$, the x-axis, and the line $x = 4$. Find the volume of Q, the solid of revolution obtained by rotating this region about the x-axis. The region and the solid are drawn in Figure 16. That part of the parabola in the upper half-plane is the set of all points (x, y) such that $y^2 = 4x$ and $y \geq 0$. This set is the graph of the equation $y = \sqrt{4x}$, and we therefore take $f(x) = \sqrt{4x}$ in formula (4.2), obtaining

$$vol(Q) = \pi \int_0^4 [f(x)]^2 dx = \pi \int_0^4 4x\, dx$$
$$= 4\pi \left.\frac{x^2}{2}\right|_0^4 = 32\pi.$$

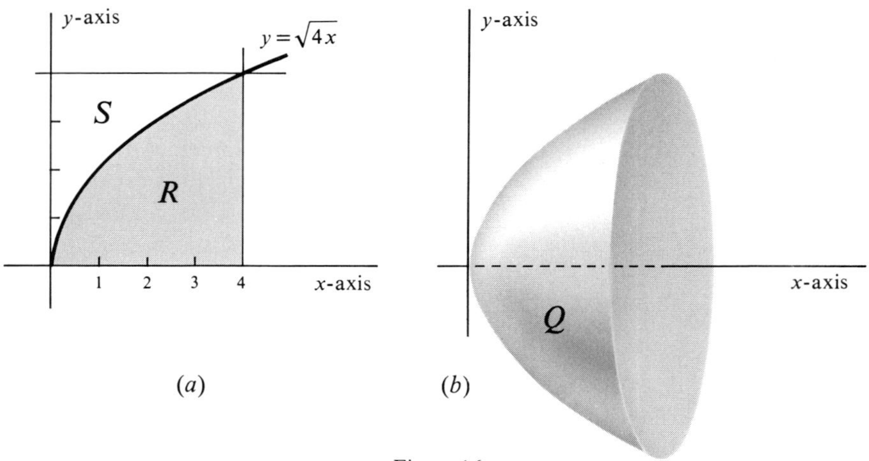

Figure 16

EXAMPLE 4. Let R be the region in Example 3, and let S be the region bounded by the parabola $y^2 = 4x$, the y-axis, and the line $y = 4$. Both regions are shown in Figure 16(a). Find the volumes of the two solids of revolution Q_1 and Q_2, obtained by rotating S and R, respectively, about the y-axis. The two solids are illustrated in Figure 17. The easiest way to find the volume of Q_1 is to use the counterpart of formula (4.2) for functions of y. The equation $y^2 = 4x$ is equivalent to the equation $x = \frac{y^2}{4}$, and the parabola is therefore the graph of the function of y defined by $f(y) = \frac{y^2}{4}$. By symmetry therefore,

$$
\begin{aligned}
vol(Q_1) &= \pi \int_0^4 [f(y)]^2 dy = \pi \int_0^4 \frac{y^4}{16}\, dy \\
&= \frac{\pi}{16} \frac{y^5}{5}\bigg|_0^4 = \frac{64\pi}{5}.
\end{aligned}
$$

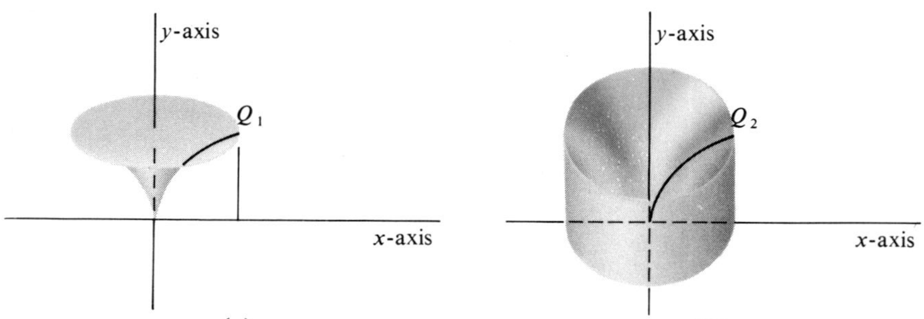

Figure 17

The union of Q_1 and Q_2 is a right circular cylinder of radius 4 and height 4. Hence

$$vol(Q_1 \cup Q_2) = (\pi 4^2) \cdot 4 = 64\pi.$$

It is obvious that

$$vol(Q_1 \cup Q_2) = vol(Q_1) + vol(Q_2).$$

Hence

$$\begin{aligned} vol(Q_2) &= vol(Q_1 \cup Q_2) - vol(Q_1) \\ &= 64\pi - \frac{64\pi}{5} = \frac{256\pi}{5}. \end{aligned}$$

Another way of computing the volumes of certain solids of revolution is the method of **cylindrical shells.** It may be used, for example, to find the volume of the solid Q_2 in Figure 17(b) directly, and it constitutes another interesting application of the integral as a limit of Riemann sums. Let f be a function which is nonnegative and continuous at every point of a closed interval $[a, b]$, where $a \geq 0$. Let R be the region bounded by the graph of f, the x-axis, and the lines $x = a$ and $x = b$. We shall derive a formula for the volume of P, the solid of revolution obtained by revolving the region R about the y-axis (see Figure 18).

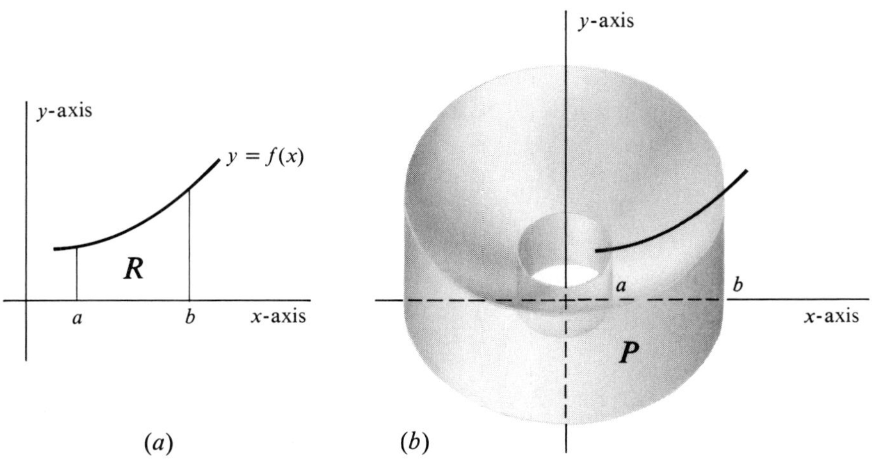

Figure 18

Consider a partition $\sigma = \{x_0, \ldots, x_n\}$ of the interval $[a, b]$ such that

$$a = x_0 \leq x_1 \leq \cdots \leq x_n = b,$$

and choose numbers $x_1^*, \ldots, x_n^*$ such that $x_{i-1} \leq x_i^* \leq x_i$ for each $i = 1, \ldots, n$. Then $\pi(x_i^2 - x_{i-1}^2)$ is the area of the shaded annulus shown in Figure 19(a), and $\pi f(x_i^*)(x_i^2 - x_{i-1}^2)$ is the volume of the cylindrical shell of height $f(x_i^*)$ and thickness $x_i - x_{i-1}$ shown in Figure 19(b). The essential idea in the derivation is that, if $x_i - x_{i-1}$ is small, then the volume of this

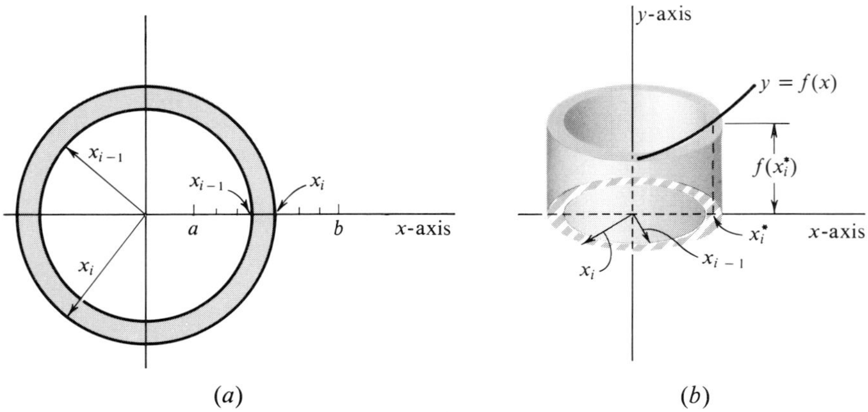

Figure 19

shell should be a good approximation to the volume of that part of the solid P that lies between the two cylinders of radii x_{i-1} and x_i. Hence if the mesh of the partition σ is small, then the sum of the volumes of the shells, i.e., the sum

$$\sum_{i=1}^{n} \pi f(x_i^*)(x_i^2 - x_{i-1}^2),$$

should be a good approximation to the volume of P. Specifically, we shall assume that

$$vol(P) = \lim_{\|\sigma\|\to 0} \sum_{i=1}^{n} \pi f(x_i^*)(x_i^2 - x_{i-1}^2). \tag{1}$$

On the basis of this assumption we now prove the theorem

(4.3)
$$vol(P) = 2\pi \int_a^b x f(x)\, dx.$$

Proof. Since each x_i^* may be any number in the subinterval $[x_{i-1}, x_i]$, let us for convenience choose it to be the midpoint. This means that $x_i^* = \dfrac{x_i + x_{i-1}}{2}$ or, equivalently, that $x_i + x_{i-1} = 2x_i^*$. Hence

$$\begin{aligned} x_i^2 - x_{i-1}^2 &= (x_i + x_{i-1})(x_i - x_{i-1}) \\ &= 2x_i^*(x_i - x_{i-1}), \end{aligned}$$

and it follows that

$$\sum_{i=1}^{n} \pi f(x_i^*)(x_i^2 - x_{i-1}^2) = \sum_{i=1}^{n} 2\pi x_i^* f(x_i^*)(x_i - x_{i-1}).$$

By our assumption (1), the left side of this equation approaches $vol(P)$ as a limit as the mesh $\|\sigma\|$ approaches zero. The right side is a Riemann sum for the continuous function $2\pi x f(x)$ and therefore approaches the integral $\int_a^b 2\pi x f(x)\,dx$ as a limit as $\|\sigma\|$ approaches zero. This completes the proof.

The region R in Figure 16(a), when rotated about the y-axis, generates the solid of revolution Q_2 illustrated in Figure 17(b). Let us compute the volume of Q_2 by the method of cylindrical shells. The function f in the formula is in this case the one defined by $f(x) = \sqrt{4x} = 2x^{1/2}$. In addition, $a = 0$ and $b = 4$. We therefore obtain

$$\begin{aligned} vol(Q_2) &= 2\pi \int_0^4 x\, 2x^{1/2} dx \\ &= 4\pi \int_0^4 x^{3/2} dx = 4\pi \cdot \tfrac{2}{5} \cdot x^{5/2}\Big|_0^4 \\ &= 4\pi \cdot \tfrac{2}{5} \cdot 32 = \frac{256\pi}{5}, \end{aligned}$$

which agrees with the value obtained in Example 4.

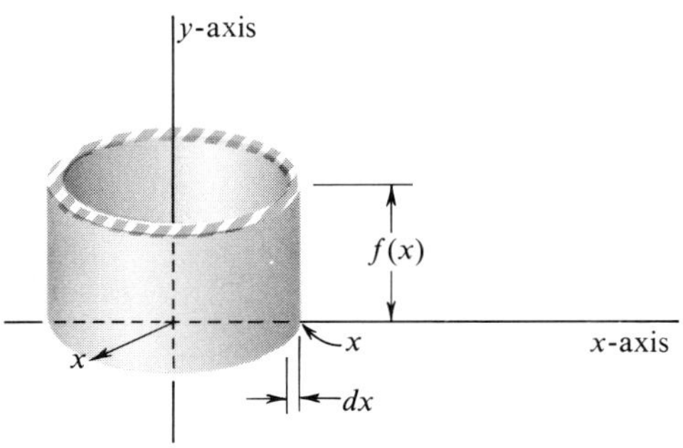

Figure 20

It was pointed out in Section 6 of Chapter 4 that the dx which appears in $\int f(x)\,dx$ can be legitimately regarded as a differential. Moreover, it was remarked in Section 6 of Chapter 2 that the traditional attitude toward a differential is that it represents an infinitesimally small quantity. These ideas are a good aid to the imagination here, too. Thus, as the meshes of the subdivisions in the Riemann sums $\sum_{i=1}^{n} f(x_i^*)(x_i - x_{i-1})$ become infinitesimally small, the differences $x_i - x_{i-1}$ become dx and the summation

$\sum$ becomes the integral $\int$. Consider the following rough-and-ready derivation of the formula used in the method of cylindrical shells. We imagine a shell of infinitesimal thickness at each x in the interval $[a, b]$. The radius is x, the thickness dx, and the height $f(x)$ (see Figure 20). The circumference of the shell is $2\pi x$, and the area of the edge is the product of the circumference and the thickness, $2\pi x\, dx$. Multiplying by the height, we get the infinitesimal volume $2\pi x f(x)\, dx$. The total volume is obtained by adding all the infinitesimal volumes. We get $\sum 2\pi x f(x)\, dx$ or, actually,

$$\int_a^b 2\pi x f(x)\, dx = 2\pi \int_a^b x f(x)\, dx$$

for the answer.

The derivation of integral formulas by this process of "summing infinitesimals" is extremely useful both as a guide to memory and in helping one to discover the right formula in the first place. Of course, any such heuristic approach is justified mathematically only if it can be supplemented by careful analysis.

PROBLEMS

1. A solid Q has a flat base which is the region in the plane bounded by the parabola $y^2 = x$ and the line $x = 4$. Each cross-section perpendicular to the x-axis is a square with one edge lying in the base. Find the volume of Q.
2. The solid P has the same base as Q in Problem 1, but each cross-section perpendicular to the x-axis is a semicircular disk with diameter lying in the base. Compute $vol(P)$.
3. A tetrahedron is a solid with four vertices and four flat triangular faces. Let T be a tetrahedron which has three mutually perpendicular edges of lengths 3, 4, and 10 meeting at a vertex. Draw a picture of T and compute its volume.
4. The graph of the function $f(x) = \sqrt{a^2 - x^2}$ is a semicircle of radius a. Use this function and an integral formula for the volume of a solid of revolution to compute the volume of a sphere of radius a.
5. Find the volume of the ellipsoid of revolution obtained by rotating about the x-axis the region bounded by the ellipse $\dfrac{x^2}{a^2} + \dfrac{y^2}{b^2} = 1$.
6. Sketch and find the volume of each of the solids of revolution obtained by rotating about the x-axis the region bounded by the indicated curves and lines.

   (a) $y = x^2 - 1$, $x = 1$, $x = 2$, and the x-axis.

   (b) $y = \frac{1}{2}x$, $x = 8$, and the x-axis.

   (c) $y = 1 + 2x - x^2$, $x = 0$, $y = 0$, and $x = 2$.

   (d) $y = \dfrac{1}{x^2}$, $x = 1$, $x = 2$, and the x-axis.

   (e) $y = 1 - x^2$ and the x-axis.

7. Find the volume of a right circular cone of height h and with a base of radius a.
8. (a) Find the volume of the solid of revolution obtained by rotating about the y-axis the region bounded by the x-axis, and the graphs of $y = x^2 - 1$ and $y = 3$.

   (b) Using (a), find the volume generated by rotating the region in Problem 6(a) about the y-axis. (Use Example 4 as a model.)
9. Using the method of cylindrical shells, find the volume of the solid of revolution obtained by rotating each of the regions in Problem 6 about the y-axis.
10. Sketch the region R in the plane which is bounded by the parabola $(y - 1)^2 = x$, the line $y = 2$, and the x-axis and y-axis. Find the volume of the solid of revolution obtained by rotating R about the x-axis, using

    (a) formula (4.2) twice, i.e., $\pi\int_a^b y^2dx$ once with $y - 1 = \sqrt{x}$ and again with $y - 1 = -\sqrt{x}$.

    (b) the counterpart of formula (4.3), i.e., the method of cylindrical shells, for functions of y.
11. Since $(x - h)^2 + y^2 = a^2$ is an equation of the circle with radius a and center at $(h, 0)$, it follows by solving for y in terms of x that the graph of the function $f(x) = \sqrt{a^2 - (x - h)^2}$ is a semicircle.

    (a) Assuming that $h > a$ and using the method of cylindrical shells, write a definite integral for the volume of the solid torus (doughnut) with radii h and a.

    (b) Evaluate the integral in (a) by making the change of variable $y = x - h$ [see Theorem (6.6), page 215], and using the fact that $\int_{-a}^{a} \sqrt{a^2 - y^2}\, dy = \dfrac{\pi a^2}{2}$ (area of a semicircle).
12. In a solid mass of material, the infinitesimal mass dm of an infinitesimal amount of volume dv located at an arbitrary point is given by

$$dm = \rho\, dv,$$

    where ρ is the density of the material at that point.

    Consider a cylindrical container of radius a filled to a depth h with a liquid whose density is greater at the bottom and less at the top. Specifically, at a point a distance x below the surface the density is given by $\rho = 2 + x$. What is the total mass of liquid in the container?
13. Same as Problem 12, but this time the container is a right circular cone (apex at the bottom) of height h and base of radius a which is filled to the top.

5. Work. The concept in physics of the work done by a force acting on an object as it moves a given distance provides another important application of the definite integral.

Throughout this section we shall consider only those situations in which the object moves in a straight line L, and in which the direction of the force

F is also along L. Mathematically, F is a function, which may or may not be constant. We shall assume that L is a coordinate axis and that, for every number x on L, the value of the force F at x is equal to $F(x)$. The sign convention will be as follows: $F(x) > 0$ means that the direction of the force at x is in the direction of increasing numbers on L, and $F(x) < 0$ means that the direction of the force is in the direction of decreasing numbers.

We first consider the special case in which the force F is constant as the object moves along L from a to b. Thus $F(x) = k$ for all x such that $a \leq x \leq b$. Then the **work done by the force** denoted by W, is defined by the simple equation

$$W = k(b - a). \tag{1}$$

Frequently, we wish to speak of the **work done against the force** which we shall denote by W_*. This is just the negative of W. Hence

$$W_* = -W = (-k)(b - a). \tag{2}$$

EXAMPLE 1. Compute the work done in raising a 10-pound weight a distance of 50 feet against the force of gravity. We choose a coordinate axis as shown in Figure 21 with the origin at the initial position of the object. The mag-

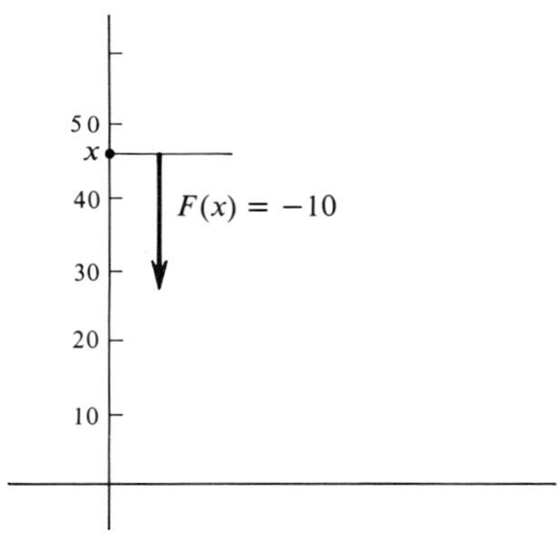

Figure 21

nitude of the force is constant and equal to 10 pounds. Thus $|F(x)| = 10$. By our sign convention, however, $F(x)$ is negative, and so $F(x) = -10$. The work done against the force of gravity in raising the weight is therefore given by

$$\begin{aligned} W_* &= (-F(x))(b - a) = (-(-10))(50 - 0) \\ &= 500 \text{ foot-pounds.} \end{aligned}$$

Next, let us consider the problem of defining the work done by a force F which is not necessarily constant. We shall assume that the function F is integrable over the closed interval having endpoints a and b (it may be that $a \leq b$ or that $b < a$). Then the **work done by the force** F as the object moves from a to b will be denoted by $W(F, a, b)$ and is defined by

$$\boxed{W(F, a, b) = \int_a^b F(x)\,dx.} \tag{3}$$

Thus work depends on the function F and the numbers a and b, and hence is a function of these three quantities. As in (1), we frequently abbreviate $W(F, a, b)$ as simply W.

Is this definition of work a reasonable one? The answer is yes only if $W(F, a, b)$ has the properties which correspond to the physical concept we are trying to describe. For example, we should expect that the work done by a force in moving an object from a to b plus the work done in moving it from b to c should equal the work done in moving it from a to c. This property is expressed in the equation

(5.1) $$W(F, a, b) + W(F, b, c) = W(F, a, c),$$

which is an immediate corollary of the definition of $W(F, a, b)$ and the fundamental additive property of the integral [see Proposition (4.2), page 191]. Second, the work done by a greater force acting on an object as it moves from a to b in the direction of the force should certainly be larger than the work done by a smaller force. This is expressed in the proposition

(5.2) *If $F_1(x) \leq F_2(x)$ for every x such that $a \leq x \leq b$, then*

$$W(F_1, a, b) \leq W(F_2, a, b).$$

This is also simply a restatement of one of the fundamental properties of the definite integral [see Proposition (4.3), page 191]. Finally, we note that the definition is consistent with the earlier one in equation (1). That is, if the force is constant, then the work is simply the product of the constant value and the change in position. Thus

(5.3) *If $F(x) = k$ for every x in the closed interval with endpoints a and b, then*

$$W(F, a, b) = k(b - a).$$

The proof is just the elementary fact that $\int_a^b k\,dx = k(b - a)$ [see Proposition (4.1), page 191].

We have just shown that work, as we have defined it, has three natural and apparently quite basic properties. This suggests that the definition is reasonable. Actually, we can conclude much more than that. We shall now show that our definition of $W(F, a, b)$ as a definite integral is the *only* one which has these three properties. That is, we have proved that the definition implies the properties, and we shall now prove, conversely, that the properties imply the definition. This is such an important fact that we state it as a theorem:

(5.4) THEOREM. *Let W be a function which assigns to every function F and any interval $[a, b]$ over which F is integrable a real number $W(F, a, b)$ such that* (5.1), (5.2), *and* (5.3) *hold. Then*

$$W(F, a, b) = \int_a^b F(x)\,dx.$$

Proof. Let F be a function, and $[a, b]$ an interval over which F is integrable. We shall first show that, for every partition σ of $[a, b]$, the upper and lower sums, U_σ and L_σ, satisfy the inequalities

$$L_\sigma \leq W(F, a, b) \leq U_\sigma. \tag{4}$$

To do this, we let $\sigma = \{x_0, \ldots, x_n\}$ and assume the usual ordering:

$$a = x_0 \leq x_1 \leq \cdots \leq x_n = b.$$

As we have done in the past, we denote by M_i the least upper bound of the values of F on the ith subinterval $[x_{i-1}, x_i]$, and by m_i the greatest lower bound. Then

$$m_i \leq F(x) \leq M_i, \qquad \text{whenever } x_{i-1} \leq x \leq x_i.$$

The two constant functions with values M_i and m_i, respectively, are certainly integrable over the subinterval $[x_{i-1}, x_i]$. Following the common practice of denoting a constant function by its value, we know, as a result of (5.2), that

$$W(m_i, x_{i-1}, x_i) \leq W(F, x_{i-1}, x_i) \leq W(M_i, x_{i-1}, x_i).$$

Using (5.3), we obtain

$$W(m_i, x_{i-1}, x_i) = m_i(x_i - x_{i-1}),$$
$$W(M_i, x_{i-1}, x_i) = M_i(x_i - x_{i-1}).$$

Hence

$$m_i(x_i - x_{i-1}) \leq W(F, x_{i-1}, x_i) \leq M_i(x_i - x_{i-1}).$$

Adding these inequalities for $i = 1, \ldots, n$, we get

$$\sum_{i=1}^{n} m_i(x_i - x_{i-1}) \leq \sum_{i=1}^{n} W(F, x_{i-1}, x_i) \leq \sum_{i=1}^{n} M_i(x_i - x_{i-1}).$$

The left and right sides of the inequalities in the preceding equation are precisely L_σ and U_σ, respectively. It follows from repeated use of (5.1) that

$$\sum_{i=1}^{n} W(F, x_{i-1}, x_i) = W(F, a, b),$$

and we have therefore proved that the inequalities (4) do hold.

The proof of Theorem (5.4) is now essentially complete. Let σ and τ be two arbitrary partitions of $[a, b]$. The union $\sigma \cup \tau$ is the partition which is the common refinement of both. It is shown in the last line of the proof of Proposition (1.1), page 168, that

$$L_\sigma \leq L_{\sigma\cup\tau} \leq U_{\sigma\cup\tau} \leq U_\tau.$$

It follows from equation (4) that

$$L_{\sigma\cup\tau} \leq W(F, a, b) \leq U_{\sigma\cup\tau},$$

and we conclude that

$$L_\sigma \leq W(F, a, b) \leq U_\tau, \tag{5}$$

for any two partitions σ and τ of $[a, b]$. But, by assumption, F is integrable over $[a, b]$, and that means that there is only one number, $\int_a^b F(x)\,dx$, which lies between all lower sums and all upper sums. Hence

$$W(F, a, b) = \int_a^b F(x)\,dx,$$

and the proof is complete.

The significance of Theorem (5.4) is more than just its present application to the definition of work. We can infer from this theorem another, and perhaps more basic, description of the definite integral. This is a description, or characterization, of the integral in terms of three of its properties. The theorem states that the integral $\int_a^b F$ is the only function which has these properties. Hence they may be regarded as a set of axioms for the integral. As such, they are sometimes called a set of **characteristic properties.**

In the remainder of the section we shall give a few examples of the work done by, and also against, a nonconstant force.

EXAMPLE 2. The physical principle known as Hooke's Law states that the force necessary to stretch a spring a distance d from its rest position is proportional to d. The stretched spring exerts a restoring force which is equal in magnitude, but opposite in direction, to the force required to stretch it. Consider the spring shown in Figure 22, which is 1 foot long when under no tension. A 5-pound load B attached to the end of the spring has stretched it to a length of 2 feet (i.e., an additional 1 foot from rest position). How much work is done by the restoring force of the spring if the load is raised $\frac{1}{2}$ foot?

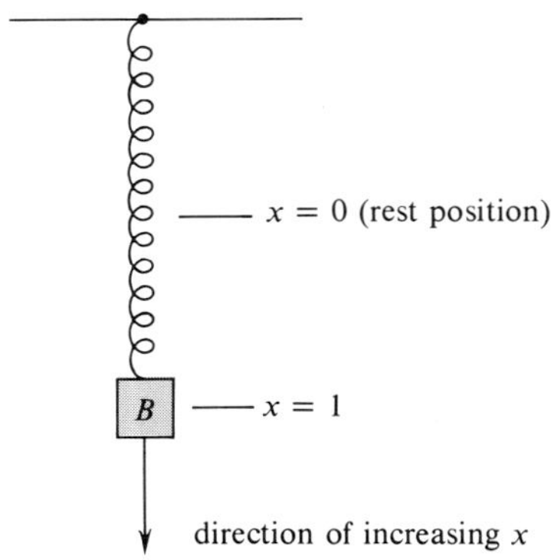

Figure 22

We choose a vertical x-axis with increasing values of x pointing down, and for convenience take the origin to be the rest position. For $x \geq 0$, the restoring force $F(x)$ of the spring is upward and, therefore, $F(x) \leq 0$. It follows by Hooke's Law that

$$F(x) = -kx,$$

for some positive number k. To find the constant k, we use the fact that the 5-pound load B has stretched the spring 1 foot. Hence $F(1) = -5$, from which it follows that $-5 = -k \cdot 1$ and so $k = 5$. Thus

$$F(x) = -5x.$$

In raising B a distance of $\frac{1}{2}$ foot, the movement is from $x = 1$ to $x = \frac{1}{2}$. Hence the work W done by the restoring force F is given by

$$\begin{aligned} W &= \int_1^{1/2} F(x)\, dx = \int_1^{1/2} (-5x)\, dx \\ &= \int_{1/2}^{1} 5x\, dx = 5\frac{x^2}{2}\bigg|_{1/2}^{1} \\ &= 5(\tfrac{1}{2} - \tfrac{1}{8}) = \tfrac{15}{8} \text{ foot-pounds.} \end{aligned}$$

The **work done against the force** F as the object moves from a to b will be denoted by $W_*(F, a, b)$, or simply by W_* as before, and is by definition the negative of the work done by the force F. Thus

$$W_*(F, a, b) = -\int_a^b F(x)\,dx. \tag{6}$$

EXAMPLE 3. Consider the spring in Example 2 loaded as shown in Figure 22. How many foot-pounds of work are required to pull the load B down an additional 1 foot (i.e., so that the spring is stretched to a total length of 3 feet)? The total, or resultant, force $F(x)$ acting at x is the sum of two forces: The first is the restoring force of the spring, which we have computed to be $-5x$, and the second is the force of gravity on B, which is equal to 5. (We ignore the weight of the spring.) Hence

$$F(x) = 5 - 5x.$$

The work required to pull the load B down an additional 1 foot, i.e., to move from $x = 1$ to $x = 2$, will be the work done against the resultant force F. Thus

$$\begin{aligned} W_* &= -\int_1^2 F(x)\,dx = -\int_1^2 (5 - 5x)\,dx \\ &= \left(\frac{5x^2}{2} - 5x\right)\Bigg|_1^2 = \tfrac{5}{2} \text{ foot-pounds.} \end{aligned}$$

EXAMPLE 4. According to Newton's Law of Gravitation, two bodies of masses M and m are attracted to each other by a force equal in magnitude to $G\dfrac{Mm}{r^2}$, where r is the distance between them and G is a universal constant. If the earth has mass M, find the work done in projecting a missile of mass m radially outward 500 miles from the surface of the earth. Let the center of the earth be fixed at the origin of an axis along which the missile is projected in the direction of increasing x, as shown in Figure 23. By Newton's Law, the gravitational force $F(x)$ acting on the missile at x is toward the origin and equal in magnitude to $G\dfrac{Mm}{x^2}$. By our sign convention, therefore,

$$F(x) = -G\frac{Mm}{x^2}.$$

Let $a = 4000$ miles, the radius of the earth, and let $b = 4500$. The work required to project the missile from a to b is equal to the work W_* done against the force F in moving from a to b.

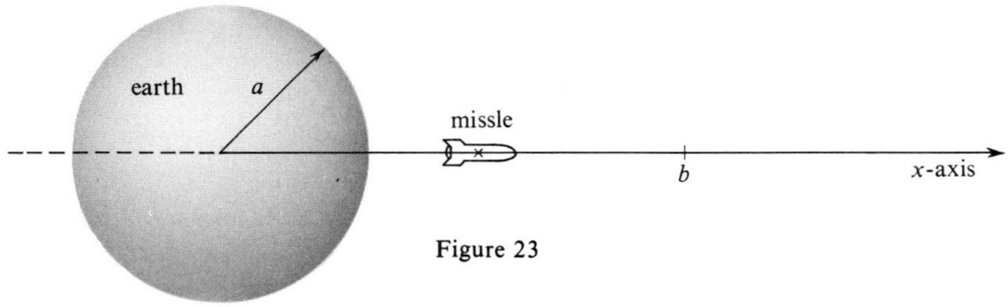

Figure 23

Hence

$$\begin{aligned} W_* &= -\int_a^b F(x)\,dx \\ &= -\int_a^b -G\frac{Mm}{x^2}\,dx = GMm\int_a^b \frac{1}{x^2}\,dx \\ &= GMm\left(-\frac{1}{x}\right)\Big|_a^b = GMm\left(\frac{1}{a}-\frac{1}{b}\right) \\ &= GMm\left(\frac{1}{4000}-\frac{1}{4500}\right). \end{aligned}$$

Suppose that we consider the work required to project the missile from the surface of the earth to points successively farther and farther away. We find that

$$\begin{aligned} \lim_{b\to\infty} W_* &= \lim_{b\to\infty} GMm\left(\frac{1}{a}-\frac{1}{b}\right) \\ &= \frac{GMm}{a}. \end{aligned}$$

This number can be regarded as the work necessary to carry the missile completely out of the earth's gravitational field. In this example we have, of course, ignored the gravitational forces which exist because of the presence of other bodies in the universe.

PROBLEMS

1. Compute the work in foot-pounds done by the force of gravity when a 50-pound rock falls 200 feet off a vertical cliff.
2. Compute the work in foot-pounds done against the force of gravity in raising a 10-pound weight vertically 6 feet from the ground.
3. A car on a horizontal track is attached to a fixed point by a spring, as shown in Figure 24. The spring has been stretched 2 feet beyond its rest position, and the car is held there by a force of 10 pounds. If the car is released, how many foot-pounds of work are done by the restoring force of the spring in moving the car 2 feet back to the rest position?

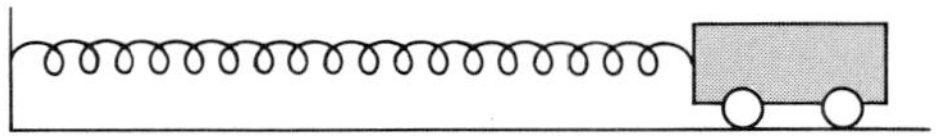

Figure 24

4. An electron is attracted to a nucleus by a force which is inversely proportional to the square of the distance r between them; i.e., $\frac{k}{r^2}$. If the nucleus is fixed, compute the work done by the attractive force in moving the electron from $r = 2a$ to $r = a$.
5. A container holding water is raised vertically a distance of 10 feet at the constant rate of 10 feet per minute. Simultaneously water is leaking from the container at the constant rate of 15 pounds per minute. If the empty container weighs 1 pound and if it holds 15 pounds of water at the beginning of the motion, find the work done against the force of gravity.
6. Suppose that a straight cylindrical hole is bored from the surface of the earth through the center and out the other side. An object of mass m inside the hole and at a distance r from the center of the earth is attracted to the center by a gravitational force equal in absolute value to $\frac{mgr}{R}$, where g is constant and R is the radius of the earth. Compute the work done by this force of gravity in terms of m, g, and R as the object falls

   (a) from the surface to the center of the earth,

   (b) from the surface of the earth through the center to a point halfway between the center and surface on the other side,

   (c) all the way through the hole from surface to surface.

   [*Hint:* Let the x-axis be the axis of the cylinder, and the origin the center of the earth. Define the gravitational force $F(x)$ acting on the object at x so that: (i) its absolute value agrees with the above prescription, and (ii) its sign agrees with the convention given at the beginning of this section.]

7. Consider a cylinder and piston as shown in Figure 25. The inner chamber, which contains gas, has radius a and length b. According to the simplest gas law, the expansive force of the gas on the piston is inversely proportional to the volume v of gas; i.e., $F = \frac{k}{v}$ for some constant k. Compute the work done against this force in compressing the gas to half its initial volume by pushing in the piston.

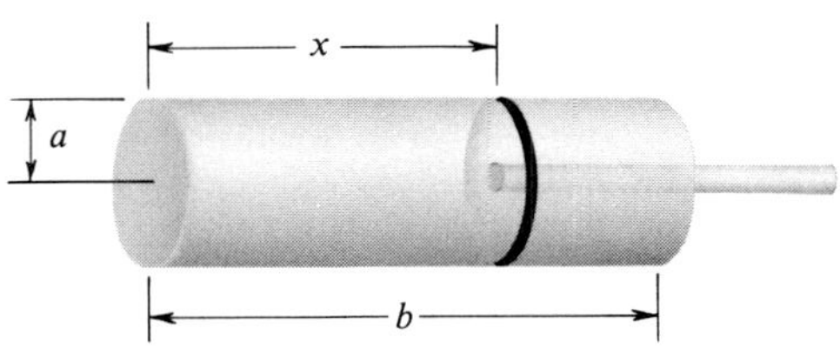

Figure 25

8. A rocket of mass m is on its way from the earth to the moon along a straight line joining their centers. Two gravitational forces act simultaneously on the rocket and in opposite directions. One is the gravitational pull toward earth, equal in absolute value to $\dfrac{GM_1m}{r_1^2}$, where G is the universal gravitational constant, M_1 the mass of the earth, and r_1 the distance between the rocket and the center of the earth. The other is the analogous gravitational attraction toward the moon, equal in absolute value to $\dfrac{GM_2m}{r_2^2}$, where M_2 is the mass of the moon and r_2 is the distance between the rocket and the center of the moon. Denote the radii of the earth and moon by a and b, respectively, and let d be the distance between their centers.

   (a) Take the path of the rocket for the x-axis with the centers of earth and moon at 0 and d, respectively, and compute $F(x)$, the resultant force acting on the rocket at x.

   (b) Set up the definite integral for the work done against the force F as the rocket moves from the surface of the earth to the surface of the moon.

6. Integration of Discontinuous Functions. If a function f is continuous at every point of an interval $[a, b]$, then we know that f is integrable over $[a, b]$ [see Theorem (5.1), page 199]. Continuity is certainly the most important criterion for integrability that we have. For example, in the fundamental theorem of calculus it is assumed that the integrand is continuous over the interval of integration. However, it is important to realize that a function does not have to be continuous to be integrable and that there are many simple discontinuous functions which can be integrated.

We begin with the following theorem:

(6.1) *If f is bounded on $[a, b]$ and is continuous at every point of $[a, b]$ except possibly at the endpoints, then f is integrable over $[a, b]$.*

Proof. If $a = b$, the conclusion follows at once since $\int_a^b f = \int_a^a f = 0$. Hence we shall assume that $a < b$. To be specific, we shall furthermore assume that f is continuous at every point of $[a, b]$ except at a. The necessary modification in the argument if a discontinuity occurs at b (or at both a and b) should be obvious. According to the definition of integrability (page 168), it is sufficient to prove that there exist partitions σ and τ of $[a, b]$ such that $U_\sigma - L_\tau$, the difference between the corresponding upper and lower sums, is arbitrarily small. For this purpose, we choose an arbitrary positive number ϵ. Since f is bounded on $[a, b]$, there exists a positive number k such that $|f(x)| \leq k$, for every x in $[a, b]$. We next pick a point a' which is in $[a, b]$ and sufficiently close to a that

$$0 < a' - a < \frac{\epsilon}{3k}$$

(see Figure 26). Since f is continuous on the smaller interval $[a', b]$, we know that f is integrable over it. Hence there exist partitions σ' and τ' of $[a', b]$ such that the upper sum $U_{\sigma'}$ and lower sum $L_{\tau'}$ for f satisfy

$$|U_{\sigma'} - L_{\tau'}| < \frac{\epsilon}{3}. \tag{1}$$

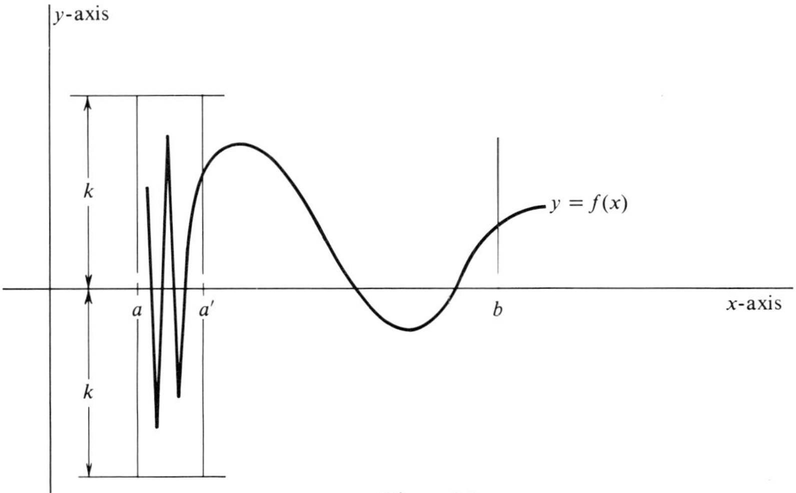

Figure 26

Let σ and τ be the partitions of $[a, b]$ obtained from σ' and τ', respectively, by adjoining the point a; i.e., $\sigma = \sigma' \cup \{a\}$ and $\tau = \tau' \cup \{a\}$. Since the maximum value of $|f(x)|$ on the subinterval $[a, a']$ is less than or equal to k, it follows that

$$|U_\sigma - U_{\sigma'}| \leq k(a' - a) < k \cdot \frac{\epsilon}{3k} = \frac{\epsilon}{3}. \tag{2}$$

By the same argument, we have

$$|L_{\tau'} - L_\tau| \leq k(a' - a) < k \cdot \frac{\epsilon}{3k} = \frac{\epsilon}{3}. \tag{3}$$

Next, consider the algebraic identity

$$U_\sigma - L_\tau = (U_\sigma - U_{\sigma'}) + (L_{\tau'} - L_\tau) + (U_{\sigma'} - L_{\tau'}).$$

The sum of three numbers is always less than or equal to the sum of their absolute values. Using this fact and the inequalities (1), (2), and (3), we obtain

$$\begin{aligned} U_\sigma - L_\tau &\leq |U_\sigma - U_{\sigma'}| + |L_{\tau'} - L_\tau| + |U_{\sigma'} - L_{\tau'}| \\ &< \frac{\epsilon}{3} + \frac{\epsilon}{3} + \frac{\epsilon}{3} = \epsilon. \end{aligned}$$

Thus there exist upper and lower sums lying arbitrarily close to each other, and the proof is complete.

EXAMPLE 1. Let f be the function defined by

$$f(x) = \begin{cases} \sin \dfrac{\pi}{x}, & \text{if } x \neq 0, \\ 0 & \text{if } x = 0. \end{cases}$$

This function is continuous everywhere except at 0, and its values oscillate wildly as x approaches 0. The graph, for values of x in the interval $[0, 2]$, is shown in Figure 27. Since $|f(x)| \leq 1$ for every x, the function is bounded on every interval. It therefore follows by Theorem (6.1) that f is integrable over $[0, 2]$.

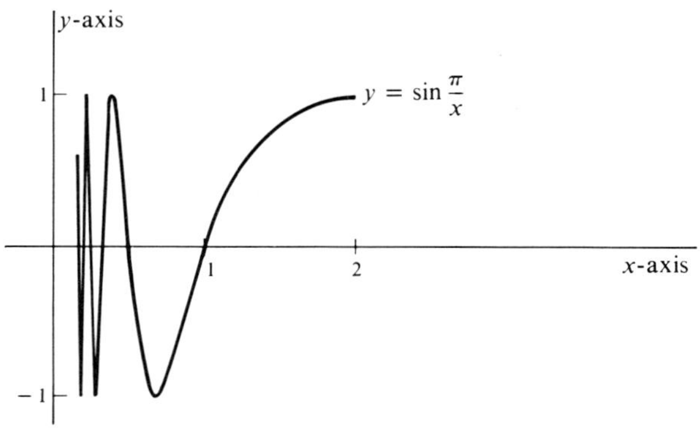

Figure 27

An important extension of Theorem (6.1) is the following:

(6.2) *If f is bounded on $[a, b]$ and is continuous at all but a finite number of points in the interval, then f is integrable over $[a, b]$. Furthermore, if $a_1, \ldots, a_n$ are the points of discontinuity and if $a \leq a_1 \leq \cdots \leq a_n \leq b$, then*

$$\int_a^b f = \int_a^{a_1} f + \int_{a_1}^{a_2} f + \cdots + \int_{a_n}^b f.$$

Proof. It is a direct corollary of (6.1) that f is integrable over each subinterval $[a, a_1]$, $[a_1, a_2], \ldots, [a_n, b]$. By repeated applications of Theorem (4.2), page 191, we may then conclude that f is integrable over $[a, b]$ and that

$$\int_a^b f = \int_a^{a_1} f + \cdots + \int_{a_n}^b f.$$

This completes the proof.

Consider the function f, whose graph is shown in Figure 28, and which is defined by

$$f(x) = \begin{cases} 0 & -\infty < x < -1, \\ 2 & -1 \le x \le 2, \\ 3 & 2 < x \le 3, \\ -1 & 3 < x < \infty. \end{cases}$$

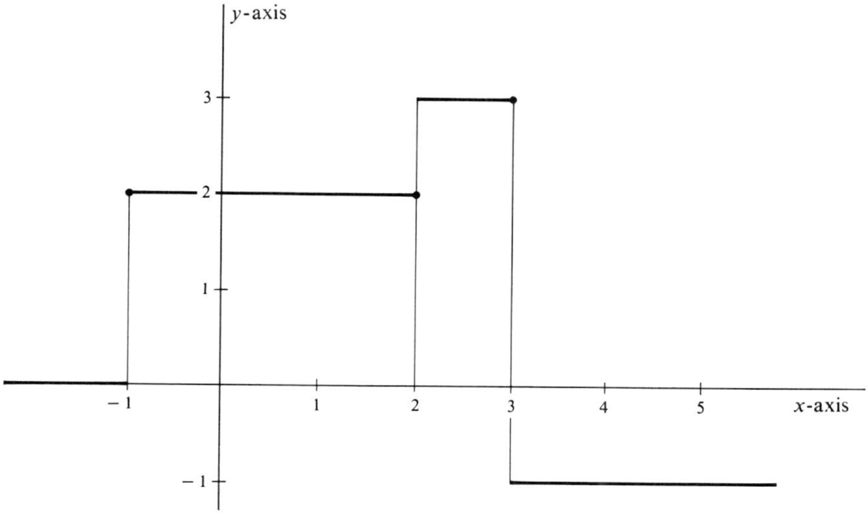

Figure 28

This function, which is constant over certain intervals, is an example of a step function. A function whose domain is the entire set of real numbers is a **step function** if every bounded interval is the union of a finite number of subintervals on each of which the function is a constant. A step function is bounded on any bounded interval and is continuous there at all but possibly a finite number of points. In the present example the only discontinuities occur at -1, 2, and 3. Hence, Theorem (6.2) implies that f is integrable over any interval $[a, b]$. In particular,

$$\int_0^4 f = \int_0^2 f + \int_2^3 f + \int_3^4 f.$$

For each of the three integrals on the right side of the preceding equation, the integrand f is constant on the interval of integration except possibly at the endpoints. If we think of an integral as area or as an average value, we shall almost certainly support the conjecture that the value of an integral is not affected by isolated discontinuities in the integrand. Thus we expect that

$$\begin{aligned} \int_0^4 f &= \int_0^2 f + \int_2^3 f + \int_3^4 f \\ &= 2 \cdot (2 - 0) + 3 \cdot (3 - 2) + (-1)(4 - 3) = 6. \end{aligned}$$

This conjecture is correct (hence, so is the preceding computation), and is implied by the next theorem.

(6.3) *Let $[a, b]$ be a subset of the domains of two functions f and g, and let $f(x) = g(x)$ for all but a finite number of values of x in $[a, b]$. If f is integrable over $[a, b]$, then so is g and $\int_a^b f = \int_a^b g$.*

Proof. It is sufficient to prove this theorem under the assumption that the values of f and g differ at only a single point c in the interval $[a, b]$ (because the result can then be iterated). To be specific, we shall assume that $f(c) < g(c)$. The proof is completed if we can show that there exist upper and lower sums for g which differ from the integral $\int_a^b f$ by an arbitrarily small amount. For this purpose, we choose an arbitrary positive number ϵ. Since f is, by hypothesis, integrable over $[a, b]$, there exists a partition τ of $[a, b]$ such that the corresponding lower sum for f, which we denote by $L_\tau(f)$, satisfies

$$\int_a^b f - L_\tau(f) < \epsilon.$$

However, every lower sum for f is also a lower sum for g. Hence we may substitute $L_\tau(g)$ for $L_\tau(f)$ in the preceding inequality and obtain

$$\int_a^b f - L_\tau(g) < \epsilon. \tag{4}$$

We next derive a similar inequality involving an upper sum for g. The integrability of f also implies the existence of a partition σ' of $[a, b]$ such that the corresponding upper sum for f satisfies

$$U_{\sigma'}(f) - \int_a^b f < \frac{\epsilon}{2}.$$

By possibly adjoining to σ' a point on either side of c, we can assure ourselves of getting a partition $\sigma = \{x_0, \ldots, x_n\}$ of $[a, b]$ with the property that if c lies in the ith subinterval $[x_{i-1}, x_i]$, then

$$x_i - x_{i-1} < \frac{\epsilon}{2[g(c) - f(c)]}. \tag{5}$$

We have already shown [see the proof of (1.1), page 168] that if one partition σ is a refinement of another σ' (i.e., if σ' is a subset of σ), then $U_\sigma \leq U_{\sigma'}$. Thus $U_\sigma(f)$ is, if anything, a better approximation to $\int_a^b f$ than $U_{\sigma'}(f)$. Hence

$$U_\sigma(f) - \int_a^b f < \frac{\epsilon}{2}. \tag{6}$$

Let M_i and N_i be the least upper bounds of the values of f and g, respectively, on $[x_{i-1}, x_i]$. Since $f(x) = g(x)$ except at c, it follows that

$$U_\sigma(g) - U_\sigma(f) = (N_i - M_i)(x_i - x_{i-1}).$$

But the difference $N_i - M_i$ can be no more than $g(c) - f(c)$. Hence

$$U_\sigma(g) - U_\sigma(f) \leq [g(c) - f(c)](x_i - x_{i-1}),$$

and this inequality combined with (5) yields

$$U_\sigma(g) - U_\sigma(f) < \frac{\epsilon}{2}\cdot \tag{7}$$

Finally, adding the inequalities (6) and (7), we obtain

$$U_\sigma(g) - \int_a^b f < \epsilon.$$

This is the analogue of (4) and completes the proof.

EXAMPLE 2. Let f be the function defined by

$$f(x) = \begin{cases} x^3 & -\infty < x \leq 0, \\ 2 - x^2 & 0 < x \leq 2, \\ 2x - 5 & 2 < x < \infty. \end{cases}$$

The graph of f is drawn in Figure 29. The function is clearly continuous except at 0 and at 2, and is bounded on any bounded interval. It follows by Theorem (6.2) that f is integrable over the interval $[-1, 3]$ and that

$$\int_{-1}^3 f = \int_{-1}^0 f + \int_0^2 f + \int_2^3 f.$$

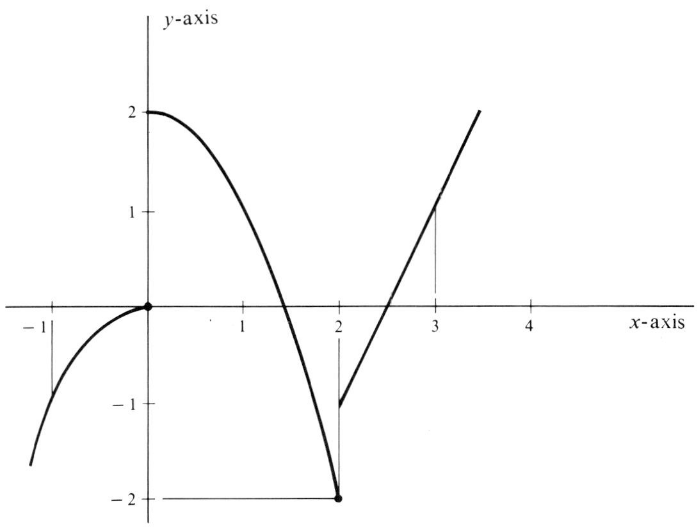

Figure 29

For every x in $[-1, 0]$, we have $f(x) = x^3$, and so

$$\int_{-1}^{0} f = \int_{-1}^{0} x^3 dx = \left.\frac{x^4}{4}\right|_{-1}^{0} = -\tfrac{1}{4}.$$

For every x in $[0, 2]$, we have $f(x) = 2 - x^2$ except that $f(0) = 0$. Hence, by Theorem (6.3),

$$\int_{0}^{2} f = \int_{0}^{2} (2 - x^2)\, dx = \left.\left(2x - \frac{x^3}{3}\right)\right|_{0}^{2} = \tfrac{4}{3}.$$

Similarly, $f(x) = 2x - 5$ for every x in $[2, 3]$ except that $f(2) = -2$. Again, by Theorem (6.3),

$$\int_{2}^{3} f = \int_{2}^{3} (2x - 5)\, dx = \left.(x^2 - 5x)\right|_{2}^{3} = -6 - (-6) = 0.$$

Hence

$$\int_{-1}^{3} f = -\tfrac{1}{4} + \tfrac{4}{3} + 0 = \tfrac{13}{12}.$$

EXAMPLE 3. Is each of the following integrals defined?

$$\text{(a)} \quad \int_{0}^{1} \sin \frac{1}{x}\, dx,$$

$$\text{(b)} \quad \int_{1}^{2} \frac{\ln x}{1 - x}\, dx,$$

$$\text{(c)} \quad \int_{0}^{\pi/2} \tan x\, dx.$$

This is the same as asking whether or not each function is integrable over its proposed interval of integration. Strictly speaking, the answer is no in every case, because each function fails to be defined at one of the endpoints of the interval. However, Theorem (6.3) shows that this answer is based on a technicality and misses the real point of the question. If a function f is bounded on an open interval (a, b) and if $f(a)$ and $f(b)$ are any real numbers whatever, then f is also bounded on the closed interval $[a, b]$. Let us suppose, therefore, that f is bounded and continuous on the open interval (a, b). We may choose values $f(a)$ and $f(b)$ completely arbitrarily, and the resulting function will be integrable over $[a, b]$ as a result of Theorem (6.1). Furthermore, by Theorem (6.3), the integral $\int_a^b f$ is independent of the choice of $f(a)$ and $f(b)$. Hence, if f is bounded and continuous on (a, b), we shall certainly adopt the point of view that f is integrable over $[a, b]$ and, equivalently, that $\int_a^b f$ is defined.

Following this convention, we see that the function $\sin \frac{1}{x}$ is bounded and continuous on $(0, 1)$, and so $\int_0^1 \sin \frac{1}{x}\, dx$ is defined. Using L'Hôpital's Rule (page 123), one can easily show that

$$\lim_{x \to 1+} \frac{\ln x}{1 - x} = -1.$$

Hence, $\frac{\ln x}{1 - x}$ is bounded and continuous on $(1, 2)$, and so $\int_1^2 \frac{\ln x}{1 - x}\, dx$ exists. On the other hand,

$$\lim_{x \to (\pi/2)-} \tan x = \infty,$$

and we therefore conclude that $\tan x$ is not integrable over the interval $\left[0, \frac{\pi}{2}\right]$.

PROBLEMS

1. Determine whether or not each of the following functions is integrable over the proposed interval (see Example 3). Give a reason for your answer.

(a) $\cos \frac{1}{x}$, over $[0, 1]$

(b) $\frac{x^2 + x - 2}{x - 1}$, over $[0, 1]$

(c) $\frac{x^2 + x - 2}{x - 1}$, over $[0, 2]$

(d) $\frac{x^2 + x + 2}{x - 1}$, over $[0, 2]$.

2. Is each of the following integrals defined? (See Example 3.) Give a reason for your answer.

(a) $\int_0^1 \frac{\sin x}{x}\, dx$

(b) $\int_0^{1/2} \frac{\tan 2x}{x}\, dx$

(c) $\int_0^1 \frac{1}{x}\, dx$

(d) $\int_0^{1/e} \frac{1}{\ln x}\, dx$

(e) $\int_0^e \ln x\, dx$.

3. Draw the graph of f, and evaluate $\int_a^b f(x)\, dx$ in each of the following examples.

(a) $f(x) = \begin{cases} 1 & \text{if } -\infty < x \le 0, \\ 5 & \text{if } 0 < x < 2, \\ 3 & \text{if } 2 \le x < \infty, \end{cases}$

and $[a, b] = [-3, 3]$.

(b) $f(x) = \begin{cases} x^2 & \text{if } -\infty < x < 0, \\ 2 - x^2 & \text{if } 0 \leq x < \infty, \end{cases}$

and $[a, b] = [-2, 2]$.

(c) $f(x) = n$ if $n \leq x < n + 1$ where n is any integer, and $[a, b] = [0, 5]$.

4. Prove that if a function f is bounded on an open interval (a, b) and, if $f(a)$ and $f(b)$ are any two real numbers, then f is also bounded on the closed interval $[a, b]$.
5. Compute

(a) $\lim_{t \to 0+} \int_t^1 \frac{1}{x} \, dx$ (b) $\lim_{t \to 1-} \int_0^t \tan \frac{\pi}{2} x \, dx.$

How does the result give insight into the fact that neither integrand is integrable over the interval $[0, 1]$?

6. A function f is said to be **piecewise continuous** on an interval $[a, b]$ if it is continuous at all but possibly a finite number of points of the interval, and if, for every point c of discontinuity in the interval, there exist numbers k and l such that

$$\lim_{x \to c+} f(x) = k \quad \text{and} \quad \lim_{x \to c-} f(x) = l.$$

Using Theorems (6.2) and (6.3), prove that if f is piecewise continuous on $[a, b]$, then it is integrable over $[a, b]$.

7. Improper Integrals. It is assumed in the definition of integrability (pages 168f) that if a function is integrable over an interval, then it is necessarily bounded on that interval. Hence the function f defined by $f(x) = \frac{1}{x^2}$ is not integrable over $[0, 1]$ because it satisfies neither condition for boundedness: The number 0 is not in the domain of f, and there is no upper bound for f [for values of x near zero, $f(x)$ becomes arbitrarily large]. The fact that $f(0)$ is not defined is not a serious difficulty because, as was proved in Section 6, the values of a function at any finite set of points can be defined arbitrarily without affecting the integrability of the function. Thus we could set

$$f(x) = \begin{cases} \frac{1}{x^2} & \text{if } x \neq 0, \\ 3 & \text{if } x = 0, \end{cases}$$

and thereby satisfy the first condition of boundedness. However, there is no way to make f bounded on $[0, 1]$ by changing a finite number of its values.

In this section we shall show that it is possible to extend the concept of integrability to include many functions which are not bounded on their intervals of integration. In addition, the extensions will allow the possibility of integrating over intervals which are not bounded. These integrals are called improper integrals. Two examples are

$$\int_0^1 \frac{1}{\sqrt{x}}\,dx \qquad \text{unbounded integrand,}$$

$$\int_0^\infty e^{-x}dx \qquad \text{unbounded interval.}$$

Let $(a, b]$ be a half-open interval (containing b but not a), and let f be a function which is integrable over the closed interval $[t, b]$ for every number t in $(a, b]$. The integral $\int_t^b f$ is thus defined if $a < t \leq b$, and our definition will concern the limit

$$\lim_{t \to a+} \int_t^b f. \tag{1}$$

We consider the following three cases:

(i) *The function f is bounded on* $(a, b]$. It is not difficult in this case to prove that f is integrable over the closed interval $[a, b]$ and, in addition, that the limit (1) exists and is equal to $\int_a^b f$. [If a is not in the domain of f, we define $f(a)$ arbitrarily.]

(ii) *The function f is not bounded on* $(a, b]$, *but the limit* (1) *exists.* In this case f is not integrable over $[a, b]$ according to our original definition. Hence we define the **improper integral,** which is still denoted by $\int_a^b f$, to be the limit (1).

(iii) *The limit* (1) *does not exist.* In this case the integral is not defined.

Thus, if the limit exists, we have the equation

$$\boxed{\int_a^b f = \lim_{t \to a+} \int_t^b f.}$$

If f is bounded on $(a, b]$, the integral is called **proper.** If f is not bounded on $(a, b]$, the improper integral exists only if the limit exists. However, the traditional terminology, which we shall adopt, is that the improper integral is **convergent** if the limit exists and **divergent** if it does not.

In spite of the improper integrals defined in this section, we emphasize that whenever we say that f is integrable over $[a, b]$ we mean it in the sense of the original definition of integrability, in which $[a, b]$ is a bounded interval and f is bounded on it.

The situation is analogous if $[a, b)$ is a half-open interval and f is integrable over $[a, t]$ for any t in $[a, b)$. We have

$$\int_a^b f = \lim_{t \to b-} \int_a^t f.$$

If f is bounded on $[a, b)$, then f is integrable over $[a, b]$, and $\int_a^b f$ is a proper integral. If f is not bounded, the integral is improper, and it is convergent or divergent according as the limit does or does not exist.

EXAMPLE 1. Classify each of the following integrals as proper or improper. If improper, determine whether convergent or divergent, and, if convergent, evaluate it.

(a) $\displaystyle\int_0^1 \frac{1}{\sqrt{x}}\,dx,$ (c) $\displaystyle\int_0^2 \frac{1}{2-x}\,dx,$

(b) $\displaystyle\int_0^1 \frac{1}{x^2}\,dx,$ (d) $\displaystyle\int_0^1 \sin\frac{1}{x}\,dx.$

Since $\dfrac{1}{\sqrt{x}}$ takes on arbitrarily large values near 0, we know that $\displaystyle\int_0^1 \frac{1}{\sqrt{x}}\,dx$ is not a proper integral. For every t in $(0, 1]$,

$$\int_t^1 \frac{1}{\sqrt{x}}\,dx = 2\sqrt{x}\Big|_t^1 = 2(1 - \sqrt{t}).$$

Since $\lim_{t \to 0+} 2(1 - \sqrt{t})$ exists, we get

$$\int_0^1 \frac{1}{\sqrt{x}}\,dx = \lim_{t \to 0+} 2(1 - \sqrt{t}) = 2.$$

Hence (a) is a convergent improper integral with value 2.

The values of $\dfrac{1}{x^2}$ also increase without bound as x approaches zero, and (b) is therefore not a proper integral. For every t in $(0, 1]$,

$$\int_t^1 \frac{1}{x^2}\,dx = -\frac{1}{x}\Big|_t^1 = \frac{1}{t} - 1.$$

However,

$$\lim_{t \to 0+} \int_t^1 \frac{1}{x^2}\,dx = \lim_{t \to 0+} \frac{1}{t} - 1 = \infty,$$

and, since the limit does not exist, the improper integral is divergent.

The function $\dfrac{1}{2 - x}$ is not bounded on $[0, 2)$, and so (c) is also an improper integral. For every t such that $0 \le t < 2$, we have

$$\begin{aligned}\int_0^t \frac{1}{2 - x}\,dx &= -\ln|2 - x|\Big|_0^t \\ &= -\ln(2 - t) + \ln 2 = \ln\frac{2}{2 - t}.\end{aligned}$$

Hence

$$\lim_{t \to 2-} \int_0^t \frac{1}{2 - x}\,dx = \lim_{t \to 2-} \ln\frac{2}{2 - t} = \infty,$$

and we conclude that $\displaystyle\int_0^2 \frac{1}{2 - x}\,dx$ is a divergent improper integral.

Since $\left|\sin\dfrac{1}{x}\right| \le 1$ for all nonzero x, the function f defined by $f(x) = \sin\dfrac{1}{x}$ is bounded on $(0, 1]$. It is also continuous at every point of that interval. We now assign a value, say 0, to $f(0)$, and it follows by Theorems (6.1) and (6.3) that f is integrable over $[0, 1]$, and the value of $\int_0^1 f(x)\,dx$ is independent of the choice of $f(0)$. As in Section 6, we therefore consider $\displaystyle\int_0^1 \sin\frac{1}{x}\,dx$ to be a proper integral.

We next define improper integrals over unbounded intervals. Let a be a given real number and f a function which, for every $t \ge a$, is integrable over $[a, t]$. If the limit $\lim_{t \to \infty} \int_a^t f$ exists, we define it to be the value of the **convergent improper integral** $\int_a^\infty f$. Thus

$$\boxed{\int_a^\infty f = \lim_{t \to \infty} \int_a^t f.}$$

If the limit does not exist, the integral of f over $[a, \infty)$ also does not exist. Although it is not defined, we follow tradition and say that the improper integral is **divergent.**

As before, the analogous definition is given for the unbounded interval $(-\infty, a]$. We have

$$\boxed{\int_{-\infty}^{a} f = \lim_{t \to -\infty} \int_{t}^{a} f,}$$

and the improper integral $\int_{-\infty}^{a} f$ is convergent if the limit exists, and divergent if it does not.

EXAMPLE 2. Test the following improper integrals for convergence or divergence, and evaluate the convergent ones.

(a) $\displaystyle\int_0^{\infty} e^{-x} dx,$ (c) $\displaystyle\int_2^{\infty} \frac{1}{x^2}\, dx,$

(b) $\displaystyle\int_1^{\infty} \frac{1}{\sqrt{x}}\, dx,$ (d) $\displaystyle\int_{-\infty}^{t} \frac{1}{1 + x^2}\, dx.$

For (a) we have

$$\int_0^t e^{-x} dx = -e^{-x}\Big|_0^t = 1 - e^{-t}.$$

Hence

$$\int_0^{\infty} e^{-x} dx = \lim_{t \to \infty} (1 - e^{-t}) = 1 - 0 = 1,$$

and so the integral is convergent and equal to 1.

Similarly, for (b),

$$\int_1^t \frac{1}{\sqrt{x}}\, dx = 2\sqrt{x}\Big|_1^t = 2\sqrt{t} - 2.$$

However, since $\lim_{t \to \infty} (2\sqrt{t} - 2) = \infty$, we conclude that $\displaystyle\int_1^{\infty} \frac{1}{\sqrt{x}}\, dx$ is a divergent integral.

For (c) we obtain

$$\int_2^t \frac{1}{x^2}\, dx = -\frac{1}{x}\Big|_2^t = \frac{1}{2} - \frac{1}{t}.$$

From this it follows that $\int_2^\infty \frac{1}{x^2}\,dx$ is convergent and equal to $\frac{1}{2}$, since

$$\int_2^\infty \frac{1}{x^2}\,dx = \lim_{t\to\infty}\left(\frac{1}{2} - \frac{1}{t}\right) = \frac{1}{2}.$$

If the integral in (d) exists, its value depends on t. Hence in testing for convergence, we use another variable.

$$\int_s^t \frac{1}{1+x^2}\,dx = \arctan x\Big|_s^t = \arctan t - \arctan s.$$

Since $\lim_{s\to-\infty} \arctan s = -\frac{\pi}{2}$, we conclude that

$$\int_{-\infty}^t \frac{1}{1+x^2}\,dx = \lim_{s\to-\infty} (\arctan t - \arctan s)$$
$$= \arctan t + \frac{\pi}{2},$$

and the integral is convergent for all real values of t.

We next enlarge the class of improper integrals to include integrands which are unbounded near both endpoints of the interval of integration. Let (a, c) be an open interval (we include the possibility that $a = -\infty$, or $c = \infty$, or both), and let f be a function which is integrable over every closed subinterval $[s, t]$ of (a, c). Choose an arbitrary point b in (a, c), and consider the two integrals $\int_a^b f$ and $\int_b^c f$. If either of these is proper, their sum is equal to $\int_a^c f$, and we need no new definition. If both $\int_a^b f$ and $\int_b^c f$ are improper integrals, then we define the improper integral $\int_a^c f$ to be their sum. Furthermore, $\int_a^c f$ is defined to be convergent if and only if *both* $\int_a^b f$ and $\int_b^c f$ are convergent; otherwise, $\int_a^c f$ is defined to be divergent. Thus, in all cases, we have the equation

$$\boxed{\int_a^c f = \int_a^b f + \int_b^c f.} \tag{2}$$

For the definition to be a valid one, it is necessary to know that $\int_a^c f$, as defined in (2), is independent of the choice of b. Hence, we need the following simple theorem, whose proof is left as an exercise.

(7.1) *If f is integrable over every closed subinterval $[s, t]$ of (a, c), and if b_1 and b_2 belong to (a, c), then*

$$\int_a^{b_1} f + \int_{b_1}^c f = \int_a^{b_2} f + \int_{b_2}^c f.$$

EXAMPLE 3. Classify each of the following improper integrals as convergent or divergent. Evaluate, if convergent.

(a) $\displaystyle\int_0^\infty \frac{1}{x^2}\,dx$, (b) $\displaystyle\int_{-\infty}^\infty \frac{1}{1+x^2}\,dx$.

For (a) we write

$$\int_0^\infty \frac{1}{x^2}\,dx = \int_0^1 \frac{1}{x^2}\,dx + \int_1^\infty \frac{1}{x^2}\,dx.$$

We have already shown in Example 1(b) that $\displaystyle\int_0^1 \frac{1}{x^2}\,dx$ is divergent, and it follows that $\displaystyle\int_0^\infty \frac{1}{x^2}\,dx$ is divergent. For (b) we have

$$\int_{-\infty}^\infty \frac{1}{1+x^2}\,dx = \int_{-\infty}^0 \frac{1}{1+x^2}\,dx + \int_0^\infty \frac{1}{1+x^2}\,dx,$$

and

$$\begin{aligned}\int_{-\infty}^0 \frac{1}{1+x^2}\,dx &= \lim_{t\to-\infty} \arctan x\Big|_t^0 \\ &= \lim_{t\to-\infty} (0 - \arctan t) = -\left(-\frac{\pi}{2}\right) = \frac{\pi}{2}\cdot\end{aligned}$$

Similarly,

$$\int_0^\infty \frac{1}{1+x^2}\,dx = \lim_{t\to\infty} \arctan t = \frac{\pi}{2}\cdot$$

Hence $\displaystyle\int_{-\infty}^\infty \frac{1}{1+x^2}\,dx$ is a convergent integral equal to $\dfrac{\pi}{2} + \dfrac{\pi}{2} = \pi$.

As a final extension of the class of improper integrals, we include the possibility that the integrand may be unbounded near a finite number of points in the interior of the interval of integration. Let (a, b) be an open interval (including possibly $a = -\infty$, or $b = \infty$, or both), let $a_1, \ldots, a_n$

be points of (a, b) such that $a_1 < \cdots < a_n$, and let f be a function which is integrable over every closed bounded subinterval of (a, b) which contains none of the points $a_1, \ldots, a_n$. Then the equation

$$\boxed{\int_a^b f = \int_a^{a_1} f + \cdots + \int_{a_n}^b f} \tag{3}$$

is either a consequence of the theory so far developed, or is taken as the definition of the improper integral $\int_a^b f$. As before, $\int_a^b f$ is divergent if any one of the integrals on the right is divergent, and is otherwise either convergent or proper.

EXAMPLE 4. Classify each of the following integrals, and evaluate any which are not divergent.

$$\text{(a)} \quad \int_{-1}^{1} \frac{1}{x^{1/3}}\, dx,$$

$$\text{(b)} \quad \int_{-1}^{1} \frac{1}{x}\, dx,$$

$$\text{(c)} \quad \int_{0}^{3} \frac{1}{(x-1)(x-3)}\, dx.$$

Since each integrand has arbitrarily large values near one or more points of the interval of integration, we conclude that none of the integrals is proper.

For (a) we first observe that

$$\int \frac{1}{x^{1/3}}\, dx = \tfrac{3}{2}x^{2/3} + c,$$

from which we obtain

$$\begin{aligned} \int_{-1}^{0} \frac{1}{x^{1/3}}\, dx &= \lim_{t \to 0-} \int_{-1}^{t} \frac{1}{x^{1/3}}\, dx \\ &= \lim_{t \to 0-} \tfrac{3}{2}t^{2/3} - \tfrac{3}{2} = -\tfrac{3}{2}, \end{aligned}$$

and, in the same way,

$$\begin{aligned} \int_{0}^{1} \frac{1}{x^{1/3}}\, dx &= \lim_{t \to 0+} \int_{t}^{1} \frac{1}{x^{1/3}}\, dx \\ &= \tfrac{3}{2} - \lim_{t \to 0+} \tfrac{3}{2}t^{2/3} = \tfrac{3}{2}. \end{aligned}$$

From the definition in (3) it follows that

$$\int_{-1}^{1} \frac{1}{x^{1/3}}\,dx = \int_{-1}^{0} \frac{1}{x^{1/3}}\,dx + \int_{0}^{1} \frac{1}{x^{1/3}}\,dx$$
$$= -\tfrac{3}{2} + \tfrac{3}{2} = 0.$$

Hence (a) is a convergent improper integral equal to 0.

If (b) is convergent, it follows from the definition that

$$\int_{-1}^{1} \frac{1}{x}\,dx = \int_{-1}^{0} \frac{1}{x}\,dx + \int_{0}^{1} \frac{1}{x}\,dx,$$

and that both integrals on the right are convergent. However,

$$\int_{0}^{1} \frac{1}{x}\,dx = \lim_{t \to 0+} \int_{t}^{1} \frac{1}{x}\,dx$$
$$= \lim_{t \to 0+} (\ln 1 - \ln t) = \infty,$$

and $\int_{-1}^{0} \frac{1}{x}\,dx$ is similarly divergent. We conclude that $\int_{-1}^{1} \frac{1}{x}\,dx$ is divergent. (*Warning:* Failure to note the discontinuity of the function $\frac{1}{x}$ at 0 can result in the following incorrect computation:

$$\int_{-1}^{1} \frac{1}{x}\,dx = \ln|x|\,\Big|_{-1}^{1} = 0 - 0 = 0.)$$

If the integral (c) is convergent, then it is given by

$$\int_{0}^{3} \frac{1}{(x-1)(x-3)}\,dx = \int_{0}^{1} \frac{1}{(x-1)(x-3)}\,dx + \int_{1}^{3} \frac{1}{(x-1)(x-3)}\,dx,$$

and both integrals on the right are convergent. However, it is easy to show that neither is convergent. A partial-fractions decomposition yields

$$\frac{1}{(x-1)(x-3)} = -\frac{1}{2}\,\frac{1}{x-1} + \frac{1}{2}\,\frac{1}{x-3},$$

and so

$$\int \frac{1}{(x-1)(x-3)}\,dx = -\tfrac{1}{2}\ln|x-1| + \tfrac{1}{2}\ln|x-3| + c$$
$$= \tfrac{1}{2}\ln\left|\frac{x-3}{x-1}\right| + c.$$

In particular, therefore,

$$\int_0^1 \frac{1}{(x-1)(x-3)}\,dx = \lim_{t\to 1-}\int_0^t \frac{1}{(x-1)(x-3)}\,dx$$
$$= \left(\lim_{t\to 1-} \tfrac{1}{2}\ln\left|\frac{t-3}{t-1}\right|\right) - \tfrac{1}{2}\ln 3 = \infty,$$

which is sufficient to establish that (c) is divergent.

We conclude this section with a theorem which gives a convenient test for the convergence or divergence of improper integrals. Called the Comparison Test for Integrals, it can frequently be used to classify an improper integral whose integrand has no simple antiderivative, such as $\int_0^\infty e^{-x^2}\,dx$.

(7.2) COMPARISON TEST FOR INTEGRALS. *Let f and g be integrable over every bounded closed subinterval of a not necessarily bounded interval (a, b). If $|f(x)| \le g(x)$ for every x in (a, b) and if $\int_a^b g$ is either convergent or proper, then $\int_a^b f$ is also either convergent or proper.*

Since an open interval can be split into two pieces, this theorem also holds for half-open intervals. For simplicity, we shall prove it for the interval $(a, b]$.

Proof. We first prove that $\int_a^b |f|$ is either convergent or proper. We shall assume without proof the theorem which states that if a function f is integrable over an interval, then so is $|f|$. [In most applications of (7.2) the function f is continuous at every point of $(a, b]$. In this case, $|f|$ is also continuous and the problem does not arise.] Since

$$0 \le |f(x)| \le g(x), \tag{4}$$

for every x in $(a, b]$, it follows that

$$\int_t^b |f| \le \int_t^b g,$$

for every t in $(a, b]$. Since g has nonnegative values, $\int_t^b g$ increases as t approaches a from the right. Hence

$$\int_t^b g \le \lim_{t\to a+}\int_t^b g = \int_a^b g,$$

and therefore

$$\int_t^b |f| \le \int_a^b g,$$

for every t in $(a, b]$. But $\int_t^b |f|$ also increases as t approaches a from the right, and the preceding inequality shows that it is bounded from above by the number $\int_a^b g$. An increasing bounded function must approach a limit. Hence $\lim_{t \to a+} \int_t^b |f|$ exists, and therefore $\int_a^b |f|$ is either convergent or proper.

Since $-f(x) \leq |f(x)|$, it follows that

$$0 \leq f(x) + |f(x)| \leq 2g(x), \tag{5}$$

for every x in $(a, b]$. In this part of the proof, the inequalities (5) are the analogues of (4). In exactly the same way as in the preceding paragraph they imply that

$$\int_t^b (f + |f|) \leq 2\int_a^b g$$

and thence that the integral $\int_a^b (f + |f|)$ is either proper or convergent. Finally, therefore, we have

$$\lim_{t \to a+} \int_t^b f = \lim_{t \to a+} \int_t^b (f + |f|) - \lim_{t \to a+} \int_t^b |f|.$$

Since both limits on the right exist, so does the one on the left. We conclude that $\int_a^b f$ is either proper or convergent, and the proof is complete.

EXAMPLE 5. Prove that $\int_0^\infty e^{-x^2}dx$ is convergent. Since $x^2 \geq x$ whenever $x \geq 1$, it follows that $e^{-x^2} \leq e^{-x}$ for $x \geq 1$. An exponential is never negative, so $e^{-x^2} = |e^{-x^2}|$, and therefore

$$|e^{-x^2}| \leq e^{-x}, \qquad \text{for } x \geq 1.$$

The convergence of $\int_0^\infty e^{-x}dx$, shown in Example 2, implies the convergence of $\int_1^\infty e^{-x}dx$. It follows by the comparison test, i.e., by Theorem (7.2), that $\int_1^\infty e^{-x^2}dx$ is a convergent integral. This, in turn, implies the convergence of $\int_0^\infty e^{-x^2}dx$.

PROBLEMS

1. Classify each of the following integrals as proper, improper and convergent, or improper and divergent. Evaluate any which are convergent if an indefinite integral can be found.

(a) $\displaystyle\int_0^2 \frac{3}{x^{2/3}}\,dx$

(b) $\displaystyle\int_1^2 \frac{1}{\sqrt{2 - x}}\,dx$

(c) $\displaystyle\int_2^3 \frac{1}{(x - 2)^2}\,dx$

(d) $\displaystyle\int_0^1 \frac{1}{x^2 + x + 1}\,dx$

(e) $\displaystyle\int_0^{\pi/2} \tan x\,dx$

(f) $\displaystyle\int_0^1 \frac{\sin x}{x}\,dx.$

2. Classify each of the following integrals and evaluate any which are not divergent.

(a) $\int_2^\infty \frac{1}{x^3}\,dx$

(b) $\int_0^2 \frac{1}{x^3}\,dx$

(c) $\int_{-1}^\infty (x^2 - x + 1)\,dx$

(d) $\int_0^\infty xe^{-x^2}\,dx$

(e) $\int_1^\infty \sin x\,dx$

(f) $\int_{-\infty}^1 e^x\,dx$

(g) $\int_0^\infty \frac{1}{(x+2)^2}\,dx$

(h) $\int_0^1 x \ln x\,dx.$

3. Show that the integral $\int_0^1 \frac{1}{x^s}\,dx$ is

(a) proper if $-\infty < s \leq 0$.

(b) improper and convergent if $0 < s < 1$.

(c) improper and divergent if $1 \leq s < \infty$.

4. For what values of s is the integral $\int_1^\infty \frac{dx}{x^s}$ convergent, and for what values is it divergent? Give reasons for your answers.

5. Classify each of the following integrals, and evaluate any which are not divergent if an indefinite integral can be found.

(a) $\int_{-1}^1 \frac{1}{x^{2/3}}\,dx$

(b) $\int_0^3 \frac{1}{(x-1)^{1/3}}\,dx$

(c) $\int_0^1 \frac{\tan x}{x}\,dx$

(d) $\int_0^\infty \frac{1}{x^s}\,dx$

(e) $\int_{-\infty}^0 \frac{1}{(x-2)^2}$

(f) $\int_{-\infty}^\infty e^{-|x|}\,dx$

(g) $\int_2^\infty \frac{1}{\sqrt{x-2}}\,dx$

(h) $\int_0^2 \frac{1}{(x+1)(x-1)}\,dx$

(i) $\int_1^\infty \frac{\ln x}{x}\,dx$

(j) $\int_1^\infty \frac{\ln x}{x^2}\,dx.$

6. Prove Theorem (7.1).

7. Using the Comparison Test for Integrals if necessary, classify each of the following integrals.

(a) $\displaystyle\int_{-\infty}^{0} e^{-x^2}dx$

(b) $\displaystyle\int_{1}^{\infty} \frac{1}{x^2}\sin x\,dx$

(c) $\displaystyle\int_{0}^{\infty} e^{-x}\sin x\,dx$

(d) $\displaystyle\int_{3}^{\infty} \frac{1}{\sqrt{(x-1)(x-2)}}\,dx$

(e) $\displaystyle\int_{0}^{1} x\sin\frac{1}{x}\,dx$

(f) $\displaystyle\int_{0}^{1} \frac{1}{\sqrt{(x-1)(x-2)}}\,dx.$

8. If $F(t) = \int_{-\infty}^{t} e^{-x^2}dx$, find $F'(0)$ and $F'(1)$.

9. (a) Show that the area of the region P bounded by the x-axis, the line $x = 1$, and the curve $y = \dfrac{1}{x}$ is infinite.

(b) Show that the volume of the solid of revolution obtained by rotating the region P about the x-axis is finite.

10. Prove that if f is bounded on $(a, b]$ and integrable over $[t, b]$ for every t in $(a, b]$, then f is integrable over $[a, b]$ and $\lim_{t\to a+} \int_t^b f = \int_a^b f$. [*Hint:* The argument is essentially the same as that in the proof of Theorem (6.1).]

CHAPTER 9

Infinite Series

Addition of real numbers is basically a binary operation: Given any *two* real numbers a and b, there is defined a real number denoted by $a + b$ and called their sum. The sum of numbers $a_1, \ldots, a_n$, where $n \geq 3$, is then defined by repeated applications of the binary operation. For example, one way of grouping the terms is given by

$$(\cdots(((a_1 + a_2) + a_3) + a_4) + \cdots + a_{n-1}) + a_n.$$

The Associative Law of Addition implies that the sums obtained by all the different possible groupings are the same; so we can discard the parentheses and write

$$\sum_{i=1}^{n} a_i = a_1 + \cdots + a_n.$$

Thus addition of any finite number of terms is defined. However, without further definitions, the sum of an infinite number of terms makes no sense at all. In this chapter we shall make the necessary definitions and develop the theory of infinite series. Two examples are

$$\sum_{i=0}^{\infty} \frac{1}{2^i} = 1 + \tfrac{1}{2} + \tfrac{1}{4} + \tfrac{1}{8} + \tfrac{1}{16} + \cdots,$$

$$\sum_{i=1}^{\infty} (-1)^i(2i + 1) = -3 + 5 - 7 + 9 - 11 + \cdots.$$

Later in the chapter we shall consider infinite series in which each term contains the power of an independent variable. An example is the series

$$1 + x + \frac{x^2}{2} + \frac{x^3}{3} + \frac{x^4}{4} + \cdots,$$

which, for every real number x, is an infinite series of real numbers. We shall see that many functions can be defined by these power series, and this fact is of fundamental importance in mathematics and its applications.

1. Sequences and Their Limits. Infinite series are defined in terms of limits of infinite sequences, and make sense only in these terms. We therefore begin by reviewing the ideas of sequences which were introduced in Section 2 of Chapter 4. Following that, we shall develop some additional facts about the limits of infinite sequences. The definition of infinite series, i.e., of the sum of an infinite number of terms, will be given in Section 2.

An **infinite sequence** is a function whose domain consists of all integers greater than or equal to some integer m. In the normal terminology of functions the value of a sequence s at an integer i in its domain would be denoted by $s(i)$. However, it is customary with sequences to denote this value by s_i. Thus

$$s_i = s(i), \qquad \text{for every integer } i \geq m.$$

The sequence s itself is frequently denoted by $\{s_i\}$ or by an indicated enumeration of its values: $s_m, s_{m+1}, s_{m+2}, \ldots$. In the majority of examples m is either 0 or 1, and the first term of the sequence is then s_0 or s_1, respectively.

An infinite sequence s of real numbers is said to **converge** to a real number L, or, alternatively, the number L is called the **limit** of the sequence s, written

$$\lim_{n \to \infty} s_n = L,$$

if the difference $s_n - L$ is arbitrarily small in absolute value for every sufficiently large integer n. The formal definition is therefore: $\lim_{n \to \infty} s_n = L$ if, for every positive real number ϵ, there exists an integer N such that $|s_n - L| < \epsilon$ for every integer $n > N$.

Geometrically, a sequence s of real numbers is an indexed set of points on the real line. If the sequence converges to L, then the points s_n of the sequence cluster ever more closely about L as n increases (see Figure 1). That is, s_n lies arbitrarily close to L if n is sufficiently large.

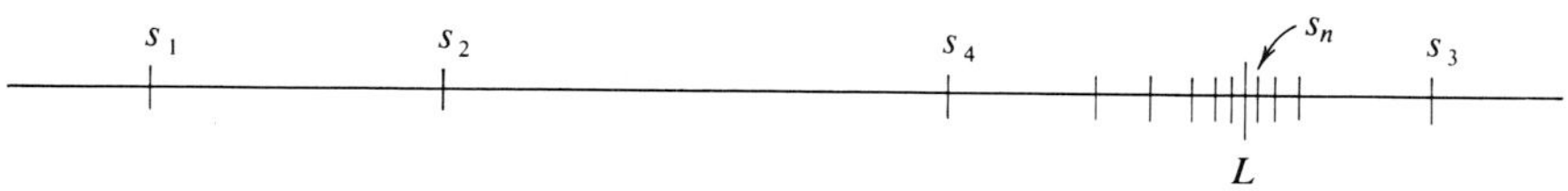

Figure 1

If the numbers s_n become arbitrarily large as n increases, then the sequence does not converge and no limit exists. In the special case that, for every real number B, the values s_n are all greater than B for sufficiently large n, we shall write

$$\lim_{n \to \infty} s_n = \infty.$$

The complete definition is: $\lim_{n \to \infty} s_n = \infty$ if, for every real number B, there exists an integer N such that $s_n > B$ for every integer $n > N$. A simple example of a sequence which "converges to infinity" in this way is the sequence of positive integers 1, 2, 3, 4, 5, ... By reversing the single inequality $s_n > B$ in the above definition, we obtain the analogous definition of

$$\lim_{n\to\infty} s_n = -\infty.$$

It should not be supposed that if an infinite sequence s of real numbers fails to converge, then it follows that $\lim_{n\to\infty} s_n = \pm\infty$. For example, the oscillating sequence

$$1, -1, 1, -1, 1, -1, \ldots$$

is bounded and does not converge. Another example is the sequence

$$0, 1, 0, 2, 0, 3, 0, 4, \ldots,$$

defined, for every integer $n \geq 1$, by

$$s_n = \begin{cases} \dfrac{n}{2} & \text{if } n \text{ is even,} \\ 0 & \text{if } n \text{ is odd.} \end{cases}$$

This sequence is not bounded and does not converge, since, as n increases, there exist arbitrarily large values s_n. However, because of the regular recurrence of the value 0, it also does not satisfy $\lim_{n\to\infty} s_n = \infty$.

The basic algebraic properties of limits of real-valued functions of a real variable, which are summarized in Theorem (4.1), page 32, also hold for infinite sequences of real numbers. We have

(1.1) THEOREM. *If sequences $\{s_n\}$ and $\{t_n\}$ converge and if c is a real number, then*

(i) $\lim_{n\to\infty} (s_n + t_n) = \lim_{n\to\infty} s_n + \lim_{n\to\infty} t_n.$

(ii) $\lim_{n\to\infty} (cs_n) = c \lim_{n\to\infty} s_n.$

(iii) $\lim_{n\to\infty} (s_n t_n) = (\lim_{n\to\infty} s_n)(\lim_{n\to\infty} t_n).$

(iv) $\lim_{n\to\infty} \dfrac{s_n}{t_n} = \dfrac{\lim_{n\to\infty} s_n}{\lim_{n\to\infty} t_n}$, *provided* $\lim_{n\to\infty} t_n \neq 0.$

We give the proof of (i). Let $L_1 = \lim_{n\to\infty} s_n$ and $L_2 = \lim_{n\to\infty} t_n$, and choose an arbitrary number ϵ. To prove (i) we use the fact that there exist integers N_1 and N_2 such that

$$|s_i - L_1| < \frac{\epsilon}{2} \qquad \text{and} \qquad |t_j - L_2| < \frac{\epsilon}{2},$$

for every integer $i > N_1$ and $j > N_2$. If we set N equal to the larger of N_1 and N_2, then, for every integer $n > N$, we have

$$|s_n - L_1| < \frac{\epsilon}{2} \quad \text{and} \quad |t_n - L_2| < \frac{\epsilon}{2} \cdot$$

Since $(s_n + t_n) - (L_1 + L_2) = (s_n - L_1) + (t_n - L_2)$ and since $|a + b| \leq |a| + |b|$ for any two numbers a and b, it follows that

$$\begin{aligned}|(s_n + t_n) - (L_1 + L_2)| = |(s_n - L_1) + (t_n - L_2)| &\leq |s_n - L_1| + |t_n - L_2| \\ &< \frac{\epsilon}{2} + \frac{\epsilon}{2} = \epsilon,\end{aligned}$$

for $n > N$. This completes the proof of (i). The proofs of the other parts of the theorem are similar, and the methods are exactly the same as those used in Appendix A to prove (4.1), page 32.

Similar to (1.1) is the following result, whose proof we omit.

(1.2) *If* $\lim_{n\to\infty} s_n$ *is the real number L and if* $\lim_{n\to\infty} t_n = \pm\infty$, *then*

(i) $\lim_{n\to\infty} \frac{s_n}{t_n} = 0.$

(ii) $\lim_{n\to\infty} \frac{t_n}{s_n} = \begin{cases} \pm\infty & \text{if } L > 0, \\ \mp\infty & \text{if } L < 0. \end{cases}$

Actually we have already used (1.1) and (1.2) in Chapter 4 in evaluating definite integrals as the limits of upper and lower sums. The following example is included primarily as a review.

EXAMPLE 1. Determine whether or not each of the following sequences converges, and evaluate the limit if it does.

(a) $\{a_n\}$ defined by $a_n = \dfrac{2n^2 + 5n + 2}{3n^2 - 7}$,

(b) $\{b_i\}$ defined by $b_i = \dfrac{2^{i+1}i}{(i + 1)2^i 3}$,

(c) $\{c_k\}$ defined by $c_k = \dfrac{k + 1}{k^2 + 1}$,

(d) $\{d_k\}$ defined by $d_k = (-1)^k \dfrac{k^2 + 1}{k + 1}$.

Note that the definition of each of the above sequences is incomplete because we have neglected to specify the domain. However, the omission does not matter, since we are concerned only with the question of the limit of each sequence. It follows immediately from the definition of convergence that the limit of an infinite sequence is unaffected by dropping or adding a finite number of terms at the beginning.

For (a), after dividing numerator and denominator by n^2, we get

$$\frac{2n^2 + 5n + 2}{3n^2 - 7} = \frac{2 + \frac{5}{n} + \frac{2}{n^2}}{3 - \frac{7}{n^2}}.$$

Using (1.1) and (1.2), we conclude that

$$\lim_{n\to\infty} a_n = \lim_{n\to\infty} \frac{2 + \frac{5}{n} + \frac{2}{n^2}}{3 - \frac{7}{n^2}}$$

$$= \frac{2 + 0 + 0}{3 - 0} = \frac{2}{3}.$$

So the sequence $\{a_n\}$ converges to $\frac{2}{3}$.

The ith term of the sequence $\{b_i\}$ can be written

$$b_i = \frac{2^{i+1}i}{(i+1)2^i 3} = \frac{2}{3\left(1 + \frac{1}{i}\right)}.$$

Since $\lim_{i\to\infty} \left(1 + \frac{1}{i}\right) = 1 + 0 = 1$, we have

$$\lim_{i\to\infty} b_i = \frac{2}{3 \lim_{i\to\infty} \left(1 + \frac{1}{i}\right)} = \frac{2}{3}.$$

Hence the sequence $\{b_i\}$ also converges to the limit $\frac{2}{3}$.

For large values of k, the number $k + 1$ is approximately equal to k, and the number $k^2 + 1$ is approximately equal to k^2. Thus the behavior of the ratio $\frac{k+1}{k^2+1}$, as k increases, is the same as that of $\frac{k}{k^2} = \frac{1}{k}$, which approaches zero. We conclude that the sequence $\{c_k\}$ converges to zero. A more systematic analysis is obtained by writing

$$\frac{k+1}{k^2+1} = \frac{k\left(1 + \frac{1}{k}\right)}{k^2\left(1 + \frac{1}{k^2}\right)} = \frac{1}{k} \frac{\left(1 + \frac{1}{k}\right)}{\left(1 + \frac{1}{k^2}\right)},$$

from which it follows by (1.1) and (1.2) that

$$\lim_{k\to\infty} c_k = \lim_{k\to\infty} \frac{k+1}{k^2+1} = 0\frac{(1+0)}{(1+0)} = 0.$$

The sequence obtained by taking the absolute value of each term in (d) is one which increases without bound. That is, it is clear that $|d_k| = \dfrac{k^2+1}{k+1}$ and that

$$\lim_{k\to\infty} |d_k| = \lim_{k\to\infty} \frac{k^2+1}{k+1} = \infty.$$

However, the factor $(-1)^k$ implies that the terms of the sequence $\{d_k\}$ alternate in sign, and for this sequence we can conclude only that no limit exists.

A sequence s of real numbers is said to be an **increasing sequence** if

$$s_{i+1} \geq s_i, \tag{1}$$

for every integer i in the domain of s. If the inequality (1) is reversed so that $s_{i+1} \leq s_i$, for every i in the domain of s, then we say that s is a **decreasing sequence.** A sequence is **monotonic** if it is either increasing or decreasing. Note that, just as in the analogous definitions for functions, we use "increasing" and "decreasing" in the weak sense. That is, an increasing sequence is one which is strictly speaking nondecreasing, and a decreasing sequence is one which is literally nonincreasing.

The following two theorems will form the basis of some fundamental conclusions about infinite series. Both are statements about increasing sequences, and corresponding to each there is an obvious analogous theorem about decreasing sequences.

(1.3) *Let s be an infinite sequence of real numbers. If s is increasing and if $\lim_{n\to\infty} s_n = L$, then $s_n \leq L$ for every n in the domain of s.*

Proof. Suppose that the conclusion is false. Then there exists an integer N such that $s_N > L$. Let a be the positive number $s_N - L$. Since s is an increasing sequence, we know that $s_n \geq s_N$ for all $n \geq N$. It follows that

$$s_n - L \geq s_N - L = a,$$

for all $n \geq N$. Since the difference $s_n - L$ is greater than or equal to the positive constant a, it cannot be arbitrarily small. Hence the sequence cannot approach L as a limit, contradicting the premise that $\lim_{n\to\infty} s_n = L$. This completes the proof.

A sequence s of real numbers is said to be **bounded above** by a real number B if $s_n \leq B$ for every n in the domain of s. If the inequality is reversed to read $B \leq s_n$, we obtain the analogous definition of a sequence s which is **bounded below** by B. The second theorem is:

(1.4) *Let s be an infinite sequence of real numbers. If s is increasing and bounded above by B, then s converges and* $\lim_{n\to\infty} s_n \leq B$.

It is easy to see geometrically that (1.4) must be true. Because the sequence is increasing, each point s_n on the real line lies at least as far to the right as its predecessor s_{n-1} (see Figure 2). In addition, we are given that no points lie to the right of B. Hence the points of the sequence must "pile up" or cluster at some point less than or equal to B. The proof which follows serves to make these intuitive ideas precise.

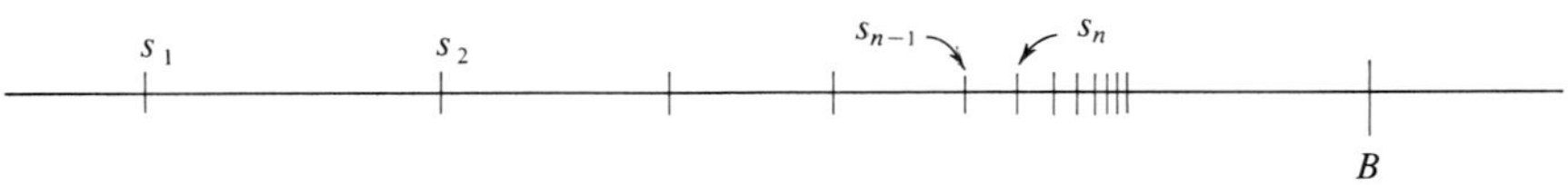

Figure 2

Proof. The range of s, which is the set of all numbers s_n, has the number B as an upper bound. By the Least Upper Bound Property (see page 7), this set has a least upper bound, which we denote by L. Obviously,

$$L \leq B. \tag{2}$$

We contend that $\lim_{n\to\infty} s_n = L$. Since L is an *upper* bound, we have $s_n \leq L$ for every n in the domain of s. Since L is a *least* upper bound, there must exist values s_n of the sequence which are arbitrarily close to L. That is, for any $\epsilon > 0$, there exists an integer N such that $|L - s_N| = L - s_N < \epsilon$. Since the sequence is increasing, we have $s_n \geq s_N$ for all $n \geq N$. Hence $-s_n \leq -s_N$ and so

$$|L - s_n| = L - s_n \leq L - s_N < \epsilon,$$

for every integer $n \geq N$. This proves that $\lim_{n\to\infty} s_n = L$, and this fact, together with the inequality (2), completes the proof.

It should be remarked that the essential ideas of Theorems (1.3) and (1.4) are not limited to sequences. For example, by making only trivial changes in the proofs, we obtain the following analogous results about an arbitrary real-valued function f defined on an interval $[a, \infty)$:

(1.3′) *If f is increasing and if* $\lim_{x\to\infty} f(x) = L$, *then $f(x) \leq L$ for every x in* $[a, \infty)$.

(1.4′) *If f is increasing and if $f(x) \leq B$ for some number B and for every x in $[a, \infty)$, then* $\lim_{x\to\infty} f(x)$ *exists and, furthermore,* $\lim_{x\to\infty} f(x) \leq B$.

The latter asserts that every increasing bounded function must approach a limit, a result which we assumed without proof in the proof of the Comparison Test for Integrals on page 469.

PROBLEMS

1. Determine whether or not each of the following sequences converges, and evaluate the limit if it does.

(a) $a_n = \dfrac{n^2 - 1}{n^2 + 1}$

(b) $b_k = \dfrac{k^2}{(k+1)^2}$

(c) $s_n = \dfrac{n+1}{n^2}$

(d) $s_n = (-1)^n \dfrac{n+1}{n^2}$

(e) $a_i = \left(1 + \dfrac{1}{i}\right)^i$

(f) $s_i = i \sin \dfrac{1}{i}$.

2. Evaluate each of the following limits.

(a) $\lim_{n\to\infty} \dfrac{(n+1)(n+3)}{n^2+3}$

(b) $\lim_{i\to\infty} e^{1/i}$

(c) $\lim_{k\to\infty} \dfrac{k}{\sqrt{k+1}}$

(d) $\lim_{n\to\infty} \cos n$

(e) $\lim_{n\to\infty} \dfrac{1}{n} \cos n$

(f) $\lim_{k\to\infty} (\sqrt{k} - \sqrt{k+1})$.

3. Determine whether or not each of the following sequences $\{s_n\}$ converges, and, if it does, evaluate the limit.

(a) $s_n = (-1)^n, \quad n = 1, 2, \ldots.$

(b) $s_n = \begin{cases} 1 + \dfrac{1}{n}, & \text{for every integer } n \text{ such that } 1 \leq n \leq 10, \\ 1, & \text{for every integer } n > 10. \end{cases}$

(c) $s_n = \begin{cases} 1 + \dfrac{1}{n}, & \text{if } n \text{ is a positive even integer,} \\ 1, & \text{if } n \text{ is a positive odd integer.} \end{cases}$

(d) $s_n = \begin{cases} 1 + \dfrac{1}{n}, & \text{for every integer } n \text{ such that } 1 \leq n \leq 10, \\ 2, & \text{for every integer } n > 10. \end{cases}$

4. Let s be the sequence defined by $s_n = \frac{1}{n}$, for every positive integer n. Draw an x-axis, and plot on it the first ten points of the sequence. What is $\lim_{n \to \infty} s_n$?

5. Let r be a real number, and consider the sequence $1, r, r^2, r^3, \ldots$. Show that the sequence converges if and only if $-1 < r \leq 1$, and that

$$\lim_{n \to \infty} r^n = \begin{cases} 0 & \text{if } -1 < r < 1, \\ 1 & \text{if } r = 1, \\ \infty & \text{if } r > 1. \end{cases}$$

What is the behavior of the sequence for $r = -1$ and for $r < -1$? (*Hint:* Let $r^n = e^{n \ln r}$, for $r > 0$.)

6. Finish the proof of Theorem (1.1):

   (a) Prove part (ii).

   (b) Prove part (iii).

   (c) Prove part (iv).

7. Let s and t be two infinite sequences and a a real number such that

$$s_n = a + t_n,$$

for every integer n greater than or equal to some integer k. Prove that

$$\lim_{n \to \infty} s_n = a + \lim_{n \to \infty} t_n.$$

[*Suggestion:* It is easy to prove this result directly from the definition of convergence. Alternatively, one may consider a constant sequence with the single value a, and obtain the result as a corollary of Theorem (1.1), part (i).]

8. Consider the sequence $\{s_n\}$ defined by $s_n = n + (-1)^n$, for every integer $n \geq 0$.

   (a) Write the first ten terms of the sequence.

   (b) Show that $\lim_{n \to \infty} s_n = \infty$, but that $\{s_n\}$ is not an increasing sequence.

   (c) Give another example of a sequence $\{s_n\}$ which is not monotonic but for which $\lim_{n \to \infty} s_n = \infty$.

2. Infinite Series: Definition and Properties. We are now ready to define infinite series. Consider an infinite sequence of real numbers a_m, a_{m+1},

$a_{m+2}, \ldots$. From this sequence $\{a_i\}$ we construct another sequence $\{s_n\}$ with the same domain, called the **sequence of partial sums** and defined by

$$\begin{aligned} s_m &= a_m, \\ s_{m+1} &= a_m + a_{m+1}, \\ s_{m+2} &= a_m + a_{m+1} + a_{m+2}, \\ &\vdots \end{aligned}$$

That is, for every integer $n \geq m$, the number s_n is given by

$$s_n = \sum_{i=m}^{n} a_i = a_m + \cdots + a_n. \tag{1}$$

If the sequence $\{s_n\}$ of partial sums converges, we define its limit to be the value of the **infinite series** determined by the original sequence $\{a_i\}$, and we write

$$\boxed{\sum_{i=m}^{\infty} a_i = \lim_{n\to\infty} s_n.} \tag{2}$$

EXAMPLE 1. Show that

$$\sum_{i=0}^{\infty} \frac{1}{2^i} = 1 + \tfrac{1}{2} + \tfrac{1}{4} + \tfrac{1}{8} + \cdots = 2.$$

For this series the sequence of partial sums is given by

$$\begin{aligned} s_0 &= 1, \\ s_1 &= 1 + \tfrac{1}{2}, \\ s_2 &= 1 + \tfrac{1}{2} + \tfrac{1}{4}, \\ &\vdots \end{aligned}$$

and, more generally, by

$$s_n = 1 + \frac{1}{2} + \cdots + \frac{1}{2^n} \cdot$$

Note that $s_0 = 2 - 1$, $s_1 = 2 - \tfrac{1}{2}$, and $s_2 = 2 - \tfrac{1}{4}$. It is not hard to show that $s_n = 2 - \dfrac{1}{2^n}$ for every positive integer n. Hence

$$\lim_{n\to\infty} s_n = \lim_{n\to\infty} \left(2 - \frac{1}{2^n}\right) = 2,$$

and it then follows from the above definition that $\displaystyle\sum_{i=0}^{\infty} \frac{1}{2^i} = 2$.

If, for a given sequence of real numbers $a_m, a_{m+1}, \ldots$, it happens that the corresponding sequence of partial sums does not converge, then the value of the infinite series is not defined. In this case we shall follow the customary terminology and say that the infinite series $\sum_{i=m}^{\infty} a_i$ **diverges.** On the other hand, if the sequence of partial sums does converge, we shall say that the infinite series $\sum_{i=m}^{\infty} a_i$ **converges.** Summarizing the above definitions (1) and (2) in a single formula, we obtain the equation

$$\boxed{\sum_{i=m}^{\infty} a_i = \lim_{n\to\infty} \sum_{i=m}^{n} a_i,} \qquad (3)$$

in which the series on the left converges if and only if the limit on the right exists.

Our first theorem states that if an infinite series $\sum_{i=m}^{\infty} a_i$ converges, then the sequence $\{a_i\}$ must converge to zero:

(2.1) THEOREM. *If* $\sum_{i=m}^{\infty} a_i$ *converges, then* $\lim_{n\to\infty} a_n = 0$.

Proof. Let $s = \{s_n\}$ be the sequence of partial sums. Since the infinite series converges, there exists a real number L such that

$$\sum_{i=m}^{\infty} a_i = \lim_{n\to\infty} s_n = L.$$

Let s' be the sequence defined by $s'_n = s_{n-1}$, for every integer $n \geq m + 1$. The range of the function s' is the same as that of s, and the order is the same. That is, enumeration of the terms of both sequences gives the same list of numbers: $s_m, s_{m+1}, \ldots$. We conclude that

$$\lim_{n\to\infty} s'_n = \lim_{n\to\infty} s_n.$$

We next observe that, for every integer $n \geq m + 1$,

$$a_n = s_n - s_{n-1} = s_n - s'_n.$$

Since the limit of the sum or difference of two convergent sequences is the sum or difference of their limits [see Theorem (1.1), page 475], we have

$$\lim_{n\to\infty} a_n = \lim_{n\to\infty} s_n - \lim_{n\to\infty} s'_n = L - L = 0,$$

and the proof is complete.

As a result of Theorem (2.1) we see at once that both infinite series

$$\sum_{i=0}^{\infty} (-1)^i 2 = 2 - 2 + 2 - 2 + \cdots,$$

$$\sum_{i=1}^{\infty} \left(2 + \frac{1}{i^2}\right) = 3 + 2\tfrac{1}{4} + 2\tfrac{1}{9} + 2\tfrac{1}{16} + \cdots$$

are divergent. For the first, $\lim_{n\to\infty} a_n = \lim_{n\to\infty} (-1)^n 2$, which does not exist, and for the second, $\lim_{n\to\infty} a_n = \lim_{n\to\infty} \left(2 + \frac{1}{n^2}\right) = 2$.

[*Warning:* The converse of Theorem (2.1) is false. That is, it is not true that if $\lim_{n\to\infty} a_n = 0$, then $\sum_{i=m}^{\infty} a_i$ converges. A well-known counterexample is the series discussed in the following example.]

EXAMPLE 2. Show that the infinite series

$$\sum_{k=1}^{\infty} \frac{1}{k} = 1 + \tfrac{1}{2} + \tfrac{1}{3} + \tfrac{1}{4} + \cdots$$

diverges. This series is called the **harmonic series** and is particularly interesting because it diverges in spite of the fact that $\lim_{n\to\infty} \frac{1}{n} = 0$. To prove divergence, we first observe that s_n, the nth partial sum of the series, is given by

$$s_n = 1 + \frac{1}{2} + \cdots + \frac{1}{n}.$$

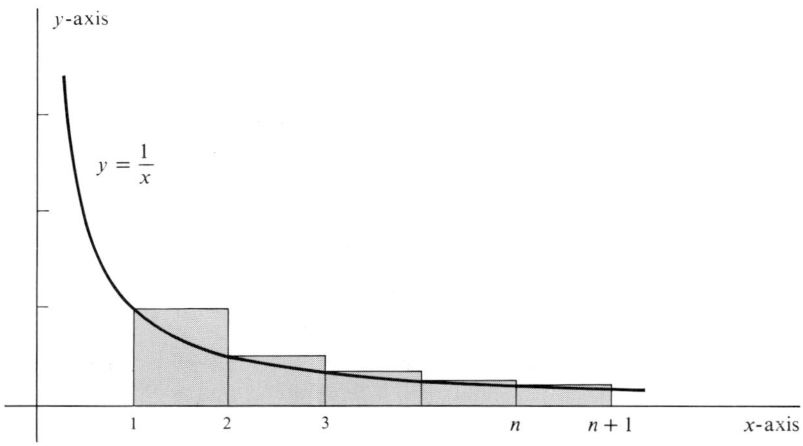

Figure 3

Next, consider Figure 3, which shows the graph of the function $\frac{1}{x}$ between $x = 1$ and $x = n + 1$. With respect to the partition $\sigma = \{1, 2, \ldots, n + 1\}$, the upper sum U_σ is equal to the sum of the areas of the shaded rectangles and is given by

$$U_\sigma = 1 + \frac{1}{2} + \cdots + \frac{1}{n}.$$

Thus $U_\sigma = s_n$. Since every upper sum is greater than or equal to the corresponding definite integral, we obtain

$$s_n = U_\sigma \geq \int_1^{n+1} \frac{1}{x}\,dx = \ln(n+1).$$

We know that $\ln(n + 1)$ increases without bound as n increases, hence the same is true of s_n. Thus

$$\lim_{n\to\infty} s_n = \infty,$$

which completes the proof that the harmonic series diverges.

The next theorem states that infinite series have what is commonly called the property of **linearity.** The result is a useful one because it shows that convergent infinite series may be added in the natural way and also multiplied by real numbers. Note that we have come across the property of linearity before. It is one of the basic features of finite series and also of definite integrals.

(2.2) *If $\sum_{i=m}^{\infty} a_i$ and $\sum_{i=m}^{\infty} b_i$ are convergent infinite series and if c is a real number, then the series $\sum_{i=m}^{\infty} (a_i + b_i)$ and $\sum_{i=m}^{\infty} ca_i$ are also convergent, and*

$$\text{(i)} \quad \sum_{i=m}^{\infty} (a_i + b_i) = \sum_{i=m}^{\infty} a_i + \sum_{i=m}^{\infty} b_i.$$

$$\text{(ii)} \quad \sum_{i=m}^{\infty} ca_i = c\sum_{i=m}^{\infty} a_i.$$

Proof. The proofs of (i) and (ii) are direct corollaries of the corresponding parts of Theorem (1.1), page 475. Let $\{s_n\}$ and $\{t_n\}$ be the two convergent sequences of partial sums corresponding to $\sum_{i=m}^{\infty} a_i$ and $\sum_{i=m}^{\infty} b_i$, respectively. That is,

$$s_n = \sum_{i=m}^{n} a_i, \qquad t_n = \sum_{i=m}^{n} b_i,$$

$$\sum_{i=m}^{\infty} a_i = \lim_{n\to\infty} s_n, \qquad \sum_{i=m}^{\infty} b_i = \lim_{n\to\infty} t_n. \tag{4}$$

By part (i) of Theorem (1.1), we have

$$\lim_{n\to\infty} (s_n + t_n) = \lim_{n\to\infty} s_n + \lim_{n\to\infty} t_n, \tag{5}$$

which shows, first of all, that $\{s_n + t_n\}$ is a convergent sequence. The linearity property of finite sums implies that

$$s_n + t_n = \sum_{i=m}^{n} a_i + \sum_{i=m}^{n} b_i = \sum_{i=m}^{n} (a_i + b_i),$$

from which we conclude that $\{s_n + t_n\}$ is the sequence of partial sums corresponding to the series $\sum_{i=m}^{\infty} (a_i + b_i)$. Hence

$$\sum_{i=m}^{\infty} (a_i + b_i) = \lim_{n\to\infty} (s_n + t_n). \tag{6}$$

Substituting from equations (6) and (4) into equation (5), we obtain

$$\sum_{i=m}^{\infty} (a_i + b_i) = \sum_{i=m}^{\infty} a_i + \sum_{i=m}^{\infty} b_i,$$

and this completes the proof of part (i). Part (ii) is proved in the same way, and we omit the details.

As an application of Theorem (2.2) we may conclude that *if a series* $\sum_{i=m}^{\infty} a_i$ *diverges and if* $c \neq 0$, *then* $\sum_{i=m}^{\infty} ca_i$ *also diverges.* For if the latter series converges, we know from part (ii) of (2.1) that

$$\frac{1}{c}\sum_{i=m}^{\infty} ca_i = \sum_{i=m}^{\infty} \frac{1}{c} ca_i = \sum_{i=m}^{\infty} a_i.$$

and that the series on the right converges, contrary to assumption. For example, since the harmonic series $\sum_{i=1}^{\infty} \frac{1}{i}$ diverges, it follows at once that the series

$$\sum_{i=1}^{\infty} \frac{1}{5i} = \frac{1}{5} + \frac{1}{10} + \frac{1}{15} + \cdots$$

also diverges.

It is an important corollary of the next theorem that the convergence or divergence of an infinite series is unaffected by the addition or deletion of any finite number of terms at the beginning.

(2.3) *If* $m < l$, *then the series* $\sum_{i=m}^{\infty} a_i$ *converges if and only if* $\sum_{i=l}^{\infty} a_i$ *converges. If either converges, then*

$$\sum_{i=m}^{\infty} a_i = \sum_{i=m}^{l-1} a_i + \sum_{i=l}^{\infty} a_i.$$

Proof. Let $\{s_n\}$ and $\{t_n\}$ be the sequences of partial sums for $\sum_{i=m}^{\infty} a_i$ and $\sum_{i=l}^{\infty} a_i$, respectively. Then

$$s_n = \sum_{i=m}^{n} a_i, \qquad \text{for every integer } n \geq m,$$

$$t_n = \sum_{i=l}^{n} a_i, \qquad \text{for every integer } n \geq l.$$

If n is any integer greater than or equal to l, then obviously

$$\sum_{i=m}^{n} a_i = \sum_{i=m}^{l-1} a_i + \sum_{i=l}^{n} a_i.$$

Hence

$$s_n = \sum_{i=m}^{l-1} a_i + t_n, \qquad \text{for every integer } n \geq l.$$

The number $\sum_{i=m}^{l-1} a_i$ does not depend on n, and is fixed throughout the proof. Thus, for $n \geq l$, the sequences $\{s_n\}$ and $\{t_n\}$ differ by a constant. It follows that one converges if and only if the other does and that

$$\lim_{n \to \infty} s_n = \sum_{i=m}^{l-1} a_i + \lim_{n \to \infty} t_n,$$

(see Problem 7, page 481). This completes the proof, since by definition,

$$\lim_{n \to \infty} s_n = \sum_{i=m}^{\infty} a_i \qquad \text{and} \qquad \lim_{n \to \infty} t_n = \sum_{i=l}^{\infty} a_i.$$

As an illustration, consider an infinite series $\sum_{i=1}^{\infty} a_i$ whose first thousand terms we know nothing about, but which has the property that $a_n = \frac{1}{2^n}$ for every integer $n > 1000$. We have shown in Example 1 that the series $\sum_{i=0}^{\infty} \frac{1}{2^i}$ converges, and it follows by Theorem (2.3) that $\sum_{i=1001}^{\infty} \frac{1}{2^i}$ also converges. Since the latter series is precisely the series $\sum_{i=1001}^{\infty} a_i$, a second application of (2.3) establishes the convergence of the original series $\sum_{i=1}^{\infty} a_i$.

An **infinite geometric series** is one of the form

$$\sum_{i=0}^{\infty} ar^i = a + ar + ar^2 + \cdots,$$

in which a and r are arbitrary real numbers. For example, by taking $a = 1$ and $r = \frac{1}{2}$, we obtain the convergent series $\sum_{i=0}^{\infty} \frac{1}{2^i}$. In studying the question of the convergence or divergence of geometric series, it is sufficient to take $a = 1$ and consider the simpler series

$$\sum_{i=0}^{\infty} r^i = 1 + r + r^2 + \cdots. \qquad (7)$$

For if this series converges, then so does $\sum_{i=0}^{\infty} ar^i$, and

$$\sum_{i=0}^{\infty} ar^i = a\sum_{i=0}^{\infty} r^i.$$

On the other hand, if (7) diverges and $a \neq 0$, then $\sum_{i=0}^{\infty} ar^i$ also diverges. The principal result about the convergence of geometric series is the following:

(2.4) *The geometric series* (7) *converges if and only if* $-1 < r < 1$. *If it converges, then*

$$\sum_{i=0}^{\infty} r^i = \frac{1}{1-r}.$$

Proof. If $r = 1$, the series (7) is the divergent series $1 + 1 + 1 + \cdots$. Hence, in what follows, we shall assume that $r \neq 1$. The sequence $\{s_n\}$ of partial sums is defined by

$$s_n = \sum_{i=0}^{n} r^i = 1 + r + \cdots + r^n,$$

for every integer $n \geq 0$. Observe that

$$\begin{aligned} 1 + rs_n &= 1 + r(1 + r + \cdots + r^n) \\ &= 1 + r + r^2 + \cdots + r^{n+1} = s_{n+1}. \end{aligned}$$

On the other hand, we have the equation

$$s_n + r^{n+1} = s_{n+1}.$$

It follows that $1 + rs_n = s_n + r^{n+1}$, whence $1 - r^{n+1} = s_n(1 - r)$, and so

$$s_n = \frac{1 - r^{n+1}}{1 - r}.$$

The proof is completed by considering two cases. First of all, suppose that $-1 < r < 1$. Then $\lim_{n\to\infty} r^{n+1} = \lim_{n\to\infty} r^n = 0$ (see Problem 5, page 481), and therefore

$$\sum_{i=0}^{\infty} r^i = \lim_{n\to\infty} s_n = \frac{1-0}{1-r} = \frac{1}{1-r}.$$

Second, suppose that $r \leq -1$ or $r > 1$. For neither of these possibilities does $\lim_{n\to\infty} r^{n+1}$ exist (again, see Problem 5, page 481). It follows that $\lim_{n\to\infty} s_n$ also does not exist, and hence the series $\sum_{i=0}^{\infty} r^i$ diverges. This completes the proof.

PROBLEMS

1. Determine whether or not each of the following infinite series converges, and evaluate it if it does.

(a) $\displaystyle\sum_{i=0}^{\infty} \frac{7}{5^i}$

(b) $\displaystyle\sum_{k=1}^{\infty} \frac{a}{5^k}$

(c) $\displaystyle\sum_{n=1}^{\infty} \left(\frac{1}{2^n} + \frac{1}{n}\right)$

(d) $\displaystyle\sum_{j=0}^{\infty} \left(\frac{1}{2^j} - \frac{1}{3^j}\right)$

(e) $\displaystyle\sum_{i=1}^{\infty} \frac{5 \cdot 2^i + 6i}{i2^i}$

(f) $\displaystyle\sum_{k=0}^{\infty} \left(3 + \frac{1}{3^k}\right)$

(g) $\displaystyle\sum_{i=1}^{\infty} \frac{i^2 - 1}{i^2 + 1}$

(h) $\displaystyle\sum_{k=0}^{\infty} \frac{2^k + 3^k}{6^k}.$

2. Consider the infinite series $\sum_{i=0}^{\infty} a_i$ defined by

$$a_{2i} = \frac{1}{2^i}, \qquad i = 0, 1, 2, \ldots,$$

$$a_{2i+1} = 0, \qquad i = 0, 1, 2, \ldots.$$

Write out the sum of the first ten terms. Does the series converge? If so, to what value?

3. Using Theorem (2.2), show that if $\sum_{i=m}^{\infty} a_i$ converges and if $\sum_{i=m}^{\infty} b_i$ diverges, then $\sum_{i=m}^{\infty} (a_i + b_i)$ must diverge.

4. Is it true that if the series $\sum_{i=m}^{\infty} (a_i + b_i)$ converges, then both $\sum_{i=m}^{\infty} a_i$ and $\sum_{i=m}^{\infty} b_i$ must also converge? Give a reason for your answer.

5. Prove that the harmonic series $\displaystyle\sum_{k=1}^{\infty} \frac{1}{k}$ diverges using the following elementary argument. Begin by grouping the terms of the series:

$$\sum_{k=1}^{\infty} \frac{1}{k} = 1 + (\tfrac{1}{2} + \tfrac{1}{3}) + (\tfrac{1}{4} + \tfrac{1}{5} + \tfrac{1}{6} + \tfrac{1}{7})$$
$$+ (\tfrac{1}{8} + \tfrac{1}{9} + \tfrac{1}{10} + \tfrac{1}{11} + \tfrac{1}{12} + \tfrac{1}{13} + \tfrac{1}{14} + \tfrac{1}{15}) + \cdots,$$

and observe that

$$\tfrac{1}{2} + \tfrac{1}{3} > \tfrac{1}{4} + \tfrac{1}{4} = \tfrac{1}{2},$$
$$\tfrac{1}{4} + \tfrac{1}{5} + \tfrac{1}{6} + \tfrac{1}{7} > \tfrac{1}{8} + \tfrac{1}{8} + \tfrac{1}{8} + \tfrac{1}{8} = \tfrac{1}{2}, \qquad \text{etc.}$$

6. Consider the infinite series $\displaystyle\sum_{k=1}^{\infty} \left(\frac{1}{k} - \frac{1}{k+1}\right)$. By writing out a few terms of the sequence of partial sums, show that the series converges, and give its value.

7. An infinite series of the form $\sum_{i=m}^{\infty} (a_i - a_{i+1})$ is called a **telescoping series** (see Problem 6). Prove that it converges if and only if the sequence $\{a_n\}$ converges. If it does converge, what is its value?

8. Determine whether or not each of the following infinite series converges, and evaluate it if it does.

(a) $\sum_{k=1}^{\infty} \frac{1}{k(k+1)}$.

(b) $\sum_{i=2}^{\infty} \frac{2i+1}{i^2(i^2+2i+1)}$.

(c) $\sum_{k=1}^{\infty} \ln\left(\frac{k+1}{k}\right) = \ln\left(\frac{2}{1}\right) + \ln\left(\frac{3}{2}\right) + \ln\left(\frac{4}{3}\right) + \cdots$.

(*Hint:* Look at Problems 6 and 7.)

3. Nonnegative Series. The theory of convergence of infinite series is in many respects simpler for those series which do not contain both positive and negative terms. A series which contains no negative terms is called **nonnegative.** Thus $\sum_{i=m}^{\infty} a_i$ is nonnegative if and only if $a_i \geq 0$ for every integer $i \geq m$. In this section we shall study two convergence criteria for such series: The Integral Test and the Comparison Test.

Let $\sum_{i=m}^{\infty} a_i$ be an arbitrary infinite series (not necessarily nonnegative), and let $\{s_n\}$ be the corresponding sequence of partial sums. We recall that $s_n = \sum_{i=m}^{n} a_i$, for every integer $n \geq m$, and that, if $\{s_n\}$ converges, then $\sum_{i=m}^{\infty} a_i = \lim_{n\to\infty} s_n$. We shall extend the convention regarding the symbol ∞ and write

$$\sum_{i=m}^{\infty} a_i = \infty \qquad (\text{or } -\infty)$$

if and only if $\lim_{n\to\infty} s_n = \infty$ (or $-\infty$).

It may very well happen that a series neither converges nor satisfies $\sum_{i=m}^{\infty} a_i = \pm\infty$. For example, the divergent series

$$\sum_{i=0}^{\infty} (-1)^i = 1 - 1 + 1 - 1 + 1 - \cdots$$

has for its sequence of partial sums the oscillating sequence 1, 0, 1, 0, 1, ... However, for nonnegative series, there are only two alternatives:

(3.1) *Every nonnegative series* $\sum_{i=m}^{\infty} a_i$ *either converges or satisfies* $\sum_{i=m}^{\infty} a_i = \infty$.

The proof of this fact follows directly from the following two lemmas:

(3.2) *If* $\sum_{i=m}^{\infty} a_i$ *is a nonnegative series, then the corresponding sequence* $\{s_n\}$ *of partial sums is an increasing sequence.*

Proof. For every integer $n \geq m$, we have $s_{n+1} = s_n + a_{n+1}$. Since the series is nonnegative, it follows that $s_{n+1} - s_n = a_{n+1} \geq 0$. Hence

$$s_{n+1} \geq s_n, \qquad \text{for every integer } n \geq m,$$

which is the definition of an increasing sequence.

(3.3) *If $\{s_n\}$ is an increasing infinite sequence of real numbers, then either it is bounded above and therefore converges or else* $\lim_{n\to\infty} s_n = \infty$.

Proof. If the sequence is bounded above, then it is proved in Theorem (1.4), page 479, that it must converge. Suppose it is not so bounded. Then, for every real number B, there exists an integer N such that $s_N > B$. Since the sequence is increasing, it follows that $s_n \geq s_N$ for every $n > N$. Hence $s_n > B$, for every integer $n > N$, and this is precisely the definition of the expression $\lim_{n\to\infty} s_n = \infty$.

We come now to the first of the tests for convergence of nonnegative infinite series. It is a generalization of the method used in Section 2 to prove the divergence of the harmonic series.

(3.4) INTEGRAL TEST. *Let f be a function which is nonnegative and decreasing on the interval $[m, \infty)$. Then the infinite series $\sum_{i=m}^{\infty} a_i$ defined by*

$$a_i = f(i), \qquad \textit{for every integer } i \geq m,$$

is convergent if and only if the improper integral $\int_m^\infty f(x)\,dx$ is convergent.

Proof. The series $\sum_{i=m}^{\infty} a_i$ is nonnegative, and its corresponding sequence $\{s_n\}$ of partial sums is therefore increasing. Figure 4 illustrates the graph of the

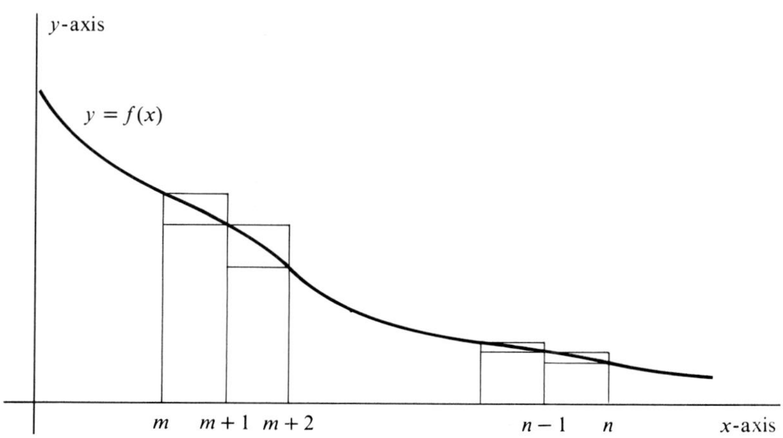

Figure 4

function f over an interval $[m, n]$, where n is an arbitrary integer greater than m. Since f is decreasing, its maximum value on each subinterval of the partition $\sigma = \{m, m + 1, \ldots, n\}$ occurs at the left endpoint. Moreover, each subinterval has length 1. Hence the upper sum U_σ, which is equal to the sum of the areas of the rectangles lying above the graph in the figure, is given by

$$U_\sigma = \sum_{i=m}^{n-1} f(i) = \sum_{i=m}^{n-1} a_i = s_{n-1}.$$

Similarly, the lower sum L_σ is equal to

$$L_\sigma = \sum_{i=m+1}^{n} f(i) = \sum_{i=m+1}^{n} a_i = s_n - a_m.$$

As always, we have

$$L_\sigma \leq \int_m^n f(x)\,dx \leq U_\sigma,$$

and it follows that

$$s_n - a_m \leq \int_m^n f(x)\,dx \leq s_{n-1}. \tag{1}$$

The crux of the proof of the Integral Test is in the inequalities (1). In completing the argument, we consider the "if" and "only if" parts of the theorem separately.

"*If.*" Let $\int_m^\infty f(x)\,dx$ be a convergent improper integral. That is, $\lim_{b\to\infty} \int_m^b f(x)\,dx$ exists. Since f is nonnegative on $[m, \infty)$, the integral $\int_m^b f(x)\,dx$ is an increasing function of b. Hence

$$\int_m^n f(x)\,dx \leq \int_m^\infty f(x)\,dx, \qquad \text{for every integer } n \geq m,$$

[see (1.3′), page 479]. From (1) it follows that

$$s_n \leq a_m + \int_m^n f(x)\,dx \leq a_m + \int_m^\infty f(x)\,dx,$$

for every integer $n > m$. Hence the increasing sequence $\{s_n\}$ is bounded above and therefore converges. The convergence of the sequence of partial sums is equivalent to the convergence of the corresponding infinite series, so we conclude that $\sum_{i=m}^\infty a_i$ converges.

"*Only if.*" Suppose that $\sum_{i=m}^\infty a_i$ converges. Then

$$s_{n-1} \leq s_n \leq \sum_{i=m}^\infty a_i,$$

for every integer $n > m$ [see (1.3), page 478]. For any real number b in $[m, \infty)$, choose an integer $n > b$. Since $\int_m^b f(x)\,dx$ is an increasing function of b, we obtain from (1)

$$\int_m^b f(x)\,dx \leq \int_m^n f(x)\,dx \leq s_{n-1}.$$

Hence

$$\int_m^b f(x)\,dx \leq \sum_{i=m}^{\infty} a_i.$$

The integral is therefore bounded above, and it follows that $\lim_{b\to\infty} \int_m^b f(x)\,dx$ exists [see (1.4′), page 480]. Hence $\int_m^\infty f(x)\,dx$ is a convergent improper integral, and the proof of the Integral Test is complete.

The convergence or divergence of many infinite series can be determined easily by the Integral Test. Important among these are series of the form

$$\boxed{\sum_{i=1}^{\infty} \frac{1}{i^p} = 1 + \frac{1}{2^p} + \frac{1}{3^p} + \cdots,}$$

where p is a positive real number. Such a series is called a ***p*-series.** An example is the divergent harmonic series, for which $p = 1$. The basic convergence theorem is

(3.5) *The p-series* $\sum_{i=1}^{\infty} \frac{1}{i^p}$ *converges if and only if* $p > 1$.

Proof. The function f defined by $f(x) = \frac{1}{x^p}$ is nonnegative on the interval $[1, \infty)$, and is also decreasing on that interval since we have made the assumption that $p > 0$. Moreover, it is obvious that $f(i) = \frac{1}{i^p}$ for every positive integer i. If $p \neq 1$, then

$$\int \frac{1}{x^p}\,dx = \frac{1}{1-p} x^{1-p} + c.$$

Hence we have the three computations:

$$0 < p < 1\colon \int_1^\infty \frac{1}{x^p}\,dx = \frac{1}{1-p} \lim_{b\to\infty} (b^{1-p} - 1) = \infty,$$

$$p = 1\colon \int_1^\infty \frac{1}{x^p}\,dx = \int_1^\infty \frac{1}{x}\,dx = \lim_{b\to\infty} \ln b = \infty,$$

$$p > 1\colon \int_1^\infty \frac{1}{x^p}\,dx = \frac{1}{1-p} \lim_{b\to\infty} \left(\frac{1}{b^{p-1}} - 1\right) = \frac{1}{p-1}.$$

It follows that $\int_1^\infty \frac{1}{x^p}\,dx$ is convergent if and only if $p > 1$, and the Integral Test therefore completes the proof.

Thus the first of the following three p-series diverges, and the last two converge:

$$\sum_{i=1}^{\infty} \frac{1}{\sqrt{i}} = 1 + \frac{1}{\sqrt{2}} + \frac{1}{\sqrt{3}} + \cdots,$$

$$\sum_{k=1}^{\infty} \frac{1}{k^{3/2}} = 1 + \frac{1}{2\sqrt{2}} + \frac{1}{3\sqrt{3}} + \cdots,$$

$$\sum_{j=1}^{\infty} \frac{1}{j^2} = 1 + \frac{1}{2^2} + \frac{1}{3^2} + \cdots.$$

EXAMPLE 1. Determine whether the series $\sum_{k=1}^{\infty} \frac{1}{2k^2+1}$ converges or diverges. The function f defined by $f(x) = \frac{1}{2x^2+1}$ is nonnegative and decreasing on the interval $[1, \infty)$, and obviously $f(k) = \frac{1}{2k^2+1}$. Since

$$\int \frac{1}{2x^2+1}\,dx = \frac{1}{\sqrt{2}} \arctan \sqrt{2}\,x + c,$$

we have

$$\int_1^\infty \frac{1}{2x^2+1}\,dx = \frac{1}{\sqrt{2}} \lim_{b\to\infty} (\arctan \sqrt{2}\,b - \arctan \sqrt{2})$$

$$= \frac{1}{\sqrt{2}}\left(\frac{\pi}{2} - \arctan \sqrt{2}\right).$$

Hence the integral is convergent, and therefore so is the series.

We come now to the second of our convergence tests.

(3.6) COMPARISON TEST. *If $\sum_{i=m}^{\infty} a_i$ is a nonnegative series and if $\sum_{i=m}^{\infty} b_i$ is a convergent series with $a_i \le b_i$ for every $i \ge m$, then $\sum_{i=m}^{\infty} a_i$ converges and $\sum_{i=m}^{\infty} a_i \le \sum_{i=m}^{\infty} b_i$.*

Proof. Let $\{s_n\}$ and $\{t_n\}$ be the sequences of partial sums for $\sum_{i=m}^{\infty} a_i$ and $\sum_{i=m}^{\infty} b_i$, respectively. The hypotheses $0 \le a_i \le b_i$ imply that

$$s_n \le t_n, \qquad \text{for every integer } n \ge m, \tag{2}$$

and also that both series are nonnegative. Hence both $\{s_n\}$ and $\{t_n\}$ are increasing sequences. The convergence of $\sum_{i=m}^{\infty} b_i$ means that $\{t_n\}$ converges and that $\lim_{n\to\infty} t_n = \sum_{i=m}^{\infty} b_i$. It follows that

$$t_n \le \sum_{i=m}^{\infty} b_i, \qquad \text{for every integer } n \ge m$$

[see (1.3), page 478]. Hence, by (2),

$$s_n \le \sum_{i=m}^{\infty} b_i, \qquad \text{for every integer } n \ge m.$$

Thus $\{s_n\}$ is increasing and bounded above by $\sum_{i=m}^{\infty} b_i$. The sequence therefore converges and

$$\lim_{n\to\infty} s_n \le \sum_{i=m}^{\infty} b_i$$

[see (1.4), page 479]. The convergence of s_n implies the convergence of $\sum_{i=m}^{\infty} a_i$ and that the value of the series is $\lim_{n\to\infty} s_n$. Hence

$$\sum_{i=m}^{\infty} a_i \le \sum_{i=m}^{\infty} b_i,$$

and the proof is complete.

EXAMPLE 2. Use the Comparison Test to prove that the series $\sum_{i=0}^{\infty} \frac{1}{2i^2 - 7}$ converges. We first observe that the first two terms of the series are negative. However, $2i^2 - 7 > 0$ provided $i \ge 2$, and so the series $\sum_{i=2}^{\infty} \frac{1}{2i^2 - 7}$ is nonnegative. It is sufficient to prove the latter series convergent because of the important fact that the convergence of an infinite series is unaffected by any finite number of terms at the beginning. As our test series we take the convergent p-series $\sum_{i=1}^{\infty} \frac{1}{i^2}$. To use the Comparison Test, we wish to show that

$$\frac{1}{2i^2 - 7} \le \frac{1}{i^2}, \qquad \text{for every integer } i \ge 2. \tag{3}$$

This inequality is equivalent to $i^2 \le 2i^2 - 7$, which in turn is equivalent to $i^2 \ge 7$. The last is clearly true provided $i \ge 3$. Thus we have proved

$$\frac{1}{2i^2 - 7} \le \frac{1}{i^2}, \qquad \text{for every integer } i \ge 3, \tag{4}$$

which is slightly weaker than (3). However, (4) is certainly sufficient. We know that the series $\sum_{i=3}^{\infty} \frac{1}{i^2}$ converges. It follows from (4) by the Comparison Test that $\sum_{i=3}^{\infty} \frac{1}{2i^2 - 7}$ converges and, as a result, the original series $\sum_{i=0}^{\infty} \frac{1}{2i^2 - 7}$ does also.

Example 2 illustrates a useful extension of the Comparison Test: *The series $\sum_{i=m}^{\infty} a_i$ converges if there exists a convergent series $\sum_{i=m}^{\infty} b_i$ such that $0 \leq a_i \leq b_i$ eventually.* The assertion that $0 \leq a_i \leq b_i$ **eventually** means simply that there exists an integer N such that $0 \leq a_i \leq b_i$ for every integer $i \geq N$. The justification for this extension is Theorem (2.3), page 486. A similar observation should be made about the Integral Test. It may be necessary to drop a finite number of terms from the beginning of the series under consideration before a convenient function f can be found which satisfies the conditions of the test.

The Comparison Test is as useful for proving divergence as convergence. It is an immediate corollary that *if the nonnegative series $\sum_{i=m}^{\infty} a_i$ diverges and if $a_i \leq b_i$ for every $i \geq m$, then $\sum_{i=m}^{\infty} b_i$ also diverges.* For if $\sum_{i=m}^{\infty} b_i$ converges, the Comparison Test implies that $\sum_{i=m}^{\infty} a_i$ converges, which is contrary to assumption.

EXAMPLE 3. Determine whether the series $\sum_{k=1}^{\infty} \frac{1}{(k^2 + 5)^{1/3}}$ converges or diverges. If we use the Comparison Test, we must decide whether to look for a convergent test series with larger terms to prove convergence, or a divergent test series with smaller terms to prove divergence. To decide which, observe that for large values of k, the number $k^2 + 5$ is not very different from k^2, and therefore $\frac{1}{(k^2 + 5)^{1/3}}$ is approximately equal to $\frac{1}{k^{2/3}}$. Stated more formally, we have

$$\begin{aligned} \lim_{k \to \infty} \frac{(k^2 + 5)^{1/3}}{k^{2/3}} &= \lim_{k \to \infty} \left(\frac{k^2 + 5}{k^2} \right)^{1/3} \\ &= \lim_{k \to \infty} \left(1 + \frac{5}{k^2} \right)^{1/3} = 1. \end{aligned}$$

This comparison, together with the divergence of the p-series $\sum_{k=1}^{\infty} \frac{1}{k^{2/3}}$, leads us to believe that the series $\sum_{k=1}^{\infty} \frac{1}{(k^2 + 5)^{1/3}}$ diverges. Hence we shall try a divergent test series. The most obvious candidate, $\sum_{k=1}^{\infty} \frac{1}{k^{2/3}}$, fails to be useful, since the necessary inequality,

$$\frac{1}{k^{2/3}} \leq \frac{1}{(k^2 + 5)^{1/3}},$$

is clearly false for every value of k. However, the series $\sum_{k=1}^{\infty} \frac{1}{2k^{2/3}}$ is also divergent, and we may ask whether or not it is true that

$$\frac{1}{2k^{2/3}} \leq \frac{1}{(k^2 + 5)^{1/3}}. \tag{5}$$

This inequality is equivalent to $8k^2 \geq k^2 + 5$, and hence to $7k^2 \geq 5$, which is certainly true for every positive integer k. Hence (5) holds for every integer $k \geq 1$, and it therefore follows by the Comparison Test that the series $\sum_{k=1}^{\infty} \frac{1}{(k^2 + 5)^{1/3}}$ diverges.

PROBLEMS

1. Test the following infinite series for convergence or divergence.

(a) $\sum_{i=1}^{\infty} \frac{1}{7i - 2}$

(b) $\sum_{i=1}^{\infty} \frac{1}{7i^2 - 2}$

(c) $\sum_{k=1}^{\infty} \frac{1}{\sqrt{k}}$

(d) $\sum_{k=1}^{\infty} \frac{1}{\sqrt{k + 7}}$

(e) $\sum_{k=1}^{\infty} \frac{1}{k^{3/2}}$

(f) $\sum_{k=1}^{\infty} \frac{1}{\sqrt{k^2 + 1}}$

(g) $\sum_{n=1}^{\infty} \frac{1}{\sqrt{n^3 + 2}}$

(h) $\sum_{i=0}^{\infty} \frac{1}{1 + i^2}$

(i) $\sum_{i=4}^{\infty} \frac{1}{e^i}$

(j) $\sum_{i=0}^{\infty} \frac{1}{i^2 - 3i + 1}$.

2. Using the Integral Test for infinite series and the Comparison Test for integrals (page 469), determine whether each of the following series converges or diverges.

(a) $\sum_{k=1}^{\infty} e^{-k^2}$

(b) $\sum_{i=1}^{\infty} \frac{1}{i^2} \sin \frac{1}{i^2}$.

3. Using the Integral Test, prove the theorem that, for positive r, the geometric series $\sum_{i=0}^{\infty} r^i$ converges if and only if $r < 1$.

4. Let $\sum_{i=m}^{\infty} a_i$ and $\sum_{i=m}^{\infty} b_i$ be two convergent nonnegative series. Using the Comparison Test, prove that the series $\sum_{i=m}^{\infty} a_i b_i$ also converges.
5. Prove the following theorem, which is hinted at in Example 3. *If* $\sum_{i=m}^{\infty} b_i$ *is a positive series* (i.e., $b_i > 0$ *for every integer* $i \geq m$) *and if* $\lim_{n\to\infty} \frac{a_n}{b_n} = 1$, *then the series* $\sum_{i=m}^{\infty} a_i$ *converges if and only if* $\sum_{i=m}^{\infty} b_i$ *does.*

4. Alternating Series. Special among infinite series which contain both positive and negative terms are those whose terms alternate in sign. More precisely, we define the series $\sum_{i=m}^{\infty} a_i$ to be **alternating** if $a_i a_{i+1} < 0$ for every integer $i \geq m$. It follows from this definition that an alternating series is one which can be written in one of the two forms

$$\sum_{i=m}^{\infty} (-1)^i b_i \qquad \text{or} \qquad \sum_{i=m}^{\infty} (-1)^{i+1} b_i,$$

where $b_i > 0$ for every integer $i \geq m$. An example is the **alternating harmonic series**

$$\sum_{i=1}^{\infty} (-1)^{i+1} \frac{1}{i} = 1 - \tfrac{1}{2} + \tfrac{1}{3} - \tfrac{1}{4} + \tfrac{1}{5} - \cdots.$$

An alternating series converges under surprisingly weak conditions. The next theorem gives two simple hypotheses whose conjunction is sufficient to imply convergence.

(4.1) THEOREM. *The alternating series* $\sum_{i=m}^{\infty} a_i$ *converges if:*

(i) $|a_{n+1}| \leq |a_n|$, *for every integer* $n \geq m$, *and*
(ii) $\lim_{n\to\infty} a_n = 0$ (or, equivalently, $\lim_{n\to\infty} |a_n| = 0$).

Proof. We shall assume for convenience and with no loss of generality that $m = 0$ and that $a_i = (-1)^i b_i$ with $b_i > 0$ for every integer $i \geq 0$. The series is therefore $\sum_{i=0}^{\infty} (-1)^i b_i$, and the hypotheses (i) and (ii) become

(i′) $b_{n+1} \leq b_n$, for every integer $n \geq 0$, and

(ii′) $\lim_{n\to\infty} b_n = 0$.

The proof is completed by showing the convergence of the sequence $\{s_n\}$ of partial sums, which is defined recursively by the equations

$$s_0 = (-1)^0 b_0 = b_0,$$
$$s_n = s_{n-1} + (-1)^n b_n, \qquad n = 1, 2, 3, \ldots.$$

The best proof that $\lim_{n\to\infty} s_n$ exists is obtained by an illustration. In Figure 5 we first plot the point $s_0 = b_0$, and then the point $s_1 = s_0 - b_1$. Next we

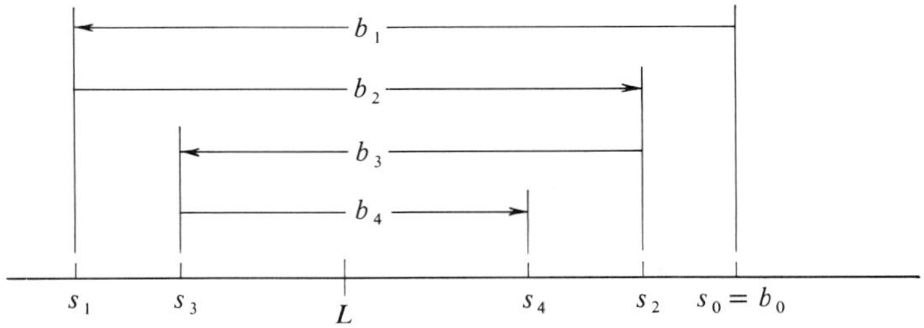

Figure 5

plot $s_2 = s_1 + b_2$ and observe that, since $b_2 \leq b_1$, we have $s_2 \leq s_0$. After that comes $s_3 = s_2 - b_3$ and, since $b_3 \leq b_2$, it follows that $s_1 \leq s_3$. Continuing in this way, we see that the odd-numbered points of the sequence $\{s_n\}$ form an increasing subsequence:

$$s_1 \leq s_3 \leq s_5 \leq \cdots \leq s_{2n-1} \leq \cdots, \tag{1}$$

and the even-numbered points form a decreasing subsequence:

$$s_0 \geq s_2 \geq s_4 \geq \cdots \geq s_{2n} \geq \cdots.$$

Furthermore, every odd-numbered partial sum is less than or equal to every even-numbered one. Thus the increasing sequence (1) is bounded above by any one of the numbers s_{2n}, and it therefore converges [see (1.4), page 479]. That is, there exists a real number L such that

$$\lim_{n\to\infty} s_{2n-1} = L.$$

For every integer $n \geq 1$, we have

$$s_{2n} = s_{2n-1} + b_{2n},$$

and, since it follows from (ii′) that $\lim_{n\to\infty} b_{2n} = 0$, we conclude that

$$\begin{aligned}\lim_{n\to\infty} s_{2n} &= \lim_{n\to\infty} s_{2n-1} + \lim_{n\to\infty} b_{2n} \\ &= L - 0 = L.\end{aligned}$$

We have shown that both the odd-numbered subsequences $\{s_{2n-1}\}$ and the even-numbered subsequence $\{s_{2n}\}$ converge to the same limit L. This implies that $\lim_{n\to\infty} s_n = L$. For, given an arbitrary real number $\epsilon > 0$, we have proved that there exist integers N_1 and N_2 such that

$$\begin{aligned}|s_{2n-1} - L| &< \epsilon, &&\text{whenever } 2n - 1 > N_1, \\ |s_{2n} - L| &< \epsilon, &&\text{whenever } 2n > N_2.\end{aligned}$$

Hence, if n is any integer (odd or even) which is greater than both N_1 and N_2, then $|s_n - L| < \epsilon$. Thus

$$L = \lim_{n\to\infty} s_n = \sum_{i=0}^{\infty} (-1)^i b_i,$$

and the proof is complete.

As an application of Theorem (4.1) consider the alternating harmonic series

$$\sum_{i=1}^{\infty} (-1)^{i+1}\frac{1}{i} = 1 - \tfrac{1}{2} + \tfrac{1}{3} - \tfrac{1}{4} + \cdots.$$

The hypotheses of the theorem are obviously satisfied:

$$\text{(i)} \quad \frac{1}{n+1} \leq \frac{1}{n}, \qquad \text{for every integer } n \geq 1, \text{ and}$$

$$\text{(ii)} \quad \lim_{n\to\infty} (-1)^{n+1}\frac{1}{n} = \lim_{n\to\infty} \frac{1}{n} = 0.$$

Hence it follows that the alternating harmonic series is convergent. It is interesting to compare this series with the ordinary harmonic series $\sum_{i=1}^{\infty} \frac{1}{i} = 1 + \frac{1}{2} + \frac{1}{3} + \frac{1}{4} + \cdots$, which we have shown to be divergent. We see that the alternating harmonic series is a convergent infinite series $\sum_{i=m}^{\infty} a_i$ for which the corresponding series of absolute values $\sum_{i=m}^{\infty} |a_i|$ diverges.

For practical purposes, the value of a convergent infinite series $\sum_{i=m}^{\infty} a_i$ is usually approximated by a partial sum $\sum_{i=m}^{n} a_i$. The **error** in the approximation, denoted by E_n, is the absolute value of the difference between the true value of the series and the approximating partial sum; i.e.,

$$E_n = \left| \sum_{i=m}^{\infty} a_i - \sum_{i=m}^{n} a_i \right|.$$

In general, it is a difficult problem to know how large n must be chosen to ensure that the error E_n be less than a given size. However, for those alternating series which satisfy the hypotheses of Theorem (4.1), the problem is an easy one.

(4.2) *If the alternating series* $\sum_{i=m}^{\infty} a_i$ *satisfies hypotheses* (i) *and* (ii) *of Theorem* (4.1), *then the error* E_n *is less than or equal to the absolute value of the first omitted term. That is,*

$$E_n \leq |a_{n+1}|, \qquad \textit{for every integer } n \geq m.$$

Proof. We shall use the same notation as in the proof of (4.1). Thus we assume that $m = 0$ and that $a_i = (-1)^i b_i$, where $b_i > 0$ for every integer $i \geq 0$. The value of the series is the number L, and the error E_n is therefore given by

$$E_n = \left| \sum_{i=0}^{\infty} a_i - \sum_{i=0}^{n} a_i \right| = |L - s_n|.$$

Since $|a_{n+1}| = b_{n+1}$, the proof is completed by showing that

$$|L - s_n| \le b_{n+1}, \qquad \text{for every integer } n \ge 0.$$

Geometrically, $|L - s_n|$ is the distance between the points L and s_n, and it can be seen immediately from Figure 5 that the preceding inequality is true. To arrive at the conclusion formally, we recall that $\{s_{2n-1}\}$ is an increasing sequence converging to L, and that $\{s_{2n}\}$ is a decreasing sequence converging to L. Thus if n is odd, then $n + 1$ is even and

$$s_n \le L \le s_{n+1}.$$

On the other hand, if n is even, then $n + 1$ is odd and

$$s_{n+1} \le L \le s_n.$$

In either case, we have $|L - s_n| \le |s_{n+1} - s_n|$. Hence, for every integer $n \ge 0$,

$$E_n = |L - s_n| \le |s_{n+1} - s_n| = |a_{n+1}|,$$

and the proof is complete.

In Table 1 we have computed some partial sums which approximate the value of the alternating harmonic series. Each entry in the second column is an approximation, and the corresponding entry in the third column is the upper bound on the error provided by Theorem (4.2).

TABLE 1

Alternating harmonic series: $\sum_{i=1}^{\infty} (-1)^{i+1} \frac{1}{i}$.

Partial sums: $s_n = 1 - \frac{1}{2} + \frac{1}{3} - \cdots + (-1)^{n+1} \frac{1}{n}$.

n	s_n = approximation	$\lvert a_{n+1}\rvert = \frac{1}{n+1}$ = upper bound for error
1	1	$\frac{1}{2}$
2	$\frac{1}{2}$	$\frac{1}{3}$
3	$\frac{5}{6}$	$\frac{1}{4}$
4	$\frac{7}{12}$	$\frac{1}{5}$
10	0.646	0.091
100	0.6882	0.0099
1000	0.6926	0.0010
10,000	0.6931	0.0001

PROBLEMS

1. Determine whether each of the following alternating series converges or diverges. Give the reasons for your answers.

(a) $\sum_{i=1}^{\infty} (-1)^i \frac{1}{\sqrt{i}}$

(b) $\sum_{i=1}^{\infty} (-1)^i \frac{1}{i^2 + 1}$

(c) $\sum_{k=1}^{\infty} (-1)^k \frac{k^2 - 1}{k^2 + 1}$

(d) $\sum_{k=1}^{\infty} (-1)^k \frac{1}{(k^2 + 1)^{1/3}}$

(e) $\sum_{n=2}^{\infty} (-1)^n e^{-n}$

(f) $\sum_{i=1}^{\infty} (-1)^i \frac{1}{\sqrt{2i^3 - 1}}$

(g) $\sum_{i=0}^{\infty} \cos(i\pi)$

(h) $\sum_{k=1}^{\infty} \frac{\cos(k\pi)}{k^2}$.

2. Prove that, for any infinite sequence $\{a_n\}$ of real numbers, $\lim_{n\to\infty} a_n = 0$ if and only if $\lim_{n\to\infty} |a_n| = 0$. (*Hint:* The proof is simple and straightforward. Go directly to the definition of convergence of an infinite sequence.)
3. For each of the series $\sum_{i=m}^{\infty} a_i$ in Problem 1, determine whether or not the corresponding series of absolute values $\sum_{i=m}^{\infty} |a_i|$ converges.
4. Give an example of an alternating series $\sum_{i=m}^{\infty} a_i$ which you can show converges, but which fails to satisfy condition (i) of the Convergence Test (4.1).
5. The first of the following examples comes from the formula for a geometric series, and the last two follow from the theory developed later in this chapter:

(a) $\frac{2}{3} = \frac{1}{1 + \frac{1}{2}} = \sum_{i=0}^{\infty} (-\frac{1}{2})^i = 1 - \frac{1}{2} + \frac{1}{4} - \cdots$.

(b) $\ln 2 = \sum_{i=1}^{\infty} (-1)^{i+1} \frac{1}{i} = 1 - \frac{1}{2} + \frac{1}{3} - \frac{1}{4} + \cdots$.

(c) $\pi = 4 \arctan 1 = \sum_{i=0}^{\infty} (-1)^i \frac{4}{2i + 1} = 4 - \frac{4}{3} + \frac{4}{5} - \frac{4}{7} + \cdots$.

If the value of each of these series is approximated by a partial sum $\sum_{i=m}^{n} a_i$, how large must n be taken to ensure an error no greater than 0.1, 0.01, 0.001, 10^{-6}?

5. Absolute and Conditional Convergence. An infinite series $\sum_{i=m}^{\infty} a_i$ is said to be **absolutely convergent** if the corresponding series of absolute values $\sum_{i=m}^{\infty} |a_i|$ is convergent. If a series $\sum_{i=m}^{\infty} a_i$ converges, but $\sum_{i=m}^{\infty} |a_i|$ does not, then we say that $\sum_{i=m}^{\infty} a_i$ is **conditionally convergent.** An example of a conditionally convergent series is the alternating harmonic series: We have shown that

$$\sum_{i=1}^{\infty} a_i = \sum_{i=1}^{\infty} (-1)^{i+1} \frac{1}{i} = 1 - \frac{1}{2} + \frac{1}{3} - \frac{1}{4} + \cdots$$

converges, but that

$$\sum_{i=1}^{\infty} |a_i| = \sum_{i=1}^{\infty} \frac{1}{i} = 1 + \tfrac{1}{2} + \tfrac{1}{3} + \tfrac{1}{4} + \cdots$$

diverges.

There are many examples of series for which both $\sum_{i=m}^{\infty} a_i$ and $\sum_{i=m}^{\infty} |a_i|$ converge, and also many where both diverge. (In particular, for nonnegative series, the two are the same.) There is the remaining possibility that $\sum_{i=m}^{\infty} |a_i|$ might converge, and $\sum_{i=m}^{\infty} a_i$ diverge. However, the following theorem shows that this cannot happen.

(5.1) THEOREM. *If the infinite series $\sum_{i=m}^{\infty} a_i$ is absolutely convergent, then it is convergent.*

Proof. Since $|a_i| \geq -a_i$, we have $a_i + |a_i| \geq 0$, for every integer $i \geq m$. Hence the series $\sum_{i=m}^{\infty} (a_i + |a_i|)$ is nonnegative. Since $a_i \leq |a_i|$, we also have

$$a_i + |a_i| \leq |a_i| + |a_i| = 2|a_i|, \tag{1}$$

for every integer $i \geq m$. The assumption that $\sum_{i=m}^{\infty} a_i$ is absolutely convergent means that the series $\sum_{i=m}^{\infty} |a_i|$ converges, and, hence, so does the series $\sum_{i=m}^{\infty} 2|a_i|$. It therefore follows from (1) by the Comparison Test that the nonnegative series $\sum_{i=m}^{\infty} (a_i + |a_i|)$ is convergent. We conclude from Theorem (2.2), page 485, that

$$\sum_{i=m}^{\infty} a_i = \sum_{i=m}^{\infty} (a_i + |a_i|) - \sum_{i=m}^{\infty} |a_i|$$

and that $\sum_{i=m}^{\infty} a_i$ converges. This completes the proof.

Thus the only possibilities for a given series are those illustrated by the following scheme:

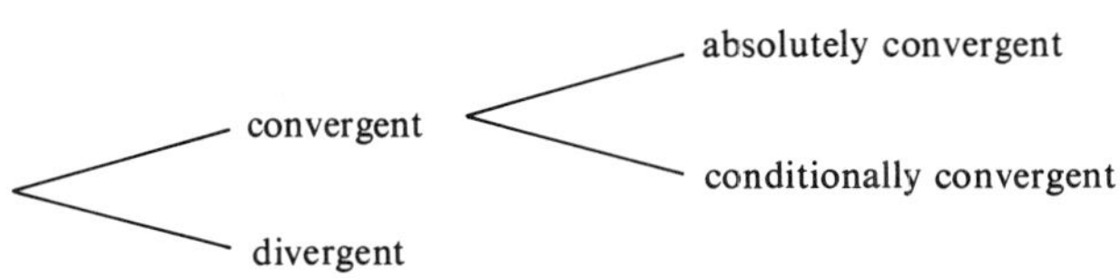

EXAMPLE 1. Classify each of the following infinite series as absolutely convergent, conditionally convergent, or divergent.

(a) $\displaystyle\sum_{k=1}^{\infty} (-1)^k \frac{1}{\sqrt{k+1}}$, (b) $\displaystyle\sum_{k=1}^{\infty} (-1)^k \frac{1}{2k^2 - 15}$.

If we let $a_k = (-1)^k \dfrac{1}{\sqrt{k+1}}$, the alternating series in (a) will converge if:

(i) $|a_{k+1}| \leq |a_k|$, for every integer $k \geq 1$, and

(ii) $\lim_{k\to\infty} |a_k| = 0$.

[See Theorem (4.1), page 498.] We have

$$|a_k| = \frac{1}{\sqrt{k+1}} \quad \text{and} \quad |a_{k+1}| = \frac{1}{\sqrt{k+2}}\cdot$$

Hence condition (i) becomes

$$\frac{1}{\sqrt{k+2}} \leq \frac{1}{\sqrt{k+1}}, \quad \text{for every integer } k \geq 1,$$

which is certainly true. Condition (ii) is also satisfied, since

$$\lim_{k\to\infty} \frac{1}{\sqrt{k+1}} = 0,$$

and it follows that the series $\sum_{k=1}^{\infty} a_k$ converges. However, it is easy to show that $\sum_{k=1}^{\infty} |a_k|$ diverges by either the Comparison Test or the Integral Test. Using the latter, we consider the function f defined by $f(x) = \dfrac{1}{\sqrt{x+1}}$, which is nonnegative and decreasing on the interval $[1, \infty)$. We have $f(k) = \dfrac{1}{\sqrt{k+1}} = |a_k|$ and

$$\int_1^{\infty} f(x)\,dx = \int_1^{\infty} \frac{1}{\sqrt{x+1}}\,dx = \lim_{b\to\infty} [2\sqrt{x+1}\,\Big|_1^b]$$
$$= \lim_{b\to\infty} [2\sqrt{b+1} - 2\sqrt{2}] = \infty.$$

The divergence of the integral implies the divergence of the corresponding series $\sum_{k=1}^{\infty} |a_k|$, and we conclude that the series (a) is conditionally convergent.

For the series in (b), we might apply the same technique: Test first for convergence and then for absolute convergence. However, if we suspect that the series is absolutely convergent, we may save a step by first testing for absolute convergence. In this particular case, the corresponding series of absolute values is $\sum_{k=1}^{\infty} \dfrac{1}{|2k^2 - 15|}\cdot$ The latter can be shown to be convergent by the Comparison Test. For a test series we choose the convergent series $\sum_{k=1}^{\infty} \frac{2}{k^2}$. The condition of the test is that the inequality

$$\frac{1}{|2k^2 - 15|} \leq \frac{2}{k^2}$$

must be true eventually. We shall consider only integers $k \geq 3$, since, for these values, $2k^2 \geq 18$ and hence $|2k^2 - 15| = 2k^2 - 15$. For those integers for which $k \geq 3$, the inequality

$$\frac{1}{2k^2 - 15} \leq \frac{2}{k^2}$$

is equivalent to $k^2 \leq 4k^2 - 30$, which in turn is equivalent to $k^2 \geq 10$. The last is true for every integer $k \geq 4$. Hence

$$\frac{1}{|2k^2 - 15|} \leq \frac{2}{k^2}, \qquad \text{for every integer } k \geq 4.$$

It follows that $\sum_{k=1}^{\infty} \frac{1}{|2k^2 - 15|}$ converges, and therefore that the series (b) is absolutely convergent.

(5.2) RATIO TEST. Let $\sum_{i=m}^{\infty} a_i$ *be an infinite series for which* $\lim_{n \to \infty} \frac{|a_{n+1}|}{|a_n|} = q$ (*or* ∞).

(i) *If* $q < 1$, *then the series is absolutely convergent.*

(ii) *If* $q > 1$ (*including* $q = \infty$), *then the series is divergent.*

(iii) *If* $q = 1$, *then the series may either converge or diverge;* i.e., *the test fails.*

Proof. Suppose, first of all, that $\lim_{n \to \infty} \frac{|a_{n+1}|}{|a_n|} = q < 1$. This implies that the ratio $\frac{|a_{n+1}|}{|a_n|}$ is arbitrarily close to q if n is sufficiently large. Hence if we pick an arbitrary number r such that $q < r < 1$, then there exists an integer $N \geq m$ such that

$$\frac{|a_{n+1}|}{|a_n|} \leq r, \qquad \text{for every integer } n \geq N. \tag{2}$$

We shall show by mathematical induction that (2) implies that

$$|a_{N+i}| \leq r^i |a_N|, \qquad \text{for every integer } i \geq 0. \tag{3}$$

If $i = 0$, then the inequality in (3) becomes $|a_{N+0}| \leq r^0 |a_N|$, which is true. In the second part of an inductive proof we need to show that, if the inequality

(3) is true for $i = k$, then it is also true for $i = k + 1$. The assumption, then, is that

$$|a_{N+k}| \leq r^k |a_N|, \tag{4}$$

and we want to prove that

$$|a_{N+k+1}| \leq r^{k+1} |a_N|.$$

If we multiply both sides of inequality (4) by the positive number r, we get

$$r|a_{N+k}| \leq r^{k+1} |a_N|. \tag{5}$$

But, inequality (2) tells us that

$$\frac{|a_{N+k+1}|}{|a_{N+k}|} \leq r,$$

and hence that

$$|a_{N+k+1}| \leq r|a_{N+k}|. \tag{6}$$

Combining inequalities (5) and (6) we have

$$|a_{N+k+1}| \leq r^{k+1} |a_N|,$$

completing the inductive proof. Since $|r| < 1$, the geometric series $\sum_{i=0}^{\infty} |a_N| r^i$ converges, and it follows from (3) by the Comparison Test that the series $\sum_{i=0}^{\infty} |a_{N+i}|$ converges. However,

$$\sum_{i=0}^{\infty} |a_{N+i}| = \sum_{i=N}^{\infty} |a_i|,$$

and the convergence of $\sum_{i=N}^{\infty} |a_i|$ implies the convergence of $\sum_{i=m}^{\infty} |a_i|$. Hence the series $\sum_{i=m}^{\infty} a_i$ converges absolutely, and the proof of part (i) of the theorem is complete.

We next assume that $\lim_{n \to \infty} \frac{|a_{n+1}|}{|a_n|} = q > 1$, and let r be an arbitrary number such that $1 < r < q$. Then there exists an integer $N \geq m$ such that

$$\frac{|a_{n+1}|}{|a_n|} \geq r, \qquad \text{for every integer } n \geq N.$$

In the same way in which we proved that (2) implies (3), it follows by induction from the preceding inequality that

$$|a_{N+i}| \geq r^i |a_N|, \qquad \text{for every integer } i \geq 0.$$

Since $r > 1$, we know that $\lim_{i \to \infty} r^i = \infty$ (see Problem 5, page 481), and therefore also that

$$\lim_{n \to \infty} |a_n| = \lim_{i \to \infty} |a_{N+i}| = \infty.$$

However, if the series $\sum_{i=m}^{\infty} a_i$ converges, then it necessarily follows that $\lim_{n \to \infty} |a_n| = \lim_{n \to \infty} a_n = 0$. [See Theorem (2.1), page 483, and Problem 2, page 502.] Hence $\sum_{i=m}^{\infty} a_i$ diverges, and part (ii) is proved.

Part (iii) is proved by giving an example of an absolutely convergent series and one of a divergent series such that $q = 1$ for each of them. Consider the convergent p-series $\sum_{i=1}^{\infty} \frac{1}{i^2}$, which, being nonnegative, is also absolutely convergent. Setting $a_n = \frac{1}{n^2}$, we obtain

$$a_{n+1} = \frac{1}{(n+1)^2} = \frac{1}{n^2 + 2n + 1}$$

and

$$\frac{|a_{n+1}|}{|a_n|} = \frac{a_{n+1}}{a_n} = \frac{n^2}{n^2 + 2n + 1} = \frac{1}{1 + \frac{2}{n} + \frac{1}{n^2}}.$$

Hence

$$\lim_{n\to\infty} \frac{|a_{n+1}|}{|a_n|} = \lim_{n\to\infty} \frac{1}{1 + \frac{2}{n} + \frac{1}{n^2}} = 1.$$

For the second example, we take the divergent harmonic series $\sum_{i=1}^{\infty} \frac{1}{i}$. If we let $a_n = \frac{1}{n}$, then $a_{n+1} = \frac{1}{n+1}$ and

$$\frac{|a_{n+1}|}{|a_n|} = \frac{a_{n+1}}{a_n} = \frac{n}{n+1} = \frac{1}{1 + \frac{1}{n}}.$$

For this series we also get

$$\lim_{n\to\infty} \frac{|a_{n+1}|}{|a_n|} = \lim_{n\to\infty} \frac{1}{1 + \frac{1}{n}} = 1.$$

The Ratio Test is therefore inconclusive if $q = 1$, and this completes the proof.

If n is an arbitrary positive integer, the product $n(n-1)\cdots 3 \cdot 2 \cdot 1$ is called ***n* factorial** and is denoted by $n!$ Thus $3! = 3 \cdot 2 \cdot 1 = 6$ and $5! = 5 \cdot 4 \cdot 3 \cdot 2 \cdot 1 = 120$. Although it may seem strange, $0!$ is also defined and has the value 1. A convenient recursive definition of the factorial is given by the formulas

$$0! = 1,$$
$$(n+1)! = (n+1)n!, \qquad \text{for every integer } n \geq 0.$$

EXAMPLE 2. Prove that the following series converges:

$$\sum_{n=0}^{\infty} \frac{1}{n!} = 1 + 1 + \frac{1}{2!} + \frac{1}{3!} + \cdots.$$

We write the series as $\sum_{i=0}^{\infty} a_n$ by defining $a_n = \dfrac{1}{n!}$ for every integer $n \geq 0$. Then

$$\frac{|a_{n+1}|}{|a_n|} = \frac{\dfrac{1}{(n+1)!}}{\dfrac{1}{n!}} = \frac{n!}{(n+1)!}$$

$$= \frac{n!}{(n+1)n!} = \frac{1}{n+1}.$$

Hence

$$q = \lim_{n \to \infty} \frac{|a_{n+1}|}{|a_n|} = \lim_{n \to \infty} \frac{1}{n+1} = 0.$$

Since $q < 1$, it follows from the Ratio Test that the series is absolutely convergent. But absolute convergence implies convergence [Theorem (5.1)], and we conclude that the series $\displaystyle\sum_{n=0}^{\infty} \frac{1}{n!}$ converges.

EXAMPLE 3. Show that the infinite series

$$\sum_{i=1}^{\infty} ir^{i-1} = 1 + 2r + 3r^2 + 4r^3 + \cdots$$

converges absolutely if $|r| < 1$ and diverges if $|r| \geq 1$. This series is related to the geometric series $\sum_{i=0}^{\infty} r^i = 1 + r + r^2 + \cdots$, and in a later section we shall make use of the relationship. To settle the immediate question of convergence, however, we set $a_i = ir^{i-1}$ for every positive integer i, and write the series as $\sum_{i=1}^{\infty} a_i$. Observe, first of all, that if $|r| \geq 1$, then $|a_n| = n|r|^{n-1}$ and

$$\lim_{n \to \infty} |a_n| = \lim_{n \to \infty} n|r|^{n-1} = \infty.$$

Hence, if $|r| \geq 1$, the series must diverge, since convergence would imply $\lim_{n \to \infty} |a_n| = 0$. This proves the second part of what is asked, and we now assume that $|r| < 1$. If $r = 0$, the series is absolutely convergent with value 1, so we further assume that $r \neq 0$. Then

$$\frac{|a_{n+1}|}{|a_n|} = \frac{(n+1)|r|^n}{n|r|^{n-1}} = \frac{n+1}{n}|r| = \left(1 + \frac{1}{n}\right)|r|,$$

and so

$$\lim_{n \to \infty} \frac{|a_{n+1}|}{|a_n|} = \lim_{n \to \infty} \left(1 + \frac{1}{n}\right)|r| = |r|.$$

Thus $q = |r| < 1$, and the Ratio Test therefore implies that the series is absolutely convergent.

The next theorem, with which we conclude the section, establishes a useful inequality.

(5.3) *If the series* $\sum_{i=m}^{\infty} a_i$ *converges, then* $|\sum_{i=m}^{\infty} a_i| \leq \sum_{i=m}^{\infty} |a_i|$.

The result is true even if $\sum_{i=m}^{\infty} a_i$ is not absolutely convergent, for in that case $\sum_{i=m}^{\infty} |a_i| = \infty$, and the inequality becomes $|\sum_{i=m}^{\infty} a_i| \leq \infty$.

Proof. In view of the preceding remark, we shall assume throughout the proof that $\sum_{i=m}^{\infty} |a_i|$ converges. Let $\{s_n\}$ be the sequence of partial sums corresponding to the series $\sum_{i=m}^{\infty} a_i$. Then

$$s_n = \sum_{i=m}^{n} a_i, \qquad \text{for every integer } n \geq m,$$

and the assumption that $\sum_{i=m}^{\infty} a_i$ converges means that the sequence $\{s_n\}$ converges and that

$$\lim_{n\to\infty} s_n = \sum_{i=m}^{\infty} a_i. \tag{7}$$

The general fact that $|a + b| \leq |a| + |b|$, for any two real numbers a and b, can be extended to any finite number of summands, and we therefore have

$$|s_n| = \left|\sum_{i=m}^{n} a_i\right| \leq \sum_{i=m}^{n} |a_i|.$$

Furthermore,

$$\sum_{i=m}^{n} |a_i| \leq \sum_{i=m}^{\infty} |a_i|$$

[see (3.2), page 490, and (1.3), page 478). Hence

$$|s_n| \leq \sum_{i=m}^{\infty} |a_i|, \qquad \text{for every integer } n \geq m. \tag{8}$$

It follows from (8) that

$$|\lim_{n\to\infty} s_n| \leq \sum_{i=m}^{\infty} |a_i|. \tag{9}$$

[It is easy to see that (8) implies (9) if we regard the numbers s_n and $\sum_{i=m}^{\infty} |a_i|$ as points on the line. The geometric statement of (8) is that all the points s_n lie in the closed interval whose endpoints are $-\sum_{i=m}^{\infty} |a_i|$ and $\sum_{i=m}^{\infty} |a_i|$. If

(9) were false, it would mean that $\lim_{n\to\infty} s_n$ lay outside this interval, a positive distance away from it. But this cannot happen, since s_n is arbitrarily close to $\lim_{n\to\infty} s_n$ for n sufficiently large.] Combining (7) and (9), we obtain the inequality which was to be proved.

PROBLEMS

1. Classify each of the following infinite series as absolutely convergent, conditionally convergent, or divergent. Show how you obtain your answer starting from a standard test or series.

(a) $\sum_{i=0}^{\infty} (-1)^i \frac{1}{2i-3}$

(b) $\sum_{k=1}^{\infty} \frac{1}{(k^3+1)^{1/2}}$

(c) $\sum_{k=1}^{\infty} (-1)^k \frac{1}{(k+1)^{2/3}}$

(d) $\sum_{i=1}^{\infty} \frac{i2^i}{3^{i+1}}$

(e) $\sum_{n=1}^{\infty} (-1)^n \frac{5^n}{4^{n+1}}$

(f) $\sum_{k=0}^{\infty} \frac{100^k}{k!}$

(g) $\sum_{k=1}^{\infty} (-1)^k \frac{k!}{100k}$

(h) $\sum_{i=1}^{\infty} (-1)^i e^{-i^2}$

2. (a) Prove that the series $\sum_{n=0}^{\infty} \frac{n}{2^n}$ is absolutely convergent.

(b) Prove that $\lim_{n\to\infty} \frac{n}{2^n} = 0$.

3. Classify each of the following series as absolutely convergent, conditionally convergent, or divergent. Show how you obtain your answer.

(a) $\sum_{n=1}^{\infty} \frac{\ln n}{n}$

(b) $\sum_{n=1}^{\infty} (-1)^n \frac{\ln n}{n}$

(c) $\sum_{n=1}^{\infty} (-1)^n \frac{\ln n}{n^3}$

(d) $\sum_{n=1}^{\infty} \frac{\ln n}{n^2}$.

4. (a) Prove that, for every positive number a, the series $\sum_{i=0}^{\infty} \frac{a^i}{i!}$ is absolutely convergent.

(b) Prove that $\lim_{n\to\infty} \frac{a^n}{n!} = 0$ for every positive number a.

5. The infinite series

$$\sum_{n=0}^{\infty} a_n = 1 + \frac{1}{2} + \frac{1}{2\cdot 3} + \frac{1}{2^2 3} + \frac{1}{2^2 3^2} + \frac{1}{2^3 3^2} + \frac{1}{2^3 3^3} + \cdots$$

is defined, for every integer $n \geq 0$, by the two equations:

$$a_{2n} = \frac{1}{2^n 3^n}$$

$$a_{2n+1} = \frac{1}{2^{n+1} 3^n}.$$

(a) Show that $\sum_{n=0}^{\infty} a_n$ is absolutely convergent.

(b) What is $\lim_{n \to \infty} \frac{|a_{n+1}|}{|a_n|}$?

6. As a corollary of (5.1), prove the following extension of the Comparison Test: *The series $\sum_{i=m}^{\infty} a_i$ is absolutely convergent if there exists an absolutely convergent series $\sum_{i=l}^{\infty} b_i$ such that $|a_i| \leq |b_i|$ eventually.*

6. Power Series. Associated with every infinite sequence of real numbers $a_0, a_1, a_2, \ldots$ and every real number x, there is the infinite series

$$\boxed{\sum_{i=0}^{\infty} a_i x^i = a_0 + a_1 x + a_2 x^2 + \cdots.}$$

Such a series is called a **power series in x**. As a general rule, it will converge for some values of x, but not all. For example, the geometric series

$$1 + x + x^2 + x^3 + \cdots$$

converges and has the same value as $\frac{1}{1 - x}$ for every real number x in absolute value less than 1, but it diverges for all other values of x. A power series which converges for some real number c, i.e., which converges if $x = c$, is commonly said to converge at c. Note that every power series in x converges at 0, since, if $x = 0$, then

$$\sum_{i=0}^{\infty} a_i x^i = a_0 + a_1 0 + a_2 0^2 + \cdots = a_0.$$

The following proposition is the basic theorem in studying the convergence of power series:

(6.1) *If a power series $\sum_{i=0}^{\infty} a_i x^i$ converges for some real number c, then it is absolutely convergent for every real number x such that $|x| < |c|$.*

Proof. If $c = 0$, the result is vacuously true, so we shall assume that $c \neq 0$. The fact that the series $\sum_{i=0}^{\infty} a_i c^i$ converges implies that $\lim_{n \to \infty} a_n c^n = 0$. Hence there exists a nonnegative integer N such that $|a_n c^n| \leq 1$, for every integer $n \geq N$. Since $|a_n c^n| = |a_n|\, |c|^n$, it follows that

$$|a_n| \leq \frac{1}{|c|^n},$$

and thence that

$$|a_n x^n| = |a_n|\, |x|^n \leq \frac{|x|^n}{|c|^n},$$

for every real number x and for every integer $n \geq N$. We now impose the restriction that $|x| < |c|$, and set $r = \dfrac{|x|}{|c|}$. Then $r < 1$, and

$$|a_n x^n| \leq r^n, \qquad \text{for every integer } n \geq N.$$

That is, we have shown that $|a_n x^n| \leq r^n$ eventually. Since the geometric series $\sum_{i=0}^{\infty} r^i$ converges if $|r| < 1$, it follows by the Comparison Test that $\sum_{i=0}^{\infty} |a_i x^i|$ converges. This completes the proof.

We shall derive three corollaries of (6.1). The first asserts that *the set of all real numbers x at which a power series $\sum_{i=0}^{\infty} a_i x^i$ converges is a nonempty interval on the real line*. The set is nonempty because, as is remarked above, it contains the number 0. A set of real numbers is an interval if, whenever it contains two numbers, it contains every number in between those two. Thus we must prove that if the series converges at a and at c and if $a < b < c$, then it also converges at b. This is quickly done. Suppose first that $b \geq 0$. Then

$$|b| = b < c = |c|,$$

and (6.1) implies that the series converges at b. On the other hand, if $b < 0$, then

$$|b| = -b < -a = |a|,$$

and it again follows from (6.1) that the series converges at b. This completes the proof, and, as a result, we call the set of all numbers at which a power series converges the **interval of convergence** of the power series.

A number a is called an **interior point** of a set S of real numbers if there exists an open interval which contains a and which is a subset of S. The set of all interior points of S is called **the interior** of S. For example, if S is itself an open interval, then all its points are interior points and hence S equals its own interior. More generally, the interior of an arbitrary interval consists of the interval with its endpoints deleted.

The second corollary states that *a power series converges absolutely at every interior point of its interval of convergence*. The proof is virtually the same as that of the first corollary. Let b be an arbitrary interior point of the interval of converence. Because it is an interior point, we know there exist real numbers a and c which also lie in the interval and for which $a < b < c$. As before, if $b \geq 0$, then $|b| < |c|$, but if $b < 0$, then $|b| < |a|$. In either case it follows from (6.1) that the series converges absolutely at b, and this completes the argument.

The third corollary is the following: *The interior of the interval I of convergence of a power series $\sum_{i=0}^{\infty} a_i x^i$ is symmetric about the origin.* That is, if b is an interior point of I, then so is $-b$. Again, there exist numbers a and c in I such that $a < b < c$. Now consider the open interval $(-c, -a)$. It certainly contains $-b$, and, if we can show that $(-c, -a)$ is a subset of I, then we shall have proved that $-b$ is an interior point of I. Let x be an arbitrary number in $(-c, -a)$, that is, $-c < x < -a$. There are the, by now familiar, two possibilities: If $x \geq 0$, then $-a > 0$ and

$$|x| = x < -a = |a|.$$

If $x < 0$, then $-c < 0$, whence $c > 0$, and

$$|x| = -x < c = |c|.$$

For either possibility, the convergence of the series at x is implied by (6.1), and so the third corollary is proved.

If a power series converges for every real number, then its interval of convergence is the set of all real numbers. The only other possibility, according to the third corollary above, is that the interval of convergence is bounded with symmetrically located endpoints $-\rho$ and ρ. We define the **radius of convergence** of a power series $\sum_{i=0}^{\infty} a_i x^i$ to be infinite if the interval of convergence is the set $(-\infty, \infty)$ of all real numbers, and to be the right endpoint of the interval of convergence if the interval is bounded. The preceding results are then summarized in Figure 6 below and in the following theorem:

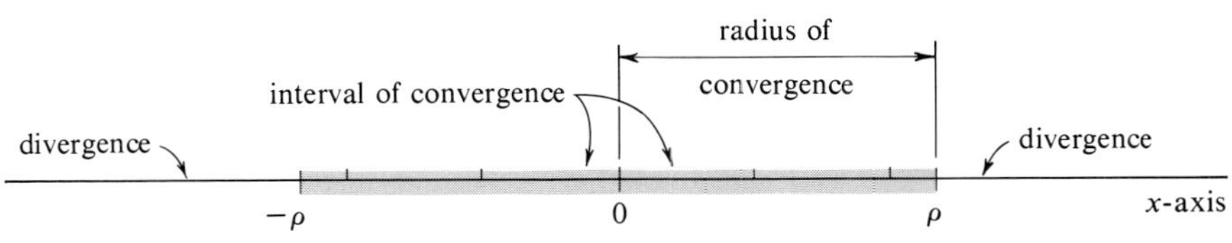

Figure 6

(6.2) THEOREM. *If a power series $\sum_{i=0}^{\infty} a_i x^i$ has radius of convergence ρ, then the series converges absolutely at every x in the open interval $(-\rho, \rho)$. If ρ is not infinite, then $-\rho$ and ρ are the endpoints of the interval of convergence.*

It is important to realize that, when ρ is finite, we have made no prediction as to whether the series converges or diverges at ρ and at $-\rho$. All we know is that it converges absolutely in the open interval $(-\rho, \rho)$ and diverges outside the closed interval $[-\rho, \rho]$.

EXAMPLE 1. Find the interval and radius of convergence of each of the following power series:

$$\text{(a)} \quad \sum_{i=0}^{\infty} \frac{x^i}{i!} = 1 + x + \frac{x^2}{2!} + \frac{x^3}{3!} + \cdots.$$

$$\text{(b)} \quad \sum_{k=1}^{\infty} (-1)^{k-1} \frac{x^k}{k} = x - \frac{x^2}{2} + \frac{x^3}{3} - \cdots.$$

$$\text{(c)} \quad \sum_{n=0}^{\infty} n!x^n = 1 + x + 2x^2 + 3!x^3 + \cdots.$$

In many examples the radius of convergence can be found easily using the Ratio Test.

For the series in (a), we set $u_i = \dfrac{x^i}{i!}$ and form the ratio

$$\frac{|u_{n+1}|}{|u_n|} = \left|\frac{x^{n+1}}{(n+1)!}\right| \left|\frac{n!}{x^n}\right| = |x| \frac{n!}{(n+1)!}.$$

Since $(n + 1)! = (n + 1)n!$,

$$\frac{|u_{n+1}|}{|u_n|} = |x| \frac{1}{n+1}.$$

Hence

$$\lim_{n \to \infty} \frac{|u_{n+1}|}{|u_n|} = \lim_{n \to \infty} \frac{|x|}{n+1} = 0.$$

It follows from the Ratio Test that the series $\sum_{i=0}^{\infty} \frac{x^i}{i!}$ is absolutely convergent for every real number x. Hence the interval of convergence is the entire real line, and the radius of convergence is infinite.

Let $u_k = (-1)^{k-1} \dfrac{x^k}{k}$ for the series in (b). For every integer $k \geq 1$, we have $|u_k| = \dfrac{|x|^k}{k}$. Hence

$$\frac{|u_{k+1}|}{|u_k|} = \frac{|x|^{k+1}}{k+1} \frac{k}{|x|^k} = |x| \frac{k}{k+1}.$$

Since

$$\lim_{k \to \infty} \frac{k}{k+1} = \lim_{k \to \infty} \frac{1}{1 + \frac{1}{k}} = 1,$$

we obtain

$$\lim_{k\to\infty} \frac{|u_{k+1}|}{|u_k|} = |x| \lim_{k\to\infty} \frac{k}{k+1} = |x|.$$

The Ratio Test therefore implies that the series $\sum_{k=0}^{\infty} (-1)^{k-1} \frac{x^k}{k}$ converges absolutely if $|x| < 1$ and diverges if $|x| > 1$. It follows that the endpoints of the interval of convergence are the numbers -1 and 1 and that the radius of convergence is 1. If $x = 1$, the series becomes

$$\sum_{k=1}^{\infty} (-1)^{k-1} \frac{1}{k} = 1 - \tfrac{1}{2} + \tfrac{1}{3} - \tfrac{1}{4} + \cdots,$$

which is the convergent alternating harmonic series. On the other hand, if $x = -1$, the series becomes

$$\sum_{k=1}^{\infty} (-1)^{k-1} \frac{(-1)^k}{k} = \sum_{k=1}^{\infty} (-1)^{2k-1} \frac{1}{k}.$$

Since $2k - 1$ is always an odd integer, we have $(-1)^{2k-1} = -1$, and so

$$\sum_{k=1}^{\infty} (-1)^{2k-1} \frac{1}{k} = -\sum_{k=1}^{\infty} \frac{1}{k},$$

which diverges. Hence the interval of convergence of the power series $\sum_{k=1}^{\infty} (-1)^{k-1} \frac{x^k}{k}$ is the half-open interval $(-1, 1]$.

For the series in (c), let $u_n = n!x^n$. We then get

$$\frac{|u_{n+1}|}{|u_n|} = \frac{|(n+1)!x^{n+1}|}{|n!x^n|} = (n+1)|x|.$$

It follows that

$$\lim_{n\to\infty} \frac{|u_{n+1}|}{|u_n|} = \lim_{n\to\infty} (n+1)|x| = \begin{cases} \infty, & \text{if } x \neq 0, \\ 0, & \text{if } x = 0. \end{cases}$$

From the Ratio Test we conclude that the series $\sum_{n=0}^{\infty} n!x^n$ converges only at $x = 0$. The radius of convergence is therefore equal to 0, and the interval of convergence contains the one number 0.

A significant generalization of the definition of power series can be made as follows: Consider an arbitrary real number a and an infinite sequence of real numbers $a_0, a_1, a_2, \ldots$. For every real number x, a **power series in $x - a$** is defined by

$$\sum_{i=0}^{\infty} a_i(x-a)^i = a_0 + a_1(x-a) + a_2(x-a)^2 + \cdots \tag{1}$$

The power series in x studied earlier in this section are simply instances of the present definition for which $a = 0$.

Fortunately, it is not necessary to start from the beginning again to develop the theory of convergence of power series in $x - a$. Consider the power series in y obtained by making the substitution $x - a = y$ in the series (1). We obtain

$$\sum_{i=0}^{\infty} a_i y^i = a_0 + a_1 y + a_2 y^2 + \cdots. \tag{2}$$

Let I be the set of all real numbers y for which (2) converges, i.e., the interval of convergence of the power series (2). Similarly, let J be the set of all real numbers x for which (1) converges. Since $x - a = y$, or equivalently, $x = y + a$, a number b will belong to I if and only if $b + a$ belongs to J. Thus the set J consists of all numbers of the form $b + a$, where b belongs to I. Symbolically, we write

$$J = I + a.$$

Figure 7

Geometrically, therefore, the set J is obtained by translating the interval I along the real line a distance $|a|$, translating to the right if $a > 0$, and to the left if $a < 0$ (see Figure 7). Hence, the set J of all real numbers x for which $\sum_{i=0}^{\infty} a_i(x-a)^i$ converges is also an interval, and is called the **interval of convergence** of that series. The corresponding **radius of convergence** is equal to the radius of convergence of the series (2), but this time it should be interpreted (if it is not infinite) as the distance from the number a to the right endpoint of the interval of convergence J. The interior of J is, of course, symmetric about the number a.

A power series $\sum_{i=0}^{\infty} a_i(x - a)^i$ is frequently called a power series about the number a. If we recall that $|x - a|$ is geometrically equal to the distance between x and a, we see that the appropriate analogue of theorem (6.2) is:

(6.3) THEOREM. *If the power series $\sum_{i=0}^{\infty} a_i(x - a)^i$ has radius of convergence ρ, then the series converges absolutely at every x such that $|x - a| < \rho$. If ρ is not infinite, the series diverges at every x such that $|x - a| > \rho$.*

EXAMPLE 2. Find the radius and interval of convergence of the power series

$$\sum_{i=0}^{\infty} \frac{1}{3^i}(x + 2)^i = 1 + \frac{x + 2}{3} + \frac{(x + 2)^2}{3^2} + \cdots.$$

Observe, first of all, that this is a power series about the number -2. That is, to put it in the standard form (1), we must write the series as

$$\sum_{i=0}^{\infty} \frac{1}{3^i}(x - (-2))^i.$$

Let us set $u_n = \frac{1}{3^n}(x + 2)^n$ and apply the Ratio Test. For every integer $n \geq 0$, we obtain

$$\frac{|u_{n+1}|}{|u_n|} = \left|\frac{(x + 2)^{n+1}}{3^{n+1}}\right| \left|\frac{3^n}{(x + 2)^n}\right| = \frac{|x + 2|}{3}.$$

Hence

$$q = \lim_{n\to\infty} \frac{|u_{n+1}|}{|u_n|} = \frac{|x + 2|}{3}.$$

The series is therefore absolutely convergent if $q < 1$, or, equivalently, if $|x + 2| < 3$, and it is divergent if $q > 1$, or, equivalently, if $|x + 2| > 3$. Remember that $|x + 2|$ is the distance between x and -2. Thus the interior of the interval of convergence is the set of all real numbers whose distance from -2 is less than 3. Hence the radius of convergence is equal to 3, and the endpoints of the interval of convergence are the numbers -5 and 1. If $x = -5$, the series becomes

$$\sum_{i=0}^{\infty} \frac{1}{3^i}(-5 + 2)^i = \sum_{i=0}^{\infty} \frac{(-3)^i}{3^i}$$

$$= \sum_{i=0}^{\infty} (-1)^i = 1 - 1 + 1 - 1 + 1 - \cdots,$$

which diverges. If $x = 1$, we also obtain a divergent series,

$$\sum_{i=0}^{\infty} \frac{1}{3^i}(1+2)^i = \sum_{i=0}^{\infty} 1 = 1 + 1 + 1 + 1 \cdots.$$

It follows that the interval of convergence of the power series

$$\sum_{i=0}^{\infty} \frac{1}{3^i}(x+2)^i$$

is the open interval $(-5, 1)$ (see Figure 8).

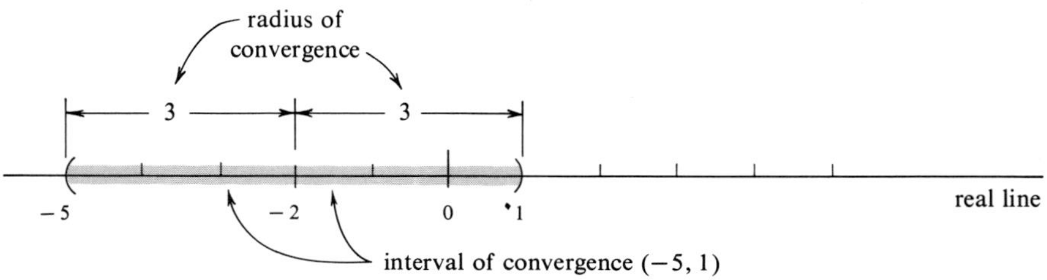

Figure 8

PROBLEMS

1. Find the radius of convergence of each of the following power series.

(a) $\sum_{i=0}^{\infty} \frac{x^i}{2^i}$

(b) $\sum_{i=1}^{\infty} \frac{x^i}{i^2}$

(c) $\sum_{k=1}^{\infty} \frac{x^k}{\sqrt{k}}$

(d) $\sum_{k=0}^{\infty} x^k$

(e) $\sum_{k=0}^{\infty} (-1)^k x^k$

(f) $\sum_{i=0}^{\infty} 2^i y^i$

(g) $\sum_{i=1}^{\infty} i x^{i-1}$

(h) $\sum_{n=0}^{\infty} \frac{y^n}{n(n+1)3^n}$.

2. Find the interval of convergence of each of the power series in Problem 1.
3. Is the following statement true or false: Every power series $\sum_{i=0}^{\infty} a_i x^i$ converges absolutely only in the interior of its interval of convergence? Why?

4. Find the radius of convergence of each of the following power series.

(a) $\sum_{i=1}^{\infty} \frac{(x-2)^i}{i}$

(b) $\sum_{i=0}^{\infty} \frac{(x-2)^i}{i!}$

(c) $\sum_{k=0}^{\infty} \frac{k}{k+1}(x+2)^k$

(d) $\sum_{n=0}^{\infty} \frac{n!}{2^n}(x-1)^n$

(e) $\sum_{k=1}^{\infty} \frac{k^2}{5^k}(y+1)^k$

(f) $\sum_{k=0}^{\infty} (-1)^{k-1}\frac{x^{2k+1}}{(2k+1)!}$.

5. Find the interval of convergence of each of the power series in Problem 4.

6. Prove the following: If $\lim_{n\to\infty} \frac{|a_{n+1}|}{|a_n|} = \rho$, then the radius of convergence of the power series $\sum_{i=0}^{\infty} a_i(x-a)^i$ is equal to $\frac{1}{\rho}$. (Assume that $\frac{1}{0} = \infty$ and that $\frac{1}{\infty} = 0$.)

7. Functions Defined by Power Series. For every power series

$$\sum_{i=0}^{\infty} a_i(x-a)^i,$$

the **function defined by the power series** is the function f which, to every real number c at which the power series converges, assigns the real number $f(c)$ given by

$$f(c) = \sum_{i=0}^{\infty} a_i(c-a)^i.$$

The domain of f is obviously equal to the interval of convergence of the power series. Speaking more casually, we say simply that the function f is defined by the equation

$$f(x) = \sum_{i=0}^{\infty} a_i(x-a)^i.$$

As an example, let f be the function defined by

$$f(x) = \sum_{i=0}^{\infty} \frac{x^i}{i!} = 1 + x + \frac{x^2}{2!} + \frac{x^3}{3!} + \cdots.$$

This power series was studied in Example 1 of Section 6 and was shown to converge for all values of x. Thus the domain of the function which the series defines is the set of all real numbers.

Functions defined by power series have excellent analytic properties. One of the most important is the fact that every such function is differentiable and that its derivative is the function defined by the power series obtained by differentiating the original series term by term. That is, if

$$f(x) = \sum_{i=0}^{\infty} a_i(x - a)^i = a_0 + a_1(x - a) + a_2(x - a)^2 + \cdots,$$

then

$$f'(x) = \sum_{i=1}^{\infty} ia_i(x - a)^{i-1} = a_1 + 2a_2(x - a) + 3a_3(x - a)^2 + \cdots.$$

This is not a trivial result. To prove it, we begin with the following theorem:

(7.1) *A power series $\sum_{i=0}^{\infty} a_i(x - a)^i$ and its derived series $\sum_{i=1}^{\infty} ia_i(x - a)^{i-1}$ have the same radius of convergence.*

In Section 6 we showed that the essential difference between the power series $\sum_{i=0}^{\infty} a_i(x - a)^i$ and the corresponding series $\sum_{i=0}^{\infty} a_ix^i$ is that the interval of convergence of one is obtained from that of the other by translation. In particular, both series have the same radius of convergence. To prove (7.1), it is therefore sufficient (and notationally easier) to prove the same result for power series about the origin 0. We shall therefore prove the following: *If the power series $\sum_{i=0}^{\infty} a_ix^i$ has radius of convergence ρ and if the derived series $\sum_{i=1}^{\infty} ia_ix^{i-1}$ has radius of convergence ρ', then $\rho = \rho'$.*

Proof. Suppose that $\rho < \rho'$, and let c be an arbitrary real number such that $\rho < c < \rho'$. Then the series $\sum_{i=0}^{\infty} a_ic^i$ diverges, whereas the series $\sum_{i=1}^{\infty} ia_ic^{i-1}$ converges absolutely. Since c is positive,

$$c \sum_{i=1}^{\infty} |ia_ic^{i-1}| = \sum_{i=1}^{\infty} |ia_ic^i|,$$

and it follows that the series $\sum_{i=1}^{\infty} ia_ic^i$ is also absolutely convergent. However, it is obvious that, for every positive integer i,

$$|a_ic^i| \leq i|a_ic^i| = |ia_ic^i|.$$

The Comparison Test therefore implies that the series $\sum_{i=1}^{\infty} |a_ic^i|$ converges, and this fact implies the convergence of $\sum_{i=0}^{\infty} a_ic^i$, which is a contradiction. Hence the original assumption is false, and we conclude that

$$\rho' \leq \rho. \tag{1}$$

Next, suppose that $\rho' < \rho$. We shall derive a contradiction from this assumption also, which, together with the inequality (1), proves that $\rho' = \rho$. Let b and c be any two real numbers such that $\rho' < b < c < \rho$. It follows from the definition of ρ' that the series $\sum_{i=1}^{\infty} ia_i b^{i-1}$ diverges. Similarly, from the definition of ρ, we know that the series $\sum_{i=0}^{\infty} a_i c^i$ converges, and therefore $\lim_{i\to\infty} a_i c^i = 0$. Because c is positive, it follows that there exists a positive integer N such that, for every integer $i \geq N$,

$$|a_i c^i| < c.$$

But $|a_i c^i| = |a_i| c^i$, and so the preceding inequality becomes $|a_i| c^i < c$, or, equivalently,

$$|a_i| < \frac{1}{c^{i-1}}.$$

Hence, since b is also positive, we obtain

$$|ia_i b^{i-1}| = ib^{i-1}|a_i| < i\frac{b^{i-1}}{c^{i-1}} = i\left(\frac{b}{c}\right)^{i-1},$$

for every integer $i \geq N$. Let us set $\frac{b}{c} = r$. Then $0 < r < 1$, and we have shown that

$$|ia_i b^{i-1}| < ir^{i-1}, \qquad \text{for every integer } i \geq N.$$

However, it is shown in Example 3, page 508, that the series $\sum_{i=1}^{\infty} ir^{i-1}$ converges if $|r| < 1$. Hence the preceding inequality and the Comparison Text imply that the series $\sum_{i=1}^{\infty} |ia_i b^{i-1}|$ converges, and this contradicts the above conclusion that $\sum_{i=1}^{\infty} ia_i b^{i-1}$ diverges. This completes the proof that $\rho' = \rho$, and, as we have remarked, also proves (7.1).

Note that Theorem (7.1) does not state that a power series $\sum_{i=0}^{\infty} a_i(x - a)^i$ and its derived series have the same *interval* of convergence, but only that they have the same *radius* of convergence. For example, in Example 1(b), page 514, the interval of convergence of the power series $\sum_{k=1}^{\infty} (-1)^{k-1} \frac{x^k}{k}$ is shown to be the half-open interval $(-1, 1]$. However, the derived series is

$$\sum_{k=1}^{\infty} (-1)^{k-1} k \frac{x^{k-1}}{k} = \sum_{k=1}^{\infty} (-1)^{k-1} x^{k-1}$$
$$= 1 - x + x^2 - x^3 + \cdots,$$

which does not converge for $x = 1$. It is a geometric series having the open interval of convergence $(-1, 1)$.

Let $\sum_{i=0}^{\infty} a_i(x - a)^i$ be a power series, and let f and g be the two functions defined respectively by $f(x) = \sum_{i=0}^{\infty} a_i(x - a)^i$ and by $g(x) = \sum_{i=1}^{\infty} ia_i(x - a)^{i-1}$. We have proved that there is an interval, which, with the possible exception of its endpoints, is the common domain of f and g. However, we have not yet proved that the function g is the derivative of the function f. This fact is the content of the following theorem.

(7.2) THEOREM. *If the radius of convergence* ρ *of the power series* $\sum_{i=0}^{\infty} a_i(x-a)^i$ *is not zero, then the function* f *defined by* $f(x) = \sum_{i=0}^{\infty} a_i(x-a)^i$ *is differentiable at every* x *such that* $|x-a| < \rho$ *and*

$$f'(x) = \sum_{i=1}^{\infty} ia_i(x-a)^{i-1}.$$

Proof. It is a direct consequence of the Chain Rule that if (7.2) is proved for $a = 0$, then it is true in general. We shall therefore assume that $f(x) = \sum_{i=0}^{\infty} a_i x^i$. Let g be the function defined by $g(x) = \sum_{i=1}^{\infty} ia_i x^{i-1}$, and let c be an arbitrary number such that $|c| < \rho$. We must prove that $f'(c)$, which can be defined by

$$f'(c) = \lim_{x \to c} \frac{f(x) - f(c)}{x - c},$$

exists and is equal to $g(c)$. Hence the proof is complete when we show that

$$\lim_{x \to c} \left(\frac{f(x) - f(c)}{x - c} - g(c) \right) = 0. \qquad (2)$$

Let d be an arbitrary real number such that $|c| < d < \rho$ (see Figure 9).

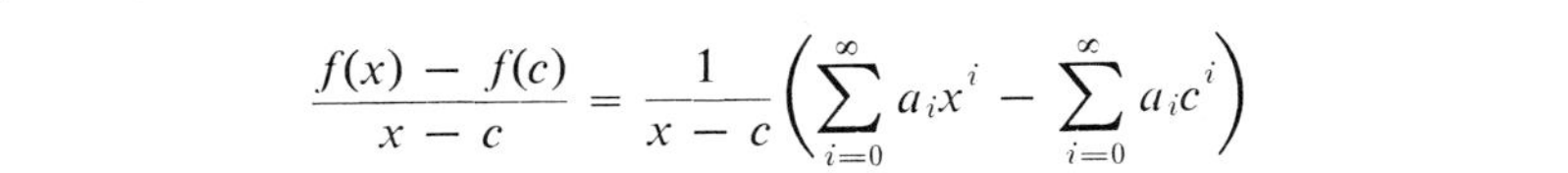

Figure 9

Henceforth, we shall consider only values of x which lie in the closed interval $[-d, d]$. For every such x other than c, we have

$$\begin{aligned} \frac{f(x) - f(c)}{x - c} &= \frac{1}{x - c} \left(\sum_{i=0}^{\infty} a_i x^i - \sum_{i=0}^{\infty} a_i c^i \right) \\ &= \sum_{i=1}^{\infty} a_i \left(\frac{x^i - c^i}{x - c} \right). \end{aligned}$$

For each integer $i \geq 1$, we apply the Mean Value Theorem to the function x^i, whose derivative is ix^{i-1}. The conclusion is that, for each i, there exists a real number c_i in the open interval whose endpoints are c and x such that $x^i - c^i = ic_i^{i-1}(x - c)$. Hence

$$\frac{x^i - c^i}{x - c} = ic_i^{i-1},$$

and so

$$\frac{f(x) - f(c)}{x - c} = \sum_{i=1}^{\infty} ia_i c_i^{i-1}.$$

From this it follows that

$$\begin{aligned}\frac{f(x) - f(c)}{x - c} - g(c) &= \sum_{i=1}^{\infty} ia_i c_i^{i-1} - \sum_{i=1}^{\infty} ia_i c^{i-1} \\ &= \sum_{i=2}^{\infty} ia_i(c_i^{i-1} - c^{i-1}).\end{aligned}$$

For each integer $i \geq 2$, we now apply the Mean Value Theorem to the function x^{i-1}, whose derivative is $(i - 1)x^{i-2}$. We conclude that there exists a real number b_i in the open interval whose endpoints are c and c_i such that

$$c_i^{i-1} - c^{i-1} = (i - 1)b_i^{i-2}(c_i - c).$$

Hence

$$\frac{f(x) - f(c)}{x - c} - g(c) = \sum_{i=2}^{\infty} i(i - 1)a_i b_i^{i-2}(c_i - c).$$

Since $|c_i - c| \leq |x - c|$, for every i, we obtain, using Theorem (5.3), page 509,

$$\left|\frac{f(x) - f(c)}{x - c} - g(c)\right| \leq |x - c| \sum_{i=2}^{\infty} |i(i - 1)a_i b_i^{i-2}|. \tag{3}$$

Two applications of (7.1) imply that the power series $\sum_{i=2}^{\infty} i(i - 1)a_i x^{i-2}$, which is the derived series of $\sum_{i=1}^{\infty} ia_i x^{i-1}$, also has radius of convergence equal to ρ, and it is therefore absolutely convergent for $x = d$. Moreover, $|b_i| < d$ for each i, and so

$$|i(i - 1)a_i b_i^{i-2}| \leq |i(i - 1)a_i d^{i-2}|,$$

for every integer $i \geq 2$. It follows from the Comparison Test that

$$\sum_{i=2}^{\infty} |i(i - 1)a_i b_i^{i-2}| \leq \sum_{i=2}^{\infty} |i(i - 1)a_i d^{i-2}|. \tag{4}$$

The value of the series on the right in (4) does not depend on x, and we denote it by M. Combining (3) and (4), we therefore finally obtain

$$\left|\frac{f(x) - f(c)}{x - c} - g(c)\right| \leq |x - c|\, M.$$

The left side of this inequality can be made arbitrarily small by taking $|x - c|$ sufficiently small. But this is precisely the meaning of the assertion in (2), and so the proof is finished.

EXAMPLE 1. (a) Show that

$$e^x = \sum_{i=0}^{\infty} \frac{x^i}{i!} = 1 + x + \frac{x^2}{2!} + \frac{x^3}{3!} + \cdots,$$

for every real number x, and (b) show that

$$\ln(1 + x) = \sum_{k=1}^{\infty} (-1)^{k-1} \frac{x^k}{k} = x - \frac{x^2}{2} + \frac{x^3}{3} - \cdots,$$

for every real number x such that $|x| < 1$.

For (a), let f be the function defined by $f(x) = \sum_{i=0}^{\infty} \frac{x^i}{i!}$. We have already shown that the domain of f is the set of all real numbers; i.e., the radius of convergence is infinite. It follows from the preceding theorem that

$$f'(x) = \sum_{i=1}^{\infty} i \frac{x^{i-1}}{i!}, \qquad \text{for every real number } x.$$

Since $\frac{i}{i!} = \frac{1}{(i-1)!}$, we obtain

$$\sum_{i=1}^{\infty} \frac{ix^{i-1}}{i!} = \sum_{i=1}^{\infty} \frac{x^{i-1}}{(i-1)!} = \sum_{k=0}^{\infty} \frac{x^k}{k!},$$

where the last equation is obtained by replacing $i - 1$ by k. Thus we have proved that

$$f'(x) = f(x) \qquad \text{for every real number } x.$$

The function f therefore satisfies the differential equation $\frac{dy}{dx} = y$, whose general solution is $y = ce^x$. Hence $f(x) = ce^x$ for some constant c. But it is obvious from the series which defines f that $f(0) = 1$. It follows that $c = 1$, and (a) is proved.

A similar technique is used for (b). Let $f(x) = \sum_{k=1}^{\infty} (-1)^{k-1} \frac{x^k}{k}$. The domain of f is the half-open interval $(-1, 1]$, and the radius of convergence is 1. Hence

$$\begin{aligned} f'(x) &= \sum_{k=1}^{\infty} (-1)^{k-1} k \frac{x^{k-1}}{k} = \sum_{k=1}^{\infty} (-1)^{k-1} x^{k-1} \\ &= 1 - x + x^2 - x^3 + \cdots, \end{aligned}$$

for $|x| < 1$. The latter is a geometric series with sum equal to $\frac{1}{1+x}$. Hence we have shown that

$$f'(x) = \frac{1}{1+x}, \qquad \text{for every } x \text{ such that } |x| < 1.$$

Integration yields

$$f(x) = \int \frac{dx}{1+x} = \ln|1 + x| + c.$$

Since $|x| < 1$, we have $|1 + x| = (1 + x)$. From the series which defines f we see that $f(0) = 0$. Hence

$$0 = f(0) = \ln(1 + 0) + c = 0 + c = c.$$

It follows that $f(x) = \ln(1 + x)$ for every real number x such that $|x| < 1$, and (b) is therefore established.

Example 1 illustrates an important point. The domain of the function f defined by $f(x) = \sum_{k=1}^{\infty} (-1)^{k-1} \frac{x^k}{k}$ is the half-open interval $(-1, 1]$. On the other hand, the domain of the function $\ln(1 + x)$ is the unbounded interval $(-1, \infty)$. It is essential to realize that the equation

$$\ln(1 + x) = \sum_{k=1}^{\infty} (-1)^{k-1} \frac{x^k}{k}$$

has been shown to hold *only for values of x which are interior points of the interval of convergence of the power series.* It certainly does not hold at points outside the interval of convergence where the series diverges. As far as the endpoints of the interval are concerned, it can be proved that a function defined by a power series is continuous at every point of its interval of convergence. Hence the above equation does in fact hold for $x = 1$, and we therefore obtain the following formula for the sum of the alternating harmonic series:

$$\boxed{\ln 2 = \sum_{k=1}^{\infty} (-1)^{k-1} \frac{1}{k} = 1 - \tfrac{1}{2} + \tfrac{1}{3} - \tfrac{1}{4} + \cdots.}$$

Let $\sum_{i=0}^{\infty} a_i(x - a)^i$ be a power series with a nonzero radius of convergence ρ, and let f be the function defined by the power series

$$f(x) = \sum_{i=0}^{\infty} a_i(x - a)^i,$$

for every x in the interval of convergence. By iterated applications of Theorem (7.2), i.e., first to the series, then to the derived series, then to the derived series of the derived series, etc., we may conclude that f has derivatives of arbitrarily high order within the radius of convergence. The formula for the nth derivative is easily seen to be

(7.3) $$f^{(n)}(x) = \sum_{i=n}^{\infty} i(i-1)\cdots(i-n+1)a_i(x-a)^{i-n},$$

for every x such that $|x - a| < \rho$.

Is it possible for a function f to be defined by two different power series about the same number a? The answer is no, provided the domain of f is not just the single number a. The reason, as the following corollary of Theorem (7.3) shows, is that the coefficients of any power series about a which defines f are uniquely determined by the function f.

(7.4) *If $f(x) = \sum_{i=0}^{\infty} a_i(x-a)^i$ and if the radius of convergence of the power series is not zero, then, for every integer $n \geq 0$,*

$$a_n = \frac{1}{n!} f^{(n)}(a).$$

[By the zeroth derivative $f^{(0)}$ we mean simply f itself. Hence, for $n = 0$, the conclusion is the obviously true equation $a_0 = f(a)$.]

Proof. The radius of convergence ρ is not zero, and so the formula in (7.3) holds. Since $i(i-1)\cdots(i-n+1) = \dfrac{i!}{(i-n)!}$, we obtain

$$f^{(n)}(x) = \sum_{i=n}^{\infty} \frac{i!}{(i-n)!} a_i(x-a)^{i-n}$$

$$= n!a_n + (n+1)!a_{n+1}(x-a) + \frac{(n+2)!}{2!} a_{n+2}(x-a)^2 + \cdots,$$

for every x such that $|x - a| < \rho$. Setting $x = a$, we obtain

$$f^{(n)}(a) = n!a_n,$$

from which the desired conclusion follows.

PROBLEMS

1. Let f be the function defined by

$$f(x) = \sum_{i=1}^{\infty} \frac{x^i}{i^2}.$$

(a) Find the radius of convergence of the power series and also the domain of f.

(b) Write the derived series, and find its radius of convergence directly [i.e., verify Theorem (7.1) for this particular series].

(c) Find the domain of the function defined by the derived series.

2. Let f be the function defined by $f(x) = \sum_{i=1}^{\infty} \frac{1}{i2^i}(x - 2)^i$, and follow the same instructions as in Problem 1.

3. Let f be the function defined by $f(x) = \sum_{i=1}^{\infty} \frac{1}{\sqrt{i}}(x - 2)^i$, and follow the same instructions as in Problem 1.

4. Let f be the function defined by $f(x) = \sum_{i=0}^{\infty} \frac{(x - 1)^i}{\sqrt{i + 1}}$, and follow the same instructions as in Problem 1.

5. Find the domains of the functions f and g defined by the following power series.

(a) $f(x) = \sum_{i=1}^{\infty} \frac{(-1)^{i-1}}{(2i - 1)!} x^{2i-1}$ (b) $g(x) = \sum_{i=0}^{\infty} \frac{(-1)^i}{(2i)!} x^{2i}$.

6. If f and g are the two functions defined in Problem 5, show that

(a) $f'(x) = g(x)$ (b) $g'(x) = -f(x)$.

7. Let f and g be the two functions defined in Problem 5 (see also Problem 6).

(a) Evaluate $f(0)$, $f'(0)$, $g(0)$, and $g'(0)$.

(b) Show that f and g are both solutions of the differential equation $\frac{d^2y}{dx^2} + y = 0$.

(c) Write the general solution of the differential equation in (b), and thence, using the results of part (a), show that $f(x) = \sin x$ and that $g(x) = \cos x$.

8. Show, as is claimed at the beginning of the proof of Theorem (7.2), that it is a direct consequence of the Chain Rule that if this theorem is proved for $a = 0$, then it is true for an arbitrary real number a.

9. Prove that every power series can be integrated, term by term. Specifically, prove the following two theorems.

(a) *A power series $\sum_{i=0}^{\infty} a_i(x - a)^i$ and its integrated series*

$$\sum_{i=0}^{\infty} \frac{a_i}{i + 1}(x - a)^{i+1}$$

have the same radius of convergence.

(b) *If the radius of convergence ρ of the power series $\sum_{i=0}^{\infty} a_i(x - a)^i$ is not zero and if f and F are the functions defined, respectively, by*

$$f(x) = \sum_{i=0}^{\infty} a_i(x - a)^i \quad \text{and} \quad F(x) = \sum_{i=0}^{\infty} \frac{a_i}{i + 1}(x - a)^{i+1},$$

then

$$F(x) = \int f(x)\,dx + c.$$

10. Starting from the geometric series

$$\sum_{i=0}^{\infty} (-1)^i x^{2i} = 1 - x^2 + x^4 - x^6 + \cdots$$

and the results of Problem 9, show that

$$\arctan x = \sum_{i=0}^{\infty} \frac{(-1)^i}{2i+1} x^{2i+1},$$

for every x such that $|x| < 1$.

11. (a) Show that the interval of convergence of the integrated series in Problem 10 is the closed interval $[-1, 1]$.

(b) True or false?

$$\frac{\pi}{4} = \arctan 1 = \sum_{i=0}^{\infty} \frac{(-1)^i}{2i+1}.$$

8. Taylor Series. The subject of Section 7 was the function defined by a given power series. In contrast, in this section we start with a given function and ask whether or not there exists a power series which defines it. More precisely, if f is a function containing the number a in its domain, then does there exist a power series $\sum_{i=0}^{\infty} a_i(x - a)^i$ with nonzero radius of convergence which defines a function equal to f on the interval of convergence of the power series? If the answer is yes, then the power series is uniquely determined. Specifically, it follows from Theorem (7.4), page 526, that $a_i = \frac{1}{i!} f^{(i)}(a)$, for every integer $i \geq 0$. Hence

$$\begin{aligned} f(x) &= \sum_{i=0}^{\infty} \frac{1}{i!} f^{(i)}(a)(x - a)^i \\ &= f(a) + f'(a)(x - a) + \frac{1}{2!} f''(a)(x - a)^2 + \cdots, \end{aligned}$$

for every x in the interval of convergence of the power series.

Let f be a function which has derivatives of every order at a. The power series $\sum_{i=0}^{\infty} \frac{1}{i!} f^{(i)}(a)(x - a)^i$ is called the **Taylor series** of the function f about the number a. The existence of this series, whose definition is motivated by the preceding paragraph, requires only the existence of every derivative $f^{(i)}(a)$. However, the natural inference that the existence of the Taylor series for a given function implies the convergence of the series to the values of the function is false. In a later theorem we shall give an additional condition which makes the inference true. Two examples of Taylor series are the series for e^x and the series for $\ln(1 + x)$ developed in Example 1 of Section 7.

The value of a convergent infinite series can be approximated by its partial sums. For a Taylor series $\sum_{i=0}^{\infty} \frac{1}{i!} f^{(i)}(a)(x - a)^i$, the nth partial sum is a polynomial in $(x - a)$, which we shall denote by T_n. The definition is as follows: Let n be a nonnegative integer and f a function such that $f^{(i)}(a)$ exists for every integer $i = 0, \ldots, n$. Then the **nth Taylor approximation** to the function f about the number a is the polynomial T_n given by

$$\begin{aligned} T_n(x) &= \sum_{i=0}^{n} \frac{1}{i!} f^{(i)}(a)(x - a)^i \\ &= f(a) + f'(a)(x - a) + \cdots + \frac{1}{n!} f^{(n)}(a)(x - a)^n, \end{aligned} \tag{1}$$

for every real number x.

For each integer $k = 0, \ldots, n$, direct computation of the kth derivative at a of the Taylor polynomial T_n shows that it is equal to $f^{(k)}(a)$. Thus we have the simple but important result:

(8.1) *The* nth *Taylor approximation* T_n *to the function* f *about* a *satisfies*

$$T_n^{(k)}(a) = f^{(k)}(a), \qquad \textit{for each } k = 0, \ldots, n.$$

EXAMPLE 1. Let f be the function defined by $f(x) = \frac{1}{x}$. For $n = 0, 1, 2,$ and 3, compute the Taylor polynomial T_n for the function f about the number 1, and superimpose the graph of each on the graph of f. The derivatives are:

$$\begin{aligned} f'(x) &= -\frac{1}{x^2}, && \text{whence } f'(1) = -1, \\ f''(x) &= \frac{2}{x^3}, && \text{whence } f''(1) = 2, \\ f'''(x) &= -\frac{6}{x^4}, && \text{whence } f'''(1) = -6. \end{aligned}$$

From the definition in (1), we therefore obtain

$$\begin{aligned} T_0(x) &= f(1) = 1, \\ T_1(x) &= 1 - (x - 1), \\ T_2(x) &= 1 - (x - 1) + (x - 1)^2, \\ T_3(x) &= 1 - (x - 1) + (x - 1)^2 - (x - 1)^3. \end{aligned}$$

These equations express the functions T_n as polynomials in $x - 1$ rather than as polynomials in x. The advantage of this form is that it exhibits most clearly the behavior of each approximation in the vicinity of the number 1. Each one can, of course, be expanded to get a polynomial in x. If we do this, we obtain

$$\begin{aligned} T_0(x) &= 1, \\ T_1(x) &= -x + 2, \\ T_2(x) &= x^2 - 3x + 3, \\ T_3(x) &= -x^3 + 4x^2 - 6x + 4. \end{aligned}$$

The graphs are shown in Figure 10. The larger the degree of the approximating polynomial, the more closely its graph "hugs" the graph of f for values of x close to 1.

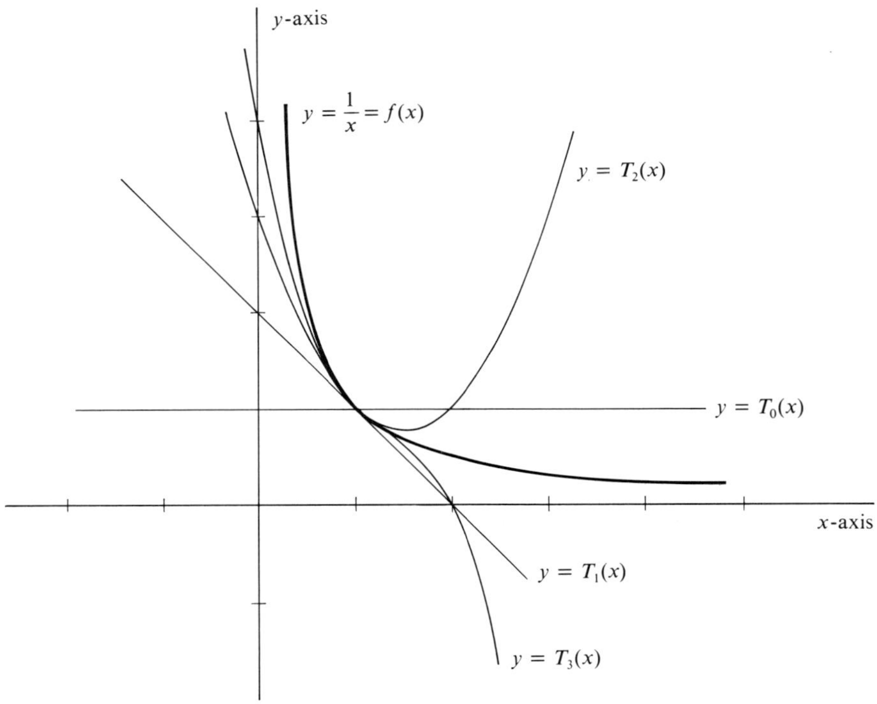

Figure 10

The basic relation between a function f and the approximating Taylor polynomials T_n will be presented in Theorem (8.3). In proving it, we shall use the following lemma, which is an extension of Rolle's Theorem (see pages 111 and 112).

(8.2) *Let F be a function with the property that the* $(n + 1)$st *derivative* $F^{(n+1)}(t)$ *exists for every t in a closed interval* $[a, b]$, *where* $a < b$. *If*

(i) $F^{(i)}(a) = 0,$ *for* $i = 0, 1, \ldots, n$, *and*

(ii) $F(b) = 0,$

then there exists a real number c in the open interval (a, b) *such that* $F^{(n+1)}(c) = 0$.

Proof. The idea of the proof is to apply Rolle's Theorem over and over again, starting with $i = 0$ and finishing with $i = n$. (In checking the continuity requirements of Rolle's Theorem, remember that if a function has a derivative at a point, then it is continuous there.) A proof by induction on n proceeds as follows: If $n = 0$, the result is a direct consequence of Rolle's Theorem. We must next prove from the assumption that if the lemma is true for $n = k$, then it is also true for $n = k + 1$. Thus we assume that there exists a real number c in the open interval (a, b) such that $F^{(k+1)}(c) = 0$ and shall prove that there exists another real number c' in (a, b) such that $F^{(k+2)}(c') = 0$. The hypotheses of (8.2) with $n = k + 1$ assure us that $F^{(k+2)}(t)$ exists for every t in $[a, b]$ and that $F^{(k+1)}(a) = 0$. The function $F^{(k+1)}$ satisfies the premises of Rolle's Theorem, since it is continuous on $[a, c]$, differentiable on (a, c), and $F^{(k+1)}(a) = F^{(k+1)}(c) = 0$. Hence there exists a real number c' in (a, c) with the property that $F^{(k+2)}(c') = 0$. Since (a, c) is a subset of (a, b), the number c' is also in (a, b), and this completes the proof.

We come now to the main theorem of the section:

(8.3) TAYLOR'S THEOREM. *Let f be a function with the property that the* $(n + 1)$st *derivative* $f^{(n+1)}(t)$ *exists for every t in the closed interval* $[a, b]$, *where* $a < b$. *Then there exists a real number c in the open interval* (a, b) *such that*

$$f(b) = \sum_{i=0}^{n} \frac{1}{i!} f^{(i)}(a)(b - a)^i + R_n,$$

where

$$R_n = \frac{1}{(n + 1)!} f^{(n+1)}(c)(b - a)^{n+1}.$$

Using the approximating Taylor polynomials, we can write the conclusion of this theorem equivalently as

$$f(b) = T_n(b) + \frac{1}{(n + 1)!} f^{(n+1)}(c)(b - a)^{n+1}. \tag{2}$$

Note that the particular value of c depends not only on the function f and the numbers a and b but also on the integer n.

Proof. Let the real number K be defined by the equation

$$f(b) = T_n(b) + K(b - a)^{n+1}.$$

The proof of Taylor's Theorem is completed by showing that

$$K = \frac{1}{(n+1)!} f^{(n+1)}(c),$$

for some real number c in (a, b). For this purpose, we define a new function F by setting

$$F(t) = f(t) - T_n(t) - K(t - a)^{n+1},$$

for every t in $[a, b]$. From the equation which defines K, it follows at once that

$$f(b) - T_n(b) - K(b - a)^{n+1} = 0,$$

and hence that $F(b) = 0$. In computing the derivatives of the function F, we observe that any derivative of $K(t - a)^{n+1}$ of order less than $n + 1$ will contain a factor of $t - a$, and therefore

$$\left.\frac{d^i}{dt^i} K(t - a)^{n+1}\right|_{t=a} = 0, \qquad \text{for } i = 0, \ldots, n.$$

Since $f^{(i)}(a) = T_n^{(i)}(a)$, for every integer $i = 0, \ldots, n$, [see (8.1)], we conclude that

$$F^{(i)}(a) = f^{(i)}(a) - T_n^{(i)}(a) - 0 = 0, \qquad i = 0, \ldots, n.$$

Hence, by Lemma (8.2), there exists a real number c in (a, b) such that

$$F^{(n+1)}(c) = 0.$$

Finally, we compute $F^{(n+1)}(t)$ for an arbitrary t in $[a, b]$. Since the degree of the polynomial T_n is at most n, its $(n + 1)$st derivative must be zero. Moreover, the $(n + 1)$st derivative of $K(t - a)^{n+1}$ is equal to $(n + 1)!K$. Hence

$$F^{(n+1)}(t) = f^{(n+1)}(t) - (n + 1)!K.$$

Letting $t = c$, we obtain

$$0 = F^{(n+1)}(c) = f^{(n+1)}(c) - (n + 1)!K,$$

from which it follows that $K = \dfrac{1}{(n+1)!} f^{(n+1)}(c)$. This completes the proof.

It has been assumed in the statement and proof of Taylor's Theorem that $a < b$. However, if $b < a$, the same statement is true except that the $(n + 1)$st derivative exists in $[b, a]$ and the number c lies in (b, a). Except for the obvious modifications, the proof is identical to the one given. Suppose now that we are given a function f such that $f^{(n+1)}$ exists at every point of some interval I containing the number a. Since Taylor's Theorem holds for *any* choice of b in I other than a, we may regard b as the value of a variable. If we denote the variable by x, we have:

(8.4) ALTERNATIVE FORM OF TAYLOR'S THEOREM. *If $f^{(n+1)}(t)$ exists for every t in an interval I containing the number a, then, for every x in I other than a, there exists a real number c in the open interval with endpoints a and x such that*

$$f(x) = f(a) + f'(a)(x-a) + \cdots + \frac{1}{n!} f^{(n)}(a)(x-a)^n + R_n,$$

where

$$R_n = \frac{1}{(n+1)!} f^{(n+1)}(c)(x-a)^{n+1}.$$

The conclusion of this theorem is frequently called **Taylor's Formula** and R_n is called the **Remainder.** As before, using the notation for the approximating Taylor polynomials, we can write the formula succinctly as

$$\boxed{f(x) = T_n(x) + R_n.} \tag{3}$$

EXAMPLE 2. (a) Compute Taylor's Formula with the Remainder where $f(x) = \sin x$, $a = 0$, and n is arbitrary. (b) Draw the graphs of $\sin x$ and of the polynomials $T_n(x)$, for $n = 1, 2$, and 3. (c) Prove that, for every real number x,

$$\sin x = \sum_{i=1}^{\infty} (-1)^{i-1} \frac{x^{2i-1}}{(2i-1)!} = x - \frac{x^3}{3!} + \frac{x^5}{5!} - \frac{x^7}{7!} + \cdots.$$

The first four derivatives are given by

$$\frac{d}{dx} \sin x = \cos x,$$

$$\frac{d^2}{dx^2} \sin x = -\sin x,$$

$$\frac{d^3}{dx^3} \sin x = -\cos x,$$

$$\frac{d^4}{dx^4} \sin x = \sin x.$$

Thus the derivatives of $\sin x$ follow a regular cycle which repeats after every fourth derivation. In general, the even-order derivatives are given by

$$\frac{d^{2i}}{dx^{2i}} \sin x = (-1)^i \sin x, \qquad i = 0, 1, 2, \ldots,$$

and the odd-order derivatives by

$$\frac{d^{2i-1}}{dx^{2i-1}} \sin x = (-1)^{i-1} \cos x, \qquad i = 1, 2, 3, \ldots.$$

If we set $f(x) = \sin x$, then

$$f^{(2i)}(0) = (-1)^i \sin 0 = 0, \qquad i = 0, 1, 2, \ldots,$$
$$f^{(2i-1)}(0) = (-1)^{i-1} \cos 0 = (-1)^{i-1}, \qquad i = 1, 2, 3, \ldots.$$

Hence the nth Taylor approximation is the polynomial

$$T_n(x) = \sum_{i=0}^{n} \frac{1}{i!} f^{(i)}(0)x^i,$$

in which the coefficient of every even power of x is zero. To handle this alternation, we define the integer k by the rule

$$k = \begin{cases} \dfrac{n}{2}, & \text{if } n \text{ is even,} \\ \dfrac{n+1}{2}, & \text{if } n \text{ is odd.} \end{cases} \tag{4}$$

It then follows that

$$T_n(x) = \sum_{i=1}^{k} \frac{1}{(2i-1)!} (-1)^{i-1} x^{2i-1} = \sum_{i=1}^{k} (-1)^{i-1} \frac{x^{2i-1}}{(2i-1)!}. \tag{5}$$

[If $n = 0$, we have the exception $T_0(x) = 0$.] For the remainder, we obtain

$$\begin{aligned} R_n &= \frac{1}{(n+1)!} f^{(n+1)}(c)\, x^{n+1} \\ &= \begin{cases} \dfrac{x^{n+1}}{(n+1)!} (-1)^k \cos c, & \text{if } n \text{ is even,} \\ \dfrac{x^{n+1}}{(n+1)!} (-1)^k \sin c, & \text{if } n \text{ is odd,} \end{cases} \end{aligned} \tag{6}$$

for some real number c (which depends on both x and n) such that $|c| < |x|$. The Taylor formula for $\sin x$ about the number 0 is therefore given by

$$\sin x = \sum_{i=1}^{k} (-1)^{i-1} \frac{x^{2i-1}}{(2i-1)!} + R_n,$$

where k is defined by equation (4), and the remainder R_n by (6).

For part (b), the approximating polynomials T_1, T_2, and T_3 can be read directly from equation (5) [together with (4)]. We obtain

$$T_1(x) = x,$$
$$T_2(x) = x,$$
$$T_3(x) = x - \frac{x^3}{3!} = x - \frac{x^3}{6}.$$

Their graphs together with the graph of $\sin x$ are shown in Figure 11.

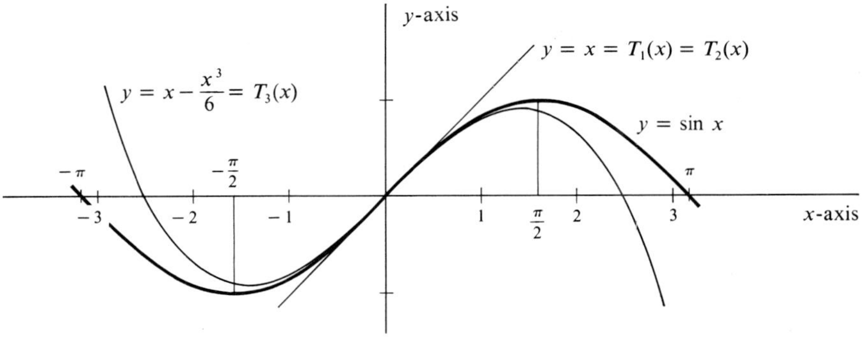

Figure 11

To prove that sin x can be defined by the infinite power series given in (c), we must show that, for every real number x,

$$\begin{aligned}\sin x &= \lim_{n\to\infty} T_n(x)\\ &= \lim_{k\to\infty} \sum_{i=1}^{k} (-1)^{i-1} \frac{x^{2i-1}}{(2i-1)!},\end{aligned}$$

where k is the integer defined in (4). Since $\sin x = T_n(x) + R_n$, an equivalent proposition is

$$\lim_{n\to\infty} R_n = 0.$$

To prove the latter, we use the important fact that the absolute values of the functions sine and cosine are never greater than 1. Hence, in the expression for R_n in (6), we know that $|\cos c| \leq 1$ and $|\sin c| \leq 1$. It therefore follows from (6) that

$$|R_n| \leq \frac{|x|^{n+1}}{(n+1)!}.$$

It is easy to show by the Ratio Test [see Problem 4(b), page 510] that $\lim_{n\to\infty} \frac{|x|^{n+1}}{(n+1)!} = 0$. Hence $\lim_{n\to\infty} R_n = 0$, and we have proved that

$$\sin x = x - \frac{x^3}{3!} + \frac{x^5}{5!} - \frac{x^7}{7!} + \cdots.$$

The form of the remainder in Taylor's Theorem provides one answer to the question posed at the beginning of the section, which, briefly stated, was: When can a given function be defined by a power series? The answer provided in the following theorem is obtained by a direct generalization of the method used to establish the convergence of the Taylor series for sin x.

(8.5) *Let f be a function which has derivatives of every order at every point of an interval I containing the number a. If the derivatives are uniformly bounded on I,* i.e., *if there exists a real number B such that $|f^{(n)}(t)| \leq B$, for every integer $n \geq 0$ and every t in I, then*

$$f(x) = \sum_{i=0}^{\infty} \frac{1}{i!} f^{(i)}(a)(x - a)^i,$$

for every real number x in I.

Proof. Since $f(x) = T_n(x) + R_n$ [see Theorem (8.4) and formula (3)], we must prove that $f(x) = \lim_{n\to\infty} T_n(x)$, or, equivalently, that $\lim_{n\to\infty} R_n = 0$. Generally speaking, the number c which appears in the expression for the remainder R_n will be different for each integer n and each x in I. But since the number B is a bound for the absolute values of *all* derivatives of f *everywhere* in I, we have $|f^{(n+1)}(c)| \leq B$. Hence

$$|R_n| = \left| \frac{1}{(n+1)!} f^{(n+1)}(c)(x - a)^{n+1} \right|$$

$$= \frac{|x - a|^{n+1}}{(n+1)!} |f^{(n+1)}(c)| \leq \frac{|x - a|^{n+1}}{(n+1)!} B.$$

However [see Problem 4(b), page 510],

$$\lim_{n\to\infty} \frac{|x - a|^{n+1}}{(n+1)!} B = 0 \cdot B = 0,$$

from which it follows that $\lim_{n\to\infty} R_n = 0$. This completes the proof.

It is an important fact, referred to at the beginning of the section, that the convergence of the Taylor series to the values of the function which defines it cannot be inferred from the existence of the derivatives alone. In Theorem (8.5), for example, we added the very strong hypothesis that all the derivatives of f are uniformly bounded on I. The following function defined by

$$f(x) = \begin{cases} 0 & \text{if } x = 0, \\ e^{-1/x^2} & \text{if } x \neq 0, \end{cases}$$

has the property that $f^{(n)}(x)$ exists for every integer $n \geq 0$ and every real number x. Moreover, it can be shown that $f^{(n)}(0) = 0$, for every $n \geq 0$. It follows that the Taylor series about 0 for this function is the trivial power series $\sum_{i=0}^{\infty} 0 \cdot x^i$. This series converges to 0 for every real number x, and does not converge to $f(x)$, except for $x = 0$.

When a Taylor polynomial or series is computed about the number zero, as in Example 2, there is a tradition for associating with it the name of the mathematician Maclaurin instead of that of Taylor. Thus the **Maclaurin series** for a given function is a power series in x, and is simply the special case of the Taylor series in which $a = 0$.

Suppose that, for a given n, we replace the values of a function f by those of the nth Taylor approximation to the function about some number a. How good is the resulting approximation? The answer depends on the interval (containing a) over which we wish to use the values of the polynomial T_n. Since $f(x) - T_n(x) = R_n$, the problem becomes one of finding a bound for $|R_n|$ over the interval in question.

EXAMPLE 3. (a) Compute the first three terms of the Taylor series of the function $f(x) = (x + 1)^{1/3}$ about $x = 7$. That is, compute

$$T_2(x) = f(7) + f'(7)(x - 7) + \frac{1}{2!}f''(7)(x - 7)^2.$$

(b) Show that $T_2(x)$ approximates $f(x)$ to within $\dfrac{5}{3^4 \cdot 2^8} = 0.00024$ (approximately) for every x in the interval $[7, 8]$.

Taking derivatives, we get

$$f'(x) = \tfrac{1}{3}(x + 1)^{-2/3} = \frac{1}{3(x + 1)^{2/3}},$$

$$f''(x) = -\tfrac{2}{9}(x + 1)^{-5/3} = -\frac{2}{9(x + 1)^{5/3}},$$

$$f'''(x) = \frac{2 \cdot 5}{9 \cdot 3}(x + 1)^{-8/3} = \frac{2 \cdot 5}{3^3(x + 1)^{8/3}}.$$

Hence

$$f(7) = 8^{1/3} = 2,$$

$$f'(7) = \frac{1}{3 \cdot 2^2} = \frac{1}{12},$$

$$f''(7) = -\frac{2}{9 \cdot 2^5} = -\frac{1}{3^2 \cdot 2^4} = -\frac{1}{144},$$

and the polynomial approximation to $f(x)$ called for in (a) is therefore

$$T_2(x) = 2 + \frac{1}{12}(x - 7) - \frac{1}{288}(x - 7)^2.$$

For part (b), we have $|f(x) - T_2(x)| = |R_2|$ and

$$R_2 = \frac{1}{3!}f'''(c)(x - 7)^3,$$

for some number c which is between x and 7. To obtain a bound for $|R_2|$ over the prescribed interval $[7, 8]$, we observe that in that interval the maximum value of $(x - 7)$ occurs when $x = 8$ and the maximum value of $|f'''|$ occurs when $x = 7$. Hence

$$|R_2| \leq \frac{1}{3!}|f'''(7)|\,|8 - 7|^3.$$

Since $f'''(7) = \dfrac{2 \cdot 5}{3^3 \cdot 2^8}$, we get

$$|R_2| \leq \frac{1}{3 \cdot 2}\,\frac{2 \cdot 5}{3^3 \cdot 2^8} \cdot 1^3 = \frac{5}{3^4 \cdot 2^8}.$$

Hence for every x in the interval $[7, 8]$, the difference in absolute value between $(x + 1)^{1/3}$ and the quadratic polynomial $T_2(x)$ is less than 0.00025.

PROBLEMS

1. For each of the values of n indicated, compute the Taylor polynomial T_n which approximates the function f near the number a.

   (a) $f(x) = \dfrac{1}{x + 2}$, $a = 0$, $n = 0$, 1, and 2.

   (b) $f(x) = \dfrac{1}{1 + x^2}$, $a = 0$, $n = 0$, 2, and 4.

   (c) $f(x) = \dfrac{1}{1 + x^2}$, $a = 1$, $n = 0$, 1, and 2.

   (d) $f(x) = \sqrt{x + 1}$, $a = 3$, $n = 1$, 2, and 3.

   (e) $f(x) = \sin x$, $a = \dfrac{\pi}{4}$, $n = 0$, 1, and 2.

2. Compute the formula, for an arbitrary nonnegative integer n, for the approximating Taylor polynomial to the function f about the number a.

   (a) $f(x) = \cos x$, $a = 0$ (b) $f(x) = \ln x$, $a = 1$.

3. For $n = 0$, 1, and 2, compute the Taylor polynomial T_n which approximates the function f near 0. Draw the graphs of the three polynomials together with the graph of f.

   (a) $f(x) = e^x$ (b) $f(x) = \cos x$.

4. Let p be a polynomial in x of degree $\leq m$; i.e., the function p is defined by an equation

$$p(x) = a_0 + a_1x + \cdots + a_mx^m,$$

   and let T_n be the Taylor polynomial which approximates p near an arbitrary real number a. Prove, as a simple consequence of Taylor's formula with the remainder, that $p(x) = T_n(x)$, for every real number x, provided $n \geq m$.

5. For each of the values of n indicated, compute the approximation T_n to the polynomial p near the number a.

   (a) $p(x) = x^2 + 3x - 1$, $a = 2$, $n = 1$, 2, and 3.

(b) $p(x) = 2x^3 - 5x^2 + 3$, $a = 0$, $n = 1$, 2, and 3.

(c) $p(x) = x^4 + 3x^2 + x + 2$, $a = 0$, $n = 3$, 4, and 17.

(d) $p(x) = x^3 - 1$, $a = 1$, $n = 2$, 3, and 4.

6. Prove that, for every real number x,

$$\cos x = \sum_{i=0}^{\infty} (-1)^i \frac{x^{2i}}{(2i)!}$$
$$= 1 - \frac{x^2}{2} + \frac{x^4}{4!} - \frac{x^6}{6!} + \cdots.$$

7. For each of the following functions, compute the Taylor series about a.

(a) e^x, if $a = 0$ (b) $\frac{e^x}{e^2}$, if $a = 2$ (c) $\arctan x$, if $a = 0$.

8. (a) Compute the cubic Taylor polynomial $p(x)$ which approximates the function $\frac{1}{x+2}$ for values of x near the number 1.

(b) Show that, for every x in the interval $[0, 2]$, the approximation $p(x)$ differs in absolute value from $\frac{1}{x+2}$ by less than 0.04.

9. Show that $\sin x$ differs in absolute value from the approximation $x - \frac{x^3}{6}$ by no more than $\frac{\pi^5}{15 \cdot 2^8} = 0.025$ (approximately) for every x in the interval $\left[-\frac{\pi}{2}, \frac{\pi}{2}\right]$.

10. The Taylor approximation T_n to a function f about the number a is frequently called the *best polynomial approximation of degree* $\leq n$ to the function f near a because it can be shown that T_n is the only polynomial of degree $\leq n$ with the property that, as x approaches a, the difference $f(x) - T_n(x)$ approaches zero faster than $(x - a)^n$.

Prove the following part of the above assertion: If f has continuous $(n + 1)$st derivative in an open interval containing a, then $\lim_{x \to a} \frac{f(x) - T_n(x)}{(x - a)^n} = 0$.

11. What cubic polynomial best approximates $x^4 - 2x^3 + 3x - 3$ near $x = 2$?

12. Another statement of Taylor's Theorem which gives a different form for the remainder is the following: *Let f be a function with continuous $(n + 1)$st derivative at every point of the interval $[a, b]$. Then*

$$f(b) = f(a) + f'(a)(b - a) + \cdots + \frac{1}{n!} f^{(n)}(a)(b - a)^n$$
$$+ \int_a^b \frac{(b - t)^n}{n!} f^{(n+1)}(t)\, dt.$$

(a) Using integration by parts, show that

$$\int_a^b \frac{(b-t)^n}{n!} f^{(n+1)}(t)\,dt$$

$$= -\frac{1}{n!} f^{(n)}(a)(b-a)^n + \int_a^b \frac{(b-t)^{n-1}}{(n-1)!} f^{(n)}(t)\,dt.$$

(b) Using induction on n and the result of part (a), prove the above form of Taylor's Theorem in which the remainder appears as an integral.

13. Let f have a continuous second derivative at every point of an interval containing the number a in its interior, and let $f'(a) = 0$. Show that f has a local maximum value at a if $f''(a) < 0$, and a local minimum value at a if $f''(a) > 0$. [*Hint:* Use the Taylor Formula $f(x) = T_1(x) + R_1$ and the fact that, if a continuous function is positive (or negative) at a, then it is positive (or negative) near a.]

CHAPTER 10

Geometry in the Plane

Suppose that we are concerned with the motion of a particle as it moves in a plane. At any time t during the motion, the position of the particle is given by its two coordinates, which depend on time, and may therefore be denoted by $x(t)$ and $y(t)$, respectively. The set of points traced out by the particle as it moves during a given interval of time is a curve. The function which describes the position of the particle is called a parametrization, and a curve described by such a function is said to be parametrized. In the first sections of this chapter we shall develop the mathematical theory of parametrized curves, abstracting from the picture of a physical particle in motion. Later we shall return to this application and define the notions of velocity and acceleration of such particles.

Parametrized curves represent an important generalization of the curves encountered thus far as the graphs of functions. As we shall see, a parametrized curve is not necessarily the graph of an equation $y = f(x)$.

1. Parametrically Defined Curves. When we speak of the plane in this book, we assume, unless otherwise stated, that a pair of coordinate axes has been chosen. As a result, we identify the set of points in the plane with the set R^2 of all ordered pairs of real numbers. A convenient notation for a function P whose domain is an interval I of real numbers and whose range is a subset of the plane is $P: I \to R^2$. Every function $P: I \to R^2$ defines two **coordinate functions,** the functions which assign to every t in I the two coordinates of the point $P(t)$. If we denote the first coordinate function by f, and the second one by g, then they are defined by the equation

$$P(t) = \big(f(t), g(t)\big), \qquad \text{for every } t \text{ in } I. \tag{1}$$

Conversely, of course, every ordered pair of real-valued functions f and g with an interval I as common domain defines a function $P: I \to R^2$ by equation (1).

Since the first and second coordinates of an element of R^2 are usually the x- and y-coordinates, respectively, we may alternatively define a function

$P: I \to R^2$ by a pair of equations

$$\begin{cases} x = f(t), \\ y = g(t), \end{cases}$$

where f and g are real-valued functions with domain I. Then, for every t in I, we have $P(t) = (x, y) = \big(f(t), g(t)\big)$. It is also common practice to denote the coordinate functions themselves by x and y. When this is done, we do not hesitate to write the equations $x = x(t)$ and $y = y(t)$, and the function $P: I \to R^2$ is defined by

$$P(t) = \big(x(t), y(t)\big), \qquad \text{for every } t \text{ in } I.$$

A function $P: I \to R^2$ is said to be **continuous at** t_0 if both coordinate functions are continuous at t_0. If the coordinate functions are denoted by x and y, then we define

$$\lim_{t \to t_0} P(t) = \big(\lim_{t \to t_0} x(t), \lim_{t \to t_0} y(t)\big).$$

As a result, the definition of continuity for P is entirely analogous to that for a real-valued function: P is continuous at t_0 if t_0 is in the domain of P and if $\lim_{t \to t_0} P(t) = P(t_0)$. As before, the function P is simply said to be **continuous** if it is continuous at every number in its domain.

A **curve** in the plane is by definition a subset of R^2 which is the range of some continuous function $P: I \to R^2$. Every curve is the range of many such functions, and, as a result, it is necessary to choose our terminology carefully. We shall call a continuous function $P: I \to R^2$ a **parametrization** of the curve C which is the range of P, and we shall say that C is **parametrically defined** by $P: I \to R^2$. The points of the curve C obviously consist of the set of all points $P(t)$, for every t in I. By a **parametrized curve** we shall mean the range of a specified continuous function $P: I \to R^2$. Speaking more casually, we shall refer to the curve defined parametrically by

$$P(t) = \big(x(t), y(t)\big),$$

or, equivalently, to the curve defined parametrically by the equations

$$\begin{cases} x = x(t), \\ y = y(t), \end{cases}$$

for every t in some interval I which is the common domain of the continuous functions x and y. If t is regarded as an independent variable, it is called the **parameter** of the parametrized curve.

EXAMPLE 1. Draw the curve defined parametrically by

$$P(t) = (t^2, t), \qquad -\infty < t < \infty.$$

This is, of course, also the curve defined by the equations

$$\begin{cases} x = t^2, \\ y = t, \qquad -\infty < t < \infty. \end{cases}$$

It is plotted in Figure 1. Since the set of all points (x, y) which satisfy the above two equations is equal to the set of all points (x, y) such that $x = y^2$, we recognize the curve as a parabola.

TABLE 1

t	(x, y)
0	(0, 0)
1	(1, 1)
−1	(1, −1)
2	(4, 2)
−2	(4, −2)

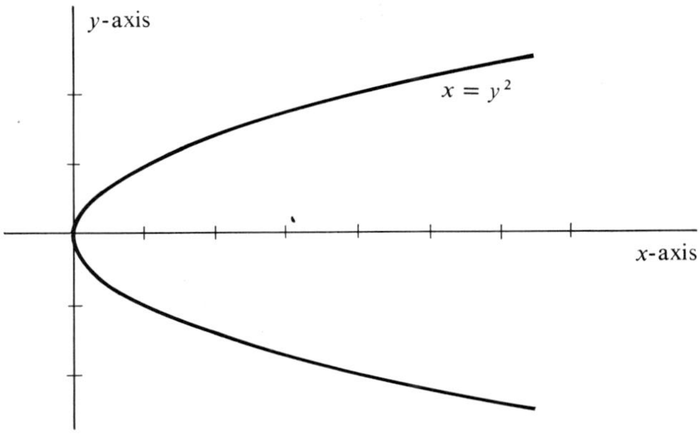

Figure 1

It is worth noting that every curve which we have previously encountered as the graph of a continuous function f can be defined parametrically. The graph is the set of all points (x, y) such that x is in the domain of f and such that $y = f(x)$. This set is obviously equal to the set of all points (x, y) such that

$$\begin{cases} x = t, \\ y = f(t), \qquad \text{and } t \text{ is in the domain of } f. \end{cases} \tag{2}$$

Hence the graph of f is defined parametrically by equations (2).

A function $P: I \to R^2$ is **differentiable at** t_0 if the derivatives of both coordinate functions exist at t_0. Moreover, following the usual style, we say that P is a **differentiable function** if it is differentiable at every number in its domain. This terminology is also applied to parametrized curves. That is, a curve defined parametrically by $P: I \to R^2$ is said to be differentiable at t_0, or simply differentiable, according as P is differentiable at t_0, or is a differentiable function.

EXAMPLE 2. Draw and identify the curve C defined parametrically by

$$P(t) = \big(x(t), y(t)\big) = (4\cos t, 3\sin t),$$

for every real number t. If (x, y) is an arbitrary point on the curve, then

$$\begin{cases} x = 4\cos t, \\ y = 3\sin t, \end{cases}$$

for some value of t. Hence, $\frac{x}{4} = \cos t$ and $\frac{y}{3} = \sin t$, and, consequently,

$$\frac{x^2}{16} + \frac{y^2}{9} = \cos^2 t + \sin^2 t = 1.$$

Thus for every point (x, y) on the curve, we have shown that

$$\frac{x^2}{16} + \frac{y^2}{9} = 1. \tag{3}$$

The latter is an equation of the ellipse shown in Figure 2, and it follows that the curve C is a subset of the ellipse. Conversely, let (x, y) be an arbitrary point on the ellipse. Then $|x| \leq 4$, and so there exists a number t such that

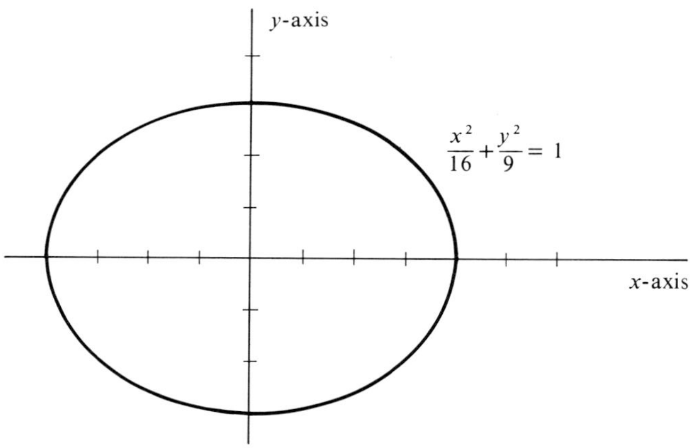

Figure 2

$x = 4\cos t$. Since $\cos t = \cos(-t)$ and $\sin t = -\sin(-t)$, we may choose t so that $\sin t$ and y have the same sign. Then, solving equation (3) for y and setting $x = 4\cos t$, we obtain

$$\begin{aligned} y^2 &= 9\left(1 - \frac{x^2}{16}\right) = 9\left(1 - \frac{16\cos^2 t}{16}\right) = 9(1 - \cos^2 t) \\ &= 9\sin^2 t. \end{aligned}$$

Since y and $\sin t$ have the same sign, it follows that $y = 3 \sin t$. We have therefore proved that, if (x, y) is an arbitrary point on the ellipse, then there exists a real number t such that

$$(x, y) = (4 \cos t, 3 \sin t) = P(t).$$

That is, every point on the ellipse also lies on C. We have already shown that the converse is true, and we therefore conclude that the parametrized curve C is equal to the ellipse.

Consider a curve C defined parametrically by a differentiable function $P: I \rightarrow R^2$, and let t_0 be an interior point of the interval I. A typical example is shown in Figure 3. Generally it will not be the case that the whole curve is a function of x, since there may be distinct points on C with the same x-coordinate. However, it can happen that a subset of C containing the point $P(t_0)$ is a differentiable function. Such a subset is shown in Figure 3, drawn

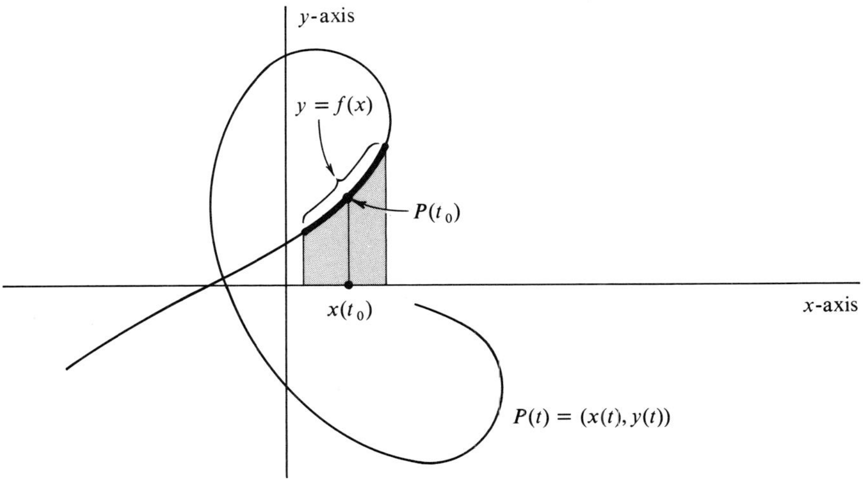

Figure 3

with a heavy line. Thus if $P(t) = \big(x(t), y(t)\big)$ for every t in I, then there may exist a differentiable function f such that

$$y(t) = f\big(x(t)\big), \tag{4}$$

for every t in some subinterval of I containing t_0 in its interior. If such a function does exist, we shall say that *y is a differentiable function of x on the parametrized curve $P(t) = \big(x(t), y(t)\big)$ in a neighborhood of the point $P(t_0)$.* Applying the Chain Rule to equation (4), we obtain

$$y'(t) = f'\big(x(t)\big)x'(t).$$

Hence

$$f'(x(t)) = \frac{y'(t)}{x'(t)}, \tag{5}$$

for every t in the subinterval, for which $x'(t) \neq 0$. If we write $y = f(x)$ and use the differential notation for the derivative, formula (5) becomes

$$\frac{dy}{dx} = \frac{\frac{dy}{dt}}{\frac{dx}{dt}}. \tag{6}$$

It should be apparent that $f'(x(t))$, or, equivalently, $\frac{dy}{dx}$ at t, is equal to the slope of the curve C at the point $P(t)$.

EXAMPLE 3. Find the slope, when $t = \frac{\pi}{3}$, of the parametrized ellipse in Example 2. The parametrization is defined by the equations

$$\begin{cases} x = 4 \cos t, \\ y = 3 \sin t. \end{cases}$$

We shall assume the analytic result that y is defined as a differentiable function of x in a neighborhood of the point

$$\left(4 \cos \frac{\pi}{3}, 3 \sin \frac{\pi}{3}\right).$$

Since

$$\left(4 \cos \frac{\pi}{3}, 3 \sin \frac{\pi}{3}\right) = \left(4 \cdot \frac{1}{2}, 3 \cdot \frac{\sqrt{3}}{2}\right) = \left(2, \frac{3\sqrt{3}}{2}\right),$$

one can see by simply looking at Figure 2 that this should certainly be true since the curve passes smoothly through the point and, in the immediate vicinity of the point, does not double back on itself. We have

$$\frac{dx}{dt} = \frac{d}{dt} 4 \cos t = -4 \sin t,$$

$$\frac{dy}{dt} = \frac{d}{dt} 3 \sin t = 3 \cos t,$$

and so

$$\left.\frac{dx}{dt}\right|_{t=\pi/3} = -4 \sin \frac{\pi}{3} = -4 \frac{\sqrt{3}}{2} = -2\sqrt{3},$$

$$\left.\frac{dy}{dt}\right|_{t=\pi/3} = 3 \cos \frac{\pi}{3} = \frac{3}{2}.$$

Hence, by formula (6), the slope is equal to

$$\left.\frac{dy}{dx}\right|_{t=\pi/3} = \frac{\left.\frac{dy}{dt}\right|_{t=\pi/3}}{\left.\frac{dx}{dt}\right|_{t=\pi/3}} = \frac{\frac{3}{2}}{-2\sqrt{3}} = -\frac{3}{4\sqrt{3}}.$$

The problem of giving analytic conditions which imply that y is a differentiable function of x on a parametrized curve in the neighborhood of a point is akin to the problem of determining when an equation $F(x, y) = c$ implicitly defines y as a differentiable function of x in a neighborhood of a point. As mentioned on page 81, the latter is solved by the Implicit Function Theorem, and the techniques needed here are similar.

As a final example, let us consider the curve traced by a point fixed on the circumference of a wheel as the wheel rolls along a straight line. We take the x-axis for the straight line. The radius of the wheel we denote by a, and the point on the circumference by (x, y). If we assume that the point passes through the origin as the wheel rolls by to the right, then the curve is defined parametrically by the equations

$$\begin{cases} x = a(\theta - \sin \theta), \\ y = a(1 - \cos \theta), \qquad -\infty < \theta < \infty, \end{cases}$$

where the parameter θ is the radian measure of the angle with vertex the center of the wheel, initial side the half-line pointing vertically downward, and terminal side the half-line through (x, y) (see Figure 4). (An alternative

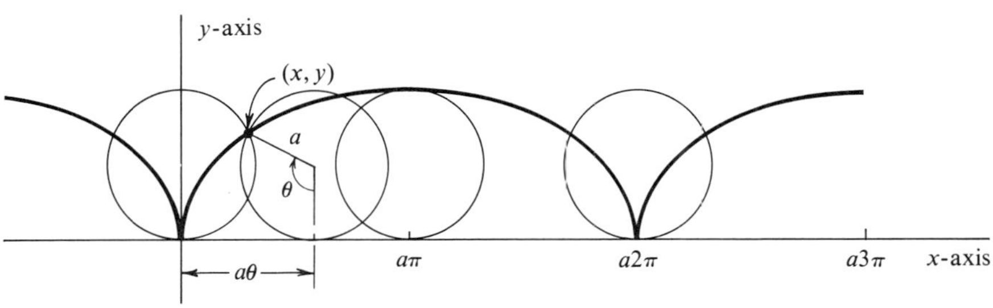

Figure 4

geometric interpretation of the parameter is that $a\theta$ is the coordinate of the point of tangency of the wheel on the x-axis.) The curve is called a **cycloid.** Note that the parametric equations are quite simple, whereas it would be difficult to express y as a function of x.

PROBLEMS

1. Draw and identify each of the curves defined by the following parametrizations.

(a) $P(t) = (t, t^2), \quad -\infty < t < \infty.$

(b) $P(t) = (t - 1, t^2), \quad -\infty < t < \infty.$

(c) $P(t) = (t^2 - 1, t + 1), \quad -\infty < t < \infty.$

(d) $P(t) = (2t^{1/3}, 3t^{1/3}), \quad -\infty < t < \infty.$

(e) $P(t) = (t - 1, t^3), \quad -\infty < t < \infty.$

(f) $P(t) = (3 \cos t, 3 \sin t), \quad 0 \leq t \leq \pi.$

(g) $P(s) = (\sin s, 2), \quad -\infty < s < \infty.$

(h) $Q(r) = (2 \sin r, 3 \cos r), \quad -\infty < r < \infty.$

2. Draw and identify each of the following parametrized curves.

(a) $\begin{cases} x = t - 1, \\ y = 2t + 3, \quad -\infty < t < \infty. \end{cases}$

(b) $\begin{cases} x = t^2, \\ y = t - 3, \quad -\infty < t < \infty. \end{cases}$

(c) $\begin{cases} x = 2 \cos t, \\ y = \cos t, \quad -\infty < t < \infty. \end{cases}$

(d) $\begin{cases} x = 3 \sec \theta, \\ y = 2 \tan \theta, \quad -\dfrac{\pi}{2} < \theta < \dfrac{\pi}{2}. \end{cases}$

3. Each of the following parametrized curves is a function f of x. [To put it another way, each is the graph of an equation $y = f(x)$.] Find $f(x)$.

(a) $P(t) = (t - 1, t^2 + 1), \quad -\infty < t < \infty.$

(b) $\begin{cases} x = t, \\ y = e^{t^2}, \quad -\infty < t < \infty. \end{cases}$

(c) $\begin{cases} x = 2 \cos t, \\ y = 3 \sin t, \quad 0 \leq t \leq \pi. \end{cases}$

(d) $P(t) = (e^t, t), \quad -\infty < t < \infty.$

4. For each of the following parametrizations, find an equation $F(x, y) = c$ whose graph is the parametrized curve

(a) $P(t) = (t^2, t), \quad -\infty < t < \infty.$

(b) $\begin{cases} x = e^{3t}, \\ y = e^t, \quad -\infty < t < \infty. \end{cases}$

(c) $\begin{cases} x = e^t + e^{-t}, \\ y = e^t - e^{-t}, \quad -\infty < t < \infty. \end{cases}$

[For (b) and (c), you will need in addition to the equation $F(x, y) = c$, the inequality $x > 0$.]

5. For the ellipse in Example 2 parametrized by the equations $x = 4 \cos t$ and $y = 3 \sin t$, interpret t geometrically. (*Hint:* See Figure 5.)

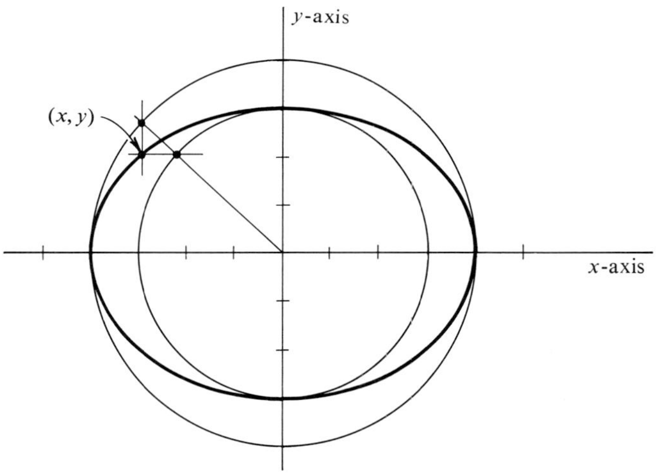

Figure 5

6. Sketch the curve defined by the parametrization

$$\begin{cases} x = a\theta - b \sin \theta, \\ y = a - b \cos \theta, \quad -\infty < \theta < \infty. \end{cases}$$

This curve is traced by a point, fixed on a radius of a wheel of radius a at a distance b from the center, as the wheel rolls along a straight line. There are two cases.

(a) The **curtate cycloid,** where $a > b$. (Think of a point on the spoke of a wheel.)

(b) The **prolate cycloid,** where $a < b$. (Think of a point on the flange of a railway wheel.)

7. For each of the following parametrized curves, assume that y is defined as a differentiable function of x in a neighborhood of the points indicated, and find $\frac{dy}{dx}$ at the point.

(a) $P(t) = (2t + 1, t^2)$, when $t = 2$.

(b) $\begin{cases} x = 5 \cos s, \\ y = 3 \sin s, \text{ when } s = \frac{\pi}{4}. \end{cases}$

(c) $\begin{cases} x = e^t, \\ y = t, \text{ when } t = 0. \end{cases}$

(d) $\begin{cases} x = e^t, \\ y = t, \text{ when } t = \ln 5. \end{cases}$

8. Find the slope of each of the following parametrized curves at the point indicated.

(a) $P(t) = (t - 1, t^3 - 3t^2 + 3t - 1)$, at $P(1)$.

(b) $\begin{cases} x = 3 \cos t, \\ y = 3 \sin t, \text{ when } t = \frac{\pi}{4}. \end{cases}$

(c) $\begin{cases} x = t^3 - t + 1, \\ y = t^2 + t + 1, \text{ at } (1, 3). \end{cases}$

(d) $Q(t) = (t^2 - t + 1, e^t + 1)$, at $Q(0)$.

9. (a) Assuming the necessary differentiability conditions on the parametrization $P(t) = \big(x(t), y(t)\big)$, derive a formula for $\frac{d^2y}{dx^2}$.

(b) For the curve parametrized by the equations $x = t^2 + t + 1$ and $y = t^3 + 3t$, find $\frac{d^2y}{dx^2}$ when $t = 1$.

2. Arc Length of a Parametrized Curve. The straight-line distance in the plane between two points $P = (x_1, y_1)$ and $Q = (x_2, y_2)$ is defined in Section 2 of Chapter 1 by the formula

$$distance(P, Q) = \sqrt{(x_2 - x_1)^2 + (y_2 - y_1)^2}. \qquad (1)$$

In this section we shall consider the harder problem of defining distance, or arc length, along a parametrized curve.

Let C be a curve parametrically defined by a continuous function $P: I \to R^2$, and let a and b be two numbers in the interval I such that $a < b$. As we have seen in Section 1, C is the set of all points

$$P(t) = \big(x(t), y(t)\big), \qquad \text{for every } t \text{ in } I.$$

Consider a partition $\sigma = \{t_0, \ldots, t_n\}$ of the closed interval $[a, b]$ such that

$$a = t_0 \leq t_1 \leq \cdots \leq t_n = b,$$

and set $P(t_i) = P_i$, for every $i = 0, \ldots, n$. We shall take the number L_σ defined by

$$L_\sigma = \sum_{i=1}^{n} distance\ (P_{i-1}, P_i) \tag{2}$$

as an approximation to the arc length along C from $P(a)$ to $P(b)$. In Figure 6, the number L_σ is the sum of the lengths of the straight-line segments joining the points along the curve. Using (1), we may also write

$$L_\sigma = \sum_{i=1}^{n} \sqrt{[x(t_i) - x(t_{i-1})]^2 + [y(t_i) - y(t_{i-1})]^2}.$$

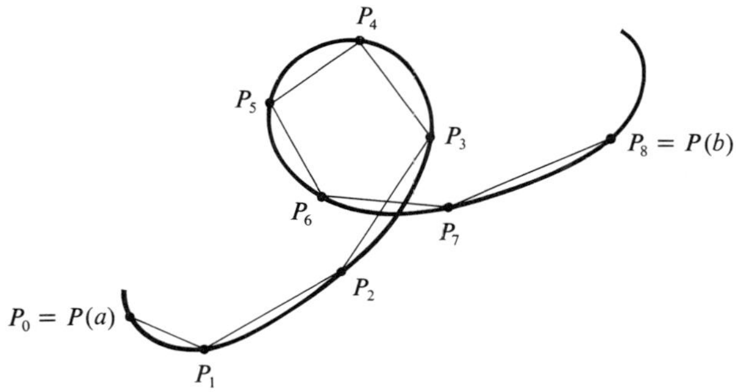

Figure 6

The principle which motivates the definition of arc length is the fact that if one partition σ of $[a, b]$ is a subset of another partition τ, then L_τ is in general a better approximation than L_σ. The basic reason for this is simply that the finer partition τ determines more points on the curve. As an example, consider Figure 7, in which L_σ is the sum of the lengths of the solid straight-

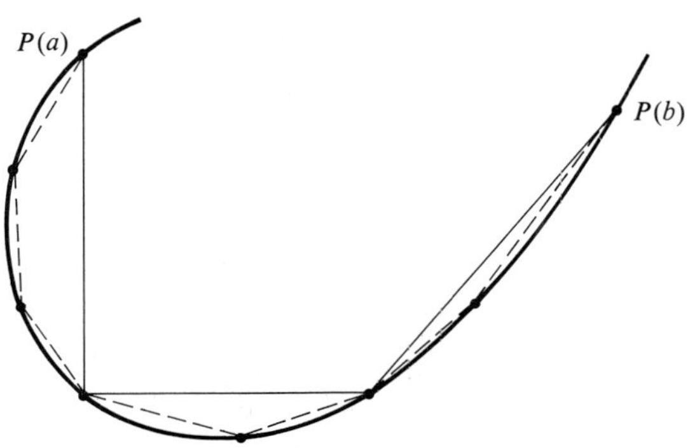

Figure 7

line segments, and L_τ is the sum of the lengths of the dashed-line segments. It is clear from the picture that L_τ is closer than L_σ to our intuitive idea of the arc length of the curve. Moreover,

(2.1) *If one partition σ of $[a, b]$ is a subset of another τ, then $L_\sigma \leq L_\tau$.*

Proof. This fact is geometrically apparent from Figure 7. The argument can be reduced to consideration of a single triangle by the realization that, since τ can be obtained from σ by adjoining one new point at a time, it will be sufficient to prove the result under the assumption that τ differs from σ by the inclusion of only one additional point, which we denote by t_*. Let $P_* = P(t_*)$. For some integer i, we have $t_{i-1} < t_* < t_i$. Then the expressions for L_σ and L_τ are obviously the same except that the term $distance(P_{i-1}, P_i)$, in L_σ is replaced by the sum $distance(P_{i-1}, P_*) + distance(P_*, P_i)$ in L_τ. Hence

$$L_\tau - L_\sigma = distance(P_{i-1}, P_*) + distance(P_*, P_i) - distance(P_{i-1}, P_i).$$

It is clear from the triangle in Figure 8 that the right side of the preceding equation cannot be negative. We conclude that $L_\tau - L_\sigma \geq 0$, or, equivalently, that $L_\tau \geq L_\sigma$, and the proof is complete.

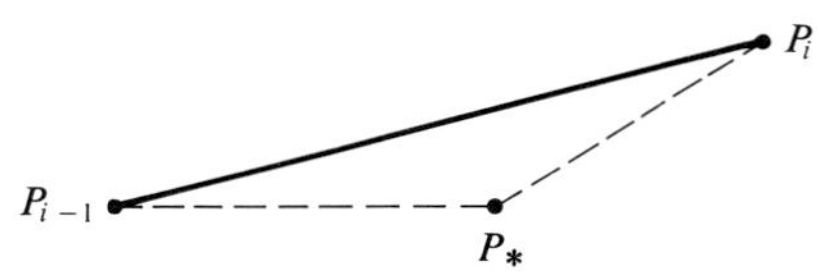

Figure 8

Thus partitions with more points result in approximations at least as large. This brings us to the definition of arc length: Let C be a curve parametrized by a continuous function $P: I \rightarrow R^2$, and let a and b be two numbers in I with $a \leq b$. We consider the set of all real numbers L_σ formed from all partitions σ of the interval $[a, b]$. This set, denoted by $\{L_\sigma\}$, either has an upper bound or it does not. The **arc length** of the parametrized curve C from $P(a)$ to $P(b)$ will be denoted by $L_a{}^b$ and is defined by

$$L_a{}^b = \begin{cases} \text{the least upper bound of the set } \{L_\sigma\}, & \text{if an upper bound exists,} \\ \infty, & \text{if } \{L_\sigma\} \text{ has no upper bound.} \end{cases}$$

A curve parametrized by a continuous function $P:[a, b] \rightarrow R^2$ is said to be **rectifiable** if its arc length $L_a{}^b$ is finite.

The main difficulty with the above definition, like that of the definite integral, is that it is by no means immediately apparent how to use it to compute the arc lengths of even very simple rectifiable curves. We shall now show that if the parametrization satisfies a simple differentiability condition, then the arc length is given by a definite integral.

(2.2) THEOREM. *Consider a parametrization defined by* $P(t) = (x(t), y(t))$, *for every t in an interval* $[a, b]$. *If the derivatives* x' *and* y' *are continuous functions, then the curve C parametrized by P is rectifiable and its arc length is given by*

$$L_a^b = \int_a^b \sqrt{x'(t)^2 + y'(t)^2}\, dt.$$

(Note that we have used $x'(t)^2$ for $[x'(t)]^2$ and $y'(t)^2$ for $[y'(t)]^2$. This is a common abbreviation, which we shall not hesitate to use whenever it causes no ambiguity.)

Proof. If $a = b$, then L_a^b is certainly equal to zero, as is the integral; so we assume that $a < b$. Let $\sigma = \{t_0, \ldots, t_n\}$ be a partition of $[a, b]$ with

$$a = t_0 < t_1 < \cdots < t_n = b,$$

and set $P_i = P(t_i)$. In each open subinterval (t_{i-1}, t_i) there exist, by the Mean Value Theorem, numbers t_{i1} and t_{i2} such that

$$x(t_i) - x(t_{i-1}) = x'(t_{i1})(t_i - t_{i-1}),$$

$$y(t_i) - y(t_{i-1}) = y'(t_{i2})(t_i - t_{i-1}).$$

Hence

$$\begin{aligned} distance(P_{i-1}, P_i) &= \sqrt{[x(t_i) - x(t_{i-1})]^2 + [y(t_i) - y(t_{i-1})]^2} \\ &= \sqrt{x'(t_{i1})^2 + y'(t_{i2})^2}\,(t_i - t_{i-1}), \end{aligned}$$

and so

$$L_\sigma = \sum_{i=1}^{n} \sqrt{x'(t_{i1})^2 + y'(t_{i2})^2}\,(t_i - t_{i-1}). \tag{3}$$

The conclusion of the theorem should now seem a natural one. Since $x'(t)$ and $y'(t)$ are continuous functions, so is $\sqrt{x'(t)^2 + y'(t)^2}$. We know that continuous functions are integrable. It is therefore very reasonable to suppose that, for successively finer and finer partitions, the right side of equation (3) approaches the integral $\int_a^b \sqrt{x'(t)^2 + y'(t)^2}\, dt$. If this is so, it follows in a straightforward manner from (2.1) that the set $\{L_\sigma\}$, for all partitions σ, must have the integral as a least upper bound, and the proof is then finished.

To complete the argument, it therefore remains to prove that

$$\lim_{\|\sigma\| \to 0} L_\sigma = \int_a^b \sqrt{x'(t)^2 + y'(t)^2}\, dt.$$

We recall that the fineness of a partition σ is measured by its mesh $\|\sigma\|$, defined on page 413 to be the length of a subinterval of maximum length.

Unfortunately, the preceding equation does not follow directly from the theory of Riemann sums because L_σ, as it is given by equation (3), is not a Riemann sum for the function $\sqrt{x'(t)^2 + y'(t)^2}$. It fails to be one because, in each subinterval of the partition, we have chosen *two* numbers t_{i1} and t_{i2} instead of one. To overcome this difficulty, we shall use a theorem about continuous functions of two variables, whose proof, although not deep, requires the concept of uniform continuity and we shall omit. From equation (3), we write the identity

$$L_\sigma = \sum_{i=1}^{n} \sqrt{x'(t_{i1})^2 + y'(t_{i1})^2}\,(t_i - t_{i-1})$$

$$+ \sum_{i=1}^{n} \left[\sqrt{x'(t_{i1})^2 + y'(t_{i2})^2} - \sqrt{x'(t_{i1})^2 + y'(t_{i1})^2}\right](t_i - t_{i-1}).$$

The first expression on the right *is* a Riemann sum for $\sqrt{x'(t)^2 + y'(t)^2}$ relative to σ, and we shall abbreviate it by R_σ. Next, let F be the function of two variables defined by

$$F(t, s) = \sqrt{x'(t)^2 + y'(s)^2} - \sqrt{x'(t)^2 + y'(t)^2},$$

for every t and s in the interval $[a, b]$. As a result, we can write the above expression for L_σ as

$$L_\sigma = R_\sigma + \sum_{i=1}^{n} F(t_{i1}, t_{i2})(t_i - t_{i-1}).$$

Hence

$$L_\sigma - R_\sigma = \sum_{i=1}^{n} F(t_{i1}, t_{i2})(t_i - t_{i-1}),$$

which implies that

$$|L_\sigma - R_\sigma| \le \sum_{i=1}^{n} |F(t_{i1}, t_{i2})|(t_i - t_{i-1}). \tag{4}$$

The function F is continuous, and, as is obvious from its definition, $F(t, t) = 0$ for every t in $[a, b]$. As a result, it can be proved that $|F(t, s)|$ is arbitrarily small provided the difference $|t - s|$ is sufficiently small. This is the theorem which we shall assume without proof. It follows that, for any positive number ϵ, there exists a positive number δ such that, if σ is any partition with mesh less than δ, then

$$|F(t_{i1}, t_{i2})| < \epsilon, \qquad \text{for every } i.$$

Hence, by the inequality (4), we obtain

$$|L_\sigma - R_\sigma| \le \sum_{i=1}^{n} \epsilon(t_i - t_{i-1}) = \epsilon(b - a),$$

for every partition σ for which $\|\sigma\| < \delta$. Since ϵ can be chosen so that $\epsilon(b - a)$ is arbitrarily small, we may conclude that $\lim_{\|\sigma\| \to 0} (L_\sigma - R_\sigma) = 0$. The proof is now virtually finished. We write

$$L_\sigma = R_\sigma + (L_\sigma - R_\sigma).$$

Since R_σ is a Riemann sum, we know that

$$\lim_{\|\sigma\| \to 0} R_\sigma = \int_a^b \sqrt{x'(t)^2 + y'(t)^2}\, dt.$$

Hence

$$\begin{aligned} \lim_{\|\sigma\| \to 0} L_\sigma &= \lim_{\|\sigma\| \to 0} R_\sigma + \lim_{\|\sigma\| \to 0} (L_\sigma - R_\sigma) \\ &= \int_a^b \sqrt{x'(t)^2 + y'(t)^2}\, dt + 0, \end{aligned}$$

and Theorem (2.2) is proved.

EXAMPLE 1. Compute the arc length of the curve C defined parametrically by $P(t) = \big(x(t), y(t)\big)$, where

$$\begin{cases} x(t) = a(t - \sin t), \\ y(t) = a(1 - \cos t), \qquad a > 0, \end{cases}$$

between $P(0) = (0, 0)$ and $P(2\pi) = (2\pi a, 0)$. The curve C is the cycloid discussed at the end of Section 1 and illustrated in Figure 4. We have

$$\begin{aligned} x'(t) &= a(1 - \cos t), \\ y'(t) &= a \sin t. \end{aligned}$$

These are obviously continuous functions, and the arc length is therefore given by the integral formula. We obtain

$$\begin{aligned} x'(t)^2 + y'(t)^2 &= a^2[(1 - \cos t)^2 + \sin^2 t] \\ &= a^2[1 - 2 \cos t + \cos^2 t + \sin^2 t] \\ &= a^2[1 - 2 \cos t + 1] = 2a^2[1 - \cos t]. \end{aligned}$$

However, we have the trigonometric identities

$$\begin{aligned} \cos t &= \cos 2 \cdot \frac{t}{2} = \cos^2 \frac{t}{2} - \sin^2 \frac{t}{2}, \\ 1 &= \qquad\qquad \cos^2 \frac{t}{2} + \sin^2 \frac{t}{2}, \end{aligned}$$

from which it follows that

$$1 - \cos t = 2 \sin^2 \frac{t}{2}.$$

Hence

$$x'(t)^2 + y'(t)^2 = 4a^2 \sin^2 \frac{t}{2} \cdot$$

Since $\sin \frac{t}{2}$ is nonnegative for every t in the interval $[0, 2\pi]$, we conclude that

$$\sqrt{x'(t)^2 + y'(t)^2} = 2a \sin \frac{t}{2}, \qquad \text{for } 0 \leq t \leq 2\pi.$$

Thus the arc length $L = L_0^{2\pi}$ is given by

$$L = \int_0^{2\pi} \sqrt{x'(t)^2 + y'(t)^2}\, dt = \int_0^{2\pi} 2a \sin \frac{t}{2}\, dt$$

$$= -4a \cos \frac{t}{2}\Big|_0^{2\pi} = -4a \cos \pi + 4a \cos 0 = 8a.$$

Suppose that a curve is given as the graph of a continuously differentiable function. In more detail: Let the derivative f' of a function f be continuous at every x in some interval $[a, b]$. The graph of the equation $y = f(x)$ is a curve which can be parametrically defined by

$$\begin{cases} x(t) = t, \\ y(t) = f(t), \qquad a \leq t \leq b. \end{cases}$$

Then $x'(t) = 1$ and $y'(t) = f'(t)$. Since x' is a constant function, it is certainly continuous. Since f' is by assumption continuous on $[a, b]$ and since $y' = f'$, the function y' is also continuous. Hence the arc length L_a^b is given by

$$L_a^b = \int_a^b \sqrt{x'(t)^2 + y'(t)^2}\, dt = \int_a^b \sqrt{1 + f'(t)^2}\, dt.$$

The variable of integration which appears in a definite integral is a dummy variable (see page 171), and we may therefore replace t by x in the right integral. Thus, we have proved, as a corollary of Theorem (2.2),

(2.3) *If the derivative of a function f is continuous at every x in an interval $[a, b]$, then the graph of $y = f(x)$ is a rectifiable curve and its arc length L_a^b is given by*

$$L_a^b = \int_a^b \sqrt{1 + f'(x)^2}\, dx.$$

EXAMPLE 2. Find the arc length L of the graph of the equation $y = x^2$ from the point (0, 0) to the point (2, 4). The curve is the familiar parabola shown in Figure 9. Using the result of the preceding theorem, we have

$$L = \int_0^2 \sqrt{1 + \left(\frac{dy}{dx}\right)^2}\, dx.$$

Since $\dfrac{dy}{dx} = 2x$,

$$L = \int_0^2 \sqrt{1 + 4x^2}\, dx.$$

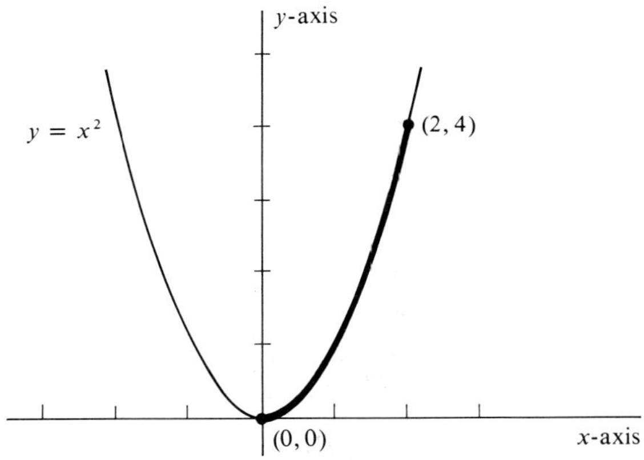

Figure 9

This integral can be evaluated by means of the trigonometric substitution $x = \frac{1}{2}\tan\theta$, or, equivalently, $2x = \tan\theta$. If $x = 0$, then $\theta = 0$, and, similarly, if $x = 2$, then $\theta = \arctan 4$. For convenience we shall set $\arctan 4 = \theta_0$. The substitution yields $\sqrt{1 + 4x^2} = \sqrt{1 + \tan^2\theta} = \sec\theta$, and $dx = \frac{1}{2}\sec^2\theta\, d\theta$. Hence using the Change of Variable Theorem for Definite Integrals, we obtain

$$\int_0^2 \sqrt{1 + 4x^2}\, dx = \frac{1}{2}\int_0^{\theta_0} \sec^3\theta\, d\theta.$$

The reduction formula on page 369 gives

$$\int \sec^3\theta\, d\theta = \frac{\sec\theta\tan\theta}{2} + \frac{1}{2}\int \sec\theta\, d\theta,$$

and also on page 369 we have

$$\int \sec\theta\, d\theta = \ln|\sec\theta + \tan\theta| + c.$$

It follows that

$$L = \tfrac{1}{2}\int_0^{\theta_0} \sec^3\theta\, d\theta = \left[\frac{\sec\theta\tan\theta}{4} + \tfrac{1}{4}\ln|\sec\theta + \tan\theta|\right]_0^{\theta_0}$$
$$= \tfrac{1}{4}[\sec\theta_0 \tan\theta_0 + \ln|\sec\theta_0 + \tan\theta_0|].$$

Since $\theta_0 = \arctan 4$, we have $\tan\theta_0 = 4$ and $\sec\theta_0 = \sqrt{1 + \tan^2\theta_0} = \sqrt{17}$. Hence the arc length L is equal to

$$\begin{aligned} L &= \tfrac{1}{4}[\sqrt{17}\cdot 4 + \ln(\sqrt{17} + 4)] \\ &= \sqrt{17} + \tfrac{1}{4}\ln(\sqrt{17} + 4) \\ &= 4.64 \text{ (approximately).} \end{aligned}$$

PROBLEMS

1. Find the arc lengths of the following parametrized curves.

(a) $\begin{cases} x = t + 1, \\ y = t^{3/2}, \end{cases}$ from $(2, 1)$ to $(5, 8)$.

(b) $\begin{cases} x(t) = t^2, \\ y(t) = \frac{2}{3}(2t + 1)^{3/2}, \end{cases}$ from $\big(x(0), y(0)\big) = (0, \frac{2}{3})$ to $\big(x(4), y(4)\big) = (16, 18)$.

(c) $P(t) = (t^2, t^3)$, from $P(0)$ to $P(2)$.

(d) $\begin{cases} x(\theta) = a\cos^3\theta, & a > 0, \\ y(\theta) = a\sin^3\theta, \end{cases}$ from $\big(x(0), y(0)\big) = (a, 0)$ to $\big(x(\pi/2), y(\pi/2)\big) = (0, a)$.

2. A particle in motion in the plane has position equal to

$$P(t) = \big(t^2 + t, \tfrac{1}{6}(4t + 3)^{3/2}\big)$$

at time t. How far does the particle travel along its path from time $t = 0$ to time $t = 1$?

3. Find the arc lengths of the graphs of each of the following functions f between the points $\big(a, f(a)\big)$ and $\big(b, f(b)\big)$.

(a) $f(x) = x^{3/2}$, $a = 1$ and $b = 4$.

(b) $f(x) = \frac{2}{3}(x^2 + 1)^{3/2}$, $a = 0$ and $b = 2$.

(c) $f(x) = x^2$, $a = 0$ and $b = \frac{1}{2}$.

(d) $f(x) = \frac{1}{2}(e^x + e^{-x})$, $a = -1$ and $b = 1$.

4. Show that the circumference of an ellipse with the line segment joining $(-a, 0)$ and $(a, 0)$ as major axis and the line segment joining $(0, -b)$ and $(0, b)$ as minor axis is given by an integral

$$K\int_0^{2\pi} \sqrt{1 + k\sin^2\theta}\, d\theta.$$

Evaluate the constants K and k in terms of a and b. (Do not attempt to evaluate the integral.)

5. (a) Let g be a function which is continuously differentiable on the closed interval $[c, d]$. Prove, as a corollary of Theorem (2.2), that the arc length L_c^d of the graph of the equation $x = g(y)$ between the points $(g(c), c)$ and $(g(d), d)$ is given by the formula

$$L_c^d = \int_c^d \sqrt{1 + g'(y)^2}\, dy.$$

(b) Find the arc length of the graph of the equation $x = \frac{1}{3}(y^2 + 2)^{3/2}$ between the point $\left(\frac{2\sqrt{2}}{3}, 0\right)$ and the point $(2\sqrt{6}, 2)$.

(c) Express as a definite integral the arc length of that part of the graph of the equation $x = 2y - y^2$ for which $x \geq 0$.

6. The coordinates of a particle in motion in the plane are given by

$$\begin{cases} x = t^2, \\ y = \frac{2}{3}t^3 - \frac{1}{2}t, \end{cases}$$

at time t. What is the distance which the particle moves along its path of motion between the times $t = 0$ and $t = 2$?

7. The same curve can be defined by more than one parametrization:

(a) Draw the curve defined parametrically by

$$\begin{cases} x(t) = t, \\ y(t) = t, \qquad 0 \leq t \leq 1. \end{cases}$$

(b) Draw the curve defined parametrically by

$$\begin{cases} x(t) = \sin \pi t, \\ y(t) = \sin \pi t, \qquad 0 \leq t \leq 1. \end{cases}$$

(c) Compute the arc lengths from $t = 0$ to $t = 1$ for the parametrizations in (a) and (b).

(d) Give a geometric interpretation which explains the difference between the arc lengths obtained for the two parametrizations.

8. Let $P: [a, b] \rightarrow R^2$ and $Q: [c, d] \rightarrow R^2$ be two parametrizations of the same curve C such that all four coordinate functions are continuously differentiable. (A function is **continuously differentiable** if its derivative exists and is continuous at every number in its domain.) Then P and Q are called **equivalent parametrizations** of C if there exists a continuously differentiable function f with domain $[a, b]$ and range $[c, d]$ which has a continuously differentiable inverse function, and in addition satisfies

(i) $f(a) = c$ and $f(b) = d$,

(ii) $P(t) = Q(f(t))$, for every t in $[a, b]$.

(a) Using the Chain Rule and the Change of Variable Theorem for Definite Integrals (for the latter, see page 215), prove that equivalent parametrizations assign the same arc length to C.

(b) Show that

$$P(t) = (\cos t, \sin t), \qquad 0 \le t \le \frac{\pi}{2},$$

$$Q(s) = \left(\frac{1 - s^2}{1 + s^2}, \frac{2s}{1 + s^2}\right), \qquad 0 \le s \le 1,$$

are equivalent parametrizations of the same curve C, and identify the curve.

(c) Show that

$$P(t) = (\cos t, \sin t), \qquad 0 \le t \le 2\pi,$$

and

$$Q(s) = (\cos 5t, \sin 5t), \qquad 0 \le t \le 2\pi,$$

are nonequivalent parametrizations of the circle.

9. Prove directly from the least upper bound definition that arc length is additive, i.e., that $L_a^b + L_b^c = L_a^c$.

3. Vectors in the Plane. A **vector** in the plane is an ordered pair (P, Q) of points in the plane. The point P is called the **initial point** of the vector and Q the **terminal point.** Geometrically, the vector (P, Q) will be represented as a directed line segment, or arrow, from P to Q, as illustrated in Figure 10. We shall use boldface lower-case letters to denote vectors. For example, if $\mathbf{v}$ is the vector with initial point P and terminal point Q, then $\mathbf{v} = (P, Q)$.

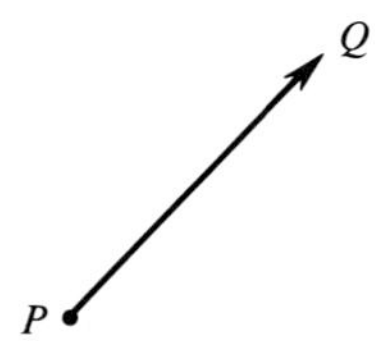

Figure 10

Having identified the plane with the set R^2 of all ordered pairs of real numbers, we see that a vector is determined by four real numbers: two coordinates of its initial point, and two of its terminal point. Let $\mathbf{v}$ be a vector with initial point $P = (a, b)$ and terminal point $Q = (c, d)$. Then the two numbers v_1 and v_2 given by the equations

$$\begin{aligned} v_1 &= c - a, \\ v_2 &= d - b, \end{aligned} \tag{1}$$

are defined to be **first** and **second coordinates,** respectively, of the vector $\mathbf{v}$ in R^2. Thus we have defined coordinates of a vector in R^2 as well as coordinates of a point in R^2. The definitions are not the same, although the concepts are certainly related.

If a vector $\mathbf{v}$ has initial point $P = (a, b)$ and coordinates v_1 and v_2, then equations (1) tell us that the terminal point $Q = (c, d)$ is given by

$$Q = (a + v_1, b + v_2).$$

It follows that a vector is completely determined by its initial point and its coordinates. Hence, another notation for a vector, which we shall use, is

$$\boxed{\mathbf{v} = (v_1, v_2)_P.} \tag{2}$$

[Although it would be consistent with this notation, we shall not write $(v_1, v_2)_{(a,b)}$ for the vector with initial point (a, b) and coordinates v_1 and v_2.]

The **length** of a vector $\mathbf{v} = (P, Q)$ in R^2 is denoted by $|\mathbf{v}|$ and defined by

$$|\mathbf{v}| = \textit{distance}(P, Q).$$

If $P = (a, b)$ and $Q = (c, d)$, then the formula for the distance between two points implies that

$$|\mathbf{v}| = \sqrt{(c - a)^2 + (d - b)^2}.$$

From equations (1) it follows that the coordinates of the vector $\mathbf{v}$ are the two numbers $v_1 = c - a$ and $v_2 = d - b$. Hence

(3.1) *The length of any vector* $\mathbf{v} = (v_1, v_2)_P$ *is given by*

$$\boxed{|\mathbf{v}| = \sqrt{v_1^2 + v_2^2}.}$$

Thus the length of a vector depends only on its coordinates.

EXAMPLE 1. Find the terminal point of each of the following vectors. Draw each one as an arrow in the xy-plane, and compute its length.

(a) $\mathbf{v} = (1, 2)_P$, where $P = (1, 1)$,

(b) $\mathbf{u} = (4, -1)_P$, where $P = (1, 1)$,

(c) $\mathbf{w} = (-2, 5)_Q$, where $Q = (0, -1)$,

(d) $\mathbf{x} = (3, -4)_O$, where $O = (0, 0)$.

We have seen that, if $P = (a, b)$, then the terminal point of the vector $(v_1, v_2)_P$ is the ordered pair $(a + v_1, b + v_2)$. It follows that

$$\text{terminal point of } \mathbf{v} = (1 + 1, 1 + 2) = (2, 3),$$
$$\text{terminal point of } \mathbf{u} = \big(1 + 4, 1 + (-1)\big) = (5, 0),$$
$$\text{terminal point of } \mathbf{w} = \big(0 + (-2), -1 + 5\big) = (-2, 4),$$
$$\text{terminal point of } \mathbf{x} = \big(0 + 3, 0 + (-4)\big) = (3, -4).$$

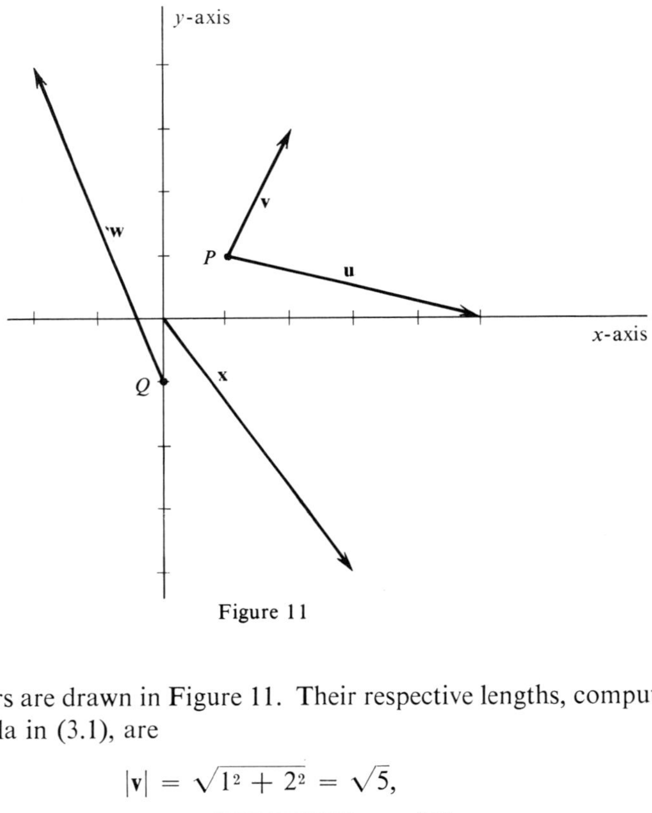

Figure 11

The vectors are drawn in Figure 11. Their respective lengths, computed from the formula in (3.1), are

$$|\mathbf{v}| = \sqrt{1^2 + 2^2} = \sqrt{5},$$
$$|\mathbf{u}| = \sqrt{4^2 + (-1)^2} = \sqrt{17},$$
$$|\mathbf{w}| = \sqrt{(-2)^2 + 5^2} = \sqrt{29},$$
$$|\mathbf{x}| = \sqrt{3^2 + (-4)^2} = \sqrt{25} = 5.$$

We shall denote the set of all vectors in R^2 by $\mathcal{V}$. For every point P in R^2, the subset of $\mathcal{V}$ consisting of all vectors with initial point P will be denoted by

$\mathcal{V}_P$. We shall now define, in each set $\mathcal{V}_P$, an operation of addition of vectors and an operation of multiplication of vectors by real numbers.

Addition in $\mathcal{V}_P$ is defined as follows: If $\mathbf{u} = (u_1, u_2)_P$ and $\mathbf{v} = (v_1, v_2)_P$ are any two vectors in $\mathcal{V}_P$, then their sum $\mathbf{u} + \mathbf{v}$ is the vector defined by

$$\boxed{\mathbf{u} + \mathbf{v} = (u_1 + v_1, u_2 + v_2)_P.} \tag{3}$$

Note that the sum of two vectors in $\mathcal{V}_P$ is again a vector in $\mathcal{V}_P$. Furthermore, if $\mathbf{u}$ is in $\mathcal{V}_P$ and $\mathbf{v}$ is in $\mathcal{V}_Q$, then their sum is not defined unless $P = Q$. That is, *the sum of two vectors is defined if and only if they have the same initial point*. For every vector $\mathbf{v} = (v_1, v_2)_P$, we denote the vector $(-v_1, -v_2)_P$ by $-\mathbf{v}$. In this way, subtraction of vectors in $\mathcal{V}_P$ is defined by the equation

$$\mathbf{u} - \mathbf{v} = \mathbf{u} + (-\mathbf{v}),$$

which implies the following companion formula to (3):

$$\boxed{\mathbf{u} - \mathbf{v} = (u_1 - v_1, u_2 - v_2)_P.} \tag{4}$$

The unique vector in $\mathcal{V}_P$ with both coordinates equal to zero is called the zero vector and will be denoted by $\mathbf{0}$. Thus

$$\mathbf{0} = (0, 0)_P = (P, P).$$

Obviously, the equations

$$\mathbf{v} + \mathbf{0} = \mathbf{v},$$

$$\mathbf{v} - \mathbf{v} = \mathbf{0}$$

are true for every vector $\mathbf{v}$ in $\mathcal{V}_P$. Geometrically the zero vector in $\mathcal{V}_P$ is represented simply by the point P. Of course, there are as many different zero vectors as there are points in the plane, and one cannot tell from the notation $\mathbf{0}$ to which set $\mathcal{V}_P$ a given zero vector belongs. It is obvious that every zero vector has length zero. Conversely, the length of a nonzero vector must be positive, since at least one of its coordinates is not zero. Hence

(3.2) *A vector* $\mathbf{v}$ *is a zero vector if and only if* $|\mathbf{v}| = 0$.

Geometrically, the sum $\mathbf{u} + \mathbf{v}$ of two nonzero vectors $\mathbf{u}$ and $\mathbf{v}$ in $\mathcal{V}_P$ is the vector in $\mathcal{V}_P$ which is a diagonal of the parallelogram which has $\mathbf{u}$ and $\mathbf{v}$ as sides. This is the famous Parallelogram Law and is illustrated in Figure 12(a). It can be verified in a straightforward way by computing the slopes of the various line segments, and we omit the details. Similarly, the vector $-\mathbf{v}$ is represented geometrically as a directed line segment lying in the same straight line as $\mathbf{v}$, but in the opposite direction, as shown in Figure 12(b). Moreover, the vectors $\mathbf{v}$ and $-\mathbf{v}$ have the same length, since

$$|\mathbf{v}| = \sqrt{v_1^2 + v_2^2} = \sqrt{(-v_1)^2 + (-v_2)^2} = |-\mathbf{v}|.$$

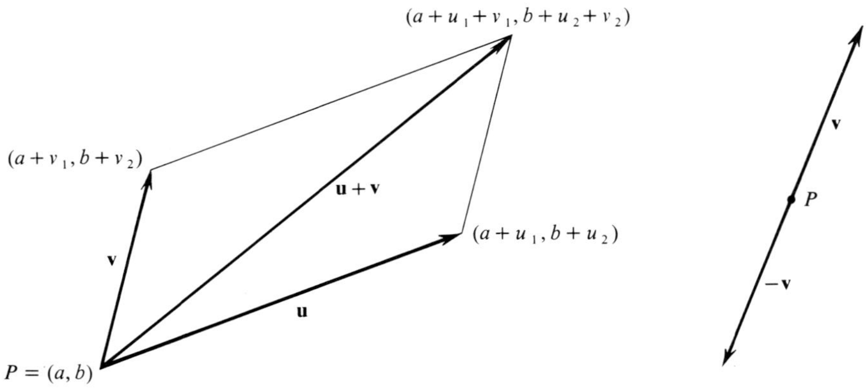

(*a*) The Parallelogram Law (*b*) The negative of a vector

Figure 12

The second algebraic operation in $\mathcal{V}_P$ is defined as follows: For every real number a and every vector $\mathbf{v} = (v_1, v_2)_P$ in $\mathcal{V}_P$, we define a vector $a\mathbf{v}$, called the **product** of a and $\mathbf{v}$, by the equation

$$a\mathbf{v} = (av_1, av_2)_P. \tag{5}$$

In traditional vector terminology, the real number a is called a **scalar.** Note that we have *not* defined a product of two vectors. If we compute the length of the vector $a\mathbf{v}$, we find that

$$\begin{aligned} |a\mathbf{v}| &= \sqrt{(av_1)^2 + (av_2)^2} = \sqrt{a^2(v_1^2 + v_2^2)} \\ &= |a|\sqrt{v_1^2 + v_2^2} = |a|\,|\mathbf{v}|, \end{aligned}$$

a result which we summarize in the statement

(3.3) $|a\mathbf{v}| = |a|\,|\mathbf{v}|$, *for every real number a and every vector* $\mathbf{v}$.

If $\mathbf{v}$ is an arbitrary nonzero vector in $\mathcal{V}_P$ and if $a \neq 0$, then the slope of the line segment joining P to the terminal point of $\mathbf{v}$ is the same as that joining P to the terminal point of $a\mathbf{v}$. Hence P and the terminal points of $\mathbf{v}$ and $a\mathbf{v}$ lie on the same straight line. In addition, it is easy to check that the arrows representing $\mathbf{v}$ and $a\mathbf{v}$ are in the same or opposite direction according as a is positive or negative.

EXAMPLE 2. Let $P = (2, -1)$, and consider the two vectors $\mathbf{u} = (1, 5)_P$ and $\mathbf{v} = (2, -1)_P$. Compute and draw each of the following vectors in the same plane with **u** and **v**.

(a) $\mathbf{u} + \mathbf{v}$,

(b) $-2\mathbf{v}$,

(c) $\mathbf{u} - 2\mathbf{v}$,

(d) $\frac{1}{4}(\mathbf{u} + \mathbf{v})$.

The computations are very simple:

$$\mathbf{u} + \mathbf{v} = (1, 5)_P + (2, -1)_P = (3, 4)_P,$$
$$-2\mathbf{v} = -2(2, -1)_P = (-4, 2)_P,$$
$$\mathbf{u} - 2\mathbf{v} = (1, 5)_P - 2(2, -1)_P = (1, 5)_P + (-4, 2)_P = (-3, 7)_P,$$
$$\tfrac{1}{4}(\mathbf{u} + \mathbf{v}) = \tfrac{1}{4}(3, 4)_P = (\tfrac{3}{4}, 1)_P.$$

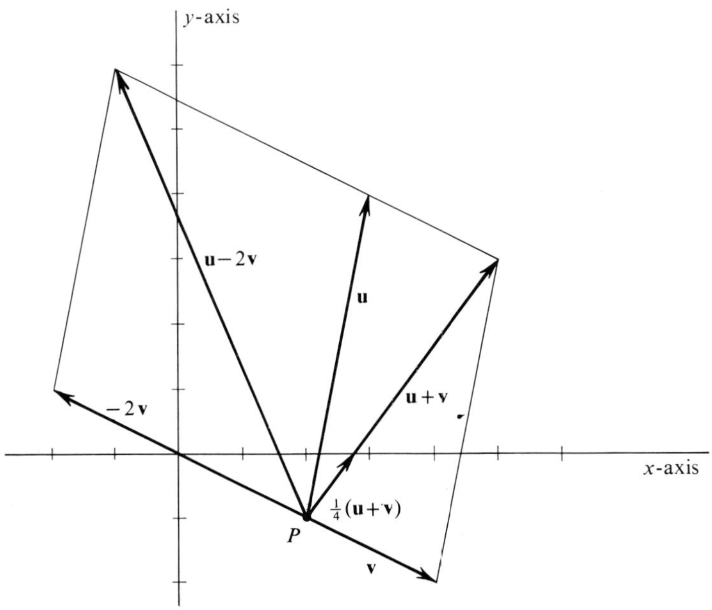

Figure 13

The directed line segments representing these vectors, as well as **u** and **v**, are shown in Figure 13. The easiest way to draw them is to make a list of their terminal points. We recall that a vector with coordinates v_1 and v_2 and initial point $P = (a, b)$ has a terminal point equal to $(a + v_1, b + v_2)$. Hence

$$\begin{aligned}
\text{terminal point of } \mathbf{u} &= (2 + 1, -1 + 5) = (3, 4),\\
\text{terminal point of } \mathbf{v} &= (2 + 2, -1 - 1) = (4, -2),\\
\text{terminal point of } \mathbf{u} + \mathbf{v} &= (2 + 3, -1 + 4) = (5, 3),\\
\text{terminal point of } -2\mathbf{v} &= (2 - 4, -1 + 2) = (-2, 1),\\
\text{terminal point of } \mathbf{u} - 2\mathbf{v} &= (2 - 3, -1 + 7) = (-1, 6),\\
\text{terminal point of } \tfrac{1}{4}(\mathbf{u} + \mathbf{v}) &= (2 + \tfrac{3}{4}, -1 + 1) = (2\tfrac{3}{4}, 0).
\end{aligned}$$

The next theorem summarizes the algebraic facts about the set $\mathcal{V}_P$ of all vectors in the plane with initial point P.

(3.4) THEOREM. *For each point P in R^2, vector addition and scalar multiplication in $\mathcal{V}_P$ have the following properties:*

(i) ASSOCIATIVITY

$$\mathbf{u} + (\mathbf{v} + \mathbf{w}) = (\mathbf{u} + \mathbf{v}) + \mathbf{w} \quad \text{and} \quad (ab)\mathbf{v} = a(b\mathbf{v}).$$

(ii) COMMUTATIVITY

$$\mathbf{u} + \mathbf{v} = \mathbf{v} + \mathbf{u}.$$

(iii) EXISTENCE OF ADDITIVE IDENTITY

There exists a vector $\mathbf{0}$ in $\mathcal{V}_P$ with the property that $\mathbf{v} + \mathbf{0} = \mathbf{v}$, for every vector $\mathbf{v}$ in $\mathcal{V}_P$.

(iv) EXISTENCE OF SUBTRACTION

For every vector $\mathbf{v}$ in $\mathcal{V}_P$, there exists a vector $-\mathbf{v}$ in *$\mathcal{V}_P$ such that* $\mathbf{v} + (-\mathbf{v}) = \mathbf{0}$.

(v) DISTRIBUTIVITY

$$a(\mathbf{u} + \mathbf{v}) = a\mathbf{u} + a\mathbf{v} \quad \text{and} \quad (a + b)\mathbf{v} = a\mathbf{v} + b\mathbf{v}.$$

(vi) EXISTENCE OF SCALAR IDENTITY

$$1\mathbf{v} = \mathbf{v}.$$

The proof of this theorem follows easily from the definitions of vector addition and scalar multiplication, from the definitions of $\mathbf{0}$ and $-\mathbf{v}$, and from the corresponding properties of addition and multiplication of real numbers given on page 2. The importance of the theorem is that every algebraic fact about vectors can be derived from the six properties listed. In fact, in abstract algebra, these properties are taken as a set of axioms: An arbitrary set **V** is called a **vector space** and its elements are called **vectors** if, for every pair of elements **u** and **v** in **V** and for every real number a, an element $\mathbf{u} + \mathbf{v}$ and an element $a\mathbf{v}$ in **V** are defined so that conditions (i) through (vi) are satisfied. This definition has proved to be of enormous value in mathematics and examples of vector spaces occur over and over again. In particular, Theorem (3.4) asserts that, for each point P in R^2, the set $\mathcal{V}_P$ is a vector space.

EXAMPLE 3. Let **v** be a nonzero vector in R^2. Then the set, which we shall denote by $R\mathbf{v}$, consisting of all products $t\mathbf{v}$, where t is a real number, is an example of a vector space. For, if P is the initial point of **v**, then **v** lies in $\mathcal{V}_P$, and it follows that every product $t\mathbf{v}$ also lies in $\mathcal{V}_P$. Hence the sum of any two vectors in $R\mathbf{v}$ is defined, and, since

$$t\mathbf{v} + s\mathbf{v} = (t + s)\mathbf{v},$$

the sum is again in $R\mathbf{v}$. Similarly, if $t\mathbf{v}$ is in $R\mathbf{v}$ and if s is any real number, then

$$s(t\mathbf{v}) = (st)\mathbf{v},$$

and $(st)\mathbf{v}$ is by definition in $R\mathbf{v}$. Thus vector addition and scalar multiplication are defined in the set $R\mathbf{v}$. Conditions (i), (ii), (v), and (vi) are automatically satisfied because they hold in the larger set $\mathcal{V}_P$. Finally, conditions (iii) and (iv) are also satisfied, since

$$\mathbf{0} = 0\mathbf{v} \quad \text{and} \quad -\mathbf{v} = (-1)\mathbf{v}.$$

This completes the proof that $R\mathbf{v}$ is a vector space. The terminal points of all the vectors in $R\mathbf{v}$ form the straight line containing the initial and terminal points of the vector **v**.

PROBLEMS

1. Find the terminal point of each of the following vectors. Draw each one as a directed line segment in the xy-plane, and compute its length.

   (a) $\mathbf{v} = (-3, 4)_P$, where $P = (1, 0)$,

   (b) $\mathbf{u} = (4, -3)_P$, where $P = (1, 0)$,

   (c) $\mathbf{x} = (3, 0)_Q$, where $Q = (-1, -1)$,

   (d) $\mathbf{a} = (4\frac{1}{2}, 3\frac{1}{2})_O$, where $O = (0, 0)$.

2. Let $P = (2, 1)$. Compute the terminal point of each of the following vectors, and draw each one as an arrow in the xy-plane. The vectors $\mathbf{u}$ and $\mathbf{v}$ in parts (b), (c), (d), and (e) are as defined in part (a).

   (a) $\mathbf{u} = (3, -2)_P$ and $\mathbf{v} = (1, 1)_P$ (d) $3\mathbf{v}$

   (b) $\mathbf{u} + \mathbf{v}$ (e) $\mathbf{u} + 3\mathbf{v}$.

   (c) $\mathbf{u} - \mathbf{v}$

3. Let $P = (0, 1)$, and consider the vectors $\mathbf{x} = (2, 5)_P$ and $\mathbf{y} = (1, 1)_P$.

   (a) Draw the vectors $\mathbf{x}$, $\mathbf{y}$, and $\mathbf{x} + \mathbf{y}$ in the xy-plane.

   (b) Compute the lengths $|\mathbf{x}|$, $|\mathbf{y}|$, and $|\mathbf{x} + \mathbf{y}|$.

4. True or false: If $P \neq Q$, then $\mathcal{V}_P$ and $\mathcal{V}_Q$ are disjoint sets?
5. Let $\mathbf{v}$ be a vector in the plane with initial point P, and let θ be the angle whose vertex is P, whose initial side is the vector $(1, 0)_P$, and whose terminal side is $\mathbf{v}$. Show that

$$\mathbf{v} = (|v| \cos \theta, |v| \sin \theta)_P.$$

   The angle θ is called the **direction** of the vector $\mathbf{v}$.
6. In physics, the force acting on a particle located at a point P in the plane is represented by a vector. The length of the vector is the magnitude of the force (e.g., the number of pounds), and the direction of the vector is the direction of the force (see Problem 5).

   (a) Draw the vector representing a force of 5 pounds acting on a particle at the point (3, 2) in a direction of $\frac{\pi}{6}$ radians.

   (b) What are the coordinates of the force vector in (a)?

7. If a particle located at a point P is simultaneously acted on by two forces $\mathbf{u}$ and $\mathbf{v}$, then the resultant force is the vector sum $\mathbf{u} + \mathbf{v}$. The fact that vectors are added geometrically by constructing a parallelogram implies a corresponding Parallelogram Law of Forces.

   Suppose that a particle at the point (1, 1) is simultaneously acted on by a force $\mathbf{v}$ of 10 pounds in the direction of $\frac{\pi}{6}$ radians and a force $\mathbf{u}$ of $\sqrt{32}$ pounds in the direction of $-\frac{\pi}{4}$ radians.

   (a) Draw the parallelogram of forces, and show the resultant force.

   (b) What are the coordinates of the resultant force on the particle?

8. Addition and scalar multiplication are defined in the set R^2 of all ordered pairs of real numbers by the equations

$$(a, b) + (c, d) = (a + c, b + d),$$

$$c(a, b) = (ca, cb).$$

Show that R^2 is a vector space with respect to these operations. This fact shows that the elements of a vector space need not necessarily be interpreted as arrows. The principal interpretation of R^2 is that of the set of points of the plane.

9. True or false?

(a) The set R of all real numbers is a vector space with respect to ordinary addition and multiplication.

(b) The set C of all complex numbers is a vector space with respect to addition and multiplication by real numbers.

(c) The set $\mathcal{V}$ of all vectors in the plane is a vector space with respect to vector addition and scalar multiplication as defined in this section.

4. The Derived Vector of a Parametrized Curve. Consider a function whose domain is a subset of the set of all real numbers and whose range is a subset of all vectors in the plane. If we denote this function by $\mathbf{v}$, then its value at each number t in the domain is the vector $\mathbf{v}(t)$. Every such vector-valued function $\mathbf{v}$ of a real variable defines two real-valued **coordinate functions** v_1 and v_2 as follows: For every t in the domain of $\mathbf{v}$, the numbers $v_1(t)$ and $v_2(t)$ are the first and second coordinates of the vector $\mathbf{v}(t)$, respectively. Hence, if the initial point of $\mathbf{v}(t)$ is $P(t)$, then $v_1(t)$ and $v_2(t)$ are defined by the equation

$$\mathbf{v}(t) = \big(v_1(t), v_2(t)\big)_{P(t)}. \tag{1}$$

Limits of vector-valued functions are defined in terms of limits of real-valued functions. Specifically, the **limit** of $\mathbf{v}(t)$, as t approaches t_0, will be denoted by $\lim_{t \to t_0} \mathbf{v}(t)$ and is defined by

$$\lim_{t \to t_0} \mathbf{v}(t) = \big(\lim_{t \to t_0} v_1(t), \lim_{t \to t_0} v_2(t)\big)_{\lim_{t \to t_0} P(t)}. \tag{2}$$

[For the definition of $\lim_{t \to t_0} P(t)$, see page 542.] There is the possibility that all the vectors $\mathbf{v}(t)$ have the same initial point P_0, i.e., that they all lie in the vector space $\mathcal{V}_{P_0}$. If this happens, (2) reduces to the simpler equation

$$\lim_{t \to t_0} \mathbf{v}(t) = \big(\lim_{t \to t_0} v_1(t), \lim_{t \to t_0} v_2(t)\big)_{P_0}.$$

Let C be a curve in the plane defined by a parametrization $P: I \to R^2$. If the coordinate functions of P are denoted by x and y, then C is the set of all points

$$P(t) = \big(x(t), y(t)\big)$$

such that t is in the interval I. A typical example is shown in Figure 14. Consider a number t_0 in I. If t is in I and distinct from t_0, then the vector $\big(P(t_0), P(t)\big)$ represents the change in the value of P from the point $P(t_0)$ to the point $P(t)$. Thus for a change in the value of the parameter from t_0 to t, the scalar product

$$\frac{1}{t - t_0}\big(P(t_0), P(t)\big) \tag{3}$$

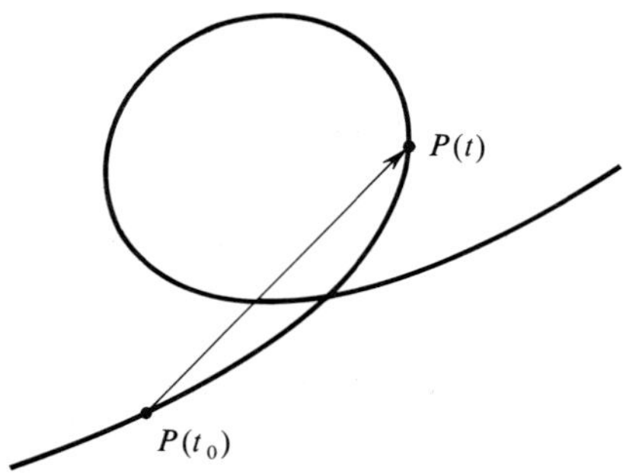

Figure 14

is the ratio of the corresponding change in the value of P to the difference $t - t_0$. Hence the vector (3) represents an average rate of change in position with respect to a change in the parameter. In analogy with the definition of the derivative of a real-valued function, we define the **derived vector** of P at t_0, denoted by $\mathbf{d}P(t_0)$, by the equation

$$\boxed{\mathbf{d}P(t_0) = \lim_{t \to t_0} \frac{1}{t - t_0}\big(P(t_0), P(t)\big).}$$

Since $P(t_0) = \big(x(t_0), y(t_0)\big)$ and $P(t) = \big(x(t), y(t)\big)$, the coordinate form of the vector $\big(P(t_0), P(t)\big)$ is given by

$$\big(P(t_0), P(t)\big) = \big(x(t) - x(t_0), y(t) - y(t_0)\big)_{P(t_0)}.$$

By the definition of the scalar product,

$$\frac{1}{t - t_0}\big(P(t_0), P(t)\big) = \left(\frac{x(t) - x(t_0)}{t - t_0}, \frac{y(t) - y(t_0)}{t - t_0}\right)_{P(t_0)},$$

and so

$$\mathbf{d}P(t_0) = \left(\lim_{t \to t_0} \frac{x(t) - x(t_0)}{t - t_0}, \lim_{t \to t_0} \frac{y(t) - y(t_0)}{t - t_0}\right)_{P(t_0)}.$$

Recall that the derivatives of the functions x and y at t_0 are by definition

$$x'(t_0) = \lim_{t \to t_0} \frac{x(t) - x(t_0)}{t - t_0},$$

$$y'(t_0) = \lim_{t \to t_0} \frac{y(t) - y(t_0)}{t - t_0},$$

provided these limits exist. It follows that

(4.1) *The parametrization defined by* $P(t) = \big(x(t), y(t)\big)$ *is differentiable at* t_0 *if and only if the derived vector* $\mathbf{d}P(t_0)$ *exists. If it does exist, then*

$$\mathbf{d}P(t_0) = \big(x'(t_0), y'(t_0)\big)_{P(t_0)}.$$

EXAMPLE 1. Consider the curve parametrized by

$$P(t) = \big(x(t), y(t)\big) = (t^2 - 1, 2t + 1), \qquad -\infty < t < \infty.$$

Compute the derived vectors of P at $t_0 = -1$, at $t_0 = 0$, and at $t_0 = 1$. Draw the curve and the three derived vectors in the xy-plane. As a result of (4.1), we have

$$\mathbf{d}P(t_0) = \big(x'(t_0), y'(t_0)\big)_{P(t_0)} = (2t_0, 2)_{P(t_0)}.$$

Hence

$$\begin{aligned} \mathbf{d}P(-1) &= (-2, 2)_{P(-1)} &\quad \text{and} \quad P(-1) &= (0, -1), \\ \mathbf{d}P(0) &= (0, 2)_{P(0)} &\quad \text{and} \quad P(0) &= (-1, 1), \\ \mathbf{d}P(1) &= (2, 2)_{P(1)} &\quad \text{and} \quad P(1) &= (0, 3). \end{aligned}$$

The terminal points of the three derived vectors are, respectively,

$$\begin{aligned} (0 - 2, -1 + 2) &= (-2, 1), \\ (-1 + 0, 1 + 2) &= (-1, 3), \\ (0 + 2, 3 + 2) &= (2, 5). \end{aligned}$$

The parametrized curve is a parabola, as can be seen by setting

$$\begin{cases} x = t^2 - 1, \\ y = 2t + 1. \end{cases}$$

Solving the second equation for t, we get $t = \dfrac{y - 1}{2}$, and substituting this value in the first, we obtain $x = \dfrac{(y - 1)^2}{4} - 1$, or, equivalently,

$$4(x + 1) = (y - 1)^2.$$

The latter is an equation of a parabola with vertex $(-1, 1)$. If $x = 0$, then $4 = (y - 1)^2$, or, equivalently, $\pm 2 = y - 1$, which implies that $y = -1$ or 3. The parametrized curve together with the three vectors is shown in Figure 15. Note that each of these vectors is tangent to the parabola.

If a parametrization $P: I \rightarrow R^2$ is differentiable at t_0, then we define a **tangent vector** to the resulting parametrized curve at t_0 to be any scalar mul-

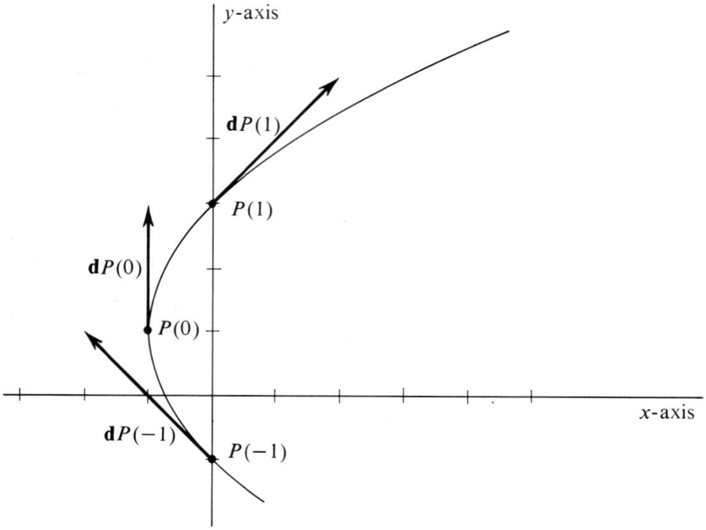

Figure 15

tiple of the derived vector $\mathbf{d}P(t_0)$. In particular, the derived vector itself is a tangent vector. The set of all tangent vectors at t_0 is a subset of $\mathcal{V}_{P(t_0)}$, since every scalar multiple of $\mathbf{d}P(t_0)$ has initial point $P(t_0)$. Moreover,

(4.2) *The set of all tangent vectors to the parametrized curve $P(t)$ at t_0 is a vector space.*

Proof. This result has nothing to do with any special properties of the derived vector, since the set of all scalar multiples of *any* vector $\mathbf{u}$ is a vector space. This result is proved, if $\mathbf{u}$ is nonzero, in Example 3 of Section 3. If $\mathbf{u}$ is a zero vector, the result is even simpler: The set of all scalar multiples of a zero vector $\mathbf{0}$ is the set having $\mathbf{0}$ as its only member, and the six conditions for a vector space are trivially satisfied. This completes the argument.

Consider a parametrization defined by $P(t) = \big(x(t), y(t)\big)$, which is differentiable at t_0 and for which the derived vector $\mathbf{d}P(t_0)$ is nonzero. If we set $x'(t_0) = d_1$ and $y'(t_0) = d_2$, then

$$\mathbf{d}P(t_0) = (d_1, d_2)_{P(t_0)},$$

where not both coordinates d_1 and d_2 are zero. The set of all tangent vectors at t_0 is the set of all scalar multiples

$$s\mathbf{d}P(t_0) = (sd_1, sd_2)_{P(t_0)},$$

where s is any real number. If $P(t_0) = (a, b)$, then the terminal point of $s\mathbf{d}P(t_0)$ is equal to

$$(sd_1 + a, sd_2 + b).$$

Hence the set of all terminal points of tangent vectors at t_0 is the set of all points (x, y) such that

$$\begin{cases} x = sd_1 + a, \\ y = sd_2 + b, \end{cases} \tag{4}$$

where s is any real number and d_1 and d_2 are not both zero. It is easy to verify that this set is a straight line (see Problem 4). We conclude that *if the derived vector* $\mathbf{d}P(t_0)$ *exists and is nonzero, then the set of all terminal points of the tangent vectors at* t_0 *to the curve parametrized by* P *is a straight line.* It is called the **tangent line** to the parametrized curve at t_0.

EXAMPLE 2. Consider the ellipse defined parametrically by

$$P(t) = \big(x(t), y(t)\big) = (4 \cos t, 2 \sin t),$$

for every real number t. Compute the derived vector at $t_0 = \frac{\pi}{6}$, and draw it and the ellipse in the xy-plane. In addition, write an equation for the tangent line at $t_0 = \frac{\pi}{6}$, and draw the tangent line in the figure. The derived vector is easily computed:

$$\begin{aligned} \mathbf{d}P(t_0) &= \big(x'(t_0), y'(t_0)\big)_{P(t_0)} = (-4 \sin t_0, 2 \cos t_0)_{P(t_0)} \\ &= \left(-4 \sin \frac{\pi}{6}, 2 \cos \frac{\pi}{6}\right)_{P(t_0)} = (-2, \sqrt{3})_{P(t_0)}, \end{aligned}$$

where

$$P(t_0) = \left(4 \cos \frac{\pi}{6}, 2 \sin \frac{\pi}{6}\right) = (2\sqrt{3}, 1).$$

The terminal point of the derived vector is therefore equal to

$$(2\sqrt{3} - 2, 1 + \sqrt{3}).$$

The parametrization P can also be written in terms of the equations

$$\begin{cases} x = 4 \cos t, \\ y = 2 \sin t, \qquad -\infty < t < \infty, \end{cases}$$

from which it follows that

$$\frac{x^2}{4^2} + \frac{y^2}{2^2} = \cos^2 t + \sin^2 t = 1.$$

Hence every point (x, y) on the parametrized curve satisfies the equation

$$\frac{x^2}{4^2} + \frac{y^2}{2^2} = 1. \tag{5}$$

Conversely, it can be shown (as in Example 2, page 544) that any ordered pair (x, y) which satisfies (5) also lies on the parametrized curve. We recognize (5) as an equation of the ellipse shown in Figure 16. The derived vector $\mathbf{d}P(t_0)$ and the tangent line at $\frac{\pi}{6}$ are also shown in the figure.

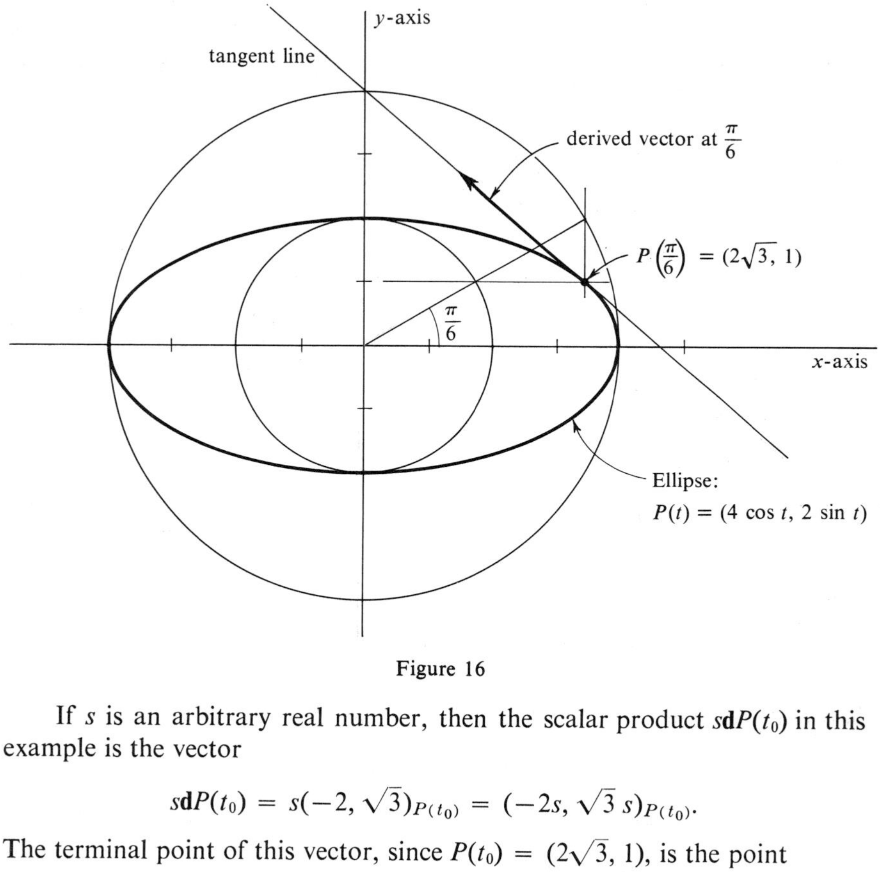

Figure 16

If s is an arbitrary real number, then the scalar product $s\mathbf{d}P(t_0)$ in this example is the vector

$$s\mathbf{d}P(t_0) = s(-2, \sqrt{3})_{P(t_0)} = (-2s, \sqrt{3}\, s)_{P(t_0)}.$$

The terminal point of this vector, since $P(t_0) = (2\sqrt{3}, 1)$, is the point

$$(-2s + 2\sqrt{3}, \sqrt{3}\, s + 1).$$

Hence the tangent line at $\frac{\pi}{6}$ is parametrically defined by the equations

$$\begin{cases} x = -2s + 2\sqrt{3}, \\ y = \sqrt{3}\, s + 1, \qquad -\infty < s < \infty. \end{cases} \tag{6}$$

Solving the first of these for s, we obtain $s = \dfrac{-x + 2\sqrt{3}}{2}$. Substitution in the second then yields

$$y = \sqrt{3}\left(\frac{-x + 2\sqrt{3}}{2}\right) + 1,$$

$$y = -\frac{\sqrt{3}}{2}x + 4. \tag{7}$$

Thus any point on the tangent line satisfies (7), and it is easy to verify that, for any x and y which satisfy (7), there is a unique s such that equations (6) hold. We conclude that (7) is an equation of the tangent line.

It is important to know that the ideas introduced in this section are consistent with related concepts developed earlier. For example, consider a differentiable parametrization defined by

$$P(t) = \big(x(t), y(t)\big), \qquad \text{for every } t \text{ in some interval } I.$$

Suppose that, for some t_0 in I, there exists a differentiable function f such that

$$y(t) = f\big(x(t)\big),$$

for every t in some subinterval of I containing t_0 in its interior. This situation was described in Section 1 and was illustrated in Figure 3 (page 545). If such a function f exists, we say that y is a differentiable function of x on the parametrized curve $P(t)$ in a neighborhood of $P(t_0)$. Formulas (5) and (6), page 546, assert that, for every t in the subinterval,

$$\frac{dy}{dx} = f'\big(x(t)\big) = \frac{y'(t)}{x'(t)},$$

provided $x'(t) \neq 0$. Hence $\dfrac{y'(t)}{x'(t)}$ is the slope of the line tangent to the graph of f at the point

$$\big(x(t), f(x(t))\big) = \big(x(t), y(t)\big) = P(t).$$

Moreover, in the vicinity of $P(t_0)$, the graph of f is the curve parametrized by P. At every t in the subinterval, the derived vector of P is equal to

$$\mathbf{d}P(t) = \big(x'(t), y'(t)\big)_{P(t)}.$$

This vector is, by definition, a tangent vector to the parametrized curve. Its initial point is $P(t) = \big(x(t), y(t)\big)$ and its terminal point is

$$Q(t) = \big(x(t) + x'(t), y(t) + y'(t)\big).$$

The slope of the line segment joining these two points is given by

$$m\big(P(t), Q(t)\big) = \frac{\big(y(t) + y'(t)\big) - y(t)}{\big(x(t) + x'(t)\big) - x(t)} = \frac{y'(t)}{x'(t)},$$

provided $x'(t) \neq 0$. We conclude that the concept of tangency, as defined in terms of the derived vector to a parametrized curve, is consistent with the earlier notion, defined in terms of the derivative.

PROBLEMS

1. For each of the following parametrizations and values of t_0, compute $P(t_0)$ and the derived vector $\mathbf{d}P(t_0)$. Draw the parametrized curve and each of the tangent vectors $\mathbf{d}P(t_0)$ to the curve.

(a) $P(t) = \big(x(t), y(t)\big) = (t - 1, t^2), \qquad -\infty < t < \infty;$
$t_0 = -1$, $t_0 = 0$, and $t_0 = 2$.

(b) $P(t) = \big(x(t), y(t)\big) = (t^2 + 1, t - 1), \qquad -\infty < t < \infty;$
$t_0 = -1$, $t_0 = 0$, and $t_0 = 1$.

(c) $P(t) = (t - 1, t^3), \qquad -\infty < t < \infty,$
$t_0 = 0$, $t_0 = 1$, and $t_0 = 2$.

(d) $P(t) = (x, y) = (e^t, t), \qquad -\infty < t < \infty;$
$t_0 = 0$ and $t_0 = \ln 2$.

(e) $P(t) = (3 \cos t, 2 \sin t), \qquad -\infty < t < \infty;$
$t_0 = 0$, $t_0 = \dfrac{\pi}{4}$, and $t_0 = \dfrac{\pi}{2}\cdot$

(f) $P(t) = (t^2, t^3), \qquad -\infty < t < \infty;$
$t_0 = -1$, $t_0 = 0$, and $t_0 = 2$.

(g) $P(t) = (t - 1, 2t + 4), \qquad -2 \le t \le 2;$
$t_0 = -1$, $t_0 = 0$, and $t_0 = 1$.

2. For each of the following parametrizations $P(t) = \big(x(t), y(t)\big)$, find the derived vector $\mathbf{d}P(t)$ for an arbitrary value of t in the domain. Draw the vectors $\mathbf{d}P(0)$, $\mathbf{d}P(1)$, and $\mathbf{d}P(2)$ in the xy-plane.

(a) $\begin{cases} x(t) = t^2 - 1, \\ y(t) = t^3, \qquad -1 \le t \le 3. \end{cases}$

(b) $\begin{cases} x(t) = \frac{1}{2}(e^t + e^{-t}), \\ y(t) = \frac{1}{2}(e^t - e^{-t}), \qquad -\infty < t < \infty. \end{cases}$

(c) $\begin{cases} x(t) = t^2, \\ y(t) = \frac{2}{3}(3t + 1)^{3/2}, \qquad -\frac{1}{3} \le t \le 5. \end{cases}$

(d) $\begin{cases} x(t) = t^2 + t + 1, \\ y(t) = \dfrac{t^3}{3} + t^2 - 1, \qquad -\infty < t < \infty. \end{cases}$

3. The cycloid shown in Figure 4 is defined by a parametrization $P(\theta) = (x, y)$ in which

$$\begin{cases} x = a(\theta - \sin\theta), \\ y = a(1 - \cos\theta), \qquad -\infty < \theta < \infty. \end{cases}$$

Compute the derived vector $\mathbf{d}P(\theta)$. Sketch the curve, and draw the tangent vectors $\mathbf{d}P(0)$, $\mathbf{d}P\left(\frac{\pi}{2}\right)$, $\mathbf{d}P(\pi)$, and $\mathbf{d}P(2\pi)$.

4. Prove that the curve defined parametrically by the equations

$$\begin{cases} x = sd_1 + a, \\ y = sd_2 + b, \qquad -\infty < s < \infty, \end{cases}$$

where not both d_1 and d_2 are zero, is a straight line. (*Note:* Check the definition of a straight line given on page 39.)

5. Converse of Problem 4: Prove that, if L is a straight line in R^2, then it can be defined by a parametrization $P(s) = (x, y)$ for which

$$\begin{cases} x = sd_1 + a, \\ y = sd_2 + b, \qquad -\infty < s < \infty, \end{cases}$$

and not both d_1 and d_2 are zero.

6. For each of the following parametrizations $P(t) = (x(t), y(t))$ and values of t_0, compute the derived vector $\mathbf{d}P(t_0)$. Draw the parametrized curve, the tangent line at t_0, and write an equation in x and y of the tangent line.

   (a) $P(t) = (t^2 + 1, t + 1)$, $-\infty < t < \infty$, and $t_0 = 2$.

   (b) $P(t) = (t^2 + 1, t + 1)$, $-\infty < t < \infty$, and $t_0 = 0$.

   (c) $P(t) = (e^t, t)$, $-\infty < t < \infty$, and $t_0 = \ln 2$.

   (d) $P(t) = (|t|, t)$, $-\infty < t < \infty$, and $t_0 = 0$.

7. Let P be the parametrization defined by $P(t) = (t^2, \frac{1}{2}t^2)$, for every real number t.

   (a) Write an equation in x and y of the tangent line at $t = 2$.

   (b) Describe the vector space of tangent vectors at $t = 2$ and at $t = 0$.

8. Let f be a real-valued function which is differentiable at a.

   (a) Write an equation of the line tangent to the graph of f at $(a, f(a))$.

   (b) Consider the parametrization

$$P(t) = (t, f(t)).$$

   Compute the derived vector $\mathbf{d}P(a)$, and write an equation of the tangent line to the parametrized curve at a.

9. Let $P: [a, b] \rightarrow R^2$ be a parametrization for which the derivatives x' and y' of the coordinate functions are continuous. Prove that the arc length of the curve parametrized by P is given by

$$L_a^b = \int_a^b |\mathbf{d}P(t)|\, dt.$$

5. Vector Velocity and Acceleration. In this section we shall consider the motion of a particle in the plane during an interval of time. We shall assume that the particle moves without jumping. As a result, if P is the function which associates to every instant of time in the interval the corresponding position of the particle in the plane, then P is continuous; i.e., it is a parametrization. The points $P(t)$ trace out the parametrized curve over which the particle moves.

Velocity is a vector concept which combines two ingredients: the number which measures how fast the particle is moving, and the direction of the motion. If the position of a particle during an interval of time I is described by a differentiable parametrization $P\colon I \to R^2$, then the **velocity** of the particle at any time t during the interval will be denoted by $\mathbf{v}(t)$ and defined to be the derived vector of P at t. Thus

$$\boxed{\mathbf{v}(t) = \mathbf{d}P(t).} \tag{1}$$

Since the derived vector is a tangent vector, the velocity is also one. Specifically, $\mathbf{v}(t)$ is a tangent vector at t to the parametrized curve defined by P. If we write $P(t) = \big(x(t), y(t)\big)$, then it follows from the formula for the derived vector [see (4.1), page 571] that the velocity vector is given by

$$\boxed{\mathbf{v}(t) = \big(x'(t), y'(t)\big)_{P(t)}.} \tag{2}$$

The **speed** of the particle at time t is defined to be the length $|\mathbf{v}(t)|$ of the velocity vector. The equation for the length of a vector in terms of its coordinates [see (3.1), page 561] implies that the speed is equal to

$$\boxed{|\mathbf{v}(t)| = \sqrt{x'(t)^2 + y'(t)^2}.} \tag{3}$$

EXAMPLE 1. A particle moves in the plane during the time interval [0, 2], and its position at any time t in this interval is given by

$$P(t) = \big(x(t), y(t)\big) = (\cos \pi t^2, \sin \pi t^2).$$

Assume that time is measured in seconds and that the unit of distance in the plane is 1 foot.

(Identify and draw the curve traced out by the particle, and describe its motion during the interval [0, 2].

(b) Compute the position, velocity, and speed of the particle at $t = 0$, $t = \frac{1}{2}$, $t = 1$, and $t = \frac{3}{2}$. Show these four positions, and draw the corresponding velocity vectors in the figure in (a).

(c) How does the speed of the particle depend on time during the entire interval of motion?

The parametrized curve over which the particle moves is the set of all points (x, y) such that

$$\begin{cases} x = \cos \pi t^2, \\ y = \sin \pi t^2, \qquad 0 \leq t \leq 2. \end{cases}$$

Hence the coordinates of every point (x, y) on the curve satisfy

$$x^2 + y^2 = (\cos \pi t^2)^2 + (\sin \pi t^2)^2 = 1.$$

The equation $x^2 + y^2 = 1$ is the familiar equation of the circle C with radius 1 and center the origin, and the particle therefore moves on this circle. In accordance with the definition of the functions sine and cosine in Chapter 6, the quantity πt^2 is the arc length along C in the counterclockwise direction from the point $(1, 0)$ to the point $P(t) = (x, y)$. As t increases from 0 to 2, the values of πt^2 increase monotonically from 0 to 4π, which is twice the circumference of the circle. We conclude that the particle starts from $(1, 0)$ at time $t = 0$, moves counterclockwise around the circle as time increases, and at $t = 2$ has gone completely around twice and has come back to its starting position at $P(2) = (\cos 4\pi, \sin 4\pi) = (1, 0)$. The curve of motion, i.e., the circle C, is shown in Figure 17.

Since $P(t) = (\cos \pi t^2, \sin \pi t^2)$, the position of the particle at each of the four values of t given in (b) is easily computed:

$$\begin{aligned} P(0) &= (\cos 0, \sin 0) = (1, 0), \\ P(\tfrac{1}{2}) &= (\cos \pi \cdot \tfrac{1}{4}, \sin \pi \cdot \tfrac{1}{4}) = \left(\frac{\sqrt{2}}{2}, \frac{\sqrt{2}}{2}\right), \\ P(1) &= (\cos \pi, \sin \pi) = (-1, 0), \\ P(\tfrac{3}{2}) &= (\cos \pi \cdot \tfrac{9}{4}, \sin \pi \cdot \tfrac{9}{4}) = \left(\frac{\sqrt{2}}{2}, \frac{\sqrt{2}}{2}\right). \end{aligned}$$

The velocity vector is

$$\begin{aligned} \mathbf{v}(t) &= \big(x'(t), y'(t)\big)_{P(t)} \\ &= (-2\pi t \sin \pi t^2, 2\pi t \cos \pi t^2)_{P(t)}. \end{aligned}$$

$$\mathbf{v}(t) = 2\pi t(-\sin \pi t^2, \cos \pi t^2)_{P(t)}. \qquad (4)$$

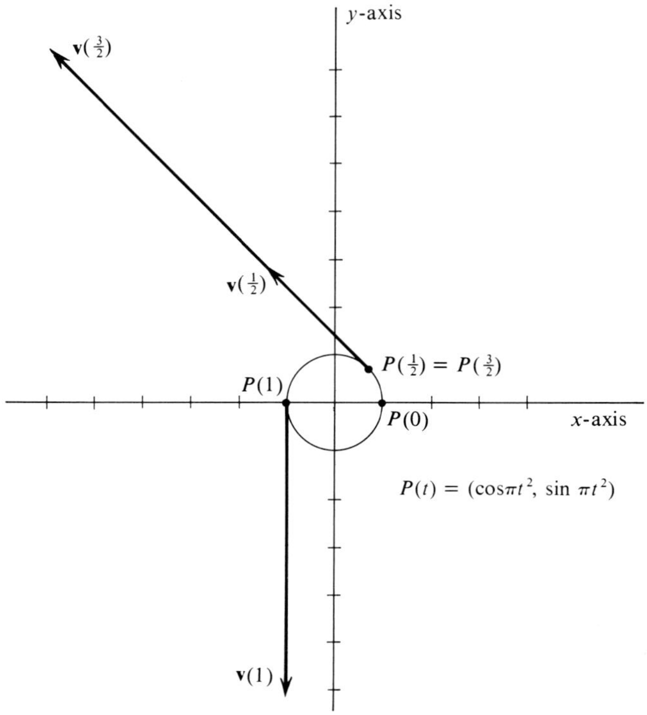

Figure 17

Hence

$$\mathbf{v}(0) = \mathbf{0} = (0, 0)_{P(0)},$$

$$\mathbf{v}(\tfrac{1}{2}) = 2\pi\tfrac{1}{2}\left(-\sin\frac{\pi}{4}, \cos\frac{\pi}{4}\right)_{P(1/2)} = \frac{\pi\sqrt{2}}{2}(-1, 1)_{P(1/2)},$$

$$\mathbf{v}(1) = 2\pi(-\sin\pi, \cos\pi)_{P(1)} = 2\pi(0, -1)_{P(1)},$$

$$\mathbf{v}(\tfrac{3}{2}) = 2\pi\tfrac{3}{2}\left(-\sin\frac{\pi 9}{4}, \cos\frac{\pi 9}{4}\right)_{P(3/2)} = \frac{3\pi\sqrt{2}}{2}(-1, 1)_{P(3/2)}.$$

The speed is by definition the length of the velocity vector. If $(b, c)_P$ is any vector, and a any real number, then the length of the scalar product $a(b, c)_P$ is given by

$$|a(b, c)_P| = |a|\sqrt{b^2 + c^2}.$$

Thus the four speeds are

$$|\mathbf{v}(0)| = 0 \text{ feet per second},$$

$$|\mathbf{v}(\tfrac{1}{2})| = \frac{\pi\sqrt{2}}{2}\sqrt{(-1)^2 + 1^2} = \pi \text{ feet per second},$$

$$|\mathbf{v}(1)| = 2\pi\sqrt{0^2 + (-1)^2} = 2\pi \text{ feet per second},$$

$$|\mathbf{v}(\tfrac{3}{2})| = \frac{3\pi\sqrt{2}}{2}\sqrt{(-1)^2 + 1^2} = 3\pi \text{ feet per second}.$$

Since each of the velocity vectors is tangent to the curve at its initial point and since we know their lengths, they can be drawn without difficulty (see Figure 17).

From the preceding computations it appears that the speed of the particle increases as time goes on. By computing the speed $|\mathbf{v}(t)|$ for an arbitrary t in the interval [0, 2], we can see that this inference is correct. Using equation (4) we get

$$|\mathbf{v}(t)| = 2\pi t\sqrt{(-\sin \pi t^2)^2 + (\cos \pi t^2)^2} = 2\pi t,$$

which shows that the speed increases linearly with time over the interval. At $t = 0$, the particle is at rest, and 2 seconds later, at $t = 2$, its speed has increased to 4π feet per second.

The motion of a particle along a straight line was studied in Section 3 of Chapter 2 and again in Section 8 of Chapter 4. When the motion is restricted to a straight line, which for convenience we may take to be the x-axis, then the velocity vector has only one nonzero coordinate, $x'(t)$. In this case velocity may be identified with $x'(t)$, and it is not necessary to consider it as a vector. In our earlier treatments $x'(t)$ was defined to be the velocity and it was denoted by $v(t)$. The distance on the line which the particle moves during the time interval $[a, b]$ was defined by the formula

$$distance\Big|_a^b = \int_a^b |v(t)|\, dt \tag{5}$$

(see page 232). We shall show that this definition is consistent with the more sophisticated notions of vector velocity and arc length of parametrized curves, which we are studying in this chapter. Consider a particle in the plane whose position is given by a parametrization P: $[a, b] \rightarrow R^2$. By the **distance** which the particle moves along the curve parametrized by P during the time interval from $t = a$ to $t = b$ we shall mean the arc length L_a^b. Let

$$P(t) = \big(x(t), y(t)\big), \qquad \text{for every } t \text{ such that } a \leq t \leq b.$$

We shall assume that the derivatives x' and y' exist and are continuous on $[a, b]$. From Theorem (2.2), page 553, it follows that

$$L_a^b = \int_a^b \sqrt{x'(t)^2 + y'(t)^2}\, dt.$$

The speed of the particle at any t in $[a, b]$ is given by

$$|\mathbf{v}(t)| = \sqrt{x'(t)^2 + y'(t)^2}.$$

Hence *the distance traveled by the particle along the parametrized curve from* $t = a$ *to* $t = b$ *is equal to*

$$L_a^b = \int_a^b |\mathbf{v}(t)|\, dt. \tag{5.1}$$

Formula (5.1) is the generalization of the distance formula (5) from rectilinear to curvilinear motion.

EXAMPLE 2. A steel ball is rolling on a plane during an interval from $t = 0$ to $t = 4$ seconds. It has an x-coordinate of velocity which is constant and equal to 2 feet per second. Its y-coordinate of velocity is $\frac{1}{2}t$ feet per second, for every t in the interval. (a) Write a definite integral equal to the distance (in feet) which the ball rolls during the interval from $t = 0$ to $t = 4$ seconds. (b) Identify and draw the curve on which the ball rolls.

The coordinates of the velocity vector $\mathbf{v}(t)$ are $x'(t)$ and $y'(t)$. Hence

$$\begin{cases} x'(t) = 2, \\ y'(t) = \frac{1}{2}t, \qquad 0 \le t \le 4. \end{cases} \tag{6}$$

It follows at once from (5.1) that the distance which the ball rolls is equal to

$$\begin{aligned} L_0^4 &= \int_0^4 \sqrt{4 + \tfrac{1}{4}t^2}\, dt \\ &= \tfrac{1}{2}\int_0^4 \sqrt{16 + t^2}\, dt. \end{aligned}$$

This answers part (a). Using a table of integrals or integration by trigonometric substitution, one can obtain

$$\begin{aligned} \int_0^4 \sqrt{16 + t^2}\, dt &= 8\sqrt{2} + 8 \ln(1 + \sqrt{2}) \\ &= 18.3 \text{ (approximately)}. \end{aligned}$$

Hence the distance the ball rolls is half this quantity, approximately 9.2 feet.

A parametrization which defines the position of the ball may be found by integrating the functions x' and y'. From equations (6), we get

$$\begin{cases} x(t) = 2t + c_1, \\ y(t) = \dfrac{t^2}{4} + c_2, \qquad 0 \le t \le 4. \end{cases}$$

Nothing in the statement of the problem specifies the position of the ball at $t = 0$, so, for simplicity, we shall choose it to be the origin. This choice is equivalent to setting $c_1 = c_2 = 0$. It follows that the parametrized curve in which the ball rolls is the set of all points (x, y) such that

$$\begin{cases} x = 2t, \\ y = \dfrac{t^2}{4}, \qquad 0 \le t \le 4. \end{cases}$$

From the first equation, we get $t = \dfrac{x}{2}$. Hence the two equations together with the inequality are equivalent to

$$y = \frac{x^2}{16}, \qquad 0 \le x \le 8.$$

The graph of this equation is the parabola shown in Figure 18, and the curve over which the ball rolls is that portion of the parabola indicated by the heavy line.

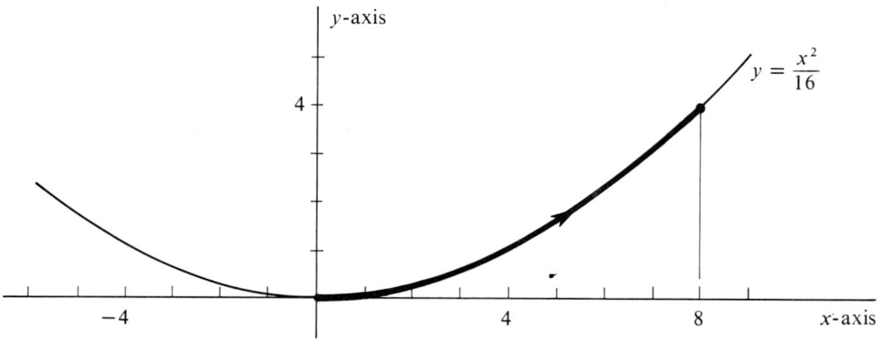

Figure 18

We next consider what is meant by the acceleration of a moving particle. The intuitive idea is that acceleration is the rate of change of the velocity vector. To be more precise: Let the position of the particle during an interval of time I be given by a differentiable parametrization $P: I \to R^2$, and let t_0 be in I. If t is a number in I distinct from t_0, then the velocity vectors $\mathbf{v}(t_0)$ and $\mathbf{v}(t)$ are tangent vectors with initial points $P(t_0)$ and $P(t)$, respectively, as illustrated in Figure 19. In defining the acceleration at t_0, we should like to form the scalar product of $\frac{1}{t - t_0}$ and the difference $\mathbf{v}(t) - \mathbf{v}(t_0)$, and to take the limit of this product as t approaches t_0. The difficulty is that, since the points $P(t)$ and $P(t_0)$ are usually distinct, the difference $\mathbf{v}(t) - \mathbf{v}(t_0)$ is generally not defined. (Recall that two vectors can be added or subtracted if and only if they have the same initial point.) It is for this reason that, before defining acceleration, we introduce the notion of parallel translation of vectors in R^2.

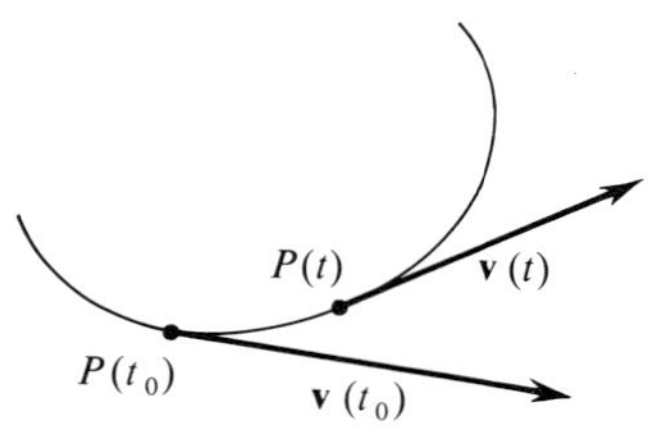

Figure 19

Let P_0 be an arbitrary point in R^2. We shall define a function T_{P_0} whose domain is the set $\mathcal{V}$ of all vectors in R^2 and whose range is the vector space $\mathcal{V}_{P_0}$ of all vectors with initial point P_0. The definition is as follows: For every vector $\mathbf{u}$ in $\mathcal{V}$, the value $T_{P_0}(\mathbf{u})$ is the vector with the same coordinates as $\mathbf{u}$, but with initial point P_0. Thus

$$\text{if } \mathbf{u} = (u_1, u_2)_Q, \qquad \text{then } T_{P_0}(\mathbf{u}) = (u_1, u_2)_{P_0}.$$

Geometrically, the vector $T_{P_0}(\mathbf{u})$ is obtained from $\mathbf{u}$ by moving the arrow representing the vector $\mathbf{u}$ parallel to itself until its initial point coincides with P_0. The process is illustrated in Figure 20, and we call the function T_{P_0} the operation of **parallel translation** of vectors to the point P_0.

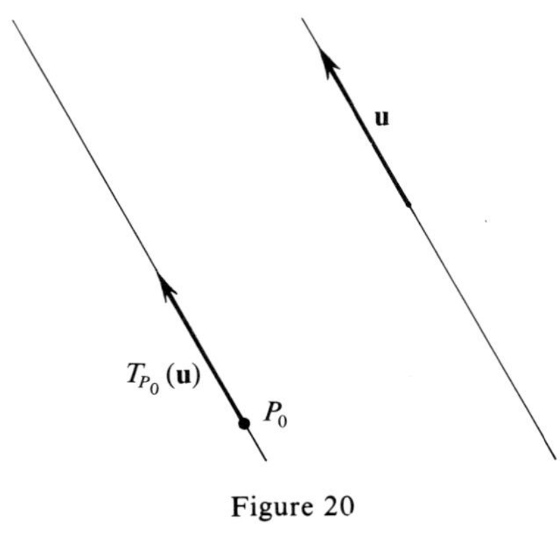

Figure 20

We can now define the acceleration of a moving particle. As before, let the position be defined by the differentiable parametrization $P: I \to R^2$. We consider t_0 in I, and set $P(t_0) = P_0$. Then the **acceleration** of the particle at t_0 is the vector $\mathbf{a}(t_0)$ defined by

$$\boxed{\mathbf{a}(t_0) = \lim_{t \to t_0} \frac{1}{t - t_0} [T_{P_0}(\mathbf{v}(t)) - \mathbf{v}(t_0)].} \tag{7}$$

Thus, acceleration, like velocity, is a vector.

We can derive a simple formula for acceleration in terms of the coordinate functions of P. Let $P(t) = (x(t), y(t))$, as usual. Then

$$\mathbf{v}(t_0) = (x'(t_0), y'(t_0))_{P(t_0)},$$
$$\mathbf{v}(t) = (x'(t), y'(t))_{P(t)},$$

and, if $P_0 = P(t_0)$, then

$$T_{P_0}(\mathbf{v}(t)) = (x'(t), y'(t))_{P(t_0)}.$$

It follows that

$$T_{P_0}(\mathbf{v}(t)) - \mathbf{v}(t_0) = (x'(t) - x'(t_0), y'(t) - y'(t_0))_{P(t_0)},$$

and thence that

$$\frac{1}{t - t_0} [T_{P_0}(\mathbf{v}(t)) - \mathbf{v}(t_0)] = \left(\frac{x'(t) - x'(t_0)}{t - t_0}, \frac{y'(t) - y'(t_0)}{t - t_0}\right)_{P(t_0)}.$$

Hence

$$\begin{aligned} \mathbf{a}(t_0) &= \lim_{t \to t_0} \frac{1}{t - t_0} [T_{P_0}(\mathbf{v}(t)) - \mathbf{v}(t_0)] \\ &= \left(\lim_{t \to t_0} \frac{x'(t) - x'(t_0)}{t - t_0}, \lim_{t \to t_0} \frac{y'(t) - y'(t_0)}{t - t_0}\right)_{P(t_0)}. \end{aligned}$$

If the two limits which are the coordinates of the preceding vector exist, they are by definition equal to the second derivatives $x''(t_0)$ and $y''(t_0)$, respectively. It follows that

(5.2) *If $P(t) = (x(t), y(t))$, then the acceleration vector $\mathbf{a}(t_0)$ exists if and only if the second derivatives $x''(t_0)$ and $y''(t_0)$ exist. If they do exist, then*

$$\mathbf{a}(t_0) = (x''(t_0), y''(t_0))_{P(t_0)}.$$

EXAMPLE 3. A particle is moving with constant speed k in a fixed circle of radius a. Show that, at any time t during the interval of motion, the acceleration vector $\mathbf{a}(t)$ has constant length equal to $\frac{k^2}{a}$ and always points directly toward the center of the circle (see Figure 21).

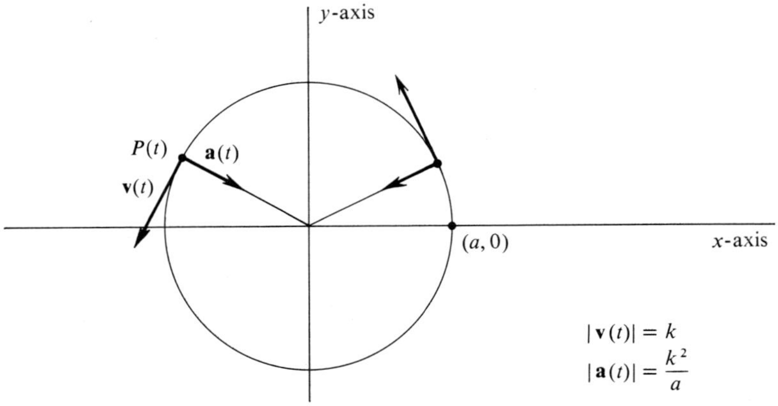

Figure 21

We shall take the center of the circle to be the origin in the xy-plane. The position of the particle can then be defined by a parametrization $P(t) = \big(x(t), y(t)\big) = (x, y)$ for which

$$\begin{cases} x = a \cos u, \\ y = a \sin u, \end{cases} \tag{8}$$

and u is some function of t having as domain the interval of time of the motion. To be specific, we shall assume that 0 is in the domain, and that, when $t = 0$, the particle is at the point $(a, 0)$ on the circle. Hence $u = 0$ when $t = 0$. We shall make the analytic assumption that the second derivative $u''(t)$ exists, for every t in the interval, and it follows that $x''(t)$ and $y''(t)$ also exist. Differentiating with respect to t in equations (8), we obtain

$$\begin{cases} x' = -au' \sin u, \\ y' = au' \cos u. \end{cases} \tag{9}$$

Thus the speed of the particle is

$$|\mathbf{v}(t)| = \sqrt{x'^2 + y'^2} = \sqrt{a^2u'^2(\sin^2 u + \cos^2 u)} = a|u'|,$$

which is assumed to be the constant k. Hence $|u'| = \frac{k}{a}$. Since u' is continuous and has constant positive absolute value, it is either always positive or always negative (depending on whether the particle is moving counterclockwise or clockwise). We shall assume the former and conclude that $u' = \frac{k}{a}$. Integrating, we obtain

$$u = \frac{k}{a}t + c.$$

Since $u = 0$, when $t = 0$, it follows that

$$u = \frac{k}{a}t.$$

Substituting this value back into equations (9), we have

$$x' = -a\frac{k}{a}\sin\frac{k}{a}t = -k\sin\frac{k}{a}t,$$

$$y' = a\frac{k}{a}\cos\frac{k}{a}t = k\cos\frac{k}{a}t.$$

Hence

$$x'' = -\frac{k^2}{a}\cos\frac{k}{a}t = -\frac{k^2}{a}\cos u,$$

$$y'' = -\frac{k^2}{a}\sin\frac{k}{a}t = -\frac{k^2}{a}\sin u,$$

or, equivalently,

$$x'' = -\frac{k^2}{a^2}a\cos u = -\frac{k^2}{a^2}x,$$

$$y'' = -\frac{k^2}{a^2}a\sin u = -\frac{k^2}{a^2}y.$$

We know from (5.2) that the acceleration vector is given by

$$\mathbf{a}(t) = (x'', y'')_{P(t)}.$$

Hence

$$\mathbf{a}(t) = \left(-\frac{k^2}{a^2}x, -\frac{k^2}{a^2}y\right)_{P(t)}$$

$$= \frac{k^2}{a^2}(-x, -y)_{P(t)}.$$

Since $P(t) = (x, y)$, the terminal point of the vector $(-x, -y)_{P(t)}$ is the point (0, 0). Thus the acceleration vector $\mathbf{a}(t)$ is a positive scalar multiple of the vector with initial point $P(t)$ and terminal point the origin. This proves that $\mathbf{a}(t)$ is always pointing directly toward the center of the circle. The length of the acceleration vector is easily computed from the preceding equation. We get

$$\begin{aligned}|\mathbf{a}(t)| &= \frac{k^2}{a^2}\sqrt{(-x)^2 + (-y^2)} = \frac{k^2}{a^2}\sqrt{x^2 + y^2}\\ &= \frac{k^2}{a^2}\cdot a = \frac{k^2}{a}\cdot\end{aligned}$$

This completes the problem. The acceleration in this example is called centripetal acceleration, and the force acting on the particle necessary to provide this acceleration is the centripetal force. In the case of a planet moving in orbit, the force is the force of gravity.

PROBLEMS

1. (a) Draw each of the following vectors.

   (i) $(0, 5)_{P_0}$, where $P_0 = (-1, 1)$.

   (ii) $(4, -1)_{P_1}$, where $P_1 = (1, -1)$.

   (iii) $(1, 3)_{P_2}$, where $P_2 = (1, 1)$.

   (iv) $(-2, -3)_{P_3}$, where $P_3 = (0, 0)$.

   (b) Let $P_0 = (-1, 1)$, and compute and draw the translated vectors $T_{P_0}(\mathbf{u})$, where $\mathbf{u}$ is taken to be each of the four vectors in (a).

2. A particle moves in the plane during the time interval from $t = 0$ to $t = 2$ seconds. Its position at any time during this interval is given by the parametrization

$$P(t) = (t, t^2 - t),$$

where it will be assumed that the unit of distance in the plane is 1 foot.

   (a) Identify and draw the curve which the particle traces out during its interval of motion.

   (b) Compute the velocity vector $\mathbf{v}(t)$. Find the position, velocity, and speed at $t = 0$, $t = 1$, and $t = 2$. Show these positions and draw the velocity vectors in the figure in part (a).

   (c) Compute the acceleration $\mathbf{a}(t)$. Find the times and corresponding positions (if any) when the acceleration and velocity vectors are perpendicular to each other.

   (d) Write a definite integral equal to the distance (in feet) which the particle moves during the interval from $t = 0$ to $t = 2$ seconds.

   (e) Evaluate the integral in (d).

3. An object is dropped from an airplane which is flying in a straight line over level ground at a constant speed of 800 feet per second and at an altitude of 10,000 feet. The horizontal coordinate of the velocity of the object is constant and equal in magnitude to the speed of the plane. The vertical coordinate of velocity is initially zero. However, the vertical component of acceleration (due to gravity) is -32 feet per second per second. (These data are realistic only if we neglect air resistance, the curvature of the earth, etc.)

   (a) Define a parametrization $P(t) = \big(x(t), y(t)\big)$ which gives the position of the particle at time t. Assume that the object was dropped when $t = 0$ and that $P(0) = (0, 0)$. Compute the velocity and acceleration vectors $\mathbf{v}(t)$ and $\mathbf{a}(t)$.

   (b) How long does it take the object to fall to the ground?

   (c) Identify and draw the curve in which the object falls.

   (d) Express the distance traveled along the curve as a definite integral.

4. Consider a particle in motion in the plane from $t = 0$ to $t = 4$ seconds. Its position at any time during this interval is given by

$$P(t) = (x, y) = \big((t - 2)^2, (t - 2)^2\big),$$

where it is assumed that the unit of distance in the plane is 1 foot.

   (a) Draw the curve in which the particle moves during the interval.

   (b) Compute the velocity $\mathbf{v}(t)$ and the speed $|\mathbf{v}(t)|$. What are the minimum and maximum speeds, and at what times are they attained?

   (c) Describe the vector space of tangent vectors to the parametrized curve at $t = 1$, and also at $t = 2$.

   (d) Compute the distance traveled by the particle during the motion.

5. The position of a particle in motion in the plane is defined by the parametrization

$$P(t) = (x, y) = (t^2, t^3), \qquad -2 \le t \le 2.$$

   (a) Draw the curve traced out by the particle during the interval $[-2, 2]$.

   (b) Compute the velocity vector $\mathbf{v}(t)$. Find the position, velocity, and speed at $t = -2$, $t = 0$, $t = 1$, and $t = 2$. Indicate these positions and draw the velocity vectors in the figure in (a).

   (c) Compute the acceleration vector $\mathbf{a}(t)$. Determine the four specific vectors $\mathbf{a}(-2)$, $\mathbf{a}(0)$, $\mathbf{a}(1)$, and $\mathbf{a}(2)$, and draw them in the figure in (a).

6. The position of a particle in the plane is defined by the parametrization

$$\begin{cases} x = a \cos kt, \\ y = b \sin kt, \qquad -\infty < t < \infty, \end{cases}$$

where a, b, and k are positive constants and $a > b$.

(a) Identify and draw the curve in which the particle moves.

(b) Prove that the particle is never at rest.

(c) Show that the acceleration vector $\mathbf{a}(t)$ always points directly toward the origin.

7. Prove that parallel translation has the following properties:

(a) $|T_{P_0}(\mathbf{u})| = |\mathbf{u}|$ for every vector $\mathbf{u}$.

(b) If $\mathbf{u}$ is any vector in $\mathcal{V}_{P_0}$, then $T_{P_0}(\mathbf{u}) = \mathbf{u}$.

(c) If $\mathbf{0}$ is any zero vector, then $T_{P_0}(\mathbf{0})$ is also a zero vector.

8. Starting at $t = 0$, a stone at the end of a string is whirled around in a fixed circle of radius a at an ever-increasing speed equal to kt for some positive constant k. The tension in the string is equal to $m|\mathbf{a}(t)|$, where m is the mass of the stone and $|\mathbf{a}(t)|$ is the length of the acceleration vector. Suppose the string breaks when the tension exceeds some value T. Compute, in terms of the constants a, k, m, and T, the moment when the string breaks.

6. Polar Coordinates. Since the set R^2 of all ordered pairs of real numbers has been identified with the set of all points in the plane, every point is uniquely determined by its x- and y-coordinates. An alternative way of specifying points in the plane is the following: To every ordered pair (r, θ) of real numbers, we assign the point $P = (x, y)$ in R^2 defined by

$$\boxed{\begin{aligned} x &= r\cos\theta, \\ y &= r\sin\theta. \end{aligned}} \tag{1}$$

The pair (r, θ) is called a pair of **polar coordinates** of the point $P = (x, y)$.

In giving the geometric interpretation of polar coordinates of a point, we distinguish three separate possibilities:

Case 1. $r > 0$. Then r is the distance between $P = (x, y)$ and the origin, since

$$\begin{aligned} distance(P, (0, 0)) &= \sqrt{x^2 + y^2} = \sqrt{r^2\cos^2\theta + r^2\sin^2\theta} \\ &= \sqrt{r^2(\cos^2\theta + \sin^2\theta)} = \sqrt{r^2} \\ &= r. \end{aligned}$$

The number θ is the radian measure of the angle which has its vertex at the origin, its initial side the positive x-axis, and its terminal side the line segment joining the origin to P. An example is shown in Figure 22.

Case 2. $r = 0$. Then P is the origin regardless of the value of θ, since

$$P = (x, y) = (0\cos\theta, 0\sin\theta) = (0, 0).$$

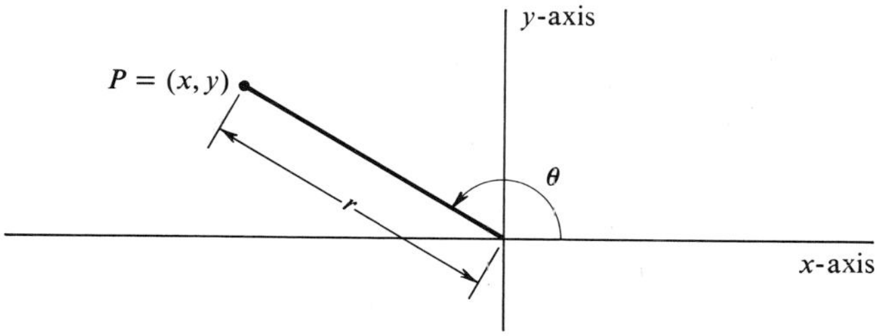

Figure 22

Case 3. $r < 0$. In this case the point $P = (x, y)$ is symmetric with respect to the origin to the point with polar coordinates $(|r|, \theta)$. This fact is illustrated in Figure 23. To prove it, we first observe that

$$\begin{aligned} -r\cos(\theta + \pi) &= -r\cos\theta\cos\pi + r\sin\theta\sin\pi = r\cos\theta, \\ -r\sin(\theta + \pi) &= -r\sin\theta\cos\pi - r\cos\theta\sin\pi = r\sin\theta. \end{aligned}$$

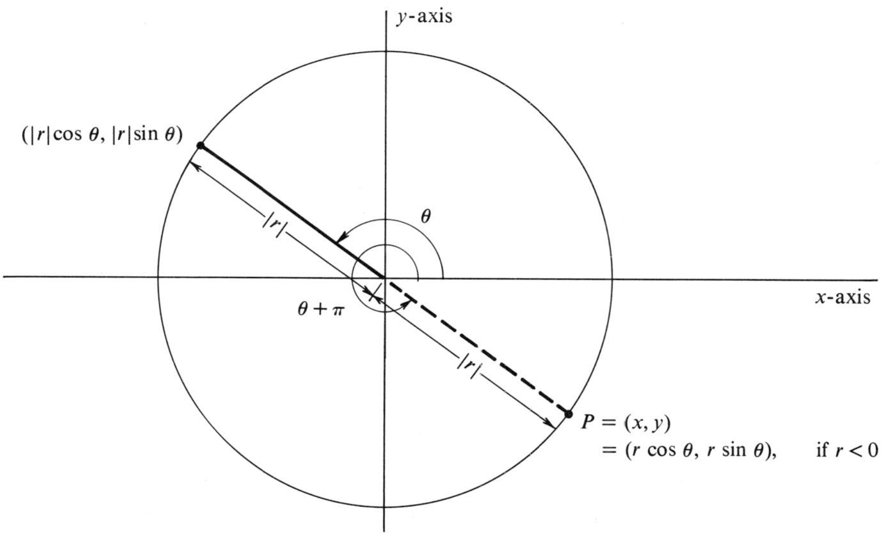

Figure 23

In the present situation r is negative. Hence $-r$ is positive, and $|r| = -r$. The preceding equations therefore imply

$$\begin{aligned} r\cos\theta &= |r|\cos(\theta + \pi), \\ r\sin\theta &= |r|\sin(\theta + \pi). \end{aligned}$$

Thus

$$\begin{aligned} P = (x, y) &= (r\cos\theta, r\sin\theta) \\ &= \big(|r|\cos(\theta + \pi), |r|\sin(\theta + \pi)\big), \qquad \text{if } r < 0, \end{aligned}$$

and this is precisely what is asserted above.

The major difference between polar coordinates and the familiar x- and y-coordinates is that if a given point $P = (x, y)$ has one pair of polar coordinates (r, θ), then it has *infinitely many:* If n is any integer, then

$$r \cos \theta = r \cos(\theta + 2\pi n),$$
$$r \sin \theta = r \sin (\theta + 2\pi n),$$

and it therefore follows that, for every integer n, the ordered pair $(r, \theta + 2\pi n)$ is a pair of polar coordinates for the one point $P = (x, y) = (r \cos \theta, r \sin \theta)$. In addition, as shown in Case 3, we have

$$-r \cos(\theta + \pi) = r \cos \theta,$$
$$-r \sin(\theta + \pi) = r \sin \theta.$$

Hence $(-r, \theta + \pi)$ is also a pair of polar coordinates of P. We conclude that there is no such thing as *the* polar coordinates of a point.

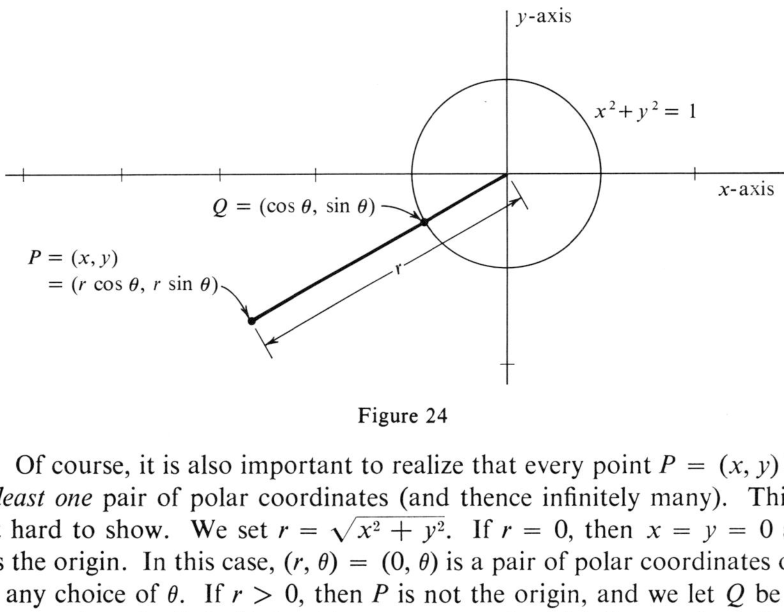

Figure 24

Of course, it is also important to realize that every point $P = (x, y)$ has *at least one* pair of polar coordinates (and thence infinitely many). This is not hard to show. We set $r = \sqrt{x^2 + y^2}$. If $r = 0$, then $x = y = 0$ and P is the origin. In this case, $(r, \theta) = (0, \theta)$ is a pair of polar coordinates of P for any choice of θ. If $r > 0$, then P is not the origin, and we let Q be the point on the unit circle (defined by $x^2 + y^2 = 1$) which lies on the half-line emanating from the origin and passing through P (see Figure 24). From our definition of the functions sine and cosine, we know that $Q = (\cos \theta, \sin \theta)$ for some number θ. It follows that

$$P = (r \cos \theta, r \sin \theta),$$

and so (r, θ) is a pair of polar coordinates of P.

EXAMPLE 1. Determine the x- and y-coordinates of the points with the following polar coordinates, and plot the points in the plane.

$$P_1\colon (r, \theta) = \left(2, \frac{\pi}{6}\right),$$

$$P_2\colon (r, \theta) = \left(-2, \frac{\pi}{6}\right),$$

$$P_3\colon (r, \theta) = \left(2, -\frac{\pi}{6}\right),$$

$$P_4\colon (r, \theta) = \left(2, \frac{7\pi}{6}\right),$$

$$P_5\colon (r, \theta) = \left(3, \frac{10\pi}{3}\right),$$

$$P_6\colon (r, \theta) = (0, \pi).$$

If (r, θ) is a pair of polar coordinates of a point P, then the x- and y-coordinates of P are given by

$$(x, y) = (r \cos \theta, r \sin \theta).$$

Hence, for each of the above, we get

$$P_1 = (x, y) = \left(2 \cos \frac{\pi}{6}, 2 \sin \frac{\pi}{6}\right) = (\sqrt{3}, 1),$$

$$P_2 = (x, y) = \left(-2 \cos \frac{\pi}{6}, -2 \sin \frac{\pi}{6}\right) = (-\sqrt{3}, -1),$$

$$\begin{aligned} P_3 = (x, y) &= \left(2 \cos \left(-\frac{\pi}{6}\right), 2 \sin \left(-\frac{\pi}{6}\right)\right) \\ &= \left(2 \cos \frac{\pi}{6}, -2 \sin \frac{\pi}{6}\right) = (\sqrt{3}, -1), \end{aligned}$$

$$\begin{aligned} P_4 = (x, y) &= \left(2 \cos \frac{7\pi}{6}, 2 \sin \frac{7\pi}{6}\right) \\ &= \left(-2 \cos \frac{\pi}{6}, -2 \sin \frac{\pi}{6}\right) = (-\sqrt{3}, -1), \end{aligned}$$

$$\begin{aligned} P_5 = (x, y) &= \left(3 \cos \frac{10\pi}{3}, 3 \sin \frac{10\pi}{3}\right) \\ &= \left(3 \cos \frac{4\pi}{3}, 3 \sin \frac{4\pi}{3}\right) = \left(-\frac{3}{2}, -\frac{3\sqrt{3}}{2}\right), \end{aligned}$$

$$P_6 = (x, y) = (0 \cos \pi, 0 \sin \pi) = (0, 0).$$

The points are plotted in Figure 25. Note that $P_2 = P_4$ even though the polar coordinates defining them are different. Although we have found the rectangular coordinates of each point, these are not necessary for plotting. For example, the point P_1 with polar coordinates $\left(2, \frac{\pi}{6}\right)$ is most easily plotted by drawing from the origin the line segment of length 2, which makes an angle of $\frac{\pi}{6}$ radians with the positive x-axis.

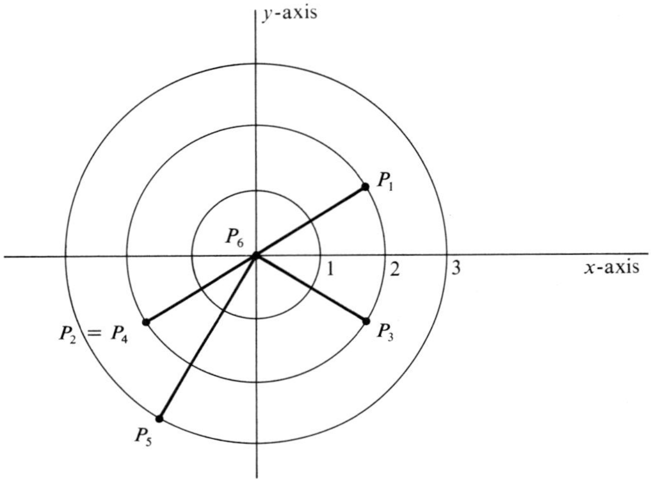

Figure 25

We next study curves in the plane defined by equations in polar coordinates. Let F be a real-valued function of two real variables. The set of all points in the plane whose polar coordinates satisfy the equation

$$\boxed{F(r, \theta) = 0} \tag{2}$$

will be called the **graph in polar coordinates,** or simply, the **polar graph** of the equation. More formally: The polar graph of equation (2) is the set of all points $P = (x, y)$ for which there exists a pair (r, θ) such that $F(r, \theta) = 0$ and

$$\begin{cases} x = r\cos\theta, \\ y = r\sin\theta. \end{cases}$$

Frequently, r is explicitly defined as a function of θ. This means that there is a real-valued function f of one real variable, and we consider the equation

$$\boxed{r = f(\theta).} \tag{3}$$

The **polar graph** of this equation is that of the equation $r - f(\theta) = 0$, which is a special case of equation (2) in the preceding paragraph. It follows that the polar graph of $r = f(\theta)$ is the set of all points (x, y) such that

$$\begin{cases} x = f(\theta) \cos \theta, \\ y = f(\theta) \sin \theta, \end{cases} \tag{4}$$

for every θ in the domain of f. If f is a continuous function with domain an interval of real numbers, then equations (4) constitute a parametrization, and so the polar graph of $r = f(\theta)$ is a parametrized curve.

EXAMPLE 2. Identify and draw the curve defined by the equation

$$r = 4 \cos \theta \tag{5}$$

in polar coordinates. By the curve defined by an equation $r = f(\theta)$ in polar coordinates we mean, of course, the polar graph of $r = f(\theta)$. A partial list of values of θ and r which satisfy equation (5) is given in Figure 26, and the points which have these pairs as polar coordinates are plotted. *Since the cosine is an even function, i.e., since* $\cos(-\theta) = \cos \theta$, *it follows that the resulting curve is symmetric about the x-axis.* The periodicity of the cosine implies that we may limit values of θ to an interval of length 2π. However, we can do considerably better than that. The fact that

$$\cos(\theta + \pi) = -\cos \theta$$

implies that if $r = 4 \cos \theta$, then $-r = 4 \cos(\theta + \pi)$. But we have already observed that (r, θ) and $(-r, \theta + \pi)$ are polar coordinates of the same point. It follows that all the points of the curve will be included even if the values of θ are limited to the interval $\left[-\frac{\pi}{2}, \frac{\pi}{2}\right]$. Finally, in view of the symmetry about the x-axis, it is sufficient in plotting points to consider only values in $\left[0, \frac{\pi}{2}\right]$.

It appears from Figure 26 that the polar graph of $r = 4 \cos \theta$ is the circle of radius 2 with center at the point $P = (2, 0)$. This can be verified as follows: An equation with the same polar graph is

$$r^2 = 4r \cos \theta. \tag{6}$$

Since $x = r \cos \theta$, $y = r \sin \theta$, and $x^2 + y^2 = r^2$, the polar graph of (6) is the same as the graph of the equation

$$x^2 + y^2 = 4x$$

in rectangular coordinates. This equation is equivalent to

$$x^2 - 4x + 4 + y^2 = 4,$$

θ	$r = 4 \cos \theta$
0	4
$\frac{\pi}{6}$	$2\sqrt{3}$
$\frac{\pi}{4}$	$2\sqrt{2}$
$\frac{\pi}{3}$	2
$\frac{\pi}{2}$	0

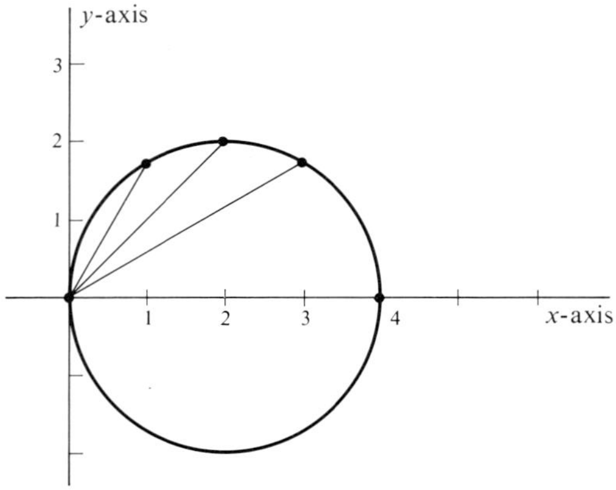

Figure 26

which is the same as

$$(x - 2)^2 + y^2 = 2^2.$$

The latter is the standard form of the equation of the circle of radius 2 with center at (2, 0).

EXAMPLE 3. A parabola is by definition the set of all points in the plane which are equidistant from a fixed line and a fixed point not on the line (see page 136). The line is called the directrix, and the point the focus. Find an equation in polar coordinates of the parabola whose focus is the origin and whose directrix is the vertical line cutting the x-axis in the point $(-1, 0)$. The parabola is drawn in Figure 27.

An equation of a curve in polar coordinates means an equation $F(r, \theta) = 0$ whose polar graph is the given curve. In the present example, let (r, θ), with $r > 0$, be a pair of polar coordinates of an arbitrary point on the parabola. Then r is the distance from the point to the focus, and, as can be seen from the figure, the distance from the point to the directrix is $1 + r \cos \theta$. Hence the geometric condition which defines the parabola is expressed in polar coordinates by the equation

$$r = 1 + r \cos \theta,$$

which is equivalent to

$$r(1 - \cos \theta) = 1,$$

and thence to

$$\boxed{r = \frac{1}{1 - \cos \theta}.} \qquad (7)$$

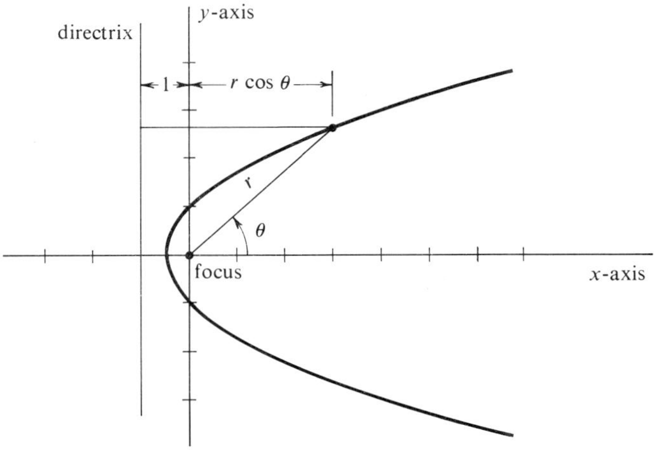

Figure 27

Conversely, if r and θ satisfy equation (7), it is clear that they are the polar coordinates of a point which satisfies the conditions for lying on the parabola.

EXAMPLE 4. Draw the curve defined by the equation

$$r = a\theta, \qquad -\infty < \theta < \infty,$$

in polar coordinates, where a is some positive constant. The polar graph of this equation is called an **Archimedean spiral.** If $\theta = 0$, then $r = a \cdot 0 = 0$, and we conclude that the origin is a point on the curve. As θ increases from zero, so does r, and it follows that a spiral is traced out in the counterclockwise direction. This part of the curve is drawn with a solid line in Figure 28. Thus the curve drawn with the solid line is the polar graph of the equation

$$r = a\theta, \qquad 0 \leq \theta < \infty.$$

For negative θ, the polar graph of $r = a\theta$ is obtained by reflecting the graph for positive θ about the y-axis. This part of the curve, indicated by a dashed line in the figure, is a spiral in the clockwise direction. One way to verify this assertion of symmetry about the y-axis is to write the equations which define the curve parametrically. They are

$$\begin{cases} x = r\cos\theta = a\theta\cos\theta, \\ y = r\sin\theta = a\theta\sin\theta, \qquad -\infty < \theta < \infty. \end{cases}$$

That is, we have a parametrization

$$P(\theta) = \big(x(\theta), y(\theta)\big) = (a\theta\cos\theta, a\theta\sin\theta),$$

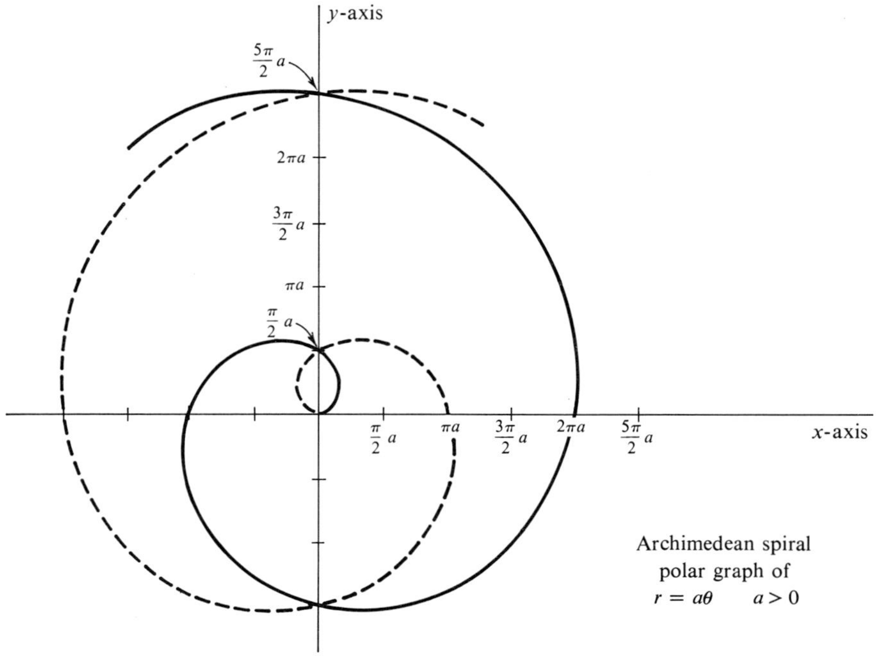

Figure 28

for every real number θ. Since the functions cosine and sine are, respectively, even and odd, we obtain

$$x(-\theta) = a(-\theta)\cos(-\theta) = -a\theta\cos\theta = -x(\theta),$$
$$y(-\theta) = a(-\theta)\sin(-\theta) = -a\theta\,(-\sin\theta) = y(\theta).$$

It follows that the point $P(-\theta) = \big(-x(\theta), y(\theta)\big)$ is situated symmetrically across the y-axis from the point $P(\theta) = \big(x(\theta), y(\theta)\big)$, and thus the curve is symmetric about the y-axis.

EXAMPLE 5. Draw the curve, called a **lemniscate,** defined by the equation

$$r^2 = 2a^2\cos 2\theta, \qquad a \neq 0,$$

in polar coordinates.

We first observe that the polar graph of this equation is symmetric about the origin; i.e., if the pair (r, θ) satisfies the equation, then so does $(-r, \theta)$. This fact is a consequence of the equations

$$r^2 = 2a^2\cos 2\theta = (-r)^2.$$

In addition, if (r, θ) satisfies the equation, then so does $(r, -\theta)$, since the cosine is an even function. Thus the polar graph is also symmetric about the x-axis. It follows from these observations of symmetry that the entire curve is obtained from that part which lies in the first quadrant of the xy-plane (those points for which $x \geq 0$ and $y \geq 0$) by reflecting about both the x-axis and the y-axis. Moreover, any point on the curve in the first quadrant has polar coordinates for which $r \geq 0$ and $0 \leq \theta \leq \frac{\pi}{2}$. Finally, if (r, θ) satisfies the equation, then

$$\cos 2\theta = \frac{r^2}{2a^2} \geq 0.$$

However, for values of θ in the interval $\left[0, \frac{\pi}{2}\right]$, $\cos 2\theta$ is nonnegative only if $0 \leq \theta \leq \frac{\pi}{4}$. We conclude that the entire curve is obtained by symmetry from those points which have polar coordinates (r, θ) with $r \geq 0$ and $0 \leq \theta \leq \frac{\pi}{4}$. A partial list of such pairs is given in Table 2, and the corresponding points are shown on the curve in Figure 29.

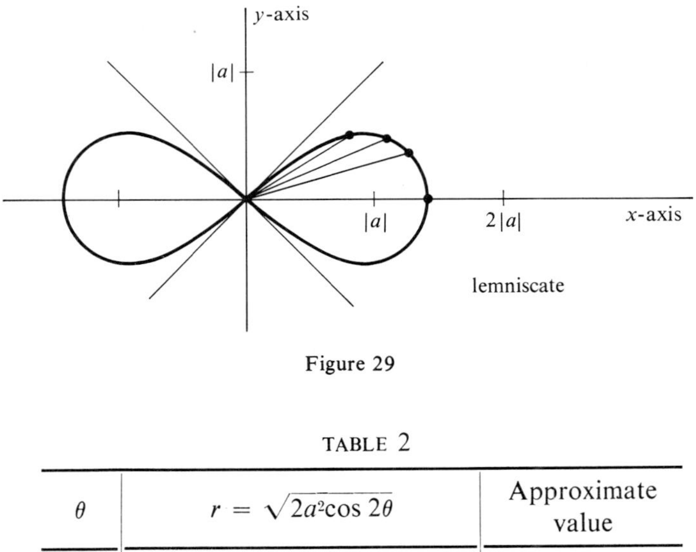

Figure 29

TABLE 2

θ	$r = \sqrt{2a^2 \cos 2\theta}$	Approximate value
0	$\sqrt{2}\,\lvert a\rvert$	$1.4\lvert a\rvert$
$\frac{\pi}{12}$	$\sqrt{2}\,\lvert a\rvert \left(\frac{\sqrt{3}}{2}\right)^{1/2} = \sqrt[4]{3}\,\lvert a\rvert$	$1.3\lvert a\rvert$
$\frac{\pi}{8}$	$\sqrt{2}\,\lvert a\rvert \left(\frac{\sqrt{2}}{2}\right)^{1/2} = \sqrt[4]{2}\,\lvert a\rvert$	$1.2\lvert a\rvert$
$\frac{\pi}{6}$	$\sqrt{2}\,\lvert a\rvert \left(\frac{1}{2}\right)^{1/2} = \lvert a\rvert$	$\lvert a\rvert$
$\frac{\pi}{4}$	0	0

PROBLEMS

1. (a) For each of the following values of θ, find the value of r such that $r = 4 \sin \theta$:
$$\theta = 0, \frac{\pi}{6}, \frac{\pi}{4}, \frac{\pi}{2}, \frac{3\pi}{4}, \frac{5\pi}{6}, \pi.$$
   (b) Plot the seven points with the polar coordinates (r, θ) found in part (a).
   (c) Draw and identify the curve defined by the equation $r = 4 \sin \theta$ in polar coordinates.
2. (a) For each of the following values of θ, find the value of r such that $r = 2(1 + \cos \theta)$:
$$\theta = 0, \frac{\pi}{4}, \frac{\pi}{3}, \frac{\pi}{2}, \frac{2\pi}{3}, \frac{5\pi}{6}, \pi.$$
   (b) Plot the seven points with the polar coordinates (r, θ), found in part (a).
   (c) What symmetry property is possessed by the curve defined by the equation $r = 2(1 + \cos \theta)$ in polar coordinates?
   (d) Draw the curve in part (c).
3. Using a figure and the geometric interpretation of polar coordinates, show that $r = \dfrac{5}{\cos \theta}$ is an equation in polar coordinates of the vertical line cutting the x-axis in the point $(5, 0)$.
4. Using a figure and the geometric interpretation of polar coordinates, find an equation in polar coordinates of the horizontal line cutting the y-axis in the point $(0, 5)$.
5. Assume the well-known fact that, if one side of a triangle inscribed in a circle is a diameter, then the triangle is a right triangle. Using this fact and the geometric interpretation of polar coordinates, show that $\cos \theta = \dfrac{r}{2a}$ is an equation of the circle which passes through the origin and has radius $a > 0$ and center on the x-axis.
6. Identify and draw the polar graphs of the two equations

   (a) $r = 7$ (b) $\theta = \dfrac{\pi}{6}$.
7. Consider the curves defined by each of the following equations in polar coordinates. Write each curve as the graph of an equation in x- and y-coordinates. Identify and draw the curve in the xy-plane.

   (a) $r \cos \theta = -2$ (d) $r = \dfrac{2}{\sin \theta - 2 \cos \theta}$

   (b) $r \sin \theta = 4$ (e) $r = \dfrac{1}{1 - \cos \theta}$ (see Example 3)

   (c) $r = -4 \cos \theta$ (f) $r = 5$

(g) $\theta = \arcsin \dfrac{3}{\sqrt{10}}$ (h) $r = \dfrac{1}{2 - \sqrt{3} \cos \theta}$.

8. Let f be a real-valued function of a real variable. Prove that:

   (a) If f is an even function, then the polar graph of the equation $r = f(\theta)$ is symmetric about the x-axis.

   (b) If f is an odd function, then the polar graph of the equation $r = f(\theta)$ is symmetric about the y-axis.

9. Let F be a real-valued function of two real variables. Prove that the polar graph of the equation $F(r^2, \theta) = 0$ is symmetric about the origin.

10. Draw the curve defined by each of the following equations in polar coordinates (the number a is an arbitrary positive constant).

    (a) $r = a(1 + \cos \theta)$ (a **cardioid**).

    (b) $r = a(2 + \cos \theta)$ (a **limaçon**).

    (c) $r = a(\frac{1}{2} + \cos \theta)$ (a **limaçon**).

    (d) $r^2 = 2a^2 \sin 2\theta$ (a **lemniscate**).

    (e) $r\theta = 2$ (a **hyperbolic spiral**).

11. Consider the Archimedean spiral defined by the equation $r = a\theta$ and discussed in Example 4. Describe the space of tangent vectors to this curve at $\theta = 0$, and also at $\theta = \dfrac{\pi}{2}$.

12. (a) Show that the equations $y = 4 \cos x$ and $y^2 = 4y \cos x$ are not equivalent.

    (b) In spite of part (a), the polar graphs of $r = 4 \cos \theta$ and of $r^2 = 4r \cos \theta$ are the same. Explain.

13. (a) If f is a real-valued function of a real variable, prove that the polar graph of the equation $r = f(\sin \theta)$ is symmetric about the y-axis.

    (b) Draw the curve (a cardioid) defined by the equation $r = 2(1 + \sin \theta)$ in polar coordinates.

    (c) Draw the curve (a limaçon) defined by the equation $r = 1 + 2 \sin \theta$ in polar coordinates.

7. Area and Arc Length in Polar Coordinates. This section is divided into two parts. In the first, which is the longer of the two, we shall study the problem of finding the areas of regions bounded by curves defined by equations in polar coordinates. To solve this problem, an integral formula for area in polar coordinates will be derived. The second part is concerned with the computation of the arc lengths of polar curves by applying the methods developed in Section 2.

Let f be a continuous function which contains the closed interval $[a, b]$ in its domain. We have already observed that the polar graph of the equation

$$r = f(\theta), \tag{1}$$

where θ takes on all values in the interval $[a, b]$, is the parametrized curve defined by the equations

$$\begin{cases} x(\theta) = r\cos\theta = f(\theta)\cos\theta, \\ y(\theta) = r\sin\theta = f(\theta)\sin\theta, \qquad a \leq \theta \leq b. \end{cases} \tag{2}$$

For the area problem, we shall assume to begin with that the interval $[a, b]$ has length no greater than 2π, i.e., that

$$b - a \leq 2\pi, \tag{3}$$

and also that f is nonnegative on $[a, b]$:

$$f(\theta) \geq 0, \qquad \text{for every number } \theta \text{ in } [a, b]. \tag{4}$$

Let R be the subset of the plane consisting of all points which have polar coordinates (r, θ) such that $a \leq \theta \leq b$ and $0 \leq r \leq f(\theta)$. An example is the shaded region R shown in Figure 30. The problem is to compute the area of R. The effect of the two assumptions (3) and (4) is that every point of R has precisely one pair of polar coordinates (r, θ) with $a \leq \theta \leq b$ (except, if $b - a = 2\pi$, for those points of R along the line defined by $\theta = a$).

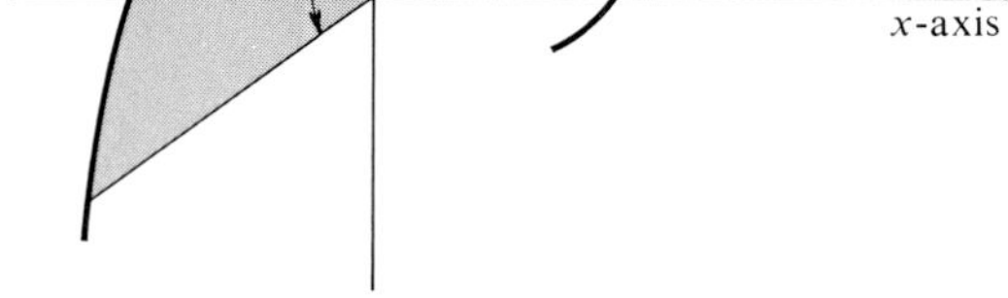

Figure 30

To derive a formula for the area of R, we consider an arbitrary partition $\sigma = \{\theta_0, \ldots, \theta_n\}$ of $[a, b]$ with the property that

$$a = \theta_0 \leq \theta_1 \leq \cdots \leq \theta_n = b.$$

For each $i = 1, \ldots, n$, let m_i and M_i be, respectively, the minimum and maximum values of the function f in the subinterval $[\theta_{i-1}, \theta_i]$. In addition, let R_i be the subset of R consisting of all points with polar coordinates (r, θ) such that $\theta_{i-1} \leq \theta \leq \theta_i$ and $0 \leq r \leq f(\theta)$, as illustrated in Figure 31. It follows from the preceding paragraph that except for their boundaries the sets $R_1, \ldots, R_n$ are pairwise disjoint. Hence

$$area(R) = area(R_1) + \cdots + area(R_n). \tag{5}$$

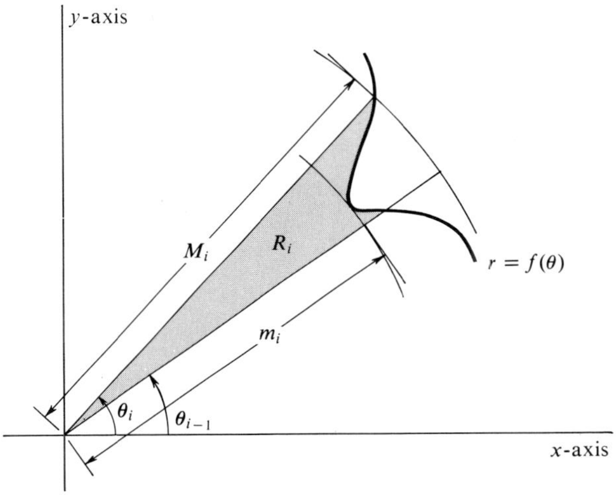

Figure 31

Each set R_i is contained in a sector of the circle of radius M_i and center the origin with angle equal to $\theta_i - \theta_{i-1}$ radians, and it contains a sector of the circle of radius m_i with the same center and the same angle. Since any sector of a circle of radius ρ and angle α radians has area equal to $\frac{1}{2}\rho^2\alpha$, we conclude that

$$\tfrac{1}{2}m_i^2(\theta_i - \theta_{i-1}) \leq area(R_i) \leq \tfrac{1}{2}M_i^2(\theta_i - \theta_{i-1}),$$

for each $i = 1, \ldots, n$. Adding inequalities and using equation (5), we get

$$\sum_{i=1}^{n} \frac{m_i^2}{2}(\theta_i - \theta_{i-1}) \leq area(R) \leq \sum_{i=1}^{n} \frac{M_i^2}{2}(\theta_i - \theta_{i-1}).$$

However, $\sum_{i=1}^{n} \frac{m_i^2}{2}(\theta_i - \theta_{i-1})$ and $\sum_{i=1}^{n} \frac{M_i^2}{2}(\theta_i - \theta_{i-1})$ are, respectively, the lower and upper sums for the function $\frac{f^2}{2}$ relative to the partition σ (see page 165). Denoting them by $L_\sigma\left(\frac{f^2}{2}\right)$ and $U_\sigma\left(\frac{f^2}{2}\right)$, respectively, we have proved that

$$L_\sigma\left(\frac{f^2}{2}\right) \leq area(R) \leq U_\sigma\left(\frac{f^2}{2}\right), \tag{6}$$

for every partition σ of $[a, b]$. Since f is continuous, so is $\frac{f^2}{2}$, and every function which is continuous on a closed bounded interval is integrable over that interval [see Theorem (5.1), page 199]. Hence the function $\frac{f^2}{2}$ is integrable over $[a, b]$, and it therefore follows immediately from the inequalities (6) and the definition of integrability on page 168 that

$$area(R) = \int_a^b \frac{f^2}{2}.$$

Summarizing, we have proved:

(7.1) *If the function f is continuous and nonnegative at every point of the closed interval $[a, b]$ and if $b - a \leq 2\pi$, then the area of the region R bounded by the polar graphs of the equations $r = f(\theta)$, $\theta = a$, and $\theta = b$ is given by*

$$area(R) = \tfrac{1}{2}\int_a^b f(\theta)^2 d\theta = \tfrac{1}{2}\int_a^b r^2 d\theta.$$

EXAMPLE 1. The curve defined by the equation $r = 1 + \cos\theta$ in polar coordinates, and drawn in Figure 32, is a cardioid. Compute the area of the

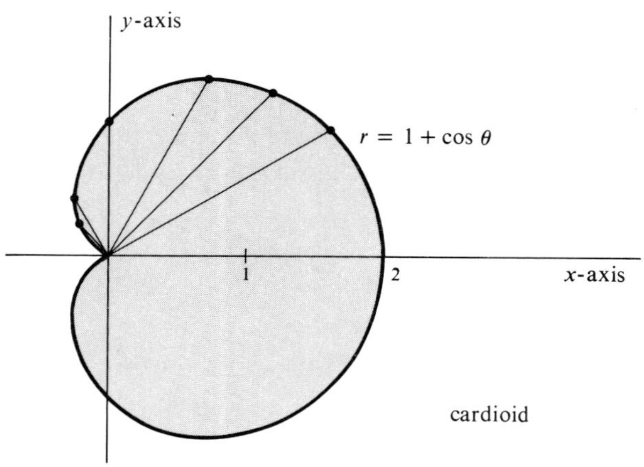

Figure 32

region R which it bounds. Since this curve is symmetric about the x-axis, it is sufficient (but in this example no easier) to find the area of that part of R lying on or above the x-axis and to multiply the result by 2. The function f defined by

$$f(\theta) = 1 + \cos\theta, \qquad 0 \leq \theta \leq \pi,$$

is both continuous and nonnegative. It follows from (7.1) that

$$\begin{aligned}
area(R) &= 2\left[\tfrac{1}{2}\int_0^{\pi} (1 + \cos\theta)^2 d\theta\right] \\
&= \int_0^{\pi} (1 + 2\cos\theta + \cos^2\theta)\, d\theta \\
&= \int_0^{\pi} [1 + 2\cos\theta + \tfrac{1}{2}(1 + \cos 2\theta)]\, d\theta \\
&= \int_0^{\pi} (\tfrac{3}{2} + 2\cos\theta + \tfrac{1}{2}\cos 2\theta)\, d\theta \\
&= (\tfrac{3}{2}\theta + 2\sin\theta + \tfrac{1}{4}\sin 2\theta)\Big|_0^{\pi} \\
&= \tfrac{3}{2}\pi.
\end{aligned}$$

If f is negative on the interval $[a, b]$, the integral $\frac{1}{2}\int_a^b f(\theta)^2\, d\theta$ is also equal to an area. Specifically, let us assume that f is continuous on $[a, b]$, that $b - a \leq 2\pi$, and that $f(\theta) \leq 0$ for every θ in $[a, b]$. Let R be the set of all points which have polar coordinates (r, θ) such that $a \leq \theta \leq b$ and $f(\theta) \leq r \leq 0$ (see Figure 33). Then the following formula is still valid:

(7.2) $$area(R) = \frac{1}{2}\int_a^b f(\theta)^2 d\theta.$$

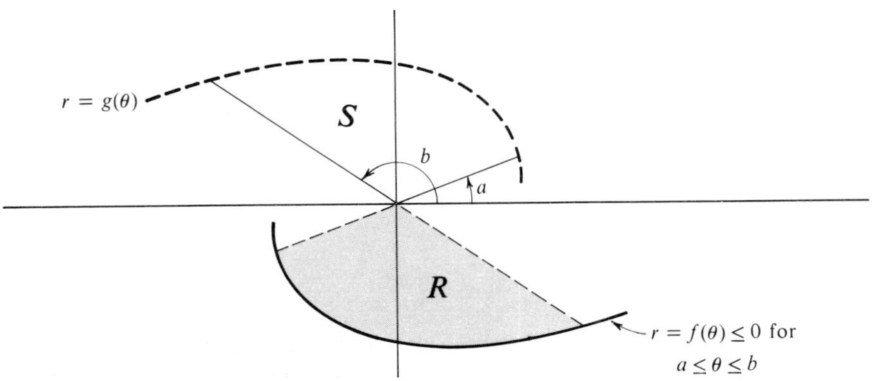

Figure 33

Proof. Let g be the function defined by $g(\theta) = -f(\theta)$, and let S be the set of all points with polar coordinates (r, θ) such that $a \leq \theta \leq b$ and $0 \leq r \leq g(\theta)$. The set S is symmetric about the origin to the set R, and we therefore conclude that

$$area(R) = area(S).$$

But, by (7.1),

$$area(S) = \frac{1}{2}\int_a^b g(\theta)^2 d\theta = \frac{1}{2}\int_a^b [-f(\theta)]^2 d\theta$$

$$= \frac{1}{2}\int_a^b f(\theta)^2 d\theta,$$

which completes the proof.

If the function f can take on both positive and negative values in the interval $[a, b]$ or if $b - a > 2\pi$ (or both), then the integral $\frac{1}{2}\int_a^b f(\theta)^2\, d\theta$ will in general give the sum of the areas of nondisjoint (i.e., overlapping) regions. It is frequently necessary to subdivide the interval $[a, b]$ into subintervals and to compute the integrals of $\frac{f^2}{2}$ over these subintervals separately to find a desired area.

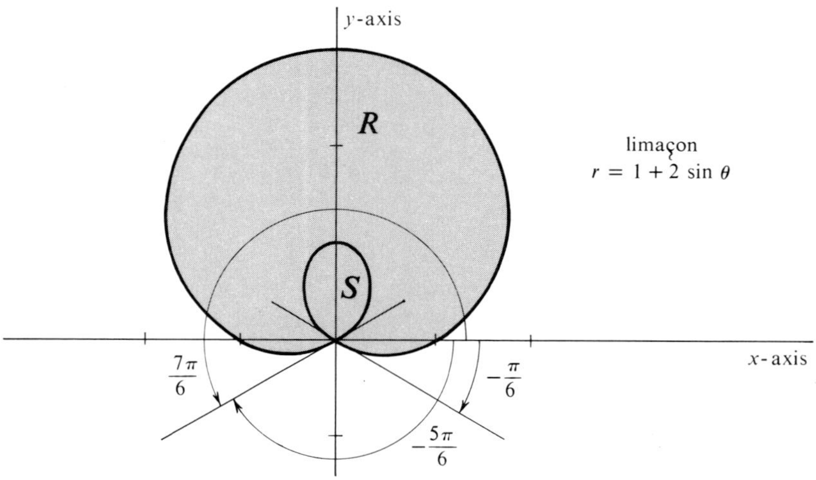

Figure 34

EXAMPLE 2. The polar graph of the equation $r = 1 + 2 \sin \theta$ is the limaçon shown in Figure 34. The function f defined by $f(\theta) = 1 + 2 \sin \theta$ satisfies the inequalities

$$f(\theta) \geq 0 \quad \text{if} \quad -\frac{\pi}{6} \leq \theta \leq \pi + \frac{\pi}{6},$$

$$f(\theta) \leq 0 \quad \text{if} \quad \frac{\pi}{6} - \pi \leq \theta \leq -\frac{\pi}{6}.$$

Let R and S be, respectively, the regions bounded by the outer and inner loops of the curve, as shown in the figure. Then

$$area(R) = \frac{1}{2}\int_{-\pi/6}^{7\pi/6} (1 + 2\sin\theta)^2 d\theta, \tag{7}$$

$$area(S) = \frac{1}{2}\int_{-5\pi/6}^{-\pi/6} (1 + 2\sin\theta)^2 d\theta. \tag{8}$$

If we integrate $\frac{1}{2}f(\theta)^2$ from 0 to 2π, the result will be equal to the area of R plus the area of S. That is, we will pick up the area of S twice and get

$$area(R) + area(S) = \frac{1}{2}\int_{0}^{2\pi} (1 + 2\sin\theta)^2 d\theta. \tag{9}$$

The consistency of equations (7), (8), and (9) can be checked as follows: From (7), (8), and the additivity of the definite integral, we get

$$\begin{aligned} area(S) + area(R) &= \frac{1}{2}\left[\int_{-5\pi/6}^{-\pi/6} (1 + 2\sin\theta)^2 d\theta + \int_{-\pi/6}^{7\pi/6} (1 + 2\sin\theta)^2 d\theta\right] \\ &= \frac{1}{2}\int_{-5\pi/6}^{7\pi/6} (1 + 2\sin\theta)^2 d\theta. \end{aligned}$$

Since the function $(1 + 2\sin\theta)^2$ has period 2π, its definite integral over every interval of length 2π will be the same. In particular,

$$\begin{aligned} area(S) + area(R) &= \frac{1}{2}\int_{-5\pi/6}^{7\pi/6} (1 + 2\sin\theta)^2 d\theta \\ &= \frac{1}{2}\int_{0}^{2\pi} (1 + 2\sin\theta)^2 d\theta, \end{aligned}$$

in agreement with (9). Evaluation of the integrals is left to the reader. The results are

$$area(R) = \frac{1}{2}\int_{-\pi/6}^{7\pi/6} (1 + 2\sin\theta)^2 d\theta = 2\pi + \frac{3}{2}\sqrt{3},$$

$$area(S) = \frac{1}{2}\int_{-5\pi/6}^{-\pi/6} (1 + 2\sin\theta)^2 d\theta = \pi - \frac{3}{2}\sqrt{3}.$$

It follows that the area of the region between the two loops of the limaçon is equal to the difference, $\pi + 3\sqrt{3}$.

EXAMPLE 3. Find the area A of the region bounded by the positive y-axis and the Archimedean spiral $r = a\theta\ (a > 0)$, where $0 \leq \theta \leq \dfrac{5\pi}{2}$. The region, shown in Figure 35, is the union of two subsets R_1 and R_2. The set R_1 consists of all points with polar coordinates (r, θ) which satisfy the inequalities $0 \leq \theta \leq 2\pi$ and $0 \leq r \leq a\theta$; i.e., it is the region bounded by the positive x-axis and that part of the spiral for which $0 \leq \theta \leq 2\pi$. We find

$$area(R_1) = \tfrac{1}{2}\int_0^{2\pi} r^2 d\theta = \tfrac{1}{2}\int_0^{2\pi} a^2\theta^2 d\theta$$

$$= \frac{a^2}{2}\frac{\theta^3}{3}\bigg|_0^{2\pi} = \frac{4a^2\pi^3}{3}.$$

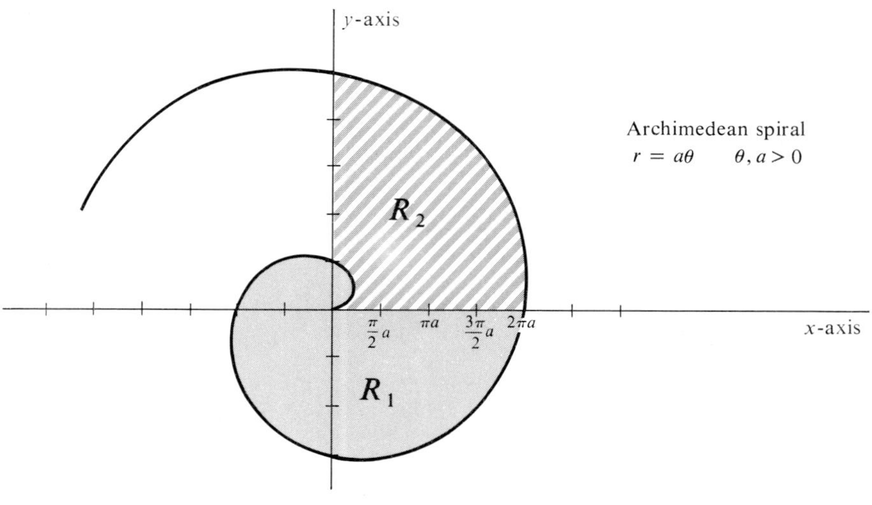

Figure 35

The set R_2 consists of all points with polar coordinates (r, θ) which satisfy the inequalities $2\pi \leq \theta \leq \dfrac{5\pi}{2}$ and $a(\theta - 2\pi) \leq r \leq a\theta$ (see Figure 35). This region can be equivalently described as that bounded by the lines $\theta = 0$ and $\theta = \dfrac{\pi}{2}$ and the two polar curves:

$$\begin{cases} r_1 = a\theta, \\ r_2 = a(\theta + 2\pi), \qquad 0 \leq \theta \leq \dfrac{\pi}{2}. \end{cases}$$

Since $0 \le r_1(\theta) \le r_2(\theta)$ for every θ on the interval $\left[0, \frac{\pi}{2}\right]$, the area of R_2 is obviously given by the formula

$$\begin{aligned} area(R_2) &= \tfrac{1}{2}\int_0^{\pi/2} (r_2^2 - r_1^2)d\theta \\ &= \tfrac{1}{2}\int_0^{\pi/2} [a^2(\theta + 2\pi)^2 - a^2\theta^2]d\theta \\ &= \frac{a^2}{2}\int_0^{\pi/2} (4\pi\theta + 4\pi^2)d\theta \\ &= 2\pi a^2\left(\frac{\theta^2}{2} + \pi\theta\right)\Bigg|_0^{\pi/2} = 2\pi a^2\left(\frac{\pi^2}{8} + \frac{\pi^2}{2}\right) \\ &= \frac{5a^2\pi^3}{4}. \end{aligned}$$

We conclude that

$$\begin{aligned} A &= area(R_1) + area(R_2) \\ &= \frac{4a^2\pi^3}{3} + \frac{5a^2\pi^3}{4} = \frac{31a^2\pi^3}{12}. \end{aligned}$$

An alternative way of finding the answer is to realize that the integral

$$\tfrac{1}{2}\int_0^{5\pi/2} r^2 d\theta = \tfrac{1}{2}\int_0^{5\pi/2} a^2\theta^2 d\theta$$

is equal to the area A except for the fact that it counts twice the area bounded by the lines $\theta = 0$ and $\theta = \frac{\pi}{2}$ and the curve $r = a\theta$ for $0 \le \theta \le \frac{\pi}{2}$. Hence we also obtain

$$\begin{aligned} A &= \tfrac{1}{2}\int_0^{5\pi/2} a^2\theta^2 d\theta - \tfrac{1}{2}\int_0^{\pi/2} a^2\theta^2 d\theta \\ &= \frac{31a^2\pi^3}{12}. \end{aligned}$$

The second topic of this section is the computation of the arc length of a curve defined by an equation in polar coordinates. No new methods are needed, since the problem is simply a special case of the more general one of finding the arc length of a parametrized curve. As noted in the second paragraph of this section, if f is a continuous function containing the interval $[a, b]$ in its domain, then the polar graph of the equation

$$r = f(\theta), \qquad \text{with } a \leq \theta \leq b,$$

is a parametrized curve [see equations (2)]. Specifically, the curve is the range of the parametrization $P: [a, b] \to R^2$ defined by

$$P(\theta) = \big(x(\theta), y(\theta)\big) = \big(f(\theta)\cos\theta, f(\theta)\sin\theta\big),$$

for every θ in $[a, b]$. We shall make the assumption that the derivative f' is a continuous function on $[a, b]$, and this implies that the derivatives x' and y' are also continuous. It then follows directly from Theorem (2.2), page 553, that the arc length of the curve from $P(a)$ to $P(b)$ is given by

$$L_a^b = \int_a^b \sqrt{x'(\theta)^2 + y'(\theta)^2}\, d\theta.$$

Since

$$x'(\theta) = f'(\theta)\cos\theta - f(\theta)\sin\theta,$$

$$y'(\theta) = f'(\theta)\sin\theta + f(\theta)\cos\theta,$$

we find that

$$\begin{aligned} x'(\theta)^2 + y'(\theta)^2 &= f'(\theta)^2\cos^2\theta - 2f'(\theta)f(\theta)\sin\theta\cos\theta + f(\theta)^2\sin^2\theta \\ &\quad + f'(\theta)^2\sin^2\theta + 2f'(\theta)f(\theta)\sin\theta\cos\theta + f(\theta)^2\cos^2\theta \\ &= f'(\theta)^2 + f(\theta)^2. \end{aligned}$$

Thus we obtain the following integral formula for the arc length L_a^b of the polar graph of the equation $r = f(\theta)$, in which $a \leq \theta \leq b$:

(7.3)
$$L_a^b = \int_a^b \sqrt{f'(\theta)^2 + f(\theta)^2}\, d\theta.$$

Alternatively, if we set $r = f(\theta)$ in the formula and write $f'(\theta) = \dfrac{dr}{d\theta}$, we have

(7.3′)
$$L_a^b = \int_a^b \sqrt{\left(\frac{dr}{d\theta}\right)^2 + r^2}\, d\theta.$$

EXAMPLE 4. Find the arc length of the cardioid defined in polar coordinates by the equation $r = 1 + \cos\theta$. This curve is shown in Figure 32, and the area of the region which it bounds is computed in Example 1. We have

$$\frac{dr}{d\theta} = -\sin\theta,$$

$$r^2 = (1 + \cos\theta)^2 = 1 + 2\cos\theta + \cos^2\theta.$$

Hence

$$\begin{aligned} r^2 + \left(\frac{dr}{d\theta}\right)^2 &= 1 + 2\cos\theta + \cos^2\theta + \sin^2\theta \\ &= 2(1 + \cos\theta). \end{aligned}$$

The trigonometric identity

$$\cos^2\frac{\theta}{2} = \frac{1}{2}(1 + \cos\theta)$$

implies that

$$r^2 + \left(\frac{dr}{d\theta}\right)^2 = 4\cos^2\frac{\theta}{2},$$

and it follows from the integral formula (7.3′) that the arc length L of the cardioid is given by

$$L = \int_0^{2\pi} \sqrt{4\cos^2\frac{\theta}{2}}\, d\theta = 2\int_0^{2\pi} \left|\cos\frac{\theta}{2}\right| d\theta.$$

However, because the cardioid is symmetric about the x-axis, we conclude that

$$L = 2\int_0^{\pi} \sqrt{4\cos^2\frac{\theta}{2}}\, d\theta = 4\int_0^{\pi} \left|\cos\frac{\theta}{2}\right| d\theta.$$

If $0 \le \theta \le \pi$, then $\cos\frac{\theta}{2} \ge 0$ and so $\left|\cos\frac{\theta}{2}\right| = \cos\frac{\theta}{2}$. Hence the arc length L is equal to

$$L = 4\int_0^{\pi} \cos\frac{\theta}{2}\, d\theta = 8\sin\frac{\theta}{2}\bigg|_0^{\pi} = 8.$$

PROBLEMS

1. In each of the following, draw the curve defined by the equation $r = f(\theta)$ in polar coordinates. Show the region R bounded by the curve and the lines $\theta = a$ and $\theta = b$, and compute its area.

   (a) $r = 4\cos\theta$, $a = 0$ and $b = \dfrac{\pi}{2}\cdot$

   (b) $r = 3(1 + \cos\theta)$, $a = 0$ and $b = \pi$.

   (c) $r = 3(1 + \sin\theta)$, $a = 0$ and $b = \dfrac{\pi}{2}\cdot$

   (d) $r = \dfrac{2}{\cos\theta}$, $a = -\dfrac{\pi}{4}$ and $b = \dfrac{\pi}{4}\cdot$

2. For each of the following equations $r = f(\theta)$ and pairs of numbers a and b, draw the region R consisting of all points with polar coordinates (r, θ) such that $a \leq \theta \leq b$ and $0 \leq r \leq f(\theta)$. Compute $area(R)$.

   (a) $r = 4 \sin \theta$, $a = 0$ and $b = \pi$.

   (b) $r = \dfrac{4}{\sin \theta}$, $a = \dfrac{\pi}{4}$ and $b = \dfrac{3\pi}{4}$.

   (c) $r = 2\theta$, $a = \pi$ and $b = 2\pi$.

   (d) $r = \dfrac{1}{2 \cos \theta + 3 \sin \theta}$, $a = 0$ and $b = \dfrac{\pi}{2}$.

   (e) $r = \sqrt{2 \cos 2\theta}$, $a = 0$ and $b = \dfrac{\pi}{4}$. (See Example 5, Section 6.)

3. Identify and draw the curve defined by the equation $r = 4 \sin \theta$ in polar coordinates, and show the region R bounded by the curve. Is it true in this case that

$$area(R) = \tfrac{1}{2}\int_0^{2\pi} r^2\, d\theta?$$

   Explain your answer.

4. Each of the following curves, defined by an equation $r = f(\theta)$ in polar coordinates, bounds a region R in the plane. Draw the curve and find the area of R.

   (a) $r = a(1 + \cos \theta)$, $a > 0$

   (b) $r = a(1 + \sin \theta)$, $a > 0$

   (c) $r = 5$

   (d) $r = 2 + \cos \theta$

   (e) $r = 4 \sin \theta$

   (f) $r = -4 \cos \theta$.

5. The curve defined by the equation $r = \dfrac{1}{1 + \cos \theta}$ in polar coordinates is a parabola similar to the one discussed in Example 3, Section 6.

   (a) Draw the parabola, and show the region R bounded by this curve and the line $\theta = \dfrac{\pi}{2}$.

   (b) Express $area(R)$ as a definite integral using the integral formula for area in polar coordinates.

   (c) Evaluate the integral in part (b) using the trigonometric substitution $z = \tan \dfrac{\theta}{2}$ (see page 397) and the Change of Variable Theorem for Definite Integrals.

   (d) Write this curve as the graph of an equation in x- and y-coordinates, and thence compute $area(R)$.

6. Find the area of the region which lies between the two loops of the limaçon $r = 1 + 2\cos\theta$.
7. Find the area of the region bounded by the lemniscate $r^2 = 2a^2\cos 2\theta$.
8. Find the area A of the region which lies inside the cardioid $r = 2(1 + \cos\theta)$ and outside the circle $r = 3$.
9. The region R bounded by the cardioid $r = 4(1 + \sin\theta)$ is cut into two regions R_1 and R_2 by the polar graph of the equation $r = \dfrac{3}{\sin\theta}$. Compute the areas of R, R_1, and R_2.
10. Find the arc length of the cardioid defined by the equation $r = a(1 + \cos\theta)$, where a is an arbitrary positive constant.
11. Consider the spiral defined in polar coordinates by the equation $r = e^{2\theta}$. Compute the arc length of this curve from $\theta = 0$ to $\theta = \ln 10$.
12. (a) Using the integral formula for arc length in polar coordinates, compute the arc length of the polar graph of the equation $r = 2\sec\theta$ from $\theta = -\dfrac{\pi}{4}$ to $\theta = \dfrac{\pi}{4}$.

    (b) Identify and draw the curve in part (a), and verify from the geometry the value obtained for the arc length.
13. Consider the curve defined by the equation $r = 2\cos^2\dfrac{\theta}{2}$ in polar coordinates.

    (a) Find the arc length of this curve from $\theta = 0$ to $\theta = \pi$.

    (b) Find the arc length of this curve from $\theta = 0$ to $\theta = 2\pi$.

CHAPTER 11

Differential Equations

1. Review. This section is primarily a review of the differential equations studied in Section 5 of Chapter 5 and also in Section 8 of Chapter 6. We begin by recalling the definition of a first-order differential equation (see page 272): Consider an equation $F(x, y, z) = 0$ in which not all the variables need occur, but at least z does. The equation

$$F\left(x, y, \frac{dy}{dx}\right) = 0, \tag{1}$$

obtained by substituting $\frac{dy}{dx}$ for z, is a first-order differential equation. By a solution of (1) is meant any differentiable function f for which the equation

$$F(x, f(x), f'(x)) = 0$$

is true for every x in the domain of f. If f is a solution, we write

$$y = f(x).$$

The general problem, given a differential equation, is to find all its solutions. A more specialized problem is to find a particular solution $y = f(x)$ which has a specified value b at some specified number a, i.e., a solution for which $b = f(a)$.

The simplest first-order differential equations are those of the type $\frac{dy}{dx} = f(x)$, where f is some given function (not to be confused with the solutions f discussed in the preceding paragraph). Every solution of this differential equation can be written

$$y = \int f(x)\, dx + c,$$

for some real number c. Hence if c is left as an arbitrary undetermined constant of integration, we call $\int f(x)\, dx + c$ the general solution.

We next considered differential equations of the form $\frac{dy}{dx} = \frac{f(x)}{g(y)}$, in which f and g are given functions. Equations of this type are called separable, since, if we use the fact that the derivative is equal to the ratio of two differentials, we can "separate" the expression containing x from that containing y by writing the equivalent differential equation

$$g(y)\,dy = f(x)\,dx.$$

Integrating both sides, we get the equation

$$\int g(y)\,dy = \int f(x)\,dx + c,$$

which defines the general solution y implicitly as a function of x. Note that the differential equation $\frac{dy}{dx} = f(x)$ discussed in the preceding paragraph is a separable equation in which $g(y) = 1$.

Of special interest among separable equations is the first-order linear differential equation $\frac{dy}{dx} + ky = 0$, in which k is a constant. This is the type of differential equation which describes the rate of decay of a radioactive substance and also the rate of growth of bacteria in a culture. It can be solved without difficulty as a separable differential equation (see pages 276 and 277). However, this equation occurs sufficiently often and has such an obvious general solution that most people recognize it at sight. The general solution is

$$y = ce^{-kx}.$$

EXAMPLE 1. Find the general solution of each of the following differential equations:

$$\text{(a)} \quad \frac{dy}{dx} = \tan^4 x \sec^2 x,$$

$$\text{(b)} \quad \frac{dy}{dx} = e^{x+y},$$

$$\text{(c)} \quad \frac{dy}{dx} + 14y = 0.$$

In (b) find the particular solution $y = f(x)$ such that $f(0) = -\ln 2$, and in (c) find the particular solution which has value 5 when $x = 0$.

The general solution of (a) can be obtained directly by integrating:

$$\begin{aligned} y &= \int \tan^4 x \sec^2 x\,dx + c \\ &= \tfrac{1}{5}\tan^5 x + c. \end{aligned}$$

Equation (b) is separable, since $\frac{dy}{dx} = e^{x+y} = e^x e^y$. Hence we may write

$$e^{-y}dy = e^x dx.$$

Integrating both sides, we obtain the equation

$$-e^{-y} = e^x + c,$$

which defines y implicitly as a function of x. In this case, it is not difficult to solve for y explicitly. We first get $e^y = \frac{1}{-c - e^x}$. Replacing the arbitrary constant $-c$ by simply c, and taking logarithms, we then obtain

$$y = \ln \frac{1}{c - e^x} \tag{2}$$

as the general solution. To find the particular solution $y = f(x)$ for which $f(0) = -\ln 2$, we substitute these values in equation (2) and solve the resulting equation for c. Thus

$$-\ln 2 = \ln \frac{1}{c - 1} \cdot$$

Since $-\ln 2 = \ln \frac{1}{2}$, it follows that $2 = c - 1$, and so $c = 3$. Hence the particular solution required is

$$y = \ln \frac{1}{3 - e^x} \cdot$$

The general solution of (c) can be written down on inspection. It is

$$y = ce^{-14x}.$$

The particular solution which has value 5 when x equals 0 is obtained by writing

$$5 = ce^{-14 \cdot 0} = c.$$

Hence the particular solution is

$$y = 5e^{-14x}.$$

The definition of an nth-order differential equation, $n \geq 1$, is entirely analogous to that of a first-order equation. Let $F(x, y_0, y_1, \ldots, y_n) = 0$ be an equation in $n + 2$ variables in which not all the variables occur, but at least y_n does. Then the equation

$$F\left(x, y, \frac{dy}{dx}, \frac{d^2y}{dx^2}, \ldots, \frac{d^ny}{dx^n}\right) = 0, \tag{3}$$

obtained by substituting the ith derivative $\frac{d^i y}{dx^i}$ for y_i $\left(\text{where it is understood that } \frac{d^0 y}{dx^0} = y\right)$, is an **$n$th-order differential equation.** A **solution** is any n-times-differentiable function f such that the equation

$$F(x, f(x), f'(x), f''(x), \ldots, f^{(n)}(x)) = 0$$

is true for every x in the domain of f.

Our study of higher-order differential equations has thus far been limited to those of the type

$$\frac{d^2y}{dx^2} + a\frac{dy}{dx} + by = 0, \tag{4}$$

where a and b are constants. Such an equation is a second-order, linear, homogeneous differential equation with constant coefficients (see page 344). It is called "linear" because y and its derivatives occur to no power higher than the first, "homogeneous," because the right side is zero, and "with constant coefficients," because a and b are constants.

You will recall that the form of the general solution of the differential equation (4) is determined by the nature of the roots of its characteristic equation $t^2 + at + b = 0$. The roots of this equation are given by the quadratic formula

$$r_1, r_2 = \frac{-a \pm \sqrt{a^2 - 4b}}{2},$$

and there are three cases depending on the discriminant $a^2 - 4b$.

Case 1. If $a^2 - 4b$ is positive, then there are two distinct real roots r_1 and r_2. In this case the general solution of (4) is

$$y = c_1 e^{r_1 x} + c_2 e^{r_2 x},$$

where c_1 and c_2 are arbitrary constants.

Case 2. If $a^2 - 4b = 0$, then $r_1 = r_2 = r$ and the general solution of the differential equation (4) is

$$y = (c_1 x + c_2)e^{rx},$$

where c_1 and c_2 are arbitrary constants.

Case 3. If $a^2 - 4b$ is negative, then r_1 and r_2 are distinct conjugate complex numbers, i.e., $r_1 = \alpha + i\beta$, $r_2 = \alpha - i\beta$, and $\beta \neq 0$. In this case the general solution of (4) is

$$y = e^{\alpha x}(c_1 \cos \beta x + c_2 \sin \beta x),$$

where c_1 and c_2 are arbitrary constants.

The above statements imply that, if y is any solution of the differential equation (4), then there exist real numbers c_1 and c_2 such that

$$y = c_1e^{r_1x} + c_2e^{r_2x} \qquad \text{if } a^2 - 4b > 0,$$

$$y = (c_1x + c_2)e^{rx} \qquad \text{if } a^2 - 4b = 0,$$

$$y = e^{\alpha x}(c_1 \cos \beta x + c_2 \sin \beta x) \qquad \text{if } a^2 - 4b < 0.$$

This fact, first stated in Section 8 of Chapter 6, has not yet been proved, but will be in Section 4.

Although we have thus far not used the letter D to denote the derivative, this notation is quite useful in the study of differential equations. We write Dy for $\frac{dy}{dx}$ and D^2y for $\frac{d^2y}{dx^2}$. We then combine these symbols and the conventions of algebra to write $(D^2 + aD + b)y$ for $D^2y + aDy + by$. In so doing we have defined a function, denoted by $D^2 + aD + b$, which has the set of twice-differentiable functions as its domain and a set of functions as its range. This function assigns to each function y in its domain the function

$$(D^2 + aD + b)y = \frac{d^2y}{dx^2} + a\frac{dy}{dx} + by$$

as value. Such a function is an example of a **differential operator.** Using it, the differential equation $\frac{d^2y}{dx^2} + a\frac{dy}{dx} + by = 0$ can be written

$$(D^2 + aD + b)y = 0. \tag{4'}$$

Note the similarity between the operator and the characteristic equation of the differential equation. The latter is the equation obtained by replacing D in the operator by t and setting the resulting quadratic polynomial equal to zero.

EXAMPLE 2. Find the general solution of each of the following differential equations:

(a) $\frac{d^2y}{dx^2} - 5\frac{dy}{dx} + 6y = 0,$

(b) $(D^2 + 6D + 9)y = 0,$

(c) $(D^2 - 6D + 10)y = 0.$

For the first, the characteristic equation is $t^2 - 5t + 6 = 0$, which is equivalent to $(t - 2)(t - 3) = 0$. Hence the two roots are 2 and 3, and the general solution is given by

$$y = c_1e^{2x} + c_2e^{3x}.$$

In (b), the characteristic equation is $t^2 + 6t + 9 = 0$, which is equivalent to $(t + 3)^2 = 0$. Thus there is only one root, -3. The solutions of the differential equation are therefore all functions

$$y = (c_1x + c_2)e^{-3x},$$

where c_1 and c_2 are arbitrary constants.

The characteristic equation for (c) is $t^2 - 6t + 10 = 0$ and, since its discriminant is equal to -4, the roots are not real. Solving the quadratic equation, we find that the roots are $3 + i$ and $3 - i$. Hence the general solution is

$$y = e^{3x}(c_1 \cos x + c_2 \sin x).$$

EXAMPLE 3. Find the particular solution of the differential equation $D(D - 5)y = 0$ which has value equal to 2 and derivative equal to -15 when $x = 0$. The characteristic equation is $t(t - 5) = 0$, whose roots are obviously 0 and 5. The general solution is therefore

$$y = c_1e^{0x} + c_2e^{5x} = c_1 + c_2e^{5x}.$$

The derivative is

$$y' = 5c_2e^{5x}.$$

When $x = 0$, we are told that $y = 2$ and $y' = -15$. Substituting these values in the preceding equations, we obtain

$$\begin{aligned} 2 &= c_1 + c_2e^{5\cdot 0} = c_1 + c_2, \\ -15 &= 5c_2e^{5\cdot 0} \quad\;\; = 5c_2. \end{aligned}$$

It follows that $c_2 = -3$ and thence that $c_1 = 5$. Hence the required solution is

$$y = 5 - 3e^{5x}.$$

It is extremely useful to recognize alternative forms of the general solution of the differential equation $(D^2 + aD + b)y = 0$ in the case where the roots of the characteristic equation are the complex numbers $\alpha + i\beta$ and $\alpha - i\beta$. In particular, it is easy to verify that the functions

$$y = ce^{\alpha x}\sin(\beta x + k), \tag{5}$$

$$y = ce^{\alpha x}\cos(\beta x + k), \tag{6}$$

where c and k are arbitrary real numbers, are both solutions. To see that this is so, we expand (5) using the trigonometric identity for the sine of the sum of two numbers. The result is

$$\begin{aligned} y &= ce^{\alpha x}(\sin \beta x \cos k + \cos \beta x \sin k) \\ &= e^{\alpha x}[(c \sin k) \cos \beta x + (c \cos k) \sin \beta x]. \end{aligned}$$

Setting $c_1 = c \sin k$ and $c_2 = c \cos k$, we obtain $y = e^{\alpha x}(c_1 \cos \beta x + c_2 \sin \beta x)$, which we know to be a solution. The proof for (6) is analogous.

Conversely, any solution $y = e^{\alpha x}(c_1 \cos \beta x + c_2 \sin \beta x)$ can be written in the forms (5) and (6). For if both $c_1 = c_2 = 0$, then $y = 0$, and we need only set $c = 0$ in (5) and (6). If c_1 and c_2 are not both zero, then $\sqrt{c_1^2 + c_2^2} \neq 0$, and we can write

$$y = \sqrt{c_1^2 + c_2^2}\, e^{\alpha x} \left[\frac{c_1}{\sqrt{c_1^2 + c_2^2}} \cos \beta x + \frac{c_2}{\sqrt{c_1^2 + c_2^2}} \sin \beta x \right].$$

To put this equation in the form of (5), we set $c = \sqrt{c_1^2 + c_2^2}$ and observe that, since

$$\left(\frac{c_2}{\sqrt{c_1^2 + c_2^2}} \right)^2 + \left(\frac{c_1}{\sqrt{c_1^2 + c_2^2}} \right)^2 = 1,$$

it follows from our definition of the functions sine and cosine that there exists a real number k such that $\cos k = \dfrac{c_2}{\sqrt{c_1^2 + c_2^2}}$ and $\sin k = \dfrac{c_1}{\sqrt{c_1^2 + c_2^2}}$.

Hence we get

$$\begin{aligned} y &= ce^{\alpha x}(\sin k \cos \beta x + \cos k \sin \beta x) \\ &= ce^{\alpha x} \sin(\beta x + k). \end{aligned}$$

Again, by an analogous argument, the solution can also be written in the form of equation (6).

An advantage in using the forms (5) and (6) for the general solution is that it is easy to see what the graphs of such functions look like. As the next example illustrates, they are all sinusoidal curves lying between the graphs of $y = ce^{\alpha x}$ and $y = -ce^{\alpha x}$.

EXAMPLE 4. Find and draw the graph of the particular solution of the differential equation $(D^2 + 2D + \pi^2 + 1)y = 0$ which has value $\sqrt{2}$ and derivative equal to $(\pi - 1)\sqrt{2}$ when $x = 0$. The characteristic equation is $t^2 + 2t + \pi^2 + 1 = 0$, which has roots $-1 + i\pi$ and $-1 - i\pi$. Hence one form of the general solution is

$$y = ce^{-x} \sin(\pi x + k).$$

Its derivative is

$$\frac{dy}{dx} = -ce^{-x} \sin(\pi x + k) + c\pi e^{-x} \cos(\pi x + k).$$

Substituting the given values of y and $\frac{dy}{dx}$ when $x = 0$ into the preceding two equations, we get

$$\sqrt{2} = c \sin k,$$
$$(\pi - 1)\sqrt{2} = -c \sin k + c\pi \cos k.$$

Hence, $(\pi - 1)\sqrt{2} = -\sqrt{2} + c\pi \cos k$, from which we obtain

$$\sqrt{2} = c \cos k.$$

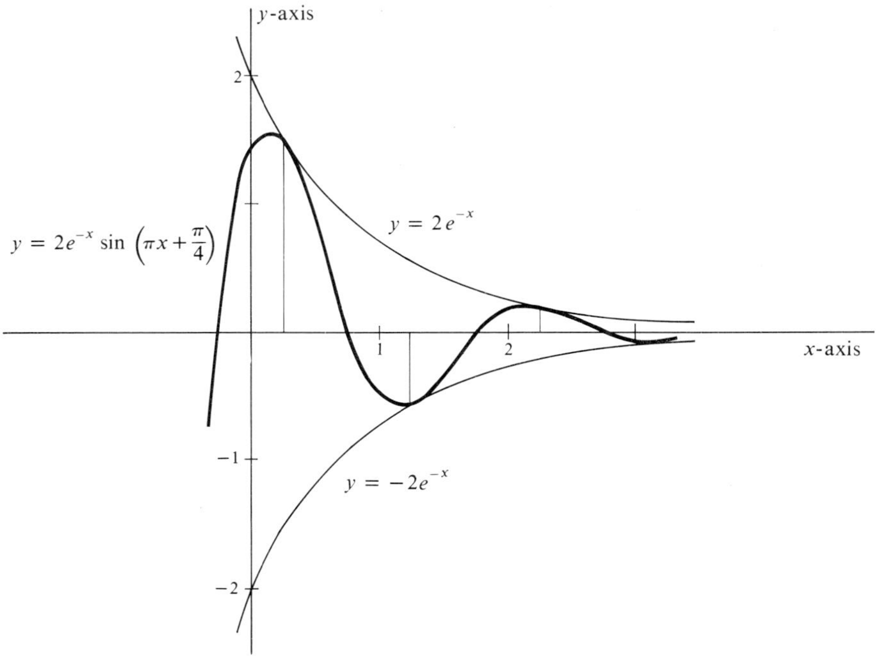

Figure 1

Since $c \cos k$ does not equal zero, it follows that

$$\tan k = \frac{c \sin k}{c \cos k} = \frac{\sqrt{2}}{\sqrt{2}} = 1,$$

and so $k = \frac{\pi}{4}$. This implies that $c = 2$, and we conclude that the particular solution is

$$y = 2e^{-x} \sin\left(\pi x + \frac{\pi}{4}\right).$$

The graph of this equation is drawn in Figure 1. Such a curve is frequently called an "exponentially damped sine wave."

PROBLEMS

1. Find the general solution of each of the following differential equations.

(a) $\dfrac{dy}{dx} = x^3 + 2e^x$

(b) $x\dfrac{dy}{dx} = 6x^3 + 5x + 1$

(c) $\dfrac{dy}{dx} = (y^2 + 1)(2x + 3)$

(d) $\dfrac{dy}{dx} = xy + x$

(e) $2xy^2 + \dfrac{dy}{dx} - 4x^3y^2 = 0$

(f) $y\dfrac{dy}{dx} = \ln x$

(g) $x\dfrac{dy}{dx} = \ln x$

(h) $\dfrac{dy}{dx} + 16y = 0$

(i) $\dfrac{d^2y}{dx^2} + 16y = 0$

(j) $\dfrac{d^2y}{dx^2} = 16y$

(k) $y'' - 19y' - 20y = 0$

(l) $(D^2 + 10D + 16)y = 0$

(m) $2\dfrac{d^2y}{dx^2} - 14\dfrac{dy}{dx} = -20y$

(n) $\dfrac{d^2y}{dx^2} + a^2y = 2a\dfrac{dy}{dx}$

(o) $(D^2 + 4D + 29)y = 0$

(p) $(y + 5)\dfrac{dy}{dx} = 7x - e^{-x}$

(q) $\dfrac{dy}{dx} = \dfrac{x}{y}$

(r) $\dfrac{dy}{dx} = \dfrac{y}{x}$

(s) $\dfrac{dy}{dx} = -\dfrac{x}{y}$

(t) $\dfrac{1}{y}\dfrac{d^2y}{dx^2} = 49$

(u) $(3x + 4)\,dt + (4t + 3)\,dx = 0$

(v) $\dfrac{dy}{dx} = \cot y$

(w) $\dfrac{1}{t}\dfrac{dy}{dt} = e^{3t^2+4}$

(x) $\dfrac{dy}{dx} = 3\sin^2 x\cos^2 x$

(y) $\dfrac{dy}{dx} = 3\sin^2 x\cos^2 y$

(z) $\dfrac{d^2y}{dx^2} = 6x^2 - 4x + 2.$

2. Find the particular solution of each of the following differential equations which satisfies the given conditions.

(a) $\dfrac{dy}{dx} = 3y, \quad y = 5$ when $x = 0$.

(b) $\dfrac{d^2y}{dx^2} = 12x^2 + 1$, graph passes through the point $(1, -1)$ with a slope of 3.

(c) $y\dfrac{dy}{dx} = -x, \quad$ graph passes through the point $(-3, -4)$.

(d) $\dfrac{d^2s}{dt^2} = -g$ (constant), when $t = 0$, $\dfrac{ds}{dt} = v_0$ and $s = s_0$.

(e) $(D^2 - 2D - 3)y = 0$, $y = 7$ and $\dfrac{dy}{dx} = 1$ when $x = 0$.

(f) $(D^2 - 4D + 13)y = 0$, graph passes through (0, 5) with a slope of 2.

(g) $(x + 2)\dfrac{dy}{dx} = 1$, $y = \ln 9$ when $x = 1$.

(h) $(D^2 - 12D + 36)y = 0$, $y = 3$ and $\dfrac{dy}{dx} = 7$ when $x = 0$.

3. (a) Sketch the graph of $y = e^{x/2}\cos\left(x + \dfrac{\pi}{4}\right)$.

(b) Find a second-order, linear, homogeneous differential equation with constant coefficients of which the function in (a) is a solution.

4. (a) Find the general solution of the differential equation $(4D^2 + 8D + 5)y = 0$.

(b) Find the particular solution of the differential equation in (a) whose graph passes through the point $\left(0, \dfrac{\sqrt{2}}{2}\right)$ with a slope of $-\dfrac{3\sqrt{2}}{4}$.

(c) Sketch the graph of the function in (b).

5. (a) Find the general solution of the differential equation $(4D^2 - 8D + 5)y = 0$.

(b) Find the particular solution of the differential equation in (a) whose graph passes through the point $\left(0, \dfrac{\sqrt{2}}{2}\right)$ with a slope of $\dfrac{\sqrt{2}}{4}$.

(c) Sketch the graph of the function in (b).

6. (a) Find the general solution of the differential equation $(D^2 + \frac{1}{4})y = 0$.

(b) Find the particular solution of the differential equation in (a) whose graph passes through the point $\left(0, \dfrac{\sqrt{2}}{2}\right)$ with a slope of $-\dfrac{\sqrt{2}}{4}$.

(c) Sketch the graph of the function in (b).

7. Find the general solution of the differential equation $(D^2 - 2\alpha D + \alpha^2 + 1)y = 0$, and sketch the graph for

(a) $\alpha > 0$ (b) $\alpha = 0$ (c) $\alpha < 0$.

[See equations (5) and (6).]

2. First-Order Linear Differential Equations. A differential equation which can be written in the form

$$a_1(x)\frac{dy}{dx} + a_0(x)y = b(x),$$

where a_1, a_0, and b are given functions of x and where a_1 is not the zero function, is a **first-order linear differential equation.** In this section we shall show

how to obtain the general solution of equations of this type. Since a_1 is by assumption not the zero function, we can divide both sides of the above equation by $a_1(x)$. Setting $\frac{a_0(x)}{a_1(x)} = P(x)$ and $\frac{b(x)}{a_1(x)} = Q(x)$, we therefore obtain the differential equation

$$\boxed{\frac{dy}{dx} + P(x)y = Q(x),} \tag{1}$$

which is the form we shall use in deriving the solution. We shall assume that the functions P and Q are continuous, thus assuring ourselves that they have antiderivatives.

Let us suppose that the function $y = f(x)$ is a solution to the differential equation (1). We shall derive a formula which expresses y in terms of P and Q and a constant c of integration. Conversely, it will be a simple matter to verify that any function y defined by this formula is a solution to (1). Hence the formula gives the general solution to the differential equation.

The derivative of the product of y and a function φ is given by

$$\frac{d}{dx}\big(\varphi(x)y\big) = \varphi(x)\frac{dy}{dx} + \varphi'(x)y. \tag{2}$$

Note that the first term on the right has $\frac{dy}{dx}$ as a factor and the second has y as a factor, and that the same is true of the two terms on the left side of equation (1). This fact suggests seeking a function φ which has the property that, if both sides of (1) are multiplied by $\varphi(x)$, then the left side of the resulting equation is the derivative of the product $\varphi(x)y$. If both sides of (1) are multiplied by an arbitrary $\varphi(x)$, the result is

$$\varphi(x)\frac{dy}{dx} + \varphi(x)P(x)y = \varphi(x)Q(x). \tag{3}$$

Comparison of this equation with (2) shows that its left side is equal to $\frac{d}{dx}\big(\varphi(x)y\big)$ provided $\varphi(x)P(x)y = \varphi'(x)y$, which will in turn be true provided

$$\varphi(x)P(x) = \varphi'(x). \tag{4}$$

However, it is easy to find a function φ which satisfies (4), since, as a differential equation with φ the unknown function, it is separable. Solving it, we obtain

$$\frac{\varphi'(x)}{\varphi(x)} = P(x).$$

Whence

$$\int \frac{\varphi'(x)}{\varphi(x)}\,dx = \int P(x)\,dx,$$

which implies

$$\ln|\varphi(x)| = \int P(x)\,dx + c,$$

and so

$$|\varphi(x)| = e^{\int P(x)\,dx + c}.$$

Since we are only seeking *a* solution to (4), and not the most general form of the solution, we may assume that $\varphi(x)$ is positive and also ignore the constant of integration. We conclude that *if*

$$\boxed{\varphi(x) = e^{\int P(x)\,dx},} \tag{5}$$

then the left side of equation (3) *is equal to* $\dfrac{d}{dx}\big(\varphi(x)y\big)$.

With (5), equation (3) therefore becomes

$$\frac{d}{dx}\big(\varphi(x)y\big) = \varphi(x)Q(x).$$

Integration yields

$$\varphi(x)y = \int \varphi(x)Q(x)\,dx + c,$$

and so

$$y = \frac{1}{\varphi(x)}\left[\int \varphi(x)Q(x)\,dx + c\right],$$

for some real number c. Replacing $\varphi(x)$ by $e^{\int P(x)dx}$, we obtain the promised formula:

(2.1)

$$\boxed{y = e^{-\int P(x)\,dx}\left[\int e^{\int P(x)\,dx}Q(x)\,dx + c\right].}$$

Suppose next that c is an arbitrary constant and that the function y is defined by (2.1). Then

$$ye^{\int P(x)\,dx} = \int e^{\int P(x)\,dx}Q(x)\,dx + c.$$

The derivative of the right side of this equation is $e^{\int P(x)\,dx}Q(x)$ and that of the left side is

$$\frac{dy}{dx}e^{\int P(x)\,dx} + yP(x)e^{\int P(x)\,dx}.$$

Hence

$$e^{\int P(x)\,dx}\left[\frac{dy}{dx} + P(x)y\right] = e^{\int P(x)\,dx}Q(x),$$

which implies at once that $\frac{dy}{dx} + P(x)y = Q(x)$; i.e., y is a solution to (1). We conclude that formula (2.1) gives the general solution to the differential equation (1).

We strongly recommend that no one memorize (2.1). The important fact to remember is that, if the first-order linear differential equation $\frac{dy}{dx} + P(x)y = Q(x)$ is multiplied through by $\varphi(x) = e^{\int P(x)\,dx}$, then the left side of the resulting equation is equal to the derivative of the product $\varphi(x)y$. Consequently, the new equation can be integrated to give

$$\varphi(x)y = \int \varphi(x)Q(x)\,dx + c,$$

which can then be solved for y. This function $\varphi(x) = e^{\int P(x)\,dx}$, which enables us to change a seemingly nonintegrable sum into the derivative of a product by multiplication, is called an **integrating factor.**

EXAMPLE 1. Solve the differential equation

$$x^2\frac{dy}{dx} - 3xy - 2x^2 = 4x^4.$$

To put this in the form of (1), we add $2x^2$ to both sides and then divide by x^2. The result is

$$\frac{dy}{dx} - \frac{3}{x}y = 4x^2 + 2, \tag{6}$$

where $P(x) = -\frac{3}{x}$ and $Q(x) = 4x^2 + 2$. An antiderivative of $P(x)$ is given by

$$\int P(x)\,dx = \int -\frac{3}{x}\,dx = -3\ln|x| = \ln|x^{-3}|,$$

and it follows that the function

$$\varphi(x) = e^{\int P(x)\,dx} = e^{\ln|x^{-3}|} = |x^{-3}|$$

is an integrating factor. Equation (4) shows that if $\varphi(x)$ is an integrating factor, then so also is $-\varphi(x)$. Hence we may drop the absolute values and write simply

$$\varphi(x) = x^{-3}.$$

Multiplying both sides of (6) by x^{-3}, we obtain

$$x^{-3}\frac{dy}{dx} - 3x^{-4}y = 4x^{-1} + 2x^{-3}.$$

It is easy to see that the left side of this equation is equal to $\dfrac{d}{dx}(x^{-3}y)$. Hence

$$\frac{d}{dx}(x^{-3}y) = 4x^{-1} + 2x^{-3},$$

and so

$$\begin{aligned} x^{-3}y &= \int (4x^{-1} + 2x^{-3})\,dx + c \\ &= 4\ln|x| + 2\frac{x^{-2}}{-2} + c \\ &= 4\ln|x| - \frac{1}{x^2} + c, \end{aligned}$$

where c is an arbitrary constant. It follows that

$$y = 4x^3\ln|x| - x + cx^3$$

is the general solution.

EXAMPLE 2. Find the general solution of the differential equation

$$\frac{dy}{dx} + 3y = 2\sin x.$$

Note that this is a first-order linear differential equation with constant coefficients, but that it is not homogeneous, because the right side is not the zero function. In this example we have $P(x) = 3$ and $Q(x) = 2\sin x$. Hence

$$\int P(x)\,dx = \int 3\,dx = 3x,$$

and an integrating factor is

$$\varphi(x) = e^{\int P(x)\,dx} = e^{3x}.$$

It follows that

$$\frac{d}{dx}(e^{3x}y) = 2e^{3x}\sin x,$$

and so

$$e^{3x}y = \int 2e^{3x}\sin x\,dx + c. \tag{7}$$

To evaluate $\int 2e^{3x}\sin x\,dx = 2\int e^{3x}\sin x\,dx$, we use integration by parts twice:

$$\begin{aligned}\int e^{3x}\sin x\,dx &= -\int e^{3x}\,d\cos x\\ &= -e^{3x}\cos x + \int \cos x\,de^{3x}\\ &= -e^{3x}\cos x + 3\int e^{3x}\cos x\,dx.\end{aligned}$$

$$\begin{aligned}\int e^{3x}\cos x\,dx &= \int e^{3x}\,d\sin x\\ &= e^{3x}\sin x - \int \sin x\,de^{3x}\\ &= e^{3x}\sin x - 3\int e^{3x}\sin x\,dx.\end{aligned}$$

Combining these results, we get

$$\int e^{3x}\sin x\,dx = -e^{3x}\cos x + 3e^{3x}\sin x - 9\int e^{3x}\sin x\,dx,$$

whence

$$10\int e^{3x}\sin x\,dx = e^{3x}(3\sin x - \cos x),$$

and so

$$2\int e^{3x}\sin x\,dx = \frac{e^{3x}}{5}(3\sin x - \cos x).$$

Returning to (7), we have

$$e^{3x}y = \frac{e^{3x}}{5}(3\sin x - \cos x) + c,$$

and consequently the general solution of the differential equation is given by

$$y = \tfrac{1}{5}(3\sin x - \cos x) + ce^{-3x}, \tag{8}$$

where c is an arbitrary constant.

Note that the above solution (8) of the differential equation of Example 2 is the sum of two terms. The second, which is ce^{-3x}, is the general solution of the homogeneous differential equation $\frac{dy}{dx} + 3y = 0$. The first term, $\frac{1}{5}(3 \sin x - \cos x)$, is one particular solution of the nonhomogeneous differential equation $\frac{dy}{dx} + 3y = 2 \sin x$. As we shall see, this situation is typical of the solutions of linear differential equations.

PROBLEMS

1. Find the general solution of each of the following differential equations.

(a) $\frac{dy}{dx} - \frac{2}{x}y = 3x^2 + 4$

(b) $x\frac{dy}{dx} + 3y + x = 0$

(c) $7y + 2x\frac{dy}{dx} = x^7 + 2$

(d) $\frac{dy}{dx} + 2xy = 5x$

(e) $\frac{dy}{dx} - 8y = e^{2x} + 4$

(f) $6x^2y + \frac{dy}{dx} = x^2$

(g) $y \cos x + \frac{dy}{dx} = \cos x$

(h) $\frac{dy}{dx} + (2x + 3)y = 8x + 12$

(i) $\frac{dy}{dx} + 2y = 3 \cos x$

(j) $\frac{dy}{dx} + \frac{y}{x} = 2e^{-x}$

(k) $11y + x\frac{dy}{dx} = ax^2 + bx + c$

(l) $(D + 9)y = \pi$

(m) $\frac{dy}{dx} + \frac{3}{x}y = \frac{e^{2x}}{x^3}$

(n) $x^2\frac{dy}{dx} + 5xy = \frac{\cos x}{x^3}$.

2. (a) Find the general solution y_h of the homogeneous differential equation $\frac{dy}{dx} + 2xy = 0$.

(b) Show that the general solution of the nonhomogeneous equation $\frac{dy}{dx} + 2xy = 3xe^{-x^2}$ is equal to the solution y_h in part (a) plus a particular solution to the nonhomogeneous equation.

3. This problem is the general version of the preceding one. Let P and Q be continuous functions of x.

(a) Find the general solution y_h of the homogeneous differential equation $\frac{dy}{dx} + Py = 0$.

(b) Show that the general solution of the nonhomogeneous equation $\frac{dy}{dx} + Py = Q$ is equal to the solution y_h in part (a) plus a particular solution to the nonhomogeneous equation.

3. Linear Differential Operators. This section is divided into three parts. In the first, we shall systematically develop and extend the differential operators $D^2 + aD + b$ which were introduced in Section 1. In the second part we shall use these operators to obtain directly the general solutions of certain linear differential equations with constant coefficients. Finally, we shall show how these methods can be used to solve any linear differential equation with constant coefficients (whether homogeneous or not) provided we extend our range of functions to include those whose values may be complex numbers.

By a **linear operator** we shall mean any function L whose domain and range are sets of numerical-valued functions and which satisfies the equations

$$L(y_1 + y_2) = L(y_1) + L(y_2), \tag{1}$$

$$L(ky) = kL(y), \tag{2}$$

for every real number k and every y, y_1, and y_2 in the domain of L. [The function $L(y)$ is frequently written simply Ly.] An important example is the function D, which, to every differentiable function y, assigns its derivative $Dy = \dfrac{dy}{dx}$. Another example is the operation of multiplication by a real number. That is, for any real number a, the function L defined by

$$Ly = ay$$

obviously satisfies (1) and (2) and hence is a linear operator.

If L_1 and L_2 are linear operators, then their **sum** is the function $L_1 + L_2$ defined by

$$(L_1 + L_2)y = L_1y + L_2y, \tag{3}$$

for every y which is in the domains of both L_1 and L_2. It is easy to show that

(3.1) *If L_1 and L_2 are linear operators, then the sum $L_1 + L_2$ is also a linear operator.*

Proof. We shall show that $L_1 + L_2$ satisfies equation (1) by using successively the definition of $L_1 + L_2$, the linearity of L_1 and L_2 separately, the commutative law of addition for functions, and finally the definition again. Thus

$$\begin{aligned}(L_1 + L_2)(y_1 + y_2) &= L_1(y_1 + y_2) + L_2(y_1 + y_2)\\ &= L_1y_1 + L_1y_2 + L_2y_1 + L_2y_2\\ &= (L_1y_1 + L_2y_1) + (L_1y_2 + L_2y_2)\\ &= (L_1 + L_2)y_1 + (L_1 + L_2)y_2.\end{aligned}$$

The proof that $L_1 + L_2$ satisfies (2) is similar:

$$\begin{aligned}(L_1 + L_2)(ky) &= L_1(ky) + L_2(ky)\\ &= kL_1y + kL_2y\\ &= k(L_1y + L_2y)\\ &= k(L_1 + L_2)y,\end{aligned}$$

and this completes the proof.

If L_1 and L_2 are linear operators, then the **composition** of L_2 followed by L_1 is the function denoted by L_1L_2 and defined by

$$(L_1L_2)y = L_1(L_2y), \tag{4}$$

for every y for which the right side is defined. The proof of the following proposition is entirely analogous to that of (3.1) and is left to the reader as an exercise.

(3.2) *If L_1 and L_2 are linear operators, then the composition L_1L_2 is also a linear operator.*

The composition L_1L_2 is also called the **product** of L_1 and L_2. There is no reason to suppose from the definition that the commutative law of multiplication holds, and, for linear operators in general, $L_1L_2 \neq L_2L_1$. However, the distributive laws hold:

(3.3)
$$\begin{cases} L_1(L_2 + L_3) = L_1L_2 + L_1L_3,\\ (L_1 + L_2)L_3 = L_1L_3 + L_2L_3. \end{cases}$$

Proof. The first of these is proved as follows:

$$\begin{aligned}(L_1(L_2 + L_3))y &= L_1((L_2 + L_3)y)\\ &= L_1(L_2y + L_3y)\\ &= L_1(L_2y) + L_1(L_3y)\\ &= (L_1L_2)y + (L_1L_3)y\\ &= (L_1L_2 + L_1L_3)y.\end{aligned}$$

The proof of the second is similar and is left as an exercise.

An important example of the product of linear operators is the composition of a linear operator L followed by the operation of multiplication by a real number a. This product, denoted aL, assigns to every y in the domain of L the value $(aL)y$ which is equal to the product of a with the function Ly. That is,

(3.4)
$$(aL)y = a(Ly).$$

The composition in the other order is the product La. Here we have $(La)y = L(ay)$, and the latter quantity, by the linearity of L, is equal to $a(Ly)$. Combining this with (3.4), we obtain the equation $(La)y = (aL)y$. Thus the operators La and aL are equal, and we have proved the following special case of the commutative law:

(3.5) $$aL = La.$$

Another example of the product, already encountered, is the operator D^2, which is the composition $D^2 = DD$ of D with itself. More generally, for every integer $n > 1$, we define the operator D^n inductively by

$$D^n = DD^{n-1}.$$

The domain of D^n is the set of all n-times differentiable functions, and, for each such function y, we have

$$D^n y = \frac{d^n y}{dx^n}\cdot$$

By repeated applications of (3.1) and (3.2), we may conclude that any function formed in a finite number of steps by taking sums and products of linear operators is itself a linear operator. As an example, consider a polynomial $p(t)$ of degree n; i.e.,

$$p(t) = a_n t^n + a_{n-1}t^{n-1} + \cdots + a_1 t + a_0,$$

where $a_0, \ldots, a_n$ are real numbers and $a_n \neq 0$. Then the function

$$p(D) = a_n D^n + a_{n-1}D^{n-1} + \cdots + a_1 D + a_0$$

is a linear operator. To every n-times differentiable function y, it assigns as value the function

$$\begin{aligned} p(D)y &= a_n D^n y + a_{n-1}D^{n-1}y + \cdots + a_1 Dy + a_0 y \\ &= a_n \frac{d^n y}{dx^n} + a_{n-1}\frac{d^{n-1}y}{dx^{n-1}} + \cdots + a_1 \frac{dy}{dx} + a_0 y. \end{aligned}$$

We call $p(D)$ a **linear differential operator of order *n*.** It is the natural generalization of the differential operators of order 2, of the form $D^2 + aD + b$, which were discussed in Section 1. [Linear differential operators of types more general than $p(D)$ certainly exist; e.g., see Problem 9. They are of importance in more advanced treatments of differential equations, but we shall not study them here.]

The polynomial differential operators $p(D)$ can be added and multiplied just like ordinary polynomials. In particular, the following theorem follows from the distributive laws (3.3) and the commutative law (3.5):

(3.6) *If $p(t)$ and $q(t)$ are polynomials and if $p(t)q(t) = r(t)$, then*

$$p(D)q(D) = r(D).$$

As an illustration, observe how (3.3) and (3.5) are used to prove the special case of this theorem in which $p(t) = at + b$ and $q(t) = ct + d$. First of all, we have

$$\begin{aligned} r(t) = p(t)q(t) &= (at + b)(ct + d) \\ &= act^2 + bct + adt + bd. \end{aligned}$$

Then

$$\begin{aligned} p(D)q(D) &= (aD + b)(cD + d) \\ &= (aD + b)cD + (aD + b)d \\ &= aDcD + bcD + aDd + bd \\ &= acD^2 + bcD + adD + bd \\ &= r(D). \end{aligned}$$

The proof is the same in principle for arbitrary polynomials $p(t)$ and $q(t)$.

It is a corollary of (3.6) that polynomial differential operators satisfy the commutative law of multiplication. Thus

(3.7) $$p(D)q(D) = q(D)p(D).$$

For, since $p(t)q(t) = q(t)p(t) = r(t)$, both sides of (3.7) are equal to $r(D)$.

We begin the second part of the section by considering the differential equation

$$\frac{d^2y}{dx^2} - 2\frac{dy}{dx} - 3y = e^{-x},$$

which, with the notation of differential operators, can be written

$$(D^2 - 2D - 3)y = e^{-x}. \qquad (5)$$

We have thus far defined the characteristic equation only for homogeneous, second-order, linear differential equations with constant coefficients. The generalization to nonhomogeneous and higher-order equations is: For any polynomial $p(t)$ and function $F(x)$, the **characteristic equation** of the differential equation

$$p(D)y = F(x)$$

is the equation $p(t) = 0$, and the polynomial $p(t)$ is its **characteristic polynomial.**

Returning to (5), we see that the characteristic polynomial, which is $t^2 - 2t - 3$, factors into the product $(t - 3)(t + 1)$. It follows from (3.6) that $D^2 - 2D - 3 = (D - 3)(D + 1)$, and (5) can therefore be written

$$(D - 3)(D + 1)y = e^{-x}.$$

Let us define the function u by setting $(D + 1)y = u$. Then (5) becomes equivalent to the pair of first-order linear equations

$$\begin{cases} (D - 3)u = e^{-x}, & (6) \\ (D + 1)y = u. & (7) \end{cases}$$

To solve (6), we use the technique developed in Section 2. For this equation, $P(x) = -3$ and $Q(x) = e^{-x}$. Hence an integrating factor is $e^{\int P(x)\,dx} = e^{-3x}$, and therefore

$$\frac{d}{dx}(e^{-3x}u) = e^{-3x}e^{-x} = e^{-4x}.$$

Integrating, we obtain

$$e^{-3x}u = \int e^{-4x}dx + c_1 = -\tfrac{1}{4}e^{-4x} + c_1,$$

whence

$$u = e^{3x}(-\tfrac{1}{4}e^{-4x} + c_1) = -\tfrac{1}{4}e^{-x} + c_1e^{3x}.$$

We now substitute this value for u in equation (7) to obtain the first-order linear equation

$$(D + 1)y = -\tfrac{1}{4}e^{-x} + c_1e^{3x}.$$

Here, $P(x) = 1$ and the integrating factor is e^x. Accordingly, we have

$$\begin{aligned} \frac{d}{dx}(e^x y) &= e^x(-\tfrac{1}{4}e^{-x} + c_1e^{3x}) \\ &= -\tfrac{1}{4} + c_1e^{4x}. \end{aligned}$$

Integration yields

$$e^x y = -\tfrac{1}{4}x + \frac{c_1}{4}e^{4x} + c_2.$$

Replacing $\frac{c_1}{4}$ by c_1, and multiplying both sides by e^{-x}, we get finally

$$y = -\tfrac{1}{4}xe^{-x} + c_1e^{3x} + c_2e^{-x}.$$

This, where c_1 and c_2 are arbitrary real constants, is the general solution to the differential equation

$$\frac{d^2y}{dx^2} - 2\frac{dy}{dx} - 3y = e^{-x}.$$

This example illustrates the fact that we can in principle solve any second-order, linear differential equation with constant coefficients provided the characteristic polynomial is the product of linear factors. Thus, if we are given

$$(D^2 + aD + b)y = F(x),$$

and if $t^2 + at + b = (t - r_1)(t - r_2)$, then the differential equation can be written

$$(D - r_1)(D - r_2)y = F(x).$$

If u is defined by setting $(D - r_2)y = u$, then the original second-order equation is equivalent to the two first-order linear differential equations

$$\begin{cases}(D - r_1)u = F(x),\\ (D - r_2)y = u,\end{cases}$$

and these can be solved successively to find first u and then y.

The same technique can be applied to higher-order equations. Consider an arbitrary polynomial

$$p(t) = t^n + a_{n-1}t^{n-1} + \cdots + a_1 t + a_0,$$

where $n > 1$ and $a_0, \ldots, a_{n-1}$ are real constants. In addition, we assume that $p(t)$ is the product of linear factors; i.e.,

$$p(t) = (t - r_1)(t - r_2) \cdots (t - r_n).$$

Let $F(x)$ be given and consider the differential equation

$$p(D)y = F(x), \tag{8}$$

which is the same as

$$\frac{d^n y}{dx^n} + a_{n-1}\frac{d^{n-1}y}{dx^{n-1}} + \cdots + a_1\frac{dy}{dx} + a_0 y = F(x).$$

Since the factorization of $p(t)$ is assumed, the differential equation can also be written

$$(D - r_1)(D - r_2) \cdots (D - r_n)y = F(x).$$

The functions $u_1, \ldots, u_{n-1}$ are defined by

$$\begin{aligned} u_1 &= (D - r_2) \cdots (D - r_n)y,\\ u_2 &= (D - r_3) \cdots (D - r_n)y,\\ &\vdots\\ u_{n-1} &= (D - r_n)y.\end{aligned}$$

Then (8) is equivalent to the following set of first-order linear differential equations

$$\begin{cases}(D - r_1)u_1 = F(x),\\ (D - r_2)u_2 = u_1,\\ \quad\vdots\\ (D - r_n)y = u_{n-1},\end{cases}$$

which can be solved successively to finally obtain y.

In Section 4 of Chapter 7 use was made of the fact that any polynomial with real coefficients and degree at least 1 can be written as the product of linear and irreducible quadratic factors (see page 386). Suppose $ct^2 + dt + e$ is irreducible. This is equivalent to the assertion that the discriminant $d^2 - 4ce$ is negative. According to the quadratic formula, the two roots of the equation $ct^2 + dt + e = 0$ are equal to $r_1 = \alpha + i\beta$ and $r_2 = \alpha - i\beta$, where $\alpha = -\dfrac{d}{2c}$ and $\beta = \dfrac{\sqrt{4ce - d^2}}{2c}$. By multiplying and substituting these values, one can then easily verify the equation

$$c(t - r_1)(t - r_2) = ct^2 + dt + e.$$

Thus any irreducible quadratic polynomial with real coefficients is the product of two linear factors with complex coefficients. It follows that, *for any polynomial*

$$p(t) = a_n t^n + a_{n-1}t^{n-1} + \cdots + a_1 t + a_0,$$

with real coefficients a_i *and with* $n \geq 1$ *and* $a_n \neq 0$, *we have*

$$p(t) = a_n(t - r_1)(t - r_2)\cdots(t - r_n),$$

where roots which are complex occur in conjugate pairs.

It is this fact which introduces the third part of this section. It is very natural to ask the following: If the class of possible solutions is enlarged to include complex-valued functions of a real variable, can we proceed to solve linear differential equations with constant coefficients just as before, but with the added knowledge that now the characteristic polynomial can always be factored into linear factors? The answer is yes!

To justify this answer, we must of course know the definition of the derivative. Let f be a function whose domain Q is a subset of the real numbers and whose range is a subset of the complex numbers. Then two real-valued functions f_1 and f_2 with domain Q are defined by

$$f_1(x) = \text{real part of } f(x),$$

$$f_2(x) = \text{imaginary part of } f(x).$$

That is, we have $f(x) = f_1(x) + if_2(x)$, for every x in Q. The **derivative** f' is defined simply by the equation

$$f'(x) = f_1'(x) + if_2'(x),$$

for every x for which both $f_1'(x)$ and $f_2'(x)$ exist. Alternatively, if we write $y = f(x)$, $u = f_1(x)$, and $v = f_2(x)$, then $y = u + iv$, and we also use the notations

$$\begin{aligned} f'(x) &= \frac{dy}{dx} = \frac{du}{dx} + i\frac{dv}{dx} \\ &= Dy = Du + i\,Dv. \end{aligned}$$

Logically, we must now go back and check that all the formal rules for differentiation and finding antiderivatives are still true for complex-valued functions, and the same applies to several theorems (see, for example, Problems 10 and 11). Much of this work is purely routine, and, to avoid an interruption of our study of differential equations, we shall omit it.

It now follows, by factoring the operator $p(D)$ into linear factors, that any linear differential equation

$$p(D)y = F(x)$$

with constant coefficients can be solved. That is, it can first be replaced by an equivalent set of first-order linear differential equations. For each of these an explicit integrating factor $e^{\int P(x)\,dx}$ exists, and by solving them successively, we can eventually obtain the general solution y.

EXAMPLE 1. Solve the differential equation $(D^2 + 1)y = 2x$. Since $t^2 + 1 = (t + i)(t - i)$, we have

$$(D + i)(D - i)y = 2x.$$

Let $(D - i)y = u$, and consider the first-order equation

$$(D + i)u = 2x.$$

Since $P(x) = i$, an integrating factor is e^{ix}, and we obtain

$$\frac{d}{dx}(e^{ix}u) = e^{ix}2x,$$

from which it follows by integrating that

$$e^{ix}u = 2\int xe^{ix}dx + c_1.$$

By integration by parts it can be verified that

$$\int xe^{ax}dx = \frac{xe^{ax}}{a} - \frac{e^{ax}}{a^2}. \tag{9}$$

In this case, $a = i$ and we know that $\frac{1}{i} = -i$ and that $i^2 = -1$. Hence

$$e^{ix}u = -2ixe^{ix} + 2e^{ix} + c_1,$$

and so

$$u = -2ix + 2 + c_1e^{-ix}.$$

It therefore remains to solve the differential equation

$$(D - i)y = -2ix + 2 + c_1e^{-ix}.$$

This time, an integrating factor is e^{-ix}. Hence

$$\frac{d}{dx}(e^{-ix}y) = -2ixe^{-ix} + 2e^{-ix} + c_1e^{-2ix}.$$

Integration [with a second application of (9)] yields

$$e^{-ix}y = 2xe^{-ix} - \frac{c_1}{2i}e^{-2ix} + c_2.$$

Replacing the constant $-\frac{c_1}{2i}$ by simply c_1, and multiplying both sides by e^{ix}, we obtain

$$y = 2x + c_1e^{-ix} + c_2e^{ix}.$$

If the function y is real-valued, then it is easy to prove that c_1 and c_2 are complex conjugates [see (4.3), page 644]. In this case $c_1e^{-ix} + c_2e^{ix}$ may be replaced by $c_1 \cos x + c_2 \sin x$, where now the constants c_1 and c_2 denote arbitrary real numbers. We conclude that

$$y = 2x + c_1 \cos x + c_2 \sin x$$

is the general real-valued solution to the original differential equation

$$\frac{d^2y}{dx^2} + y = 2x.$$

The computations in this section were long and involved. The important fact we have shown is that the equations can be solved by an iteration of routine steps. As a practical matter, however, it is clear that some general computationally simple techniques are badly needed. These will be developed in the next two sections by breaking the problem into a homogeneous part and a nonhomogeneous part and attacking each one separately.

PROBLEMS

1. Find the general solution of each of the following differential equations.

(a) $\frac{d^2y}{dx^2} + \frac{dy}{dx} - 2y = 5e^{-x}$

(b) $(D + 2)(D - 1)y = 6e^{-2x}$

(c) $(D^2 - 3D + 2)y = 4x + 3$

(d) $\frac{d^2y}{dx^2} + y = e^x$

(e) $(D^2 + 1)y = x^2 + 1.$

2. Using equations (1) and (2), prove that, if L is a linear operator, then
$$L(y_1 - y_2) = L(y_1) - L(y_2).$$
3. Show that equations (1) and (2) can be replaced by a single equation. That is, prove that a function L is a linear operator if and only if
$$L(ay_1 + by_2) = aLy_1 + bLy_2.$$
4. Prove (3.2); i.e., if L_1 and L_2 are linear operators, then the composition L_1L_2 is also a linear operator.
5. Prove the second equation in (3.3), i.e., the distributive law $(L_1 + L_2)L_3 = L_1L_3 + L_2L_3$.
6. It might at first seem more natural to define the product of two linear operators L_1 and L_2 by the equation
$$(L_1L_2)y = (L_1y)(L_2y).$$
(This *is* the way the product of two real-valued functions is defined.) Using this definition, show that, if D is the derivative, then D^2 is not a linear operator.
7. Let $f(x)$ be a given function and L a linear operator. Define $f(x)L$ by the equation
$$(f(x)L)y = f(x)(Ly).$$
Show that $f(x)L$ satisfies equations (1) and (2) and hence is a linear operator.
8. (a) Show that the operation of multiplication by a given function $f(x)$ is a linear operator. That is, prove that, if M is defined by
$$My = f(x)y,$$
then M is a linear operator.

   (b) Show that the composition of a linear operator L followed by the operation of multiplication by $f(x)$ is just the operator $f(x)L$ defined in Problem 7.
9. (See Problems 7 and 8.) If $f(x)$ is a differentiable function and if D is the derivative, then both linear operators $f(x)D$ and $Df(x)$ are examples of **linear differential operators** more general than the type discussed in the text. Show that
$$xD \neq Dx,$$
by applying both sides to the function $y = x$. Thus the commutative law of multiplication fails.
10. Let f and g be differentiable complex-valued functions of a real variable. Show that the ordinary product rule for differentiation is still valid; i.e., prove that
$$\frac{d}{dx}\big(f(x)g(x)\big) = \left(\frac{d}{dx}f(x)\right)g(x) + f(x)\left(\frac{d}{dx}g(x)\right).$$
[*Hint:* Let $f(x) = f_1(x) + if_2(x)$ and $g(x) = g_1(x) + ig_2(x)$, and apply the definitions of the derivative and of multiplication of complex numbers.]
11. (a) Let f be a complex-valued function of a real variable which is differentiable at every point x of an interval I. Show that if $f'(x) = 0$, for every x in I, then $f(x)$ is a constant on I.

(b) Let f and g be two complex-valued functions of a real variable with $f'(x) = g'(x)$ at every x of some interval I. Show that there exists a complex number c such that $f(x) = g(x) + c$, for every x in I.

12. Find the general solution of each of the following differential equations.

(a) $(D - 1)^2(D + 2)y = 0$ (b) $\dfrac{d^3y}{dx^3} - \dfrac{d^2y}{dx^2} - 4\dfrac{dy}{dx} + 4y = 0.$

4. Homogeneous Differential Equations. For a given function $F(x)$ and a given polynomial

$$p(t) = t^n + a_{n-1}t^{n-1} + \cdots + a_1t + a_0,$$

let us consider the differential equation

$$p(D)y = F(x). \tag{1}$$

The simplification of the theory gained by enlarging the set of possible solutions to include complex-valued functions of a real variable was demonstrated in Section 3, and we shall continue to use this technique. Nevertheless, our primary concern is still that of finding real-valued solutions to real differential equations. For this reason, we shall assume throughout that the coefficients $a_0, \ldots, a_{n-1}$ of the polynomial $p(t)$ are real numbers and that $F(x)$ is a real-valued function. Associated with the differential equation (1) is the homogeneous differential equation

$$p(D)y = 0, \tag{2}$$

called the **associated homogeneous equation** of $p(D)y = F(x)$. A theorem of basic importance is the following:

(4.1) THEOREM. *If y_0 is any particular solution of* (1) *and if y is the general solution of* (2), *then $y + y_0$ is the general solution of* (1).

Proof. Once the statement of this theorem is understood, its proof becomes almost a triviality. First, one should realize that, strictly speaking, the general solution of a differential equation is the set of all its solutions. Referring to a function y as the general solution is actually a common and very convenient misuse of language. What it really means is that y depends not only on x, but also on one or more other variables which are arbitrary constants of integration and can take on any real, or complex, values. That is, we have a function $\varphi(x, u_1, \ldots, u_n)$, and, for every set of real (or complex) numbers $c_1, \ldots, c_n$, the function y defined by

$$y = \varphi(x, c_1, \ldots, c_n)$$

is a solution of the differential equation. Conversely, corresponding to every solution $f(x)$, there exist numbers $c_1, \ldots, c_n$ such that $f(x) = \varphi(x, c_1, \ldots, c_n)$.

Thus y, as expressed in the above equation, does exhibit the set of all solutions.

With this understanding, it follows that (4.1) is equivalent to the following proposition. *Let y_0 be an arbitrary solution of equation* (1). *Then:*

(i) *If y_1 is any solution of* (1), *then there exists a solution y_2 of* (2) *such that $y_1 = y_2 + y_0$.*

(ii) *If y_2 is any solution of* (2), *then $y_2 + y_0$ is a solution of* (1).

The proofs use only the fact that $p(D)$ is a linear operator. To prove (i), we set $y_2 = y_1 - y_0$ and check that y_2 is a solution of (2). We get

$$\begin{aligned} p(D)y_2 = p(D)(y_1 - y_0) &= p(D)y_1 - p(D)y_0 \\ &= F(x) - F(x) = 0. \end{aligned}$$

For (ii), we need only verify that $y_2 + y_0$ is a solution of (1). We have

$$\begin{aligned} p(D)(y_2 + y_0) &= p(D)y_2 + p(D)y_0 \\ &= 0 + F(x) = F(x), \end{aligned}$$

and the proof of (4.1) is complete.

As a result of this theorem, our approach to the problem of solving the differential equation $p(D)y = F(x)$ will be divided into two parts. We shall first concentrate on finding the general solution of the associated homogeneous equation $p(D)y = 0$, and then consider methods of finding a particular solution to the original nonhomogeneous equation. The remainder of this section will be devoted to the first part.

We begin with the second-order linear homogeneous differential equation with constant coefficients:

$$(D^2 + aD + b)y = 0. \tag{3}$$

The general solution of this equation has been presented earlier (see page 617), but without proof. We shall supply the proof now by factoring the linear operator $D^2 + aD + b$ and solving the equation by the iterative technique of Section 3. The characteristic polynomial can be written as the product

$$t^2 + at + b = (t - r_1)(t - r_2),$$

where the roots r_1 and r_2 are either both real or distinct conjugate complex numbers. Equation (3) can therefore be written

$$(D - r_1)(D - r_2)y = 0. \tag{4}$$

(4.2) THEOREM. *The general solution of the differential equation* (3) [or equivalently, of (4)] is:

$$\text{(i)} \quad y = c_1e^{r_1x} + c_2e^{r_2x}, \textit{ if } r_1 \neq r_2, \textit{ or}$$

$$\text{(ii)} \quad y = (c_1x + c_2)e^{rx}, \textit{ if } r_1 = r_2 = r,$$

where c_1 and c_2 are arbitrary complex numbers.

Note that these solutions include all the real-valued ones, since the set of all real numbers is a subset of the set of all complex numbers.

Proof. Let y be an arbitrary solution of (4). We define the function u by setting $u = (D - r_2)y$. Then (4) is equivalent to the two first-order linear equations:

$$\begin{cases}(D - r_1)u = 0, \\ (D - r_2)y = u.\end{cases}$$

An integrating factor for the first of these is $e^{-r_1 x}$, because, in the notation of first-order linear equations, we have $P(x) = -r_1$. It follows that

$$\frac{d}{dx}(e^{-r_1 x}u) = 0.$$

Integration yields $e^{-r_1 x}u = c_1$, whence

$$u = c_1 e^{r_1 x}, \qquad \text{for some complex number } c_1.$$

Substituting the expression for u into the second differential equation above, we obtain

$$(D - r_2)y = c_1 e^{r_1 x}.$$

This time an integrating factor is $e^{-r_2 x}$, and so

$$\frac{d}{dx}(e^{-r_2 x}y) = c_1 e^{(r_1 - r_2)x}. \tag{5}$$

We now distinguish two cases.

Case 1. $r_1 \neq r_2$. Integration of (5) yields

$$e^{-r_2 x}y = \frac{1}{r_1 - r_2} c_1 e^{(r_1 - r_2)x} + c_2,$$

for some complex number c_2. Multiplying both sides by $e^{r_2 x}$ and replacing $\frac{c_1}{r_1 - r_2}$ by simply c_1, we get

$$y = c_1 e^{r_1 x} + c_2 e^{r_2 x}.$$

Case 2. $r_1 = r_2 = r$. Then $e^{(r_1 - r_2)x} = e^0 = 1$, and (5) reduces to

$$\frac{d}{dx}(e^{-rx}y) = c_1.$$

Integrating, we obtain $e^{-rx}y = c_1 x + c_2$, for some complex number c_2, and it follows that

$$y = (c_1 x + c_2)e^{rx}.$$

We have now proved that, if y is an arbitrary solution of the original differential equation (4), then there exist complex numbers c_1 and c_2 (either or both of which may perfectly well be real—every real number is a special case of a complex number) such that y is of form (i) if $r_1 \neq r_2$ and of form (ii) if $r_1 = r_2 = r$. Conversely, it is a simple matter to check by substitution that, for any complex numbers c_1 and c_2, the function $c_1e^{r_1x} + c_2e^{r_2x}$ is a solution if $r_1 \neq r_2$ and the function $(c_1x + c_2)e^{rx}$ is a solution if $r_1 = r_2 = r$. This completes the proof of the theorem.

How can we use Theorem (4.2) to obtain the general real-valued solution of the differential equation $(D^2 + aD + b)y = (D - r_1)(D - r_2)y = 0$? Suppose, to begin with, that r_1 and r_2 are both real and that $r_1 \neq r_2$. It follows from part (i) of Theorem (4.2) that the function defined by

$$y = c_1e^{r_1x} + c_2e^{r_2x}, \qquad \text{for any two real numbers } c_1 \text{ and } c_2, \tag{6}$$

is a solution, and it is certainly real-valued. There is only one obstacle in the way of the conclusion that (6) is the general real-valued solution. This is the *a priori* possibility that there might exist complex numbers c_1 and c_2, which are not both real, but are such that $c_1e^{r_1x} + c_2e^{r_1x}$ is a real-valued function. This, in fact, cannot happen, as the following argument shows: Let $c_1 = \gamma_1 + i\delta_1$ and $c_2 = \gamma_2 + i\delta_2$. Since

$$(\gamma_1 + i\delta_1)e^{r_1x} + (\gamma_2 + i\delta_2)e^{r_2x}$$

is by assumption real-valued, then so is

$$(\gamma_1 + i\delta_1)e^{r_1x} + (\gamma_2 + i\delta_2)e^{r_2x} - \gamma_1e^{r_1x} - \gamma_2e^{r_2x} = i(\delta_1e^{r_1x} + \delta_2e^{r_2x}).$$

Hence

$$\delta_1e^{r_1x} + \delta_2e^{r_2x} = 0,$$

and so

$$\delta_1e^{(r_1-r_2)x} = -\delta_2.$$

This equation holds for all real values of x. But, since $r_1 - r_2 \neq 0$, the left side has constant value only if $\delta_1 = 0$, which in turn immediately implies that $\delta_2 = 0$. Hence $\delta_1 = \delta_2 = 0$, and the argument is complete. With this problem disposed of, it now follows from (4.2)(i) that, if r_1 and r_2 are real and unequal, then the general real-valued solution of the differential equation is given by (6).

A similar situation arises if $r_1 = r_2 = r$. In this case r must be a real number, and it is a corollary of part (ii) of Theorem (4.2) that the function defined by

$$y = (c_1x + c_2)e^{rx}, \qquad \text{for any two real numbers } c_1 \text{ and } c_2, \tag{7}$$

is a solution, and, of course, it is real-valued. Again, we must show that it is not possible to have complex numbers c_1 and c_2, not both real, such that $(c_1x + c_2)e^{rx}$ is a real-valued function. The proof of this fact is similar to that of the analogous result in the preceding paragraph, and we leave it as an exercise. It then follows from (4.2)(ii) that the general real-valued solution is given by (7).

The third and final possibility is that the roots r_1 and r_2 of the characteristic polynomial are distinct conjugate complex numbers. In this case, we need the lemma:

(4.3) *If* $r_1 = \alpha + i\beta$, $r_2 = \alpha - i\beta$, *and* $\beta \neq 0$, *then the function defined by*

$$y = c_1 e^{r_1 x} + c_2 e^{r_2 x}, \qquad \textit{for arbitrary complex numbers } c_1 \textit{ and } c_2,$$

is real-valued if and only if c_1 *and* c_2 *are complex conjugates. Moreover, if* $c_1 = \gamma + i\delta$ *and* $c_2 = \gamma - i\delta$, *then*

$$y = e^{\alpha x}(2\gamma \cos \beta x - 2\delta \sin \beta x).$$

A proof in the "if" direction is given in detail in (8.3) on page 347. In addition, the above equation giving y in terms of α, β, γ, and δ is also derived there. The "only if" direction can be proved in the same direct manner as the analogous results for the other two cases: Let $c_1 = \gamma_1 + i\delta_1$ and $c_2 = \gamma_2 + i\delta_2$, substitute these values into $c_1 e^{r_1 x} + c_2 e^{r_2 x}$, and impose the condition that y is real-valued. It will then follow that $\gamma_1 = \gamma_2$ and that $\delta_1 = -\delta_2$. Again, we leave this task as an exercise.

Let us replace the real constants 2γ and -2δ which appear in the equation in the last line of (4.3) by c_1 and c_2, respectively. It is then a corollary of (4.3) and (4.2)(i) that the general real-valued solution of the differential equation $(D^2 + aD + b)y = (D - r_1)(D - r_2)y = 0$ is

$$y = e^{\alpha x}(c_1 \cos \beta x + c_2 \sin \beta x), \quad \text{for any two real numbers } c_1 \text{ and } c_2, \tag{8}$$

provided $r_1 = \alpha + i\beta$, $r_2 = \alpha - i\beta$, and $\beta \neq 0$.

This completes the proof that second-order, homogeneous, linear differential equations with real constant coefficients have the general solutions first described in Section 8 of Chapter 6 and again in Section 1 of this chapter.

The higher-order homogeneous equations can be solved in the same way. If

$$p(t) = t^n + a_{n-1}t^{n-1} + \cdots + a_1 t + a_0,$$

then the general solution of the differential equation $p(D)y = 0$ can be obtained by first factoring $p(D)$ to obtain an equivalent set of n first-order linear differential equations which are then solved successively to find y. As an illustration, we shall solve a third-order equation by this method. Following this example, we shall give (without proof) the form of the general real-valued solution for arbitrary order n.

EXAMPLE 1. Find the general solution of the differential equation

$$\frac{d^3y}{dx^3} - 3\frac{d^2y}{dx^2} + 4y = 0.$$

The characteristic polynomial is $p(t) = t^3 - 3t^2 + 4$. Substituting -1 for t, we obtain $p(-1) = 0$, from which it follows that $(t + 1)$ is a factor of $p(t)$. Dividing, we find that

$$t^3 - 3t^2 + 4 = (t + 1)(t^2 - 4t + 4) = (t + 1)(t - 2)^2.$$

Hence the differential equation can be written

$$(D + 1)(D - 2)^2y = 0.$$

We set $u_1 = (D - 2)^2y$ and $u_2 = (D - 2)y$ and, by so doing, obtain the equivalent set of three first-order equations

$$\begin{cases} (D + 1)u_1 = 0, \\ (D - 2)u_2 = u_1, \\ (D - 2)y = u_2. \end{cases}$$

The general solution of the first of these is $u_1 = c_1e^{-x}$, and the second equation is therefore

$$(D - 2)u_2 = c_1e^{-x}.$$

An integrating factor is e^{-2x}, and so

$$\frac{d}{dx}(e^{-2x}u_2) = e^{-2x}c_1e^{-x} = c_1e^{-3x}.$$

Hence

$$e^{-2x}u_2 = -\frac{c_1}{3}e^{-3x} + c_2,$$

from which it follows by multiplying both sides by e^{2x} and replacing $-\dfrac{c_1}{3}$ by simply c_1 that

$$u_2 = c_1e^{-x} + c_2e^{2x}.$$

The third equation is now seen to be

$$(D - 2)y = c_1e^{-x} + c_2e^{2x}.$$

Again, e^{-2x} is an integrating factor, and we have

$$\frac{d}{dx}(e^{-2x}y) = e^{-2x}(c_1e^{-x} + c_2e^{2x}) = c_1e^{-3x} + c_2.$$

Integration yields

$$e^{-2x}y = -\frac{c_1}{3}e^{-3x} + c_2x + c_3.$$

Multiplying both sides by e^{2x} and replacing $-\frac{c_1}{3}$ by simply c_1 again, we have

$$y = c_1e^{-x} + (c_2x + c_3)e^{2x},$$

where c_1, c_2, and c_3 are arbitrary real constants. This is the general real-valued solution and completes the example.

We now give the general solution for arbitrary order n. Let

$$p(t) = t^n + a_{n-1}t^{n-1} + \cdots + a_1t + a_0,$$

and suppose that factorization into real-valued irreducible factors yields the product

$$p(t) = (t - r_1)^{m_1} \cdots (t - r_k)^{m_k}(t^2 + c_1t + d_1)^{n_1} \cdots (t^2 + c_lt + d_l)^{n_l},$$

where $m_1, \ldots, m_k$ and $n_1, \ldots, n_l$ are positive integers, the factors $t - r_i$ are all distinct, and the factors $t^2 + c_jt + d_j$ are all distinct. For each factor $(t - r_i)^{m_i}$, define the function

$$f_i(x) = (C_{i1}x^{m_i-1} + C_{i2}x^{m_i-2} + \cdots + C_{im_i})e^{r_ix}, \quad \textit{for arbitrary real numbers } C_{i1}, \ldots, C_{im_i}. \tag{9}$$

For each factor $(t^2 + c_jt + d_j)^{n_j}$, let $\alpha_j + i\beta_j$ and $\alpha_j - i\beta_j$ be the roots of $t^2 + c_jt + d_j$, and define the function

$$\begin{aligned} g_j(x) = {} & (A_{j1}x^{n_j-1} + A_{j2}x^{n_j-2} + \cdots + A_{jn_j})e^{\alpha_jx}\cos \beta_jx \\ & +(B_{j1}x^{n_j-1} + B_{j2}x^{n_j-2} + \cdots + B_{jn_j})e^{\alpha_jx}\sin \beta_jx, \\ & \textit{for arbitrary real numbers } A_{j1}, \ldots, A_{jn_j} \\ & \textit{and } B_{j1}, \ldots, B_{jn_j}. \end{aligned} \tag{10}$$

Then it can be proved that

(4.4) *The general real-valued solution of the homogeneous differential equation* $p(D)y = 0$ *is the sum*

$$y = f_1(x) + \cdots + f_k(x) + g_1(x) + \cdots + g_l(x).$$

Note that, since $m_1 + \cdots + m_k + 2n_1 + \cdots + 2n_l = n$, the number of arbitrary constants in the general solution is equal to n, the order of the differential equation.

EXAMPLE 2. Find the general solution of the differential equation

$$(D + 2)(D - 5)^3(D^2 + D + 1)^2y = 0.$$

This is an equation of order 8. The polynomial $t^2 + t + 1$ is irreducible with roots equal to $-\frac{1}{2} + i\frac{\sqrt{3}}{2}$ and $-\frac{1}{2} - i\frac{\sqrt{3}}{2}$. It follows directly from (4.4) that the general real-valued solution is

$$y = C_1e^{-2x} + (C_2x^2 + C_3x + C_4)e^{5x}$$

$$+(C_5x + C_6)e^{-(1/2)x}\cos\frac{\sqrt{3}}{2}x + (C_7x + C_8)e^{-(1/2)x}\sin\frac{\sqrt{3}}{2}x,$$

for any set of real numbers $C_1, C_2, \ldots, C_8$.

PROBLEMS

1. Find the general real-valued solution of each of the following differential equations.

(a) $\frac{d^2y}{dx^2} - 6\frac{dy}{dx} - 7y = 0$

(b) $\frac{d^2y}{dx^2} + 6\frac{dy}{dx} + 9y = 0$

(c) $(D^2 + 6D + 5)y = 0$

(d) $(D^2 - 2D + 10)y = 0$

(e) $(4D^2 + 4D - 3)y = 0$

(f) $\frac{d^2y}{dx^2} + 6\frac{dy}{dx} = 0$

(g) $(D^2 + 2D + 6)y = 0$

(h) $4\frac{d^2y}{dx^2} + 4\frac{dy}{dx} + y = 0$

(i) $\frac{d^2y}{dt^2} + 4\frac{dy}{dt} + 5y = 0$

(j) $\frac{d^2x}{dt^2} + \frac{dx}{dt} + x = 0.$

2. Show by substitution that the function defined by $y = (c_1x + c_2)e^{rx}$ is a solution of the differential equation $(D - r)^2y = 0$.
3. Let r be a real number, and c_1 and c_2 complex numbers. Prove that, if $(c_1x + c_2)e^{rx}$ is a real-valued function, then c_1 and c_2 must both be real.
4. Let $r_1 = \alpha + i\beta$ and $r_2 = \alpha - i\beta$, where α and β are real numbers and $\beta \neq 0$. Prove that, for any two complex numbers c_1 and c_2, if the function

$$c_1e^{r_1x} + c_2e^{r_2x}$$

is real-valued, then c_1 and c_2 are complex conjugates of each other.
5. For each of the following differential equations, find the general real-valued solution by first finding an equivalent set of first-order linear differential equations and then solving these successively to find y.

(a) $(D + 1)(D - 2)(D - 3)y = 0$.

(b) $(D - 2)(D^2 - 6D + 9)y = 0$.

(c) $(D - a)(D - b)(D - c)y = 0$, where a, b, and c are distinct real numbers.

6. Find the general real-valued solution of the differential equation

$$(D^2 + 4)(D - 3)y = 0$$

by solving an equivalent pair of equations. Use the fact that we have already derived the general real-valued solution of the second-order, homogeneous, linear differential equation with constant coefficients.

7. Using Theorem (4.4), which gives the general real-valued solution of the nth-order differential equation $p(D)y = 0$, solve each of the following.

(a) $(D - 2)(D + 1)^2 y = 0$

(b) $\dfrac{d^3y}{dx^3} - 7\dfrac{dy}{dx} + 6y = 0$

(c) $(D - 3)^2(D + 1)(D - 5)y = 0$

(d) $D(D^2 + 3D - 4)y = 0$

(e) $(D + 2)^3(D - 1)y = 0$

(f) $(D + 3)^2(D^2 + 3)y = 0$

(g) $\dfrac{d^3y}{dx^3} + \dfrac{d^2y}{dx^2} - 2\dfrac{dy}{dx} = 0$

(h) $(D^2 + 2D + 2)^2 y = 0$

(i) $(D + 1)(D^2 + 2D + 2)^2 y = 0$

(j) $D^2(D^2 + 2D + 2)^2 y = 0$

(k) $\dfrac{d^4y}{dx^4} - 81y = 0$

(l) $\dfrac{d^3y}{dx^3} + \dfrac{d^2y}{dx^2} + \dfrac{dy}{dx} + y = 0.$

5. Nonhomogeneous Equations. We continue to consider a given real-valued function $F(x)$, a given polynomial

$$p(t) = t^n + a_{n-1}t^{n-1} + \cdots + a_1 t + a_0,$$

with real coefficients, and the resulting differential equation

$$p(D)y = F(x). \tag{1}$$

In this section our objective is to develop techniques for solving many examples of (1) quickly. This is in contrast to Section 3, where it is demonstrated that (1) can always be solved by successively solving first-order linear equations. The task of solving all these first-order equations can be extremely tedious, and we therefore look for a simpler method.

The technique to be discussed is based on two premises. The first is the fact, demonstrated in Section 4, that one can write down the general solution of the associated homogeneous equation

$$p(D)y = 0 \tag{2}$$

immediately, once $p(t)$ has been factored into irreducible polynomials. The second is Theorem (4.1), page 640, which asserts that the general solution of (1) is equal to the general solution of (2) plus any particular solution of (1).

Hence the problem of solving (1) reduces to that of finding *any one* solution.

As an introductory example, consider the differential equation

$$(D^2 + 4D + 3)y = 3x^2 + 2x - 6. \tag{3}$$

The characteristic polynomial is $t^2 + 4t + 3$, which factors into the product $(t + 1)(t + 3)$. The associated homogeneous equation is therefore

$$(D + 1)(D + 3)y = 0,$$

and its general solution, which we shall denote by y_h, is given by

$$y_h = c_1e^{-x} + c_2e^{-3x},$$

for arbitrary real numbers c_1 and c_2. To obtain the general solution of (3), it remains to find a particular solution y_p, and any one is as good as any other. If we can find one, it follows by Theorem (4.1) that $y_h + y_p$ is the general solution of (3).

Since the derivatives of polynomials are polynomials and since the right side of (3) is the polynomial $3x^2 + 2x - 6$, it is natural to seek a polynomial solution. Let us set

$$y_p = A_nx^n + A_{n-1}x^{n-1} + \cdots + A_0, \qquad \text{with } A_n \neq 0,$$

and try to find n and coefficients $A_n, \ldots, A_0$ so that $(D^2 + 4D + 3)y_p = 3x^2 + 2x - 6$. Since Dy_p is a polynomial of degree $n - 1$, and D^2y_p is a polynomial of degree $n - 2$, it follows that $(D^2 + 4D + 3)y_p$ is a polynomial of degree n. If this polynomial is to equal $3x^2 + 2x - 6$, for every x, then it must be the case that $n = 2$. Hence we let

$$y_p = Ax^2 + Bx + C.$$

Then

$$Dy_p = 2Ax + B,$$
$$D^2y_p = 2A,$$

and so

$$\begin{aligned}(D^2 + 4D + 3)y_p &= 2A + 4(2Ax + B) + 3(Ax^2 + Bx + C)\\ &= 3Ax^2 + (8A + 3B)x + 2A + 4B + 3C.\end{aligned}$$

The right side of the preceding equation is equal to $3x^2 + 2x - 6$, for all real numbers x, if and only if

$$\begin{aligned}3 &= 3A,\\ 2 &= 8A + 3B,\\ -6 &= 2A + 4B + 3C.\end{aligned}$$

Solving these equations, we get $A = 1$, $B = -2$, and $C = 0$. The function

$$y_p = x^2 - 2x$$

is therefore a particular solution of (3). It follows from Theorem (4.1) that

$$y = y_h + y_p = c_1e^{-x} + c_2e^{-3x} + x^2 - 2x$$

is the general solution, where c_1 and c_2 are arbitrary real numbers.

A second example is the differential equation

$$(D^2 + 4)y = 3e^{5x}. \tag{4}$$

The characteristic polynomial $t^2 + 4$ is irreducible with roots $2i$ and $-2i$, and the general solution y_h of the associated homogeneous equation $(D^2 + 4)y = 0$ is therefore given by

$$y_h = c_1 \cos 2x + c_2 \sin 2x,$$

for arbitrary real numbers c_1 and c_2. A particular solution y_p of (4) will be any function with the property that its second derivative plus four times itself is equal to $3e^{5x}$. Since the derivative of an exponential function is again an exponential function, an intelligent guess is that a particular solution might be a function of the form

$$y_p = Ae^{5x}.$$

Trying this, we obtain

$$Dy_p = 5Ae^{5x},$$

$$D^2y_p = 25Ae^{5x},$$

and so

$$(D^2 + 4)y_p = 25Ae^{5x} + 4Ae^{5x} = 29Ae^{5x}.$$

Obviously, $29Ae^{5x} = 3e^{5x}$ if and only if $A = \frac{3}{29}$. Hence a particular solution of the differential equation $(D^2 + 4)y = 3e^{5x}$ is

$$y_p = \tfrac{3}{29}e^{5x},$$

and it is a consequence of Theorem (4.1) that the general solution is

$$y = y_h + y_p = c_1 \cos 2x + c_2 \sin 2x + \tfrac{3}{29}e^{5x},$$

where c_1 and c_2 are arbitrary real constants.

The method of finding particular solutions used in the above two examples is sometimes called the *method of undetermined coefficients*. For a third example, consider the differential equation

$$(D^2 + 4)y = 7 \sin 2x. \tag{5}$$

The associated homogeneous equation $(D^2 + 4)y = 0$ is the same as for equation (4), and its general solution is

$$y_h = c_1 \cos 2x + c_2 \sin 2x.$$

In attempting to find a particular solution of (5), one might reason from the experience of the preceding examples as follows: The right side is the function $7 \sin 2x$. Since the derivatives of any function which is a linear combination of sines and cosines are functions of the same type, a reasonable candidate for a particular solution is some function y_p of the form

$$y_p = A \sin 2x + B \cos 2x.$$

However, when we try to determine values of the coefficients A and B which will make y_p a solution, we find that $(D^2 + 4)y_p = 0$. This is actually not surprising, since any function of this type has already been shown to be a solution of the associated homogeneous equation. Hence we must try some other form for y_p.

With some ingenuity and willingness to experiment, it is not at all impossible to discover a particular solution to (5). Nevertheless, this example serves to illustrate the desirability of analyzing our technique to reduce the amount of inspiration necessary. For this purpose, we again consider the differential equation (1); i.e.,

$$p(D)y = F(x) \tag{1}$$

with given function $F(x)$ and polynomial $p(t)$ of degree n. To apply the method of undetermined coefficients, it is necessary that the right side of (1) is itself a solution of a homogeneous linear differential equation with constant coefficients. Hence in the discussion which follows, *we make the assumption that there exists a polynomial $q(t)$ of degree m such that* $q(D)F(x) = 0$.

Such a linear differential operator $q(D)$ is sometimes called an **annihilator** of the right side of (1). For the differential equation (3), a suitable annihilator is the operator D^3, since

$$D^3(3x^2 + 2x - 6) = 0.$$

For equation (5), whose right side is the function $7 \sin 2x$, we have $D(7 \sin 2x) = 14 \cos 2x$ and $D^2(7 \sin 2x) = D(14 \cos 2x) = -4(7 \sin 2x)$. Hence

$$(D^2 + 4)7 \sin 2x = 0,$$

and thus $D^2 + 4$ is an annihilator of the right side. Similarly, it is easy to see that

$$(D - 5)3e^{5x} = 0,$$

from which it follows that $D - 5$ is an annihilator of the right side of equation (4).

Returning to the general case, we first observe that, if y is an arbitrary solution of the differential equation (1), then

$$q(D)p(D)y = q(D)F(x) = 0.$$

That is, every solution of (1) is also a solution of the equation

$$q(D)p(D)y = 0, \tag{6}$$

which is homogeneous and of order $m + n$. Let us denote by y_* the general solution of (6), and by y_h the general solution of the associated homogeneous equation of (1), i.e., of the equation $p(D)y = 0$. It is clear that y_h is also a solution of (6). We know that y_* contains $m + n$ arbitrary constants and that y_h contains n. It follows from the form of the general solution of a homogeneous linear differential equation with constant coefficients, as presented in Theorem (4.4), page 646, that we can write

$$y_* = y_h + u, \tag{7}$$

where u contains m arbitrary constants. It will follow that these are the "undetermined coefficients" of the particular solution we are seeking.

Let y_1 be a solution of (1); i.e., y_1 is some function with the property that $p(D)y_1 = F(x)$. Then y_1 is also a solution of (6), and so there exists a set of values for the n constants in y_h and for the m constants in u such that, with these values substituted, we have

$$y_1 = y_h + u.$$

Hence

$$\begin{aligned} F(x) = p(D)y_1 &= p(D)(y_h + u) \\ &= p(D)y_h + p(D)u \\ &= 0 + p(D)u \\ &= p(D)u. \end{aligned}$$

Thus we have proved that there exists a set of values for the m constants in u such that, with these values substituted, the resulting function u is a solution of the differential equation (1). Moreover, it can be proved that there is only one such set of values. Hence, as the following examples will illustrate, these "undetermined coefficients" are uniquely determined by the equation

$$p(D)u = F(x). \tag{8}$$

We take for the particular solution y_p the function u specified by equations (7) and (8).

EXAMPLE 1. Find the general solution of the differential equation (5), i.e., of

$$(D^2 + 4)y = 7 \sin 2x.$$

As indicated earlier, the general solution of the associated homogeneous equation $(D^2 + 4)y = 0$ is

$$y_h = c_1 \cos 2x + c_2 \sin 2x.$$

Moreover, we have observed that $(D^2 + 4)7 \sin 2x = 0$, and therefore the operator $D^2 + 4$ is an annihilator of the right side. Hence we consider the homogeneous fourth-order equation

$$(D^2 + 4)(D^2 + 4)y = (D^2 + 4)^2 y = 0.$$

The general solution of this equation is given by

$$\begin{aligned} y_* &= (Ax + B) \cos 2x + (Cx + D) \sin 2x \\ &= B \cos 2x + D \sin 2x + Ax \cos 2x + Cx \sin 2x, \end{aligned}$$

for arbitrary real numbers, A, B, C, and D. It is clear that

$$y_h = B \cos 2x + D \sin 2x,$$

and we therefore set

$$u = Ax \cos 2x + Cx \sin 2x.$$

It follows that

$$Du = A \cos 2x - 2Ax \sin 2x + C \sin 2x + 2Cx \cos 2x,$$

and

$$\begin{aligned} D^2u &= -2A \sin 2x - 2A \sin 2x - 4Ax \cos 2x \\ &\quad +2C \cos 2x + 2C \cos 2x - 4Cx \sin 2x \\ &= (4C - 4Ax) \cos 2x + (-4A - 4Cx) \sin 2x. \end{aligned}$$

Hence

$$\begin{aligned} (D^2 + 4)u &= (4C - 4Ax + 4Ax) \cos 2x + (-4A - 4Cx + 4Cx) \sin 2x \\ &= 4C \cos 2x - 4A \sin 2x. \end{aligned}$$

Setting $(D^2 + 4)u = 7 \sin 2x$, we obtain

$$4C \cos 2x - 4A \sin 2x = 7 \sin 2x.$$

Since this equation is to be true for all real values of x, we conclude that $4C = 0$ and $-4A = 7$. Thus $C = 0$ and $A = -\frac{7}{4}$. It follows that the function u, with these values substituted for the constants, is a solution of the given differential equation. We therefore set

$$y_p = -\tfrac{7}{4}x \cos 2x,$$

and obtain

$$y = y_h + y_p = c_1 \cos 2x + c_2 \sin 2x - \tfrac{7}{4}x \cos 2x$$

as the general solution.

EXAMPLE 2. Find the general solution of the differential equation

$$\frac{d^2y}{dx^2} + \frac{dy}{dx} - 2y = 5e^{-2x}.$$

The characteristic polynomial is $t^2 + t - 2 = (t + 2)(t - 1)$, and the differential equation can therefore be written

$$(D^2 + D - 2)y = (D + 2)(D - 1)y = 5e^{-2x}.$$

The general solution of the associated homogeneous equation

$$(D + 2)(D - 1)y = 0$$

is given by

$$y_h = c_1e^{-2x} + c_2e^{x}.$$

The right side of the nonhomogeneous equation is the function $5e^{-2x}$. Since $D(5e^{-2x}) = -2(5e^{-2x})$, it follows that

$$(D + 2)5e^{-2x} = 0,$$

and so $D + 2$ is an annihilator. We therefore consider the third-order homogeneous equation

$$(D + 2)(D + 2)(D - 1)y = (D + 2)^2(D - 1)y = 0,$$

whose general solution is

$$y_* = (Ax + B)e^{-2x} + Ce^{x},$$

for any real numbers A, B, and C. Recognizing that $Be^{-2x} + Ce^{x} = y_h$, we set

$$u = Axe^{-2x}.$$

The constant A is evaluated by setting $(D^2 + D - 2)u = 5e^{-2x}$. Differentiating to obtain the left side, we get

$$\begin{aligned} Du &= Ae^{-2x} - 2Axe^{-2x}, \\ D^2u &= -2Ae^{-2x} - 2Ae^{-2x} + 4Axe^{-2x} \\ &= -4Ae^{-2x} + 4Axe^{-2x}. \end{aligned}$$

Hence

$$\begin{aligned} (D^2 + D - 2)u &= -4Ae^{-2x} + 4Axe^{-2x} + Ae^{-2x} - 2Axe^{-2x} - 2Axe^{-2x} \\ &= -3Ae^{-2x}. \end{aligned}$$

We therefore obtain the equation $-3Ae^{-2x} = 5e^{-2x}$, which implies that $A = -\frac{5}{3}$. Hence the function u obtained by substituting this value for A is a particular solution. Thus we take

$$y_p = -\tfrac{5}{3}xe^{-2x},$$

and it follows that the general solution is given by

$$y = y_h + y_p = c_1e^{-2x} + c_2e^{x} - \tfrac{5}{3}xe^{-2x},$$

for arbitrary real numbers c_1 and c_2.

EXAMPLE 3. Solve the differential equation

$$D^3(D + 2)y = 8x + 1.$$

The characteristic polynomial is $t^3(t + 2)$, whose roots 0 and -2 occur with multiplicities three and one, respectively. It follows that the general solution of the associated homogeneous equation is

$$\begin{aligned} y_h &= (c_1x^2 + c_2x + c_3)e^{0x} + c_4e^{-2x} \\ &= c_1x^2 + c_2x + c_3 + c_4e^{-2x}. \end{aligned}$$

The right side of the given nonhomogeneous equation is $8x + 1$, and the operator D^2 is an annihilator, since $D^2(8x + 1) = 0$. Hence we consider the sixth-order homogeneous equation

$$D^2D^3(D + 2)y = D^5(D + 2)y = 0,$$

the general solution of which is

$$y_* = Ax^4 + Bx^3 + Cx^2 + Dx + E + Fe^{-2x}.$$

It is obvious that $y_h = Cx^2 + Dx + E + Fe^{-2x}$, and we set

$$u = Ax^4 + Bx^3.$$

It follows that

$$\begin{aligned} Du &= 4Ax^3 + 3Bx^2, \\ D^2u &= 12Ax^2 + 6Bx, \\ D^3u &= 24Ax + 6B, \\ D^4u &= 24A, \end{aligned}$$

and so

$$\begin{aligned} D^3(D + 2)u &= D^4u + 2D^3u \\ &= 24A + 48Ax + 12B \\ &= 48Ax + 24A + 12B. \end{aligned}$$

Setting $D^3(D + 2)u = 8x + 1$, we obtain the equation

$$48Ax + 24A + 12B = 8x + 1,$$

which is true for all real values of x if and only if $A = \frac{1}{6}$ and $B = -\frac{1}{4}$. It follows that a particular solution of the differential equation

$$D^3(D + 2)y = 8x + 1$$

is defined by

$$y_p = \tfrac{1}{6}x^4 - \tfrac{1}{4}x^3,$$

and the general solution is, therefore,

$$y = y_h + y_p = c_1x^2 + c_2x + c_3 + c_4e^{-2x} + \tfrac{1}{6}x^4 - \tfrac{1}{4}x^3,$$

for arbitrary real numbers c_1, c_2, and c_3, and c_4.

The method of undetermined coefficients which we have studied in this section is not applicable to all linear differential equations with constant coefficients. For example, it will not work for the equation $(D^2 + 2)y = \tan x$, because there is no polynomial $q(t)$ with the property that $q(D)\tan x = 0$. Of course, this equation can be solved by replacing it by two first-order linear equations and solving these successively as in Section 3. It can also be solved by another well-known technique, called the method of *variation of parameters*, which we shall not discuss in this book. Finally, it is important to realize that there exist tables in which particular solutions of the equation $p(D)y = F(x)$ are tabulated for a variety of functions $F(x)$. In particular, see pages 112 to 114 of the book by E. J. Cogan and R. Z. Norman, *Handbook of Calculus, Difference and Differential Equations*, Prentice-Hall, 2nd ed., 1963.

PROBLEMS

1. For each of the following differential equations, a particular solution can be found by inspection. Obtain such a solution y_p, and also find the general solution.

   (a) $(D^2 + 3D - 10)y = 5$

   (b) $(D^2 + 1)y = 2x$

   (c) $\dfrac{d^2y}{dx^2} - 4y = 12x - 20$

   (d) $\dfrac{d^2y}{dx^2} + \dfrac{dy}{dx} - 2y = -2x^2 + 6x - 4$

   (e) $(D^2 - 2D - 3)y = e^x$

   (f) $\dfrac{d^2y}{dx^2} - 2\dfrac{dy}{dx} + y = 8e^{3x}$

   (g) $D(D^2 - 9)y = 2e^{-x}$

   (h) $(D^2 + 4)y = 3\sin x$

(i) $(D^2 + 4)y = 3 \sin x + 4x + 8$

(j) $\dfrac{d^2y}{dx^2} + 3\dfrac{dy}{dx} = 5 \cos x - 5 \sin x$

(k) $(D^2 + 3)y = 5 \cos 3x$

(l) $(D^2 + 2D - 2)y = 13 \cos 2x.$

2. Find the particular solution $f(x)$ of the differential equation $(D^2 + 1)y = 2x$ which has the property that $f(0) = 3$ and $f'(0) = 2$. (*Hint:* Find the general solution first and then apply the given boundary conditions to find the values of the constants.)
3. Find the particular solution $y(x)$ of the differential equation

$$\frac{d^2y}{dx^2} + 2\frac{dy}{dx} + 2y = 5e^{2x}$$

with the property that $y(0) = \frac{1}{2}$ and $y'(0) = \frac{1}{2}$.
4. The current i in a given alternating-current circuit is a function of time t and is governed by the differential equation

$$\frac{d^2i}{dt^2} + 2\frac{di}{dt} + 5i = 12e^{-t}.$$

Find i as a function of t, if $i = 0$ and $\dfrac{di}{dt} = 6$ when $t = 0$.

5. Find the general solution of each of the following differential equations.

(a) $(D - 2)^2y = 4x^2 - 5$

(b) $(D^2 - 3D + 2)y = 4x + 3$

(c) $\dfrac{d^2y}{dx^2} + \dfrac{dy}{dx} - 2y = 5e^{-x}$

(d) $D(D - 2)y = 6x^2 + 2x - 6$

(e) $(D^2 + D - 2)y = 6e^{-2x}$

(f) $(D^2 + D - 2)y = 6e^{-2x} + 2x - 4$

(g) $(D^2 + D - 2)y = 6e^{-2x} + 15e^x$

(h) $D^2(D + 3)y = 5x - 2$

(i) $\dfrac{d^2y}{dx^2} + 4y = 5 \cos 3x$

(j) $\dfrac{d^2y}{dx^2} + 9y = 2 \sin 3x$

(k) $(D^2 + 1)y = 10 \sin x + 3e^{-x}$

(l) $(D^2 + 1)y = 4 \sin x + 8 \cos x$

(m) $(D^2 - 2D + 1)y = 3e^x \sin x$

(n) $(D^2 + 2D + 2)y = 3e^x \cos x.$

6. Hyperbolic Functions. In solving linear differential equations, we have encountered many combinations of e^{r_1x} and e^{r_2x}. Among these, two particular linear combinations occur sufficiently often that they have been given special names. These are the two functions $\frac{1}{2}e^x + \frac{1}{2}e^{-x}$ and $\frac{1}{2}e^x - \frac{1}{2}e^{-x}$.

Let us look at some of the properties of these two functions, which motivate their names. First, we observe that each is the derivative of the other:

$$\frac{d}{dx}\left(\tfrac{1}{2}e^x + \tfrac{1}{2}e^{-x}\right) = \tfrac{1}{2}e^x - \tfrac{1}{2}e^{-x},$$

$$\frac{d}{dx}\left(\tfrac{1}{2}e^x - \tfrac{1}{2}e^{-x}\right) = \tfrac{1}{2}e^x + \tfrac{1}{2}e^{-x}.$$

This fact implies, of course, that each function is its own second derivative. There is a clear analogy here with the trigonometric functions cosine and sine, each of which is, up to sign, the derivative of the other and each of which is the negative of its own second derivative.

The result of squaring these two functions is

$$\left(\tfrac{1}{2}e^x + \tfrac{1}{2}e^{-x}\right)^2 = \tfrac{1}{4}e^{2x} + \tfrac{1}{2} + \tfrac{1}{4}e^{-2x},$$
$$\left(\tfrac{1}{2}e^x - \tfrac{1}{2}e^{-x}\right)^2 = \tfrac{1}{4}e^{2x} - \tfrac{1}{2} + \tfrac{1}{4}e^{-2x},$$

from which it follows that

$$\left(\tfrac{1}{2}e^x + \tfrac{1}{2}e^{-x}\right)^2 - \left(\tfrac{1}{2}e^x + \tfrac{1}{2}e^{-x}\right)^2 = 1. \qquad (1)$$

Thus the difference of their squares is equal to 1, and this fact is analogous to the trigonometric identity $\cos^2 x + \sin^2 x = 1$. It is a consequence of equation (1) that, for every real number t, the ordered pair

$$(x, y) = \left(\tfrac{1}{2}e^t + \tfrac{1}{2}e^{-t}, \tfrac{1}{2}e^t - \tfrac{1}{2}e^{-t}\right)$$

satisfies the equation $x^2 - y^2 = 1$ of an equilateral hyperbola. Similarly, we know that, for every real number t, the ordered pair

$$(x, y) = (\cos t, \sin t)$$

is a point on the unit circle $x^2 + y^2 = 1$. With this motivation, we define the **hyperbolic cosine,** abbreviated cosh, and the **hyperbolic sine,** abbreviated sinh, by setting

$$\begin{aligned} \cosh x &= \tfrac{1}{2}e^x + \tfrac{1}{2}e^{-x}, \\ \sinh x &= \tfrac{1}{2}e^x - \tfrac{1}{2}e^{-x}, \qquad \textit{for every real number } x. \end{aligned} \qquad (2)$$

It is trivial to verify that

(6.1)

$$\begin{aligned} \cosh(-x) &= \cosh x, \\ \sinh(-x) &= -\sinh x, \qquad \textit{for every real number } x. \end{aligned}$$

Thus, like their respective trigonometric counterparts, the hyperbolic cosine is an even function, and the hyperbolic sine is an odd function.

Equation (1) now becomes the identity

(6.2) $$\boxed{\cosh^2 x - \sinh^2 x = 1, \quad \textit{for every real number } x,}$$

and we have also already established the two derivative formulas

(6.3) $$\boxed{\frac{d}{dx}\cosh x = \sinh x,}$$

and

(6.4) $$\boxed{\frac{d}{dx}\sinh x = \cosh x.}$$

Since each of the two functions, cosh and sinh, is equal to its own second derivative, each is a solution of the differential equation $(D^2 - 1)y = 0$. More generally, the functions $\cosh kx$ and $\sinh kx$, where k is an arbitrary real constant, are both solutions of the differential equation

$$(D^2 - k^2)y = 0. \tag{3}$$

From the linearity of the differential operator $D^2 - k^2$ it follows that the function defined by

$$\boxed{y = c_1 \cosh kx + c_2 \sinh kx,} \tag{4}$$

for any two real numbers c_1 and c_2, is also a solution. In fact, (4) *is an alternative form of the general solution of the differential equation* (3).

To prove this fact, let y_0 be an arbitrary solution of (3). The characteristic polynomial is $t^2 - k^2$, which equals the product $(t - k)(t + k)$. Hence there exist real numbers A and B such that

$$y_0 = Ae^{kx} + Be^{-kx}.$$

However, we have

$$\cosh kx + \sinh kx = \frac{e^{kx}}{2} + \frac{e^{-kx}}{2} + \frac{e^{kx}}{2} - \frac{e^{-kx}}{2} = e^{kx},$$

$$\cosh kx - \sinh kx = \frac{e^{kx}}{2} + \frac{e^{-kx}}{2} - \frac{e^{kx}}{2} + \frac{e^{-kx}}{2} = e^{-kx}.$$

It follows that

$$\begin{aligned} y_0 &= A(\cosh kx + \sinh kx) + B(\cosh kx - \sinh kx) \\ &= (A + B)\cosh kx + (A - B)\sinh kx, \end{aligned}$$

which is of the form of (4). This completes the proof.

In drawing the graphs of the hyperbolic functions, we make use of the fact that cosh is an even function, and sinh is an odd function. In addition, each of the following simple results follows quickly from the definition of the relevant function:

(6.5)
$$\begin{cases} \text{(i)} & \cosh 0 = 1, \\ \text{(ii)} & \sinh 0 = 0, \\ \text{(iii)} & \cosh x > 0, \textit{ for every real number } x, \\ \text{(iv)} & \sinh x = 0 \textit{ if and only if } x = 0, \\ \text{(v)} & \sinh x > 0, \textit{ for every } x > 0. \end{cases}$$

Applying these facts to the first and second derivatives, we conclude that the graph of $\sinh x$ has positive slope everywhere, has therefore no local maximum or minimum points, and passes through the origin with slope 1. Moreover, it is concave upward if x is positive, is concave downward if x is negative, and as a result has one point of inflection at the origin. Similarly, the graph of $\cosh x$ has positive slope if x is positive, negative slope if x is negative, and one critical point at (0, 1). It is concave upward everywhere, from which it follows that there are no points of inflection and the critical point at (0, 1) is a local minimum. The graphs of the two functions are drawn in the same xy-plane in Figure 2.

The curve which is the graph of the equation $y = \cosh x$ is called a **catenary.** More generally, a catenary is the graph of an equation of the form $y = a \cosh \left(\frac{x}{a}\right)$, where a is a nonzero constant. The word comes from the latin word meaning "chain," and it can be shown that, if a chain or cable with a uniform weight per unit length is suspended between two points, then it hangs in the shape of a catenary.

In a manner completely analogous to that for defining the other four trigonometric functions from the sine and cosine, we define four other hyperbolic functions. They are the **hyperbolic tangent,** denoted by tanh; the **hyperbolic secant,** denoted by sech; the **hyperbolic cosecant,** denoted by csch; and the **hyperbolic cotangent,** denoted by coth. The definitions are

$$\boxed{\begin{aligned} \tanh x &= \frac{\sinh x}{\cosh x}, & \operatorname{sech} x &= \frac{1}{\cosh x}, \\ \coth x &= \frac{\cosh x}{\sinh x}, & \operatorname{csch} x &= \frac{1}{\sinh x}. \end{aligned}} \qquad (5)$$

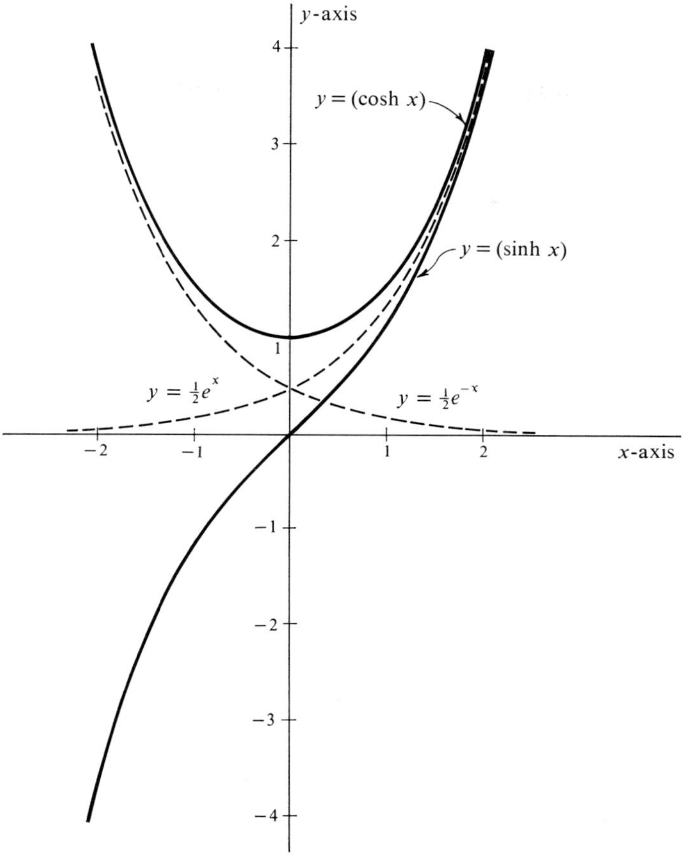

Figure 2

In the problems at the end of the section you are asked to find the derivatives of these functions. These derivative formulas, and also the many identities among the hyperbolic functions, are all closely akin to those for the trigonometric functions.

The inverse hyperbolic functions are also defined. For example, y is the inverse hyperbolic cosine of x if and only if x is the hyperbolic cosine of y. That is,

$$y = \text{arccosh } x \qquad \textit{if and only if } x = \cosh y.$$

The domain of the inverse hyperbolic cosine *arccosh* is the set of all real numbers greater than or equal to 1, and the range is chosen to be the set of all nonnegative real numbers. The definitions of the other inverse hyperbolic functions follow the same pattern.

PROBLEMS

1. Find the following derivatives.

(a) $\dfrac{d}{dx}\cosh 5x$

(b) $\dfrac{d}{dx}(\cosh^2 3x + \sinh^2 3x)$

(c) $\dfrac{d}{dx}\ln\cosh(x^2 + 1)$

(d) $\dfrac{d}{dx}\sinh\sqrt{x^2 + 1}$

(e) $\dfrac{d}{dx}\tanh x$

(f) $\dfrac{d}{dx}\operatorname{sech} x$

(g) $\dfrac{d}{dx}\operatorname{csch} x$

(h) $\dfrac{d}{dx}\coth x$

(i) $\dfrac{d}{dt}\tanh\dfrac{1}{1 + t^2}$

(j) $\dfrac{d}{dx}a\cosh\left(\dfrac{x}{a}\right)$.

2. Find the following integrals.

(a) $\int \sinh 7x\, dx$

(b) $\int \cosh\frac{t}{2}\, dt$

(c) $\int \sinh 3x \cosh^3 3x\, dx$

(d) $\int \tanh x\, dx$

(e) $\int \dfrac{\operatorname{sech}^2 x}{\tanh x}\, dx$

(f) $\int 2x\sinh(2x^2 + 1)\, dx$

(g) $\int \tanh^5 2x \operatorname{sech}^2 2x\, dx$

(h) $\int \coth x \ln\sinh x\, dx$

(i) $\int \cosh^2 x\, dx$

(j) $\int \sinh^2 x\, dx$.

3. Prove the following identities.

(a) $\cosh 2x = \cosh^2 x + \sinh^2 x$

(b) $\sinh 2x = 2\sinh x\cosh x$

(c) $1 - \tanh^2 x = \operatorname{sech}^2 x$

(d) $\coth^2 x - 1 = \operatorname{csch}^2 x$.

4. Prove that $\cosh x$ is an even function and that $\sinh x$ is an odd function.

5. Find the general solution of each of the following differential equations in terms of the hyperbolic functions.

(a) $\dfrac{d^2y}{dx^2} = 4y$

(b) $(D^2 - 7)y = 0$

(c) $\dfrac{d^2y}{dx^2} - 9y = 5e^{2x}$

(d) $(D^2 - k^2)y = x + \sin x$

(e) $(D^2 - 16)y = 5\sinh 8x$

(f) $(D^2 - 16)y = 5\cosh 4x$.

6. Prove that $\sinh x = 0$ if and only if $x = 0$.

7. Find $\lim\limits_{x \to \infty} \dfrac{\sinh x}{\cosh x}$.

8. Identify and draw the curve defined parametrically by

$$\begin{cases} x(t) = \cosh t, \\ y(t) = \sinh t, \qquad -\infty < t < \infty. \end{cases}$$

9. (a) Draw the region R bounded by the x-axis, the hyperbola $x^2 - y^2 = 1$, and the straight line joining the origin to the point (x, y) on the hyperbola defined by $x = \cosh t$ and $y = \sinh t$, for an arbitrary $t > 0$.

   (b) Compute the area of the region R.

10. Find the arc length of the graph of the equation $y = 3 \cosh\left(\dfrac{x}{3}\right)$ from the point $(0, 3)$ to the point $(6, 3 \cosh 2)$.

11. Compute the following derivatives.

    (a) $\dfrac{d}{dx} \operatorname{arctanh} x$ (b) $\dfrac{d}{dx} \operatorname{arcsinh} x$.

12. Sketch the graph of the following equations.

    (a) $y = \tanh x$ (b) $y = \operatorname{arccosh} x$.

APPENDIX A

Properties of Limits

In this appendix we shall establish the fundamental properties of limits stated without proof in Theorem (4.1), page 32. Before restating the theorem and giving the proof, we recall one of the basic facts about inequalities and absolute values, which we shall use. Called the **triangle inequality,** it asserts that, for any two real numbers a and b,

$$|a \pm b| \leq |a| + |b|.$$

This result is stated and proved for $a + b$ in (1.4), page 7. It holds equally well for $a - b$, since

$$|a - b| = |a + (-b)| \leq |a| + |-b| = |a| + |b|.$$

The theorem, which we shall prove, is the following:

THEOREM. *If* $\lim\limits_{x \to a} f(x) = b_1$ *and* $\lim\limits_{x \to a} g(x) = b_2$, *then*

(i) $\lim\limits_{x \to a} [f(x) + g(x)] = b_1 + b_2$.

(ii) $\lim\limits_{x \to a} cf(x) = cb_1$.

(iii) $\lim\limits_{x \to a} f(x)g(x) = b_1 b_2$.

(iv) $\lim\limits_{x \to a} \dfrac{f(x)}{g(x)} = \dfrac{b_1}{b_2}$ *provided* $b_2 \neq 0$.

According to the definition of limit, the hypotheses tell us that, for any positive number ϵ, there exist positive numbers δ_1 and δ_2 such that if x is in the domains of both f and g and if $0 < |x - a| < \delta_1$ and $0 < |x - a| < \delta_2$, then $|f(x) - b_1| < \epsilon$ and $|g(x) - b_2| < \epsilon$. Where it is relevant in the proofs which follow, we shall assume without explicitly stating it the condition that x lies in the appropriate domain of f or g (or both).

Proof of (i). Let ϵ be an arbitrary positive number. Then there exist positive numbers δ_1 and δ_2 such that $|f(x) - b_1| < \dfrac{\epsilon}{2}$, whenever $0 < |x - a| < \delta_1$, and $|g(x) - b_2| < \dfrac{\epsilon}{2}$, whenever $0 < |x - a| < \delta_2$. (It is legitimate to write

$\frac{\epsilon}{2}$ in these inequalities, since the definition specifies the existence of δ's for *any* positive number ϵ. Given a choice of ϵ, we can then take $\frac{\epsilon}{2}$ to be the number which implies the existence of δ_1 and δ_2.) We set

$$\delta = \text{minimum } \{\delta_1, \delta_2\}.$$

Let us now suppose that $0 < |x - a| < \delta$. It follows that $0 < |x - a| < \delta_1$ and $0 < |x - a| < \delta_2$, and thence that $|f(x) - b_1| < \frac{\epsilon}{2}$ and $|g(x) - b_2| < \frac{\epsilon}{2}$. Clearly,

$$|[f(x) + g(x)] - [b_1 + b_2]| = |[f(x) - b_1] + [g(x) - b_2]|.$$

Hence, using the triangle inequality, we obtain

$$|[f(x) + g(x)] - [b_1 + b_2]| \leq |f(x) - b_1| + |g(x) - b_2| < \frac{\epsilon}{2} + \frac{\epsilon}{2} = \epsilon.$$

Thus we have shown that, for any $\epsilon > 0$, there exists a $\delta > 0$ such that, whenever $0 < |x - a| < \delta$, then $|[f(x) + g(x)] - [b_1 + b_2]| < \epsilon$. By the definition of limit we have therefore proved that

$$\lim_{x \to a} [f(x) + g(x)] = b_1 + b_2,$$

which is the result (i).

Proof of (ii). Suppose first that $c = 0$. Then cf is the constant function with value 0, and $cb_1 = 0$. Hence

$$|cf(x) - cb_1| = |0 - 0| = 0,$$

for every x in the domain of f. Thus, for any two positive numbers ϵ and δ, it is trivially true that

$$|cf(x) - cb_1| < \epsilon, \qquad \text{whenever } 0 < |x - a| < \delta,$$

and (ii) is therefore proved in this special case. We next assume that $c \neq 0$, and choose an arbitrary positive number ϵ. There then exists a positive number δ such that

$$|f(x) - b_1| < \frac{\epsilon}{|c|}, \qquad \text{whenever } 0 < |x - a| < \delta.$$

It follows immediately that

$$|cf(x) - cb_1| = |c[f(x) - b_1]| = |c|\,|f(x) - b_1| < |c|\frac{\epsilon}{|c|} = \epsilon,$$

whenever $0 < |x - a| < \delta$. This completes the proof of (ii).

Proof of (iii). Let ϵ be an arbitrary positive number. Select a positive number M such that $|b_1| < M$ and $|b_2| < M$. Then there exist positive numbers δ_1, δ_2, and δ_3 such that

$$|f(x) - b_1| < \frac{\epsilon}{2M}, \qquad \text{provided } 0 < |x - a| < \delta_1,$$

$$|g(x) - b_2| < \frac{\epsilon}{2M}, \qquad \text{provided } 0 < |x - a| < \delta_2,$$

$$|g(x) - b_2| < M - |b_2|, \qquad \text{provided } 0 < |x - a| < \delta_3.$$

We set

$$\delta = \text{minimum } \{\delta_1, \delta_2, \delta_3\},$$

and in the remainder of the argument we assume that $0 < |x - a| < \delta$. It then follows that all three of the above inequalities hold. Using the last one together with the triangle inequality, we first observe that

$$|g(x)| = |(g(x) - b_2) + b_2| \leq |g(x) - b_2| + |b_2| < (M - |b_2|) + |b_2| = M.$$

Next we obtain

$$\begin{aligned} |f(x)g(x) - b_1b_2| &= |f(x)g(x) - b_1g(x) + b_1g(x) - b_1b_2| \\ &= |g(x)[f(x) - b_1] + b_1[g(x) - b_2]| \\ &\leq |g(x)[f(x) - b_1]| + |b_1[g(x) - b_2]| \\ &= |g(x)|\,|f(x) - b_1| + |b_1|\,|g(x) - b_2|. \end{aligned}$$

Finally, therefore,

$$|f(x)g(x) - b_1b_2| < M\frac{\epsilon}{2M} + M\frac{\epsilon}{2M} = \frac{\epsilon}{2} + \frac{\epsilon}{2} = \epsilon,$$

and the proof of (iii) is finished.

Proof of (iv). We shall prove the simpler statement:

$$\lim_{x \to a} \frac{1}{g(x)} = \frac{1}{b_2}, \qquad \text{provided } b_2 \neq 0. \tag{1}$$

This fact, together with (iii), then implies

$$\lim_{x \to a} \frac{f(x)}{g(x)} = \lim_{x \to a} f(x)\frac{1}{g(x)} = b_1\frac{1}{b_2} = \frac{b_1}{b_2},$$

which is the result (iv). Since it is assumed that $b_2 \neq 0$, there exists a number m such that $0 < m < |b_2|$. Hence there exists a positive number δ_1 such that

$$|g(x) - b_2| < |b_2| - m,$$

whenever $0 < |x - a| < \delta_1$. But

$$|b_2| = |-b_2| = |(g(x) - b_2) - g(x)| \\ \le |g(x) - b_2| + |g(x)|.$$

Hence, if $0 < |x - a| < \delta_1$, we have

$$|g(x)| \geq |b_2| - |g(x) - b_2| > |b_2| - (|b_2| - m) = m.$$

Taking reciprocals, we therefore obtain

$$\frac{1}{|g(x)|} < \frac{1}{m}, \qquad \text{whenever } 0 < |x - a| < \delta_1.$$

Now let ϵ be an arbitrary positive number. There exists a positive number δ_2 such that

$$|g(x) - b_2| < m|b_2|\epsilon, \qquad \text{whenever } 0 < |x - a| < \delta_2.$$

We set

$$\delta = \text{minimum } \{\delta_1, \delta_2\}.$$

It follows that, if $0 < |x - a| < \delta$, then

$$\begin{aligned} \left|\frac{1}{g(x)} - \frac{1}{b_2}\right| &= \left|\frac{b_2 - g(x)}{b_2 g(x)}\right| \\ &= \frac{1}{|g(x)|}\frac{1}{|b_2|}|g(x) - b_2| \\ &< \frac{1}{m}\frac{1}{|b_2|}|g(x) - b_2| \\ &< \frac{1}{m|b_2|}m|b_2|\epsilon = \epsilon. \end{aligned}$$

Thus (1) is proved, and, as we have seen, (1) and (iii) imply (iv). This completes the proof of the theorem.

APPENDIX B

Properties of the Definite Integral

Five basic properties of the definite integral are listed at the beginning of Section 4 of Chapter 4. Of these, two are proved in the text and one is left as an exercise. The remaining two will be proved here.

Let f be a function which is bounded on a closed interval $[a, b]$. This implies that $[a, b]$ is contained in the domain of f and that there exists a positive number B such that $|f(x)| \leq B$ for all x in $[a, b]$. We recall that, for every partition σ of $[a, b]$, there are defined the upper and lower sums for f relative to σ, which are denoted by U_σ and L_σ, respectively. Moreover, it has been shown (see page 168) that

$$L_\sigma \leq L_{\sigma \cup \tau} \leq U_{\sigma \cup \tau} \leq U_\tau, \tag{1}$$

for any two partitions σ and τ of $[a, b]$. The function f is defined to be integrable over $[a, b]$ if there exists one and only one number, denoted $\int_a^b f$, with the property that

$$L_\sigma \leq \int_a^b f \leq U_\tau,$$

for any two partitions σ and τ of $[a, b]$. It is an immediate consequence of this definition and the inequalities (1) that f is integrable over $[a, b]$ if and only if, for any positive number ϵ, there exists a partition σ of $[a, b]$ such that $U_\sigma - L_\sigma < \epsilon$. A similar corollary, which we shall also use in the subsequent proofs, is the statement that f is integrable over $[a, b]$ and $\int_a^b f = J$ if and only if, for every positive number ϵ, there exists a partition σ of $[a, b]$ such that $|U_\sigma - J| < \epsilon$ and $|J - L_\sigma| < \epsilon$.

The first property of the definite integral, which we shall establish in this section, is presented in the following theorem:

THEOREM 1. *The function f is integrable over the intervals $[a, b]$ and $[b, c]$ if and only if it is integrable over their union $[a, c]$. Furthermore,*

$$\int_a^b f + \int_b^c f = \int_a^c f.$$

Proof. We first assume that f is integrable over $[a, b]$ and over $[b, c]$. Let ϵ be an arbitrary positive number. Then there exists a partition σ_1 of $[a, b]$, and a partition σ_2 of $[b, c]$ such that the following inequalities hold:

$$\left|U_{\sigma_1} - \int_a^b f\right| < \frac{\epsilon}{2}, \qquad \left|\int_a^b f - L_{\sigma_1}\right| < \frac{\epsilon}{2},$$

$$\left|U_{\sigma_2} - \int_b^c f\right| < \frac{\epsilon}{2}, \qquad \left|\int_b^c f - L_{\sigma_2}\right| < \frac{\epsilon}{2}.$$

It follows from these that

$$\left|(U_{\sigma_1} + U_{\sigma_2}) - \left(\int_a^b f + \int_b^c f\right)\right| < \epsilon,$$

$$\left|\left(\int_a^b f + \int_b^c f\right) - (L_{\sigma_1} + L_{\sigma_2})\right| < \epsilon.$$

Let us set $\sigma_1 \cup \sigma_2 = \sigma$. This union is a partition of $[a, c]$, and it is obvious that

$$U_{\sigma_1} + U_{\sigma_2} = U_\sigma,$$

$$L_{\sigma_1} + L_{\sigma_2} = L_\sigma.$$

Hence

$$\left|U_\sigma - \left(\int_a^b f + \int_b^c f\right)\right| < \epsilon,$$

$$\left|\left(\int_a^b f + \int_b^c f\right) - L_\sigma\right| < \epsilon.$$

These inequalities imply that f is integrable over $[a, c]$ and also that

$$\int_a^c f = \int_a^b f + \int_b^c f.$$

It remains to prove that, if f is integrable over $[a, c]$, then it is integrable over $[a, b]$ and over $[b, c]$. We choose an arbitrary positive number ϵ. Since f is integrable over $[a, c]$, there exists a partition σ of $[a, c]$ such that $U_\sigma - L_\sigma < \epsilon$. Let us form a refinement of the partition σ by adjoining the number b. That is, we set

$$\sigma' = \sigma \cup \{b\}.$$

(It is, of course, possible that σ already contains b, in which case $\sigma' = \sigma$.) Then

$$L_\sigma \le L_{\sigma'} \le U_{\sigma'} \le U_\sigma,$$

from which it follows that $U_{\sigma'} - L_{\sigma'} < \epsilon$. But, since b belongs to σ', we can write $\sigma' = \sigma_1 \cup \sigma_2$, where σ_1 is a partition of $[a, b]$ and σ_2 is a partition of $[b, c]$. Moreover,

$$U_{\sigma'} = U_{\sigma_1} + U_{\sigma_2},$$

$$L_{\sigma'} = L_{\sigma_1} + L_{\sigma_2}.$$

Hence

$$(U_{\sigma_1} - L_{\sigma_1}) + (U_{\sigma_2} - L_{\sigma_2}) = U_{\sigma'} - L_{\sigma'} < \epsilon,$$

Since $U_{\sigma_1} - L_{\sigma_1}$ and $U_{\sigma_2} - L_{\sigma_2}$ are both nonnegative, it follows that

$$U_{\sigma_1} - L_{\sigma_1} < \epsilon,$$

$$U_{\sigma_2} - L_{\sigma_2} < \epsilon.$$

The first of these inequalities implies that f is integrable over $[a, b]$, and the second that f is integrable over $[b, c]$. This completes the proof of Theorem 1.

The second result to be proved is the following:

THEOREM 2. *If f and g are integrable over $[a, b]$, then so is their sum and*

$$\int_a^b (f + g) = \int_a^b f + \int_a^b g.$$

Proof. Let ϵ be an arbitrary positive number. By taking, if necessary, the common refinement $\sigma_1 \cup \sigma_2$ of two partitions of $[a, b]$, we may select a partition σ of $[a, b]$ such that

$$\left| U_\sigma^{(f)} - \int_a^b f \right| < \frac{\epsilon}{2}, \qquad \left| \int_a^b f - L_\sigma^{(f)} \right| < \frac{\epsilon}{2},$$

$$\left| U_\sigma^{(g)} - \int_a^b g \right| < \frac{\epsilon}{2}, \qquad \left| \int_a^b g - L_\sigma^{(g)} \right| < \frac{\epsilon}{2},$$

where $U_\sigma^{(f)}$ and $L_\sigma^{(f)}$ are, respectively, the upper and lower sums for f relative to σ, and $U_\sigma^{(g)}$ and $L_\sigma^{(g)}$ are the same for g. We conclude from the above inequalities that

$$\left| (U_\sigma^{(f)} + U_\sigma^{(g)}) - \left(\int_a^b f + \int_a^b g \right) \right| < \epsilon, \tag{2}$$

$$\left| \left(\int_a^b f + \int_a^b g \right) - (L_\sigma^{(f)} + L_\sigma^{(g)}) \right| < \epsilon. \tag{3}$$

Let $[x_{i-1}, x_i]$ be the ith subinterval of the partition σ. We denote by $M_i^{(f)}$ and $M_i^{(g)}$ the least upper bounds of the values of f and of g, respectively, on

$[x_{i-1}, x_i]$, and by $m_i^{(f)}$ and $m_i^{(g)}$ the analogous greatest lower bounds. Then

$$m_i^{(f)} + m_i^{(g)} \leq f(x) + g(x) \leq M_i^{(f)} + M_i^{(g)},$$

for every x in $[x_{i-1}, x_i]$. It follows that

$$m_i^{(f)} + m_i^{(g)} \leq m_i^{(f+g)} \leq M_i^{(f+g)} \leq M_i^{(f)} + M_i^{(g)},$$

where $m_i^{(f+g)}$ and $M_i^{(f+g)}$ are, respectively, the greatest lower bound and the least upper bound of the values of $f + g$ on $[x_{i-1}, x_i]$. By multiplying each term in the preceding chain of inequalities by $(x_i - x_{i-1})$ and then summing on i, we obtain

$$L_\sigma^{(f)} + L_\sigma^{(g)} \leq L_\sigma^{(f+g)} \leq U_\sigma^{(f+g)} \leq U_\sigma^{(f)} + U_\sigma^{(g)}, \tag{4}$$

where $U_\sigma^{(f+g)}$ and $L_\sigma^{(f+g)}$ are the upper and lower sums, respectively, for $f + g$ relative to σ. The inequalities (2), (3), and (4) imply that

$$\left| U_\sigma^{(f+g)} - \left(\int_a^b f + \int_a^b g \right) \right| < \epsilon,$$

$$\left| \left(\int_a^b f + \int_a^b g \right) - L_\sigma^{(f+g)} \right| < \epsilon.$$

It follows from these two inequalities that the function $f + g$ is integrable over $[a, b]$ and that

$$\int_a^b (f + g) = \int_a^b f + \int_a^b g.$$

This completes the proof of Theorem 2.

APPENDIX C

Equivalent Definitions of the Integral

The purpose of this section is to prove that the definite integral $\int_a^b f$, defined on page 169 in terms of upper and lower sums, can be equivalently defined as the limit of Riemann sums. The fact that these two approaches to the integral are the same is stated without proof in Theorem (2.1), page 414, and we shall now supply the details of the argument. The "if" and the "only if" directions of the proof will be treated separately.

Let f be a real-valued function which is bounded on the closed interval $[a, b]$. This implies, according to our definition of boundedness, that $[a, b]$ is contained in the domain of f. Let $\sigma = \{x_0, \ldots, x_n\}$ be a partition of $[a, b]$ such that

$$a = x_0 \leq x_1 \leq \cdots \leq x_n = b.$$

If an arbitrary number x_i^* is chosen in the ith subinterval $[x_{i-1}, x_i]$, then the sum

$$R_\sigma = \sum_{i=1}^{n} f(x_i^*)(x_i - x_{i-1})$$

is a Riemann sum for f relative to σ. The fineness of a partition σ is measured by its mesh, which is denoted by $\|\sigma\|$ and defined by

$$\|\sigma\| = \underset{1 \leq i \leq n}{\text{maximum}} \{(x_i - x_{i-1})\}.$$

The first of the two theorems is:

THEOREM 1. *If f is bounded on $[a, b]$ and if $\lim_{\|\sigma\| \to 0} R_\sigma = L$, then f is integrable over $[a, b]$ and $\int_a^b f = L$.*

Proof. We assume that $a < b$, since otherwise $L = 0 = \int_a^a f$ and the result is trivial. It is a consequence of the definition of integrability that the conclusion of Theorem 1 is implied by the following proposition: For any positive number ϵ, there exists a partition σ of $[a, b]$ such that, where U_σ and L_σ are, respectively, the upper and lower sums for f relative to σ, then $|U_\sigma - L| < \epsilon$ and $|L - L_\sigma| < \epsilon$. It is this that we shall prove.

We first prove that, if U_σ is the upper sum for f relative to *any* partition σ of $[a, b]$, then there exists a Riemann sum $R_\sigma^{(1)}$ for f relative to σ such that $|U_\sigma - R_\sigma^{(1)}|$ is arbitrarily small. Let $\sigma = \{x_0, \ldots, x_n\}$ be the partition with the usual proviso that

$$a = x_0 \leq x_1 \leq \cdots \leq x_n = b,$$

and let ϵ be an arbitrary positive number. For each $i = 1, \ldots, n$, set M_i equal to the least upper bound of the values of f in the subinterval $[x_{i-1}, x_i]$. Then there exists a number x_i^* in $[x_{i-1}, x_i]$ such that

$$0 \leq M_i - f(x_i^*) \leq \frac{\epsilon}{2(b-a)} \cdot$$

Hence

$$0 \leq M_i(x_i - x_{i-1}) - f(x_i^*)(x_i - x_{i-1}) \leq \frac{\epsilon}{2(b-a)}(x_i - x_{i-1}),$$

and so

$$0 \leq \sum_{i=1}^{n} M_i(x_i - x_{i-1}) - \sum_{i=1}^{n} f(x_i^*)(x_i - x_{i-1}) \leq \frac{\epsilon}{2(b-a)} \sum_{i=1}^{n} (x_i - x_{i-1}).$$

However,

$$\sum_{i=1}^{n} M_i(x_i - x_{i-1}) = U_\sigma,$$

$$\frac{\epsilon}{2(b-a)} \sum_{i=1}^{n} (x_i - x_{i-1}) = \frac{\epsilon}{2(b-a)}(b-a) = \frac{\epsilon}{2} \cdot$$

Moreover, $\sum_{i=1}^{n} f(x_i^*)(x_i - x_{i-1})$ is a Riemann sum for f relative to σ, which we denote by $R_\sigma^{(1)}$. Thus the preceding inequalities become

$$0 \leq U_\sigma - R_\sigma^{(1)} \leq \frac{\epsilon}{2}, \tag{1}$$

which proves the assertion at the beginning of the paragraph.

In an entirely analogous manner, we can prove that, if L_σ is the lower sum relative to an arbitrary partition σ of $[a, b]$ and if ϵ is any positive number, then there exists a Riemann sum $R_\sigma^{(2)}$ such that

$$0 \leq R_\sigma^{(2)} - L_\sigma \leq \frac{\epsilon}{2} \cdot \tag{2}$$

We are now ready to use the premise of Theorem 1—the fact that $\lim_{\|\sigma\| \to 0} R_\sigma = L$. Let ϵ be an arbitrary positive number. Then there exists a positive number δ such that, if σ is any partition of $[a, b]$ with mesh less than δ, then

$$|R_\sigma - L| < \frac{\epsilon}{2},$$

for every Riemann sum R_σ. Accordingly, let σ be a partition of $[a, b]$ with $\|\sigma\| < \delta$, and let U_σ and L_σ be, respectively, the upper and lower sums for f relative to this partition. It follows from the preceding two paragraphs that there exist Riemann sums $R_\sigma^{(1)}$ and $R_\sigma^{(2)}$ for f relative to σ such that

$$|U_\sigma - R_\sigma^{(1)}| \leq \frac{\epsilon}{2},$$

$$|R_\sigma^{(2)} - L_\sigma| \leq \frac{\epsilon}{2},$$

[see inequalities (1) and (2)]. Since the mesh of σ is less than δ, we have

$$|R_\sigma^{(1)} - L| < \frac{\epsilon}{2},$$

$$|L - R_\sigma^{(2)}| = |R_\sigma^{(2)} - L| < \frac{\epsilon}{2}.$$

Hence

$$\begin{aligned} |U_\sigma - L| &= |(U_\sigma - R_\sigma^{(1)}) + (R_\sigma^{(1)} - L)| \\ &\leq |U_\sigma - R_\sigma^{(1)}| + |R_\sigma^{(1)} - L| \\ &< \frac{\epsilon}{2} + \frac{\epsilon}{2} = \epsilon, \end{aligned}$$

and, similarly,

$$\begin{aligned} |L - L_\sigma| &= |(L - R_\sigma^{(2)}) + (R_\sigma^{(2)} - L_\sigma)| \\ &\leq |L - R_\sigma^{(2)}| + |R_\sigma^{(2)} - L_\sigma| \\ &< \frac{\epsilon}{2} + \frac{\epsilon}{2} = \epsilon. \end{aligned}$$

Thus both $|U_\sigma - L|$ and $|L - L_\sigma|$ are less than ϵ, and the proof of Theorem 1 is complete.

The converse proposition is the following:

THEOREM 2. *If f is integrable over $[a, b]$, then* $\lim_{\|\sigma\| \to 0} R_\sigma = \int_a^b f$.

Proof. We assume from the outset that $a < b$. Let ϵ be an arbitrary positive number. Since f is integrable, there exist partitions of $[a, b]$ with upper and lower sums arbitrarily close to $\int_a^b f$. By taking, if necessary, the common refinement $\sigma \cup \tau$ of two partitions σ and τ (see the inequalities $L_\sigma \leq L_{\sigma \cup \tau} \leq U_{\sigma \cup \tau} \leq U_\tau$ on page 168), we may choose a partition $\sigma_0 = \{x_0, \ldots, x_n\}$ of $[a, b]$ such that

$$U_{\sigma_0} - \int_a^b f < \frac{\epsilon}{2},$$

$$\int_a^b f - L_{\sigma_0} < \frac{\epsilon}{2}.$$

The assumption of integrability implies that the function f is bounded on $[a, b]$. Thus there exists a positive number B such that $|f(x)| \leq B$ for every x in $[a, b]$. We define

$$\delta = \frac{\epsilon}{4Bn}.$$

Next, let σ be any partition of $[a, b]$ with mesh less than δ. Consider the common refinement $\sigma \cup \sigma_0$. Since

$$L_{\sigma_0} \leq L_{\sigma \cup \sigma_0} \leq \int_a^b f \leq U_{\sigma \cup \sigma_0} \leq U_{\sigma_0},$$

we have

$$U_{\sigma \cup \sigma_0} - \int_a^b f < \frac{\epsilon}{2}, \tag{3}$$

$$\int_a^b f - L_{\sigma \cup \sigma_0} < \frac{\epsilon}{2}. \tag{4}$$

The partition $\sigma \cup \sigma_0$ may be regarded as having been obtained from σ by the addition of at most $n - 1$ new points of σ_0. Hence at most $n - 1$ of the subintervals of σ have been further partitioned by the inclusion of points of σ_0 in their interiors. Each of these further partitioned subintervals has length less than δ. It follows that, on each of them, the contribution to the difference $U_\sigma - U_{\sigma \cup \sigma_0}$ is less than the product $\delta(2B)$. On those subintervals of σ which have not been hit by points of σ_0 in their interiors, the corresponding terms of U_σ and of $U_{\sigma \cup \sigma_0}$ are the same. We conclude that

$$U_\sigma - U_{\sigma \cup \sigma_0} < (n - 1)\delta(2B),$$

and, similarly,

$$L_{\sigma \cup \sigma_0} - L_\sigma < (n - 1)\delta(2B).$$

However,

$$(n - 1)\delta(2B) = (n - 1)\frac{\epsilon}{4nB}(2B) < \frac{\epsilon}{2}.$$

Hence

$$U_\sigma - U_{\sigma \cup \sigma_0} < \frac{\epsilon}{2},$$

$$L_{\sigma \cup \sigma_0} - L_\sigma < \frac{\epsilon}{2}.$$

Combining these inequalities with (3) and (4), we conclude that

$$U_\sigma - \int_a^b f < \epsilon, \tag{5}$$

$$\int_a^b f - L_\sigma < \epsilon. \tag{6}$$

for every partition σ with mesh less than δ.

Finally, let R_σ be an arbitrary Riemann sum for f relative to a partition σ of $[a, b]$ with mesh less than δ. We know that

$$L_\sigma \leq R_\sigma \leq U_\sigma$$

(see page 413). These inequalities together with those in (5) and (6) immediately imply that

$$|R_\sigma - \int_a^b f| < \epsilon.$$

Hence $\lim_{\|\sigma\| \to 0} R_\sigma = \int_a^b f$, and the proof of Theorem 2 is complete.

The conjunction of Theorems 1 and 2 is equivalent to Theorem (2.1), page 414. We have therefore proved that the definite integral defined in terms of upper and lower sums is the same as the limit of Riemann sums.

Answers to Problems

Students sometimes make the mistake of assuming that a printed answer must be correct. Unfortunately, this is not always true in first editions. The authors will appreciate being notified of any errors which are discovered.

CHAPTER 1

Section 1, pages 8–9

1. $(2, 4)$ is open, bounded; $[3, 5]$ is closed, bounded; $(-\infty, -2]$ is closed, unbounded; $[1.5, 2.5)$ is neither, but bounded; $(\sqrt{2}, \pi)$ is open, bounded.
2. (d) $x > 2$ or $x < -2$
   (e) $-3 < y < -1$ or $1 < y < 3$
   (f) $1 \leq z \leq 3$
   (g) $x \neq a$
   (h) $-4 < u < 0$ or $2 < u < 6$
8. yes

Section 2, pages 17–19

2. The point (a, b) is symmetric to the point (b, a) around the line $x = y$.
3. Use the signs of a and b.
4. $2\sqrt{5}$, $\sqrt{2}$, $2\sqrt{5}$, $2\sqrt{5}$.
8. (a), (b), (c), (d), (g), (i), (j), (k), (l) are functions.
9. (a) $f(2) = 7, f(0) = 1, f(a) = a^2 + a + 1,$
   $f(a + b) = a^2 + b^2 + 2ab + a + b + 1,$
   $f(a - b) = a^2 + b^2 - 2ab + a - b + 1.$
   (b) $g(0) = -1, g(-1) = 0, g(10) = \frac{11}{9},$
   $$g(5 + t) = \frac{t + 6}{t + 4}, \quad g(x^3) = \frac{x^3 + 1}{x^3 - 1}.$$
10. (a) $f(x) = x, g(x) = x^2$ (b) $f(x) = \sqrt{x}, g(x) = |x|$
    (c) $f(x) = x, g(x) = x + 1$ (d) $f(x) = \sqrt{x}, g(x) = \sqrt{|x|}$

11. (a) $x \neq 3$ (b) $x \neq \pm\sqrt{2}$ (c) $x \neq 3, -4$
(d) R (e) $t < 5$ (f) $x \leq -2, x \geq 10$
(g) $x \neq 0, 9$

Section 3, pages 25–26

1. (a) $\frac{1}{x^3} - \frac{4}{x^2} + \frac{5}{x} - 2 = \left(\frac{1}{x} - 2\right)\left(\frac{1}{x^2} - \frac{2}{x} + 1\right)$
(b) $x^3 - 4x^2 + 5x - 2 + \frac{1}{x}$ (c) $\frac{1}{x^3 - 4x^2 + 5x - 2}$
(d) $x^2 - 4x + 5 - \frac{2}{x}$ (e) $5x - 20 + \frac{25}{x} - \frac{10}{x^2}$
2. domain of f: R, domain of g: $x \neq 0$.
(a) $x \neq 0$ (b) $x \neq 0$ (c) $x \neq 1, 2$
(d) $x \neq 0$ (e) $x \neq 0$
6. $(3f)(x) = 3 \cdot f(x)$ for all x in the domain of f. $\sqrt{f}\,(x) = \sqrt{f(x)}$ for all x in the domain of f such that $f(x) \geq 0$.
7. (b) $v = \sqrt{g^2t^2} = g|t|$
8. (a) $x^2 + x + 1$ (b) $x^4 + 5x^2 + x + 6$
(c) $x^6 + 7x^4 + 17x^2 + 14$
9. (a) $x^6 + 3x^4 + 4x^2 + 4$ (b) $\frac{(x+1)^3}{x^3} + \frac{(x+1)}{x} + 2$
(c) $x^2 + \frac{1}{x} + 2$
12. $f + g$ is the set of all ordered pairs (a, b) such that a is in the domain of both f and g and $b = f(a) + g(a)$. fg is the set of all ordered pairs (a, b) such that a is in the domain of both f and g, and $b = f(a) \cdot g(a)$.
13. (a) the set of all real x such that x is in the domain of g and $g(x)$ is in the domain of f.
(b), (c) the set of all real x such that x is in the domains of both f and g.

Section 4, pages 38–39

1. (a) 3 (b) -1 (c) 1 (d) 2
(e) 0 (f) $\frac{1}{2}$ (g) 0 (h) 5
(i) $3a^2 + 2$ (j) $4x - 1$
2. (a) none (b) none
(c) not continuous at $x = -1, 2$ (d) not continuous at $x = 2$
(e) not continuous if $x < -3$ (f) not continuous at $x = 0$
(g) none (h) not continuous at any x
(i) not continuous at $x = 1$
3. (a), (d) and (i) have removable discontinuities.
6. (a) 0 (b) 3 (c) 4 (d) 3
7. both true

9. (a) ∞ (b) does not exist (c) 1 (d) ∞

10. all except (vi).

11. $\lim_{x \to a+} f(x) = b$ if, given $\epsilon > 0$, there exists $\delta > 0$ such that if x is in the domain of f and if $0 < x - a < \delta$, then $|f(x) - b| < \epsilon$.

12. $f(x) = \dfrac{1 + |x|}{x}$

14. (a) $\lim_{x \to a} f(x) = \infty$ if, given any $\epsilon > 0$, there exists $\delta > 0$ such that if x is in the domain of f and if $0 < |x - a| < \delta$, then $f(x) > \epsilon$.
    (b) $\lim_{x \to -\infty} f(x) = +\infty$ if, given any $\epsilon > 0$, there exists δ such that if x is in the domain of f and if $x < \delta$, then $f(x) > \epsilon$.

Section 5, pages 45–46

1. (a) $y = x + 1$ (b) $y = x + 1$ (c) $y = -2x + 1$
   (d) $y = -\frac{1}{2}x - \frac{7}{2}$ (e) $y = -2x - 3$ (f) $y = -x + 1$
   (g) $y = -19x$ (h) $y = 4$ (i) $x = 2$

2. (a) -1 (b) -1 (c) $\frac{1}{2}$
   (d) no slope (e) 0 (f) $-\frac{4}{3}$

3. (b) $y = -\frac{5}{2}x + \frac{11}{4}$
   (c) If colinear, then $y = \left(\dfrac{b_2 - a_2}{b_1 - a_1}\right) x + \dfrac{a_2 b_1 - a_1 b_2}{b_1 - a_1}$

5. (a) x-intercept $= -\frac{10}{3}$
   y-intercept $= 10$
   (b) $y = mx + b$

6. (a) $f(x) = 3x - 7$ (b) $f(x) = \frac{3}{5}$
   (c) $f(x) = -\frac{2}{3}|x| + \frac{4}{3}$ (d) $f(x) = x - 1$
   (e) $y = \pm\sqrt{-2x - 3}$ (f) $f(x) = x$
   (g) $f(x) = 3x^2 + 4x + 2$ (h) $f(x) = -\frac{5}{3}x + \frac{1}{3}$
   linear functions: (a), (b), (d), (f), (h)

7. (a) and (d) are parallel; (a) and (b), (b) and (d), and (e) and (f) are perpendicular.

8. (a) $y = -\frac{5}{2}x + \frac{11}{2}$ (b) $2y + 5x = 5$
   (c) $5y - 2x = 13$

Section 6, pages 55–57

1. (a) 6 (b) $6a$

2. (a) $y = 6x + 1$ (b) $y = 6ax - 3a^2 + 4$

3. $-\frac{6}{49}$

4. (a) $3a^2$ (b) $2a + 3$ (c) 0
   (d) $\dfrac{1}{2\sqrt{a}}$ (e) $1 - \dfrac{2}{a^3}$ (f) $3a^2 + 6a + 3$

(g) $\dfrac{a}{\sqrt{a^2+1}}$ (h) $\dfrac{-a}{(a^2+1)^{3/2}}$ (i) $\dfrac{1}{3a^{2/3}}$

5. (a) $y = 0$ (b) $y = 5x + 4$

(c) $y = 7$ (d) $y = x\left(1 - \dfrac{2}{a^3}\right) + \dfrac{3}{a^2}$

6. (a) $2x$

7. (b) $-1, 1$

9. (a) 1 (b) 27 (c) $3b^2 + 6b + 3$

(d) $\dfrac{x}{\sqrt{x^2+1}}$ (e) $2a + 3$ (f) $1 - \dfrac{2}{x^3}$

10. (a) 2 (b) 64 (c) $32t$

12. Neither $f(x) = x^{1/3}$ nor $g(x) = |x|$ has a derivative at $x = 0$.

Section 7, pages 66–68

1. (a) $6x + 4$ (b) $3x^2 + 2x$

(c) $2x^3(x+2) + 3x^2(x+2)^2$ (d) $4x^3 + 6x^2 - 2x - 8$

(e) $4x - \dfrac{1}{x^4}$ (f) $\dfrac{4}{(2-x)^2}$

(g) $\dfrac{2x+4}{(2-x)^3}$ (h) $\dfrac{-2x^7 + 3x^2}{(x^5+1)^2}$

(i) $\dfrac{-12(3-x)}{(3+x)^3}$ (j) $6x(x^2+1)^2$

(k) $\dfrac{-(2x^2+2x+1)}{(x^2+x)^2}$ (l) $\dfrac{-2x}{(x^2+1)^2}$

(m) $2\left(x - \dfrac{1}{x^3}\right)$

(n) $(x-a)(x-b) + (x-a)(x-c) + (x-b)(x-c)$

2. $y = -2x + 4$

3. $y = 2x^2 - 6x + 4$

5. (a) $(f_1f_2f_3f_4)' = f_1'f_2f_3f_4 + f_1f_2'f_3f_4 + f_1f_2f_3'f_4 + f_1f_2f_3f_4'$

(b) $(f_1f_2f_3\cdots f_n)' = f_1'f_2f_3\cdots f_n + f_1f_2'f_3\cdots f_n + \cdots + f_1f_2\cdots f_{n-1}f_n'$

6. $28x - 25y = 16$

7. (a) $7\frac{1}{2}$ (b) $4x - \dfrac{4}{x^3}$

(c) $\dfrac{-2}{(x-1)^2}$ (d) $-\frac{1}{4}$

(e) $\dfrac{-2}{a^3}$ (f) $6u + 4$

8. $a = 2, b = 8, c = 9$

9. (a) $6x + 2, 6$ (b) $5, 0$

(c) $\frac{x^3}{3} + \frac{x^2}{2} + x + 1, x^2 + x + 1$

(d) $5t^4 - 3t^2, 20t^3 - 6t$ (e) $3x^2 - 2x^{-3}, 6x + 6x^{-4}$

(f) $\frac{4s}{(s^2+1)^2}, \quad \frac{-12s^2+4}{(s^2+1)^3}$

10. $a = -1, b = 3, c = 0$

11. (a) $6x - 2$ (b) $\frac{-4}{(x+1)^3}$

(c) $6a$ (d) $-24a^{-5}$

12. $\left(1 + \sqrt{2}, \frac{(1+\sqrt{2})^3}{3} - (1+\sqrt{2})^2\right)$,

$\left(1 - \sqrt{2}, \frac{(1-\sqrt{2})^3}{3} - (1-\sqrt{2})^2\right)$

13. (a) x for $x \geq 0$, $-x$ for $x \leq 0$ (b) yes
(c) 1 for $x > 0$, -1 for $x < 0$

14. (a) $\frac{4}{3}x^{1/3}$ (b) yes (c) $\frac{4}{9x^{2/3}}$

15. (b) $g'(x) = 2x$ for $x \leq 1$, $=2$ for $x > 1$
$g''(x) = 2$ for $x < 1$, $=0$ for $x > 1$
(c) g is differentiable; g' is not

Section 8, pages 75–76

1. (a) $10x(x^2+1)^4$ (b) $4(x^2 - 2x + 3)(x - 1)$

(c) $\frac{3x^2(1-x^2)}{(x^2+1)^4}$ (d) $\frac{2x}{(x^2+1)^2}$

(e) $-\frac{2}{3}x^{-5/3}$ (f) $\frac{16x(x^2-1)^3}{(x^2+1)^5}$

(g) $\big(2(x^2 + 3x + 2) + 3\big)(2x + 3)$

2. (a) $20x(1 + x^2)^9$

(b) $6(x^4 + 3x^3 + 2x^2 + x + 4)^5(4x^3 + 9x^2 + 4x + 1)$

(c) $4t(t^2+1)^3(2t^2-3)^2(7t^2-3)$

(d) $\frac{3}{2}x^2(x^3-1)^{-1/2}$ (e) $\frac{6(x-1)^2}{(x+1)^4}$

(f) $-s(s^2+1)^{-3/2}$ (g) $2y(y^2+1)^{-3/2} - 3y^3(y^2+1)^{-5/2}$
(h) $-20(u + u^{-1/2})^{-5}(1 - \frac{1}{2}u^{-3/2})$

3. $10x + 6$

4. $[35(2x^2-6)^6 + 4(2x^2-6)](4x)$, 17,984

5. $\frac{-3t}{(t^2+5)^{5/2}}, \quad -\frac{2}{81}$

6. $\dfrac{22}{27}$

10. $\dfrac{100}{9}$

11. (a) $2 \sinh x \cosh x$ (b) $\sinh x$
(c) 1 (d) $\cosh (\cosh x) \sinh x$

13. 678

Section 9, pages 81–83

1. 0

2. $-\dfrac{\sqrt{3}}{3}, \dfrac{\sqrt{3}}{3}$

3. (a) -1 (b) three
(c) $-1, \quad -\dfrac{8+2\sqrt{5}}{5+3\sqrt{5}}, \quad -\dfrac{8-2\sqrt{5}}{5-3\sqrt{5}}$

4. (a) -2 (b) $\frac{1}{4}$
(c) $\dfrac{5}{2}$ (d) $\dfrac{1}{b(a+1)^2}$
(e) $\dfrac{2a}{b(a^2+1)^2}$ (f) $-\dfrac{8}{5}$
(g) 1 (h) $\frac{7}{16}$
(i) -1 (j) -1

5. $-\frac{1}{3}, \frac{5}{18}$

6. (a) $\sqrt{2}, -1$ (b) $-\frac{1}{2}, \frac{1}{4}$
(c) $\frac{1}{2}, -\frac{3}{8}$ (d) $\frac{2}{3}, \frac{10}{9}$

CHAPTER 2

Section 1, pages 93–94

1. (a) min $(\frac{5}{2}, -\frac{1}{4})$ (b) max $(-1, 4)$
(c) min $(\frac{3}{4}, -2\frac{1}{8})$ (d) max $(0, 5)$
(e) local max $(-1, 2)$, local min $(1, -2)$, p.i. $(0, 0)$
(f) local min $(-2, -11)$, p.i. $(-1, 0)$ and p.i. $(1, 16)$
(g) no extrema or p.i. (h) local min $(2, 12)$, p.i. $(-\sqrt[3]{16}, 0)$
(i) local min $\left(\dfrac{\sqrt{3}}{3}, \dfrac{-2\sqrt{3}}{9}\right)$, local max $\left(-\dfrac{\sqrt{3}}{3}, \dfrac{2\sqrt{3}}{9}\right)$, p.i. $(0, 0)$
(j) min $\left(\dfrac{1}{\sqrt[3]{4}}, \dfrac{-3}{\sqrt[3]{256}}\right)$

(k) max $(0, a)$, min $(\pm a, 0)$ (l) min $(0, 0)$
(m) min $(7, 0)$ (n) no extrema or p.i.

(o) local max $(0, 3)$, min $(\pm 2, -13)$, p.i. $\left(\pm \frac{2\sqrt{3}}{3}, -\frac{53}{9}\right)$

(p) local max $(-2, -\frac{9}{8})$, local min $(2, -\frac{1}{8})$

(q) local max $\left(2 - \frac{\sqrt{3}}{3}, \frac{2\sqrt{3}}{9}\right)$, local min $\left(2 + \frac{\sqrt{3}}{3}, -\frac{2\sqrt{3}}{9}\right)$, p.i. $(2, 0)$

(r) local min $(-2, -7)$, local max $(2, 9)$, p.i. $(0, 1)$

4. (b) a (c) $\dfrac{-b^2 + 4ac}{4a}$

6. $f(x) = -x + 3$

7. $f(x) = \frac{1}{2}x + \frac{5}{2}$

8. $f(x) = \dfrac{x^2}{2}$

9. $f(x) = 6$

Section 2, pages 101–103

2. 50 yd $\times$ 100 yd

3. 1

4. (a) local min $(1, 5)$, local max $(-1, 9)$
 (b) local min $(\frac{1}{2}, 3\frac{1}{4})$, local max $(-1, 10)$
 (c) local min $(|a|, 2|a|)$, local max $(-|a|, -2|a|)$
 (d) local max $(2, 27)$, local min $(5, 0)$, absolute min $(0, -25)$
 (e) local min $(2, 4)$, local max $(0, 0)$
 (f) absolute min $(-1, 41)$, local max $(0, 54)$
 (g) local min $(0, 1)$, absolute max $(2, 27)$, absolute min $(1, 0)$, local max $\left(\frac{1}{5}, \frac{2^7 3^3}{5^5}\right)$
 (h) absolute max $(-4, 2)$
 (i) local max $(1, 0)$, absolute min $(0, -4)$, absolute min $(3, -4)$

6. absolute min

7. $y = -x + 4$

8. $x = 4$, $y = 2$

9. $(3, 3)$, $(-3, -3)$

10. (b) $(1, 2)$, $(-1, -2)$
 (c) 2 units $\times$ 4 units [the vertices of the rectangle are the points $(\pm 1, \pm 2)$]

11. 6 units $\times$ 18 units

12. 2 units $\times$ 26 units

13. side of square $=$ 6 in.

14. (a) 3 in. (length of side) (b) $\dfrac{s}{6}$ in.

15. (a) Use all wire for the circle. (b) circumference of circle $= \dfrac{20\pi}{\pi + 4}$

16. Paddle directly to destination.
19. $3 + 2\sqrt{2}$
20. 350
21. no

Section 3, pages 107–108

1. (c) $\frac{5}{2}$ (d) 100 ft
2. Particle starts 100 units to left of origin ($t = 0, v = 135, a = -84$), moves toward origin, slowing down, stops at $t = \frac{5}{2}$ ($x = \frac{75}{2}, v = 0, a = -24$), moves left, reaches maximum speed toward the left at $t = \frac{7}{2}$ ($x = \frac{59}{2}$, $v = -12, a = 0$), stops at $t = \frac{9}{2}$ ($x = \frac{43}{2}, v = 0, a = 24$), and finally moves to right with ever-increasing velocity.
3. At $t = 0$ the particle is 288 units below the origin ($y = -288, v = 144$); it moves up, slowing and stopping at $t = \frac{9}{2}$ ($y = 36, v = 0$), and finally moves down with ever-increasing speed. The acceleration throughout is constant at -32.
4. Particle starts at $x = 5$ ($v = 0, a = -6$), moves left, speeding up. At $t = 1$ ($x = 3, v = -3, a = 0$) it reaches maximum speed to the left and starts slowing down, stops at $t = 2$ ($x = 1, v = 0, a = 6$), and finally moves right, at ever-increasing velocity.
5. Particle starts at $y = 10$ ($v = -9, a = 0$). It moves down until $t = 1$ ($y = 4$, $v = 0, a = 18$), then moves up at ever-increasing speed.
6. 0
8. 25 knots
9. $\dfrac{25}{12\pi}$
11. $\frac{8}{3}$ miles/hr, $\frac{20}{3}$ miles/hr
12. (a) -16 knots (b) 3:12 P.M. (c) 3 miles
13. $-\frac{4}{3}$ ft/sec
15. $\dfrac{4}{9\pi}$ ft/min

Section 4, page 111

1. (a) 1.993 (b) 4.004 (c) 1.998
   (d) 1.060 (e) 4.075 (f) 2.019
   (g) 0.062 (h) 31.875 (i) 0.850
2. 0.1008
3. 5.03
4. 4.72
5. (a) 36π in.^{3} (b) 39.6π in.^{3}
6. (a) 216 in.^{3} (b) 205.2 in.^{3}

7. 3.92

8. 31.66

9. (a) 6.93 (or $4\sqrt{3}$) in². (b) 7.21 in².

10. 0.87

Section 5, page 115

1. (a) $f(8) = f(-1) = f'(\frac{7}{2}) = 0$
   (b) $f(-2\sqrt{3}) = f(0) = f(2\sqrt{3}) = f'(-2) = f'(2) = 0$
   (c) $f(-4) = f(-1) = f(2) = f'(-1 + \sqrt{3}) = f'(-1 - \sqrt{3}) = 0$
   (d) $f(-4) = f(0) = f(4) = f'(-2\sqrt{2}) = f'(0) = f'(2\sqrt{2}) = 0$
   (e) $f(2) = f(-2) = 0$. $f'(x)$ is never zero, and $f'(0)$ does not exist.
   (f) $f(-3) = f(3) = f'(-3) = f'(0) = f'(3) = 0$

2. (a) $\frac{5}{2}$ (b) $\frac{\sqrt{3}}{3}$
   (c) $\sqrt{3}$ (d) $\frac{2}{3} \cdot 3^{2/3}$

5. (a) $2x^2 + c$, c any constant (b) $x^4 + \frac{1}{3}x^3 + 2x + c$
   (c) $-\frac{1}{x} + c$ (d) $-\frac{1}{2(x^2 + 1)^2} + c$

Section 6, pages 121–122

1. (a) $(2x + 1)\,dx$ (b) $7\,dx$
   (c) $[3x^2(5x - 1)^3 + 15(x^3 + 1)(5x - 1)^2]\,dx$
   (d) $\frac{2\,dx}{(x + 1)^2}$ (e) $7u^6\,du$
   (f) $\frac{2v^2u\,du - 2u^2v\,dv}{v^4}$ (g) $(2az + b)\,dz$
   (h) $\frac{dx}{4\sqrt{1 + \sqrt{1 + x}}\,\sqrt{1 + x}}$ (i) dx
   (j) $3v^2(u^2 + 2)\,dv + 2u(v^3 - 1)\,du$

2. $\frac{-(21x^2 + 2)\,dx}{(7x^3 + 2x + 1)^2}$

3. $\frac{-(64t + 4)\,dt}{(16t^2 + 2t)^3}$

6. $\frac{100\pi}{3}$ ft³

7. (a) 2160π in.³ (b) 162π in.³ (c) 2322π in.³

Section 7, pages 130–131

1. (a) 6 (b) $\frac{3}{20}$ (c) 0 (d) $\frac{2}{3}$

(e) $\frac{2}{3}$ (f) $\dfrac{\sqrt{3}}{6}$

(g) $\frac{3}{2}$ (h) 2

3. (a) $\dfrac{1}{n4^{n-1}}$ (b) 0 (c) $-\infty$

(d) $\frac{1}{6}$ (e) $\frac{2}{3}$ (f) -9

4. (a) 0 (b) 0 (c) 0 (d) $\frac{2}{3}$

5. (a) 0 (b) ∞

CHAPTER 3

Section 1, pages 135–136

1. (a) $x^2 + y^2 = 9$ (b) $x^2 + y^2 = \frac{100}{9}$
   (c) $x^2 + y^2 \quad 4x - 4y - 17 = 0$ (d) $x^2 - 6x + y^2 = 0$
   (e) $x^2 + y^2 + 14y = 0$ (f) $x^2 + y^2 + 6x + 6y = 0$
   (g) $x^2 + y^2 - 8x - 8y + 16 = 0$ (h) $x^2 + y^2 \pm 5y = 0$
   (i) $x^2 + y^2 - 6x \pm 4y + 9 = 0,\ x^2 + y^2 - 14x \pm 4y + 49 = 0$

2. All curves are circles, with centers and radii as follows:
   (a) $(0, 0), 8$ (b) $(0, 0), 4\sqrt{2}$
   (c) $(0, 4), 3$ (d) $(-2, 0), 4$
   (e) $(2, -7), \sqrt{19}$ (f) $(\frac{3}{2}, \frac{5}{2}), \frac{5}{4}$
   (g) $(4, 6), 5$ (h) $(\frac{5}{2}, \frac{7}{2}), 4$
   (i) $(\frac{2}{3}, -\frac{5}{3}), \frac{10}{3}$ (j) $(\frac{3}{5}, -\frac{4}{5}), \dfrac{6\sqrt{5}}{5}$

6. (a) $4y = 3x + 25$ (b) $y = 3$
   (c) $12y + 5x = 83$ (d) $3y = 7x - 43$

7. $4x + 18y + 25 = 0$

9. (a) $x^2 + y^2 = 25$ (b) $(x - 3)^2 + (y + 2)^2 = 25$
   (c) $(x + 1)^2 + (y - 4)^2 = 169$

Section 2, pages 141–142

1. (a) $y^2 = -8x$ (b) $x^2 = 12y$
   (c) $x^2 = -4y$ (d) $y^2 = 16x$
   (e) $x^2 = -16y + 32$ (f) $y^2 = -16x - 80$

5. (a) $\dfrac{2a}{y_1}$ (b) $y = \dfrac{2a}{y_1}x - \dfrac{2ax_1}{y_1} + y_1$

8. $y = \frac{1}{2}x + 3,\ y = \frac{3}{8}x + 4$

9. (a) $(-x_1, 0)$

10. (a) $y^2 - 8x - 2y + 9 = 0$ (b) $x^2 - 2x - 4y + 5 = 0$
    (c) $y^2 - 24x - 6y + 105 = 0$ (d) $x^2 + 2x + 8y - 15 = 0$

11. (a) $(3, 1)$ (b) $(1, 2)$
    (c) $(10, 3)$ (d) $(-1, 0)$

12. (a) $y^2 - 4x - 2y + 1 = 0$ (b) $x^2 - 2x - 2y + 2 = 0$
    (c) $y^2 - 12x - 6y + 21 = 0$ (d) $x^2 + 2x + 4y - 11 = 0$

13. Foci, vertices, and directrices are as follows:
 (a) $(0, \frac{1}{4}), (0, 0), y = -\frac{1}{4}$ (b) $(1, -\frac{3}{4}), (1, -1), y = -\frac{5}{4}$
 (c) $(3, -\frac{31}{4}), (3, -8), y = -\frac{33}{4}$ (d) $(0, -1), (\frac{1}{2}, -1), x = 1$
 (e) $(\frac{31}{4}, -2), (8, -2), x = \frac{33}{4}$ (f) $(-\frac{1}{2}, 0), (-\frac{1}{2}, \frac{1}{4}), y = \frac{1}{2}$

Section 3, pages 149–151

1. For each curve, the foci, the endpoints of the major axis, and the endpoints of the minor axis are as follows:
 (a) $(0, \pm\sqrt{5}), (0, \pm 3), (\pm 2, 0)$ (b) $(\pm\sqrt{5}, 0), (\pm 3, 0), (0, \pm 2)$
 (c) $(\pm 5, 0), (\pm 13, 0), (0, \pm 12)$ (d) $(\pm 6, 0), (\pm 10, 0), (0, \pm 8)$
 (e) $(\pm 1, 0), (\pm\sqrt{17}, 0), (0, \pm 4)$

2. (a) $\frac{x^2}{169} + \frac{y^2}{144} = 1$ (b) $\frac{x^2}{49} + \frac{y^2}{16} = 1$
 (c) $\frac{x^2}{12} + \frac{y^2}{4} = 1$ (d) $\frac{x^2}{25} + \frac{y^2}{9} = 1$
 (e) $\frac{4x^2}{33} + \frac{4y^2}{49} = 1$

4. All curves are ellipses; for each the center c, endpoints of major axis M.A., endpoints of minor axis m.a., and foci f are as follows:
 (a) c (2, 4), M.A. (7, 4), (−3, 4), m.a. (2, 7), (2, 1), f (6, 4), (−2, 4)
 (b) c (−3, 2), M.A. (−3, 7), (−3, 3), m.a. (1, 2), (−7, 2), f (−3, 5), (−3, −1)
 (c) c (−5, −2), M.A. (21, −2), (−31, −2), m.a. (−5, 22), (−5, −26), f (5, −2), (−15, −2)
 (d) c (0, −3), M.A. (0, 2), (0, −8), m.a. (3, −3), (−3, −3), f (0, 1), (0, −7)
 (e) c (−2, 3), M.A. (−2, 6), (−2, 0), m.a. (0, 3), (−4, 3), f $(-2, 3 \pm \sqrt{5})$

5. (a) $\frac{8}{3}$ (b) $\frac{2b^2}{a}$

6. (a) $x = \pm\frac{9\sqrt{5}}{5}, \quad e = \frac{\sqrt{5}}{3}$ (b) $y = \pm\frac{18\sqrt{5}}{5}, \quad e = \frac{\sqrt{5}}{3}$

7. (a) $\sqrt{(x + c)^2 + y^2}$ (b) $\left|x + \frac{a^2}{c}\right|$
 (c) $\frac{x^2}{a^2} + \frac{y^2}{a^2 - c^2} = 1$

9. (a) $-\frac{b^2x_1}{a^2y_1}$ (b) $y = -\frac{b^2x_1x}{a^2y_1} + \frac{b^2x_1^2}{a^2y_1} + y_1$

10. (a) $-\frac{2ax_1 + c}{2by_1 + d}$
 (b) $y = -\frac{(2ax_1 + c)x}{2by_1 + d} + \frac{2ax_1^2 + cx_1}{2by_1 + d} + y_1$

11. (a) $\frac{(x - 1)^2}{36} + \frac{(y - 2)^2}{20} = 1$ (b) $\frac{(x - 2)^2}{16} + \frac{(y + 1)^2}{25} = 1$
 (c) $\frac{(x - 7)^2}{25} + \frac{(y - 4)^2}{4} = 1$

12. line segment connecting $(-c, 0)$ and $(c, 0)$
13. $a > b$, $b > a$

Section 4, pages 159–162

2. Each curve is a hyperbola. The center c, foci f, vertices v, and asymptotes for each are as follows:
   (a) $c\ (0, 0), f\ (\pm 5, 0), v\ (\pm 3, 0), y = \pm \frac{4}{3}x$
   (b) $c\ (0, 0), f\ (0, \pm 5), v\ (0, \pm 3\sqrt{2}), y = \pm \dfrac{3\sqrt{14}}{7}x$
   (c) $c\ (0, 0), f\ (\pm 13, 0), v\ (\pm 5, 0), y = \pm \frac{12}{5}x$
   (d) $c\ (0, 0), f\ (\pm 17, 0), v(\pm 15, 0), y = \pm \frac{8}{15}x$
   (e) $c\ (0, 0), f\ (0, \pm\sqrt{34}), v\ (0, \pm 3), y = \pm \frac{3}{5}x$
   (f) $c\ (0, 0), f\ (4, 4), (-4, -4), v\ (2\sqrt{2}, 2\sqrt{2}), (-2\sqrt{2}, -2\sqrt{2})$, x- and y-axes $(y = 0, x = 0)$
   (g) $c\ (0, 0),\ f\ (2\sqrt{3}, -2\sqrt{3}), (-2\sqrt{3}, 2\sqrt{3}), v\ (\sqrt{6}, -\sqrt{6}), (-\sqrt{6}, \sqrt{6})$, x- and y-axes

4. (a) $\dfrac{x^2}{9} - \dfrac{y^2}{16} = 1$ (b) $\dfrac{x^2}{4} - \dfrac{y^2}{36} = 1$
   (c) $\dfrac{y^2}{9} - \dfrac{x^2}{3} = 1$ (d) $\dfrac{y^2}{17} - \dfrac{8x^2}{17} = 1$

6. All are hyperbolas.
   (a) $c\ (1, -3), f\ (6, -3), (-4, -3), v\ (5, -3), (-3, -3)$, $y + 3 = \pm\frac{3}{4}(x - 1)$
   (b) $c\ (1, -3), f\ (1, 2), (1, -8), v\ (1, 0), (1, -6), y + 3 = \pm\frac{3}{4}(x - 1)$
   (c) $c\ (1, -3), f\ (1 + 2\sqrt{6}, -3 - 2\sqrt{6}), (1 - 2\sqrt{6}, -3 + 2\sqrt{6})$, $v\ (1 + 2\sqrt{3}, -3 - 2\sqrt{3}), (1 - 2\sqrt{3}, -3 + 2\sqrt{3}), x = 1, y = -3$
   (d) $c\ (-4, -2), f\ (9, -2), (-17, -2), v\ (1, -2), (-9, -2)$, $y + 2 = \pm\frac{12}{5}(x + 4)$
   (e) $c\ (2, 2),\ f\ (7, 7), (-3, -3), v\left(2 + \dfrac{5\sqrt{2}}{2}, 2 + \dfrac{5\sqrt{2}}{2}\right)$, $\left(2 - \dfrac{5\sqrt{2}}{2}, 2 - \dfrac{5\sqrt{2}}{2}\right)$, $x = 2, y = 2$
   (f) $c\ (1, 1), f\ (3, 1), (-1, 1), v\ (1 + \sqrt{2}, 1), (1 - \sqrt{2}, 1), y = x$, $y = -x + 2$
   (g) $c\ (-2, -2), f\ (-2, -2 \pm 2\sqrt{5}), v\ (-2, 2), (-2, -6)$, $y + 2 = \pm 2(x + 2)$

8. (a) $\dfrac{9}{2}$ (b) $\dfrac{2b^2}{a}$

9. (a) $x = \pm\dfrac{16}{5},\quad e = \dfrac{5}{4}$ (b) $y = \pm\dfrac{25}{13},\quad e = \dfrac{13}{5}$

10. (a) $\sqrt{(x + c)^2 + y^2}$ (b) $\left|x + \dfrac{a^2}{c}\right|$
    (c) $\dfrac{x^2}{a^2} - \dfrac{y^2}{c^2 - a^2} = 1$

11. vertical distances: $10 - \sqrt{10^2 - 16}$, $100 - \sqrt{100^2 - 16}$, $1000 - \sqrt{1000^2 - 16}$, $10{,}000 - \sqrt{10{,}000^2 - 16}$

12. (a) $\dfrac{b^2x_1}{a^2y_1}$ (b) $y = \dfrac{b^2x_1x}{a^2y_1} - \dfrac{b^2x_1^{\,2}}{a^2y_1} + y_1$

13. (a) $\dfrac{2ax_1 + c}{2by_1 - d}$

(b) $y = \dfrac{(2ax_1 + c)x}{2by_1 - d} - \dfrac{2ax_1^{\,2} + cx_1}{2by_1 - d} + y_1$

16. (a) circle; $c\ (-3, -2)$, $r = 5$

(b) ellipse; $c\ (-3, -\frac{1}{2})$, $f\ (-3 \pm \sqrt{3}, -\frac{1}{2})$, directrices: $x = -3 \pm \dfrac{4\sqrt{3}}{3}$,

$e = \dfrac{\sqrt{3}}{2}$, M.A. $= 4$, m.a. $= 2$

(c) parabola; $f\ (-3, \frac{3}{4})$, directrix: $y = \frac{11}{4}$, $v\ (-3, \frac{7}{4})$, axis $x = -3$

(d) hyperbola; $c\ (-3, \frac{1}{2})$, $f\ (-3 \pm \sqrt{5}, \frac{1}{2})$, directrices: $x = -3 \pm \dfrac{4\sqrt{5}}{5}$,

$e = \dfrac{\sqrt{5}}{2}$, asymptotes: $y - \frac{1}{2} = \pm \dfrac{x + 3}{2}$, $v\ (-1, \frac{1}{2})$, $(-5, \frac{1}{2})$, t.a. $= 4$

(e) parabola; $f\ (-\frac{19}{8}, -\frac{1}{2})$, directrix: $x = -\frac{13}{8}$, $v\ (-2, -\frac{1}{2})$, axis: $y = -\frac{1}{2}$

(f) hyperbola; $c\ (-4, -6)$, $f\ (-4 + 3\sqrt{6}, -6 + 3\sqrt{6})$,
$(-4 - 3\sqrt{6}, -6 - 3\sqrt{6})$, directrices: $x + y = \pm 3\sqrt{6} - 10$, $e = \sqrt{2}$,
asymptotes: $x = -4$, $y = -6$, $v\ (-4 - 3\sqrt{3}, -6 - 3\sqrt{3})$,
$(-4 + 3\sqrt{3}, -6 + 3\sqrt{3})$, t.a. $= 6\sqrt{6}$

(g) circle; $c\ (-1, 3)$, $r = 8$

(h) hyperbola; $c\ (3, -\frac{1}{2})$, $f\left(3, -\dfrac{1}{2} \pm \dfrac{\sqrt{15}}{2}\right)$, directrices: $y = -\dfrac{1}{2} \pm \dfrac{\sqrt{15}}{10}$,

$e = \sqrt{5}$, asymptotes: $y + \frac{1}{2} = \pm\frac{1}{2}(x - 3)$, $v\left(3, -\dfrac{1}{2} \pm \dfrac{\sqrt{3}}{2}\right)$,

t.a. $= \sqrt{3}$

(i) hyperbola; $c\ (0, 1)$, $f\ (0, 1 \pm \sqrt{10})$, directrices: $y = 1 \pm \dfrac{9\sqrt{10}}{10}$,

$e = \dfrac{\sqrt{10}}{3}$, asymptotes: $y - 1 = \pm 3x$, $v\ (0, 4)$, $(0, -2)$, t.a. $= 6$

(j) ellipse; $c\ (0, 1)$, $f\ (0, 1 \pm 2\sqrt{2})$, directrices: $y = 1 \pm \dfrac{9\sqrt{2}}{4}$, $e = \dfrac{2\sqrt{2}}{3}$,

M.A. $= 6$, m.a. $= 2$

CHAPTER 4

Section 1, pages 173–174

1. (a) yes (b) yes (c) no (d) yes, $\frac{1}{2}$ or any smaller number

2. (a) 1 (b) $\frac{1}{2}$
(c) 1, $\frac{1}{2}$ (d) 1

3. (a) $\frac{11}{6}$, $\frac{13}{12}$ (b) $\frac{7}{6}$, $\frac{5}{6}$
(c) $\frac{36}{25}$, $\frac{31}{25}$ (d) $\frac{1}{2}$, $-\frac{1}{2}$

4. true

8. (a) $\frac{\pi}{2}$ (b) $\frac{1}{2}$ (c) 4 (d) 4 (e) 1

Section 2, pages 181–183

1. (a) 1 (b) $\frac{2}{5}$ (c) 1 (d) ∞
2. (a) 100 (b) -5 (c) 87 (d) $\frac{1363}{60}$ (e) $\frac{13}{12}$ (f) 62
3. (a) 85 (b) 35 (c) $1 + x + x^2 + x^3$ (d) $x + \frac{x^2}{2} + \frac{x^3}{3} + \frac{x^4}{4}$ (e) $1 - x^4$
5. (a) $\frac{1}{2}(2n^3 + 3n^2 + 5n)$ (b) $\frac{1}{6}(2n^3 - 3n^2 + n)$ (c) $1 + \frac{3}{2n} + \frac{5}{2n^2}$ (d) $1 - \frac{1}{n}$ (e) $\frac{1}{3}(n^3 + 3n^2 + 5n + 3)$
6. (a) 24 (b) $\frac{27}{8}$ (c) $\frac{n}{3}(n^2 + 2)$ (d) $\frac{5n^2 + 6n + 1}{6n}$
7. (a) 1 (b) 1 (c) 0 (d) ∞
8. (a) $\frac{1}{2}$ (b) 2
11. (a) 78 (b) 364

Section 3, page 190

1. (a) $\frac{1}{3}$ (b) 4 (c) 6 (d) 2
3. (a) 2 (b) 6 (c) 6 (d) 27

Section 4, pages 197–198

1. (a) $\int_0^1 x^2\,dx + 5\int_0^1 x\,dx$

(b) $4\int_2^3 x^5\,dx - \int_2^3 x\,dx - 2\int_2^3 dx$

(c) $3\int_1^2 t^3\,dt + 2\int_1^2 t^2\,dt + \int_1^2 t\,dt$

(d) $17\int_5^3 y^{13}\,dy - 11\int_5^3 y^7\,dy + 4\int_5^3 dy$

(e) $\int_0^1 x^4\,dx + 4\int_0^1 x^2\,dx + 4\int_0^1 dx$

2. (a) $\frac{13}{6}$ (b) $-\frac{13}{12}$ (c) 0 (d) $\frac{19}{3}$ (e) $\frac{3}{2}$

3. (a) 5 (b) $\frac{8}{3}$ (c) 14 (d) $\frac{26}{3}$
7. (a) $\geq$ (b) $\geq$ (c) $\leq$
8. (a) $\frac{2}{3}, -\frac{4}{3}, -\frac{2}{3}$ (b) $\frac{2}{3}, \frac{4}{3}, 2$
9. $4, \frac{7}{4}, \frac{23}{4}$

Section 5, pages 206–209

2. (a) $\dfrac{x^8}{8} + c$ (b) $\dfrac{x^4}{4} - \dfrac{1}{2x^2} + c$
   (c) $\frac{35}{6}y^{6/5} + c$ (d) $t^5 + t^3 + t + c$
   (e) $\frac{1}{9}(3x + 1)^3 + c$ (f) $-\dfrac{1}{x^2 + 1} + c$
3. (a) 1 (b) 4 (c) 18 (d) 18
   (e) $\frac{17}{6}$ (f) $\frac{3}{4}(2\sqrt[3]{2} - 3\sqrt[3]{3})$
   (g) $-\frac{5}{3}\sqrt[5]{2}$ (h) $\frac{13}{4}$
   (i) $\frac{8}{3}$ (j) 36
   (k) $\frac{302}{3}$ (l) $2(x^3 - x^2 + x - 21)$
   (m) $\dfrac{t^3}{3} + \dfrac{3t^2}{2} - t$ (n) $\dfrac{x^8}{4}$
   (o) $b - a$ (p) $2x(8x - 1)$
4. (a) $\dfrac{b^{n+1} - a^{n+1}}{n + 1}$ (b) $\dfrac{1}{1 - n}\left(\dfrac{1}{b^{n-1}} - \dfrac{1}{a^{n-1}}\right)$
   (c) (i) We have not yet encountered a function whose derivative is $\dfrac{1}{x}$.
   (ii) If $ab < 0$ we are integrating on an interval on which the function is not continuous.
6. $(0, \infty)$
7. (a) $F'(t) = 6t^2 - 4t + 1$, $F'(-1) = 11$, $F'(2) - 17$, $F'(x) = 6x^2 - 4x + 1$
   (b) $2t^3 - 2t^2 + t$
8. (a) $G'(x) = x + \dfrac{1}{x^2}$, $G'(2) = \frac{9}{4}$
   (b) $G(x) = \dfrac{x^2}{2} - \dfrac{1}{x} + \dfrac{1}{2}$
9. (a) $t^6 + t^2$ (b) $6t^5 + 2t$, $F'(2) = 196$
11. (a) $\sqrt{1 + t^3}$ (b) $-\dfrac{1}{1 + t^2}$
   (c) $\dfrac{1}{2t^2 + 2t + 1}$ (d) $\dfrac{2t}{t^4 + t^2 + 1} - \dfrac{1}{t^2 + t + 1}$
12. $\dfrac{1}{x^2}$ is not integrable on $[-1, 1]$.
13. (a) 6 (b) $\frac{20}{3}$ (c) 5 (d) 4 (e) $\frac{2}{3}$
14. $F'(x) = \dfrac{1}{x^2 + 1}$. No extrema; p.i. at (0, 0). Domain $= (-\infty, \infty)$.

Section 6, pages 218–219

1. (a) $\dfrac{x^3}{3} + \dfrac{x^2}{2} + x + c$ (b) $\frac{3}{4}x^4 + \dfrac{1}{6x^2} + c$
   (c) $2t^3 - t^2 + 5t + c$ (d) $\frac{2}{3}y^3 - \frac{5}{2}y^2 - 3y + c$
   (e) $\frac{1}{5}(2x-1)^{5/2} + c$ (f) $\frac{1}{54}(3x^3+2)^6 + c$
   (g) $-\frac{1}{3}(a^2 - x^2)^{3/2} + c$ (h) $\sqrt{t^2+4t+5} + c$
   (i) $\frac{1}{6}(s^3 + 3s^2 + 5)^2 + c$
   (j) $\dfrac{x^2}{2} + c$ for $x \geq 0$, $-\dfrac{x^2}{2} + c$ for $x < 0$; or $\dfrac{x|x|}{2} + c$
2. (a) cannot (b) $\frac{2}{3}x^{3/2} + 2x^{1/2} + c$ (c) $\frac{1}{15}(y^3+7)^5 + c$.
   (d) $\frac{1}{14}y^{14} + \frac{28}{11}y^{11} + \frac{147}{4}y^8 + \frac{1372}{5}y^5 + \frac{2401}{2}y^2 + c$
   (e) cannot (f) cannot
   (g) $\frac{3}{4}x^4 + 2x^3 - \dfrac{x^2}{2} - 2x + c$
   (h) $\frac{1}{10}(s^2 + 2s - 3)^5 + c$
   (i) $\dfrac{x^2}{2} - x + c$
   (j) cannot (k) cannot
3. $f(x) = 2x^4 + x^2 + 1$
4. $f(x) = x^3 + x - 1$
5. $f(x) = x^4 + x^2 + 5x - 2$
6. (a) 4 (b) 0
   (c) $\dfrac{2^{17}}{27}$ (d) $\sqrt{11} - \sqrt{2}$
   (e) $\dfrac{153{,}340}{81}$ (f) $\frac{2}{3}$
   (g) 0 (h) 12 (i) $\frac{20}{3}$ (j) $\frac{13}{3}$
7. $f'(x) = 9x^2 + 10x + 2$; $f(x) = 3x^3 + 5x^2 + 2x + 1$
8. (a) $g(x) = \frac{4}{15}x^{5/2} - \frac{2}{3}x + \sqrt{2}$ (b) $f(t) = t^3 + t^2 + t + 1$
9. 28
10. $\frac{1}{2}$

Section 7, pages 227–229

1. (a) 4; $P^+ = 4\frac{1}{2}$, $P^- = \frac{1}{2}$ (b) $\frac{3}{2}$; $P^+ = \frac{10}{3}$, $P^- = \frac{11}{6}$
   (c) $\frac{5}{6}$; $P^+ = \frac{9}{2}$, $P^- = \frac{11}{3}$ (d) 0; $P^+ = \frac{1}{4}$, $P^- = \frac{1}{4}$
2. (a) $\frac{1}{3}$ (b) 1 (c) $\frac{7}{2}$ (d) 4
5. $\frac{16}{3}$
6. $\frac{15}{2}$
7. $\frac{61}{24}$
8. $\frac{125}{24}$

9. (a) $\int_{-2}^{2} (4 - x^2)\,dx = \frac{32}{3}$ (b) $2\int_{0}^{4} \sqrt{y}\,dy = \frac{32}{3}$

10. (a) $\frac{13}{6}$ (b) $\frac{9}{2}$ (c) $\frac{16}{3}$

11. (a) yes

12. $\frac{2}{3}$

13. $\frac{a^3}{12}$

14. $\frac{2b}{a}\int_{-a}^{a} \sqrt{a^2 - x^2}\,dx = \frac{4b}{a}\int_{0}^{a} \sqrt{a^2 - x^2}\,dx,\quad \frac{2a}{b}\int_{-b}^{b} \sqrt{b^2 - y^2}\,dy$

15. $\frac{28}{3}$

Section 8, pages 235–237

1. (a) 224 miles
   (b) min. speed 0 at $t = 0$, $s = 0$; max. speed 75 at $t = 2\frac{1}{2}$, $s = 125$
2. (a) $\frac{1}{4}$ hour, $7\frac{1}{2}$ miles (b) 285 miles
3. (a) $-32t + 640$ (b) 4800 ft
   (c) 6400 ft at $t = 20$ sec (d) 8000 ft
   (e) -640 ft/sec
5. (a) to the right (positive direction on x-axis)
   (b) $s = -2t^2 + 20t - 48$ (c) 48
   (d) 68 (e) $t = 4, 6$
6. (a) $v(t) = 3t^2 - 2t - 1$; $y(t) = t^3 - t^2 - t + 1$
   (b) 16 (c) $5\frac{10}{27}$
7. (a) $3\frac{3}{4}$ miles by 20 miles (b) -22.5 miles per hour per hour
   (c) $2\frac{3}{8}$ hours
9. (a) $V = 1728\pi - 10t$ in.^{3}; 172.8π min
   (b) $r = \sqrt[3]{1728 - \frac{10t}{\pi}}$ in.
10. (a) $(\frac{1}{2}, \frac{1}{4})$, $(\frac{3}{4}, \frac{9}{16})$ (b) approaches (1, 1)
11. (a) $x(t) = v_0 t$, $y(t) = -\frac{1}{2}gt^2$ (b) $y = \frac{-g}{2v_0^2}x^2$; parabola

CHAPTER 5

Section 1, pages 247–249

1. (a) $\frac{2}{x}$ (b) $\frac{7}{7x + 2}$
   (c) $\frac{2x + 1}{2(x - 3)(x + 4)}$ (d) $\frac{2x - 9}{x^2 - 9x + 3}$
   (e) $-\frac{1}{x}$ (f) $\frac{3}{7x}$

(g) $\dfrac{2x}{x^2+3}$ (h) $\dfrac{3(\ln x)^2}{x}$

(i) $\dfrac{1}{x\ln x}$ (j) $\dfrac{1}{x}+\dfrac{1}{2(x-1)}$

(k) $\dfrac{1}{x-3}-\dfrac{1}{x+1}$ (l) $\dfrac{2(x-1)}{x^2-2x+4}-\dfrac{2x}{x^2+1}$

(m) $\dfrac{1}{x}+\dfrac{2x}{2-x^2}$

3. (a) $p+r$ (b) $-2p$ (c) $4p+q+3r$
   (d) $r-3p$ (e) $q-2(p+r)$ (f) $6p+3q$

4. (a) $\ln(x^2)+c$ (b) $\ln\sqrt{x^2+1}+c$
   (c) $\frac{1}{2}\ln|x^2-6x+2|+c$ (d) $\frac{1}{3}(x^2+3)^{3/2}+c$
   (e) $\dfrac{13}{3}\ln|x^3-6|+c$ (f) $\ln\dfrac{(x+1)^2\,|2x-1|^{3/2}}{(3x+5)^{4/3}}+c$
   (g) $x+\ln|x-1|+c$ (h) $\dfrac{-1}{2x-1}+c$
   (i) $\frac{1}{2}(\ln x)^2+c$

5. (a) no extrema or p.i. (b) no extrema or p.i.
   (c) abs. min (0, 0), p.i. at $(\pm 1, \ln 2)$
   (d) abs. min $(\pm 2, 4-4\ln 4)$, no p.i.
   (e) no extrema, p.i. at $(\pm 2, 4+4\ln 4)$

6. (a) $\ln 2$ (b) $\frac{1}{5}\ln 6$ (c) $\ln\frac{1}{2}$
   (d) $\ln\sqrt{x^2-3}$ (e) $\dfrac{x^2-1}{x^2}$ (f) $\ln\frac{1}{5}$

7. (a) $\ln\frac{7}{3}$ (b) $\dfrac{(\ln 4)^2}{2}$ (c) $\frac{2}{99}$

8. $x>\sqrt{3}$

9. (a) $x<1$ (b) The integral does not exist.

10. (a) 0 (b) 0 (c) 0 (d) -1 (e) 0

Section 2, pages 255–257

1. (a) $\dfrac{x-4}{3}$ (b) $\dfrac{1}{3x+4}$

2. The range of each must be the domain of the other.

3. (a) $7e^{7x}$ (b) e^{3x+2}
   (c) xe^x+e^x (d) $-2xe^{-x^2}$
   (e) $\dfrac{e^x}{x}+e^x\ln x$ (f) $\dfrac{e^x-e^{-x}}{2}$
   (g) $\dfrac{e^x+e^{-x}}{2}$ (h) $\dfrac{-3}{2e^x}$
   (i) $\dfrac{e^x(x-1)}{x^2}$ (j) $\dfrac{5(1-\ln x)}{x^2}$

(k) $(6x - 4)e^{3x^2-4x+5}$ (l) ae^{ax+b}

4. (a) $\frac{1}{3}e^{3x} + c$ (b) $e^{x^2} + c$
(c) $x + c$ (d) $\ln|e^x - e^{-x}| + c$
(e) $\frac{1}{8}\ln(4e^{x^2} + 5) + c$ (f) $\dfrac{-3}{8e^{4x}} + c$
(g) $\dfrac{-1}{3e^{x^3-2}} + c$ (h) $\frac{1}{2}e^{x^2+2x} + c$
(i) $e^{\pi}x + c$ (j) $\dfrac{e^{ax+b}}{a} + c$

5. (a) $\ln 2$ (b) $\frac{1}{2}(e^6 - 1)$
(c) $e^{x^2} - e^x$ (d) $e^{-4} - e^{-8}$

6. (a) no extrema, no p.i.
(b) abs. min at $\left(\dfrac{1}{e}, -\dfrac{1}{e}\right)$
(c) abs. max at $\left(1, \dfrac{1}{e}\right)$, p.i. at $\left(2, \dfrac{2}{e^2}\right)$
(d) rel. max at $\left(2, \dfrac{4}{e^2}\right)$, abs. min at $(0, 0)$; p.i. at $\left(2 \pm \sqrt{2}, \dfrac{6 \pm 4\sqrt{2}}{e^{2\pm\sqrt{2}}}\right)$
(e) abs. max at $(0, 1)$; p.i. at $\left(\dfrac{\pm\sqrt{2}}{2}, e^{-1/2}\right)$

7. (a) $\frac{1}{2}(e^8 - 1)$ (b) $\frac{1}{2}(e^{16} - e^4)$ (c) $\dfrac{1}{e^5} - \dfrac{1}{e^7}$

8. (a) 0

10. (a) 1 (b) $\frac{1}{2}$ (c) e (d) -1
(e) 0 (f) ∞ (g) 0 (h) 2

11. $g(x) = \sqrt{x}, f(x) = x^2$, domain $g = [0, \infty)$, domain $f = (-\infty, \infty)$

13. $m = 1, 2$

14. (a) e^2 (b) e^{-3}

Section 3, page 262

2. yes
3. $g(x) = x + f(x)$, where $f(x)$ is the step function defined in Problem 3, page 190.
4. $f(x) = x^3$

Section 4, pages 269–271

1. (a) $4^{x+1}\ln 4$ (b) $\dfrac{2x}{(x^2 + 1)\ln 10}$
(c) $\dfrac{\ln 4}{\ln 10}$ (d) $(2x + 1)e^{x^2+x+2}$
(e) $a^x(1 + x\ln a)$ (f) $x2^x(x\ln 2 + 2)$

(g) $e^{-x}(1 - x)$ (h) $\dfrac{2x - 4^x \ln 4}{\ln 4(x^2 - 4^x)}$

(i) $x^{x-1}\left(\dfrac{x-1}{x} + \ln x\right)$ (j) $x^{x^2+1}(1 + \ln(x^2))$

(k) $2x^{2x}(\ln x + 1)$

4. (a) $\dfrac{7^x}{\ln 7} + c$ (b) $\dfrac{2^{3x^3+4}}{9 \ln 2} + c$

(c) $\frac{1}{2}(\ln|x + 2|)^2 + c$ (d) $-\frac{1}{2}\left(\ln|x|\right)^2 + c$

(e) $\dfrac{x}{\ln 2}\left(\dfrac{7x}{2} - 5\right) + c$ (f) $\dfrac{3^{\ln|x+3|}}{\ln 3} + c$

(g) $\dfrac{5^{e^x}}{\ln 5} + c$

5. (a) $\dfrac{1}{2}\left[\dfrac{(x-1)(x+3)}{(x+2)(x-4)}\right]^{1/2}\left(\dfrac{1}{x-1} + \dfrac{1}{x+3} - \dfrac{1}{x+2} - \dfrac{1}{x-4}\right)$

(b) $x < -3$, or $-2 < x < 1$, or $x > 4$

7. (a) $\dfrac{2x^{\ln x} \ln x}{x}$ (b) $(\ln x)^x\left[\ln(\ln x) + \dfrac{1}{\ln x}\right]$

(c) $(3x^2 + 1)(e^x)^{x^2+1}$ (d) $(\ln x)^{\ln x}\left(\dfrac{\ln(\ln x) + 1}{x}\right)$

13. (a) 1 (b) 1 (c) e^2

Section 5, pages 278–280

1. (a) $y = ce^{3x}$ (b) $y = \dfrac{t^4}{4} + \dfrac{t^2}{2} + c$

(c) $xy = c$ (d) $y^2 + 1 = c(x + 1)^2$

(e) $e^y - e^t = c$ (f) $x^2c = y^2 - 2$

2. (a) $xy = 1$ (b) $y = 5e^{(3/2)x}$

(c) $y = -3\sqrt{x^4 - 7}$ (d) $x^2 + y^2 = 25$

3. $y = -e^{2x}$

4. $x^2 + y^2 = c$

5. $x^2 - y^2 = c$

6. (a) separable (b) linear (c) neither

(d) both (e) both (f) separable

(g) both (h) neither (i) neither

8. $1 - 2^{-3/2}$, or roughly $\frac{2}{3}$

9. $\dfrac{45 \ln 3}{\ln 2}$ years (roughly 70)

10. $\dfrac{-\ln 2}{16}$

11. $10^7 e^{\ln 2/3}$(roughly 1.26×10^7) bacteria

12. $\dfrac{\ln 3}{5}$ sec (roughly 0.2 sec)

13. $\frac{1}{4}$

14. $y = ce^{-ax} + \dfrac{b}{a}$

15. $y = 7e^{2x} - 3$

CHAPTER 6

Section 1, pages 288–289

1. (a) $0, -1$ (b) $\dfrac{\sqrt{2}}{2}, \dfrac{-\sqrt{2}}{2}$
   (c) $-1, 0$ (d) $\dfrac{1}{2}, \dfrac{\sqrt{3}}{2}$
   (e) $\dfrac{\sqrt{3}}{2}, \dfrac{1}{2}$ (f) $\dfrac{-\sqrt{2}}{2}, \dfrac{-\sqrt{2}}{2}$

2.

Q	cos	sin
I	+	+
II	−	+
III	−	−
IV	+	−

3. (a) $t = n\pi$, n any integer (b) $\dfrac{\pi}{2}(2n + 1)$
   (c) $\pi(2n + \frac{1}{2})$ (d) $2n\pi$
   (e) $\pi(2n + \frac{3}{2})$ (f) $\pi(2n + 1)$
   (g) no values of t (h) no values of t

4. domain: R; range: $[-1, 1]$

5. (a) ± 1 (b) 0 (c) 0 (d) ± 1

6. (a) $\dfrac{\sqrt{6} + \sqrt{2}}{4}$ (b) $\dfrac{\sqrt{6} - \sqrt{2}}{4}$
   (c) $\dfrac{\sqrt{6} - \sqrt{2}}{4}$ (d) $\dfrac{\sqrt{6} + \sqrt{2}}{4}$

8. (a) $\cos 2a = \cos^2 a - \sin^2 a$ (b) $\sin 2a = 2 \sin a \cos a$
   (c) $\cos a = \cos^2 \dfrac{a}{2} - \sin^2 \dfrac{a}{2}$ (d) $\sin a = 2 \sin \dfrac{a}{2} \cos \dfrac{a}{2}$

9. $\cos^2 a = \frac{1}{2}(1 + \cos 2a)$, $\sin^2 a = \frac{1}{2}(1 - \cos 2a)$

Section 2, pages 299–300

1. (a) 1 (b) 1
   (c) 1 (d) 2
   (e) $\frac{1}{\pi}$ (f) $\frac{2}{3}$
   (g) 1 (h) ∞
   (i) 2 (j) $\frac{1}{2}$
2. (a) $2x\cos(x^2+3)$ (b) $-e^x\sin e^x$
   (c) $\cos 2t$ (d) 0
   (e) $-\cos x\sin(\sin x)$ (f) $\dfrac{\cos x}{\sin x}$
   (g) $25x^4\sin^4 x^5\cos x^5$ (h) $\dfrac{1}{\cos^2 x}$
   (i) $-\dfrac{1}{\sin^2 x}$ (j) $e^{-x}(\cos x-\sin x)$
3. (a) $\frac{1}{7}\sin 7x+c$ (b) $\frac{1}{2}\sin 2x-\frac{1}{3}\cos 3x+c$
   (c) $\sin(e^x)+c$ (d) $-\cos(x+a)+c$
   (e) $-e^{-\sin x}+c$ (f) $\sin(\sin t)+c$
   (g) $-\ln|\cos x|+c$ (h) $-\frac{1}{7}\cos^7 x+c$
   (i) $\frac{1}{7}\sin^7 x+c$ (j) $x+c$
4. (a) $\dfrac{-\cos^{n+1}x}{n+1}+c$, if $n\neq -1$; $-\ln|\cos x|+c$, if $n=-1$
   (b) $\dfrac{\sin^{n+1}x}{n+1}+c$, if $n\neq -1$; $\ln|\sin x|+c$, if $n=-1$
   (c) $-\cos x+\dfrac{\cos^3 x}{3}+c$ (d) $\dfrac{\cos^7 x}{7}-\dfrac{\cos^5 x}{5}+c$
5. $\frac{1}{2}(1+\cos 2x)$, $\frac{1}{2}x+\frac{1}{4}\sin 2x+c$
6. $\frac{1}{2}(1-\cos 2x)$, $\frac{1}{2}x-\frac{1}{4}\sin 2x+c$
7. (a) $y=-\cos(e^x)+c$
   (b) $\sin y=-\cos x+c$
   (c) $\sin x+\frac{1}{2}\cos y^2=c$
8. (a) 1 (b) 0 (c) 1 (d) 1 (e) $\frac{1}{2}$ (f) $\frac{1}{2}$

Section 3, pages 309–311

1. even: cos, sec; odd: sin, csc, tan, cot
3. (a) $2\sec^2 x\tan x$ (b) $4x\sec^2(2x^2-1)$
   (c) $\tan x$ (d) $\cos y$
   (e) $2\sec^2 t\tan t$ (f) $-3x^2\csc(x^3-1)\cot(x^3-1)$
   (g) 0 (h) $-\dfrac{1}{\sin t\cos t}$
5. (a) $\frac{1}{5}\ln|\sec 5x|+c$ (b) $\ln|\sin x|+c$
   (c) $\tan(e^x)+c$ (d) $\tan x-x+c$

(e) $\frac{1}{5}\tan^5 x + c$ (f) $\dfrac{\sec^4 x}{4} + c$

(g) $-2\csc\sqrt{x} + c$ (h) $-\frac{1}{3}\cot^3 x - \cot x + c$

6. $\ln|\sec x + \tan x| + c$
7. $\ln|\sec x| + c$
9. (a) 1 (b) $\frac{1}{2}$
10. (a) $\frac{3}{4}$ (b) ∞ (lines are perpendicular)

(c) $-2\left(\dfrac{\sqrt{5}-1}{2}\right)^{1/2} - \left(\dfrac{2}{\sqrt{5}-1}\right)^{1/2}$

11. The functions, domains, and ranges are as follows:

sin: R, $[-1, 1]$
cos: R, $[-1, 1]$
tan: $x \neq \dfrac{\pi}{2}(2n+1)$, R
cot: $x \neq n\pi$, R
sec: $x \neq \dfrac{\pi}{2}(2n+1)$, $|x| \geq 1$
csc: $x \neq n\pi$, $|x| \geq 1$ (n is any integer)

12. (a) 1 (b) 1
13. (a) $\frac{1}{2}$ (b) $\frac{1}{24}$ (c) 3
(d) 2 (e) ∞ (f) 0

Section 4, pages 321–323

1. (a) $\dfrac{\pi}{2}$ (b) $\dfrac{\pi}{6}$

(c) $\dfrac{\pi}{3}$ (d) $-\dfrac{\pi}{3}$

(e) $\dfrac{\pi}{3}$ (f) $\dfrac{\pi}{6}$

(g) $\dfrac{\pi}{6}$ (h) $\dfrac{\pi}{6}$

(i) $(-1)^k (a + k\pi)$, k an integer such that $-\dfrac{\pi}{2} \leqq a + k\pi \leqq \dfrac{\pi}{2}$

(j) $\dfrac{\pi}{7}$ (k) $\dfrac{5\pi}{14}$

(l) $(-1)^k \left(\dfrac{\pi}{2} - (a + k\pi)\right)$, k an integer such that $0 \leqq a + k\pi \leqq \pi$

(m) -1

(n) $(-1)^k (2x + k\pi)$, k an integer such that $-\dfrac{\pi}{2} \leqq 2x + k\pi \leqq \dfrac{\pi}{2}$

(o) $\dfrac{\pi}{4}$ (p) $\dfrac{\pi}{3}$

2. (a) $\dfrac{2x}{\sqrt{1-x^4}}$ (b) $\dfrac{1}{2\sqrt{x}(1+x)}$

(c) $\dfrac{1}{\sqrt{x}(x+1)}$ (d) $\begin{cases} 1, & \text{if } \sin x > 0, \\ -1, & \text{if } \sin x < 0 \end{cases}$

(e) $\begin{cases} -1, \text{ if } \cos x > 0, \\ 1, \text{ if } \cos x < 0 \end{cases}$ (f) $\dfrac{2x}{|x|(1 + x^2)\sqrt{x^2 + 2}}$

(g) 0 (h) $\dfrac{6x^2}{1 + x^6}$

(i) $\dfrac{1}{x(1 + (\ln x)^2)}$ (j) 0

3. The domain and range for each function is as follows:
   (a) $|x| \le 1, 0 \le y \le \dfrac{\pi}{2}$ (b) $x \ge 0, 0 \le y < \dfrac{\pi}{2}$
   (c) $x \ge 0, -\dfrac{\pi}{2} \le y < \dfrac{\pi}{2}$ (d) $R, 0 \le y \le \pi$
   (e) $R, 0 \le y \le \pi$ (f) $R, 0 \le y < \dfrac{\pi}{2}$
   (g) $[-2, 0], y = \dfrac{\pi}{2}$ (h) $R, -\dfrac{3\pi}{2} < y < \dfrac{\pi}{2}$
   (i) $x > 0, -\dfrac{\pi}{2} < y < \dfrac{\pi}{2}$ (j) $|x| \ge 1, y = 0$

4. (a) $\dfrac{\sqrt{2}}{2} \arctan\left(\dfrac{x\sqrt{2}}{2}\right) + c$ (b) $\arcsin\left(\dfrac{x\sqrt{2}}{2}\right) + c$
   (c) $\frac{1}{2}\arctan y^2 + c$ (d) $\arctan(x + 1) + c$
   (e) $\arcsin(y - 1) + c$ (f) $\frac{1}{2}\arcsin(x + 1)^2 + c$
   (g) $\operatorname{arcsec} x + c$ (h) $\frac{1}{2}\arctan(x^2 + 1) + c$

7. (a) $\dfrac{\pi}{6}$ (b) $\dfrac{\pi}{6}$
   (c) $\dfrac{\pi\sqrt{3}}{9}$ (d) $\dfrac{\pi}{3}$
   (e) $\arctan x$ (f) $\dfrac{\pi}{4}$

8. (b) $R, (0, \pi)$

9. $\arctan t$

Section 5, page 326

1. (c) and (f) are polynomials

3. (a) $F(x, y) = y^2(x - 1) - (x + 1)$
   (b) $F(x, y) = y(1 + x^2) - 1$
   (c) $F(x, y) = y^2(1 - x^2) - 1$
   (d) $F(x, y) = y^2x^2(x^2 - 1) - 1$
   (e) $F(x, y) = y - (\ln 5)x$
   (f) $F(x, y) = (y - 2x)^2 - (4x^2 - 1)$

Section 6, pages 334–335

2. (a) $10 + i$ (b) -4
   (c) 6 (d) $-\frac{1}{5} - i\frac{17}{5}$
   (e) $2(a^2 + b^2)$ (f) $-6 + i29$
   (g) $-\frac{24}{25} + i\frac{7}{25}$ (h) $-\frac{19}{29} + i\frac{33}{29}$

(i) $\dfrac{a^2 - b^2}{3(a^2 + b^2)} + i\,\dfrac{2ab}{3(a^2 + b^2)}$ (j) $-\frac{3}{5} - i\frac{4}{5}$
(k) $3 + i4$ (l) $39 + i4$

4. (a) $2 + i3$ (b) $5 - i4$ (c) $7 - i$ (d) $22 + i7$
(e) $8 + i12$ (f) -7 (g) $-2i$

Section 7, pages 343–344

1. (a) $\sqrt{2}\,e^{i\pi/4}$ (b) $\sqrt{2}\,e^{i7\pi/4}$
(c) $2e^{i\pi/3}$ (d) $4e^{i5\pi/6}$
(e) $5e^{i\pi}$ (f) $e^2e^{i\pi/4}$
(g) 5 (h) $e^{i\pi/2}$

2. (a) $2e^{i\pi/6}$ (b) $\sqrt{2}\,e^{-i\pi/4}$
(c) $2\sqrt{2}\,e^{-i\pi/12}$ (d) $\sqrt{2}\,e^{i5\pi/12}$
(e) $4e^{i\pi/6}$ (f) $64e^{i\pi}$

3. (a) $(-1, 0)$ (b) $(\sqrt{2}, \sqrt{2})$
(c) $(-\sqrt{2}, -\sqrt{2})$ (d) $(\sqrt{2} - 1, \sqrt{2})$
(e) $(3, 5)$ (f) $(3, -2), (-3, -2)$

4. $|z_1|e^{-it}$

7. (a) $2e^{i(\pi/2+2k\pi)/3}$, $k = 0, 1, 2$ (b) $e^{i(\pi/4+k\pi)}$, $k = 0, 1$
(c) $\sqrt[3]{2}\,e^{i2k\pi/3}$, $k = 0, 1, 2$ (d) $e^{ik\pi/2}$, $k = 0, 1, 2, 3$
(e) $\sqrt[5]{2}\,\mathrm{Exp}\left[i\left(\dfrac{\pi/2 + 2k\pi}{5}\right)\right]$, $k = 0, 1, 2, 3, 4$
(f) $\sqrt[3]{2}\,\mathrm{Exp}\left[i\left(\dfrac{\pi/3 + 2k\pi}{3}\right)\right]$, $k = 0, 1, 2$

8. $e^{z \ln 5}$

Section 8, pages 351–353

1. (a) $y = ce^{-(3/2)x}$ (b) $y = ce^{5x}$
(c) $y = c_1e^{-5x} + c_2e^x$ (d) $y = e^x(k_1 \cos 2x + k_2 \sin 2x)$
(e) $y = (c_1x + c_2)e^{-4x}$ (f) $y = (c_1x + c_2)e^{x/2}$
(g) $y = c_1 + c_2e^{7x}$ (h) $y = k_1 \cos 2x + k_2 \sin 2x$
(i) $y = c_1e^{3x} + c_2e^{-3x}$ (j) $y = ce^{-13x}$
(k) $y = c_1 + c_2e^{-13x}$
(l) $y = e^{-(1/2)x}\left(k_1 \cos \dfrac{\sqrt{3}}{2}x + k_2 \sin \dfrac{\sqrt{3}}{2}x\right)$
(m) $y = e^{-7x}(k_1 \cos x + k_2 \sin x)$ (n) $y = (c_1x + c_2)e^{-7x}$

2. (a) $y = (c_1x + c_2)e^{4x}$ (b) $y = (x + 2)e^{4x}$

3. (a) $y = e^{-3x} + 2e^{-x}$ (b) $y = (3x + 1)e^{-(1/5)x}$
(c) $y = 5 - 2e^{-3x}$ (d) $y = 5e^{-x} \cos x$
(e) $y = e^{-(3/2)x}\left(2 \cos \dfrac{\sqrt{11}}{2}x + \dfrac{18\sqrt{11}}{11} \sin \dfrac{\sqrt{11}}{2}x\right)$

8. 0, 1, 5

9. (a) $y = c_1e^{2x} + c_2e^{6x}$ (b) $A = 2$, $B = \frac{7}{3}$

CHAPTER 7

Section 1, pages 361–362

1. (a) $x \sin x + \cos x + c$
   (b) $\left(\dfrac{2x+3}{4}\right)e^{2x} + c$
   (c) $x \arctan x - \frac{1}{2}\ln(1 + x^2) + c$
   (d) $-\dfrac{x^2}{7}\cos 7x + \dfrac{2}{49}x \sin 7x + \dfrac{2}{343}\cos 7x + c$
   (e) $\dfrac{x^4 \ln x}{4} - \dfrac{x^4}{16} + c$
   (f) $-(x^3 - 13x + 2)\cos x + (3x^2 - 13)\sin x + c$
   (g) $\frac{3}{13}e^{3x}\sin 2x - \frac{2}{13}e^{3x}\cos 2x + c$
   (h) $\frac{1}{2}(\sec x \tan x + \ln|\sec x + \tan x|) + c$
   (i) $\frac{1}{3}(x^3 + 1)\ln(x + 1) - \frac{1}{9}x^3 + \frac{1}{6}x^2 - \frac{1}{3}x + c$
   (j) $x \ln(x^2 + 2) - 2x + 2\sqrt{2}\arctan\left(\dfrac{x\sqrt{2}}{2}\right) + c$
3. (a) $\dfrac{a}{a^2 + b^2}e^{ax}\cos bx + \dfrac{b}{a^2 + b^2}e^{ax}\sin bx + c$
   (b) $\dfrac{-b}{a^2 + b^2}e^{ax}\cos bx + \dfrac{a}{a^2 + b^2}e^{ax}\sin bx + c$
5. (a) $\frac{1}{2}x + \frac{1}{4}\sin 2x + c$ (b) $\frac{1}{2}x - \frac{1}{4}\sin 2x + c$
6. (c) $\left(\dfrac{ax + b}{a}\right)(\ln|ax + b|)^n - n\displaystyle\int (\ln|ax + b|)^{n-1}\,dx + c$
7. (a) $e^x(x^5 - 5x^4 + 20x^3 - 60x^2 + 120x - 120) + c$
   (b) $-\frac{1}{4}\sin^3 x \cos x - \frac{3}{8}\sin x \cos x + \frac{3}{8}x + c$
   (c) $\frac{1}{5}\sin 5x - \frac{1}{15}\sin^3 5x + c$
   (d) $\dfrac{A}{3}[(\ln|A|)^6 - 6(\ln|A|)^5 + 30(\ln|A|)^4 - 120(\ln|A|)^3$
   $+ 360(\ln|A|)^2 - 720 \ln|A|] + 720x + c$, where $A = 3x + 7$
   (e) $\frac{2}{3}$ (f) $-2e + 6$

Section 2, pages 372–374

1. (a) $\frac{1}{6}\cos 3x - \frac{1}{14}\cos 7x + c$ (b) $\dfrac{\sin 2x}{4} + \dfrac{\sin 4x}{8} + c$
   (c) $\dfrac{\sin 2x}{4} - \dfrac{\sin 4x}{8} + c$ (d) $\dfrac{\sin 3z}{6} + \dfrac{\sin 11z}{22} + c$
   (e) $-\dfrac{1}{2}\left[\dfrac{\cos(\pi - 3)x}{\pi - 3} + \dfrac{\cos(\pi + 3)x}{\pi + 3}\right] + c$
   (f) $\dfrac{\cos \pi y}{2\pi} - \dfrac{\cos 3\pi y}{6\pi} + c$ (g) $\dfrac{\sin 5x}{10} - \dfrac{\sin 7x}{14} + c$
   (h) $-\dfrac{\cos 10w}{20} - \dfrac{\cos 24w}{48} + c$

2. (a) $-\cos\theta + \frac{1}{3}\cos^3\theta + c$
   (b) $-\frac{3}{4}\cos\theta + \frac{1}{12}\cos 3\theta + c$

3. (a) $-\frac{1}{2}\cos 2x + \frac{1}{3}\cos^3 2x - \frac{1}{10}\cos^5 2x + c$
   (b) $-\frac{1}{10}\sin^4 2x\cos 2x - \frac{2}{5}\cos 2x + \frac{2}{15}\cos^3 2x + c$

4. (a) $\frac{1}{2}\sin 2x - \frac{1}{6}\sin^3 2x + c$
   (b) $\frac{3}{8}x + \frac{3}{16}\sin 2x + \frac{1}{4}\cos^3 x\sin x + c$
   (c) $\frac{3}{8}x - \frac{1}{12}\sin 6x + \frac{1}{96}\sin 12x + c$
   (d) $\frac{1}{47}\cos^{47}x - \frac{1}{45}\cos^{45}x + c$
   (e) $\frac{1}{8}\theta - \frac{1}{32}\sin 4\theta + c$
   (f) $\frac{2}{3}\sin^{3/2}x - \frac{4}{7}\sin^{7/2}x + \frac{2}{11}\sin^{11/2}x + c$
   (g) $-\frac{1}{144}\sin^3 6x + \frac{1}{12}\sin 6x + \frac{5}{16}x + \frac{1}{64}\sin 12x + c$
   (h) $\frac{1}{5}\sin^5 y - \frac{2}{7}\sin^7 y + \frac{1}{9}\sin^9 y + c$
   (i) $\dfrac{y}{16} - \frac{1}{64}\sin 4y - \frac{1}{48}\sin^3 2y + c$
   (j) $\frac{2}{11}\cos^{11/2}x - \frac{2}{7}\cos^{7/2}x + c$

5. (a) $\frac{1}{2}\theta + \frac{1}{4}\sin 2\theta + c$
   (b) $\frac{1}{2}\cos\theta\sin\theta + \frac{1}{2}\theta + c$

6. $\frac{1}{2}\tan^2 x + c,\ \frac{1}{2}\sec^2 x + c$

7. (a) $\frac{1}{3}\tan^3 x - \tan x + x + c$ (b) $\frac{1}{8}\tan^2 4y + \frac{1}{4}\ln|\cos 4y| + c$
   (c) $\frac{1}{3}\tan\theta(\sec^2\theta + 2) + c$
   (d) $\frac{1}{4}(\tan 2x\sec 2x + \ln|\tan 2x + \sec 2x|) + c$
   (e) $\frac{1}{7}\tan^7 x + \frac{1}{5}\tan^5 x + c$ (f) $\frac{1}{5}\sec^5 x - \frac{1}{3}\sec^3 x + c$
   (g) $\frac{1}{4}\sec^3 x\tan x - \frac{1}{8}\sec x\tan x - \frac{1}{8}\ln|\sec x + \tan x| + c$
   (h) $\frac{1}{8}\tan^8 x + \frac{1}{6}\tan^6 x + c$
   (i) $\frac{2}{11}\tan^{11/2}x + \frac{4}{7}\tan^{7/2}x + \frac{2}{3}\tan^{3/2}x + c$
   (j) $-\ln|\csc x + \cot x| + \cos x + c$

8. (a) $\displaystyle\int \cot^n x\,dx = -\frac{1}{n-1}\cot^{n-1}x - \int \cot^{n-2}x\,dx$
   (b) $-\frac{1}{12}\cot^4 3\theta + \frac{1}{6}\cot^2 3\theta + \frac{1}{3}\ln|\sin 3\theta| + c$

10. (b) $-\frac{1}{5}\csc^4 y\cot y - \frac{4}{15}\csc^2 y\cot y - \frac{8}{15}\cot y + c$

11. (a) $-\frac{1}{4}\csc^3\theta\cot\theta - \frac{3}{8}\csc\theta\cot\theta - \frac{3}{8}\ln|\csc\theta + \cot\theta| + c$
   (b) $\frac{1}{3}\sin 3x + c$ (c) $-\frac{1}{3}\cot^3 y + \cot y + y + c$
   (d) $-\frac{1}{5}\cot^5 x - \frac{1}{3}\cot^3 x + c$ (e) $\frac{1}{6}\csc^3 2y - \frac{1}{10}\csc^5 2y + c$
   (f) $-\frac{1}{4}\csc^3\phi\cot\phi + \frac{1}{8}\csc\phi\cot\phi + \frac{1}{8}\ln|\csc\phi + \cot\phi| + c$

Section 3, pages 382–384

1. $\dfrac{1}{2}\left(x\sqrt{a^2 - x^2} - a^2\arccos\dfrac{x}{|a|}\right) + c$

2. (a) $x = |a|\cot\theta,\ \sqrt{a^2 + x^2} = |a|\csc\theta$
   (b) $0 < \theta < \pi$

3. $-\ln|\csc\theta + \cot\theta| + c$

4. $\left(0, \dfrac{\pi}{2}\right] \cup \left(\pi, \dfrac{3\pi}{2}\right]$

5. (a) $\sqrt{x^2 - 9} - 3 \operatorname{arcsec}\left(\frac{x}{3}\right) + c$

(b) $\frac{1}{4}x^3\sqrt{x^2 - 1} - \frac{5}{8}x\sqrt{x^2 - 1} + \frac{3}{8}\ln|x + \sqrt{x^2 - 1}| + c$

(c) $-\frac{1}{3}(16 - x^2)^{3/2} + c$

(d) $(x^2 - 4)^{3/2}[\frac{4}{3} + \frac{1}{5}(x^2 - 4)] + c$

(e) $\frac{1}{3}\ln\left|\frac{\sqrt{x^2 - 9}}{x + 3}\right| + c$

(f) $\frac{1}{4}x(x^2 + 4)^{3/2} + \frac{3}{2}x\sqrt{x^2 + 4} + 6\ln|x + \sqrt{x^2 + 4}| + c$

(g) $-(x^2 + a^2)^{-1/2} + c$

(h) $-4\sqrt{4 - x^2} + \frac{1}{3}(4 - x^2)^{3/2} + c$

(i) $\frac{1}{2}x\sqrt{x^2 - a^2} - \frac{1}{2}a^2\ln|x + \sqrt{x^2 - a^2}| + c$

(j) $\dfrac{-x}{25\sqrt{x^2 - 25}} + c$

6. $a^2\left[\frac{\pi}{2} - \arcsin\left(\frac{a - h}{a}\right)\right] - (a - h)\sqrt{2ah - h^2}$

7. (a) $\ln 3$ (b) 1 (c) $\dfrac{\pi}{12}$

(d) $\dfrac{3 - \sqrt{3}}{6}$ (e) $\dfrac{37}{3}$ (f) $\frac{1}{2}\ln\dfrac{7 + 2\sqrt{10}}{9}$

8. (a) $\dfrac{\sqrt{5}}{5}\arctan\dfrac{x\sqrt{5}}{5} + c$ (b) $\frac{1}{4}\arctan\dfrac{x}{2} + c$

(c) $\frac{1}{2}\arctan\dfrac{x + 1}{2} + c$ (d) $\frac{1}{2}\arctan(x + 3) + c$

(e) $\dfrac{x}{24(x^2 + 3)} + \dfrac{\sqrt{3}}{72}\arctan\dfrac{x\sqrt{3}}{3} + c$

(f) $\dfrac{x - 2}{8(x^2 - 4x + 8)} + \dfrac{1}{16}\arctan\dfrac{x - 2}{2} + c$

(g) $\dfrac{x}{36(x^2 + 9)^2} + \dfrac{x}{216(x^2 + 9)} + \dfrac{1}{648}\arctan\dfrac{x}{3} + c$

(h) $\dfrac{x + 1}{4(x^2 + 2x + 2)^2} + \dfrac{3x + 3}{8(x^2 + 2x + 2)} + \frac{3}{8}\arctan(x + 1) + c$

Section 4, pages 394–396

1. (a) $\dfrac{1}{x - 2} - \dfrac{1}{x + 3}$ (b) $\dfrac{3}{2x + 1} - \dfrac{1}{x + 1}$

(c) $x - \dfrac{3}{2x + 1} + \dfrac{1}{x + 1}$ (d) $\dfrac{3}{x - 4} + \dfrac{x - 1}{x^2 + 2}$

(e) $3x - 7 + \dfrac{2}{x} - \dfrac{1}{x + 4}$ (f) $1 - \dfrac{x}{x^2 + x + 1}$

2. (a) $\ln\left|\dfrac{x - 2}{x + 3}\right| + c$ (b) $\ln\dfrac{(2x + 1)^{3/2}}{|x + 1|} + c$

(c) $\dfrac{x^2}{2} + \ln\dfrac{|x + 1|}{(2x + 1)^{3/2}} + c$

(d) $\ln|(x-4)^3\sqrt{x^2+2}| - \dfrac{\sqrt{2}}{2}\arctan\dfrac{x\sqrt{2}}{2} + c$

3. (a) $\dfrac{-1}{x-3} + \dfrac{2}{x+2}$ (b) $\dfrac{-3}{x+7} + \dfrac{3}{x+1}$

(c) $\dfrac{1}{x-1} + \dfrac{2}{(x-1)^2}$ (d) $\dfrac{5}{x} + \dfrac{3}{x+5}$

(e) $\dfrac{-1}{x} + \dfrac{2}{x^2} + \dfrac{1}{x+2}$ (f) $\dfrac{4}{x+1} + \dfrac{2x-3}{x^2+4}$

(g) $\dfrac{1}{x+3} - \dfrac{2}{(x+3)^2} + \dfrac{1}{(x+3)^3}$

(h) $\dfrac{4}{2x-1} - \dfrac{2x}{(x^2+1)^2} - \dfrac{2x+1}{x^2+1}$

4. (a) $\ln\dfrac{(x+2)^2}{|x-3|} + c$ (b) $3\ln\left|\dfrac{x+1}{x+7}\right| + c$

(c) $\ln|x-1| - \dfrac{2}{x-1} + c$ (d) $5\ln|x| + 3\ln|x+5| + c$

(e) $\ln[(x+1)^4(x^2+4)] - \dfrac{3}{2}\arctan\dfrac{x}{2} + c$

(f) $\ln|x+3| + \dfrac{2}{x+3} - \dfrac{1}{2(x+3)^2} + c$

7. (a) $\ln\left|\dfrac{x}{x+1}\right| - \dfrac{2}{x+1} + c$ (b) $x + \ln\dfrac{|x-2|^5}{(x-1)^2} + c$

(c) $\frac{1}{2}\ln(x^2+1) - \ln|2x+1| + c$

(d) $3\ln|x-2| - 2\ln|x-3| - \dfrac{4}{x-2} + c$

(e) $\arctan(x+1) + c$

(f) $\ln(x^2+2x+2) - \arctan(x+1) + c$

(g) $\frac{1}{2}\ln|\tan x - 3| - \frac{1}{2}\ln|\tan x - 1| + c$

(h) $\frac{1}{3}\ln\left|\dfrac{2\sec y + 1}{\sec y + 2}\right| + c$

(i) $y - \frac{1}{2}\ln(y^2+y+1) + \dfrac{\sqrt{3}}{3}\arctan\dfrac{\sqrt{3}(2y+1)}{3} + c$

(j) $\ln\dfrac{\sqrt{z^2+z+1}}{(z+4)^2} + \dfrac{5\sqrt{3}}{3}\arctan\dfrac{\sqrt{3}(2z+1)}{3} + c$

(k) $\ln\left|\dfrac{(x-1)(x+2)}{x^2+x+1}\right| + c$ (l) $x - \dfrac{4}{x+2} + \ln\dfrac{|x+1|^3}{(x+2)^8} + c$

(m) $2\ln|2x-1| + \dfrac{1}{x^2+1} - \ln(x^2+1) - \arctan x + c$

(n) $\dfrac{2x+1}{38(x^2+x+5)^2} + \dfrac{3(2x+1)}{361(x^2+x+5)}$

$+ \dfrac{12\sqrt{19}}{6859}\arctan\left(\dfrac{\sqrt{19}(2x+1)}{19}\right) + c$

Section 5, pages 402–403

3. (a) $\frac{3}{5}(1+x)^{5/3} - \frac{3}{2}(1+x)^{2/3} + c$

(b) $3\left(\frac{x^{5/3}}{5} - \frac{x^{4/3}}{4} + \frac{x}{3} - \frac{x^{2/3}}{2} + x^{1/3} - \ln|x^{1/3}+1|\right) + c$

(c) $\sec x - \tan x + x + c$ (d) $\frac{3\sqrt{2}}{2}\ln\left|\frac{\sqrt{2} - \cos x + \sin x}{\sin x + \cos x}\right| + c$

(e) $\frac{2\sqrt{3}}{3}\arctan\left(\frac{\sqrt{3}}{3}\tan\frac{t}{2}\right) + c$ (f) $\frac{15}{13}(1+x)^{13/15} + c$

(g) $2y^{1/2} - 3y^{1/3} + 6y^{1/6} - \ln(y^{1/6}+1)^6 + c$

(h) $\frac{1}{2}\sin^2 x + c$

(i) $\frac{1}{125}(\frac{2}{5}(5x+3)^{5/2} - 4(5x+3)^{3/2} + 18(5x+3)^{1/2}) + c$

(j) $\frac{-2}{3(1+\sqrt{x})^3} + \frac{1}{2(1+\sqrt{x})^4} + c$

(k) $2(1+e^x)^{1/2} + c$ (l) $\ln\frac{\sqrt{1+e^x} - 1}{\sqrt{1+e^x}+1} + c$

4. (a) $\frac{6 - 2\sqrt{3}}{3} + \ln\left(1 + \frac{2\sqrt{3}}{3}\right)$ (b) $\frac{8}{3}$

(c) $\frac{8}{105}$ (d) $\frac{11}{2} + \frac{3}{8}\ln\frac{3}{5}$

Supplementary Problems, pages 403–404

1. $\frac{1}{4}(2x^2 - 1)\sin 2x + \frac{1}{2}x\cos 2x + c$
2. $\frac{1}{4}\arctan\frac{y}{4} + c$
3. $\frac{1}{4}\ln\left|\frac{4+y}{\sqrt{16-y^2}}\right| + c$
4. $-x + \frac{1}{2}\ln\left|\frac{1+x}{1-x}\right| + c$
5. $\frac{2}{3}e^{x^3} + c$
6. $\frac{2}{3}(x+1)^{3/2} - 2(x+1)^{1/2} + c$
7. $e^x + \frac{x^2}{2} + \sin x + c$
8. $\left(\frac{t^3}{3} + \frac{t^2}{2}\right)\ln|t| - \left(\frac{t^3}{9} + \frac{t^2}{4}\right) + c$
9. $\frac{1}{2}\left(x\sqrt{4-x^2} + 4\arcsin\frac{x}{2}\right) + c$
10. $\frac{1}{2}(x\sqrt{4+x^2} + 4\ln|x + \sqrt{4+x^2}|) + c$
11. $\frac{1}{4}\tan^4 x - \frac{1}{2}\tan^2 x + \ln|\sec x| + c$
12. $-\frac{1}{22}\cos^{11}2x + \frac{1}{13}\cos^{13}2x - \frac{1}{30}\cos^{15}2x + c$

13. $\frac{x}{2}\sqrt{4x^2+1}+\frac{1}{4}\ln|2x+\sqrt{4x^2+1}|+c$

14. $\frac{1}{2}(e^z+e^{-z})+c$

15. $2x^{1/2}-4x^{1/4}+4\ln(x^{1/4}+1)+c$

16. $-\frac{1}{4}\ln|y|+\frac{1}{8}\ln|y^2-4|+c$

17. $\frac{1}{50}e^{7t}(\sin t+7\cos t)+c$

18. $x-\ln(1+e^x)+c$

19. $-\frac{1}{2}\cot(2x+1)+\frac{1}{2}\csc(2x+1)+c$

20. $\ln|x+\sqrt{4+x^2}|+c$

21. $\ln|2+\ln x|+c$

22. $\frac{1}{4}\arctan\frac{x^2}{2}+c$

23. $\frac{1}{2}\ln(e^{2x}+e^{-2x})+c$

24. $y[(\ln y)^2-2\ln y+2]+c$

25. $-\cos x+\frac{2}{3}\cos^3x-\frac{1}{5}\cos^5x+c$

26. $\left(\frac{x^3}{3}+x\right)\ln x-\left(\frac{x^3}{9}+x\right)+c$

27. $\frac{1}{4}\ln|7+4\sec t|+c$

28. $\frac{3}{2}x-\frac{3}{10}\ln(2e^{5x}+1)+c$

29. $\frac{x^{n+1}}{n+1}\ln x-\frac{x^{n+1}}{(n+1)^2}+c$

30. $\frac{1}{2}(\ln x)^2+c$

31. $\frac{1}{4}\sec^3y\tan y+\frac{3}{8}\sec y\tan y+\frac{3}{8}\ln|\tan y+\sec y|+c$

32. $x\ln|x^2-1|-2x+\ln\left|\frac{x+1}{x-1}\right|+c$

33. $\frac{1}{2}\sin y+\frac{1}{18}\sin 9y+c$

34. $-\frac{1}{16}\cos 8x-\frac{1}{4}\cos 2x+c$

35. $\frac{1}{6}\ln\left|\frac{x-1}{x+5}\right|+c$

36. $\arctan(x+2)+c$

37. $\frac{x2^x}{\ln 2}-\frac{2^x}{(\ln 2)^2}+c$

38. $e^x(x^5-5x^4+20x^3-60x^2+120x-120)+c$

39. $-\frac{1}{2}[(1-x)\sqrt{2x-x^2}+\arcsin(1-x)]+c$

40. $\frac{-1}{6(3+2e^x)^3}+c$

41. $\frac{2}{9}\arctan\left(\frac{t^{3/2}}{3}\right)+c$

42. $\dfrac{x^3}{3} - 4x + 8\arctan\dfrac{x}{2} + c$

43. $\frac{2}{3}x - \frac{22}{9}\ln|3x + 5| + c$

44. $e^{\sec x} + c$

45. $\dfrac{5^{x^3}}{\ln 5} + c$

46. $\frac{1}{4}\ln|1 + 4\sin x| + c$

47. $\frac{1}{3}x^3\arccos x - \frac{1}{9}x^2\sqrt{1 - x^2} - \frac{2}{9}\sqrt{1 - x^2} + c$

48. $\ln|(t - 5)^2(t + 2)^3| + c$

49. $15(\frac{1}{9}x^{9/5} - \frac{1}{2}x^{8/5} + \frac{16}{7}x^{7/5} - \frac{32}{3}x^{6/5} + \frac{256}{5}x$
$- 256x^{4/5} + \frac{4096}{3}x^{3/5} - 8192x^{2/5} + 65{,}536x^{1/5}$
$- 262{,}144\ln|x^{1/5} + 1|) + c$

50. $\dfrac{x^6}{6} + \dfrac{x^5}{5} + \dfrac{x^4}{4} + \dfrac{x^3}{3} + \dfrac{x^2}{2} + x + c$

CHAPTER 8

Section 1, pages 411–412

1. (a) $\frac{1}{3}$ (b) 0 (c) $\ln 2$
   (d) $\dfrac{\ln n}{n - 1}$ (e) $\dfrac{2}{\pi}$ (f) $-1 + \frac{5}{4}\ln 5$

4. 5 sec, 80 ft/sec

5. $v(0) = 90$, $v(1) = 96$(max), $v(4) = 42$, average $= 82$, number of words $= 19{,}680$

6. $18\frac{2}{3}$ ft, $9\frac{1}{3}$ ft/sec

7. (a) $\dfrac{\sqrt{3}}{3}$ (b) $\frac{1}{2}\sqrt[3]{2}$ (c) 2
   (d) $\ln\left(\dfrac{e^2 - 1}{2}\right)$ (e) $\arccos\dfrac{2}{\pi}$

9. $x(t) = x_0e^{t\ln 2/T}$, $\dfrac{x_0}{\ln 2}$

Section 2, pages 420–422

1. (a) $\frac{1}{3}$ (b) $\frac{4}{3}$ (c) $\frac{2}{3}(2^{3/2} - 1)$ (d) $2(\sqrt{2} - 1)$

3. (a) $T_4 = \frac{43}{32} = 1.34375$, $\displaystyle\int = \frac{4}{3}$ (b) $T_4 = \frac{19}{4} = 4.75$, $\displaystyle\int = \frac{14}{3}$
   (c) $T_4 = 12$, $\displaystyle\int = 12$ (d) $T_4 = \frac{141}{200} = 0.705$, $\displaystyle\int = \frac{2}{3}$
   (e) $T_4 = \frac{1171}{1680} = 0.697024$ (f) $T_4 = \frac{5323}{6800} = 0.782794$
   (g) $T_4 = 0.742984$ (h) $T_4 = \frac{945}{1554} = 0.608108$
   (i) $T_4 = -\frac{1923}{3400} = -0.565588$ (j) $T_4 = \dfrac{3\pi + 16\sqrt{2} + 12}{24} = 1.83551$

5. (a) 16, 151, 15,001 (b) 21, 201, 20,001
(c) no error (d) 5, 41, 4083
(e) 961, 9601, 960,001 (f) 5, 41, 4083

6. (a) $T_{10} = 0.835023$, $T_{100} = 0.835642$, $T_{1000} = 0.835646$
(b) $T_{10} = 0.784981$, $T_{100} = 0.785394$, $T_{1000} = 0.785396$
(c) $T_5 = 1.11499$, $T_{50} = 1.11148$, $T_{500} = 1.11145$
(d) $T_2 = 1.7854$, $T_4 = 1.83551$, $T_8 = 1.84784$, $T_{16} = 1.85091$, $T_{100} = 1.85191$
(e) $T_{10} = 0.746211$, $T_{100} = 0.746818$, $T_{1000} = 0.746822$

Section 3, pages 431–432

1. (a) $M_4 = \frac{85}{64} = 1.32812$, $\int = \frac{4}{3}$
(b) $M_4 = 4$, $\int = 4$
(c) $M_4 = 0.648146$, $\int = \frac{2}{3}$
(d) $M_4 = 1.36616$, $\int = \ln 4 = 1.3863$
(e) $M_4 = 2.79961$ (f) $M_4 = 3.14159$
(g) $M_4 = 0.748747$ (h) $M_4 = 1.10867$

2. (a) $T_4 = 1.34375$ (b) $T_4 = 4$
(c) $T_4 = 0.705$ (d) $T_4 = 1.42809$
(e) $T_4 = 2.79306$ (f) $T_4 = 3.14159$
(g) $T_4 = 0.742984$ (h) $T_4 = 1.11699$

3. (a) $S_4 = 1.33333$ (b) $S_4 = 4$
(c) $S_4 = 0.671481$ (d) $S_4 = 1.39162$
(e) $S_4 = 2.79734$ (f) $S_4 = 4.18879$
(g) $S_4 = 0.746855$ (h) $S_4 = 1.11136$

8. (a) 301 (Trap), 214 (Midpt), 31 (Simpson)
(b) 231 (Trap), 164 (Midpt), 29 (Simpson)

Section 4, pages 442–443

1. 32
2. 4π
3. 20
4. $\frac{4}{3}\pi a^3$
5. $\frac{4}{3}\pi ab^2$
6. (a) $\dfrac{38\pi}{15}$ (b) $\dfrac{128\pi}{3}$ (c) $\dfrac{86\pi}{15}$
(d) $\dfrac{7\pi}{24}$ (e) $\dfrac{16\pi}{15}$
7. $\frac{1}{3}\pi a^2 h$

8. (a) $\frac{15\pi}{2}$ (b) $\frac{9\pi}{2}$

9. (a) $\frac{9\pi}{2}$ (b) $\frac{512\pi}{3}$ (c) $\frac{20\pi}{3}$

(d) $2\pi \ln 2$ (e) $\frac{\pi}{2}$

10. $\frac{4\pi}{3}$

11. (a) $4\pi \int_{h-a}^{h+a} x\sqrt{a^2 - (x-h)^2}\, dx$

(b) $2\pi^2 a^2 h$

12. $\pi a^2 h \left(2 + \frac{h}{2}\right)$

13. $\pi a^2 h \left(\frac{h}{12} + \frac{2}{3}\right)$

Section 5, pages 450–452

1. 10,000 ft-lb
2. 60 ft-lb
3. 10 ft-lb
4. $\frac{k}{2a}$
5. 85 ft-lb
6. $\frac{MgR}{2}, \frac{3MgR}{8}, 0$
7. $\frac{k \ln 2}{\pi a^2}$
8. (a) $Gm\left(\frac{M_2}{(d-x)^2} - \frac{M_1}{x^2}\right)$ (b) $-Gm\int_a^{d-b} \left(\frac{M_2}{(d-x)^2} - \frac{M_1}{x^2}\right) dx$

Section 6, pages 459–460

1. a, b, c are integrable
2. a, b, d are integrable
3. (a) 16 (b) 4 (c) 10
5. both limits are ∞

Section 7, pages 470–472

1. (a) improper, convergent: $9\sqrt[3]{2}$ (b) improper, convergent: 2

(c) improper, divergent (d) proper: $\frac{\pi\sqrt{3}}{9}$

(e) improper, divergent
(f) proper

2. (a) improper, convergent: $\frac{1}{8}$
(b) improper, divergent
(c) improper, divergent
(d) improper, convergent: $\frac{1}{2}$
(e) improper, divergent
(f) improper, convergent: e
(g) improper, convergent: $\frac{1}{2}$
(h) proper: $-\frac{1}{4}$

4. convergent for $s > 1$, divergent otherwise

5. (a) improper, convergent: 6
(b) improper, convergent: $\frac{3}{2}(\sqrt[3]{4} - 1)$
(c) proper,
(d) improper, divergent
(e) improper, convergent: $\frac{1}{2}$
(f) improper, convergent: 2
(g) improper, divergent
(h) improper, divergent
(i) improper, divergent
(j) improper, convergent: 1

7. (a) improper, convergent
(b) improper, convergent
(c) improper, convergent
(d) improper, divergent
(e) proper
(f) improper, convergent

8. $F'(0) = 1$, $F'(1) = \dfrac{1}{e}$

CHAPTER 9

Section 1, pages 480–481

1. (a) 1
(b) 1
(c) 0
(d) 0
(e) e
(f) 1

2. (a) 1
(b) 1
(c) ∞
(d) does not converge
(e) 0
(f) 0

3. (a) no limit
(b) 1
(c) 1
(d) 2

4. 0

8. (a) 1, 0, 3, 2, 5, 4, 7, 6, 9, 8
(c) $s_n = \begin{cases} \dfrac{n}{2}, & \text{if } n \text{ is even} \\ n, & \text{if } n \text{ is odd} \end{cases}$

Section 2, pages 489–490

1. (a) $\frac{35}{4}$
(b) $\dfrac{a}{4}$
(c) does not converge
(d) $\frac{1}{2}$
(e) does not converge
(f) does not converge
(g) does not converge
(h) $\frac{7}{2}$

2. converges to 2

6. 1

7. $a_m - \lim\limits_{n\to\infty} a_n$

8. (a) 1
(b) $\frac{1}{4}$
(c) does not converge

Section 3, pages 497–498

1. (b), (e), (g), (h), (i), (j) are convergent
2. both are convergent

Section 4, page 502

1. (a), (b), (d), (e), (f), (h) are convergent
3. (b), (e), (f), (h) are convergent
5. (a) 3, 6, 10, 20 (b) 9, 99, 999, $10^6 - 1$ (c) 20, 200, 2000, 2×10^6

Section 5, pages 510–511

1. (a) conditionally convergent (b) absolutely convergent
   (c) conditionally convergent (d) absolutely convergent
   (e) divergent (f) absolutely convergent
   (g) divergent (h) absolutely convergent
3. (a) divergent (b) conditionally convergent
   (c) absolutely convergent (d) absolutely convergent
5. (b) no limit

Section 6, pages 518–519

1. (a) 2 (b) 1 (c) 1 (d) 1
   (e) 1 (f) $\frac{1}{2}$ (g) 1 (h) 3
2. (a) $(-2, 2)$ (b) $[-1, 1]$
   (c) $[-1, 1)$ (d) $(-1, 1)$
   (e) $(-1, 1)$ (f) $(-\frac{1}{2}, \frac{1}{2})$
   (g) $(-1, 1)$ (h) $[-3, 3]$
3. false
4. (a) 1 (b) ∞ (c) 1
   (d) 0 (e) 5 (f) ∞
5. (a) $[1, 3)$ (b) $(-\infty, \infty)$ (c) $(-3, -1)$
   (d) $x = 1$ (e) $(-6, 4)$ (f) $(-\infty, \infty)$

Section 7, pages 526–528

1. (a) $r = 1$, domain $= [-1, 1]$
   (b) $\sum_{i=1}^{\infty} \frac{x^{i-1}}{i}$, $r = 1$
   (c) $[-1, 1)$
2. (a) $r = 2$, domain $= [0, 4)$
   (b) $\sum_{i=1}^{\infty} \frac{(x-2)^{i-1}}{2^i}$, $r = 2$
   (c) $(0, 4)$

3. (a) $r = 1$, domain $= [1, 3)$
   (b) $\sum_{i=1}^{\infty} \sqrt{i}\,(x-2)^{i-1}$, $r = 1$
   (c) $(1, 3)$
4. (a) $r = 1$, domain $= [0, 2)$
   (b) $\sum_{i=1}^{\infty} \frac{i(x-1)^{i-1}}{\sqrt{i+1}}$, $r = 1$
   (c) $(0, 2)$
5. $(-\infty, \infty)$ for both f and g
7. 0, 1, 1, 0
11. (b) true

Section 8, pages 538–540

1. (a) $T_0 = \frac{1}{2}$, $T_1 = \frac{1}{2} - \frac{1}{4}x$, $T_2 = \frac{1}{2} - \frac{1}{4}x + \frac{x^2}{8}$
   (b) $T_0 = 1$, $T_2 = 1 - x^2$, $T_4 = 1 - x^2 + x^4$
   (c) $T_0 = \frac{1}{2}$, $T_1 = \frac{1}{2} - \frac{1}{2}(x-1)$, $T_2 = \frac{1}{2} - \frac{1}{2}(x-1) + \frac{1}{4}(x-1)^2$
   (d) $T_1 = 2 + \frac{1}{4}(x-3)$, $T_2 = 2 + \frac{1}{4}(x-3) - \frac{1}{64}(x-3)^2$,
   $T_3 = 2 + \frac{1}{4}(x-3) - \frac{1}{64}(x-3)^2 + \frac{1}{512}(x-3)^3$
   (e) $T_0 = \frac{\sqrt{2}}{2}$, $T_1 = \frac{\sqrt{2}}{2} + \frac{\sqrt{2}}{2}\left(x - \frac{\pi}{4}\right)$,
   $T_2 = \frac{\sqrt{2}}{2} + \frac{\sqrt{2}}{2}\left(x - \frac{\pi}{4}\right) - \frac{\sqrt{2}}{4}\left(x - \frac{\pi}{4}\right)^2$
2. (a) $T_n(x) = \sum_{i=0}^{k} \frac{(-1)^i x^{2i}}{(2i)!}$, where $k = \begin{cases} \frac{n}{2}, & n \text{ even} \\ \frac{n-1}{2}, & n \text{ odd} \end{cases}$
   (b) $T_0(x) = 0$, $T_n(x) = \sum_{i=1}^{n} \frac{(-1)^{i-1}(x-1)^i}{i}$, $n \geq 1$
3. (a) $1,\ 1 + x,\ 1 + x + \frac{x^2}{2}$ (b) $1,\ 1,\ 1 - \frac{x^2}{2}$
5. (a) $T_1 = 9 + 7(x-2)$, $T_2 = 9 + 7(x-2) + (x-2)^2 = T_3$
   (b) $T_1 = 3$, $T_2 = 3 - 5x^2$, $T_3 = 3 - 5x^2 + 2x^3$
   (c) $T_3 = 2 + x + 3x^2$, $T_4 = 2 + x + 3x^2 + x^4 = T_{17}$
   (d) $T_2 = 3(x-1) + 3(x-1)^2$, $T_3 = 3(x-1) + 3(x-1)^2 + (x-1)^3 = T_4$
7. (a) $\sum_{i=0}^{\infty} \frac{x^i}{i!}$ (b) $\sum_{i=0}^{\infty} \frac{(x-2)^i}{i!}$ (c) $\sum_{i=0}^{\infty} \frac{(-1)^i x^{2i+1}}{2i+1}$
8. ( ) $\frac{1}{3} - \frac{1}{9}(x-1) + \frac{1}{27}(x-1)^2 - \frac{1}{81}(x-1)^3$
11. $3 + 11(x-2) + 12(x-2)^2 + 6(x-2)^3 = 6x^3 - 24x^2 + 35x - 19$

CHAPTER 10

Section 1, pages 548–550

1. (a), (b), (c) are parabolas (d) straight line
 (e) cubic (f) semicircle
 (g) line segment (h) ellipse

2. (a) straight line (b) parabola
 (c) line segment (d) hyperbola

3. (a) $f(x) = x^2 + 2x + 2$ (b) $f(x) = e^{x^2}$
 (c) $f(x) = \frac{3}{2}\sqrt{4 - x^2}$ (d) $f(x) = \ln x$

4. (a) $y^2 - x = 0$ (b) $y^3 - x = 0, x > 0$
 (c) $x^2 - y^2 = 4, x > 0$

7. (a) 2 (b) $-\frac{3}{5}$ (c) 1 (d) $\frac{1}{5}$

8. (a) 0 (b) -1 (c) $\frac{3}{2}$ (d) -1

9. (a) If $x' = \dfrac{dx}{dt}$, etc., then $\dfrac{d^2y}{dx^2} = \dfrac{x'y'' - y'x''}{(x')^3}$
 (b) $\frac{2}{9}$

Section 2, pages 558–560

1. (a) $\frac{8}{27}\left(10^{3/2} - \dfrac{13^{3/2}}{8}\right)$ (b) 24
 (c) $\frac{8}{27}(10^{3/2} - 1)$ (d) $\dfrac{3a}{2}$

2. 3

3. (a) $\frac{8}{27}\left(10^{3/2} - \dfrac{13^{3/2}}{8}\right)$ (b) $\frac{22}{3}$
 (c) $\frac{1}{4}(\sqrt{2} + \ln(1 + \sqrt{2}))$ (d) $e - \dfrac{1}{e}$

5. (b) $\frac{14}{3}$ (c) $\displaystyle\int_0^2 \sqrt{4y^2 - 8y + 5}\, dy$

6. $\frac{19}{3}$

7. (c) $\sqrt{2}, 2\sqrt{2}$

Section 3, pages 567–569

1. (a) $(-2, 4)$ (b) $(5, -3)$ (c) $(2, -1)$ (d) $(4\frac{1}{2}, 3\frac{1}{2})$

2. (a) $(5, -1), (3, 2)$ (b) $(6, 0)$ (c) $(4, -2)$
 (d) $(5, 4)$ (e) $(8, 2)$

3. (b) $\sqrt{29}, \sqrt{2}, 3\sqrt{5}$

4. true

6. (b) $(\frac{5}{2}\sqrt{3}, \frac{5}{2})$
7. (b) $(5\sqrt{3} + 4, 1)$
9. (a) true (b) true (c) false

Section 4, pages 576–577

1. For each value of t_0, $P(t_0)$ and the coordinates of $\mathbf{d}P(t_0)$ are as follows:
   (a) -1: $(-2, 1)$, $(1, -2)$; 0: $(-1, 0)$, $(1, 0)$; 2: $(1, 4)$, $(1, 4)$
   (b) -1: $(2, -2)$, $(-2, 1)$; 0: $(1, -1)$, $(0, 1)$; 1: $(2, 0)$, $(2, 1)$
   (c) 0: $(-1, 0)$, $(1, 0)$; 1: $(0, 1)$, $(1, 3)$; 2: $(1, 8)$, $(1, 12)$
   (d) 0: $(1, 0)$, $(1, 1)$; $\ln 2$: $(2, \ln 2)$, $(2, 1)$
   (e) 0: $(3, 0)$, $(0, 2)$; $\frac{\pi}{4}$: $(\frac{3}{2}\sqrt{2}, \sqrt{2})$, $(-\frac{3}{2}\sqrt{2}, \sqrt{2})$; $\frac{\pi}{2}$: $(0, 2)$, $(-3, 0)$
   (f) -1: $(1, -1)$, $(-2, 3)$; 0: $(0, 0)$, $(0, 0)$; 2: $(4, 8)$, $(4, 12)$
   (g) -1: $(-2, 2)$, $(1, 2)$; 0: $(-1, 4)$, $(1, 2)$; 1: $(0, 6)$, $(1, 2)$
2. The coordinates of $\mathbf{d}P(t)$ are as follows:
   (a) $(2t, 3t^2)$ (b) $(\frac{1}{2}(e^t - e^{-t}), \frac{1}{2}(e^t + e^{-t}))$
   (c) $(2t, 3(3t + 1)^{1/2})$ (d) $(2t + 1, t^2 + 2t)$
3. Coordinates of $\mathbf{d}P(\theta)$ are $(a(1 - \cos\theta), a \sin\theta)$
6. $P(t_0)$, the coordinates of $\mathbf{d}P(t_0)$, and equations of the tangent lines are as follows:
   (a) $(5, 3)$, $(4, 1)$, $x - 4y + 7 = 0$
   (b) $(1, 1)$, $(0, 1)$, $x = 1$
   (c) $(2, \ln 2)$, $(2, 1)$, $x - 2y - 2 + \ln 4 = 0$
   (d) $(0, 0)$, $\mathbf{d}P$, and a unique tangent do not exist at $t_0 = 0$.
7. (a) $y = \frac{1}{2}x$
   (b) At $t = 2$ the vector space forms the straight line $y = \frac{1}{2}x$. At $t = 0$ the vector space consists of the zero vector at the point $(0, 0)$.
8. (a) $y = xf'(a) - af'(a) + f(a)$
   (b) $\mathbf{d}P(a) = (1, f'(a))_{P(a)}$; $y - f(a) = f'(a)(x - a)$

Section 5, pages 588–590

2. (a) parabola: $y = x^2 - x$
   (b) $\mathbf{v}(t) = (1, 2t - 1)_{P(t)}$,
   $$\begin{aligned} P(0) &= (0, 0), \mathbf{v}(0) = (1, -1)_{P(0)}, & |\mathbf{v}(0)| &= \sqrt{2} \\ P(1) &= (1, 0), \mathbf{v}(1) = (1, 1)_{P(1)}, & |\mathbf{v}(1)| &= \sqrt{2} \\ P(2) &= (2, 2), \mathbf{v}(2) = (1, 3)_{P(2)}, & |\mathbf{v}(2)| &= \sqrt{10} \end{aligned}$$
   (c) $\mathbf{a}(t) = (0, 2)_{P(t)}$, perpendicular when $t = \frac{1}{2}$
   (d) $\sqrt{2}\int_0^2 \sqrt{2t^2 - 2t + 1}\,dt$
   (e) $\frac{3}{4}\sqrt{10} + \frac{\sqrt{2}}{4} + \frac{1}{4}\ln\left(\frac{3 + \sqrt{10}}{-1 + \sqrt{2}}\right)$

3. (a) $P(t) = (800t, -16t^2)$, $\mathbf{v}(t) = (800, -32t)_{P(t)}$, $\mathbf{a}(t) = (0, -32)_{P(t)}$
   (b) 25 sec
   (c) parabola: $y = \dfrac{-x^2}{40{,}000}$
   (d) $32\int_0^{25} \sqrt{625 + t^2}\,dt$
4. (b) $\mathbf{v}(t) = (2t - 4, 2t - 4)_{P(t)}$, $|\mathbf{v}(t)| = 2\sqrt{2}|t - 2|$;
   $|\mathbf{v}(0)| = |\mathbf{v}(4)| = 4\sqrt{2}$ (max), $|\mathbf{v}(2)| = 0$(min)
   (c) straight line and point (d) $8\sqrt{2}$
5. (b) $\mathbf{v}(t) = (2t, 3t^2)_{P(t)}$
   $P(-2) = (4, -8)$, $\mathbf{v}(-2) = (-4, 12)_{P(-2)}$, $|\mathbf{v}(-2)| = 4\sqrt{10}$
   $P(0) = (0, 0)$, $\mathbf{v}(0) = (0, 0)_{P(0)}$, $|\mathbf{v}(0)| = 0$
   $P(1) = (1, 1)$, $\mathbf{v}(1) = (2, 3)_{P(1)}$, $|\mathbf{v}(1)| = \sqrt{13}$
   $P(2) = (4, 8)$, $\mathbf{v}(2) = (4, 12)_{P(2)}$, $|\mathbf{v}(2)| = 4\sqrt{10}$
   (c) $\mathbf{a}(t) = (2, 6t)_{P(t)}$, $\mathbf{a}(-2) = (2, -12)_{P(-2)}$,
   $\mathbf{a}(0) = (2, 0)_{P(0)}$, $\mathbf{a}(1) = (2, 6)_{P(1)}$, $\mathbf{a}(2) = (2, 12)_{P(2)}$
6. (a) $\dfrac{x^2}{a^2} + \dfrac{y^2}{b^2} = 1$
8. $t = \dfrac{1}{k}\left(\dfrac{a^2T^2}{m^2} - a^2k^2\right)^{1/4}$

Section 6, pages 600–601

1. (a) $0, 2, 2\sqrt{2}, 4, 2\sqrt{2}, 2, 0$ (c) $x^2 + y^2 - 4y = 0$
2. (a) $4, 2 + \sqrt{2}, 3, 2, 1, 2 - \sqrt{3}, 0$
   (c) about x-axis
4. $r = \dfrac{5}{\sin\theta}$
6. (a) circle (b) straight line
7. (a) $x = -2$ (b) $y = 4$
   (c) $(x + 2)^2 + y^2 = 4$ (d) $y = 2x + 2$
   (e) $y^2 = 2x + 1$ (f) $x^2 + y^2 = 25$
   (g) $y = 3x$ (h) $(x - \sqrt{3})^2 + (2y)^2 = 4$

Section 7, pages 611–613

1. (a) 2π (b) $\dfrac{27\pi}{4}$ (c) $\dfrac{27\pi}{8} + 9$ (d) 4
2. (a) 4π (b) 16 (c) $\dfrac{14\pi^3}{3}$
   (d) $\frac{1}{12}$ (e) $\frac{1}{2}$
4. (a) $\dfrac{3a^2\pi}{2}$ (b) $\dfrac{3a^2\pi}{2}$ (c) 25π

(d) $\dfrac{9\pi}{2}$ (e) 4π (f) 4π

5. (b) $\dfrac{1}{2}\displaystyle\int_{-\pi/2}^{\pi/2} \frac{d\theta}{1 + 2\cos\theta + \cos^2\theta}$
(c) $\frac{2}{3}$ (d) $y^2 = 1 - 2x$
6. $\pi + 3\sqrt{3}$
7. $2a^2$
8. $\dfrac{9}{2}\sqrt{3} - \pi$
9. $24\pi,\ 16\pi + 9\sqrt{3},\ 8\pi - 9\sqrt{3}$
10. $8a$
11. $\dfrac{99\sqrt{5}}{2}$
12. (a) 4
13. (a) 4 (b) 8

CHAPTER 11

Section 1, pages 622–623

1. (a) $y = \frac{1}{4}x^4 + 2e^x + c$ (b) $y = 2x^3 + 5x + \ln x + c$
(c) $y = \tan(x^2 + 3x + c)$ (d) $y = ce^{(1/2)x^2} - 1$
(e) $y = \dfrac{1}{-x^4 + x^2 + c}$ (f) $y^2 = 2x\ln x - 2x + c$
(g) $y = \frac{1}{2}(\ln x)^2 + c$ (h) $y = ce^{-16x}$
(i) $y = c_1\cos 4x + c_2\sin 4x$ (j) $y = c_1e^{4x} + c_2e^{-4x}$
(k) $y = c_1e^{20x} + c_2e^{-x}$ (l) $y = c_1e^{-2x} + c_2e^{-8x}$
(m) $y = c_1e^{2x} + c_2e^{5x}$ (n) $y = (c_1x + c_2)e^{ax}$
(o) $y = e^{-2x}(c_1\cos 5x + c_2\sin 5x)$
(p) $\frac{1}{2}y^2 + 5y = \frac{7}{2}x^2 + e^{-x} + c$
(q) $y^2 - x^2 = c$ (r) $y = cx$
(s) $x^2 + y^2 = c$ (t) $y = c_1e^{7x} + c_2e^{-7x}$
(u) $\frac{1}{4}\ln|4t + 3| + \frac{1}{3}\ln|3x + 4| = c$
(v) $y = \operatorname{arcsec}(ke^x)$
(w) $y = \frac{1}{6}e^{3t^2+4} + c$ (x) $y = \dfrac{3x}{8} - \dfrac{3}{32}\sin 4x + c$
(y) $\tan y = \frac{3}{2}x - \frac{3}{4}\sin 2x + c$ (z) $y = \dfrac{x^4}{2} - \frac{2}{3}x^3 + x^2 + c_1x + c_2$

2. (a) $y = 5e^{3x}$ (b) $y = x^4 + \dfrac{x^2}{2} - 2x - \frac{1}{2}$
(c) $x^2 + y^2 = 25$ (d) $s = -\frac{1}{2}gt^2 + v_0t + s_0$
(e) $y = 2e^{3x} + 5e^{-x}$ (f) $y = e^{2x}(5\cos 3x - \frac{8}{3}\sin 3x)$
(g) $y = \ln|3x + 6|$ (h) $y = (-11x + 3)e^{6x}$

3. (b) $D^2y - Dy + \frac{5}{4}y = 0$

4. (a) $y = e^{-x}(c_1 \cos \frac{1}{2}x + c_2 \sin \frac{1}{2}x)$

(b) $y = \frac{\sqrt{2}}{2} e^{-x}(\cos \frac{1}{2}x - \sin \frac{1}{2}x)$

5. (a) $y = e^{x}(c_1 \cos \frac{1}{2}x + c_2 \sin \frac{1}{2}x)$

(b) $y = \frac{\sqrt{2}}{2} e^{x}(\cos \frac{1}{2}x - \sin \frac{1}{2}x)$

6. (a) $y = c_1 \cos \frac{1}{2}x + c_2 \sin \frac{1}{2}x$

(b) $y = \frac{\sqrt{2}}{2}(\cos \frac{1}{2}x - \sin \frac{1}{2}x)$

7. $y = e^{\alpha x}(c_1 \cos x + c_2 \sin x)$

Section 2, page 629

1. (a) $y = 3x^3 - 4x + cx^2$ (b) $y = -\frac{x}{4} + \frac{c}{x^3}$

(c) $y = \frac{x^7}{21} + \frac{2}{7} + cx^{-7/2}$ (d) $y = \frac{5}{2} + ce^{-x^2}$

(e) $y = -\frac{1}{6}e^{2x} - \frac{1}{2} + ce^{8x}$ (f) $y = \frac{1}{6} + ce^{-2x^3}$

(g) $y = 1 + ce^{-\sin x}$ (h) $y = 4 + ce^{-(x^2+3x)}$

(i) $y = \frac{3}{5}(2 \cos x + \sin x) + ce^{-2x}$

(j) $y = -2e^{-x}\left(1 + \frac{1}{x}\right) + \frac{c}{x}$ (k) $y = \frac{a}{13}x^2 + \frac{b}{12}x + \frac{c}{11} + dx^{-11}$

(l) $y = \frac{\pi}{9} + ce^{-9x}$ (m) $y = \frac{e^{2x}}{2x^3} + \frac{c}{x^3}$

(n) $y = \frac{\sin x}{x^5} + \frac{c}{x^5}$

2. (a) $y_h = ce^{-x^2}$

3. (a) $y_h = ce^{-\int P\,dx}$

Section 3, pages 638–640

1. (a) $y = c_1e^{x} + c_2e^{-2x} - \frac{5}{2}e^{-x}$ (b) $y = c_1e^{x} + c_2e^{-2x} - 2xe^{-2x}$

(c) $y = c_1e^{x} + c_2e^{2x} + 2x + \frac{9}{2}$ (d) $y = c_1 \cos x + c_2 \sin x + \frac{1}{2}e^{x}$

(e) $y = c_1 \cos x + c_2 \sin x + x^2 - 1$

12. (a) $y = c_1e^{x} + c_2xe^{x} + c_3e^{-2x}$ (b) $y = c_1e^{x} + c_2e^{2x} + c_3e^{-2x}$

Section 4, pages 647–648

1. (a) $y = c_1e^{7x} + c_2e^{-x}$ (b) $y = (c_1x + c_2)e^{-3x}$

(c) $y = c_1e^{-x} + c_2e^{-5x}$ (d) $y = e^{x}(c_1 \cos 3x + c_2 \sin 3x)$

(e) $y = c_1e^{-(3/2)x} + c_2e^{(1/2)x}$ (f) $y = c_1e^{-6x} + c_2$

(g) $y = e^{-x}(c_1 \cos \sqrt{5}\, x + c_2 \sin \sqrt{5}\, x)$

(h) $y = (c_1x + c_2)e^{-(1/2)x}$ (i) $y = e^{-2x}(c_1 \cos x + c_2 \sin x)$

(j) $y = e^{-(1/2)x}\left(c_1 \cos \frac{\sqrt{3}}{2}x + c_2 \sin \frac{\sqrt{3}}{2}x\right)$

5. (a) $y = c_1e^{-x} + c_2e^{2x} + c_3e^{3x}$ (b) $y = c_1e^{2x} + (c_2x + c_3)e^{3x}$
   (c) $y = c_1e^{ax} + c_2e^{bx} + c_3e^{cx}$

6. $y = c_1 \cos 2x + c_2 \sin 2x + c_3e^{3x}$

7. (a) $y = c_1e^{2x} + (c_2x + c_3)e^{-x}$ (b) $y = c_1e^{x} + c_2e^{2x} + c_3e^{-3x}$
   (c) $y = (c_1x + c_2)e^{3x} + c_3e^{-x} + c_4e^{5x}$
   (d) $y = c_1 + c_2e^{x} + c_3e^{-4x}$ (e) $y = c_1e^{x} + (c_2x^2 + c_3x + c_4)e^{-2x}$
   (f) $y = (c_1x + c_2)e^{-3x} + c_3 \cos \sqrt{3}\, x + c_4 \sin \sqrt{3}\, x$
   (g) $y = c_1 + c_2e^{x} + c_3e^{-2x}$
   (h) $y = (c_1x + c_2)e^{-x} \cos x + (c_3x + c_4)e^{-x} \sin x$
   (i) $y = e^{-x}[(c_1x + c_2)\cos x + (c_3x + c_4)\sin x + c_5]$
   (j) $y = e^{-x}[(c_1x + c_2)\cos x + (c_3x + c_4)\sin x] + c_5x + c_6$
   (k) $y = c_1e^{3x} + c_2e^{-3x} + c_3 \cos 3x + c_4 \sin 3x$
   (l) $y = c_1e^{-x} + c_2 \cos x + c_3 \sin x$

Section 5, pages 656–657

1. (a) $y_p = -\frac{1}{2},\ y = c_1e^{-5x} + c_2e^{2x} - \frac{1}{2}$
   (b) $y_p = 2x,\ y = c_1 \cos x + c_2 \sin x + 2x$
   (c) $y_p = -3x + 5,\ y = c_1e^{2x} + c_2e^{-2x} - 3x + 5$
   (d) $y_p = x^2 - 2x + 2,\ y = c_1e^{x} + c_2e^{-2x} + x^2 - 2x + 2$
   (e) $y_p = -\frac{1}{4}e^{x},\ y = c_1e^{3x} + c_2e^{-x} - \frac{1}{4}e^{x}$
   (f) $y_p = 2e^{3x},\ y = (c_1x + c_2)e^{x} + 2e^{3x}$
   (g) $y_p = -\frac{1}{5}e^{-x},\ y = c_1 + c_2e^{3x} + c_3e^{-3x} - \frac{1}{5}e^{-x}$
   (h) $y_p = \sin x,\ y = c_1 \cos 2x + c_2 \sin 2x + \sin x$
   (i) $y_p = \sin x + x + 2,\ y = c_1 \cos 2x + c_2 \sin 2x + \sin x + x + 2$
   (j) $y_p = \cos x + 2 \sin x,\ y = c_1 + c_2e^{-3x} + \cos x + 2 \sin x$
   (k) $y_p = -\frac{5}{6} \cos 3x,\ y = c_1 \cos \sqrt{3}\, x + c_2 \sin \sqrt{3}\, x - \frac{5}{6} \cos 3x$
   (l) $y_p = -\frac{3}{2} \cos 2x + \sin 2x,$
   $y = c_1e^{(-1+\sqrt{3})x} + c_2e^{(-1-\sqrt{3})x} - \frac{3}{2} \cos 2x + \sin 2x$

2. $y = 3 \cos x + 2x$

3. $y = \frac{1}{2}(e^{2x} - e^{-x} \sin x)$

4. $i = 3e^{-t}(-\cos 2t + \sin 2t + 1)$

5. (a) $y = (c_1x + c_2)e^{2x} + x^2 + 2x + \frac{1}{4}$
   (b) $y = c_1e^{2x} + c_2e^{3x} + 2x + \frac{9}{2}$
   (c) $y = c_1e^{-2x} + c_2e^{x} - \frac{5}{2}e^{-x}$
   (d) $y = c_1 + c_2e^{2x} - x^3 - 2x^2 + x$
   (e) $y = c_1e^{x} + c_2e^{-2x} - 2xe^{-2x}$
   (f) $y = c_1e^{x} + c_2e^{-2x} - 2xe^{-2x} - x + \frac{3}{2}$
   (g) $y = c_1e^{x} + c_2e^{-2x} - 2xe^{-2x} + 5xe^{x}$
   (h) $y = c_1e^{-x} + c_2 + c_3x - \frac{11}{18}x^2 + \frac{5}{18}x^3$
   (i) $y = c_1 \cos 2x + c_2 \sin 2x - \cos 3x$
   (j) $y = c_1 \cos 3x + c_2 \sin 3x - \frac{1}{3}x \cos 3x$
   (k) $y = c_1 \cos x + c_2 \sin x - 5x \cos x + \frac{3}{2}e^{-x}$
   (l) $y = c_1 \cos x + c_2 \sin x - 2x \cos x - 4x \sin x$
   (m) $y = (c_1x + c_2)e^{x} - 3e^{x} \sin x$
   (n) $y = e^{-x}(c_1 \cos x + c_2 \sin x) + \frac{3}{8}e^{x}(\cos x + \sin x)$

Section 6, pages 662–663

1. (a) $5 \sinh 5x$ (b) $12 \cosh 3x \sinh 3x$

(c) $2x \tanh(x^2 + 1)$ (d) $\dfrac{x \cosh \sqrt{x^2+1}}{\sqrt{x^2+1}}$

(e) $\operatorname{sech}^2 x$ (f) $-\operatorname{sech} x \tanh x$

(g) $-\operatorname{csch} x \coth x$ (h) $-\operatorname{csch}^2 x$

(i) $\dfrac{-2t \operatorname{sech}^2(1/1 + t^2)}{(1 + t^2)^2}$ (j) $\sinh\left(\dfrac{x}{a}\right)$

2. (a) $\frac{1}{7} \cosh 7x + c$ (b) $2 \sinh \dfrac{t}{2} + c$

(c) $\frac{1}{12} \cosh^4 3x + c$ (d) $\ln \cosh x + c$

(e) $\ln |\tanh x| + c$ (f) $\frac{1}{2} \cosh(2x^2 + 1) + c$

(g) $\frac{1}{12} \tanh^6 2x + c$ (h) $\frac{1}{2}(\ln \sinh x)^2 + c$

(i) $\frac{1}{4} \sinh 2x + \dfrac{x}{2} + c$ (j) $\frac{1}{4} \sinh 2x - \dfrac{x}{2} + c$

5. (a) $y = c_1 \cosh 2x + c_2 \sinh 2x$

(b) $y = c_1 \cosh \sqrt{7}\, x + c_2 \sinh \sqrt{7}\, x$

(c) $y = c_1 \cosh 3x + c_2 \sinh 3x - e^{2x}$

(d) $y = c_1 \cosh kx + c_2 \sinh kx - \dfrac{x}{k^2} - \dfrac{1}{1 + k^2} \sin x$

(e) $y = c_1 \cosh 4x + c_2 \sinh 4x + \frac{5}{48} \sinh 8x$

(f) $y = c_1 \cosh 4x + c_2 \sinh 4x + \frac{5}{8} x \sinh 4x$

7. 1

8. $x^2 - y^2 = 1$

9. (b) $\dfrac{t}{2}$

10. 3 sinh 2

11. (a) $\dfrac{1}{1 - x^2}$ (b) $\dfrac{1}{\sqrt{1 + x^2}}$

Index